$149.95
25Sep92
D0022811

Harnessing Biotechnology for the 21st Century

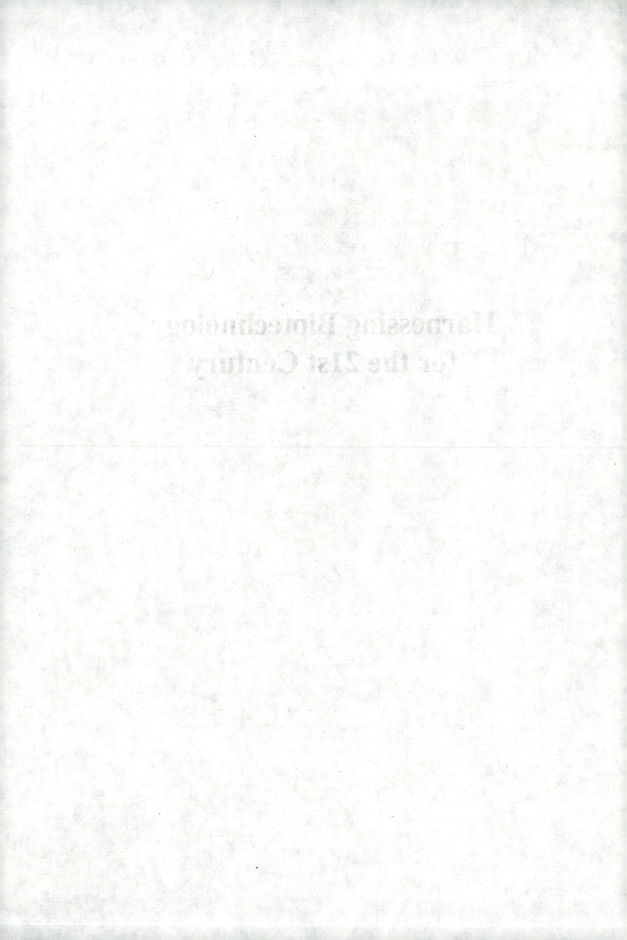

Harnessing Biotechnology for the 21st Century

EDITED BY
Michael R. Ladisch
Purdue University
and
Arindam Bose
Pfizer Central Research

Proceedings of the Ninth International Biotechnology Symposium and Exposition, Crystal City, Virginia, August 16–21, 1992

American Chemical Society, Washington, DC 1992

Library of Congress Cataloging-in-Publication Data

International Biotechnology Symposium and Exposition (9th: 1992: Crystal City, Va.)

Harnessing biotechnology for the 21st century: proceedings of the the Ninth International Biotechnology Symposium and Exposition, Crystal City, Virginia, August 19–21, 1992 / Michael R. Ladisch, editor, Arindam Bose, editor.

p. cm.—(Conference proceedings series (American Chemical Society)).

Includes bibliographical references and index.

ISBN 0–8412–2477–3

1. Biotechnology—Congresses.

I. Ladisch, Michael R., 1950– . II. Bose, Arindam, 1952– . III. Title. IV. Series

TP248.14.I56 1992
660′.6—dc20 92–21791
CIP

The paper used in this publication meets the minimum requirements of American National Standard for Information Sciences—Permanence of Paper for Printed Library Materials, ANSI Z39.48–1984. ∞

PRINTED IN THE UNITED STATES OF AMERICA

Message from the Chair

BIOTECHNOLOGY HAS CHANGED DRAMATICALLY since our first symposium (formerly named Fermentation Technology Symposium), which was held in Japan in 1960. Today, molecular biology tools are used in health care, animal nutrition, chemical production, and agriculture. These tools were not even thought of in 1960. If we compare the technical presentations from the 1960 symposium to our program in 1992, we see few similarities between the two symposia.

The Organizing and Program Committees developed 11 symposia, which contain 36 sessions. A total of 185 invited papers will be presented, representing worldwide biotechnology researchers.

I am pleased to inform you that a conference proceedings has been prepared and will be available during and after this symposium. The proceedings was edited by Michael R. Ladisch of Purdue University and by Arindam Bose of Pfizer Central Research.

I hope that you enjoy the oral presentations at the Ninth International Biotechnology Symposium and that you will obtain a copy of the symposium proceedings.

DANIEL I. C. WANG
Chevron Professor of Chemical Engineering
Massachusetts Institute of Technology
Cambridge, MA 02139

June 1, 1992

Host

American Chemical Society Division of Biochemical Technology

Sponsor

International Union of Pure and Applied Chemistry

Cosponsors

Society for Industrial Microbiology
American Society for Microbiology
American Institute of Chemical Engineers

Distinguished Corporate Sponsors

Abbott Laboratory
Alafi Capital
Amgen
Biogen
Cowen & Company
Genentech
Hoffman-LaRoche, Inc.
Johnson & Johnson
Merck, Sharp & Dohme
Miles/Cutter/Bayer
Pennie & Edwards
Pfizer Company
Sandoz
Schering–Plough

Corporate Sponsors

Archer Daniel Midland Company
Associated Bioengineering Consultants
Amoco
Bristol–Myers Squibb
Chemap
Coors Biotech
Cytogen
Genzyme
Glaxo
Jacob Engineering
Monsanto
New Brunswick Scientific
Promega
Rhone–Poulenc Rorer
SmithKline Beecham
Synergen
Teltech
U.S. Department of Agriculture

Organizing Committee

Chair

Daniel I. C. Wang, *Massachusetts Institute of Technology*

Secretary–Treasurer

Arindam Bose, *Pfizer Central Research*

James E. Bailey
California Institute of Technology

Stephen W. Drew
Merck & Company, Inc.

Robert W. Eltz
Monsanto Company

Michael C. Flickinger
University of Minnesota

Arthur E. Humphrey
Pennsylvania State University

Dianne B. Ruddy
American Chemical Society

Program Committee

Barry C. Buckland
Merck & Company, Inc.

John P. Cherry
U.S. Department of Agriculture

Douglas S. Clark
University of California—Berkeley

Charles L. Cooney
Massachusetts Institute of Technology

Arnold Hershman
Monsanto Company

Michael R. Ladisch
Purdue University

Anil S. Menawat
Tulane University

James E. Rollings
Worcester Polytechnic Institute

Dewey Y. Ryu
University of California—Davis

Gregory N. Stephanopoulos
Massachusetts Institute of Technology

James R. Swartz
Genentech, Inc.

J. Gregory Zeikus
Michigan Biotechnology Institute

Contents

MICROBIOLOGY AND PHYSIOLOGY: SYMPOSIUM III

Primary Metabolites: Session A

Secondary Metabolites: Session B

BIOCATALYSIS AND BIOTRANSFORMATION: SYMPOSIUM IV

Immobilized and Two-Phase Systems: Session A

Designed Biocatalysts: Session B

Biocatalysis in Synthesis and Biotransformation: Session C

BIOREACTOR ENGINEERING: SYMPOSIUM V

Transport Phenomena in Bioreactor Systems: Session A

Stability, Formulation, and Quality Control: Session F

BIOINSTRUMENTATION AND BIOPROCESS CONTROL: SYMPOSIUM VII

Sensors and Analytical Techniques: Session A

Modeling of Bioprocesses: Intracellular and Extracellular Events: Session B

BIOTECHNOLOGY IN DEVELOPING COUNTRIES: SYMPOSIUM XI

Suitable Bioengineering Processes: Session A

Stimulating Biotechnology for Small-Scale Farmers: Session B

INDEX

Index

Preface

THE INTERNATIONAL UNION OF PURE AND APPLIED CHEMISTRY sponsors a conference on biotechnology every four years. The intent of these conferences has been to provide a snapshot of the underlying science and the enabling technology that have allowed commercialization of biotechnology. The proceedings of the earlier conferences of this series have become valuable reference volumes for academicians, practicing scientists, and engineers in industry. We hope this volume will continue to fill that need.

The generally accepted definition of biotechnology is that it "includes any technique that uses living organisms (or parts of organisms) to make or modify products, to improve plants or animals, or to develop microorganisms for specific uses" (U.S. Congress, Office of Technology Assessment; *Commercial Biotechnology: An International Analysis;* OTA–BA–218, 1984). This definition encompasses both the "old biotechnology" (for example, production of organic acids and antibiotics by fermentation) as well as the "new biotechnology" (i.e., processes employing microorganisms, plants, or animals that have been modified by recombinant DNA or by other modern genetic manipulation techniques). Biotechnology has been contributing to human well-being for thousands of years—beginning with the use of microorganisms to ferment foods during the dawn of civilization to the wonder drugs (antibiotics, anthelmintics, and cholesterol-lowering agents) of the present era. The principal impact of the new biotechnology has so far been in the pharmaceuticals arena. However, introduction of new products and services in the areas of agriculture, food additives, energy production, and waste treatment are imminent. The Program Committee of the Ninth International Biotechnology Symposium strove to attain a balance between the coverage of the old and the new biotechnology when assembling the technical program for this conference.

The schedule of oral presentations at this symposium consists of 185 invited papers. Care was taken to ensure that approximately half of the presenters were from outside North America. Each invited lecturer was requested to submit a manuscript for inclusion in this proceedings volume. The submitted papers have been organized into 11 symposia, each of which is subdivided into two or more sessions.

The first three symposia include topics in molecular genetics, biochemistry, and microbiology and provide the scientific basis for the manipulation of organisms constituting commercially significant biotechnology production systems. The fourth symposium covers biocatalysis, including cutting-edge issues such as catalytic antibodies and the use of enzymes in novel organic synthesis tasks.

Biochemical engineering topics (microbial and animal cell bioreactors, product recovery, and monitoring and control of bioprocesses) are the subject of the next three symposia. For the new protein biopharmaceuticals, the cost of product recovery and formulation can often exceed 80% of total bulk drug production cost versus generally less than 50% of total cost for fermentation-derived antibiotics. Thus, a considerable incentive to develop new separation processes for those biopharmaceuticals exists. Most of the papers in Symposium VI address aspects of protein processing, including folding or refolding and formulation.

Within the next decade, the new biotechnology is likely to have a significant commercial impact on agriculture, treatment of wastes, and production of liquid fuels from renewable resources. The sessions on bioremediation, impacts of transgenic plants, and biomass utilization contain contributions from many leading researchers in those fields. Finally, to illustrate the potential of biotechnology in promoting economic development, descriptions of relevant projects in several developing countries are included in Symposium XI.

Any biotechnology development project, by its very nature, is an interdisciplinary effort with life scientists, engineers, and regulatory professionals working closely together. That diversity is also reflected in these proceedings. Specific sections of this book may be of interest primarily to certain groups of professionals. However, we shall consider our efforts in editing this volume worthwhile if a reader finds something useful in papers that are outside the area of his or her expertise.

We thank the numerous corporations, government agencies, and individuals (listed elsewhere in this volume) whose generous financial support made the organization of this conference possible. Special recognition is owed to our Distinguished Corporate Sponsors; their offers of financial support early in the conference planning cycle allowed us to attract prominent speakers from outside North America. We thank Norma Leuck for her assistance in ensuring the timely completion of this volume. We are also grateful to Dianne Ruddy, Latisha Best, and Michelle Hicks of the American Chemical Society's Meetings Department for handling the administrative chores and local arrangements for the conference. Finally, our heartfelt thanks are due to Cheryl Shanks and the staff of the American Chemical Society's Books Department for making this volume possible.

MICHAEL R. LADISCH
Laboratory for Renewable Resources Engineering
Purdue University
West Lafayette, IN 47907

ARINDAM BOSE
Bioprocess Research and Development
Pfizer Central Research
Groton, CT 06340

June 15, 1992

FRONTIERS IN POLYPEPTIDE PRODUCTION SYMPOSIUM I

J. R. Swartz and T. Imanaka: *Co-Chairs*

Expression in Prokaryotes: Session A
J. R. Swartz and T. Imanaka: *Co-Chairs*

Expressing, Processing, and Secretion in Eukaryotes: Session B
C. F. Goochee and W. Fiers: *Co-Chairs*

Expression in Transgenic Systems: Session C
W. H. Velander and R. Strijker: *Co-Chairs*

OPTIMIZATION OF HETEROLOGOUS GENE EXPRESSION USING GRAM NEGATIVE BACTERIA

Allan R. Shatzman
SmithKline Beecham Pharmaceuticals, King of Prussia, PA 19405

Phone: (215) 270-7732; Fax: (215) 270-7962

The last 3-4 years has seen considerable growth in the use of a wide range of gram negative bacteria (e.g. Salmonella Typhimurium, Pseudomonas putida, and Zymomonas mobilis) for the expression of heterologous gene products. However, Escherichia coli has been and continues to be the work horse of the microbial gene expression field. No other system has been developed (thus far) that can not only express virtually any gene product at levels sufficient for detailed biochemical analysis or product development but also express undefined coding sequences (open reading frames) in amounts sufficient to determine the identity of the gene product and to characterize its function. I will describe advances made over the last few years in our ability to optimize the expression of heterologous gene products in E. coli with respect to both quantity and quality of the proteins being produced with the understanding that many of these same approaches can be applied to heterologous gene expression in other gram negative bacteria.

Literally hundreds of papers are published each year in which it has been reported that genes of prokaryotic or eukaryotic origin have been cloned and expressed in E. coli. The ability to express this myriad of genes has been aided by the development of numerous expression vector systems which use highly efficient, regulated, transcriptional regulatory signals to optimize production (for a review see Brosius, 1988). We, (Shatzman and Rosenberg, 1985) and others (e.g. Schauder and McCarthy, 1989) have shown that optimization of translational regulatory signals [ribosome binding sites (RBS)] is equally important to the optimization of heterologous gene expression. For example, in our laboratories, we have seen high levels of expression of various viral antigens using a phage λ P_L promoter coupled with translation initiation using the phage λ cII RBS. In contrast, little, if any, expression of these same antigens was obtained using a pTac promoter and lacZ RBS. In fact, this result had nothing to do with the promoter chosen to drive transcription as equal levels of gene specific message were obtained following induction of both vector systems. Instead, the differences in expression were due solely to translation initiation. When mutations were introduced into the lacZ RBS to make it identical to the cII RBS, equivalent levels of antigen expression were obtained from both promoter systems.

Over the last few years, the focus on the optimization of gene expression has gone beyond the study of regulatory signals to now include alterations in the coding sequence to be expressed. In addition, major efforts have now been made to understand factors which affect not only levels of production, but also the quality (solubility, homogeneity, folding) of the products themselves. It is these newer efforts which I will now focus on.

Optimization of gene expression via alterations in the coding sequence

Initial efforts to express a heterologous gene in E. coli can often lead to less than desirable results with respect to the quantity of protein produced. In some of these cases, even the use of several different transcriptional or translational regulatory signals does not improve the level of protein production. Several reports have been published which show that production of such gene products can be dramatically improved by making alterations within the gene coding sequence. For example, Seow et al. (1989) reported that expression of human tumor necrosis factor (TNF)-β could be

1054–7487/92/0002$06.00/0

increased from undetectable levels to 34% of total cell protein (TCP) by introducing silent mutations throughout the gene, which result in optimal codon utilization for E. coli, and also mutations in the 5' end of the gene, which minimize the development of secondary structure in that region of the message. Interestingly, expression levels were quite low (1-2% of TCP) if only the changes minimizing secondary structure were made, while no expression was detected when only the codon optimization changes were present. In contrast, others have shown that a more random approach to coding sequence alterations can result in vastly improved expression levels. Devlin et al. (1988) and Sathe et al. (1990) demonstrated that expression of human granulocyte stimulating factor could be increased from undetectable levels to almost 20% of total soluble protein by altering G and C residues in the 5' end of the gene to A and T resideus without affecting the protein sequence. These changes actually greatly decreased the presence of preferred codons in this region of the gene and did not create any significant changes in mRNA secondary structure. Reduction in the GC content of the 5' end of the bovine growth hormone gene has also yielded increased expression levels in E. coli (Hsiung and MacKellar, 1987).

While we have had many successes in our laboratories using this approach, random replacement of G-C base pairs with AT proved unsuccessful in optimizing the expression of human alpha-1-antitrypsin (A1AT). Only by using a process of mutagenesis and genetic selection were we able to find mutations in the 5' end of the A1AT coding sequence which allowed us to achieve expression of A1AT at 10% of TCP in contrast to <0.1% of TCP when the native coding sequence was used (Sutiphong et al. 1987).

Improved production of soluble and active proteins in E. coli

Typically, the first concern in heterologous gene expression is "how much can be made." However, after achieving high level expression, attention then must focus on the utility of the product for, unfortunately, most heterologous gene products expressed in E. coli are not produced in an active properly folded, or soluble form. Instead, most recombinant proteins accumulate as insoluble aggregates or inclusion bodies in the cytoplasm of the bacteria.

Considerable efforts have recently been made to understand what causes these proteins to be insoluble and to find ways of expressing these proteins in more soluble, active forms. Schein (1989) compiled a list of nine suggested reasons for the formation of inclusion bodies in E. coli which includes: excessively high rates of production, which do not allow sufficient time to achieve correct folding; high local concentrations of product leading to precipitation; and lack of post-translational modifying enzymes or foldases necessary to achieve the correct conformation.

Expression of subtilisin in E. coli using a high-expression vector under standard growth and induction conditions was recently shown to lead to production of an insoluble product from which very little properly folded and active subtilisin could be obtained (Takagi, et al, 1990). However, when cells were grown at 23°C instead of 37°C, an increase in active subtilisin of up to 14-fold was achieved. Further studies showed that a reduction in the concentration of inducer used in this expression system could increase levels of active product even with growth at 37°C. When both temperature and inducer concentrations were adjusted, active product accumulated at levels 16-fold above those found under standard production conditions. These studies suggest that the rate at which a protein is made can affect its ability to fold properly. Similar effects of culture temperature on product solubility have been observed for the production of human interferon α2 (Schein, et al, 1988). In these studies, overproduction of interferon at 37°C yielded only 5% soluble, active protein while growth at 28-30°C produced 73% soluble product. Thus, although total interferon production at 37°C was three times greater than at 28-30°C, the lower production temperature actually yielded a seven-fold increase in active product.

Work in our laboratories over the last 2 years have also shown that product solubility and stability can be greatly affected by thermal parameters (Shatzman et al, in preparation). For example, expression of the first 2 domains of the human CD4 receptor in E. coli resulted in the production of completely insoluble and inactive protein when production was at 42°C. In contrast, a significant fraction of this product was soluble, properly disulfide cross-linked, and active, when produced at 28-32°C. In addition, a major degradation product observed in the 42°C sample was absent in the 28°C samples. Similar successes with a variety of human gene products expressed in E. coli have now been recorded in our laboratories using expression systems compatible with production at low temperatures.

Studies have also been initiated recently to determine whether the co-expression of foldases along with the gene product of interest can result in improved solubility and activity. This approach is based on the observations that there are numerous proteins such as thioredoxin, protein disulfide isomerase, and prolyl cis-trans isomerase that seem to be designed to aid in protein folding. Although the usefulness of these proteins in improving product solubility in E. coli in vivo has not been documented as yet, there are some initial reports that another class of foldases, known as chaperonins, do indeed aid in the production and assembly of multimeric proteins (Goloubinoff et al, 1989a). Chaperonins are a ubiquitous class of proteins that seem to facilitate the folding of other polypeptides and assist in the assembly of oligomeric structures of which they are not a part (Hemmingsen and Ellis, 1989). It has been further proposed that chaperonins prevent the formation of incorrect structures and unscramble those that do occur. Goloubinoff et al. (1989a) have shown that over-expressing the E. coli GroEL chaperonin proteins at the same time as expressing the A. nidulans ribulose-1,5-bisphosphate carboxylase/oxygenase (RuBisCo) genes in E. coli led to a five- to ten-fold increase in the level of expression of the RuBisCo gene products. In addition, the GroEL products were shown to be necessary for the formation of active dimers and correctly assembled octamers in E. coli. In a later paper, Goloubinoff et al. (1989b) have also shown that chaperonins from a number of sources, including E. coli, yeast, and chloroplasts, can be used in vitro to facilitate the reconstitution of active RuBisCo from unfolded polypeptides. While this is one of the few cases where a protein-mediated refolding has been accomplished, numerous successes have been reported in the last two years in the area of chemically mediated refolding.

In summary, while E. coli has been used for expression of heterologous gene products since the beginning of the modern molecular biology era, significant advances are still being made in our abilities to optimize the quantity and quality of the proteins produced in this system. We can fully expect such advances to continue in the future accompanied by a growing application of these advances to gene expression in other gram negative bacteria.

References

Brosius, J. In *Vectors,* Rodriguez and Denhardt Ed; Butterworths Publishers, Stoneham, MA, **1988**, pp 205-226.

Devlin, P.; Drummond, R.; Toy, P.; Mark, D.; Watt, K.; Devlin, J. *Gene* **1988**, 65:13-22.

Goloubinoff, P.; Christeller, J.; Gatenby, A.A.; Lorimer, G.H. *Nature* **1989**, 342:884-889.

Goloubinoff, P.; Gatenby, A.A.; Lorimer, G.H. *Nature* **1989**, 337:44-47.

Hemmingsen, S.M.; Ellis, R.J. *Trends Biochem Sci* **1989**, 14:339-342.

Hsiung, H.M.; MacKellar, W.C. *Methods Enzymol* **1987**, 153:390-401.

Sathe, G.; Gross, M.; Watson, F.; Lee, J. *FASEB* **1990**, 4:A1933.

Schauder, B. and McCarthy, J. *Gene* **1989**, 78:59-72.

Schein, C. *Bio/Technology* **1989**, 7:1141-1149.

Schein, C. and Notebom, M. *Bio/Technology* **1988**, 6:291-294.

Seow, H.F.; Goh, C.R.; Krishnah, I.; Porter, A. *Bio/Technology* **1989**, 7:363-368.

Shatzman, A.R. and Rosenberg, M. *Methods in Enzymology* **1987**, 252:661-673.

Sutiphong, J.; Johansen, H.; Sathe, G.; Rosenberg, M.; Shatzman, A. *Molecular Biology and Medicine* **1987**, 4:307-322.

Takagi, H.; Morinaga, Y.; Tsuchya, M.; Ikemura, H.; Inouye, M. **1990**, in press.

High Level Expression of Proteins in *Pseudomonas* sp.

Taro Iizumi*, Koichi Nakamura, Kyoichiro Sanda, and **Tetsuro Fukase,**
Kurita Water Industries Ltd, 7-1, Wakamiya, Morinosato, Atsugi 243-01 JAPAN
(FAX: 0462 48 2577)

Pseudomonas sp. KWI-56 produces an extracellular lipase with the addition of oils. A 2,916-bp DNA fragment cloned from this bacterium, containing lipase-encoding gene, its activator-encoding gene, and the promoter region, was ligated with a multiple-copy plasmid, and was introduced into the original *Pseudomonas* strain. The productivity of lipase by the recombinant increased in proportion to the gene dosage, resulting that the amount of lipase protein reached 18 g/l in a fed-batch culture. We attempted to apply the host-vector system to allow the heterologous protein to be expressed. As an example, α-amylase derived from *Bacillus subtilis* was expressed in the *Pseudomonas*, and most of the activity was detected in the culture supernatant.

There is considerable interest in developing a protein secretion system by microorganism. Secretory production is better than intracellular production in terms of the yield, purification, formation of protein molecules, and the avoidance of proteolysis.

Genus *Bacillus*, a type of typical gram-positive bacteria, is a well-known enzyme producer. These bacteria produce a large amount of enzyme in the medium. An industrial mutant is said to be capable of producing 5 to 10 g/l of amylase or protease. In addition, the available genetic information is sufficient for industrial application. As a result, attention has recently been focused on the production of foreign gene products.

In contrast to *Bacillus*, gram-negative bacteria have received little attention for protein secretion due to the facts that: (i) these bacteria have a two-layer membrane structure that consists of an inner and an outer membrane, which means that most secret protein accumulates in the periplasmic space; (ii) some types of proteins can pass through the outer membrane and be secreted into the medium by specific secretion functions, which require additional proteins, however most of these mechanisms are unclear; (iii) genetic information regarding promoter structures and regulation systems is limited except for information about *Escherichia coli.*

Genus *Pseudomonas* is known to be an extracellular lipase (triacylglycerol acylhydrolase; EC 3.1.1.3) producer. There have been many studies on the application of lipase, such as ester synthesis, trans-esterification, and the production of fatty acids because these lipases are usually stable in high temperatures and in the presence of organic solvents. However, the enzyme has not been widely used industrially, mainly because of its low productivity. We report here on the hyper-production of lipase from *Pseudomonas* sp. by application of a genetic engineering technique and on the possibility of producing foreign gene products in this bacterium.

Cloning of Lipase- and Its Activator-Encoding Genes

We isolated one *Pseudomonas* strain, which produces a thermostable lipase in the medium with oils and fatty acids, and is named *Pseudomonas* sp. KWI-56. Its taxonomical characteristics showed it to be similar to *Pseudomonas cepacia* (Iizumi et al., 1990).

A 2,916-bp *Bgl*II-*Eco*RI DNA fragment from the chromosomal DNA of KWI-56 was cloned into *Escherichia coli*, and was ligated with a *Bam*HI-*Eco*RI-digested pUC19 plasmid vector. The resulting recombinant plasmid was named pLP64 (Iizumi et al., 1991). The restriction map of this DNA fragment is shown in Fig. 1.

1054–7487/92/0005$06.00/0

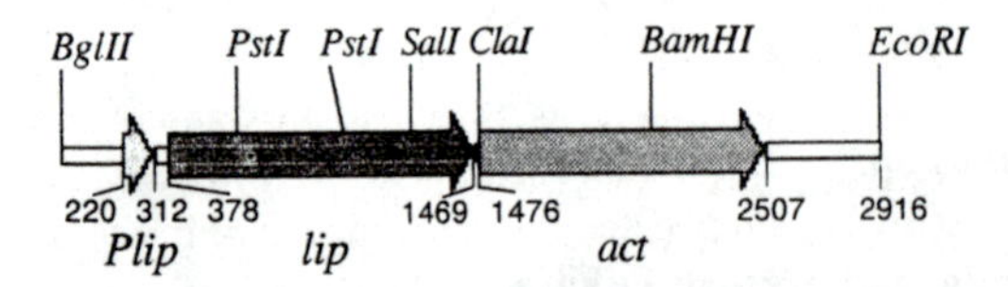

Fig. 1. Restriction map of a 2,916-bp *Bgl*II-*Eco*RI DNA fragment from *Pseudomonas* sp. KWI-56 chromosomal DNA. The DNA fragment contains the lipase-encoding gene (*lip*), its activater-encoding gene (*act*), and the promoter region (*Plip*).

The lipase was expressed in *E. coli* carrying pLP64 under the control of the *lac* promoter (*Plac*). Following cell fractionation by osmotic shock, about 95% of lipase activity could be found in the cytoplasmic fraction, while 5% was in the periplasmic fraction. Very low activity was detected in the supernatant. A Western blot analysis indicated that the lipase protein from the cytoplasm appeared in two molecular forms (33-kD and 38-kD), although the molecular weight of the periplasmic lipase was 33-kD as a single band. The molecular weight of 33-kD matched that of lipase purified from a culture supernatant of KWI-56 (Iizumi et al., 1991). These results suggest that the cloned lipase is produced as a 38-kD pro-enzyme, and cleaved by signal peptidase to be secreted into the periplasm or on the cell membrane as a 33-kD mature enzyme.

The lipase-encoding gene (*lip*) and its activator-encoding gene (*act*) which activate the lipase in *E. coli*, were identified from the sequencing experiment of the 2,916-bp DNA fragment and complementation tests (Iizumi et al., 1991). The open reading frames (ORF) of *lip* and *act* were found to contain 1,092 and 1,032 nucleotides, respectively. The ORF of *act* existed at 7-bp downstream of *lip* with the same orientation. The amino acid sequence, deduced from the nucleotide sequence of *lip* and N-terminal amino acid sequence of purified lipase, indicated that the lipase protein (pro-enzyme) consisted of 320 residues of mature protein and 44 residues of a signal peptide. The amino acid sequence deduced from *act* contained a putative signal sequence of 34 residues, suggesting that the activator seemed to be a secret protein consisting of 310 residues.

The Role of *act* in *Escherichia coli*

The *act* plays an important role in the expression of lipase activity *in trans*. When *lip* was expressed under the control of *Plac*, the lipase activity of *E. coli* carrying both *lip* and *act* was about 200 times higher than that of *E. coli* carrying only *lip*. This finding suggests that this activation process does not depend on the transcriptional signal (Iizumi et al., 1991). A similar result was reported by Jorgensen et al (1991). In their study, they cloned *lipA* and *limA* genes from *P. cepacia* DSM3959, which correspond to *lip* and *act*, respectively. In spite of the replacement of the *limA* expression signal (promoter, SD sequence, and signal peptide-encoding sequence) by a heterologous signal, the expression of *lipA* required a *limA* function in heterologous organisms. Therefore, *limA* does not affect either the transcriptional or the translational signals of *lipA*.

The results of Western blot experiments indicated that *E. coli* carrying only *lip* produced as much lipase protein as that carrying both *lip* and *act*. The lipase protein from the former also appeared in two molecular forms (33-kD and 38-kD) (unpublished data). These results suggest that the activation of lipase is a post translational event, and is proceeded by either maturation of the precursor or interaction with a cofactor. However, the lipase does not require a cofactor for the expression of the activity (Iizumi et al., 1991). Accordingly, a precursor of lipase, such as an inactive zymogen, is believed to become an active lipase by the function of *act*. We assume that *act* affects the processing or folding of lipase molecules in *E. coli*.

Introduction and Expression of Lipase Gene in *Pseudomonas*

The 2,916-bp DNA fragment containing *lip* and *act* was ligated with pSUP104 (Priefer et al., 1985), which contains a wide host range replication origin of RSF1010 (Guerry et al., 1974). The resulting plasmid, pSUP-lip64, was introduced into KWI-56M1, a high yielding mutant strain, by mobilization using *E. coli* S17-1. The plasmid was very stable in KWI-56M1 with a copy number of 15 to 20 per cell.

The lipase productivity of the recombinant strain in the flask culture was about 20-fold greater than that of KWI-56M1 in proportion to the gene dosage (Iizumi et al., 1992). The expression of *lip* on the plasmid was found to be controlled by the own promoter (*Plip*) existing upstream of *lip* (Fig. 1), which was proved by deletion analysis of 5'-flanking region of *lip* (unpublished data). However, KWI-56M1[pSUP-lip64] produced a small amount of lipase without oils, indicating that the expression of cloned lipase gene was also regulated by oils.

On the other hand, KWI-56M1 carrying pSUP-lip65, lacking *act* on the plasmid, indicated one-seventh the amount of lipase activity compared with that carrying pSUP-lip64. By a Western blot analysis of KWI-56M1[pSUP-lip65] intracellular proteins, a significant amount of lipase molecules were found to accumulate in the cell, probably in the periplasm. The accumulated molecules had been partly degraded (Iizumi et al., 1992). These results suggest that *act* is required for the secretion of lipase across the outer membrane in KWI-56M1, as well as for the activation of lipase molecules.

Generally, the translocation across the outer membrane or the expression of activity of extracellular proteins from gram-negative bacteria requires additional proteins (Pugsley, 1988; Holland et al., 1990). Many of these additional genes are known to locate near the protein-encoding gene. The *act* is assumed to be one of the additional gene. However, the secretion and activation mechanisms of lipase by the functioning of *act* have not yet been completely clarified.

High Level Production of Lipase by Fed-Batch Culture

Fed-batch culture in a 2 l jar fermenter was carried out using the basal medium (2% peptone, 0.3% yeast extract, 0.3% KH_2PO_4, 0.15% $MgSO_4 \cdot 7H_2O$) with oleic acid as a carbon source and an inducer at a constant feeding rate. As shown in Table 1, the lipase activities of KWI-56M1, KWI-56M1[pSUP-lip64], KWI56-M1[pSUP-lip642] reached 2,600, 40,100, and 90,000 U/ml, respectively (Iizumi et al., 1992). The plasmid, pSUP-lip642, used in this experiment contained two copies of the 2,916-bp DNA fragment in tandem. The productivity of KWI-56M1[pSUP-lip642] was about 35-fold greater than that of KWI-56M1. The amount of lipase protein in the KWI-56M1[pSUP-lip642] supernatant was 18 g/l calculated on the basis of the specific activity of the purified lipase (5,000 U/mg protein). The amount of lipase protein was

Table 1. Results of fed-batch cultures

Strain	Incubation Time (h)	Dry Cell Weight (g/l)	Lipase Activity (U/ml)	Concentration of Lipase Protein (g/l)
KWI-56M1	168	24.8	2,600	0.5
KWI-56M1[pSUP-lip64]	117	27.1	40,100	8.0
KWI-56M1[pSUP-lip642]	147	28.1	90,000	18.0

Fed-batch culture was performed using a basal medium (2% peptone, 0.3% yeast extract, 0.3% KH_2PO_4, and 0.15% $MgSO_4 \cdot 7H_2O$) in a 2-liter jar fermenter with a working volume of 1.2 l. Oleic acid was supplied at constant feed rates of 0.30 and 0.45 g/l·h for the plasmid-free and recombinant strains, respectively. Ammonia solution of 14% (w/v) was used for supplying a nitrogen source and controling pH. The lipase activity in the supernatant and the dry cell weight were measured over the course of the study. The maximum lipase activity is described in the above table. The concentration of lipase protein was calculated on the basis of the specific activity of the purified lipase protein (5,000 U/mg protein).

more than that of the total intracellular protein, assuming that about 50% of the dry cell weight (28.1 g-cell/l) was protein.

Such a high level production of lipase may depend on the following characteristics of this bacteria: (i) the RSF1010-derived plasmid is very stable in this strain; (ii) the expression signal of *lip* (promoter, SD-sequence, and signal peptide-encoding sequence) is very effective in KWI-56; (iii) this strain has a high protein secretion ability.

Pseudomonas sp. KWI-56 has a high potential in terms of secretory production. Our study indicates that much attention must be given to the secretory production of protein by gram-negative bacteria.

Expression of the α-Amylase-Encoding Gene from *Bacillus subtilis* in *Pseudomonas*

We attempted to produce heterologous proteins utilizing the high level productivity of protein by the *lip* expression system.

An expression vector containing *lip* expression signal (promoter, SD sequence, and signal peptide-encoding sequence), *act*, chloramphenicol acetyl transferase-encoding gene, and replication functions of RSF1010, was constructed and named pPEX50. A DNA fragment coding pro-region and mature region of α-amylase-encoding gene (*amyE*) was cloned from *Bacillus subtilis* chromosomal DNA (Yang et al., 1983), and was ligated downstream of the signal peptide-encoding region of pPEX50 to form an 'in frame' gene fusion. The resulting plasmid, pPEX1011, was introduced into KWI-56M1 by mobilization using *E. coli* S17-1.

When the recombinant strain was cultured in a PY medium (1% peptone, 0.1% yeast extract, 0.1% KH_2PO_4, 0.05% $MgSO_4 \cdot 7H_2O$, pH 6.5) containing 1% olive oil, *amyE* was successfully expressed (Fig. 2). After 72h of incubation, 90% of the total activity was detected in the medium without significant cell lysis. The secretion of amylase into the medium followed after its accumulation in the cell. In this case, lipase derived from the chromosome was also secreted into the medium, but without accumulation, indicating that the secretion pattern of amylase is a little different from that of lipase.

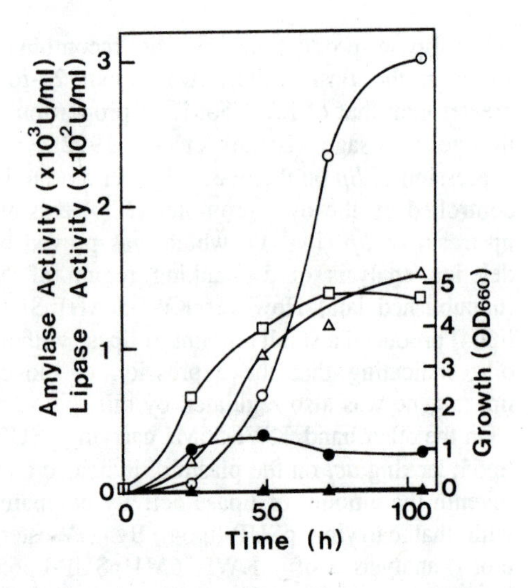

Fig. 2. Time course of α-amylase and lipase activities by *Pseudomonas* sp. KWI-56M1 [pPEX1011]. Symbols: ○, extracellular amylase activity; ●, intracellular amylase activity; △, extracellular lipase activity; ▲, intracellular lipase activity; □, bacterial growth.

We found that this system can be utilized to produce heterologous protein in the medium. However, the secretion system and the role of *act* in the production of the amylase have yet to be clarified. Further studies are needed to elucidate the secretion mechanisms of extracellular protein in this bacterium, and other attempts to express other heterologous genes should be made.

Acknowledgments

We thank Drs. Yuji Shimada, Akio Sugihara and Yoshio Tominaga of Osaka Municipal Technical Research Institute for DNA sequencing analysis, Prof. Hisao Ohtake of Hiroshima University for the expression of lipase in *Pseudomonas*, and Prof. Tadayuki Imanaka of Osaka University for cloning the *amyE*.

References

Guerry, P.; van Embdan, J.; Falkow, S. *J. Bacteriol.* **1974**, 117, 619-630.

Holland, B.; Blight, M. A.; Kenny, B. *J. Bioenergetics Biomembranes.* **1990**, 22, 473-491.

Iizumi, T.; Nakamura, K.; Fukase, T. *Agric. Biol. Chem.* **1990**, 54, 1253-1258.

Iizumi, T.; Nakamura, K.; Shimada,Y.; Sugihara, A.; Tominaga, Y.; Fukase, T. *Agric. Biol. Chem.* **1991**, 55, 2349-2357.

Iizumi, T.; Nakamura, K.; Ohtake, H.; Fukase, T. **1992**, *to be published.*

Jorgensen, S.; Skov, K. W.; Diderichsen, B. *J. Bacteriol.* **1991**, 173, 559-567.

Priefer, U. B.; Simon, R.; Puhler, A. *J. Bacteriol.* **1985**, 163, 324-330.

Pugsley, A. P. *Protein Transfer and Organelle Biogenesis;* Academic Press: New York, N.Y. **1988**; pp 607-652.

Yang, M.; Galizzi, A.; Henner, D. *Nucleic Acids Res.* **1983**, 11, 237-249.

Mono- and Bivalent Antibody Fragments Produced in *E. coli*: Binding Properties and Folding *in vivo*.

Peter Pack, Achim Knappik, Claus Krebber and Andreas Plückthun*,
Max-Planck-Institut für Biochemie, Am Klopferspitz, D-8033 Martinsried, FRG
(FAX +49 - 89 - 8589 3795)

We have designed dimeric antibody fragments of minimal size that assemble in *E. coli* and show an increase in avidity approaching a whole antibody. They are based on single-chain F_V fragments with a flexible hinge region from mouse IgG3 and an amphiphilic helix fused to the C-terminus of the antibody fragment. The sequence of the helix was taken either from that of a four-helix bundle design or a leucine zipper, optionally extended with a short cysteine containing peptide. To investigate the folding and assembly process of antibody fragments in *E. coli*, co-expression experiments with proline cis-trans-isomerase and disulfide isomerase were carried out. These folding steps do not appear to be limiting the folding process in *E. coli*.

Recent advances in the bacterial expression technology of antibodies (Plückthun, 1991), PCR cloning of antibody libraries (Orlandi et al., 1989) and library screening with the phage display technology (Marks et al., 1991) are decisive steps along a path to antibodies based almost entirely on molecular biology and biotechnology. This account will address the question of which antibody fragments are compatible with functional expression in *E. coli*, and by what folding mechanism they achieve the native state.

The bacterial expression of antibodies in the native state was first developed for the F_V fragment of the antibody (the non-covalent heterodimer of the variable domains V_L and V_H; Skerra and Plückthun, 1988) and the F_{ab} fragment (the first two domains of the heavy chain V_HC_H and the complete light chain V_LC_L; Better et al., 1988). In this approach, the principles of antibody secretion and folding in the eukaryotic cell are partially mimicked by secreting both chains into the periplasm of the same *E. coli* cell. Therefore, each of the domains can act as a folding template for the other, and the disulfide bonds can form. Without the disulfide bonds in the variable domains, no functional product can be obtained in *E. coli* in any antigen binding fragment tested (Glockshuber et al., 1992, Skerra and Plückthun, 1991).

In order to increase the stability of the F_V fragment, the V_H and V_L domain can be linked covalently *in vivo* by two strategies, which are compatible with functional expression and secretion: (i) by engineering of an intermolecular disulfide bond (Glockshuber et al., 1990) and (ii) by connecting the two domains by a genetically encoded peptide linker to give a so-called single-chain Fv fragment (scF_v) (Glockshuber et al., 1990; Huston et al. 1991). In the native expression approach, both possible designs (V_H-linker-V_L and V_L-linker V_H) have been used. Conflicting results have been obtained about the relative yields of the two designs when used on the same antibody. While Anand et al. (1991) observed dramatic differences, Knappik and Plückthun (unpublished results) found the yields to be very similar.

From Monovalent to Bivalent Fragments

All the antibody fragments so far reported as expressible in the native state in *E. coli* have been monovalent. While F_{ab}' fragments produced in *E. coli* have been linked chemically to $(F_{ab}')_2$ fragments after purification (Carter et al., 1992), we wished to investigate the question, whether an *in vivo* dimerization of antibody fragments is possible. Furthermore, our goal was to make these bivalent fragments as small as possible while achieving a maximal increase in avidity (Crothers and Metzger, 1972).

For this reason, we designed dimerizing single-chain F_V fragments, which we call miniantibodies (Fig. 1). As a model system, the phosphorylcholine antibody McPC603

(Perlmutter et al., 1984) was used. Two alternative molecular designs have been tested (Pack and Plückthun, 1992). In the first, the single chain fragment (V_H-linker-V_L) was fused to the flexible upper hinge region of mouse IgG3. This was followed by one helix taken from the the 4-helix bundle designed by Eisenberg et al. (1986). This helix was either taken as such (construct scHLX) or extended by a small hydrophilic peptide ending in a cysteine residue (scHLXc) in order to covalently link two helices. Ultracentrifugation measurements are consistent with a dimer formation *in vivo*. From the arrangement of the charged residues on the helix, an antiparallel association is anticipated.

In the other molecular design, instead of the helix from the 4-helix-bundle design, a helix from a parallel coiled-coil structure was used (O'Shea et al., 1991). Specifically, we fused the leucine zipper peptide from the yeast transcription factor GCN4 to the scF_V fragment. Again, the helix was either taken as such (scZIP) or extended with a short peptide ending in a cysteine (scZIPc).

The functionality of the fragments was investigated in several different ways. All miniantibodies can be purified by hapten affinity chromatography, illustrating that the antigen binding site forms correctly in *E. coli*. The covalently linked miniantibodies show dimer-bands on a non-reducing PAGE. In ultracentrifugation measurements, all non-covalently linked and covalently linked miniantibodies give evidence of molecular weights compatible with dimers. Only in the case of the non-covalent antiparallel helices, there is a very small amount of faster sedimenting material, perhaps consisting of tetramers. Probably the non-covalent 4-helix-bundle is not stable enough to allow persistent tetramer formation under these conditions. The linking peptide may prevent

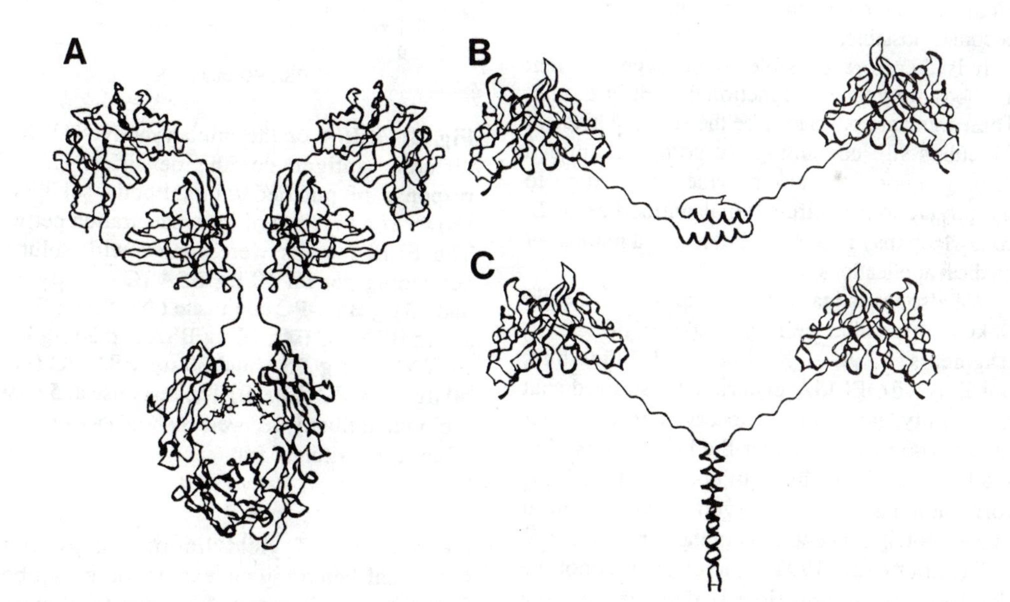

Fig. 1: (**A**) Molecular model of the human antibody KOL (**B**): Molecular model of the dimeric scHLXc miniantibody constructs derived from the single-chain F_V fragment of the mouse antibody McPC603 (Satow et al., 1986). The hinge region was modelled according to a polyproline-II helix with $\phi = -78°$ and $\psi = 149°$. A standard α-helix with $\phi = -57°$ and $\psi = -47°$ was used for the amphiphilic helix. The cys-tail peptide is presumably disordered, and the structure drawn should be taken only as a guide to the topology. (**C**): Molecular model of the scZIPc miniantibody construct. The parameters are identical as in A, except that a parallel coiled coil with about a quarter turn of the superhelix was modeled for the leucine zipper part (O'Shea et al., 1991).

tetramer formation in the case of the covalently linked variant, so that dimers are the predominant species in both cases.

Most importantly, however, all the bivalent miniantibodies show the desired gain in "avidity", an empirical measure of the increased apparent binding constant to polymeric or surface-bound antigen. The best performance was observed for the covalently linked antiparallel helix construct, whose binding properties are almost identical to those of the complete IgA, although the bivalent miniantibody has the MW of only one F_{ab} fragment. The observed differences in avidity between the different miniantibodies (Pack and Plückthun, 1992) may be due to different binding geometries between the coiled-coil and antiparallel helix structure and/or different monomer-dimer equilibria in the non-covalent forms.

The surface binding was also examined by varying the antigen density in a functional ELISA (Fig. 2), clearly showing a non-linear increase with coating density. This suggests that the avidity gain is indeed obtained when multivalent binding of the miniantibodies to the same surface becomes possible.

It is therefore possible to produce bivalent antibody fragments in functional form in *E. coli*. These miniantibodies may be the smallest bivalent structures still containing the complete antigen binding region. They also provide a direct route to the production of bifunctional antibodies in *E. coli*, which may become of interest in a number of medical applications.

We also investigated the use of the covalently linked antiparallel helix to make $(F_{ab}\text{-helix})_2$ fragments *in vivo* (Krebber and Plückthun, unpublished). ELISA experiments showed that the avidity gain can be observed in a similar fashion, no matter whether the helix is linked to the light or heavy chain. In contrast, no $(F_{ab}')_2$ formation has been observed *in vivo* even at higher protein concentration in the periplasm of *E. coli* (Carter et al., 1992). Therefore, it is not the disulfide formation itself that causes the dimerization of the fragments. Rather, a non-covalent interaction of sufficient lifetime must first occur, which can merely be made permanent by a nearby S-S bond.

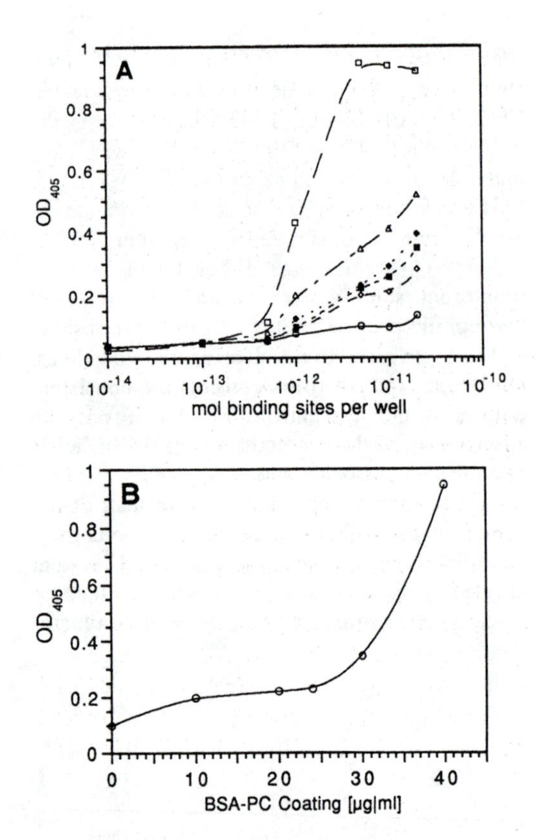

Fig. 2: ELISA of the miniantibody scHLXc at different antigen coating densities: **(A)** The response of purified miniantibody scHLXc is shown as a function of its concentration per well. The ELISA wells were coated with solutions containing per ml 40 µg BSA (O), 30 µg BSA and 10 µg BSA-PC conjugate (◇), 20 µg PC and 20 µg BSA-PC (■), 16 µg BSA and 24 µg BSA-PC (◆), 10 µg BSA and 30 µg BSA-PC (Δ) or 40 µg BSA-PC (□). **(B)** The response at $5 \cdot 10^{-12}$ mol miniantibody per well is shown as a function of antigen coating.

Folding *in vivo*

The antibodies are an ideal model system to investigate the problem of *in vivo* folding. While the question of yield limiting steps in the functional heterologous expression of antibody fragments is still unanswered, investigations on a mouse F_{ab} fragment (Skerra and Plückthun, 1991) strongly suggest that the periplasmic folding and/or assembly may be the limiting factor. Furthermore, the similarity in yields from Fv and scFv fragments is consistent with the idea that assembly is not a limiting process (Glockshuber et al., 1990). Different antibodies or those from different species may be limited at different levels (Carter et al., 1992).

It was therefore of interest to investigate whether folding catalysis in the periplasm may overcome this block. In a first approach, the periplasmic *E. coli* proline cis-trans-isomerase (Liu and Walsh, 1990) was overexpressed together with the antibody F_{ab}, F_V, and both orientations of the scF_V fragment (V_H-linker-V_L and V_L-linker-V_H) (Knappik, Walsh and Plückthun, unpublished data). As the F_{ab} fragment, whose 3-D structure is known (Satow et al., 1986), contains 5 cis-proline peptide bonds, a folding block due to the slow isomerization of any of these bonds is conceivable. However, in all cases, the effects were rather small, suggesting that proline cis-trans isomerization is not the step, at which the folding pathway diverts from its desired path.

Additionally, the recently discovered protein disulfide isomerase from *E. coli* (Bardwell et al., 1991) was co-expressed with the F_{ab} fragment, which contains a total of 5 disulfide bonds (Krebber, Beckwith and Plückthun, unpublished experiments). Again the effect was only marginal, although the functionality of the overexpressed isomerases could be clearly demonstrated. Similar results were obtained with human PDI (Pihlajaniemi et al., 1987) (Krebber, Skerra and Plückthun, unpublished data). This shows that the diversion of the folding pathway from the desired path does not occur at the step of disulfide formation or rearrangement, consistent with earlier mutant analysis (Skerra and Plückthun, 1991).

This approach of overexpressing potential folding modulators may be useful in pin-pointing bottle-necks during *in vivo* folding, and together with mutant analysis, perhaps lead to an increased understanding and handles for manipulating protein folding in the bacterial cell.

References

Anand, N. N.; Mandal, S.; MacKenzie, C. R.; Sadowska, J.; Sigurskjold, B.; Young, N. M.; Narang, S. A. *J. Biol. Chem.* **1991**, *266*, 21874-21879.

Bardwell, J. C. A.; McGovern, K.; Beckwith, J. *Cell* **1991**, *67*, 581-589.

Better, M.; Chang, C. P.; Robinson, R. R.; Horwitz, A. H. *Science* **1988**, *240*, 1041-1043.

Carter, P.; Kelley, R. F.; Rodrigues, M. L.; Snedcor, B.; Covarrubias, M.; Velligan, M. D.; Wong, W. L. T.; Rowland, A. M.; Kotts, C. E.; Carver, M. E.; Yang, M.; Bourell, J. H.; Shepard, H. M.; Henner, D. *Biotechnology* **1992**, *10*, 163-167.

Crothers, D. M.; Metzger, H. *Immunochemistry* **1972**, *9*, 341-357.

Eisenberg, D.; Wilcox, W.; Eshita, S. M.; Pryciak, P. M.; Ho, S. P.; DeGrado, W. P. *Proteins* **1986**, *1*, 16-22.

Glockshuber, R.; Malia, M.; Pfitzinger, I.; Pückthun, A. *Biochemistry* **1990**, *29*, 1362-1367.

Glockshuber, R.; Schmidt, T. Plückthun, A. *Biochemistry* **1992**, *31*, 1270-1279.

Huston, J. S; Mudgett-Hunter, M.; Tai, M. S.; McCartney, J.; Warren, F.; Haber, E.; Opperman, H. *Meth. Enzymol.* **1991**, *203*, 46-88.

Liu, C. T.; Walsh, C. *Proc. Natl. Acad. Sci. USA* **1990**, *87*, 4028-4032.

Marks, J. D.; Hoogenboom, H. R.; Bonnert, T. P.; McCafferty, J.; Griffiths, A.D.; Winter, G. *J. Mol. Biol.* **1991**, *222*, 581-597.

O'Shea, E. K.; Klemm, J. D.; Kim, P. S.; Alber, T. *Science* **1991**, *254*, 539-544.

Orlandi, R.; Güssow, D. H.; Jones, P. T.; Winter, G. *Proc. Natl Acad. Sci. USA* **1989**, *86*, 3833-3837.

Pack, P.; Plückthun, A. *Biochemistry* **1992**, *31*, 1579-1585.

Perlmutter, R. M.; Crews, S. T.; Douglas, R.; Sorensen, G.; Johnson, N.; Nivera, N.; Gearhart, P. J.; Hood, L. *Adv. Immunol.* **1984**, *35*, 1-37.

Pihlajaniemi, T.; Helaakoski, T.; Tasanen, K.; Myllylä, R., Huhtala, M. L.; Koivu, J.; Kivirikko, K. I. *EMBO J.* **1987**, *6*, 643-649.

Plückthun, A. *Biotechnology* **1991**, *9*, 545-551.

Satow, Y.; Cohen, G. H.; Padlan, E. A.; Davies, D. R. *J. Mol. Biol.* **1986**, *190*, 593-604.

Skerra, A.; Plückthun, A. *Science* **1988**, *240*, 1038-1040.

Skerra, A.; Plückthun, A. *Protein Eng.* **1991**, *4*, 971-979.

A Novel Approach For Efficient E.coli Production of Human IGF-I

Judy Chang and James Swartz, Genentech, Inc., 460 Pt. San Bruno Blvd., South San Francisco, CA 94080.

A significant amount of human insulin-like growth factor (IGF-I) peptide can be translocated into the E. coli periplasmic space. However, more than 90% of the translocated peptide accumulates as an insoluble aggregate. When the aggregate is isolated by cell lysis and centrifugation, it can be solubilized by reduction under mild chaotropic conditions. Simultaneous air oxidation results in a significant yield of properly folded, bioactive IGF-I. This provides a simple, yet effective method for the production of authentic human IGF-I. Yields were increased by optimizing urea and DTT concentrations, pH and the folding buffer. Oxidation, solubilization and folding kinetics suggest a sequential reaction series with substantial disulfide exchange. Results with higher chaotrope concentrations also suggest that the IGF-I peptide trapped in the aggregate may contain structural information which facilitates its continued folding at low chaotrope concentrations.

Human insulin-like growth factor (IGF-I) is a 70 amino acid long protein which plays an important role in the growth hormone cascade. It has formerly been called somatomedin-C because of its central role in growth regulation. It is also the only protein other than insulin known to have the ability to lower circulating blood glucose. Additionally, recent animal testing suggests that it may assist in stimulating the immune system. Thus, IGF-1 offers exciting potential as a pharmaceutical and is being tested for use in treating anabolic disorders, diabetes, and AIDS.

However, it now appears that these markets will require relatively low selling prices and high doses. Thus, the need to develop a low cost production process has prompted the exploration of E.coli as a production host. A secretion-based process offers the advantages of producing a protein without an N-terminal methionine and of facilitating protein purification. Although the IGF-1 peptide is secreted and accumulates to approximately 10% of the total cellular protein, more than 90% of the secreted IGF-1 peptide accumulates in the periplasmic space as large insoluble aggregates. These are clearly visible under a light microscope.

In exploring the nature of these IGF-1 peptide aggregates , it was discovered that relatively mild treatment not only leads to solubilization of the aggregate, but also results in significant folding of the peptide to the bioactive, natural form. This presentation describes the work that led to this discovery as well as further work to optimize and characterize the solubilization/folding process.

Isolation and Initial Characterization of IGF-I Aggregates

Because the accumulated IGF-1 peptide takes the form of large, relatively dense aggregates, it can be isolated by cell homogenization and differential centrifugation. E.coli cells grown in 10-liter fermentors and induced for IGF-I expression and secretion are harvested by centrifugation. They are then suspended in a Tris buffer containing EDTA and lysozyme and are sonicated until no intact cells are visible under a phase contrast light microscope. The resulting suspension is then lightly centrifuged to pellet the IGF-1 peptide aggregates while leaving the remainder of the cell components in suspension.

Analysis by coomassie blue stained polyacrylamide gel electrophoresis (PAGE) showed nearly quantitative recovery of of the aggregated IGF-I in the centrifuged pellet. Furthermore, very little IGF-I could be detected in the supernatant and the coomassie blue stained gel showed very little non-IGF-I protein in the pellet fraction. This preparation thus provided an excellent starting material to study solubilization of the IGF-I aggregate.

Results from initial attempts to solubilize the aggregates are shown in Table 1.

Table 1

Chaotrope	DTT (mM)	Solubility	IGF-I Activity
none	0	-	0.05
none	10	-	0.03
2M Urea	0	-	0.06
4M Urea	0	-	0.08
6M Urea	0	-	0.10
8M Urea	0	-	0.11
8M Urea	10	++++	1.11
2M GuCl	0	-	0.14
4M GuCl	0	-	0.12
6M GuCl	0	-	0.10
6M GuCl	10	++++	0.08

1054–7487/92/0014$06.00/0

High concentrations of strong chaotropes resulted in no detectable solubilization as judged by recentrifuging the aggregate-containing solutions and analyzing the supernatants on PAGE gels. By itself, the reducing agent dithiothreitol (DTT) at a 10 mM concentration also produced no significant solubilization. The combination of the two, however, resulted in complete solubilization. The supernatants were tested for IGF-I using a radio-immuno assay developed to detect only properly folded IGF-1. Surprisingly, for the 8M Urea, 10 mM DTT case, the RIA assay detected a small, but significant level of IGF-1 as compared to background levels. No significant activity was detected for the 6M guanidineHCl (GuCl), 10 mM DTT case.

These results provided convincing evidence that the periplasmic IGF-I aggregates were strongly cross-linked by intermolecular disulfide bonds. The results also suggested the possibility that properly folded IGF-1 could be obtained from the aggregates.

Initial Folding Studies and Product Characterization

In order to investigate the forces leading to and stabilizing the IGF-I peptide aggregates, further studies were undertaken to determine the minimum conditions required to solubilize the periplasmic aggregates (Bowden et al., 1991). Simultaneously, IGF-I folding was evaluated. Most solubilization and folding schemes for aggregated recombinant proteins involve high denaturant conditions and low protein concentrations (Cleland and Wang, 1990a; Marston, 1986). These approaches appeared to be too expensive for IGF-I. Thus, it was also important to test if IGF-I folding could be achieved with lower concentrations of chaotropic agent and at significant protein concentrations. For this and subsequent studies, a concentration of approximately 1.5 mg/ml IGF-1 peptide was used.

The aggregate was placed into 100 µl of solutions varying in urea and DTT concentration. The samples were the allowed to incubate with occasional vortexing at room temperature and in the presence of air. Fig. 1 presents the results expressed as the percentage of peptide that had become recognizable by the RIA antibody. As might be expected, DTT and urea concentrations have a profound effect on the efficiency of solubilization and folding of the IGF-I peptide aggregate. Optimal concentrations are suggested to be 2 M urea and 2 mM DTT. Surprisingly, these conditions resulted in complete solubilization of the aggregates.

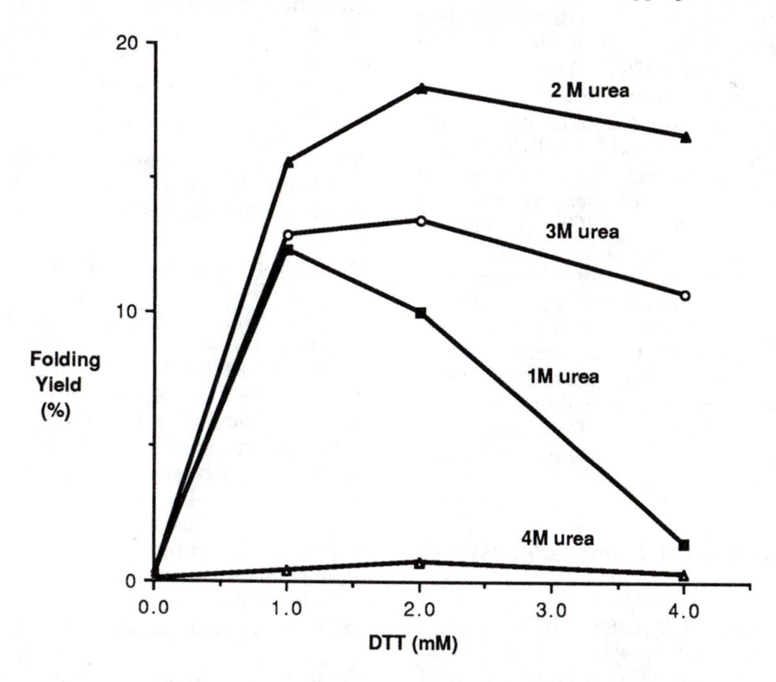

Figure 1. Initial results of the influence of urea and DTT concentration on the yield of RIA-recognizable IGF-1. These data were obtained in a 100 mM acetate buffer with 100 mM NaCl and an approximate pH of 8.5 after a three hour incubation at room temperature in the presence of air.

To determine if this one step procedure was indeed producing properly folded IGF-1, an HPLC procedure was used to analyze the product mixture. The resulting chromatogram indicated a major peak which eluted at the same retention time as authentic IGF-1. The peak area agreed well with that expected based on the RIA results. The peak was collected and further analyzed by determining the HPLC profile of a V-8 protease digest and by determining its bioactivity. Both analyses indicated that the collected peak was authentic, properly folded, and bioactive human IGF-1.

Effect of pH, Temperature, and Solvent

The necessity of using a reducing agent for solubilization and folding of the IGF-I peptide aggregate suggested the involvement of disulfide exchange. This in turn prompted an investigation of the effect of pH. As might be expected, pH has a strong effect on both the solubilization and the folding of the aggregate. Surprisingly, there were also specific buffer effects in addition to pH effects. For example, at pH 9.5, the folding yield was more than twice as great with a Ches buffer (pKa = 9.3) as with a glycylglycine buffer (pKa = 8.4). A Capso buffer (pKa = 9.6) appeared to be the most beneficial for IGF-I folding and produced the best yield at pH 10.5.

Temperature has a much less dramatic effect with virtually no effect on maximal folding yield between 15 °C and 37°C. The yield was decreased by approximately 20% at 4°C.

Previously published work had suggested the effect of various solvents and solubilizing agents on protein folding (Snyder, 1984; Cleland and Wang, 1990b). Several of these agents were tested in the folding system for IGF-I periplasmic aggregates. Most produced little or no effect, but methanol and ethanol at 20% concentration (v/v) had significant and approximately equal effects. Interestingly, these alcohols also significantly improved the ratio of the concentration of properly folded IGF-I to that of a mis-disulfided variant (the chief monomeric end product other than properly folded IGF-I). Combining 20% methanol and 1 M NaCl provided a further improvement in folding yield.

IGF-1 Concentration Effects and Kinetic Sequence

When the effect of increasing IGF-1 peptide aggregate concentration was evaluated, the results depended upon the concentration of DTT added. Results with 1 and 2 mM DTT suggested that a DTT to IGF-I molar ratio of 5-10 produced optimal folding. When the DTT/IGF-I ratio was maintained at 10, a broad optimum in IGF-1 concentration from approximately 2 to 5 mg/ml was suggested.

The relative kinetics of DTT oxidation, aggregate solubilization, and properly folded IGF-I concentration were then followed. DTT oxidation and IGF-I folding were monitored by HPLC analysis. Aggregate solubilization was monitored by centrifuging reaction tubes at discrete time intervals and measuring the amount of insoluble protein. Interestingly, at the time of full DTT oxidation, the aggregate was only approximately 50% solubilized and no properly folded IGF-1 had been formed. Continued disulfide exchange produced nearly complete solubilization and approximately a 50% yield of properly folded IGF-I.

Does the Periplasmic Aggregate contain Partially folded IGF-I ?

To investigate this important question an experiment was conducted in which IGF-I peptide aggregate was solubilized in increasing urea and DTT concentrations. The experiment was designed so that the solutions could then be quickly diluted to 2 M urea, 2 mM DTT, 20% methanol and 1.5 mg/ml IGF-1 peptide.

The results are preliminary but suggest that exposure to higher urea and DTT concentrations significantly reduce the ultimate yield obtained, even after dilution to optimal folding conditions. Interestingly, no additional formation of properly folded IGF-1 occurred after dilution of the 4M and 6M urea solutions. This type of experiment will be repeated, but these initial results suggest that exposure to high urea and DTT concentrations may be disrupting IGF-1 structure which already exists in the molecules trapped in the periplasmic aggregate.

Discussion and Summary

These results suggest an exciting new way to produce human IGF-I. Recovering the periplasmic aggregates by centrifugation provides an initial purification as well as an isolation method. The ability to solubilize and fold the IGF-I in one solution provides an approach which may not be significantly more complicated than if the IGF-I were initially produced in a soluble form. Thus, it appears that this approach offers the potential for efficient, low cost production of human IGF-I.

Work to date suggests optimal solubilization and folding conditions are: pH=10.5, 2M urea, 2 mM DTT, 1M NaCl and 20% methanol in a Capso buffer. Work is continuing in order to better understand the relevant kinetics of the reactions as well as the effect of any possible protein folding that may be resident in the IGF-I periplasmic aggregate.

References

Bowden, G.A.; Paredes, A.M.; Georgiou, G. *Bio/technology* **1991**, 9, 725-730.

Cleland, J.L.; Wang, D.I.C. *Biochemistry* **1990**, 29, 11072-11078.

Marston, F.A.O. *Biochem. J.* **1986**, 240,1-12.

Snyder, G.H. *J. Biol. Chem.* **1984**, 259, 7468-7472.

Cleland, J.L.; Wang, D.I.C. *Bio/technology* **1990**, 8, 1274-1278.

Gene Expression Using Gram-Positive Bacteria

Tadayuki Imanaka, Osaka University, Suita, Osaka 565, Japan
(FAX:81-6-876-2250)

A low copy number (one copy per chromosome) Km^r Tc^r plasmid pTB19 was isolated from thermophilic bacillus. After establishing the transformation system in B. stearothermophilus, we have constructed many derivatives from pTB19 and their copy numbers range from 1 to 214 per chromosome in B. subtilis (a mesophile) and B. stearothermophilus (a thermophile). By selecting the best combination of vector plasmid and host strain, both molecular cloning of various enzyme genes (penicillinase, α-amylase, neopullulanase, cyclodextrin glucanotransferase, neutral protease, alkaline protease, alcohol dehydrogenase, aldehyde dehydrogenase) and enhancement of enzyme production could be achieved. In addition, pleiotropic regulatory genes (degT etc.) and regulatory genes (penI and penJ) for penicillinase expression were cloned and analyzed in genus Bacillus. By constructing temperature-sensitive penicillinase repressors, the gene expression was easily controlled by the change in cultivation temperature.

Genus Bacillus is one of the most important Gram-positive bacteria which can secrete a wide variety of proteins directly into the culture medium. Therefore, Bacillus subtilis and Bacillus stearothermophilus were used as the representatives of mesophilic and thermophilic bacteria, respectively.

Construction of Low, Intermediate, and High Copy Number Plasmids in Bacillus Species

We have isolated a low copy number (one copy per chromosome) plasmid, pTB19, from a thermophilic bacillus (Imanaka et al., 1981a). The plasmid pTB19, encoding resistance to both kanamycin and tetracycline (Km^r Tc^r), could transform both B. subtilis and B. stearothermophilus (Imanaka et al., 1982). Two different replication determinants (repA and repB) were found on pTB19 (Imanaka et al., 1984). The RepA and RepB proteins function as positive and negative factors for the replication of repA and repB plasmids, and give low and high copy numbers in Bacillus, respectively (Ano et al., 1986; Imanaka et al., 1986). Although B. stearothermophilus CU21 can be transformed only by repB plasmid, strain SIC1 can be transformed by both repA and repB plasmids, (Zhang et al., 1988). We have constructed many derivatives from pTB19 and their copy numbers range from 1 to 214 per chromosome (Table 1) (Imanaka, 1987).

Importance of the Best Combination of Vector Plasmid and Host Strain

It is important to select the best combination of vector plasmid and host strain for the efficient cloning and

1054-7487/92/0018$06.00/0

Table 1. Properties of Plasmids

Plasmid	Molecular Size(MDa)	Characteristics	Rep	Copy No. B. subtilis	B. stearothermophilus
pTB19	17.2	Km^r Tc^r	A,B	1	1
pTB51	8.4	Km^r	A	8	NT
pTB52	7.0	Tc^r	A	9	NT
pTB53	11.2	Km^r Tc^r	A	8	NT
pRA1	4.9	Km^r	A	8	NT
pRAT1	2.3	Km^r	A	8	NT
pRAT11	1.7	Km^r	A	8	NT
pTB90	6.7	Km^r Tc^r	B	11	5 (18)
pTB914	5.7	Km^r Tc^r	B	13	43
pTB916	5.7	Km^r Tc^r	B	15	unstable
pTB913	2.9	Km^r	B	25	39
pTB919	3.9	Km^r	B	13	5
pTB931	3.9	Km^r	B	14	4
pTB902	2.7	Tc^r	B	24	60
pTB921	3.7	Tc^r	B	13	7
pRBT1	2.1	Km^r	B	35	NT
pRBH1	1.5	Km^r	B	42	NT
pRBH1-IR	1.5	Km^r	B	134	NT
pRBHC3	1.5	Km^r	B	134	NT
pRBHC3-IR	1.5	Km^r	B	214	NT
pRBHC7	1.5	Km^r	B	202	NT
pRBHC7-IR	1.5	Km^r	B	214	NT
pUB110	3.0	Km^r	B	48	50
pUB110dB	1.5	Km^r	B	55	NT

NT, not tested; ( ), copy number in the presence of tetracycline.

expression of specific genes. To substantiate this idea, attempts were made to clone some enzyme genes.

By using plasmid pMB9, penicillinase genes (penP and penI) from both the wild-type and constitutive strains of Bacillus licheniformis 9945a were cloned in Escherichia coli (Imanaka et al., 1981b). When a low copy number plasmid, pTB53, was used, both wild-type and constitutive penicillinase genes could be transferred into B. subtilis. However, when a high copy number plasmid, pUB110, was used, only the genes of the wild type could be transferred. These recombinant plasmids in B. subtilis could all be transferred by the protoplast transformation procedure into B. licheniformis. These results clearly show that a low copy number plasmid is desirable for gene cloning because of minimal stress to the host cells (Imanaka et al., 1981b).

Transformants of E. coli exhibited low penicillinase activities (7 U/mg of cells). In contrast, transformants of B. subtilis and B. licheniformis with constitutive penicillinase genes showed high enzyme activities (more than 10,000 U/ mg of cells).

By using plasmid pTB90, constitutive penicillinase genes were also cloned in B. stearothermophilus (Fujii et al., 1982). The recombinant plasmid carrier produced 3,300 U of penicillinase/mg of cells at 48°C, and the plasmid was stably maintained in the thermophile. However, the amount of enzyme produced at 55 and 60°C was reduced to 660 and

Table 2. Examples of gene cloning in both B. subtilis and B. stearothermophilus with the plasmids shown in Table 1.

Enzyme/protein	Gene	Origin	References
Penicillinase	penP	B. licheniformis	Imanaka et al., 1981b
			Fujii et al., 1982
Repressor of penP	penI	B. licheniformis	Himeno et al., 1986
Positive regulator	penJ	B. licheniformis	Imanaka et al., 1987
α-Amylase	amyT	B. stearothermophilus	Aiba et al., 1983
			Nakajima et al., 1985
Neutral protease	nprT	B. stearothermophilus	Fujii et al., 1983
			Takagi et al., 1985
Neutral protease	nprM	B. stearothermophilus	Kubo et al., 1988
Neutral protease	nprS	B. stearothermophilus	Nishiya et al., 1990
Activator of nprS	nprA	B. stearothermophilus	Nishiya et al., 1990
Neopullulanase	nplT	B. stearothermophilus	Kuriki et al., 1988a
			Kuriki et al., 1989
Pullulanase	pulT	B. stearothermophilus	Kuriki et al., 1988b
			Kuriki et al., 1990
Insecticidal protein	cryIA(b)	B. thuringiensis	Nakamura et al., 1989
Glucose isomerase	xylA	C. thermosulfurogenes	Lee et al., 1990
Restriction enzyme	BamHI-R	B. amyloliquefaciens	Kawakami et al., 1990
Modification enzyme	BamHI-M	B. amyloliquefaciens	Kawakami et al., 1990
Sensor protein	degT	B. stearothermophilus	Takagi et al., 1990
Alcohol dehydrogenase	adhT	B. stearothermophilus	Sakoda et al., 1992

B., Bacillus; C., Clostridium

39 U/mg of cells, respectively. The recombinant plasmid was thus getting more unstable at elevated temperatures.

Gene Cloning in Bacillus Species

By using the host-vector systems in Bacillus spp., many enzyme genes have been cloned and expressed efficiently (Table 2). The regulatory gene, degT, from B. stearothermophilus which enhanced production of extracellular alkaline protease was also cloned in B. subtilis and B. stearothermophilus with pTB53 as a vector (Takagi et al., 1990) (Table 2).

A B. subtilis strain carrying degT showed the following pleiotropic phenomena: ① enhancement of production of extracellular enzymes such as alkaline protease and levansucrase, ② repression of autolysin activity, ③ decrease of transformation efficiency for B. subtilis (competent cell procedure), ④ altered control of sporulation, ⑤ loss of flagella, and ⑥ abnormal cell division. The membrane protein DegT might function as a sensor protein and transfer the signal of environmental stimuli to the regulatory region of target genes to activate or repress transcription of the genes.

Construction of Temperature-Sensitive Penicillinase Repressors

The penicillinase gene expression system of B. licheniformis consists of the following three genes: the structural gene for penicillinase (penP), a repressor gene (penI), and a positive regulatory gene (penJ) (Himeno

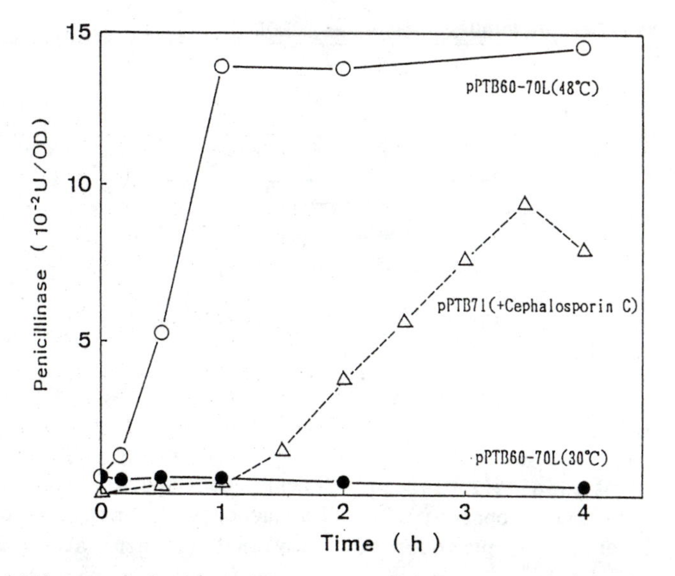

Fig. 1. Penicillinase induction by an inducer or heat. Bacillus subtilis carrying each plasmid was grown in L broth and induction started at time 0. pPTB60-70L (penIts) carrier was cultivated at 30°C (●) or 48°C (○). pPTB71 (wild type) carrier was cultivated in the presence of an inducer (cephalosporin C) (△).

et al., 1986; Imanaka et al., 1987). In the absence of an inducer (e.g., β-lactam antibiotics), the repressor protein PenI recognizes two operator sequences located between the promoter and the Shine-Dalgarno sequence of the penP gene and interrupts transcription from the penP promoter. In the presence of an inducer, the penJ gene product forms a complex with the inducer molecule and may interact with PenI to induce penP gene expression.

We constructed temperature-sensitive PenI repressors by amino acid substitutions at Pro in the predicted β-turn structure (Imanaka et al., 1992). A mutant repressor (P70L; Pro-70 is substituted with Leu) was inactive at 48°C and penP gene expression was derepressed (1,200 U/OD_{660}), although the mutant was still active at 30°C (27 U) (Fig. 1). This heat induction was much quicker than the induction with an inducer.

Heat-Inducible Secretion Vector

By using the temperature-sensitive repressor system, we constructed a useful heat inducible secretion vector plasmid for the production of extracellular enzymes (Fig. 2). A low copy number plasmid pTB52 was used as a replication unit. α-Amylase gene amyE of B. subtilis was supplied to give the SD sequence, signal sequence, and transcription terminator. The mature portion of the enzyme was deleted in amyE, and BamHI site was introduced as a cloning site. The newly constructed plasmid, $pISA^{ts}412$, can be used for the efficient gene cloning at 30°C and for the production of a large amount of extracellular enzyme at 48°C.

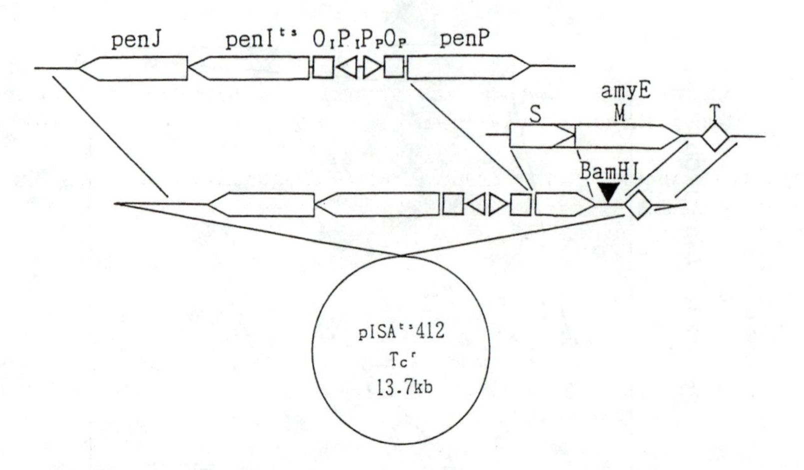

Fig. 2. Structure of pISA^{ts}412 plasmid. P_I, promoter of penI operon; O_I, operator of penI gene; P_P, promoter of penP; O_P, operator of penP; S, signal sequence; M, mature portion of α-amylase; T, terminator; ▼, BamHI site.

References

Aiba,S.;Kitai,K.;Imanaka,T. Appl.Environ.Microbiol. 1983, 46, 1059-1065.

Ano,T.;Imanaka,T.;Aiba,S. Mol.Gen.Genet 1986, 202, 416-420.

Fujii,M.;Imanaka,T.;Aiba,S. J.Gen.Microbiol. 1982, 128, 2997-3000.

Fujii,M.;Takagi,M.;Imanaka,T.;Aiba,S. J Bacteriol. 1983, 154

Himeno,T.;Imanaka,T.;Aiba,S. J.Bacteriol. 1986, 168, 1128-1132.

Imanaka,T.;Fujii,M.;Aiba,S. J.Bacteriol. 1981a, 146, 1091-1097.

Imanaka,T.;Tanaka,T.;Tsunekawa,H.;Aiba, S. J.Bacteriol. 1981b, 147, 776-786.

Imanaka,T.;Fujii,M.;Aramori,I.;Aiba,S. J.Bacteriol. 1982, 149, 824-830.

Imanaka,T.;Ano,T.;Fujii,M.;Aiba,S. J. Gen.Microbiol. 1984, 130, 1399-1408.

Imanaka,T.;Ishikawa,H.;Aiba,S. Mol.Gen. Genet. 1986, 205, 90-96.

Imanaka,T. Ann.N.Y.Acad.Sci. 1987, 506, 371-383.

Imanaka,T.;Himeno,T.;Aiba,S. J.Bacteriol. 1987, 169, 3867-3872.

Kawakami,B.;Katsuragi,N.;Maekawa,Y.; Imanaka,T. J.Ferment.Bioeng. 1990, 70, 211-214.

Kubo,M.;Imanaka,T. J.Gen.Microbiol.1988 134, 1883-1892.

Kuriki,T.;Okada,S.;Imanaka,T. J.Bacteriol. 1988a, 170, 1554-1559.

Kuriki,T.;Park,J.H.;Okada,S.;Imanaka,T. Appl.Environ.Microbiol. 1988b, 54,2881-2883.

Kuriki,T.;Imanaka,T. J. Gen. Microbiol. 1989, 135, 1521-1528.

Kuriki,T.;Park,J.H.;Imanaka,T. J.Ferment Bioeng. 1990, 69, 204-210.

Lee,C.;Bhatnagar,L.;Saha,B.C.;Lee,Y.E.; Takagi,M.;Imanaka,T.;Bagdasarian,M.; Zeikus,J.G. Appl. Environ. Microbiol. 1990, 56, 2638-2643.

Nakajima,R.;Imanaka,T.;Aiba,S. J.Bacteriol. 1985, 163, 401-406.

Nakamura,K.;Imanaka,T. Appl. Environ. Microbiol. 1989, 55, 3208-3213.

Nishiya,Y.;Imanaka,T. J.Bacteriol.1990, 172, 4861-4869.

Sakoda,H;Imanaka,T. J.Bacteriol. 1992, 174, 1397-1402.

Takagi,M.;Imanaka,T.;Aiba,S. J.Bacteriol. 1985, 163, 824-831.

Takagi,M.;Takada,H.;Imanaka,T. J.Bacteriol. 1990, 172, 411-418.

Zhang,M.;Nakai,H.;Imanaka,T. Appl.Environ. Microbiol. 1988, 54, 3162-3164

Secretion and Surface Expression in Microorganisms of Heterologous Proteins Important for Medical Research and Clinical Applications

Walter Fiers, Jan Demolder, Ronny Leemans, Marc Logghe, Marleen Maras, Nico Mertens, Johan Robbens, Lothar Steidler, Roland Contreras, and **Erik Remaut,**
Laboratory of Molecular Biology, Gent University, K.L. Ledeganckstraat 35, 9000 Gent, Belgium
(FAX:32 91 645348)

Expression systems for production of cytokines in microorganisms are discussed. High-level production in *E. coli* often leads to inclusion bodies. Some cytokines can be efficiently exported into the periplasm, and a subsequent, inducible system allows release of the heterologous proteins of interest into the medium. Alternatively, many cytokines can be efficiently secreted by yeast cells. An inducible system, such as the GAL1 promoter, can be coupled to a secretion cassette involving the prepro-sequence of the α-mating factor. Various approaches can be used to further optimalize the secretion level.

Protein hormones are known since the first quarter of this century. But it is only in the last decade, since the widespread introduction of cloning methodology, that a large plethora of signal proteins, now called cytokines, and their receptors has been identified. In order to study their biological functions, their chemical structure and their physical conformation, adequate amounts of these proteins must be produced. Moreover, quite a few of these cytokines have an important clinical potential and therefore large-scale production, purification and development processes have been developed. There are considerable advantages to produce these cytokines in bacteria or in lower eukaryotes, such as yeast. This ensures a reproducible, cheap, and fast production process.

The first expression system that can be considered is intracellular synthesis in *E. coli*. It is fairly easy to replace a eukaryotic signal sequence by a single initiating AUG codon. The various steps to optimize expression level, gene copy number, induced transcription and efficient translation have been amply studied and reviewed. For example, an elegant system involves phage T7 RNA polymerase under control of an IPTG-inducible promoter, such as Trc, and a heterologous gene transcription unit directed by a T7 promoter. High translation efficiency can then be ensured by placing the T7 gene 10 ribosome-binding site in front of the heterologous gene. The main problem, however, which is often encountered, is that the overproduced, heterologous protein is deposited in the form of inclusion bodies. This is not necessarily an unsurmountable obstacle, as inclusion bodies can fairly easily be purified from most of the bacterial debris. The protein can then be dissolved in guanidinium hydrochloride, and under appropriate conditions slowly refolded, followed by oxidation to form the correct disulfide bridges. On the other hand, one can try to retain the heterologous protein in a native conformation. For a number of proteins, this is easily possible by lowering the temperature, for example growing the cells at 20°C. We have also tried to overexpress various chaperone molecules, which we hoped would help the heterologous protein to fold properly. Such candidate assisting proteins were groEL, groES, DnaK, DnaJ, GrpE, secB, etc.; but none of these helped in a major way.

Another possibility to avoid intracellular precipitation is to have the proteins secreted into the periplasm. After proper folding, the disulfide bridges can then be formed by the

1054–7487/92/0023$06.00/0

oxidizing environment. For example, we could produce murine IL2 efficiently by fusing the gene of interest to the ompA signal sequence. The periplasmic proteins can be selectively released, and from this step on, purification of the protein of interest becomes straightforward. We have also used an inducible system which produces Kil, a protein which makes pores in the outer cell wall. Hence, after a heterologous protein has accumulated in the periplasm, one can switch on the *kil* gene, and in this way release the protein of interest in the medium. Cell separation and collection of the medium allows then directly on-stream separation of the product of interest.

For some purposes, it is important to have a protein expressed on the surface of the bacteria. This requires that the polypeptide moves from the periplasm through the outer cell wall, which is not an easy task. We have been able, however, to insert coding sequences into the pilin structural protein gene. In this way, some determinants could be exposed on the surface of the transformed bacteria.

Yeast (*Saccharomyces cerevisiae*) has several advantages as a production system. It is usually possible to make the heterologous product as a secreted protein, which considerably facilitates its subsequent purification. Furthermore, yeast cells recognize the same type of N-glycosylation recognition sequence, which means that, potentially at least, it glycosylates proteins at the same sites as eukaryotic cells. However, the glycosyl groups are of the high mannose type, such that upon injection into the circulation of mammalian species, most of the proteins are recognized by mannose receptors on various cells and removed. A convenient system for secretion in yeast, and which is often used, is the prepro-sequence of the α-mating factor. The heterologous protein is then cleaved off by the KEX protease. Sometimes it might be worthwhile to place between the pro-sequence and the heterologous gene sequence one or two Glu-Ala dipeptides, which facilitate the cleavage of the pro-sequence. As a promoter, one can either use a strong, constitutive promoter, such as the triosephosphate isomerase promoter, or an inducible promoter, such as GAL1. Using such a system, we were able to obtain about 30 mg/l of human IL6. In the case of murine IL2, the level was about 10 mg/l. It is of some interest regarding the latter cytokine that the yield could be considerably improved by removing most of the 3'-untranslated sequence. This region apparently contains a signal which makes the messenger a target for rapid degradation, and it is surprising that this same regulatory system also operates in yeast cells. Murine IL2, but not human, contains twelve consecutive glutamine codons of the CAG type, which are believed to be unfavorable for yeast; however, changing these to the more frequently used CAA codons did not affect the expression level.

It may be noted that there are several possibilities to maintain the heterologous gene in transformed yeast cells. A plasmid containing the 2μ-origin is most often used, and this then has to be maintained in a cell which contains a wild-type μ-plasmid; this requires a continuous selection pressure. Another possibility is to use a cir° yeast strain, this means a strain which does not harbor an endogenous 2μ-plasmid. Although in this case the plasmid is fairly stable, it should nevertheless be kept under selection pressure, for example when it contains an ura^+ marker in an ura^- genomic background. A third method to maintain the heterologous gene is to use an integrative plasmid, for example a plasmid linearized within a yeast sequence. We have used for this purpose plasmids cleaved in the δ-sequence. In this way, multiple copies can integrate at the many δ-sequences present in the yeast genome. One can use a dihydrofolate reductase gene as selection marker. This then allows to apply a selection pressure for a high copy number by addition of methotrexate plus sulfanilamide in the course of the transformation procedure. Under these conditions, and with plasmids linearized in the δ-segment, potentially integration could occur at a number of chromosomal sites.

Yeast cells are often difficult to open, and in order to solve the problem of heterologous proteins locked in the periplasmic space, we have expressed a plant ß-(1,3)-glucanase linked to the α-mating factor and under control of the GAL1 promoter. Apparently, this enzyme interferes with cell wall growth from within the cell. With this system, about 20% of the periplasmic proteins could be released into the medium; obviously, further improvements are necessary.

It is often possible to screen for higher producing clones by using two membranes, the second of which, nitrocellulose, retains the protein of interest. When the latter membranes are then developed by means of an appropriate antiserum, higher producing (mutant) clones can be identified. Results show that a beneficial mutation for one protein does not necessarily help the secretion of another heterologous protein. Undoubtedly, identifying the mutations which are involved in these systems, should provide a clearer picture of the processes leading to efficient secretion of heterologous proteins in yeast.

Research supported by the ETC, OOA, IUAP, and IFOB.

Hybridoma Culture -- Optimization, Characterization, Cost

Brian Maiorella, Chiron Corporation, 4560 Horton St., Emeryville, CA 94608-2916

Process development has been a critical factor in making practical the therapeutic application of human monoclonal antibodies. Antibodies act in stoichiometric relation to their antigen targets, and many therapeutic applications are expected to require doses of 100 mg/patient or more. As recently as five years ago, maximum titres of human antibodies from cell culture were reported at 30 mg/L or less, and culture media were supplemented with serum. For a significant product with a patient population of 100,000/yr, total cell culture production required at these low titres would have been expected to exceed one million liters per year, and it was not certain that these products could be manufactured economically.

Development of improved culture media has played a central role in improving culture productivity and reducing cost. We have replaced serum with defined components, and have identified synergistic combinations of nutrients which significantly increase hybridoma growth, culture longevity, and antibody production. Key ingredients of conventional media and of an optimized serum-free medium are contrasted in Table 1. Pluronic polyol replaces serum in

Table 1. Hybridoma Culture Media

Key Ingredients	Conventional Serum-Supplemented Medium	Chiron Serum-Free
Glucose Glutamine	4 gm/L 4 mM	5 → 11 gm/L 8 → mM
Amino Acids	0.8 gm/L	9 gm/L
Choline Ethanolamine	3.5 mg/L -	75 mg/L 20 mg/L
Monothioglycerol	-	10 mg/L
Trace Elements	Se	Co, Cu, Mo, Mn, Se, Zn
Chelated Iron	0.2 nM	110 nM
Insulin/Transferrin	-	10 mg/L
Pluronic F68	-	1 gm/L
Serum	50 ml/L	-

1054–7487/92/0026$06.00/0

protecting cells from hydrodynamic stress. Iron is supplied either bound to transferrin or chelated with citrate. By providing lipid precursors (choline and ethanolamine) lipoproteins are not required. The level of most amino acids is increased four to ten-fold to support increased cell and product production.

To support high density culture, the requirements for glucose and glutamine energy sources are significantly increased. Rapid metabolism of these nutrients can result in production of inhibitory levels of lactate and ammonia. By feeding these nutrients in direct response to metabolic demand (as measured by oxygen utilization) waste product production is minimized.

Figure 1 compares growth and antibody production of a human/human/murine trioma grown in a typical commercially available medium (Ventrex HL-1) and in our optimized serum-free medium. Maximum viable cell density is increased almost four-fold to over 5 million/ml, and antibody titre is increased from 30 mg/L to over 550 mg/L in our medium. The optimized medium has been useful for culture of many murine and human hybridomas (Table 2), producing antibody at 150 to 1,000 mg/L.

In implementing any change in medium or culture conditions, it is critical to assure product consistency. We have developed a battery of tests of antibody primary structure, biological activity and pharmacology. These tests support equivalence of a human IgM manufactured in two media (before and after optimization to increase product titre). Consistency of protein backbone and of glycosylation are supported by peptide maps and by high pH anion exchange chromatography of cleaved oligosaccharides (Figures 2 and 3). Clearance rate of the two materials following bolus injection in rat is identical and binding activity is identical.

The effect of product titre improvements on final cost of antibody manufacture is summarized in Figure 4. It is clear that process development has played a critical role, essential to making these products broadly available for therapeutic use.

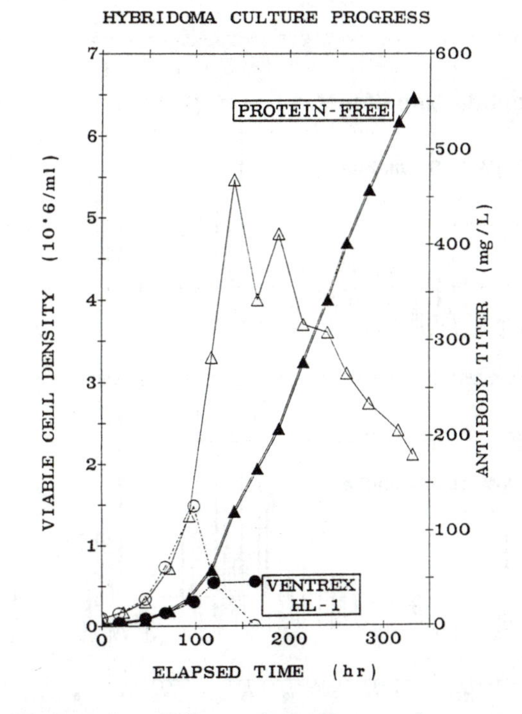

Figure 1. Effect of medium and process optimization on hybridoma culture growth and antibody production.

Table 2. Hybridoma Culture in Optimized Medium

Cell Line	Viable Cell Density	Ig Titre
Murine Hybridomas	6 x 10^6/ml	600 mg/L
	4 x 10^6/ml	150 mg/L
	6 x 10^6/ml	400 mg/L
	6 x 10^6/ml	400 mg/L
	6 x 10^6/ml	1000 mg/L
Murine Quadroma	5 x 10^6/ml	1000 mg/L
Human/Murine Triomas	5 x 10^6/ml	550 mg/L
	2 x 10^6/ml	750 mg/L
Human Lymphoblast	3 x 10^6/ml	-

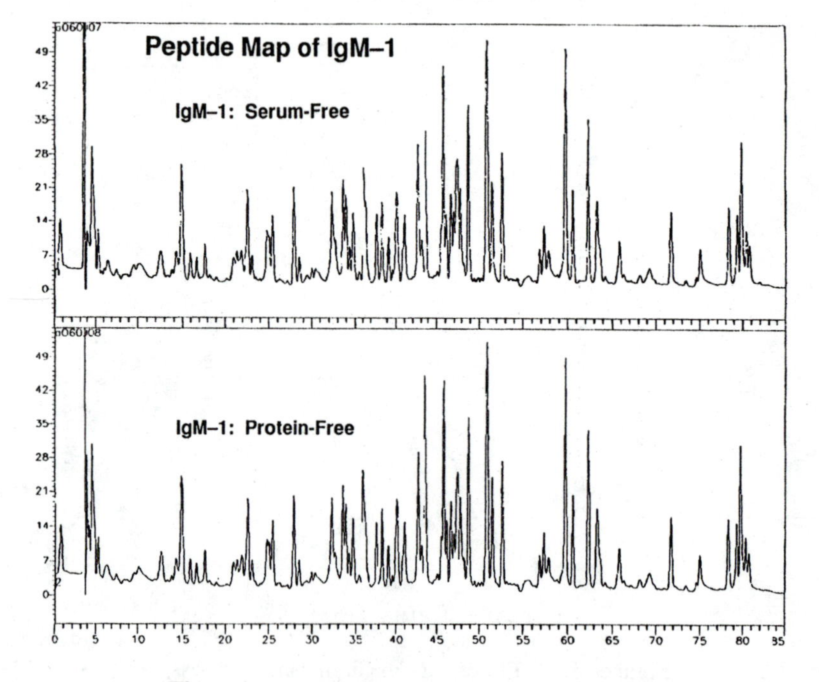

Figure 2. Tryptic map of human antibody produced in two culture media.

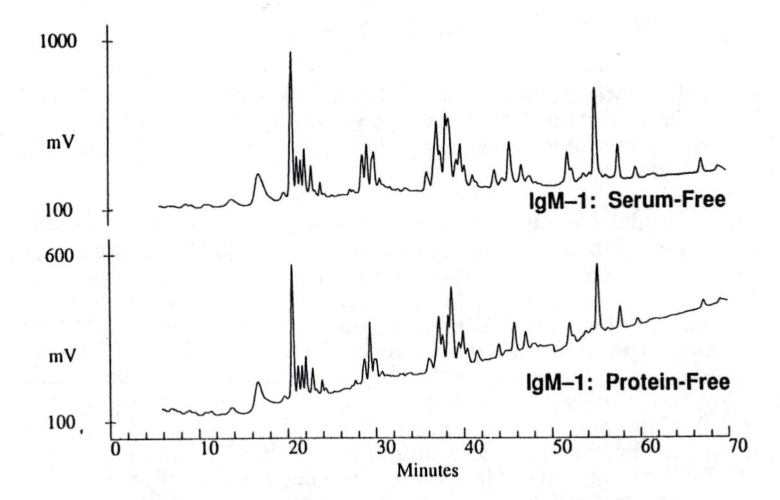

Figure 3. HPAE-PAD map of oligosaccharides from human antibody produced in two culture media.

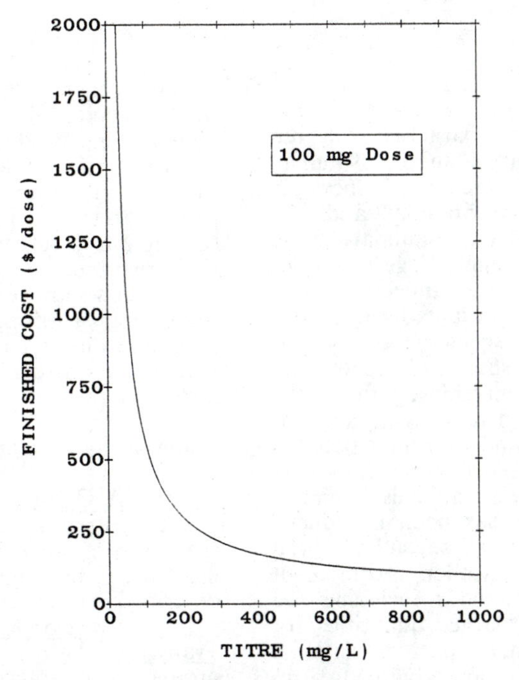

Figure 4. Effect of product titre on final cost of a therapeutic monoclonal antibody.

Production of Human Hemoglobin in Transgenic Swine

J. Kevin O'Donnell, **Mark E. Swanson**, **Michael J. Martin**, **Stephen H. Pilder**, **Kathleen M. Hoover**, **Victoria M. Huntress**, **Cynthia T. Karet**, **Carl A. Pinkert**, **William J.P. Lago** and **John S. Logan**, DNX, Inc., Princeton, New Jersey, USA, 08540

We have used transgenic technology to generate pigs which contain and express human Hb in their red blood cells. The DNA construct contained three essential elements: the human β-globin Locus Control Region (LCR) which confers copy number dependence, integration site independence, and tissue specificity; two complete copies of the human α-globin gene; and a copy of the human β-globin gene. The DNA was microinjected into the pronuclei of fertilized eggs. Three transgenic pigs were identified. The transgenic pigs are healthy, are not anemic, and grow at a rate comparable to that of their non-transgenic littermates. To date one of these animals has been analyzed in detail. In this pig human Hb accounted for 10-15% of the total Hb in the red blood cells. Purification of the human Hb from both porcine Hb and other non-hemoglobin proteins was accomplished by ion exchange chromatography and afforded essentially pure protein. The purified recombinant Hb exhibited an oxygen affinity similar to that of authentic human Hb. These experiments demonstrate that expression of functional human Hb or Hb variants in transgenic swine is technically feasible and that the Hb can be readily purified from endogenous porcine proteins.

Introduction

Crucial to the preparation of a red cell substitute is an ample supply of human hemoglobin (Hb). Recent advances in recombinant DNA technology and the production of transgenic animals have provided a means to supply large quantities of recombinant protein. Indeed, human hemoglobin and some variants have already been produced in transgenic mice (1-4). A Hb based blood substitute eliminates the need for blood type matching, reduces the risk of viral disease transmission such as AIDS and hepatitis, and also anticipated is a prolonged shelf life.

A line of transgenic animals is one in which a foreign gene has been introduced into its germ line. These animals can offer advantages over other recombinant protein production methods including the ability to produce large quantities of complex proteins where post-translational modifications such as glycosylation, phosphorylation, subunit assembly and insertion of prosthetic groups are critical for the biological activity of the molecule. Furthermore in the case of hemoglobin, production in the red blood cell (96% of soluble protein is hemoglobin) provides a good starting material for purification. The high dose levels (multigram level) and cost constraints of this product make simple purification processes an essential component. Therefore, a transgenic animal producing human Hb in its red blood cells could be an attractive alternative source of this important protein.

Materials and Methods

<u>DNA Construct</u>: The DNA construct, $LCR\alpha\alpha\beta$, used to generate the transgenic animals in this study is illustrated in Figure 1 and essentially described in Greaves *et al.* (3).

<u>Generation and Identification of Transgenic Pigs</u>: A detailed method for the production of transgenic livestock has been reviewed (5). Briefly, fertilized eggs in the form of one or two cell ova were

1054–7487/92/0030$06.00/0

removed from sexually mature gilts prior to microinjection of the vector DNA. In a typical experiment, approximately 100 copies of construct DNA (2ng/μl) were injected into one pronucleus in one cell ova or both nuclei in a two cell ova. Microinjected ova were then transferred to the uteri of suitable recipients. After birth, piglets were screened for human globin sequences by southern blotting of tail DNA as well as analysis of blood samples on isoelectric focusing (IEF) gels (Isolabs, Inc., Akron, Ohio).

Hemoglobin Purification: Hemoglobin was isolated from washed erythrocytes followed by hypotonic lysis. The clarified hemolysate containing recombinant human Hb from the transgenic pig, #6-3, was purified from porcine Hb by anion exchange chromatography using a linear salt gradient by standard chromatographic techniques (Fig. 2).

Oxygen Binding Curves: Oxygen equilibrium curves of whole blood and Hb solutions were performed on a Hemox Analyzer (TCS, Inc, Southhampton, Pennsylvania) according to the protocol recommended by the manufacturer.

Results and Discussion

The selection of pigs as the animal for the production of recombinant human Hb is based on four critical factors: 1) the relatively large size of the pig which provides sufficient Hb upon recovery from the erythrocytes; 2) the functional similarities between pig and human Hb in the regulation of oxygen binding which enables transgenic pigs to remain healthy even with high level expression of the heterologous protein in their erythrocytes;

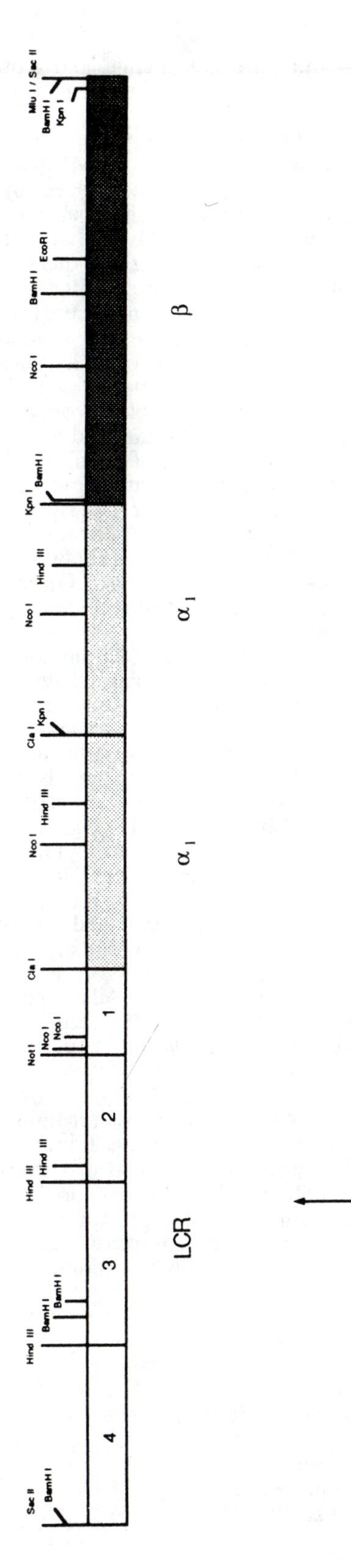

Figure 1: The LCR $\alpha\alpha\beta$ vector used for the generation of the transgenic animals. This DNA is 16.9 kb in length and contains three essential elements. First is the Locus Control Region (LCR) containing four DNase 1 super hypersensitive sites which confer copy number dependence, integration site independence and high level erythroid specific expression. Second are two copies of the human α_1-globin gene, and third is a complete copy of the human adult β-globin gene. Plasmid vector sequences are not shown. The construct is essentially as described in reference 3.

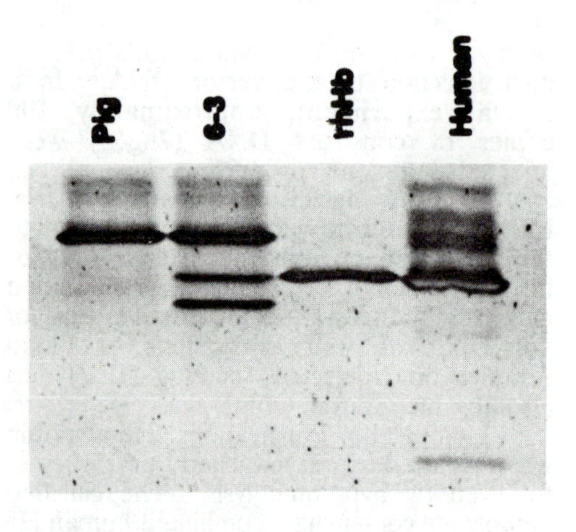

Figure 2: Isoelectric Focusing Gel of Purified Recombinant Human Hemoglobin. Aliquots of the indicated hemoglobins were analyzed by IEF gels according to the manufacturers procedure (Isolab, Inc., Akron, Ohio).

3) the size and frequency of swine litters; and 4) relative to other farm species, pigs have short periods of gestation and sexual maturation.

The expression vector, LCR$\alpha\alpha\beta$, contains three essential elements for proper globin gene expression in transgenic animals. First is the human β-globin Locus Control Region (LCR) normally located about 50 kb 5'- to the human adult β-globin gene on chromosome 11. Previous studies demonstrated that the LCR confers copy number dependence, integration site independence, and high level erythroid specific expression of human globin genes (6). The second element of the vector is a set of two complete human α_1-globin copies in tandem and lastly, a complete copy of the human β-globin gene.

The LCR $\alpha\alpha\beta$ construct was purified to remove plasmid sequences and used for microinjection experiments in pigs. Seven hundred and nine injected eggs were transferred to recipients and of a total of 112 piglets born, three were identified as transgenic. The overall efficiency was 0.42%, comparable to that obtained with many different constructs in pigs (5). These transgenic pigs are healthy, are not anemic, and grow at a rate comparable to that of their non-transgenic littermates.

One pig, #6-3, was analyzed in detail and expressed a total of 10-15% of human Hb in its erythrocytes as judged by IEF gels (Fig. 2, lane 2). IEF separates Hb dimers based on their pI and is an efficient method for screening and quantitating small aliquots of whole blood for the presence of different Hbs. Interestingly, only one hybrid band was observed in addition to the expected human α/human β and porcine α/porcine β bands. This hybrid band was isolated on a preparative IEF gel and then analyzed for globin chain composition by Triton acid-urea gel electrophoresis, identifying it as a human α/porcine β dimer. The absence of the second hybrid, porcine confirmed by reassociation experiments between human and non-transgenic pig Hbs. Once again, only one hybrid, human α/pig β, was detected.

Despite about 85% homology between human and pig globin chains, these Hbs separate readily on an anion exchange column using a linear salt gradient. Human Hb eluted first followed by the hybrid and pig Hbs. The recombinant human Hb was judged to be essentially pure as demonstrated by IEF (Fig. 2, lane 3) and SDS-PAGE gel analysis (data not shown).

Purified recombinant human Hb from transgenic pigs and human Hb A from human red blood cells exhibited equivalent P_{50} values of approximately 10mm Hg. Cooperativity values were also similar (data not shown). Additional studies to further characterize the recombinant human Hb are underway and should be completed shortly.

These results demonstrate that expression of functional recombinant human Hb or Hb variants in transgenic swine is technically feasible and that the Hb can be readily purified from endogenous porcine proteins thereby providing an attractive and economical source of human Hb for use as a potential blood substitute.

References

Behringer, R.R., J.M. Rayn, M.P. Reilly, T. Asakura, R.D. Palmiter, R.L. Brinster and T.M. Townes. (1989) *Science*, 245, 971-973.

Hancombe, O., M. Vidal, J. Kaeda, L. Luzzatto, D.R. Greaves and F. Grosveld. (1989) *Genes & Dev.*, 3, 1572-1581.

Greaves, D.R., P. Fraser, M.A. Vidal, M.J. Hedges, D. Ropers, L. Luzzatto and F. Grosvels. (1990) *Nature*, 343, 183-185.
Ryan, T.M., T.M. Townes, M.P. Reilly, T. Asakura, R.D. Palmiter, R.L. Brinster and R.R. Behringer. (1990) *Science*, 247, 566-568.
Pursel, V.G., C.A. Pinkert, K.F. Miller, D.J. Bolt, R.G. Campbell, R.D. Palmiter, R.L. Brinster and R.E. Hammer. (1989) *Science*, 244, 1281-1288.
Grosveld, F., G. Blom vanAssendelft, D.R. Greaves and G. Kollias. (1987), *Cell*, 51, 975-985.

Expression of Human Protein C in Transgenic Swine

William H. Velander*, **Anuradha Subramanian**, **Arthur W. Degener**, and **Tülin Morcöl**, Department of Chemical Engineering, Virginia Polytechnic Institute and State University, Blacksburg, VA 24061
John L. Johnson and **Tracy D. Wilkins**, Department of Anaerobic Microbiology, Virginia Polytechnic Institute and State University, Blacksburg, VA 24061
Francis C. Gwazdauskas and **Robert M. Akers**, Department of Dairy Science, Virginia Polytechnic Institute and State University, Blacksburg, VA 24061
Henryk Lubon and **William N. Drohan**, The Holland Laboratory, The American Red Cross, Rockville, MD 20855

The current understanding of gene regulation has enabled investigators to achieve high levels of recombinant protein expression in cell culture (reviewed in Grinnell et al., 1990) as well as in the mammary gland of transgenic livestock (Denman et al., 1991; Wright et al., 1991; Wall et al., 1991). The use of *in vitro* systems as bioreactors for the production of complex recombinant proteins has been limited by their inability to perform appropriate post-translational processing at high rates of synthesis (Grinnell et al., 1990). These modifications can be essential to the biological activity of many proteins having therapeutic value. We have produced transgenic swine containing the cDNA of human Protein C (hPC) in an effort to use the porcine mammary gland as a protein bioreactor (Velander et al., 1992). These studies demonstrate that regulatory elements taken from the murine Whey Acidic Protein (WAP) gene can direct expression of hPC to the mammary gland of transgenic pigs. In addition, we have documented the potential of the porcine mammary gland to perform complex post-translational modifications on recombinant hPC (rhPC) such as carboxylation of glutamic acid to form 4-carboxyglutamic acid (gla). Gla residues are necessary for the membrane-mediated activities of many plasma proteins associated with hemostasis (reviewed in Grinnell et al., 1990).

Protein C is produced by the liver and is an important regulator of hemostasis (reviewed in Grinnell et al., 1990). Therefore, hPC is being investigated as a therapeutic agent for disease states where excess blood clotting may be pathogenic. Therapeutic indications for hPC include: attenuated formation of secondary blood clots after fibrinolytic treatment, prevention of intravascular clotting that results from surgical procedures such as hip and knee replacement, and replacement therapy for congenital hPC deficiency. The relationship between the anticlotting function and the structure of hPC makes it one of the most complex of the proteins which maintain hemostasis. Human Protein C occurs naturally as a trace plasma protein at approximately 4 μg/ml, which makes large-scale isolation from human plasma difficult (see Velander et al., 1990). Only low levels of functional rhPC have been secreted from mammalian cell lines derived from human kidney, human liver, and mouse mammary epithelium (reviewed in Grinnell et al., 1990).

Human Protein C circulates in blood as a 62,000 dalton (Da) zymogen of a serine protease which contains 12 intrachain disulfide bonds, one β-hydroxylated aspartic acid residue, and 9 gamma-carboxyglutamic acid (gla) residues. These gla residues occur within the first 29 amino acid residues of the light-chain and are necessary for anticoagulant activity. Compared with other recombinant proteins which have been made in transgenic livestock, rhPC is one of the most complex.

Production of Transgenic Pigs.

A primary strategy associated with using livestock as bioreactors for recombinant proteins is utilizing the mammary gland's natural capacity to make and export high levels of protein without causing harmful physiological effects to the animal. Transgenic pigs were generated using a hybrid gene containing the human Protein C cDNA inserted into the first exon of the murine Whey Acidic Protein gene (Velander et al., 1992). No abnormalities or pathological effects were observed in transgenic pigs containing the Whey Acidic Protein-Human Protein C fusion gene. Milk was collected from

1054–7487/92/0034$06.00/0

several transgenic founder and control pigs, all of which experienced a normal 8-week lactation. In addition to nursing, each sow was milked one to five times a day which yielded about 400-1000 ml each milking with a total volume of 1-4 liters/day.

Levels of rhPC Synthesis in Porcine Mammary Tissue.

Table 1 presents the hPC antigen levels detected by Enzyme-Linked-Immunosorbent Assay (ELISA) in milk from the first lactations of transgenic founder pigs. It was somewhat unexpected to find rhPC expression at levels of 100 to 1000 μg/ml in the milk from two transgenic founder pigs (Velander et al., 1992); hPC antigen was expressed at about 1 μg/ml in the milk of transgenic founder mice containing the same Whey Acidic Protein-Human Protein C fusion gene (Velander et al., 1991). Expression of rhPC was low during the early days of lactation and increased during lactation for the founder animals studied. No immunoreactivity was detected in control whey samples (data not shown). The synthesis of milk proteins by the mammary gland does not occur at significant levels until day 3 or later. Both transgenic and control pigs exhibited a steady total protein content in their milk of about 50-70 mg/ml after day 6 of lactation (data not shown). Thus, the initiation of rhPC secretion observed for these pigs was temporally consistent with normal milk protein secretion.

The highest synthesis rates observed for rhPC in porcine mammary tissue appear similar to those reported for unamplified human kidney 293 cell line (Grinnell et al., 1990) with both systems synthesizing rhPC at rates on order of 1-10 μg/ 10^6 cells/day. The 293 cell line produced only 10-30 μg/ml/day as compared to 100-1000 μg/l/hr in the milk of transgenic pigs. Thus, an advantage of producing rhPC in the pig mammary gland appears to be a significantly higher cell density than can be obtained with kidney cells in culture. It is noted that rhPC levels from the founder pigs studied are less than 2% of the total protein level in porcine milk. Thus, the potential of the porcine mammary gland for expressing a complex recombinant protein is much higher than we have observed in these early experiments.

Evidence for Proper Vitamin K-dependent Processing in Mammary Tissue at High Rates of Synthesis.

Proper post-translational modification of the gla domain of Protein C is essential for biological activity (reviewed in Grinnell et al., 1990). In order to determine whether gamma-carboxylation had occurred properly in rhPC, its anticoagulant activity was assayed *in vitro* by delay in Activated Partial Thromboplastin (clotting) Time (APTT). This assay simulates coagulation *in vivo* by initiating clotting in a mixture containing calcium, phospholipid membrane, and the proteins associated with hemostasis. Table 1 lists the specific anticoagulant activities of immunopurified hPC from Normal human Reference Plasma Pool (NRPP) and rhPC immunopurified from milk pooled from the first lactation of transgenic pig 29-2. Immunopurification yields of >40% of the total antigen present in the milk were achieved. The theoretical specific anticoagulant activity of hPC in NRPP is 250 U/mg. The specific anticoagulant activity of 213 $\pm$ 7 U/mg obtained for hPC immunopurified from NRPP was consistent with the theoretical activity assigned to NRPP. The specific anticoagulant activity of rhPC immunopurified from pig 29-2 was determined to be 244 $\pm$ 12 U/mg. Therefore, the average specific activity of rhPC from transgenic pig 29-2 is equivalent to that of hPC from NRPP. Recombinant Protein C from the milk of another transgenic pig which expressed at similar levels possessed a similar activity. Since the anticoagulant activity of Protein C is dependent upon proper gamma-carboxylation, it appears that the porcine rhPC is sufficiently gamma-carboxylated.

We have demonstrated that pigs can be genetically engineered to produce a human protein at high levels in milk without any obvious impact upon the health of the animal. The biological activity of rhPC indicates that the membrane-binding, catalytic, and activation peptide domains are appreciably native relative to hPC.

Table 1

Average Expression Level and Anticoagulant Activities of Recombinant Human Protein C From Transgenic Swine

Immunopurified Products from Pooled Starting Material	Average hPC Antigen Level (μg/ml)	Specific Anticoagulant Activity (μ/mg)
rhPC transgenic Pig 29-2	500	$244 \pm 12^{\dagger}$
Human Plasma-derived hPC	4	213 ± 7

†immunopurified sample from pig 29-2 whey pool; > 40% antigen yield from starting whey

Table 1. Human Protein C antigen levels detected in the milk of transgenic swine. Control and transgenic whey samples (30) were diluted to an OD_{280} of 0.5 with TBS-BSA (0.5 M Tris HCl, 0.5% BSA, pH 7.5]. Immunopurified hPC was doped into control milk whey and used as reference. Samples (50 μl) were pipetted into microtiter plates which were coated with rabbit anti hPC polyclonal antisera. After washing with TBS-BSA-TWEEN (0.5 % Tween-80, pH 7.5), goat anti hPC polyclonal antisera (100 μl, 1:1000 dilution stock in TBS-BSA) was added to each sample well and incubated for 3 hours at room temperature. Sheep anti-goat IgG conjugated to horseradish peroxidase (100 μl of 1:2000 dilution stock with TBS-BSA) was then added to each well, followed by the chromophoric substrate o-phenyldiamine. The chromophore was detected at 490 nm using an EL308 Bio-Tek Microplate reader.

Activated Partial Thromboplastin Time (APTT) of porcine recombinant human Protein C. Dilutions of Normal human Reference Plasma Pool (NRPP) (American Bioproducts Inc., Parsippany, NJ) were made in Protein C depleted plasma (dhPC). NRPP was assumed to have 1 unit (U) of anticoagulant activity/ml and 4 μg hPC/ml. Thus, a theoretical specific anticoagulant activity here is defined as 250 U/mg. Sample dilutions were made in 50 mM imidazole, 100 mM NaCl, pH 7.3. Samples of immunopurified rhPC and hPC were diluted in dhPC. Identical stock solutions of APTT reagent (American Bioproducts) were prepared with or without 0.5 U/ml of snake venom activator (Protac™; American Diagnostics Inc., Greenwich, CT). Protac™ is an hPC-specific activator isolated from *Agkistrodon Contortrix* venom. Replicate samples of NRPP, hPC, and rhPC were diluted in dhPC and combined with APTT reagent. Protac™-specific anticoagulant activities were measured by the difference in APTT between sets with or without Protac™ using an Electra 750A coagulation timer. The APTT is plotted as a function of hPC antigen content. Essentially no anticoagulant activity was detected for samples not containing Protac™ (data not shown). The specific anticoagulant activities were calculated for those samples containing Protac™ (U/mg of hPC antigen ± SD).

Acknowledgements.

This research was supported in part by the Research Division of VPI&SU and by National Science Foundation grant BCS-9011098-01 to W.H.V. The authors thank Lothar Hennighausen and Christoph Pittius (Laboratory of Biochemistry and Metabolism, NIH Bethesda, MD 20892) for cloning the Whey Acidic Protein-Protein C construct.

References

Denman J, Hayes M, O'Day C, Edmunds T, Bartlett C, Hirani S, Ebert KM, Gordon K, McPherson JM: Bio/Technology 9:839, 1991.

Grinnell BW, Walls JD, Gerlitz B, Berg DT, McClure DB, Ehrlich H, Bang NU, Yan, SB: Native and Modified Recombinant Human Protein C: Function, Secretion, and Posttranslational Modifications. In *"Advances in Applied Biotechnology Series, Protein C and Related Anticoagulants,"* The Portfolio Publishing Company, 1990, pp. 29-63.

Velander WH, Morcol T, Clark DB, Gee D, Drohan WN: Technological Challenges for Large-Scale Purification of Protein C. In *"Advances in Applied Biotechnology Series, Protein C and Related Anticoagulants,"* The Portfolio Publishing Company, 1990, pp. 11-27.

Velander WH, Johnson JL, Page RL, Russell CG, Morcöl T, Wilkins TD, Canseco R, Williams BL, Gwazdauskas FC, Knight JW, Pittius C, Young JM, Drohan WN: High Level Expression in the Milk of Transgenic Swine Using the cDNA Encoding Human Protein C, in press, 1992.

Velander WH, Page RL, Morcöl T, Russell CG, Rodolfo C, Drohan WH, Gwazdauskas FC, Wilkins TD, Johnson JL: Production of Biologically Active Human Protein C in the Milk of Transgenic Mice, Annals New York Acad. Sci., in press, 1991.

Wall RJ, Pursel VG, Shamay A, McKnight RA, Pittius CW, Henninghausen L: Proc Natl Acad Sci USA 88:1696, 1991.

Wright G, Carver A, Cottom D, Reeves D, Scott A, Simons P, Wilmut I, Garner I, Colman A: Bio/Technology 9:830, 1991.

Expression of Human Lactoferrin in Milk of Transgenic Animals

Rein Strijker, Gerard Platenburg, Jan Nuyens, Frank Pieper, Paul Krimpenfort, Will Eyestone, Adriana Rademakers, Erika Kootwijk, Patricia Kooiman, Harry van Veen and Herman de Boer.
Gene Pharming Europe B.V. and Leiden University, P.O. Box 9502, 2300 RA Leiden, The Netherlands. (Fax: +31.71.215121)

Lactoferrin is the major iron binding protein in human milk. It has strong anti-bacterial properties and plays a role in iron uptake in suckling infants. We have shown that it is possible to produce high levels of human lactoferrin in the milk of transgenic mice. A transgene composed of hLF coding sequences fused to regulatory sequences of the bovine αS1-casein gene was expressed exclusively in the mammary gland and only after the onset of lactation. We have generated dairy calves transgenic for the lactoferrin fusion gene by combining microinjection technology, as developed for transgenic mice, with in vitro bovine embryology. These animals will be used to generate transgenic offspring.

Lactoferrin (LF) is the major iron binding protein in milk of many mammalian species (Reiter, 1987). Its concentration in human milk is about 1.7 mg/ml, which makes it one of the most abundant whey proteins. The ratio of iron-free and iron-saturated LF in whey is approximately 1 to 20. In human milk, two main roles have been suggested for LF. The first role would be to inhibit growth of several groups of (potentially pathogenic) bacteria both in the intestinal tract of the suckling infant and in the mammary gland of the mother. Based on 'in vitro' studies this antibiotic effect can be exerted through at least two different mechanisms viz., through direct binding to the outer membrane of gram-negative bacteria (Ellison and Giehl., 1991) and through iron deprivation thereby inhibiting growth of iron-dependent bacteria (Stuart, 1984). This latter mechanism is restricted to the iron-free form of the protein. The second role would be mediation of iron transport from the mother to the newborn (Saarinen and Siimes, 1979). This function is obviously restricted to the iron-saturated form. Beside bacterial growth inhibition and iron transport, it has also been reported that hLF can exert anti-inflammatory properties by scavenging of toxic oxygen radicals, by suppression of cytokine production, and by inhibition of complement activation. In view of this wide variety of properties, LF may have various applications in human health care.

We (Rey et al., 1990), and others (Rado et al., 1987; Powell and Ogden, 1990), have cloned the human LF cDNA and have determined its sequence. The protein contains two highly homologous domains, each with a single iron binding site and an N-glycosylation site. These domains share extensive homology with those of other members of the transferrin family. LF binds ferric-ions with high

1054–7487/92/0038$06.00/0

affinity ($K_{app}=10^{20}$), while incorporating equimolar amounts of carbonate-ions. Post-translational modification of the protein is not necessary for efficient iron binding (Anderson et al., 1990), but may play a role in the interaction with its receptor (Davidson and Lönnerdal, 1988).

The presence of both iron-free and iron-saturated LF in human milk is a prerequisite for the protein to fulfil both main functions described above. An expression system used to produce hLF should therefore also generate both forms. In addition, the protein should undergo specific post translational modifications and be available in relatively large quantities. Expression in milk of transgenic dairy animals is most probably the only way to meet these criteria. Since structure and function of the epithelial cells in the mammary gland is very conserved between different species, it is likely that post translational modifications and the degree of iron-saturation of hLF expressed in milk of transgenic animals are similar compared to human milk derived LF.

To direct expression of hLF to the mammary gland we have used regulatory sequences from the bovine αS1-casein gene. Bovine milk contains four different caseins all encoded by single-copy genes clustered in one locus (Ferreti et al., 1990; Threadgill and Womack, 1990). The proteins with the highest concentration in bovine milk are αS1- and β-casein (about 10g/l each); αS2- and κ-casein are present at lower concentrations (about 3g/l each).

As a modelsystem for the production of hLF in the milk of transgenic cows we have generated transgenic mouse lines producing hLF in their milk.

Design of mammary gland-specific expression cassettes. To direct hLF expression to the mammary gland we made several DNA constructs of which a general design is shown in Fig. 1. In these constructs we use large flanking regions of the bovine αS1-casein gene, its untranslated regions as well as its signal sequence fused to the coding sequence of the hLF cDNA. Expression levels of these fusion genes in transgenic mice are dependent on the specific design of the expressionplasmid (length of flanking regions and the origin of intervening sequences) as well as on the site of integration. The highest levels observed so far are well above 1 mg/ml.

Specificity of transgene expression. So far, the only tissue in which we were able

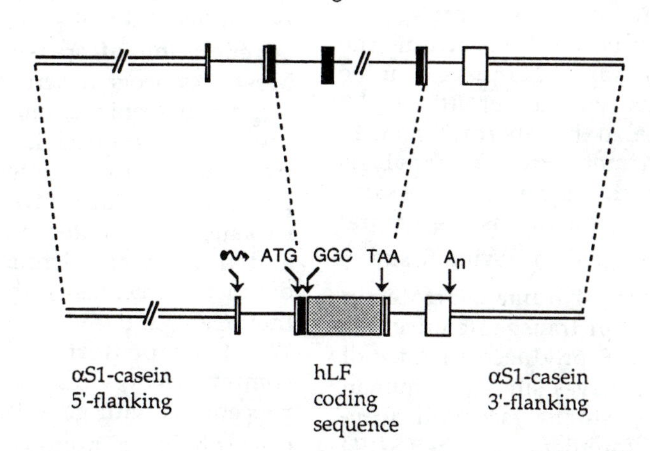

Fig.1. Construction of hLF cDNA constructs. Double lines indicate flanking regions. Single lines indicate intra-genic regions. Untranslated regions are indicated by open boxes. Translated regions are indicated by black boxes.

to detect expression of the transgene in lactating mice was the mammary gland (Platenburg et al., manuscript in preparation). Also, no expression has been observed in any tissue tested from male and virgin female mice. Tested tissues included salivary gland, muscle, thymus, kidney and liver. Around the day of parturition we observed a very strong induction of expression of the transgene. This is in contrast with endogenous murine casein genes which are already expressed abundantly during late pregnancy (Hennighausen et al., 1990), but is similar to the induction pattern observed for casein genes in the cow (Goodman and Schanbacher, 1991). Expression levels of the transgene generally increase during lactation to reach maximum levels at around mid-lactation. We have observed considerable variations in expression levels (up to three-fold) between milk samples of the same individual taken at narrow time intervals. Although physiological parameters will certainly contribute to this phenomenon it is also quite possible that other (more technical) aspects (such as he amount of interstitial fluid drawn into the milk) will play a role in this variation. It is therefore important to realize that exact expression levels of heterologous genes in milk of transgenic mice may not be very meaningful.

Transcription of the transgene. Northern analysis of RNA from the lactating mammary gland of mice transgenic for two different cDNA constructs revealed no aberrantly sized transcripts. Primer extension analysis showed that the transgene used the same transcription initiation site as the natural bovine αS1-casein gene in bovine tissue.

Characterization of human lactoferrin expressed in milk of transgenic mice. To assess whether hLF produced in milk of transgenic mice behaves similar to human milk derived LF, we analysed milk from several female founders, of their (F1) offspring and of the (F1) offspring of the male founders. Two types of radio immuno assays were developed to determine the amount of hLF present in milk. In one case monoclonal antibodies were coupled to sepharose while in the other case polyclonal antibodies were used. Serial dilutions of milk from transgenic mice were made and the amount of hLF bound to the anti-hLF antibody-sepharose was determined. Both assays showed identical levels of hLF and, more importantly, binding curves that are parallel to the standard curve. This latter result indicates that the affinities of the exposed epitopes of hLF from transgenic mouse milk to the antibodies is identical to those of hLF purified from human milk. To further characterize hLF in transgenic mouse milk, we purified the protein from transgenic mouse milk using a mono-S cation-exchange column. At the salt concentration where human milk derived hLF elutes from the column, a peak could be detected in the transgenic mouse milk as well. RIA-analysis showed that this peak contained essentially all hLF present in the original sample. Fractions eluting at lower salt concentrations contained the other whey proteins as could be judged from elution profiles of normal mouse milk. Analysis of the protein on a non-reducing 4-15% polyacrylamide-SDS gradient gel yielded the two bands that are characteristic for hLF as it occurs in human milk. We are currently comparing the two lactoferrin species in a number of functional and structural assays.

Generation of transgenic dairy cattle. Most protocols describing generation of large transgenic animals use essentially the same approach as applied for transgenic mice. However, since these protocols require two surgery steps (isolation of zygotes from donor animals and transfer of microinjected zygotes to oviducts of recipients) they are impractical, inefficient and therefore very expensive. We (Krimpenfort et al., 1991) have combined 'in vitro' embryo production procedures with gene transfer technology thus enabling non-surgical transfer of microinjected embryos that have developed from immature oocytes. Bovine oocytes were collected by aspiration of follicles present on ovaries obtained from local slaughterhouses. These oocytes were matured in vitro and subsequently fertilized with semen obtained from selected bulls. After 18-23 hours fertilized oocytes were centrifuged (to visualize pronuclei) and microinjected. After microinjection, the embryos were transferred to microdrops of medium conditioned by bovine oviduct epithelial cells. Embryo development was evaluated nine days after the onset of maturation. By that time approximately 20% of the embryos have developed to the compact morula/blastocyst stage. Embryos were transferred non-surgically to the uterus of hormonally synchronized recipient cows using standard procedures. A total of 21 pregnancies was established. During pregnancy two fetuses were lost. In both cases no intact DNA could be isolated for analysis. Nineteen calves were born after normal pregnancies. In two of these animals the transgene could be detected by Southern analysis and by PCR. In one case (a female) a rearrangement event had occurred resulting in a deletion at the 3'-end of the transgene. In addition, the calf was mosaic for integration of the transgene. In another case (a male) the transgene had integrated correctly as a head to tail concatemer of appr. 3 copies. The transgene is present in all tissues tested (including sperm). As a further improvement to this protocol of generating transgenic calves we are currently developing methods to assess integration of the transgene in pre-implantation embryos. Incorporation of these methods in our 'in vitro' program will further reduce the number of recipients needed to generate (larger numbers of) transgenic cattle on a routine basis.

References

Anderson, B.F., Baker, H.M., Noris, G.E., Rice, D.W. and Baker, E.N. *Nature* **1990**, *344*, 784-787

Davidson,L.A. and Lönnerdal,B. *Am. J. Physiol.* **1988**, 254, G580-G585

Ellison, R.T. andGiehl T.J. *J. Clin. Invest.* **1991**, *88*, 1081-1091

Ferreti, L., Leone, P. and Sgaramella, V. *Nucl. Ac. Res.* **1990**, *18*, 6829-6833

Goodman, R.E. and Schanbacher F.L. *Biochem. Biophys. Res. Comm.* **1991**, *180*, 75-84

Hennighausen, L., Westphal, C., Sankaran, L. and Pittius, C.W. in *Transgenic Animals*; Eds., N. First and F.P. Haseltine; Butterworth-Heinemann: Stoneham, Ma, **1990**; pp 65-74.

Krimpenfort, P., Rademakers,A., Eyestone W., Van der Schans, A., Van den Broek, S., Kooiman, P., Kootwijk, E., Platenburg, G., Pieper, F., Strijker, R. and De Boer, H. *Bio/Technology* **1991**, *9*, 844-847

Powel, M.J. and Ogden, J.E. *Nucl. Ac. Res.* **1990**, *18*, 4013

Rado, T.A., Wei, X. and Benz, E.J. *Blood* **1987**, *70*, 989-993

Reiter, B. *Ann. Rech. Vet.* **1987**, *9*, 205-209

Rey, M.W., Woloshuk, S.L., De Boer, H.A. and Pieper, F.R. *Nucl. Ac. Res.* **1990**, *18*, 5288

Saarinen, U.M. and Siimes, M.A. *Pediat. Res.* **1979**, *13*, 143-147.

Stuart R.J. *Int. J. Biochem.* **1984**, *16*, 1043-1048

FRONTIERS IN PRODUCTION OF METABOLITES SYMPOSIUM II

D. D. Y. Ryu and D. A. Hopwood: *Co-Chairs*

Pathway Analysis and Pathway Engineering: Session A
D. D. Y. Ryu and D. A. Hopwood: *Co-Chairs*

Designing Novel Pathways: Session B
B. D. Ensley, Jr., and K. M. Timmis: *Co-Chairs*

High Throughput Screens: Session C
R. L. Greasham and L. Nisbet: *Co-Chairs*

Nonprotein Biomaterials: Session D
A. J. Sinskey and V. Crescenzi: *Co-Chairs*

Optimization of Recombinant Fermentation Processes through the Analysis of Effects of Promoter Derepression and Metabolic Flux Change

Dewey D.Y. Ryu* **and Jeong-Yoon Kim** Department of Chemical Engineering, University of California, Davis CA 95616, USA (Fax: 916-752-3112)

Effects of derepression of a strong promoter and altering metabolic flux on the recombinant cell yield were analyzed. It was found that the turning-on of the PL promoter, irrespective of downstream protein coding sequences, reduced the cell yield. When the *par* locus from pSC101 was inserted in the same plasmid having the PL promoter, the inhibitory effect of derepressed PL promoters was partially reversed. The recombinant cells having low copy number plasmids which have the *pdc* and *adh* genes under the control of the tac or PL promoter showed the same growth rate as the host cells unless glucose is completely exhausted in rich media. In continuous culture experiments with glucose minimal media, the recombinants produced higher cell concentration than the host cells regardless of biomass yields.

One of the ways to obtain the maximum productivities of recombinant proteins is to increase cell densities, hence volumetric productivities. The metabolic burden imposed on recombinant cells actively expressing cloned genes may be divided into two categories. One is the reallocation of nutrients to cloned gene products and the other is the competition for essential protein synthesizing machinery such as RNA polymerase and ribosomes. To meet the altered metabolic requirements, recombinant cells must adapt themselves to the adverse physiological condition. Thus, it is desirable to enable the recombinant cells to overcome the metabolic constraints and obtain high cell concentration. To achieve this goal, the effects of derepression of the PL promoter, presence of the *par* locus, and alteration of metabolic flux by introducing a low copy number plasmid having the *pdc* and *adh* genes on cell growth were studied.

The Effects of Derepressed Promoter and *Par* Sequence on Cell Growth

It was reported that the concentration of functional RNA polymerase in the cytoplasm could limit transcription in *E. coli* (Churchward, *et al.*, 1982). Jensen and Pederson (1990) argued that the individual promoters compete with each other for a limited amount of free RNA polymerase. This view is based not only on the experimental evidence by Churchward *et al.* but also on the proposition that the growth behavior of *E. coli* may be considered analogous to an enzymatic reaction that shows saturation kinetics (Jensen and Pedersen, 1990). If their arguments hold true, the sequestering of functional RNA polymerase by strong promoters cloned in multicopy plasmids is expected to have a negative effect on cell growth.

We studied how the sequestering of active RNA polymerases by the strong promoters present in multicopy

1054–7487/92/0044$06.00/0

plasmids affects the growth of recombinant *E. coli*. The turning-on of the PL promoter, whether or not protein-coding sequences are present downstream of the promoter, caused a lowering of growth rate of the recombinant cell. This result suggests that the active RNA polymerase could be one of the limiting factors responsible for the decreased growth rate of recombinant cells. Interestingly, the negative effect of derepressed PL promoters on the cell growth rate was partially reversed when the *par* locus from pSC101 was inserted into the same plasmid. In all three cases, where each plasmid carried the PL promoter only, the PL promoter followed by N-terminal part of protein-coding region, or the PL promoter followed by complete sequences encoding a protein, the *par* locus showed a positive effect on the growth rate of recombinant cells which usually grew at a slower rate than the host cell.

Analysis and Implementation of Metabolic Flux Change to Recombinant Fermentaion

When carbon flux exceeds the metabolic capacity of the central metabolic pathways, *E. coli* maintains its balance between anabolism and catabolism by one of three ways: conversion of carbon source to storage polymers (e.g. glycogen), futile cycle, or excretion of acetate (El-Mansi and Holms, 1989). Acetate excretion is used to maintain the balance between the carbon flux to energy generation and the carbon fluxes to product biosynthesis and cell growth (Holms, 1986). As glucose concentration in the medium increases or specific growth rates increase, carbon flux from pyruvate to acetate increases. This coincides with the repression of NADH and succinate dehydrogenase (Doelle, *et al.*, 1982).

In order to alter the carbon flux through the acetate pathway and prevent acetate accumulation, pyruvate decarboxylase (PDC) and alcohol dehydrogenase (ADH) genes were subcloned into a low copy plasmid (5 copies per cell). *E. coli* strains carrying multicopy plasmids with the *pdc* and *adh* genes, which alter the pathway of surplus pyruvate to ethanol instead of acetate, have been grown to higher cell densities than the host strain, largely due to the relief from the inhibitory effects of low pH and acetate itself (Ingram and Conway, 1988). The expression of these two genes subcloned into a low copy plasmid (pSLL) was controlled by either the tac promoter (pSLL-T) or the PL promoter (pSLL-PL). Fig. 1 shows the growth of various strains, glucose consumption and acetate production in LB plus 0.2% glucose medium. Although the strain harboring the multicopy plasmid, pLOI295, encoding the *pdc* and *adh* genes produced a small amount of acetate compared to the pSLL-T- or the pSLL-PL-carrying strains, its growth rate was significantly lower than the strains carrying the low copy number plasmids (Fig. 1A). The slightly lower biomass yields of the pSLL-T- or pSLL-PL- containing cells as compared to the host cells may result from the depleted pyruvate pool after glucose in the medium was exhausted (Fig. 1A and 1B). The acetate accumulation was significantly lowered for both of these recombinants as compared to the host cells (Fig. 1A and 1B). As long as glucose is present in the culture media, the expression of the *pdc* and *adh* genes in the low copy plasmids seems to have no adverse effect on the cell metabolism.

To further investigate the characteristics of the new recombinant cells carrying pSLL-T or pSLL-PL, continuous culture experiments were carried out with glucose minimal medium. The recombinant strain, pSLL-T/K12ΔH1Δ*trpEA*, gave higher cell concentration than the host strain (K12ΔH1Δ*trpEA*) in the range of dilution rates tested (Fig. 2A). Acetate yields of the recombinant strain based on glucose consumed were lower by about 70% at $D=0.3h^{-1}$ and about 30% above $D=0.5h^{-1}$ compared to those of the host strain (Fig. 2B). While the host strain produced a negligible amount of ethanol, the ethanol yields for the recombinant strain were about 0.2 - 0.3 mol ethanol per mol glucose consumed (Fig. 2B). Similar

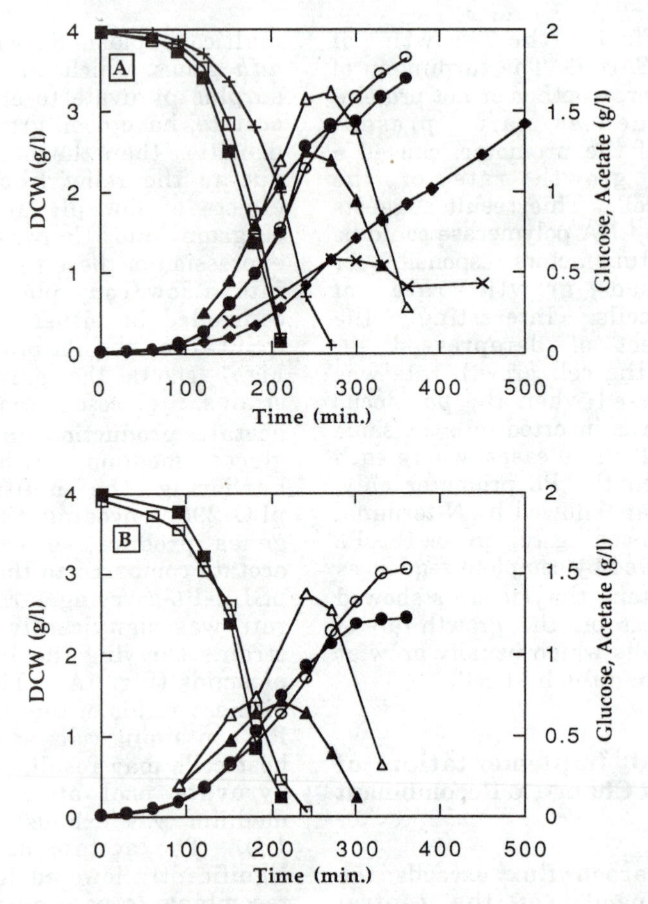

Figure 1. Growth, acetate formation, and glucose consumption in LB plus 0.2% glucose media. A. K12ΔH1Δ*trpEA* (-○-:DCW, -□-:glucose, -Δ-:acetate), pSLL-T/K12ΔH1Δ*trpEA* (-●-:DCW, -■-:glucose, -▲-:acetate), pLOI295/K12ΔH1Δ*trpEA* (-◆-:DCW, -+-:glucose, -×-:acetate). B. MV1190 (-○-:DCW, -□-:glucose, -Δ-: acetate), pSLL-PL/MV1190 (-●-: DCW, -■-: glucose, -▲-: acetate)

results were obtained with a host strain, MV1190, and a recombinant strain, pSLL-PL/MV1190 (results not shown). One interesting thing is that the expression of the *pdc* and *adh* genes has different effects on carbon fluxes depending on whether or not glucose is limiting. The input carbon flux was lower in the recombinant cell than in the host (e.g. 28.7 mmol/g/h vs. 33 mmol/g/h at D = $0.3h^{-1}$) when glucose was limiting. On the contrary, at high dilution rates where glucose was not limiting, the recombinant strain had higher input carbon flux than the host strain (e.g. 60.04 mmol/g/h vs. 57.36 mmol/g/h at D = $0.7h^{-1}$). However, more biomass was obtained with the recombinant strain regardless of biomass yields, as measured in terms of g cell per mol glucose (Fig. 2A and 2B).

This stimulating effect on the biomass formation seems to be related to the fast regeneration of oxidized pyrimidine nucleotides (NAD^+). The operation of the central metabolic pathways reduces large quantities of NAD^+ to NADH. To keep catabolic reduction charge [NADH/(NADH + NAD^+)] at a constant value (0.03-0.07), the reduction of NAD^+ to NADH should be coupled to reactions that reoxidize

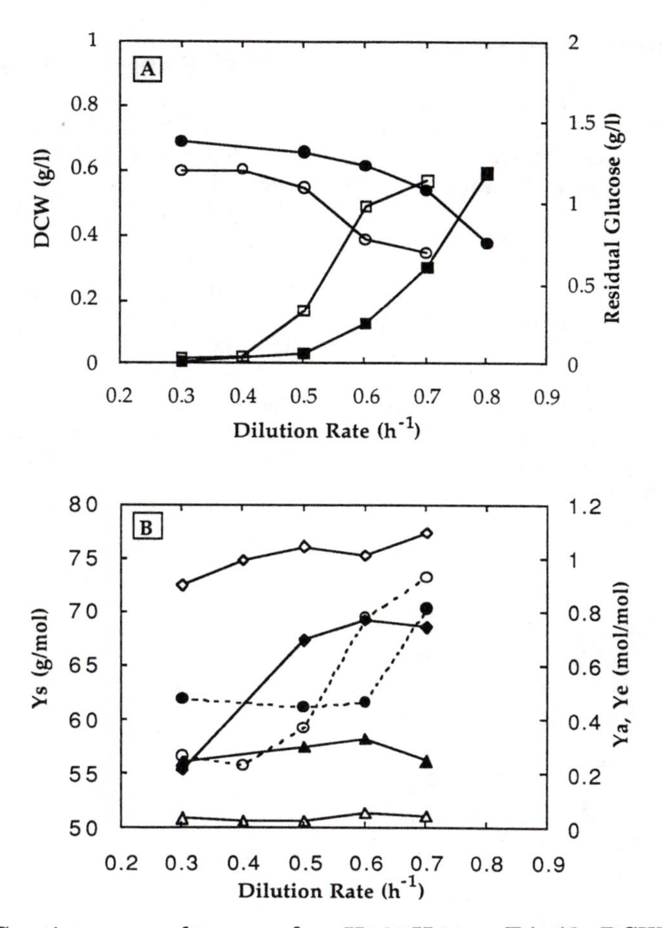

Figure 2. Continuous cultures of K12ΔH1Δ*trpEA* and pSLL-T/K12ΔH1Δ*trpEA*. A. K12ΔH1Δ*trpEA* (-○-:DCW, -□-:glucose), pSLL-T/K12ΔH1Δ*trpEA* (●-:DCW, -■-:glucose). B. K12ΔH1Δ*trpEA* (-◇-, acetate yield, Ya; -Δ-, ethanol yield, Ye; --○--, cell yield, Ys). pSLL-T/K12ΔH1Δ*trpEA* (-◆-: Ya, -▲-:Ye, --●--:Ys)

NADH, which are normally accomplished either by electron transport via the respiratory chain in aerobic conditions or by fermentation pathways in anaerobic environments. Or, transhydrogenase plays a role in balancing NADH, NAD^+, NADPH, and $NADP^+$ (Neidhardt, *et al.*, 1990). The expression of the *pdc* and *adh* genes adds the pathway of anaerobic reoxidization of NADH to the aerobic respiratory pathway so that the inhibitory effects of high concentration of NADH on the pathways leading to precursor metabolites and monomers will be decreased. The increased concentration of precursor metabolites and monomers may contribute to the formation of extra biomass, which was suggested by others (Jensen and Pedersen, 1990; Marr, 1991).

The results presented and discussed above enable us to make use of this system, namely, the *pdc* and *adh* genes cloned in a low copy plasmid, in recombinant fermentation. By cotransforming a multicopy plasmid with the genes of desired proteins and a low copy plasmid with the *pdc* and *adh* genes, replication origins of which belong to different incompatibility groups, it will be possible to increase cell densities, and reduce the accumulation of acetate in recombinant fermentation at the same time, thereby increasing productivity to cloned gene protein.

Acknowledgments

The authors are grateful to Prof. L.O. Ingram for providing the plasmid, pLOI295. This research was supported by the National Science Foundation.

References

Churchward, G.; Bremer, H.; Young, Y. *J. Bacteriol.* **1982,** 150(2), 572-581.
Doelle, H. W.; Ewings, K. N.; Hollywood, N. W. Regulation of glucose metabolism in bacterial systems. In *Advances in Biochemical Engineering* ; A. Fiechter Ed.; Springer-Verlag Berlin Heidelberg New York, **1982,** vol 23; pp. 1-35 .
El-Mansi, E. M. T.; Holms, W. H. *J. Gen. Microbiol.* **1989,** 135, 2875-2883.
Holms, W. H. *Curr. Top. Cell. Reg.* **1986,** 28, 69-105.
Ingram, L. O.; Conway, T. *Appl. Environ. Microbiol.* **1988,** 54(2), 397-404.
Jensen, K. F.; Pedersen, S. *Microbiol. Rev.* **1990,** 54(2), 89-100.
Marr, A. G. *Microbiol. Rev.* **1991,** 55(2), 316-333.
Neidhardt, F. C.; Ingraham, J. L.; Schaechter, M. *Physiology of the Bacterial Cell* ; Sinauer Associates, Inc.: Sunderland, MA; **1990.**

High level expression of a bacterial *cef*E gene in *Penicillium chrysogenum*

Stephen W. Queener, Robert J. Beckmann, Cathleen A. Cantwell, Roland L. Hodges, Deborah L. Fisher, Joe E. Dotzlaf, Wu-Kuang Yeh, and Derek McGilvray
Lilly Research Laboratories, Indianapolis, IN 46285, USA

A hybrid *cef*E gene, encoding penicillin N expandase, was constructed by fusing the promoter sequences and terminator sequences from the *Penicillium chrysogenum* *pcb*C gene to the open reading frame from the *Streptomyces clavuligerus* *cef*E gene. The resulting hybrid gene was transformed into *P. chrysogenum* on plasmid vector pRH6. Among transformants obtained with pRH6, transformant 9EN-5-1 exhibited a penicillin N expandase specific activity 4- fold higher than the activity in the *S. clavuligerus* strain used as the source of the *cef*E orf. The specific activity of the *P. chrysogenum* transformant was 75% of the specific activity observed in an industrial strain of *Cephalosporium acremonium*.

At least three clinically important oral cephalosporins-- cephalexin, cephradine, and cefadroxil (Kohler et al., 1986) -- are manufactured by adding appropriate side-chain groups to the 7-amino moiety of 7-aminodeacetoxycephalosporanic acid (7-ADCA). (Bunnell et al., 1986). The 7-ADCA is made by chemical ring expansion of penicillin G isolated from *P. chrysogenum* . For structures, see Fig.1.

In the filamentous bacterium *S. clavuligerus* , penicillin N is enzymatically ring expanded to form deacetoxycephalosporin C [DAOC] (Dotzlaf and Yeh 1989); the enzyme is encoded by the *cef*E gene [Kovacevic et al. 1989]. If penicillin N expandase could be engineered to accept penicillin G or V as substrate and the corresponding gene could be expressed efficiently in *P. chrysogenum*, then several chemical reactions in organic solvents could be replaced with a single enzymatic reaction in aqueous medium.

Here we report high level expression of a bacterial penicillin N expandase gene in *P. chrysogenum* using a hybrid *cef*E gene containing transcription and termination sequences from the fungus.

Materials and methods

Microbial strains. For all transformations of *P. chrysogenum*, strain OM6-4-16, which produces penicillin at an industrially significant level, was used. *Escherichia coli* strain RV308 was used for *E. coli* transformations in the construction of pRH6; strain DH5α was used in the construction of pRH5. Transfections of *E. coli* using m13mp18 derived bacteriophages required DH5αF' and DH5α cells (Bethesda Research Laboratories, Gaithersburg, Maryland). *C. acremonium* strains was the industrial strain 394-4 (Cantwell et al. 1990). The *S. clavuligerus* strain used was ATCC 27064 (Dotzlaf and Yeh, 1989).

Source plasmids. pUT715 was obtained from Labortoire Cayla, F-31094 Toulouse Cedex, France. pLC2 and pPS64 were constructed in this laboratory (Carr et al. 1986, Cantwell et al. 1990). M13mp18 was purchased from Boehringer Mannheim Biochemicals, Indianapolis, Indiana. pKC787, the source of an apramycin resistance gene, is a derivative of pKC222 [Kaster et al. 1983] and was obtained from Nagaraja Rao, Eli Lilly and Company, Indianapolis, Indiana.

Plasmid construction scheme. Plasmids were constructed as follows: Δ**2** : a gel purified 2.5 kb *Bam*H1 to *Eco*RI fragment from pLC2 containing the *P. chrysogenum pcb*C terminator fragment and a gel purified 2.7 kb *Xba*I to *Bam*H1 fragment from pPS64 containing the *S. clavuligerus cef*E open reading frame (orf) were ligated to a 7 kb gel purified *Eco*RI to *Xba*I fragment of m13mp18. Δ**3** : The single

1054–7487/92/0049$06.00/0

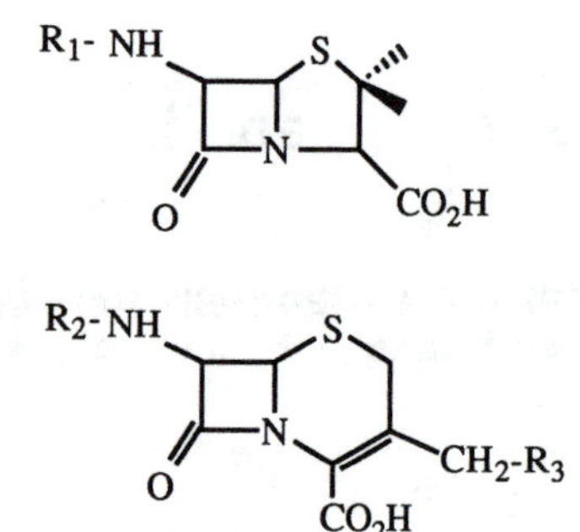

Fig. 1. Structures of β-lactam compounds and metabolites. Penicillin N, R_1 = δ-D-α-aminoadipyl-; isopenicillin N, R_1 = δ-L-α-aminoadipyl-; penicillin V, R_1 = phenoxyacetyl-; penicillin G, R_1 = phenylacetyl-; 7-aminodeacetoxycephalosporin C, R_2 = H, R_3 = H; deacetoxycephalosporin C, R_2 = δ-D-α-aminoadipyl-, R_3 = H; cephalosporin C, R_2 = δ-D-α-aminoadipyl-, R_3 = acetyl-.

stranded form of Δ2 was isolated and modified by *in vitro* mutagenesis to remove 720 bases and join the orf of the *cef*E gene in frame with the stop codon and terminator sequences of the *pcb*C gene. **pRH5**: Fragments from *Eco*RI and *Nco*I digested pLC2, *Eco*RI and *Bam*H1 digested pKC787, and *Nco*I and *Bgl*II digested pUT715, were ligated and the resulting product was transformed into *E. coli*. Transformants were first screened for apramycin resistance, ampicillin sensitivity, and lack of β-galactosidase activity. Plasmids in selected transformants were then screened by restriction endonuclease digestion analysis. A plasmid was selected that contained three fragments: a 2 kb *Eco*RI to *Nco*I *pcb*C fragment from pLC2 [source of *pcb*C promoter], a 1.2 kb *Nco*I to *Bgl*II fragment from pUT715 [source of *ble* orf and *Aspergillus nidulans* terminator sequences] and the large *Eco*RI to *BamH1* fragment of pKC787. **pRH6**: a gel isolated 3.3 kb *Xba*I to *Eco*RI fragment from Δ3 and a gel isolated 1 kb *Eco*RI to *Xba*I *pcb*C promoter fragment from pPS64 were ligated to *Eco*RI digested pRH5. The structure of pRH6 is illustrated in Figure 2.

Other methods. Isolation of DNA, Southern hybridization analysis, transformation of *E. coli*, agarose gel analysis, restriction endonuclease digestion, DNA fragment isolation, and ligations. were as described previously (Skatrud and Queener 1984; Cantwell et al. 1990.) Synthesis, purification and phosphorylation of synthetic oligonucleotides, single-stranded DNA phage isolation, in vitro single-stranded mutagenesis, plaque hybridization, genetic transformation of *P. chrysogenum*, protein extraction, penicillin N expandase assays, mitotic stability of ble^r transformants are described elsewhere (Cantwell et al., 1990, 1992 and manuscripts in preparation).

Results

Specific activity for penicillin N expandase in pRH6-P. chrysogenum transformant 9EN-5-1 approached the level observed in a commercial strain of Cephalosporium acremonium

Three pRH6-transformants that exhibited mitotically stable resistance to phleomycin and retained pRH6 DNA integrated into their high molecular weight DNA [data not shown] were analyzed for penicillin N expandase activity during fermentation. The recipient *P. chrysogenum* strain in a standard penicillin fermentation reproducibly exhibited no expandase activity in cell samples harvested at 48 hr to 120 hr of the fermentation cycle. pRH6 transformant 9EN-5-1 exhibited a maximal penicillin N expandase specific activity of 92 milliunits/mg protein at 72 hr. The expandase activity in transformant 9EN-5-1

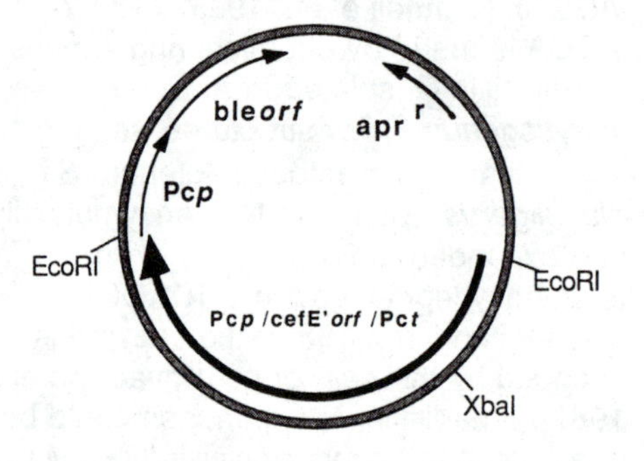

Fig. 2. Structure of plasmid pRH6, a vector for high level expression of a gene encoding *Streptomyces clavuligerus* penicillin N expandase in *P. chrysogenum*. Pc*p* = promoter from *P. chrysogenum* pcbC gene; Pc*p* = termination sequences from *P. chrysogenum* pcbC gene; cefE'*orf* = open reading frame of cefE gene, prepared as described in text; ble *orf* = open reading frame of phleomycin resistance gene attached to termination sequences of *Aspergillus nidulans* trpC gene; apr^r = apramycin resistance gene for selection in *E. coli*.

was four fold higher than that in the *S. clavuligerus* strain used as the source of the cefE $_{orf}$

The specific activity of penicillin N expandase in crude extracts of transformant 9EN-5-1 was about 75% of that observed in an industrial strain of *C. acremonium*. *C. acremonium* is a fungus used commercially to produce cephalosporin C, an important intermediate for manufacture of injectable cephalosporin antibiotics. Cephalosporin C is not used for preparation of 7-ADCA because its does not possess the desired methyl substituent at C3 in the six-membered dihydrothiazine ring. Unlike *S. clavuligerus*, *C. acremonium* produces a bifunctional expandase/hydroxylase encoded by a *cef*EF gene. The bifunctional enzyme efficiently ring expands penicillin N to DAOC and also efficiently hydroxylates DAOC at its C3 methyl. This hydroxylation activity is unwanted for manufacture of 7-ADCA.

The relative amounts of penicillin N expandase activity (at maximum expression) for the three pRH6 transformants were 9EN-5-1 > 9EN-5-2 > 9EN-5-3. pRH6-transformants did not produce DAOC.

Discussion

The construction of *P. chrysogenum* transformants that exhibit penicillin N expandase specific activities ca. equivalent to the those observed in a commercial strain of *C. acremonium* increases the likelihood that, with the further work, commercially significant amounts of cephalosporins may someday be produced in *P. chrysogenum*. Since the bacterial penicillin N expandase exhibits negligible DAOC 3'-hydroxylase activity, the cephalosporins produced in these recombinant *P. chrysogenum* strains should possess the methyl at C3 on the dihydrothiazine ring, a desirable characteristic for manufacture of 7-ADCA.

The specific activity of penicillin N expandase in pRH6 transformant 9EN-5-1 was 70-fold higher than the maximal specific activity observed in pPS65-*P. chrysogenum* transformant WG9-61L-3. WG9-61L-3 was the most active of 35 pPS65-*P. chrysogenum* transformants. Plasmid pPS65 contains a hybrid cefE gene comprised of the promoter sequence Pc_p fused to the *S. clavuligerus* *cef*E orf still attached to the *S. clavuligerus* terminator sequences.

Recently the pRH6 hybrid cefE gene *and* a hybrid bacterial *cef*D gene has been Introduced and expressed in *P. chrysogenum*. The cefD gene encodes isopenicillin N epimerase which allows conversion of isopenicillin N to penicillin N. DAOC was produced in these transformants (Cantwell et al., 1992).

References

Bunnell, C.A.; Luke, W.D.; and Perry, F.M. In: *Beta-lactam Antibiotics for Clinical Use.* Queener, S.F.; Webber, J.A.; Queener, S.W. Ed; Marcel Dekker Inc., New York, **1986**, pp 255-283.

Carr, L.G.; Skatrud, P.L.; Sheetz II, M.E.; Queener, S.W.; Ingolia, T.D. *Gene* **1986**, *48*: 257-266

Cantwell, C.A.; Beckmann, R.J.; Dotzlaf, J.E., Fisher, D.L.; Skatrud, P.L.; Yeh, W.K.; Queener, S.W. *Curr. Genet.* **1990**, 17: 213-221.

Cantwell,C.; Beckmann,R.; Whiteman,P.; Queener, S.W.; and Abraham,E.P. *Proc. Royal Soc., Ser. B* **1992**, *in press.*

Dotzlaf, J.; Yeh, W.K. *J. Bacteriol.* **1987**, *169* : 1611-1618.

Dotzlaf, J.; Yeh, W.K. *J. Biol. Chem.* **1989**, *264*: 10219-10227.

Kaster,R.K.; Burgett,S.G.; Rao,R.N, Ingolia,T.D. *Nucleic Acid Res.* **1983**, *11* : 6895-6911.

Kohler, R.B.; Wheat, L.J.; White, A.C.; In: *Beta-lactam Antibiotics for Clinical Use.* Queener, S.F.; Webber, J.A.; Queener, S.W. Ed; Marcel Dekker Inc., New York, **1986**, pp 351-370.

Kovacevic, S.; Weigel, B.; Tobin, M.B.; Ingolia, T.D.; Miller, J.R. *J. Bacteriol.* **1989**, *171*: 754-760

Kovacevic, S.; Tobin, M.B.; Miller, J.R.; *J. Bacteriol.* **1990**, *172* : 3952-3958.

Skatrud, P.L; Queener, S.W. *Curr. Genet.* **1984**, *8* : 155-163.

Production of Tropane Alkaloids by Recombinant Plant Cells and Analysis of Metabolic Pathway

Yasuyuki Yamada*, Takashi Hashimoto, and **Dae Jin Yun**
Department of Agricultural Chemistry, Faculty of Agriculture, Kyoto University, Kyoto 606-01, Japan (FAX: 81 75 753 6398)

We foresee that within several years molecular biology will be used to improve yields of useful secondary products in whole plants and in plant tissue and cell cultures. We here describe our recent work on improving plant alkaloid contents by genetic engineering. A cDNA of hyoscyamine 6β-hydroxylase, which catalyzes the first reaction in the biosynthetic pathway from hyoscyamine to scopolamine in several solanaceous plants, was cloned from *Hyoscyamus niger* and placed under the control of the cauliflower mosaic virus 35S promoter in a plant expression vector. When the engineered gene was introduced and expressed constitutively in hairy roots and in regenerated plants of normally hyoscyamine-rich *Atropa belladonna*, scopolamine synthesis was markedly enhanced in the transgenic tissues.

Hyoscyamine and scopolamine, 6,7-epoxide of hyoscyamine, are two major tropane alkaloids that accumulate in plants of the Solanaceae. Although both alkaloids are important anticholinergic agents that act on the parasympathetic nervous system, they have different effects on the central nervous system: Hyoscyamine excites the central nervous system, whereas scopolamine suppresses it. The ratio of hyoscyamine to scopolamine contents varies markedly among plant species, high scopolamine contents being found in only a few species. These differences result in a higher commercial demand for scopolamine than for hyoscyamine (and its racemic form atropine). Because several tropane alkaloid-producing species (*Datura*, *Scopolia* and *Atropa* species) accumulate hyoscyamine as the major alkaloid and scopolamine in minor quantities, it is of commercial importance to increase the scopolamine contents in these species.

Biosynthesis of Scopolamine

The epoxide oxygen of scopolamine is derived from molecular oxygen, its addition being catalyzed by a 2-oxoglutarate-dependent dioxygenase, hyoscyamine 6β-hydroxylase (H6H; EC 1.14.11.11)(Fig. 1). H6H incorporates one atom of dioxygen into the 6β-position of [*S*]-hyoscyamine as a hydroxyl group while using the other dioxygen atom for the oxidative decarboxylation of 2-oxoglutarate. This reaction requires ferrous ion and ascorbate, and probably involves a ferryl enzyme as the highly reactive reaction intermediate (Hashimoto and Yamada, 1986, 1987; Yamada and Hashimoto, 1989). In plants, the product of the enzyme reaction, 6β-hydroxyhyoscyamine (Hyos-OH) is converted to scopolamine by 7β-dehydrogenation, thus retaining the 6β-hydroxyl oxygen in scopolamine (Hashimoto and Yamada, 1989). Low enzyme activity which catalyzes this epoxidation reaction in the presence of co-factors of 2-oxoglutarate-dependent dioxygenases was found in cultured roots of *Hyoscyamus niger*, and the epoxidase was co-purified with H6H during partial purification (Hashimoto et al., 1989). Moreover, a monoclonal antibody raised against purified H6H inhibited both hydroxylase and epoxidase activities to similar extents. Therefore, the low epoxidase activity probably is catalyzed by H6H.

Tropane alkaloids mainly are synthesized in plant roots then transported to the aerial parts. Our recent immunohistochemical studies done with an H6H-specific antibody and immunogold-silver enhancement detected H6H only in the pericycle (the outermost cell layer of the young vascular cylinder in the root) in several scopolamine-producing plants (Hashimoto et al., 1991). This pericycle-specific localization of scopolamine biosynthesis provides an anatomical explanation for the root-specific biosynthesis of tropane alkaloids and may be of importance for the translocation of alkaloids from the roots through the vascular cylinder to the aerial plant parts.

1054–7487/92/0052$06.00/0

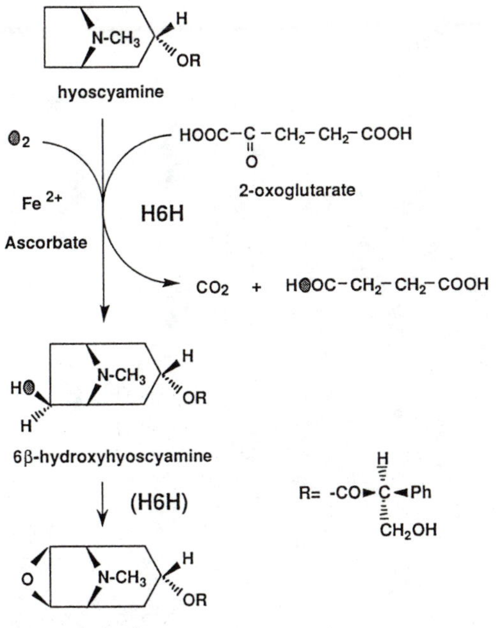

Figure 1. Hyoscyamine 6β-hydroxylase (H6H) catalyzes two consecutive oxidation reactions from hyoscyamine to scopolamine. H6H belongs to the 2-oxoglutarate-dependent dioxygenases which require 2-oxoglutarate, ferrous ions, molecular oxygen and ascorbate for catalysis. The first hydroxylation activity is about 50-fold stronger than the second.

Molecular Cloning of H6H cDNA from *H. niger*

H6H from cultured roots of *H. niger* was purified to homogeneity in a series of chromatographic steps (Yamada et al., 1990). After the homogeneous H6H protein had been digested with lysylendopeptidase, its internal peptides were separated by HPLC and analyzed in a protein sequencer. Amino acid sequences of 12 peptides were determined, three of which were used to design oligonucleotide probes. After screening a cDNA library made from the mRNA from cultured *H. niger* roots, several hybridization-positive clones were obtained. All the clones encoded a polypeptide of MW 38,999 and differed only in the lengths of their 5'-untranslated regions and their polyadenylation sites. All 12 internal peptide fragments of the purified H6H were present in the amino acid sequence deduced from the cDNA, and the predicted amino acid composition well matched the composition for the purified H6H (Matsuda et al., 1991).

To demonstrate the function of the polypeptide encoded in the cloned gene, we constructed an expression vector such that an intact protein is translated from the first ATG of the cDNA under the control of the *lacZ* promoter. The *E. coli* harboring this vector expressed a 38-kDa protein which was recognized by an H6H-specific antibody on Western blots and showed distinct H6H enzyme activity. This unequivocally demonstrates that the cDNA encoded H6H. Northern hybridization showed that H6H mRNA is abundant in cultured roots, and is present in whole plant roots but absent from plant leaves and stems, and from cultured cells of *H. niger*.

Expression of *H. niger* H6H Enhances Scopolamine Formation in Transgenic *A. belladonna*

H6H is a promising target enzyme which, when expressed highly in hyoscyamine-accumulating plants, would result in increased oxidation of hyoscyamine, and higher concentrations of scopolamine. We used two methods to introduce the *H. niger* H6H gene into *Atropa belladonna*, a typical hyoscyamine-rich plant. A plant expression vector was constructed in which the cDNA was placed under the control of the CaMV 35S promoter and the NOS terminator in pGA482 (Pharmacia)(Fig. 2).

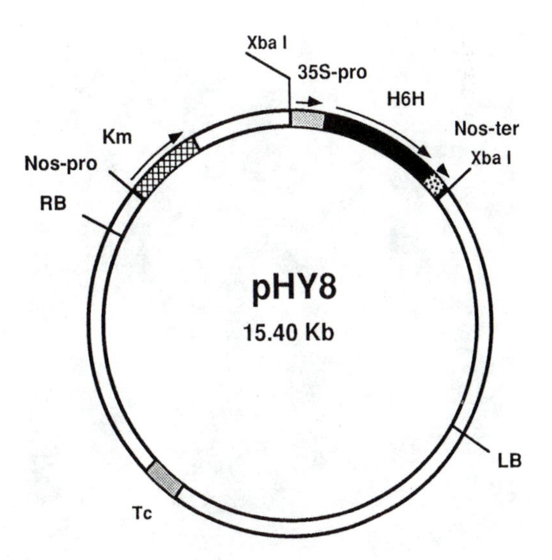

Figure 2. Structure of the binary vector pHY8. The 15.4-kb plasmid contains an NPT-II gene as well as an *XbaI* DNA fragment that encodes the chimeric 35S-H6H-Nos gene in the T-DNA region.

First, the *Agrobacterium rhizogenes*-based binary vector system was used to transform *A. belladonna* leaf explants. The hairy root clones obtained then were selected for kanamycin resistance, expected as the neomycin phosphotransferase II gene was placed next to the H6H gene in the vector. The presence of the *H. niger* H6H gene in several kanamycin-resistant root clones was confirmed by the polymerase chain reaction. These hairy root clones were further screened by Western blotting using an H6H-specific antibody for the clones with high expression of H6H protein. One of the hairy root clones obtained showed 3- to 10-fold higher H6H enzyme activity and the same fold higher contents of Hyos-OH and scopolamine than the wild-type hairy root clones.

Next, a transgenic *A. belladonna* plant having the *H. niger* H6H gene was obtained by *A. tumefaciens*-based transformation (Fig. 3). In this transgenic plant and its T_1 progeny, the introduced H6H gene (driven by the CaMV 35S promoter) was expressed in almost all the tissues (including the leaf and the stem in which H6H protein normally is absent), and the alkaloid composition in the leaf was constituted almost exclusively of scopolamine (Fig. 4). The H6H enzyme expressed in the leaf and stem probably has access to hyoscyamine during the translocation of the alkaloid from the root to the leaf. Genetic engineering of medicinal plants for better pharmaceutical constituents is now shown to be feasible.

Figure 3. A transgenic *A. belladonna* plant at flowering. The plant expresses the *H. niger* H6H constitutively in almost all its tissues and is resistant to kanamycin because of the presence of the NPT-II gene.

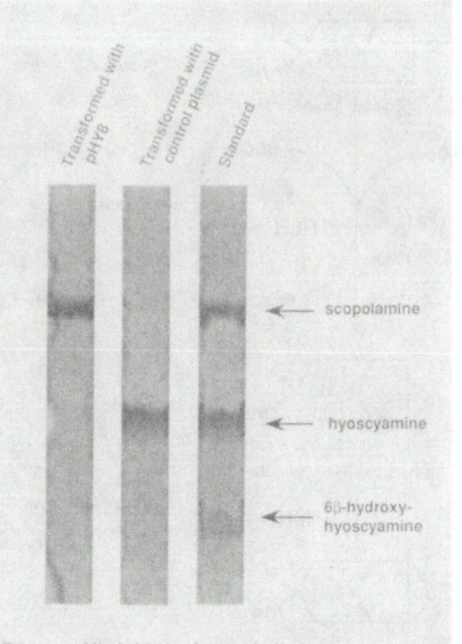

Figure 4. Tropane alkaloids in the leaves of *A. belladonna* plants analyzed by TLC together with Dragendorff's reagent. The transgenic plant expressing H6H almost exclusively contains scopolamine; whereas, the control plant transformed with the empty vector and the wild-type plants (not shown) mainly contain hyoscyamine. Right lane: positions of the alkaloid standards.

References

Hashimoto, T.; Yamada, Y. *Plant Physiol.* **1986**, 81, 619-625.

Hashimoto, T.; Yamada, Y. *Eur. J. Biochem.* **1987**,164, 277-285.

Hashimoto, T.; Yamada, Y. *Agric. Biol. Chem.* **1989**, 53, 863-864.

Hashimoto, T.; Kohno, J.; Yamada, Y. *Phytochemistry* **1989**, 28, 1077-1082.

Hashimoto, T.; Hayashi, A.; Amano, Y.; Kohno, J.; Iwanari, H.; Usuda, S.; Yamada, Y. *J. Biol. Chem.* **1991**, 266, 4648-4653.

Matsuda, J.; Okabe, S.; Hashimoto, T.; Yamada, Y. *J. Biol. Chem.* **1991**, 266, 9460-9464.

Yamada, Y.; Hashimoto, T. *Proc. Japan Acad. Sci., Ser. B* **1989**, 65, 156-159.

Yamada, Y.; Okabe, S.; Hashimoto, T. *Proc. Japan Acad. Sci., Ser. B* **1990**, 66, 73-76.

Two-Dimensional Gel Electrophoretic Approaches to the Analysis of Secondary Metabolic Pathways

Tom G. Holt*, **Karen M. Karlis**, **Richard L. Monaghan**, Merck Research Laboratories, P. O. Box 2000, Rahway, New Jersey (FAX: 908-594-5468).
Charles J. Thompson, Institut Pasteur, 25, rue du Dr. Roux, 75724 Paris Cedex 15, France.

Two-dimensional gel electrophoresis was used to analyze extracts of various *Zalerion arboricola* strains in order to identify proteins involved in the biosynthesis of the pneumocandin class of antifungal antibiotics. By comparing the two-dimensional gel patterns of crude extracts of several compositional mutants (mutants producing various ratios of the pneumocandins), proteins uniquely associated with pneumocandin A_0 biosynthesis were identified as resolved spots on two-dimensional gels. In related studies using bialaphos biosynthesis in *Streptomyces hygroscopicus* as a model system, cluster analysis was applied to quantitative two-dimensional gel electrophoretic data in order to identify families of coordinately regulated gene products.

We report the application of a two-dimensional gel electrophoretic approach to analyze secondary metabolic pathways. Two-dimensional gel electrophoretic data were used both to identify gene products essential for a given antibiotic biosynthetic pathway and to look for families of coordinately regulated functions related to antibiotic biosynthesis. Preliminary results of the approach are presented here for two biosynthetic systems. First, in a well-characterized, model system, quantitative two-dimensional gel data were analyzed using cluster analysis in order to uncover families of proteins with similar kinetics of biosynthesis. The kinetic profiles of these families were, in turn, related to growth and antibiotic biosynthesis in the culture. Second, in an essentially uncharacterized system, two-dimensional gels were compared for a series of compositional mutants in order to identify potential antibiotic biosynthetic gene products.

Model System: Bialaphos Biosynthesis in *Streptomyces hygroscopicus*

The two-dimensional gel electrophoretic approach was first applied to study the regulation of bialaphos biosynthesis in *Streptomyces hygroscopicus*. This antibiotic biosynthetic system has been well-characterized in recent years. The biosynthetic pathway was elucidated in detail (Hidaka, et al., 1990), the corresponding gene cluster was identified (Murakami, et al., 1986; Hara, et al., 1988), and a regulatory gene, *brpA*, was discovered which controls the expression of most genes in the bialaphos biosynthetic gene cluster (Raibaud, et al., 1991). The discovery of *brpA* set the stage for a two-dimensional gel electrophoresis study comparing global gene expression of the $brpA^+$ parental strain with a $brpA^-$ mutant (Holt, et al., 1992). As a result, many of the products of the

1054–7487/92/0055$06.00/0

gene cluster were identified as discrete protein spots on two-dimensional gel electropherograms.

We examined quantitative two-dimensional gel data from a series of pulse-labeled time-course experiments in an effort to uncover natural, biologically relevant groups of coordinately regulated *S. hygroscopicus* gene products. From extracts which were prepared of mycelia pulse-labeled at two-hour intervals both before and after the onset of bialaphos biosynthesis, two-dimensional gel autoradiography reproducibly resolved 346 major proteins. The integrated intensities of these two-dimensional protein spots were quantified using QUEST software (Garrels, 1989) which also served to match the coordinates of each spot with its corresponding coordinates in every gel of the time-course experiment. The QUEST analysis resulted in a quantitative measure of the relative kinetics of biosynthesis for each protein at every sampled time-point. These data were then placed into groups having similar kinetics of synthesis using the centroid clustering algorithm available on SAS/STAT®. Seventy-five clusters or groups of proteins were generated which varied in size from 1 to 57 members; the nine major kinetic patterns (accounting for 51% of the proteins clustered) are shown in Fig. 1. Products of the bialaphos gene cluster, previously identified by their positions on two-dimensional gel electropherograms, were assigned primarily into one single group. The synthesis of proteins found in this group was initiated immediately prior to the appearance of bialaphos in the medium.

Cluster analysis enabled us to identify several proteins not subject to *brpA* control but displaying similar kinetic patterns. For example, cluster 5, a group of 14 proteins, included ten *brpA*-dependent proteins and four unknown proteins with similar patterns of regulation. The suggestion that these four unknown proteins may be associated with bialaphos biosynthesis remains to be further investigated.

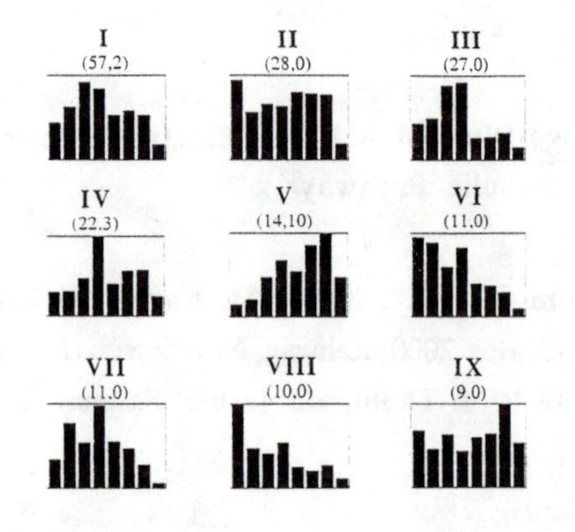

Fig. 1. Average normalized kinetic profiles of major families of *Streptomyces hygroscopicus* gene products. Cluster analysis was used to group *S. hygroscopicus* proteins, identified as spots on two-dimensional gels, into families related by their kinetics of biosynthesis. Average normalized kinetic profiles of the nine largest families are shown (ordered by size and labeled I - IX) which account for 51% of the proteins analyzed. Bar graphs show rates of synthesis in integrated intensity of labeled protein (determined by QUEST analysis (Garrels, 1989)) vs. age of culture. Bars are shown for the following time points (in order): 10 h, 12 h, 14 h, 16 h, 18 h, 20 h, 22 h, and 42 h. Two numbers are shown in parentheses above each profile: the first is the total number of proteins in the group, and the second is the number of proteins in group which are *brpA*-dependent.

The observed kinetics of expression suggested systems of coordinate control and indicated when these systems were activated with respect to the growth curve. The kinetic approach described here could also apply to studies of the regulation of secondary metabolic genes in other organisms in which genetic information is not readily available, provided a distinct onset of secondary metabolism is observed.

Unknown System: Pneumocandin Biosynthesis in *Zalerion arboricola*

The two-dimensional gel electrophoretic approach has now been applied to study biosynthesis of the pneumocandins, a system for which little is known. The pneumocandins, which are produced by the fungus, *Zalerion arboricola*, belong to the echinocandin class of acylated cyclic hexapeptide antibiotics. We became interested in the biosynthesis of pneumocandin A_0 (also known as L-671,329) with the report that the compound in animal studies was effective against *Pneumocystis carinii* pneumonia (Schmatz, et al., 1990).

The structures of pneumocandin A_0 and B_0 (also known as L-688,786) differ in the 4-position of the second of two proline residues (depicted in the 11 o'clock position in Fig. 2). Using ^{13}C-labeled precursors to pneumocandin A_0 and B_0 it was recently shown that whereas in pneumocandin B_0 the 3-hydroxyproline derives from proline, in pneumocandin A_0 the methylhydroxyproline derives from an unusual cyclization of leucine (Adefarati and et al., 1991a, 1991b). The cited study notwithstanding, little else is known regarding the biosynthesis of the pneumocandins.

The specific goal of our effort was to identify, as resolved spots on two-dimensional gels, gene products involved uniquely in the pneumocandin A_0 pathway (leucine cyclization pathway). In related studies, a series of compositional mutants were available (mutants producing various ratios of pneumocandins A_0 and B_0). This mutant series comprised a parental

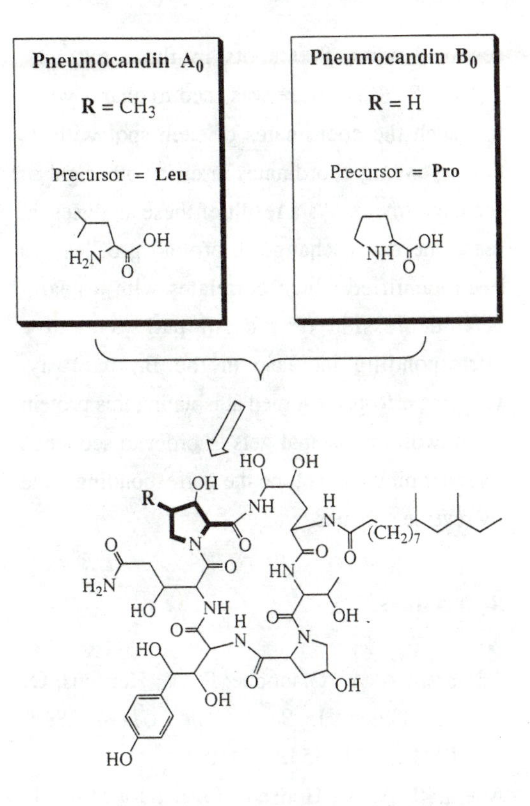

Fig. 2. Structures of pneumocandins A_0 and B_0. The two structures differ in the biosynthetic origin of the second of two proline residues (shown in boldface type at the 11 o'clock position in the cyclic peptide structure). Whereas this second proline residue derives from proline in pneumocandin B_0, in pneumocandin A_0 it derives from the cyclization of leucine (Adefarati et al., 1991a, 1991b).

strain (wild type) which produced primarily pneumocandin A_0 and two successive mutant strains which produced successively less pneumocandin A_0 and more pneumocandin B_0 (Fountoulakis, et al., 1991; Sosa, et al., 1991). The strains were pulse-labeled and protein extracts were prepared for these strains throughout the course of development of the culture. Two-dimensional gel electrophoresis resolved some 800 spots in these extracts. PDQUEST™ software was used to quantify and to match the coordinates of each spot with its corresponding coordinates in every other gel of the experiment. As a result of these analyses, at least one major change in protein profiles has been identified which correlates with a nearly total depression of the A_0 pathway and a corresponding increase in the B_0 pathway. Ongoing efforts are aimed at isolating this protein from two-dimensional gels in order to sequence its *N*-terminus and clone the corresponding gene by reverse genetics.

References

Adefarati, A. A.; Giacobbe, R. A.; Hensens, O. D.; Tkacz, J. S. *J. Am. Chem. Soc.* **1991a**, *113*, 3542-3545.

Adefarati, A. A.; Hensens, O. D.; Jones, E. T. T.; Tkacz, J. S. *Abstracts of Papers*, 202nd National Meeting of the American Chemical Society, New York, NY; American Chemical Society: Washington, DC, **1991b**; MEDI 58.

Fountoulakis, J. M.; Kaplan, L.; Masurekar, P. S. *Abstracts of Papers*, Annual Meeting of the Society for Industrial Microbiology, Philadelphia, PA; Society for Industrial Microbiology: Arlington, VA, 1991; P28.

Garrels, J. I. *J. Biol. Chem.* **1989**, *264*, 5269-5282.

Hara, O. H.; Anzai, S. I.; Kumada, Y.; Murakami, T.; Itoh, R.; Takano, E.; Satoh, A.; Nagaoka, K. *J. Antibiot.* **1988**, *41*, 538-547.

Hidaka, T.; Imai, S.; Hara, O.; Anzai, H.; Murakami, T.; Nagaoka, K.; Seto, H. *J. Bacteriol.* **1990**, *172*, 3066-3072.

Holt, T. G.; Chang, C.; Laurent-Winter, C.; Murakami, T.; Garrels, J. I.; Davies, J. E.; Thompson, C. J. *Molec. Microbiol.* **1992**, in press.

Murakami, T.; Anzai, H.; Imai, S.; Satoh, A.; Nagaoka, K.; Thompson, C. J. *Mol. Gen. Genet.* **1986**, *205*, 42-50.

Raibaud, A.; Zalacain, M.; Holt, T. G.; Tizard, R.; Thompson, C. J. *J. Bacteriol.* **1991**, *173*, 4454-4463.

Schmatz, D. M.; Romancheck, M. A.; Pittarelli, L. A.; Schwartz, R. E.; Fromtling, R. A.; Nollstadt, K. H.; Vanmiddlesworth, F. L.; Wilson, K. E.; Turner, M. J. *Proc. Natl. Acad. Sci. USA* **1990**, *87*, 5950-5954.

Sosa, M. S.; Hallada, T. C.; Kaplan, L.; Masurekar, P. S. *Abstracts of Papers*, Annual Meeting of the Society for Industrial Microbiology, Philadelphia, PA; Society for Industrial Microbiology: Arlington, VA, 1991; P27.

Genetic Manipulation of Central Carbon Metabolism in *Escherichia coli*

Neilay Dedhia, Thomas Hottiger, Wilfred Chen and James E. Bailey
Department of Chemical Engineering, 210-41, California Institute of Technology, Pasadena, CA 91125, USA

A strategy to harness the carbon flux which currently flows to harmful byproducts is discussed. Genes *glgC* and *glgA* coding for the enzymes of the glycogen biosynthesis pathway were introduced into *Escherichia coli.* We found that an *E. coli* strain deficient in the acetate synthesis pathway could grow to higher cell densities compared to control strains when glycogen was overproduced. Glycogen levels were about ten times higher and extracellular pyruvate levels were lower than in the control strain. The decrease in pyruvate levels was consistent with the increase in glycogen levels. Glycogen levels in the cells could be varied by inducible expression of *glgC* and *glgA* using a *tac* promoter.

For maximum yield of recombinant products in organisms, it is desirable to cultivate the organisms at high cell densities. Acetate production leads to a waste of carbon as well as inhibition of cell growth at high concentrations (Luli et al., 1990; Zabriskie et al., 1986). Acetate production results from an imbalance between the carbon uptake rate and the dissimilatory rate. The excess carbon overflows into acetate. To block acetate production, *E. coli* strains which are deficient in the acetate pathway can be used. We have used strain TA3476 (Levine et al., 1980) which has mutations in the genes for phosphotransacetylase (*pta*) and acetate kinase (*ackA*). This strain cannot produce acetate. Previous work in our lab (Diaz-Ricci et al., 1991) with this strain showed that other unusual products like pyruvate accumulate in the medium. This suggested that a more successful global solution would be to shift the flow of excess carbon into other products which are less harmful and potentially useful.

Redirection of Carbon Flux to Glycogen

Our strategy involves the redirection of carbon flux to glycogen, a storage polysaccharide in *E. coli.* The pathway from glucose to glycogen (Preiss, 1984) is shown in the accompanying diagram.

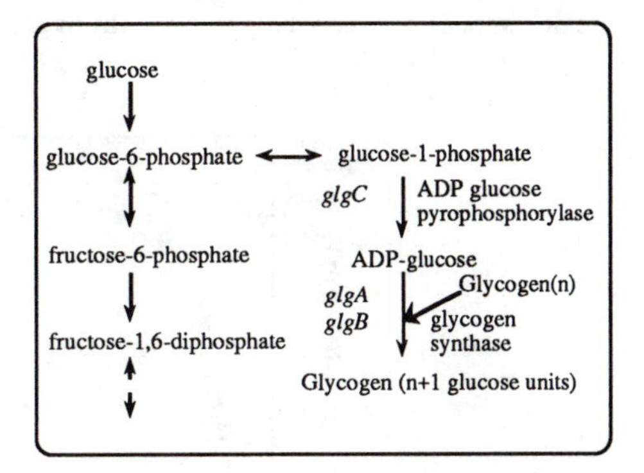

To overproduce glycogen, we used the genes *glgC* and *glgA*. The construct, pPR2, (Romeo et al., 1990) contains the *glgC*, *glgA* genes genes with their native promoter cloned into pUC19. It has been shown (Romeo et al., 1990) that the *glgC* and *glgA* are under transcriptional control. Glycogen is mainly produced only under stationery phase conditions.

1054–7487/92/0059$06.00/0

Effect of glycogen overproduction

Shake flask fermentations were done with the acetate minus strain TA3476 carrying the plasmid pPR2. The fermentations were done at 37 °C, 275 rpm using LB medium supplemented with 0.2% glucose. The control was TA3476 carrying the plasmid pUC19. TA3476:pPR2 overproduced glycogen to 10 times higher levels than in the control strain. The glycogen was produced mainly during late exponential phase. The glycogen overproducing strain grew to higher final cell densities. The total protein content was 25% higher than in the control strain (Fig. 1, Fig. 2). Pyruvate levels were lower compared to the control (Fig. 3). The decrease in pyruvate levels could be almost totally accounted for by the increase in glycogen. We speculate that the increase in final cell densities is due to the higher substrate efficiency of glycogen as compared to pyruvate. During late exponential phase, the strain with the glycogen plasmid has glycogen as a potential carbon source while the control strain has pyruvate. Glycogen depolymerization provides greater ATP production than does pyruvate dissimilation. Thus the differences in the final cell densities between the two strains can probably be explained on the basis of metabolic economics. The differences between the glycogen strain and the control strain were observed only when the LB medium was supplemented with 0.2% glucose. Differences in final cell densities were insignificant in LB medium. The differences in final cell densities for wild-type *E. coli.* strains (i.e. no mutations in the acetate pathway) with and without plasmid pPR2 were also insignificant in LB plus 0.2% glucose.

Modulation of Glycogen Synthesis

To achieve varying levels of glycogen in the cell and to study their effect on the carbon flux distribution, we cloned the *glgC* and *glgA* genes under the control of the *tac* promoter. With the resulting construct, pGT100, we were able to vary the amount of glycogen in the cell by varying the amount of inducer (IPTG) added in the medium (Fig. 4). The level of glycogen synthesis appears to be saturated at 100 μM IPTG. This system will allow us to optimize the level of glycogen production as well as to control the timing of glycogen induction.

Discussion

Glycogen is an attractive material in which to deposit the excess carbon flux. We can

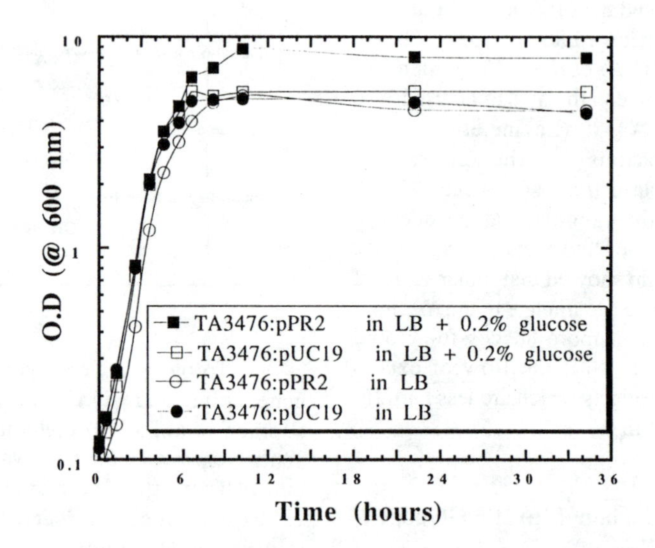

Figure 1. Growth curves of TA3476:pPR2, TA3476:pUC19 in LB+0.2% glucose. Cells are grown in shake flasks at 37°C, 275 rpm

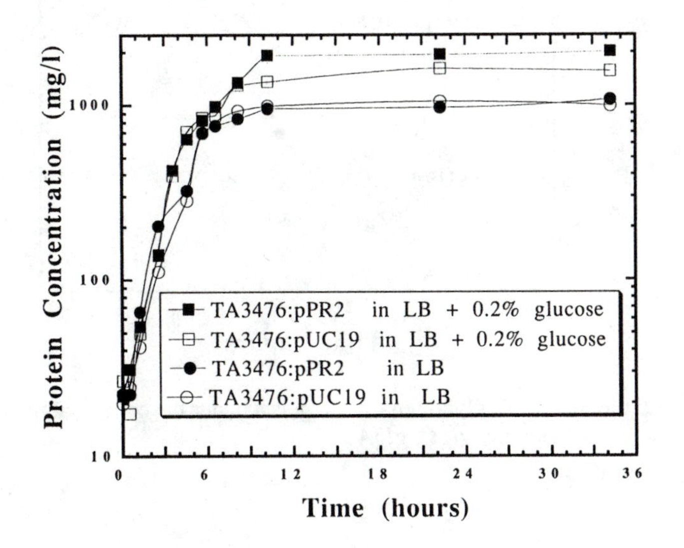

Figure 2. Total Protein Content

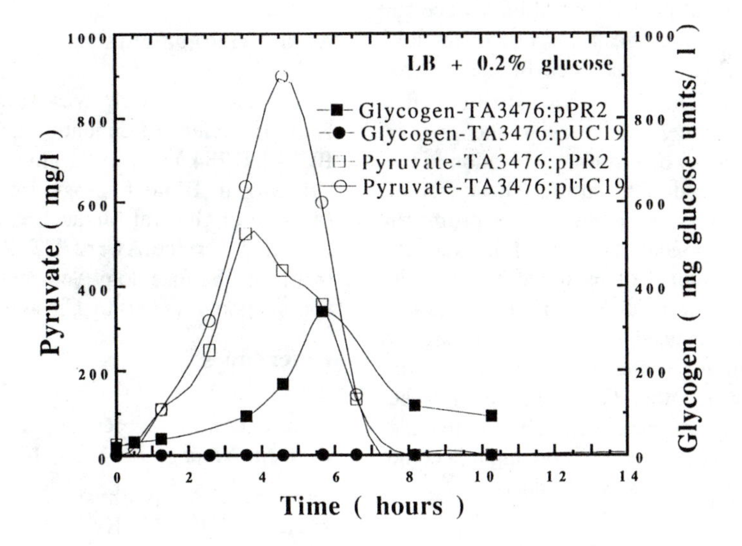

Figure 3. Glycogen and Pyruvate levels.

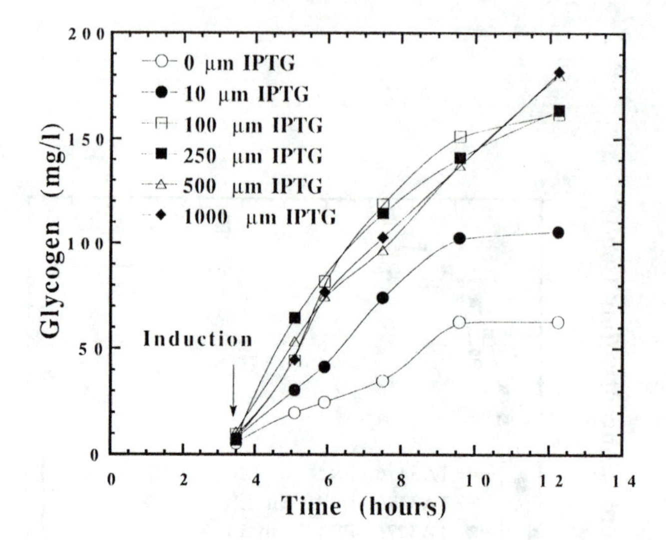

Figure 4. Varying the level of glycogen in TA3476:pGT100 which carries the *glgC*, *glgA* genes under the control of the *tac* promoter.

utilize carbon more efficiently as well as avoid end-product inhibition by organic acid byproducts. It should be noted that in our particular case end-product inhibition by acetate was probably not significant because of the low cell densities that are achieved in shake flask fermentations. Therefore, we attribute the improvement in cell densities to the greater energetic efficiency of substrate utilization in the glycogen overproducing strain. The onset of glycogen synthesis with the native promoter begins in late exponential phase. At this stage, a considerable amount of pyruvate has already accumulated in the medium. The inducible *tac* promoter can be used to vary the timing of glycogen synthesis and study its effect on pyruvate flux. Another factor that should be taken into consideration is the possibility that the rate of glycogen depolymerization might become limiting after glycogen is overproduced.

Outlook

This example demonstrates how metabolic engineering can be used to design cells to attain the goal of increasing higher cell densities and achieving greater efficiencies of substrate utilization. The strategy we have outlined could potentially be extended to channel the excess carbon flux to useful products such as amino acids.

Acknowledgements

This research was supported by the National Science Foundation (Grant No. NSF BCS- 891284) and by a grant for Predoctoral Training in Biotechnology from the National Institute of General Medical Sciences, National Research Service Award 1 T32 GM 08346-01 from the Pharmacology Sciences Program. T.H. was supported in part by Ciba-Geigy Ltd.

References

Diaz-Ricci, J.C.; Regan, L.; Bailey, J.E. *Biotech. Bioeng.* **1991,** *38,* 1318

LeVine, S.A; Ardhesir F.; Ames, G.F.A. *J. Bact.* **1980**, *143*, 1081

Luli, G.W.; Strohl W.R. *Appl Envir* **1990**, *56*, 1004

Preiss, J. *Ann. Rev. Microbiol.* **1984,** *38*, 419

Romeo, T.; Preiss J. *J. Bact.* **1989,** *171*, 2773

Zabriskie, D.W.; Arcuri, E. *J. Enz. Microb. Tech* 1986, *8(12)*, 705

The Synthesis of New Carbohydrate-Based Materials and Their Use for the Stabilization of Proteins

Matthew R. Callstrom,* Tara G. Hill, Charles A. Wartchow, Michael E. Huston, M. Bradley Smith, Department of Chemistry, The Ohio State University, Columbus, Ohio 43210 (FAX: 614 292-9128)
Mark D. Bednarski,* Lynn M. Oehler, and Peng Wang, Department of Chemistry, University of California at Berkeley, Berkeley, California 94720, and the Center for Advanced Materials, Lawrence Berkeley Laboratory, Berkeley, California 94720.

We report the synthesis of a series of carbohydrate-based polymers which contain a high-density of masked aldehyde functionality and the coupling of proteases to these materials through their ε-lysine residues by reductive amination methodology. This mild, multi-site attachment reaction does not significantly alter the active site of the enzymes. We also report the stability of these carbohydrate protein conjugates of proteins [CPC(proteins)] at elevated temperatures in distilled water and buffered solutions and their catalytic activity in organic solvents.

As part of our program for the preparation and study of biological macromolecules using chemical and enzymatic methods for their synthesis, we have discovered a series of carbohydrate-based materials that stabilize proteins. We describe here our approach to the preparation of these materials, the use of some of these macromolecules for the stabilization of proteases, and the catalytic chemistry of these carbohydrate protein conjugates of proteases [CPC(proteases)] in organic solvents for the formation of peptide bonds (Scheme I).

Our strategy is to utilize the carbohydrate as a template for the introduction of desired functionality into these polymeric materials. The combination of chemical and enzymatic syntheses of carbohydrate-based monomers and their free-radical polymerization provides complete control of the regio- and stereochemistry contained in the resultant macromolecules. Importantly for our research program, control of the structure of these materials provides a handle to manipulate the critical factors that limit protein stabilization by their covalent modification, including: (1) reactivity of the anomeric center of the carbohydrate moiety, (2) hydrophobicity/hydrophilicity of the surrounding microenvironment of the protein, (3) charge content near the surface of the protein, (4) solubility of the protein conjugate in organic solvents and organic solvent — water mixtures, and (5) molecular weight of the protein conjugate.

We have prepared a series of carbohydrate-based polymers, coupled proteases to these materials through their ε-lysine residues, examined the stability of these conjugates at elevated temperatures in low ionic strength and

1054–7487/92/0063$06.00/0

Scheme I

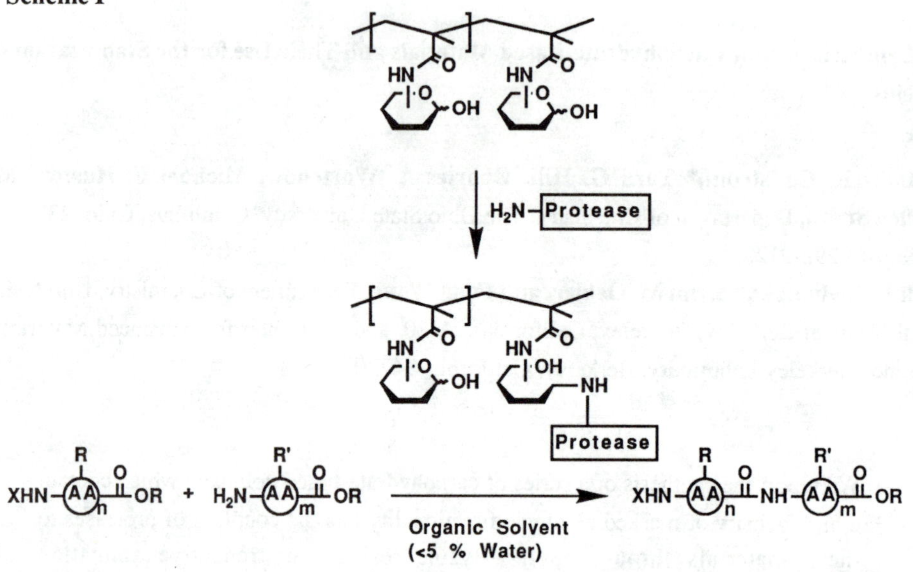

buffered water solutions (Hill, et al., 1991) and found that these carbohydrate protein conjugates of proteases [(CPC-proteases)] are extremely effective for the catalytic formation of peptide bonds in organic media (Scheme I) (Wang, et al., 1991, Wang, et al., 1992).

The preparation of these carbohydrate-derived macromolecules is efficiently accomplished by the reaction of methacryloyl chloride with an aminosugar followed by solution free-radical initiated polymerization. For example, treatment of 2-amino glucose with methacryloyl chloride in methanol gave **1** in quantitative yield. Subsequent solution polymerization (in quadruply-distilled water) of **1** with ammonium persulfate as a free-radical initiator at temperatures from 5°C - 70°C gave the carbohydrate-based macromolecule **2** in yields of >80% (Scheme II). This water-soluble macromolecule contains a high density of masked aldehyde functionality and an absolute molecular weight of >4 x 10^6 with a polydispersity of <1.4.

Incubation of **2** with the desired enzyme and sodium cyanoborohydride in borate buffer (pH 9) at 35 °C gave conjugates of chymotrypsin [CPC(CT)], trypsin [CPC(Try)], and subtilisin BPN' [CPC(BPN')] (Gray, 1974, Schwartz and Gray, 1977, Gray, et al., 1978, Gray, 1978, Roy, et al., 1984) (Scheme II). The yield of the coupling reactions were found to be greater than 40% in all cases as determined by measurement of the relative V_{max} of the native enzymes and their respective carbohydrate protein conjugates. The k_{cat} and K_m values of the native enzymes and their respective CPC-proteases did not differ by more than 30% (k_{cat}, K_m): CT (40 s^{-1}, 33 μM), **2**-CPC(CT) (46 s^{-1}, 20 μM); Try (760 s^{-1}, 0.90 μM), **2**-CPC(Try) (890 s^{-1}, 1.2 μM); BPN' (240 s^{-1}, 83 μM), **2**-CPC(BPN') (350 s^{-1}, 76 μM) (DelMar, et al., 1979, Schonbaum, et al., 1961, Erlanger, et al., 1961, Bender, et al., 1966, Thomas, et al., 1985).

The CPC materials were easily purified by dialysis of the reaction mixture against buffered solutions or by gel-filtration chromatography. Lyophilization gave white powders that could be stored indefinitely at room temperature without loss of catalytic activity. Amino-acid analysis of the CPC-proteases found that approximately 3 lysines of each enzyme are conjugated to the carbohydrate-based macromolecule.

Table I contains the results of the use of CPC(proteases) for the catalytic synthesis of peptide bonds in organic solvents. Examination of the data contained in Table I finds that **2**-CPC(CT) gives greater than 95% yields of dipeptide in tetrahydrofuran (THF), dioxane and acetonitrile (CH_3CN). **2**-CPC(Th) and **2**-CPC(BPN') also operate in acetonitrile and acetonitrile-water mixtures with high catalytic efficiency. The V_{max} for the formation of peptide bonds in acetonitrile is approximately 0.1 - 1 mmol/min/mg of **2**-CPC(CT) which is of the same order of catalytic efficiency as that for the cleavage of peptide bonds in aqueous systems.

Under identical experimental conditions [~0.1 μmol protease (~0.1 mg protease/mL) per 0.5 mmol peptide] the native enzymes exhibited little or no activity for the same coupling reactions. In some cases the use of a large excess of the protein catalyst can compensate for its low catalytic activity and high rate of decompositon in organic solvents. However, we felt that this approach will not be practical for the extension of this chemistry to the synthesis of large peptides since the isolation of the desired polypeptide from the enzyme and enzyme degradation products would be difficult. In addition, with large excesses of the native protein present in solution and the long reaction times necessary for these coupling reactions, hydrolysis of the ester substrate is competitive with peptide bond formation resulting in low yields of the desired product and further complications for its isolation.

The carbohydrate-protein conjugates of α-chymotrypsin [CPC(CT)], trypsin [CPC(Try)], and subtilisin BPN' [CPC(BPN')] exhibit enhanced stability at elevated temperatures and in low ionic strength water (distilled) solutions. α-Chymotrypsin, for example, suffers greater than 80% loss of its activity within 3 h at 45 °C in buffered solution while **2**-CPC(CT) and the 6-aminoglucose analog [**3**-CPC(CT)] retained greater than 80 % of its activity after 15 h under identical conditions (Fig. 1).

Scheme II

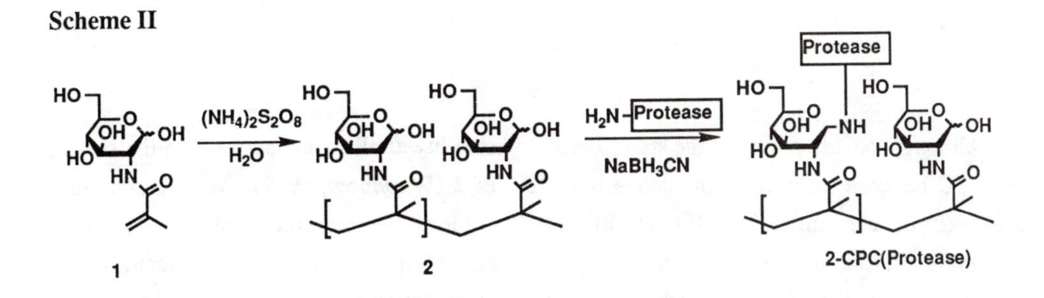

Table I. Peptide bond formation using carbohydrate protein conjugates of α-chymotrypsin [2-CPC(CT)], subtilisin BPN' [2-CPC(BPN')], thermolysin [2-CPC(Th)] and the native enzymes in organic solvents.

Enzyme	Solvent[a]	Time (h)	Acceptor Amino Acid	Donor Amino Acid[b]	Product[c]	Yield (%)
2-CPC(CT)	THF	12	Ac-F-OEt	A-NH_2	Ac-FA-NH_2	98
CT	THF	12	Ac-F-OEt	A-NH_2	Ac-FA-NH_2	2
2-CPC(CT)	Dioxane	24	Ac-F-OEt	A-NH_2	Ac-FA-NH_2	98
CT	Dioxane	24	Ac-F-OEt	A-NH_2	Ac-FA-NH_2	3
2-CPC(CT)	CH_3CN	24	Ac-F-OEt	A-NH_2	Ac-FA-NH_2	100
CT	CH_3CN	24	Ac-F-OEt	A-NH_2	Ac-FA-NH_2	4
2-CPC(BPN')	CH_3CN[d]	72	Cbz-LL-OMe	FL-O^tBu	Cbz-LLFL-O^tBu	95
BPN'	CH_3CN[d]	72	Cbz-LL-OMe	FL-O^tBu	Cbz-LLFL-O^tBu	<1
2-CPC(BPN')	CH_3CN[d]	72	Cbz-VL-OMe	FL-O^tBu	Cbz-VLFL-O^tBu	90
BPN'	CH_3CN[d]	72	Cbz-VL-OMe	FL-O^tBu	Cbz-VLFL-O^tBu	<1
2-CPC(Th)	CH_3CN	48	Cbz-F-OH	L-OMe	Cbz-FL-OMe	95
Th	CH_3CN	48	Cbz-F-OH	L-OMe	Cbz-FL-OMe	<1
2-CPC(Th)	CH_3CN	48	Cbz-F-OH	L-O^tBu	Cbz-FL-O^tBu	90
Th	CH_3CN	48	Cbz-F-OH	L-O^tBu	Cbz-FL-O^tBu	<1
2-CPC(Th)	CH_3CN	48	Boc-MLF-OMe	FL-NH_2	Boc-MLFFL-NH_2	70
Th	CH_3CN	48	Boc-MLF-OMe	FL-NH_2	Boc-MLFFL-NH_2	<1

(a) Reactions were carried out at 37 °C. Each solvent contains 5% (v/v) triethylamine and <5% (v/v) water.

(b) Two equivalents of donor amino acid, relative to the acceptor amino acid, were used in each reaction.

(c) All compounds were fully characterized by ^1H and ^{13}C NMR and high resolution mass spectrometry.

(d) The solvent was distilled from calcium hydride.

Although the differences in the stabilization at 45 °C between these two conjugates is not great (both are quite stable) at higher temperatures, the α-chymotrypsin conjugate with the greatest distance between the anomeric center and the polymer backbone, 3-CPC(CT), reproducibly has 10-15% higher thermal stability at all temperatures than the conjugates of **2** (Wartchow, 1992). We are continuing to explore the generality of the use of these carbohydrate-based macromolecules for the stabilization of enzymes and other proteins, the preparation of new carbohydrate-based macromolecules, and their applications.

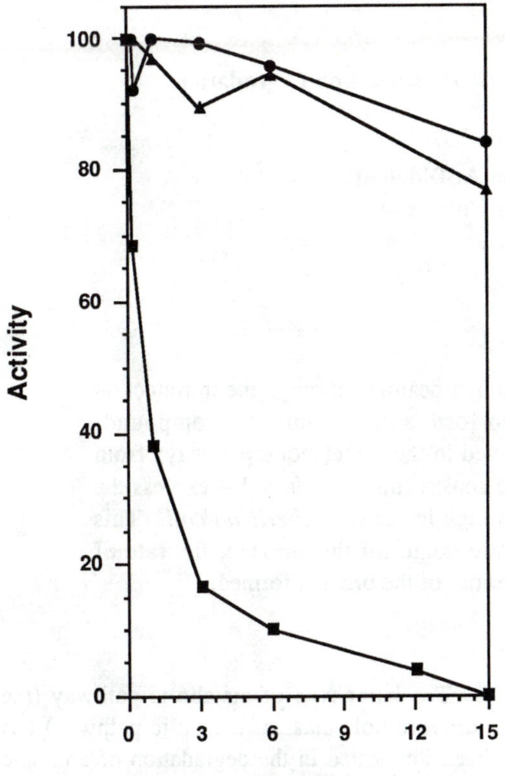

Figure 1. Activity of native α-chymotrypsin [■], 2-CPC(CT) [▲] and 3-CPC(CT) [●] in borate buffer (pH 9) at 45 °C.

Acknowledgement. We are grateful to Cargill, Incorporated (Minneapolis, MN) and to the Director, Office of Energy Research, Office of Basic Energy Sciences, Divisions of Materials Sciences and also Energy Biosciences of the U.S. Department of Energy under contract DE-AC03-76SF00098 to the Lawrence Berkeley Laboratory and grant DE-FG02-91ER45463 to The Ohio State University for their financial support of this work.

References

Bender, M. L.; Begue-Canton, M. L.; Blakeley, R. L.; Brubacher, L. J.; Feder, J.; Gunter, C. R.; Kezdy, F. J.; Killheffer, J. V., Jr.; Marshall, T. H.; Miller, C. G.; Roeske, R. W.; Stoops, J. K. *J. Am. Chem. Soc.* **1966**, *88*, 5890.

DelMar, E. G.; Largman, C.; Brodrick, J. W.; Geokas, M. C. *Anal. Biochem.* **1979**, *99*, 316.

Erlanger, B. F.; Kokowsky, N.; Cohen, W. *Arch. Biochem. Biophys.* **1961**, *95*, 271.

Gray, G. R. *Arch. Biochem. Biophys.* **1974**, *163*, 426.

Gray, G. R. *Meth. Enz.* **1978**, *50*, 155.

Gray, G. R.; Schwartz, B. A.; Kamicker, B. J. Cell Surf. Carbohydr. Biol. Recog. 1978, 583.

Hill, T. G.; Wang, P.; Oehler, L. M.; Huston, M. E.; Wartchow, C. A.; Bednarski, M. D.; Callstrom, M. R. *Tet. Lett.* **1991**, *32*, 6823.

Roy, R.; Katzenellenbogen, E.; Jennings, H. J. *Can. J. Biochem. Cell Biol.* **1984**, *62*, 270.

Schonbaum, G. R.; Zerner, B.; Bender, M. L. *J. Biol. Chem.* **1961**, *236*, 2930.

Schwartz, B. A.; Gray, G. R. *Arch. Biochem. Biophys.* **1977**, *181*, 542.

Thomas, P. G.; Russell, A. J.; Fersht, A. R. *Nature* **1985**, *318*, 375.

Wang, P.; Hill, T. G.; Bednarski, M. D.; Callstrom, M. R. *Tet. Lett.* **1991**, *32*, 6823.

Wang, P.; Hill, T. G.; Wartchow, C. A.; Huston, M. E.; Oehler, L. M.; Smith, M. B.; Bednarski, M. D.; Callstrom, M. R. *J. Am. Chem. Soc.* **1992**, *114*, 378.

Wartchow, C. A.; Huston, M. E.; Smith, M. B.; Wang, P.; Bednarski, M. D.; Callstrom, M. R. submitted.

Manipulation of the Genes for Polycyclic Aromatic Hydrocarbon Degradation

Gerben J. Zylstra*, Xiao Ping Wang, and Varsha A. Didolkar,
Center for Agricultural Molecular Biology, Rutgers University,
P.O. Box 231, New Brunswick, NJ, 08903-0231
(FAX: (908)932-6535)

Microorganisms can metabolize aromatic hydrocarbons through the introduction of oxygen into the aromatic nucleus to form a *cis*-dihydrodiol compound. Cloning and analysis of the genes involved in these metabolic pathways from different microorganisms has allowed the construction of clones that express the initial genes in the catabolic pathway to high levels in *Escherichia coli*. This allows the determination of the substrate range for the enzyme, the rate of conversion for each substrate, and the identity of the product formed.

The microbial catabolic pathways for the degradation of aromatic hydrocarbons involve the introduction of oxygen into the aromatic ring. The best studied aromatic compound in terms of the diversity of catabolic pathways for its degradation is toluene. Five metabolic pathways are known for toluene degradation, each characterized by the initial mode of enzymatic oxidation of the compound (Fig. 1). *Pseudomonas putida* strain mt-2, which harbors the TOL plasmid (Williams and Murray, 1974; Worsey and Williams, 1975), initiates metabolism of toluene through oxidation of the methyl group to form a carboxylic acid (benzoate). *P. cepacia* strain G4 (Shields et al., 1989), *P. pickettii* strain PKO1 (Kukor and Olsen, 1980), and *P. mendocina* strain KR (Whited and Gibson, 1991) initiate metabolism of toluene through the introduction of a hydroxyl group to the aromatic ring to form *o*-cresol, *m*-cresol, and *p*-cresol, respectively. *P. putida* strain F1 (Gibson et al., 1968; Gibson et al. 1970) initiates metabolism of toluene through the introduction of both atoms of molecular oxygen into the aromatic nucleus to form (+)-*cis*-1(S),2(R)-dihydroxy-3-methylcyclohexa-3,5-diene (*cis*-toluene dihydrodiol).

The latter type of metabolic pathway (the "dihydrodiol" class of catabolic pathway) has been implicated in the degradation of aromatic compounds varying in complexity from benzene (one aromatic ring) to pyrene (four aromatic rings) (Fig. 2). *P. putida* strain F1 metabolizes benzene (as well as other substituted benzenes) through a *cis*-dihydrodiol intermediate. *P. putida* strain LB400 (Nadim et al., 1987), *P. putida* strain OU83 (Khan et al., 1988), *P. testosteroni* strain B-356 (Sondossi et al., 1991), *P. paucimobilis* strain Q1 (Furukawa et al., 1983), *Alcaligenes* and *Acinetobacter* strains (Furukawa et al., 1978) and a *Beijerinckia* species (Gibson et al., 1973) metabolize biphenyl through *cis*-2,3-dihydroxy-1-phenyl-cyclohexa-4,6-diene (*cis*-biphenyl dihydrodiol). *P. putida* strain PpG7 (containing the NAH plasmid) and strain NCIB9816 metabolize naphthalene through *cis*-(1R,2S)-dihydroxy-1,2-dihydronaphthalene (*cis*-naphthalene dihydrodiol) (Jeffrey et al., 1975). A *Mycobacterium* species (Heitkamp et al., 1988) has been shown to metabolize pyrene through *cis*-4,5-dihydroxy-4,5-dihydropyrene (*cis*-pyrene dihydrodiol).

The enzymes involved in the initial steps of the catabolic pathways for aromatic

1054–7487/92/0068$06.00/0

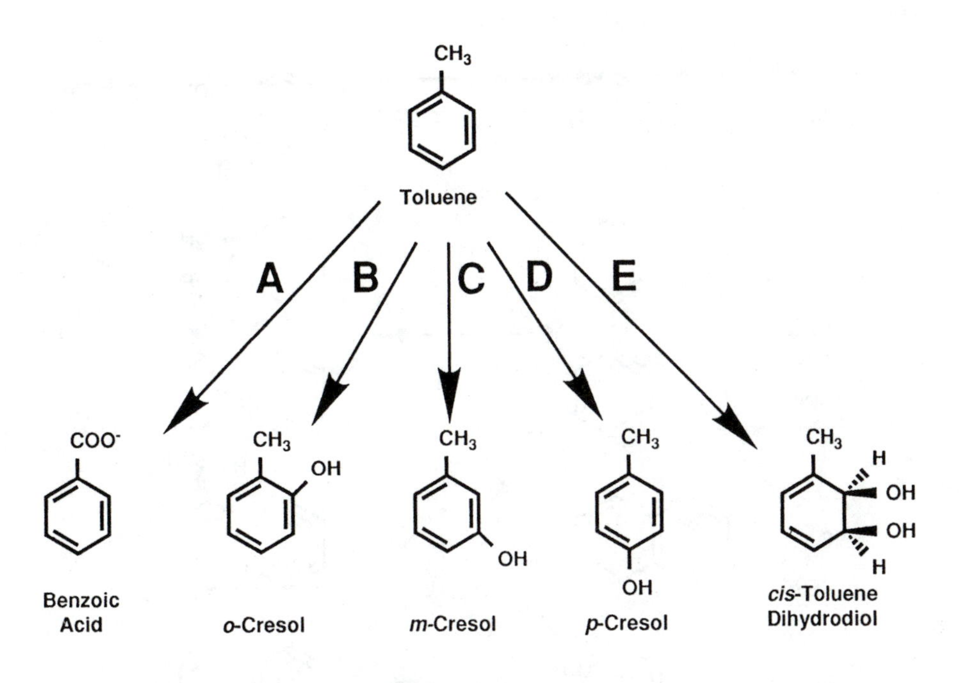

Fig. 1. Initial step in the degradation of toluene by *P. putida* mt-2 (A), *P. cepacia* G4 (B), *P. pickettii* PKO1 (C), *P. mendocina* KR (D), and *P. putida* F1 (E).

hydrocarbons often have a broad substrate range (Zylstra and Gibson, 1991; Gibson et al., 1990). However, the microbial strains producing the enzymes often have a limited substrate range for growth and/or limited range of inducers for the catabolic pathway. Through the use of appropriate mutant strains blocked in the catabolic pathway specific metabolic intermediates can be produced following induction. This allows for the biocatalytic transformation of aromatic hydrocarbons to specific, oxygenated products. In the case of the *cis*-dihydrodiols these compounds are also chiral molecules. The genes for several of these dihydrodiol pathway enzymes have been cloned from microorganisms capable of growing on particular aromatic hydrocarbons as carbon and energy sources. The genes for toluene degradation (Zylstra et al., 1988), naphthalene degradation (Serdar and Gibson, 1989; Yen and Gunsalus, 1982), biphenyl degradation (Ahmad et al., 1990; Furukawa and Miyazaki, 1986; Khan and Walia, 1991; Mondello, 1989), and pyrene degradation (Wang and Zylstra, unpublished data) have all been cloned and in some cases the nucleotide sequences are known. Manipulation of these genes in *Escherichia coli* has allowed the construction of clones allowing high levels of expression (Zylstra and Gibson, 1991; Furukawa and Suzuki, 1988). Particularly useful in this case are clones that express the dioxygenase enzyme that catalyzes the transformation of the aromatic hydrocarbon to a *cis*-dihydrodiol molecule. Expression of the genes (and therefore induction of the biocatalytic activity) in *E. coli* has allowed a detailed analysis of the range of substrates that can be oxygenated by a specific dioxygenase enzyme (Zylstra et al., 1990; Zylstra and Gibson, 1991). Accumulation of large amounts of the *cis*-dihydrodiol in the culture medium has aided in the identification of the precise dihydrodiol formed. This type of approach has shown that in addition to oxidizing monocyclic aromatic hydrocarbons, toluene dioxygenase from *P. putida* strain F1 can also oxidize the polycyclic

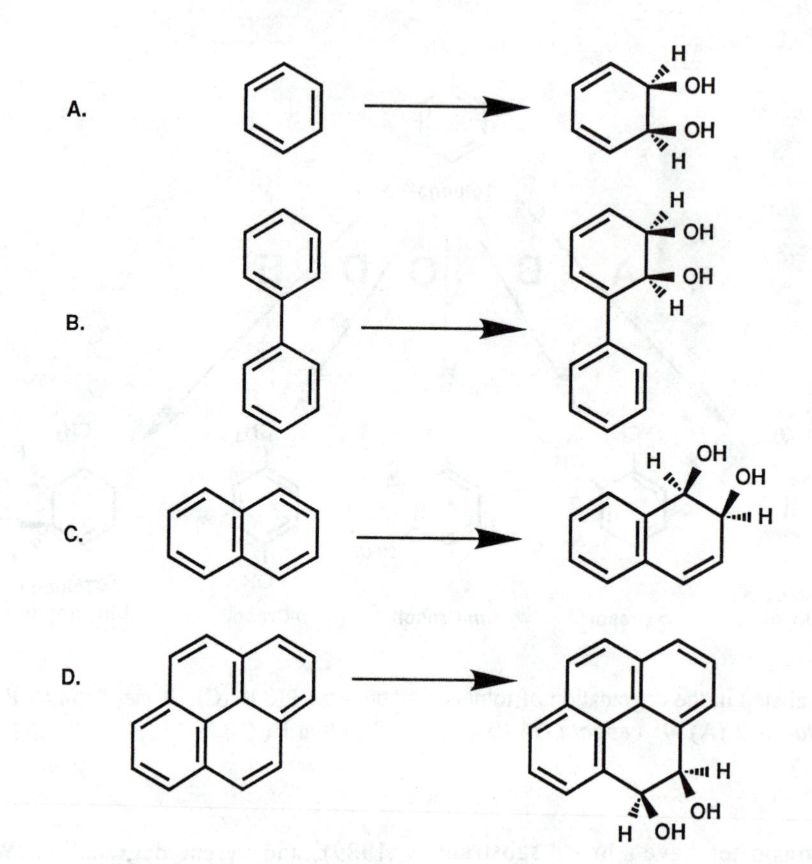

Fig. 2. Initial enzymatic reaction in the degradation of benzene (A), biphenyl (B), naphthalene (C), and pyrene (D) by the dihydrodiol pathway.

aromatic hydrocarbons biphenyl and naphthalene to *cis*-dihydrodiol compounds. Newly cloned dioxygenases from other microbial strains have similar wide substrate ranges. The genes from microorganisms that metabolize polycyclic aromatic hydrocarbons were manipulated to obtain expression in *E. coli*. This allowed a comparison of their substrate range and their rate of biotransformation.

Acknowledgements

This research was supported in part by a grant from Envirogen, Inc., and by the sponsorship of the U.S. Department of Energy, Environmental Restoration and Waste Management Young Faculty Award Program administered by Oak Ridge Associated Universities.

References

Ahmad, D.; Masse, R.; Sylvestre, M. *Gene* **1990**, 86, 53-61.

Furukawa, K.; Matsumura, F.; Tonomura, K. *Agric. Biol. Chem.* **1978**, 42, 543-548.

Furukawa, K.; Miyazaki, T. *J. Bacteriol.* **1986**, 166, 392-398.

Furukawa, K.; Simon, J. R.; Chakrabarty, A. M. *J. Bacteriol.* **1983**, 154, 1356-1362.

Furukawa, K.; Suzuki, H. *Appl. Microbiol. Biotechnol.* **1988**, 29, 363-369.

Gibson, D. T.; Koch, J. R.; Kallio, R. E. *Biochemistry* **1968**, 7, 2653-2662.

Gibson, D. T.; Roberts, R. L.; Wells, M. C.; Kobal, V. M. *Biochem. Biophys. Res. Commun.* **1973**, 50, 211-219.

Gibson, D. T.; Yoshioka, H.; Mabry, T. J. *Biochemistry* **1970**, 9, 1626-1630.

Gibson, D. T.; Zylstra, G. J.; Chauhan, S. In *Pseudomonas: Biotransformations, Pathogenesis, and Evolving Biotechnology*; Silver, S.; Chakrabarty, A. M.; Iglewski, B.; Kaplan, S., Ed.; American Society for Microbiology, Washington, D.C., **1990**; pp. 121-140.

Heitkamp, M. A.; Freeman, J. P.; Miller, D. W.; Cerniglia, C. E. *Appl. Environ. Microbiol.* **1988**, 54, 2556-2565.

Jeffrey, A. M.; Yeh, H. J. C.; Jerina, D. M.; Patel, T. R.; Davey, J. F.; Gibson, D. T. *Biochemistry* **1975**, 14, 575-584.

Khan, A. A.; Walia, S. K. *Appl. Environ. Microbiol.* **1991**, 57, 1325-1332.

Khan, A.; Tewari, R.; Walia, S. *Appl. Environ. Microbiol.* **1988**, 54, 2664-2671.

Kukor, J. J.; Olsen, R. H. *J. Bacteriol.* **1990**, 172, 4624-4630.

Mondello, F. J. *J. Bacteriol.* **1989**, 171, 1725-1732.

Nadim, L. M.; Schocken, M. J.; Higson, F. K.; Gibson, D. T.; Bedard, D. L.; Bopp, L. H.; Mondello, F. J. *Proceedings of the Thirteenth Annual Hazardous Waste Research Symposium: Remedial Action, Treatment, and Disposal of Hazardous Waste*; Environmental Protection Agency: Cincinnati, OH; **1987**; pp. 395-402.

Serdar, C. M.; Gibson, D. T. *Biochem. Biophys. Res. Commun.* **1989**, 164, 772-779.

Shields, M. S.; Montgomery, S. O.; Chapman, P. J.; Cuskey, S. M.; Pritchard, P. H. *Appl. Environ. Microbiol.* **1989**, 55, 1624-1629.

Sondossi, M.; Sylvestre, M.; Ahmad, D.; Massé, R. *J. Industrial Microbiol.* **1991**, 7, 77-88.

Whited, G. M.; Gibson, D. T. *J. Bacteriol.* **1991**, 173, 3010-3016.

Williams, P. A.; Murray, K. *J. Bacteriol.* **1974**, 120, 416-423.

Worsey, M. J.; Williams, P. A. *J. Bacteriol.* **1975**, 124, 7-13.

Yen, K.-M; Gunsalus, I. C. *Proc. Natl. Acad. Sci. USA* **1982**, 79, 874.

Zylstra, G. J.; Chauhan, S.; Gibson, D. T. *Proceedings of the Sixteenth Annual Hazardous Waste Research Symposium: Remedial Action, Treatment, and Disposal of Hazardous Waste*; Environmental Protection Agency: Cincinnati, OH; **1990**; pp. 290-302.

Zylstra, G. J.; Gibson, D. T. In *Genetic Engineering*; Setlow, J. K., Ed.; Plenum Press: New York, NY, **1991**, Vol. 13; pp. 183-203.

Zylstra, G. J.; McCombie, W. R.; Gibson, D. T.; Finette, B. A. *Appl. Environ. Microbiol.* **1988**, 54, 1498-1503.

Mechanism Based Screens for Natural Product Leads as Sources for Antitumor Drugs

Hirofumi Nakano * and Tatsuya Tamaoki
Tokyo Research Laboratories, Kyowa Hakko Co. Ltd.
3-6-6 Asahimachi, Machida, Tokyo, Japan (FAX : 81-427-26-8330)
Fusao Tomita, Hokkaido University, Sapporo, Hokkaido, Japan

This review focuses on our screening in the 1980's to identify new antitumor agents. Based on mechanism of action of established antitumor drugs, we screened cultures of actinomyces and fungi for new drugs which target DNA and DNA topoisomerases. A microbial prescreen (DC screening) with supersensitive strains of *Bacillus* was successful to discover 10 new antitumor agents from which duocarmycins and leinamycin are selected as lead compounds. New topoisomerase-active drugs were discovered by a biochemical assay in vitro using mammalian topoisomerases. New prescreening methods to detect inhibitors of oncogene and signal transduction were developed. Among new protein kinase inhibitors isolated, UCN-01 is evaluated as a new antitumor agent for clinical study.

Introduction ; methods to discover new antitumor drugs from fermentation.

Since the 1950's, antitumor drugs had been discovered from fermentation in antimicrobial screening programs accompanied with mammalian cell toxicity assay and murine tumor models. While clinical studies revealed that these drugs, mitomycin C, actinomycin D, bleomycin and anthracyclines, are useful in treatment of several human tumors, they have been less effective for common human solid tumors. In contrast to the successful development of semisynthetic derivatives of β-lactams and other antimicrobial agents, analog design of established antitumor drugs has been of limited success, whereas introduction of new cytotoxic agents such as cis-platin and etoposide expanded the field of cancer chemotherapy. Thus, screening of new antitumor agents for the treatment of common human solid tumors remains an active area of drug development. The successful findings of new microbial metabolites as new drugs in the fields of immunomodulators (cyclosporin A and FK-506), hypocholesteremic drugs (HMG-coA reductase inhibitor, mevalotin, lovastatin) and anthelmintics (avermectin) have stimulated efforts to find new lead compounds from microbes. Regardless of the target of screening, the methods of detection of active compounds are critical to the discovery of new microbial metabolites.

This review describes our screening programs to identify new antitumor agents from fermentation over the past decade, focusing on the method of detection and new microbial metabolites discovered. [Fig. 1] A microbial prescreen (DC screening), which we developed as a selective, sensitive and simple assay, has been successful to detect new cytotoxic agents which target DNA. In our screening for topoisomerase active drugs, new antitumor drugs were detected by enzyme assays *in vitro* with purified mammalian topoisomerases but not by several methods with special microbes. We have also discovered potent and selective inhibitors of protein kinases using enzyme assay. Based on the findings that many oncogenes and signal transduction pathway are conserved in yeast and mammalian cells, we have developed a new microbial prescreen to detect drugs which can suppress the growth defect of yeast strains under restrictive conditions. These methods have been applied to culture broth and appear to be promising.

1054–7487/92/0072$06.00/0

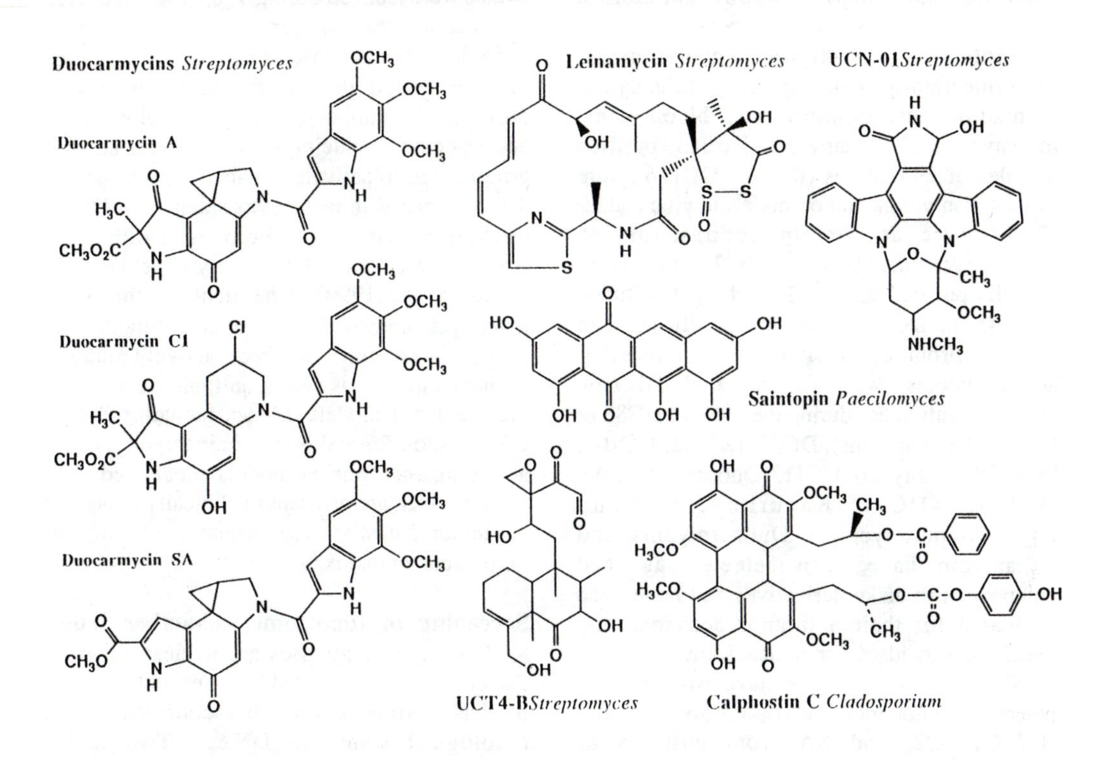

Figure 1. Structures of new microbial antitumor agents ; duocarmycins, leinamycin, UCN-01, calphostin C, saintopin and UCT4-B.

Screening for new cytotoxic drugs which target DNA (DC screening)

With the knowledge that the most useful antitumor drugs in the clinic function by initiating DNA damage, several detection methods to find these DNA-interactive compounds have been developed. After we screened microbial cultures with different systems in late the 1970's, we selected a method using supersensitive strains of *Bacillus subtilis*, which are deficient in enzymes for recombination (rec$^-$) or DNA repair (polA$^-$). During the 1980's, the following procedure was most conveniently used : Melted agar medium prewarmed at 60°C was inoculated with spore suspensions of wild type strain at 10^3 spores/ml and a supersensitive strain at 10^6 spores/ml, and was poured into plates on a level surface. After solidification, paper disks impregnated with samples were applied on the agar plates and the plates were incubated. Satellite zones surrounding the test disks indicate the presence of antibiotics which inhibit growth of the supersensitive strain selectively. This assay has been useful to screen large numbers of cultures in a highly sensitive and selective manner for cytotoxic antibiotics which initiate DNA damage.

During the 1980's, we mainly screened cultures of *Streptomyces* (>$3x10^4$ strains) and fungi (>10^4 strains) which were newly isolated from soils and plants of Japan. 7 % of cultures of *Streptomyces* were active in the DC screening, whereas yield of anti-*Bacillus* activity, which are not selective to the strain supersensitive to DNA-active drugs, was 15 %. Thus a very high percentage of freshly isolated *Streptomyces* produced DNA damaging metabolites, and these results provide an interesting question whether the DNA-active metabolites of *Streptomyces* are responsible for the own genome instability and for their high ability to produce a great variety of antibiotics.

Most of these active compounds were

classified into groups of known antibiotics: xanthomycins, griseoluteins, benzodiazepine antibiotics, streptonigrins, chartreusins, azaserine family, anthracyclines, kinamycin, anthraquinones, quinoxalines, bleomycins, mitomycins and saframycins. Partially purified samples of 29 cultures (DC88 - DC116) were evaluated in murine tumor models in vivo and 22 (75%) were active. (In contrast to the *Streptomyces* metabolites, only 2 out of 11 of partially purified samples from fungal cultures, DC1043 and DC1149, were marginally active in vivo.) From 22 in vivo active cultures of actinomycetes, we have isolated 10 new antitumor antibiotics during the 1980's : DC88 & DC89 (Duocarmycins), DC92, DC102, DC105, DC107(Leinamycin), DC113(DuocarmycinSA), DC114, DC115(Kapurimycin) and DC116(Sapurimycin). Duocarmycins and leinamycin have been selected as lead compounds to make derivatives and have been evaluated for their antitumor activities and toxicity as candidates for clinical study.

Duocarmycins : We have isolated new potent antitumor antibiotics duocarmycin A, B1, B2, C1, C2, and SA from cultures of *Streptomyces*(Ichimura et. al., 1988 & 1990, Takahashi et.al., 1988). Duocarmycin A and SA have a cyclopropylhexadienone moiety which is known to be the DNA alkylating group of the "left hand segment " of CC-1065. Duocarmycin B1, B2, C1, C2 possess five or six-membered ring which were generated via attacking of bromide or chloride ion at different carbon positions on the cyclopropane ring of duocarmycin A. Duocarmycin A is unstable in aqueous solution. On the other hand duocarmycin SA is a more stable compound and one of the most cytotoxic agents with IC_{50} of pM. Although duocarmycins have not shown the irreversible delayed hepatotoxicity which was caused by CC-1065, several derivatives of duocarmycins were synthesized to improve antitumor activity in murine and human xenografted tumor models. KW-2189, a 8-O-carbamoyl analog, has been selected for further evaluation toward clinical trial (Saito H. et al., 1992).

Leinamycin (DC107) : Leinamycin is a newly discovered potent antitumor antibiotic which contains an unusual 1,3-dioxo-1,2-dithiolane structure. Five different producing strains were isolated during 1985-1988 and were assigned as *Streptomyces* (Hara et al.,1989). In vitro, leinamycin causes single-strand cleavage of supercoiled double strand DNA in the presence of a thiol co-factor such as glutathione and cysteine. Reducing agents without an -SH group failed to activate leinamycin. Because S-deoxy leinamycin lacks cytotoxicity and DNA-cleaving activity, the 1,3-dioxo-1,2-dithiolane structure is essential for activity of leinamycin (Hara et al., 1990). Chemical synthesis of leinamycin and compounds with 1,3-dioxo-1,2-dithiolane structure has been actively studied. Leinamycin shows potent antitumor activity in murine tumor models *in vivo,* including S-180, B-16, Colon 26 and P388. Leinamycin is also active against murine models inoculated with tumors which are resistant to clinically important antitumor drugs, such as cisplatin, adriamycin, mitomycin or endoxan.

Screening of topoisomerase-active drugs

DNA topoisomerases are nuclear enzymes that catalyze the concerted breaking and rejoining of DNA strands, thereby controlling the topological states of DNA. Two major topoisomerases have been detected in all eukaryotic cells; topoisomerase I catalyzes the passage of DNA strand through transient single-strand break and topoisomerase II catalyzes the passage of DNA double strands. In addition to their many important functions in DNA metabolism, both topoisomerases have generated extensive clinical interest in chemotherapy. In the 1980's it has been shown that topoisomerases I and II are the principal intracellular targets for a number of clinically important antitumor drugs including m-AMSA, VP-16 and camptothecin. Despite their apparent structural diversity, these drugs have the common properties of stabilizing a key covalent reaction intermediate of topoisomerases, termed the cleavable complex, which upon exposure to denaturant results in the induction of DNA cleavage. Since 1985, according to this attractive model, we have screened cultures of actinomycetes and fungi for their ability to induce topoisomerase -mediated DNA cleavage *in vitro*. In our assay condition in which unlabeled DNA was used as substrate DNA, anthracyclines and actinomycin-D could not induce an increase of topoisomerase -mediated DNA cleavage. The early findings in the course of screening were the effects of ingredients of the media ; 5 mM Ca^{++} suppressed the mammalian topoisomerase II-mediated DNA cleavage induced by m-AMSA. Flavonoids such as genistein involved in natural nitrogen source of media induced the topoisomerase II mediated DNA cleavage. Using fermentation media excluding the above interfering factors, we found that the following metabolites induce the DNA cleavable complex with topoisomerase II: streptonigrin, terpentecin, clerocidin and UCT4-B(Yamashita et al., 1990 ; Kawada et al., 1991). Saintopin (UCT1003), a new antitumor antibiotic, was isolated from *Paecilomyces*, which induces DNA cleavable complex with both topoisomerase I & II (Yamashita et al.,1991). We also have identified several microbial metabolites which induce topoisomerase I -mediated DNA cleavage.

Thus, biochemical assay in vitro with purified mammalian topoisomerases have been successful to identify new topoisomerase-active metabolites. Although we applied several microbial prescreening methods to detect topoisomerase-active agents in culture broth, they were low throughput.

Inhibitors of oncogene products and signal transduction pathway

Since the early 1980's, rapid progress in the studies on human oncogenes and tumor cell biology have provided attractive targets for screening of new antitumor drugs. A number of lead compounds have been discovered from microbial metabolites. Using enzyme assays in vitro, we identified potent inhibitors of protein kinases, staurosporine and UCN-01, from *Streptomyces* and discovered specific inhibitors of protein kinase C, calphostins, from *Cladosporium* (Review, Tamaoki and Nakano, 1990). UCN-01 shows antitumor activity in several murine and human tumor xenografted models in vivo and studies on toxicology are in progress to evaluate UCN-01 as a candidate for clinical study. While UCN-01 seems to be a selective inhibitor of protein kinase C, it remains for further studies to identify the critical target in tumor cells of this new antitumor drug.

References

Hara, M., Takahashi, I., Yoshida, M., Asano, K., Kawamoto, I., Morimoto, M. and Nakano, H. ; *J. Antibiotics*. **1989**, 42, 333-335.

Hara, M., Saitoh, Y. and Nakano. H.; *Biochemistry*. **1990**, 29, 5666-5681.

Ichimura, M., Muroi K., Asano, K., Kawamoto, I., Tomita, F., Morimoto, M. and Nakano H.; *J. Antibiotics*. **1988,** 41,1285-1288.

Ichimura M., Ogawa, T., Takahashi, K., Kobayashi, E., Kawamoto, I., Yasuzawa, T., Takahashi, I. and Nakano, H. ; *J. Antibiotics*. **1990**, 43, 1037-1038

Kawada, S., Yamashita, Y., Fujii, N. and Nakano, H.; *Cancer Research*. **1991**, 51 :2922-2925

Takahashi, I., Takahashi, K., Ichimura, M., Morimoto, M., Asano, K., Kawamoto, K., Tomita, F. and Nakano, H.; *J. Antibiotics*. **1988**, 41, 1915-1917.

Tamaoki, T and Nakano, H.; *Biotechnology*. **1990**. 8: 732-735

Yamashita, Y., Kawada,S. and Nakano, H.; *Cancer Research*. **1990**, 50 , 5841-5844

Yamashita Y., Kawada, S., Fujii, N. and Nakano, H.; *Biochemistry*. **1991**, 30 : 5853-5845

Emerging Technologies and Future Prospects for Industrialization of Microbially Derived Cellulose

R. Malcolm Brown, Jr.* Department of Botany, The University of Texas, Austin, Texas 78712 (FAX:512 471 3573)

This presentation will describe a new application of biotechnology to the production of one of the most widely used biopolymers, cellulose. The new application exploits microbially derived cellulose on an industrial scale. The gram negative bacterium, *Acetobacter xylinum* is the microorganism of choice since it synthesizes copious amounts of cellulose of unsurpassed purity, mechanical strength and absorbability. The kinetics and potential efficiency of microbial cellulose production as well as goals, obstacles, and future strategies will be presented. Optimal fermentation pathways in relation to the final cellulose product characteristic will be described. The importance of molecular biology and genetic engineering for intermediate and long term strategies of the cellulose industry will be discussed.

Cellulose is the most abundant macromolecule on the planet (Brown, 1985). Over 10^{11} tons is made and destroyed annually. Thus, it is of vital interest to applied science to understand how this vast polymeric material is synthesized, for if we can control the biosynthesis of cellulose, we are in a unique position to beneficially alter and promote our global climate as well as produce a major industrial raw material. Considering that cellulose is a sink for atmospheric CO_2, the more gaseous CO_2 we can "fix" into solidified mass, the better we can control the global warming trend.

How should we tackle this problem? The first idea might be to simply stop using our forests and let them grow! We all know that this is not feasible in view of the continuing demand for cellulose from forests and from cotton. Thus, we must turn to other sources of cellulose, and this is where *microbially derived cellulose* can provide a new purpose and direction. Cellulose produced under a directed and controlled *fermentation process* is a biotechnological ideal. Since the gram negative bacterium, *Acetobacter xylinum* is the most prolific synthesizer of microbial cellulose (Brown, 1989), it is the focus of current studies.

What is microbially derived cellulose and what are its advantageous characteristics?

Microbially derived cellulose is *pure* cellulose, without any lignin or other cell wall contaminating material. *Acetobacter xylinum* synthesizes cellulose in the form of a composite ribbon of microfibrils which exit the cell envelope from a distinct row of pores or "spinnerets" (Fig. 1). Since the microfibrils are spun into the growth medium, all cells in the medium contribute to the gross morphology of the cellulose which is in the form of a gelatinous membrane or *pellicle*. This membrane contains entrapped cells and is formed at the gas/liquid interface of the culture medium in static cultures. Typical nutrients are glucose, sucrose, proteose peptone, and yeast extract. Culture medium is inoculated with cells and a visible pellicle usually forms within 24 hrs. Static cultivation for one week or more results in a thick, strong, hydrophilic pellicle.

Cellulose production is active only in a narrow zone of cells at the upper surface of the pellicle. Cells become entrapped within the pellicle as new cells produce overlaying cellulose. Those entrapped cells will eventually die due to oxygen deprivation from above and nutrient deprivation from below. In spite of this limitation, cellulose yields of 45% are not uncommon.

Two properties of the microbial cellulose membrane far exceed those typical of celluloses of vascular plants: (a) the cellulose is very hydrophilic with a water absorption capacity of over 100 times the weight of cellulose; (b) the cellulose has great mechanical strength, far surpassing that of pulp paper and cotton textiles. Given these important characteristics, it is reasonable to expect that microbial cellulose can find a useful place in the industrial market. This is one of the major goals of our microbial cellulose research and development program at The University of Texas at Austin.

1054–7487/92/0076$06.00/0 © 1992 American Chemical Society

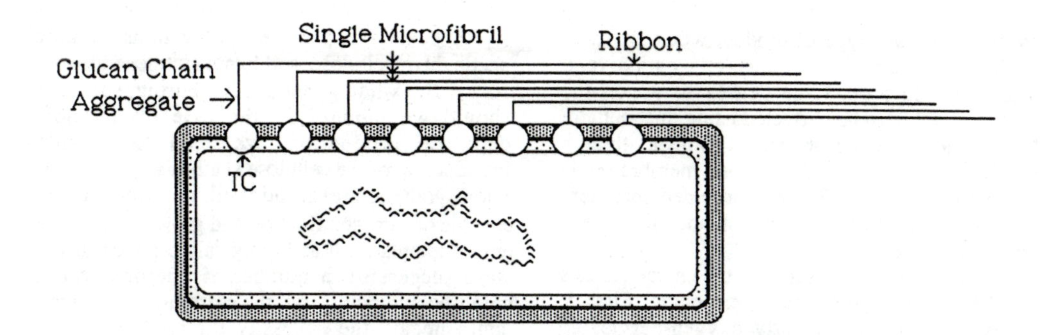

Figure 1. Idealized schematic diagram of an *Acetobacter* cell showing the linear arrangement of the cellulose synthesizing complexes (=TC). Each TC synthesizes a glucan chain aggregates consisting of approximate 6 -8 glucan chains. Three TCs are required to synthesize their respective glucan chain aggregates which are required to form a crystalline microfibril. Microfibrils aggregate via H-bonding to form the cellulose ribbon which is secreted into the medium. The matrix of interwoven ribbons constitutes the bacterial cellulose membrane or pellicle.

Kinetics and efficiency of microbially derived cellulose production

In contrast to cellulose production from a crop such as cotton, microbially derived cellulose has a significantly larger yield. Consider the average yield of one bale of cotton per acre per year as a standard of comparison. Since microbially derived cellulose can be synthesized by living bacterial cells in shallow pans requiring only adequate nutrition and oxygenation, the yield of cellulose from *a single tray* only several mm deep with the corresponding surface area of one acre, has the potential to produce more than 26,000 lb of cellulose per year culture time! This comes to more than 10^8 glucose molecules incorporated into cellulose each hour per cell.

Microbially derived cellulose has yet another unique feature in that it can be substantially modified *during synthesis*. This means that the physical properties of the cellulose can be predetermined by the circumstances of the fermentation, a tremendous advantage over the absence of control of the characteristics of the cellulose produced within a cotton boll or forest trees.

Goals, obstacles, and future strategies for microbially derived cellulose production

No large scale production of microbially derived cellulose has been attempted to date. Thus critical data needed for verification of scale up parameters are lacking. The economy of scale will most certainly determine the ultimate production price for the product, but this price cannot be accurately set until additional fermentation research is completed. Considering these limitations and the reserve of large companies to engage in potentially high risk enterprises, what are the prospects of microbially derived cellulose as a useful commodity product? The future depends on several factors, among them the commitment of major R & D support for fermentation research, the discovery of more efficient strains, and the results of fundamental research into the nature of the cellulose synthase system -- its regulation and its potential for genetic manipulation.

Optimal fermentation pathways in relation to final product characteristics

If the industrial production of microbially derived cellulose required only a deep tank fermentation of a suspended product, the fermentation engineering would be straightforward inasmuch as parameters such as pH, oxygen, viscosity, nutrient input, and batch outflow, all could be controlled with fermentation technology similar to that currently used for such industrially important polysaccharides as xanthan gum. In fact, Cetus and Weyerhaeuser have used this approach to produce a slurry of partially dried microbially derived cellulose known as "Cellulon". Current uses for this product are rather limited in view of product applications which can be adequately fulfilled by conventional sources of cellulose or cellulose derivatives.

Fermentation designed to produce intact microbial cellulose membranes, on the other hand, leads to a product with unmatched strength and water absorption properties (White and Brown,

1989). While this type of product would be more readily appreciated from the market perspective, it is difficult to predict the degree to which the cellulose yield can be perfected using this method. Nevertheless, for the present, it looks as though passive fermentation of cellulose membranes is reasonable and cost effective, provided sufficient space is available for fermentation and the space is efficiently utilized.

What would be some of the characteristics of such a passive membrane fermentor? First, it needs to have a sufficiently oxygen-permeable surface to permit optimal cellulose production. Too much or too little oxygen can be detrimental to cellulose production. Second, the fermentor needs to be automated with respect to the operations of nutrient delivery, inoculation, cellulose production, and cellulose harvesting. Third, this passive fermentor must be modular to protect against loss due to accidental contamination. All of these characteristics can be incorporated using present technology, and our laboratory is committed through a State of Texas grant to the development of fermentation technology for microbial cellulose.

Molecular biology and designer genes for cellulose production

Short term strategies will most likely inaugurate the era of intermediate and large scale fermentation of microbial cellulose; however, for this technology to be truly revolutionary, we must utilize genetic engineering (Saxena, Lin and Brown, 1990, 1991). It is here that the potential for improvement of cellulose biosynthesis is greatest; however, we need to know what to improve! We need to develop, over many years, a thorough well-integrated understanding of the *natural process of cellulose biosynthesis*. What nature has done over the millennia is to perfect adaptation to niches. If we can understand the overall process, the task for the genetic engineer becomes much easier. For example, the movement of eukaryotic cells from the sea to the land required adaptation to a less buoyant environment. This meant that extensive cell wall structural modifications were required. Along with the cellulose came the requirement for other polymers to integrate into a composite material for maximum strength at the minimal cost for the energy budget of the cell. For example, cells did not evolve using chitin as the fibrillar material in the wall, because chitin contains nitrogen, and for the equivalent structural performance, cellulose with only H, O, and C would work equally well. The evolutionary strategies were really quiet complicated when one thinks of the various biochemical pathways required to assemble a typical eukaryotic plant cell wall with its cellulose, pectin, hemicellulose, and lignin components.

So what genetic engineering strategies should we pursue? In the case of microbial cellulose production, we need to consider introduction of the cellulose synthesis operon into a photosynthetic prokaryotic cell, one which is also capable of nitrogen fixation and growth in extreme environments such as highly saline ponds. If this were successful, a number of onerous criteria could be avoided. First, the extreme environment could negate the necessity for maintenance of sterile conditions, a great potential cost saving. Second, the energy budget could be reduced significantly without need for nitrogen-based fertilizers and carbon-based substrates. Certainly, this is a theoretical consideration at this point, but it is important to define the ultimate goals which, if sought and achieved, could dramatically change the agricultural practices throughout the world.

One final thought is that a better understanding of the molecular biology and biochemistry of microbial cellulose can result in more productive and efficient cellulose from conventional sources such as cotton and trees. The earth and its inhabitants can benefit not only from an industrial transition to a microbial cellulose economy thus reducing deforestation and freeing additional arable land for food production but also from more efficient cellulose production on land because the overall CO_2 balance will continue to depend on land plants. Increasing land plant cellulose production efficiency will also help achieve the goal of providing for human needs without disastrous consequences on our environment.

Acknowledgements

I would like to express my appreciation to Richard Santos for helpful suggestions and to the Texas Advanced Technology Program for support of this research through grant TATP-121.

References

Brown, Jr. R. M. The microbial synthesis of cellulose. In *Bioexpo85*, Kahners Exposition Group, Boston, MA. **1985** pp325-335.

Brown, Jr. R. M. "Bacterial Cellulose" In *Cellulose: Structural and Functional Aspects* Ed. Kennedy, Phillips, & Williams. Ellis Horwood Ltd. **1989** pp145-151.

Saxena, I. M., F. C. Lin, and R. M. Brown, Jr. *Plant Mol. Biol* **1990,** 15, 673-683.

Saxena, I. M., F. C. Lin, and R. M. Brown, Jr. *Plant Mol. Biol* **1991,** 16, 947-954.

White, D.G. and R. M. Brown, Jr. Prospects for the commercialization of microbial cellulose. In *Cellulose and Wood-Chemistry and Technology* Ed. C.S. Schuerch, John Wiley and Sons, Inc., N.Y. **1989**, pp573-590

Rheological Characteristics of Microbially Derived Sodium Hyaluronate

L.P. Yu; **J.W. Burns**; **A. Shiedlin**, Biopolymers Department, Genzyme Corporation, Cambridge, MA 02139
Y. Guo; **T. Jankowski**; **P. Pradipasena**; **C. Rha**, Biomaterials Science & Engineering Laboratory, MIT, Cambridge, MA 02139

We have investigated the solution viscosity of the naturally occuring biopolymer sodium hyaluronate (NaHA) with Mw range from 3.0 x 10^6 D to 1.0 x 10^5 D at concentrations of 0.1 to 25 g/dl. At high molecular weights, the low shear rate viscosity increased with about 3.7 power of the Mw which is in agreement with theroretical predication of entanglement (Bueche 1956). Moreover, a relation of viscosity, Mw, and concentration for NaHA solutions was constructed for a viscosity range from 1 to 5 x 10^5 cps. The shear rate (0.001 to 1000 s^{-1}) and frequency (0.005 to 20 hz) dependence of NaHA solution rheology was also determined by employing a Bohlin Rheometer.

Hyaluronic acid, or its sodium salt sodium hyaluronate (NaHA), is a naturally occurring biopolymer found throughout various physiological systems including the vitreous humor, joints, organs, and the extracellular matrix. Structurally, sodium hyaluronate is a linear glycosaminoglycan chain composed of alternating disaccharide units of β(1,4) linked D-glucuronic acid and N-acetyl glucosamine, joined through a β(1,3) glucosidic linkage (Structure 1). NaHA has one acidic group per repeating disaccharide unit. Although there exists regions of local order along the backbone chain, NaHA behaves hydrodynamically as a random coil of overall spherical shape.

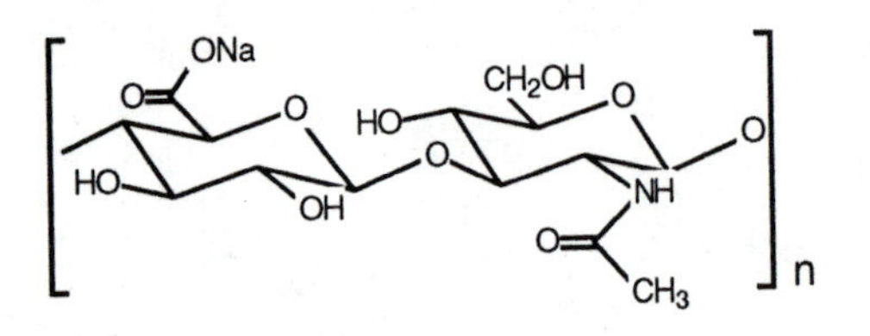

Structure 1: Sodium Salt of Hyaluronic Acid

A number of biocompatible and bioresorbable viscoelastic surgical products have been developed based on the inherent biocompatibility and viscous nature of NaHA solutions. In particular, NaHA solutions have been developed to protect corneal tissues during ophthalmic surgery (Balazs, E.A., 1983), to reduce joint inflammation due to arthritis (Namihi, O., et al., 1982), and to protect surgical tissues during surgery in order to prevent adhesion formation (Burns, J.W., et al., 1991). In each of these applications the ability of NaHA solutions to function as an effective product is dependent upon its rheological characteristics.

In order to better understand the properties of NaHA solutions and their potential applications as viscoelastic surgical aids, we have evaluated the steady and dynamic shear rheological characteristics of NaHA solutions over a range of NaHA molecular weights and concentrations.

Experimental

NaHA Solution Preparation

Highly purified sodium hyaluronate (NaHA) was obtained from Genzyme Corporation (Cambridge, MA). NaHA solutions for all rheological evaluations were prepared in physiological phosphate buffered saline (PBS) and adjusted to pH 7.0. NaHA of varying molecular weight was prepared by the controlled degradation of high molecular weight polymer, generally greater than 2.5 million D, employing high temperature or ultrasonic energy.

1054–7487/92/0080$06.00/0

Rheology Measurements

The apparent viscosity of NaHA solutions was measured with a Brookfield cone and plate viscometer (LV, RV & HB models) at shear rates ranging from 0.6 s^{-1} to 200 s^{-1} and at a sample temperature of 25°C. A Bohlin Rheometer (Lund Sweden) was employed for measuring the zero shear rate viscosity (η_o), of NaHA solutions at shear rates from 0.001 s^{-1} to 1000 s^{-1}. The storage modulus and loss modulus were determined in the frequency range of 0.005 Hz to 20 Hz.

Molecular Weight Measurement

The weight average molecular weight (Mw) of NaHA was measured employing a size exclusion chromatography system in line with a multi-angle laser light scattering detector (SEC/MALLS). The chromatographic system consisted of an LKB 2150 pump, a WISP 710B autoinjecter (Waters), and two µ-Bondagel E-high Å and E-500Å columns connected in series. Sample elutions were carried out isocratically at a flow rate of 0.3 ml/min. The mobile phase consisted of 0.15M Na_2SO_4 and 0.01 M NaH_2PO_4 pH adjusted to 3.5-4.5 with phosphoric acid. Fifty µl of a 0.2 mg/ml NaHA solution was injected for each sample. The mass of injected sample was determined with a refractive index detector (HP 1037A Hewlett Packard) in-line with the SEC columns and the MALLS molecular weight detector (DAWN-F, Wyatt Technologies). Data from both detectors was collected and processed through a personal computer to determine the weight average molecular weight (Mw) of NaHA.

Results

The shear rate dependency of NaHA solutions was strongly affected by NaHA molecular weight and concentration (Fig. 1). Non-Newtonian flow occurred at progressively lower shear rates as the molecular weight and concentration of NaHA increased. One percent and 2.0% NaHA solutions of high molecular weight (>2 million D) still exhibited non-Newtonian flow at a shear rate of 1 s^{-1}. The shear thinning behavior which is a result of mechanical disentangling of the chains as the shear rate increases was much more pronounced for high molecular weight NaHA solutions. As shown in Fig. 1, the 2% 0.65 million NaHA solution had a lower viscosity than the 1% 1.5 million NaHA solution at low shear rates (<10 s^{-1}); it was less shear rate dependent as the shear rate increased than the 1.5 million NaHA solution.

Figure 2 shows the relationship between apparent viscosity at 1 s^{-1} and molecular weight of NaHA solutions with a molecular weight range from 100,000 D to 3 million D. For each NaHA concentration studied there was a biphasic relationship of viscosity and molecular weight represented by $\eta \propto Mw^{\beta}$. Below a critical molecular weight, Mc, the exponent β was typically between 1.2 and 1.5. At higher NaHA molecular weights β ranged from 3.3 to 3.8, which is in agreement with Bueche's theoretical prediction (Bueche, 1956) of β =3.5 based on the effect of molecular entanglements. According to Bueche's theory (Bueche, F., 1962), Mc is related to the average molecular weight spacing between entanglement points, Me, which is approximated by Mc = 2 Me. Additionally, as shown in Fig. 2, Mc decreased as NaHA concentration increased.

Figure 3 shows the effect of NaHA concentration on viscosity for several NaHA molecular weights. The apparent viscosity at shear rate = 1 s^{-1} (η_1) was used to construct the plots for the 100,000 D, 500,000 D, 1.2 million D HA, and 1.8 million D NaHA. The zero shear rate viscosity (η_o) was used for the 2.5 million D NaHA. As we showed for the concentration-viscosity relationship of NaHA solutions there were two different molecular interactive regimes with slopes of 1.3 - 1.6 and 3.8 - 4.1. As expected, at NaHA concentrations where the NaHA molecules overlap and entangle the viscosity increased much more rapidly than in the dilute solutions regime. We observed that the slopes of curves for NaHA of different molecular weights were almost identical in the entanglement region of Fig. 3.

Figure 4 shows the relationship between viscosity (η_1) and the log C*Mw (product of concentration and molecular weight) over a considerable range of NaHA concentrations and molecular weights. The observed biphasic relationship, as in Fig. 3, is due to a critical C*Mw value where the NaHA chains overlap. The slope of the curve is 3.75 in the entanglement region and 1.42 in the non-entanglement region. The critical transition point for C*Mw is 400,000 which correlates to about 40 cps viscosity. The NaHA molecules start to have strong intermolecular interactions as the solution viscosity reaches 40 cps (or C*Mw = 400,000). In that highly interactive region the relationship between viscosity (η_1), molecular weight, and concentration is given by $\log \eta_1 = -19.5 + 3.75 \log C{*}Mw$. Below the transition point, the relationship is given by $\log \eta_1 = -6.45 + 1.42 \log C{*}Mw$. The zero shear rate viscosity

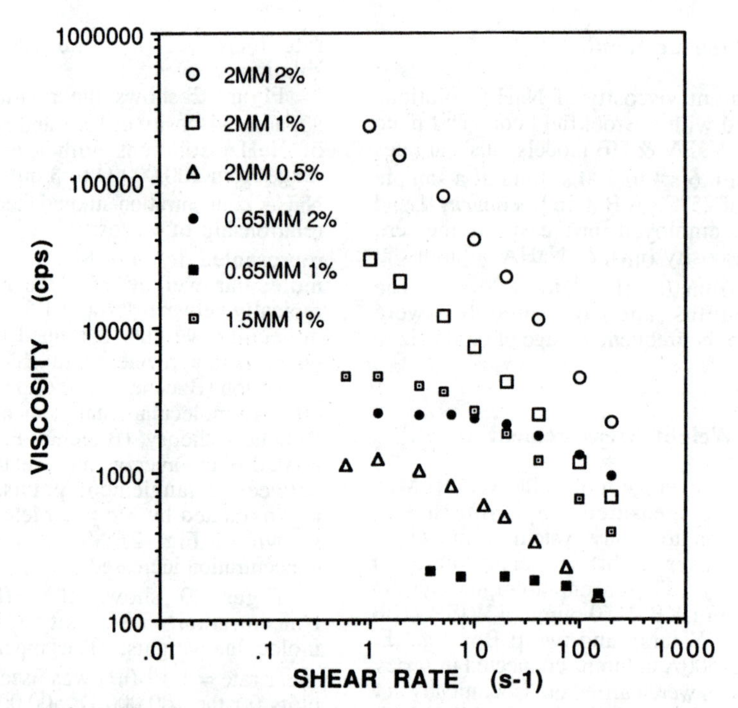

Fig. 1. Apparent viscosity plotted logarithmically against shear rate for six NaHA solutions with different molecular weights and concentrations.

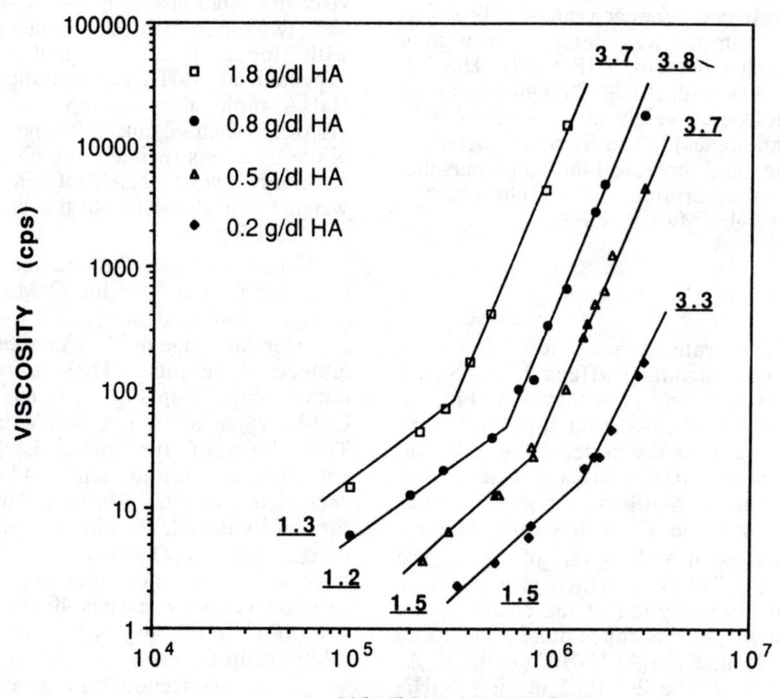

Fig. 2. Viscosity at shear rate 1 s^{-1} (η_1) plotted logarithmically against NaHA molecular weight for 0.2, 0.5, 0.8 and 1.8 g/dl. The slope of each line is underlined.

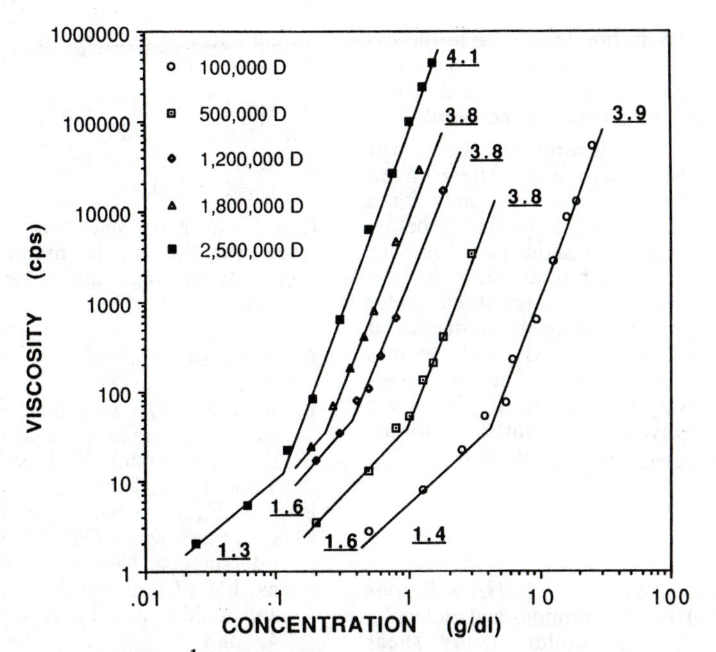

Fig. 3. Viscosity at shear rate 1 s^{-1} (η_1) plotted logarithmically against NaHA concentration for four NaHA molecular weights, 100,000 D, 500,000 D, 1.2 million D and 1.8 million D. The fifth curve, with 2.5 million D molecular weight, the zero shear rate viscosity (η_o) was employed for the plot. The slope of each line is underlined.

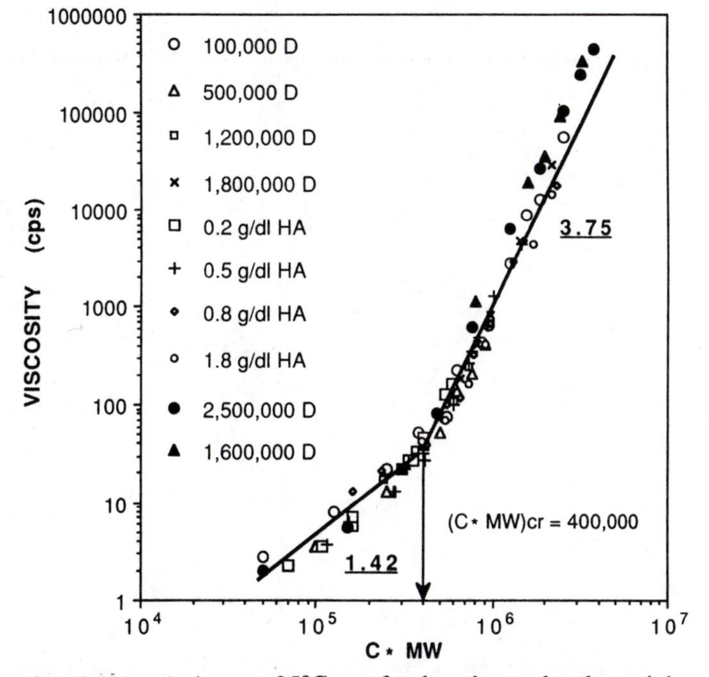

Fig. 4. Viscosity for NaHA solutions at 25°C plotted logarithmically against C*Mw of NaHA. The zero shear viscosity (η_o) was used for 2.5 million D and 1.6 million D NaHA and the viscosity at shear rate 1 s^{-1} (η_1) was employed for the other molecular weights. A master curve with two straight lines was constructed employing η_1 data for all NaHA molecular weights and concentrations. The slopes of the lines are underlined. The magnitude of C*Mw at which the slope changes is about 400,000.

(η_o) for Mw of 2.5 million D and 1.6 million D are also shown in Fig. 4 with solid symbols. As expected, the discrepancy between η_o and η_1 was more pronounced as the viscosity increased.

Through dynamic testing, the crossover frequency of NaHA solutions have been investigated. The crossover frequency for a polymer solution is the critical frequency below which a solution exhibits viscous flow behavior and above which it exhibits elastic behavior. This critical frequency is determined under dynamic testing by measuring the frequency at which the storage modulus (G') and the loss modulus (G") are equal. The crossover frequency for NaHA solutions of Mw = 2.5 million D was inversely proportional to the square of NaHA concentration: $W_c = 0.12/C^2$.

Summary

The rheological properties of NaHA solutions with varying NaHA concentrations and molecular weights were studied under steady shear conditions. For shear rates at or approaching zero shear rate, the relationship between apparent viscosity, and the product of NaHA molecular weight and concentration was biphasic and was represented by:

$\log \eta_1 = - 19.5 + 3.75 \log C*Mw$
(for C*Mw>400,000)
$\log \eta_1 = - 6.45 + 1.45 \log C*Mw$
(for C*Mw<400,000)

The crossover frequency for the storage and loss moduli was shown to be inversely proportional to the polymer concentration for a given molecular weight.

References

Balazs, E.A., *In Healon (sodium hyaluronate): a guide to its use in ophthalmic surger;* ; Miller, D. and Stegmann, R., Eds; Wiley, New York, **1983**; 5-28.

Bueche, F., *J. Chem. Phys.*, **1956,** 5, 271.

Bueche, F., *Physical Properties of Polymers*, Interscience, New York, **1962.**

Burns, J.W., Choodnovskiy, A., Yaacobi, Y., and Goldberg, E.P., *J. Invest. Surg.*, **1991, 4, 388.**

Namiki, O., Toyoshima, H., and Morisaki, N. *Int. J. Clin. Pharmacol. Ther. Toxicol.*, **1982**, 5-28.

Production and Properties of Microbial Poly(hydroxyalkanoates)

Yoshiharu Doi*, Yasushi Kawaguchi, Shigeo Nakamura, and Naoyuki Koyama
Research Laboratory of Resources Utilization, Tokyo Institute of Technology, Nagatsuta, Midori-ku, Yokohama 227, Japan
(**FAX** 81-45-921-0897)

Three types of copolymers, poly(3-hydroxybutyrate-co-3-hydroxyvalerate), poly(3-hydroxybutyrate-co-4-hydroxybutyrate) and poly(3-hydroxybutyrate-co-3-hydroxypropionate), are produced by *Alcaligenes eutrophus* from various carbon substrates. These microbial polyesters are thermoplastics with biodegradable properties, and the physical properties can be regulated by varying the compositions of copolymers.

A wide variety of microorganisms accumulate an optically active polymer of (R)-3-hydroxybutyric acid, P(3HB), as an intracellular storage material of carbon and energy (Anderson and Dawes, 1990; Doi, 1990). Many prokaryotic organisms such as bacteria and cyanobacteria have been found to accumulate P(3HB) up to 80% of their cellular dry weight, when growth is limited by the depletion of an essential nutrient such as nitrogen, oxygen, phosphorus or magnesium. Recently, some bacteria have been found to accumulate copolymers containing (R)-3-hydroxyalkanoate units other than (R)-3-hydroxybutyrate. More recently, copolymers containing 4-hydroxybutyrate unit have been produced by several bacterial strains. The general class of microbial polyesters is called poly(hydroxyalkanoates) (PHA) (Anderson and Dawes, 1990; Doi, 1990).

These microbial polyesters have attracted much attention as environmentally degradable thermo-plastics for a wide range of agricultural, marine and medical applications (Holmes, 1988). The microbial polyesters are degradable in soil, sludge or sea water. Some microorganisms such as bacteria and fungi secrete extracellular P(3HB) depolymerase to degrade environmental microbial polyesters and utilize the decomposed compounds as nutrient.

This paper surveys the PHA copolymers produced by *Alcaligenes eutrophus* from various carbon sources and discusses the pathway and its regulation of PHA synthesis. In addition, the biodegradability of PHA products is studied.

Production of P(3HB-co-3HV) Copolymer

The copolymers of (R)-3-hydroxybutyrate and (R)-3-hydroxyvalerate P(3HB-co-3HV) have been commercially produced by ICI, UK in a large scale, two-stage, fed-batch fermentation of *A.eutrophus*, feeding propionic acid and glucose as the carbon sources (Byrom, 1987). In the first stage, *A.eutrophus* cells grow and multiply in a glucose-salts medium under conditions of carbon and nutrient excess. In the second stage, the phosphate supply becomes depleted and propionic acid is fed. The P(3HB-co-3HV) copolymers are accumulated up to 75% of total dry cell weight in the second stage of phosphate limitation. The total fermentation time is in the order of 110 - 120 hours. The copolyester compositions vary from 0 to 47 mol% 3HV, depending on the ratio of propionic acid and glucose supplied. The biosynthetic pathway of P(3HB-co-3HV) in *A.eutrophus* has been investigated by using ^{13}C-labeled acetic and propionic acids as the carbon sources (Doi et al., 1987).

The P(3HB-co-3HV) copolymers with a wide range of compositions from 0 to 90 mol% 3HV are accumulated in

A.eutrophus by using butyric and pentanoic acids as the carbon sources (Doi et al.,1988). The butyric and pentanoic acids are respectively incorporated into 3HB and 3HV units via acetoacetyl-CoA in the β-oxidation cycle. The regulation of P(3HB-co-3HV) biosynthesis from butyric and pentanoic acids is of interest because the key regulatory enzyme of polymer synthesis from acetyl-CoA, 3-ketothiolase, is not involved.

The P(3HB-co-3HV) copolymers were produced at 30°C from various carbon sources by *A.eutrophus* under fed-batch growth conditions. The production of P(3HB-co-3HV) from butyric and pentanoic acids was effective under nitrogen-limited conditions, and the conversion of carbon sources into copolymers was as high as 56 wt% at the C/N mole ratio of 40. In contrast, under nitrogen-excess conditions (C/N<10), the cell growth was good, while the production of P(3HB-co-3HV) from fructose and propionic acid was almost completely inhibited under nitrogen-excess conditions.

Production of P(3HB-co-4HB) Copolymer

The copolymers of (R)-3-hydroxybutyrate and 4-hydroxybutyrate P(3HB-co-4HB) are produced by *A.eutrophus* from 4-hydroxybutyric acid, 1,4-butanediol or γ-butyrolactone (Kunioka et al., 1988; 1989). When 4-hydroxybutyric acid was used as the sole carbon source, a P(3HB-co-33%4HB) was produced. The addition of butyric acid in the 4-hydroxybutyric acid culture solution resulted in a decrease in the 4HB fraction. Thus, the copolymer compositions were varied from 0 to 33 mol% 4HB, depending on the carbon substrates supplied in the feed. Recently, we have found that the P(3HB-co-4HB) copolymers with a wide range of compositions from 0 to 100 mol% 4HB are produced by *A.eutrophus* from 4-hydroxybutyric acid in the presence of some additives. When 4-hydroxybutyric acid, citrate and ammo nium sulfate were fed as the mixed substrates, P(3HB-co-4HB) copolymers with compositions of 70 - 100 mol% 4HB were produced.

4-Hydroxybutyryl-CoA is first formed from 4-hydroxybutyric acid in the *A. eutrophus* cells. A portion of 4-hydroxybutyryl-CoA is then metabolized into (R)-3-hydroxybutyryl-CoA via acetoacetyl-CoA in the β-oxidation cycle. A random copolymer of 3HB and 4HB units is synthesized by the copolymerization of (R)-3-hydroxybutyryl-CoA with 4-hydroxy-butyryl-CoA under the actions of P(3HB) polymerase. When $(NH_4)_2SO_4$ and citrate are added to *A.eutrophus*, acetoacetyl-CoA from 4-hydroxybutyryl- CoA is metabolized into acetyl-CoA rather than into (R)-3-hydroxybutyryl-CoA under growth conditions, resulting in an increase in the 4HB fraction.

The copolymers of (R)-3-hydroxybutyrate and 3-hydroxypropionate P(3HB- co-3HP) are produced by *A.eutrophus*, when 3-hydroxypropionic acid is used as the carbon source (Nakamura et al., 1991). The 3HP content is still limited to the range 0 - 7 mol%. The P(3HB-co-3HP) copolymers were also produced from the alkanediols of odd carbon numbers such as 1,5-pentanediol, 1,7-heptanediol and 1,9-nonanediol. In contrast, P(3HB-co-4HB) copolymers were produced from the alkanediols of even carbon numbers such as 1,4-butanediol, 1,6-hexanediol, 1,8-octanediol, 1,10-decanediol and 1,12-dodecanediol.

These microbial polyesters are thermoplastics with biodegradable properties (Kunioka et al., 1989; Scandola et al., 1990), and the mechanical properties can be regulated by varying the compositions of copolymers (Table 1).

Biodegradation of Microbial Polyesters

A remarkable characteristic of microbial polyesters is that they are thermoplastic with environmentally degradable properties. The biodegradability of PHA products has been studied in environments such as soil, aerobic sewage and sea water. The processes of biodegradation were analyzed by monitoring the time-dependent changes in weight loss (erosion), molecular weights and mechanical strength of films, plates and fibers of microbial polyesters. All the samples exposed in environments were degraded via surface erosion. The rates of surface erosion of P(3HB) film in various environments at 25°C are given in Table 2. The rate of

Table 1 Physical and mechanical properties of P(3HB-co-4HB) films at 23 °C

Composition, (mol%) 3HB	4HB	Crystal-linity (%)	Density (g/cm^3)	Stress at yield (MPa)	Elonga-tion at yield (%)	Tensile strength (MPa)	Elonga-tion to break (%)
100	0	60 ± 5	1.250	-	-	43	5
97	3	55 ± 5	n.d.	34	4	28	45
90	10	45 ± 5	1.232	28	5	24	242
84	16	45 ± 5	1.234	19	7	26	444
56	44	15 ± 5	n.d.	-	-	10	511

Table 2 Decrease in the thickness of P(3HB) films in various environments at 25 °C

Environment	Rate of degradation (μm/week)
Aerobic sewage	7
Soil	5
Sea water	5

surface erosion in sea water was almost independent of the copolymer compositions of P(3HB-co-3HV) and P(3HB-co-4HB) samples. In contrast, the erosion rates in soil and aerobic sewage were strongly dependent on copolymer compositions and decreased in the order P(3HB-co-4HB) > P(3HB) > P(3HB-co-3HV). These results suggests that extracellular P(3HB) depolymerases from various bacteria have different specificities on the degradation of microbial polyesters.

An extracellular P(3HB) depolymerase was purified from *Alcaligenes faecalis* which had been isolated in aerobic sewage (Tanio et al., 1982). In a previous paper (Doi et al., 1990), we showed that the rate of enzymatic degradation of PHA films was faster by two or three orders of magnitude than the rate of simple hydrolytic degradation. The enzymatic degradation occurred at the surface of PHA film and the rate of surface erosion decreased in the order P(3HB-co-4HB) > P(3HB) > P(3HB-co-3HV).

References

Anderson, A.J.; Dawes, E.A. *Microbiol. Rev.* **1990**, *54*, 450-472.

Byrom, D. *Trends Biotechnol.* **1987**, *5*, 246-250.

Doi, Y.; Kunioka, M.; Nakamura, Y.; Soga, K. *Macromolecules* **1987**, *20*, 2988-2991.

Doi, Y.; Tamaki, A.; Kunioka, M.; Soga, K.; *Appl. Microbiol. Biotechnol.* **1988**, *28*, 330-334.

Doi, Y.; Kanesawa, Y.; Kunioka, M. *Macromolecules* **1990**, *23*, 26-31.

Doi, Y. *Microbial Polyesters*; VCH Pub., New York, 1990.

Holmes, P.A. In *Development in Crystalline Polymers-2*; Bassett, D.C. Ed.; Elsevier, London, **1988**, pp 1-65.

Kunioka, M.; Nakamura, Y.; Doi, Y. *Polym. Commun.* **1988**, *29*, 174-176.

Kunioka, M.; Kawaguchi, Y.; Doi, Y. *Appl. Microbiol. Biotechnol.* **1989**, *30*, 569-573.

Kunioka, M.; Tamaki, A.; Doi, Y. *Macromolecules* **1989**, *22*, 694-697.

Nakamura, S.; Kunioka, M.; Doi, Y. *Macromol. Rep.* **1991**, *A28*, 15-24.

Scandola, M.; Ceccoruli, G.; Doi, Y. *Int. J. Biol. Macromol.* **1990**, *12*, 112-117.

Tanio, T.; Fukui, T.; Shirakura, Y.; Saito, T.; Tomita, K.; Kaiho, T.; Masamune, S. *Eur. J. Biochem.* **1982**, *124*, 71-77.

POTENTIAL BIOMEDICAL APPLICATIONS OF PARTIAL ESTERS OF THE POLYSACCHARIDE GELLAN

Vittorio Crescenzi, Mariella Dentini,
Department of Chemistry, University "La Sapienza", 00185 Rome, (FAX : +39-6-445.7112);
Lanfranco Callegaro,
Fidia Research Laboratories, Abano Terme (PD), Italy.

A mild, non-degrading chemical procedure can give good yields of partial or total alkyl or alkylaryl esters (including mixed esters) of natural polyuronans. The physico-chemical properties in bulk and in solution of said esters markedly depend, for each given starting polyuronan, on the degree of esterification (DE) and on the nature of the alkyl or alkylaryl residues.
In the case of the exocellular bacterial polyuronan gellan, different partial esters have thus been obtained and characterized. Microspheres of gellan benzyl esters (DE 70-80 %) have been shown to provide convenient means for the controlled release of incapsulated simple drugs. Prednisolone and methylprednisolone partial esters of gellan have also been prepared and studied as potential nasal drug delivery systems.

The background

A rational route to truly biocompatible polymeric and/or bioactive materials consists in synthesising appropriate partial or total esters starting from a variety of natural biocompatible polyuronans.

For the production of inert/biocompatible polymeric products soluble in organic solvents and useful for the preparation - according to standard procedures - of films, threads, non-woven tissues, microspheres, etc., the carboxyl groups along the polysaccharide chains may be esterified with simple alkyl or alkylaryl residues, e.g. ethyl or benzyl groups, or a mixture of such residues.

The final products may actually incorporate bioactive compounds (for instance drug molecules can be incapsulated in the microspheres) and may thus be useful in controlled drug delivery formulations.

Alternatively, whenever feasible and convenient, a given drug may be directly linked *via* ester bonds onto the polysaccharide backbone.

In addition, since depending on the degree of esterification and the nature of alkyl/alkylaryl residues nature the product may be water soluble, different formulations may be realised by changing the nature of the counterions (e.g. bioactive base cations) neutralising the non-esterified carboxylate groups.

All these products, when in contact with tissues or biological fluids, more or less rapidly hydrolise and liberate the original, biocompatible polysaccharide - which may be itself an active component as in the case of hyaluronic acid - together with inert small molecules and/or pharmacologically beneficial species (esterified drugs or bioactive bases).

Realisation of these products for biomedical applications is, however, heavily dependent on our skills in devising processes for

1054-7487/92/0088$06.00/0

the synthesis of esters of polyuronans having the following essential prerequisites:

1 - to involve a minimum number of steps and easily accessible reagents;
2 - to leave the polysaccharide chains basically undegraded;
3 - to allow control of the degree of esterification even in the case of mixed ester formation;
4 - to lead to thoroughly purified products.

Recent results (Crescenzi et al., 1992) show that points 1-4 can be fulfilled by a relatively simple procedure based, essentially, on the reaction of a given polyuronan once solubilised in an aprotic organic solvent (e.g. anhydrous methylsulphoxide) with the appropriate alkyl or alkylaryl halide.

Motivations for the choice of polyuronans of the "gellan family" as starting materials.

The bacterial polysaccharide gellan (Fig. 1) , commercialised with the trade name "gelrite" by Kelco-Merck , is used in the food and in the biomedical industries for its biocompatibility and excellent solution and gelling properties.

Other polysaccharides structurally very closely related to gellan (e.g. welan and rhamsan : Fig. 1), but of strikingly different solution behaviour, are also commercialised by Kelco for different end uses, in particular as rheological or suspending agents.

The results of detailed physicochemical investigations (Crescenzi et al., 1991) clearly point out the extreme importance of the effective "charge density" and hydration of gellan-like chains in determining their accessible conformations and hence their peculiar solution/gelling properties.

We then reasoned that modulating the hydrophilic/hydrophobic character and the effective charge density of gellan, welan and rhamsan by esterifying to controlled extents the carboxylate functions along their chains with simple alkyl or alkylaryl groups should lead to :

1 - samples with markedly different solubilities and conformational properties, of interest in basic physico-chemical studies;
2 - a range of novel polymeric derivatives with diversified physico-chemical bulk/solution properties and with potential applications in different industrial sectors (e.g. : gellan esters in the biomedical field).

Another good, practical reason for the choice of the "gellan family" polysaccharides for our type of work is that both welan and rhamsan (sodium salts) are soluble in methylsulphoxide, a frequently used esterification reaction solvent.

According to our findings of relevance in the context of this presentation, the novel derivatives of gellan, welan and rhamsan do indeed display interesting physico-chemical properties and should maintain at least the biocompatibility of the parent polymers, reconversion into which only involves hydrolysis of ester bonds, a process taking place easily in contact with biological fluids.

A discussion of the main physico-chemical properties and of the potential industrial interest of welan and rhamsan derivatives will be presented elsewhere.

Gellan esters as potential biomaterials

For the preparation of gellan esters a simple and convenient procedure is based on the treatment of quaternary ammonium salts of the polysaccharide with conventional alkylating agents in organic solvents, preferably aprotic, such as methylsulphoxide (Crescenzi et al., 1992).

The procedure allows the production of a large variety of gellan esters, including mixed esters, above all the esters of monovalent n-aliphatic alcohols and the esters of arylaliphatic, alicyclic

$$\text{GELLAN: } \rightarrow3)\text{-}\overset{A}{\beta\text{-D-Glc}}\text{-}(1\rightarrow4)\text{-}\overset{B}{\beta\text{-D-GlcA}}\text{-}(1\rightarrow4)\text{-}\overset{C}{\beta\text{-D-Glc}}\text{-}(1\rightarrow4)\text{-}\overset{D}{\alpha\text{-L-Rha}}\text{-}(1\rightarrow$$

$$\text{WELAN: } \rightarrow3)\text{-}\beta\text{-D-Glc-}(1\rightarrow4)\text{-}\beta\text{-D-GlcA-}(1\rightarrow4)\text{-}\beta\text{-D-Glc-}(1\rightarrow4)\text{-}\alpha\text{-L-Rha-}(1\rightarrow$$

$$\text{(side chain on C-3 of residue C: } \overset{E}{\alpha\text{-L-Rha (or L-Man)}}\ 1\rightarrow3)$$

$$\text{RHAMSAN: } \rightarrow3)\text{-}\beta\text{-D-Glc-}(1\rightarrow4)\text{-}\beta\text{-D-GlcA-}(1\rightarrow4)\text{-}\beta\text{-D-Glc-}(1\rightarrow4)\text{-}\alpha\text{-L-Rha-}(1\rightarrow$$

$$\text{(side chain on C-6 of residue A: } \overset{E}{\alpha\text{-D-Glc}}\text{-}(6\leftarrow1)\text{-}\overset{F}{\beta\text{-D-Glc}}\ 1\rightarrow6)$$

Fig. 1. Chemical repeat of unbranched gellan and branched welan and rhamsan. The residues marked A, B, C, and D in the main chain of gellan are common to all the polymers. The side-chain residues are marked E and F in the branched polymers.

and heterocyclic monovalent alcohols.

One variant of the procedure allows the intra/intermolecular esterification (macrocyclic lactone formation) of gellan with the production of highly water swellable products.

Still another variant of the process allows the bridging of gellan chains with appropriate bifunctional derivatives of alcohols yielding, depending on the experimental conditions, insoluble derivatives cross-linked to different extents.

For the sake of brevity, attention will be limited here, however, to the case of soluble gellan esters, in particular of partial gellan esters containing :

a) - benzyl groups ;
b) - methylprednisolone and prednisolone groups.

a) - A number of benzyl esters of gellan, of different degree of esterification (DE), have been prepared and partially characterised. The derivatives have been primarily used for the production of microspheres of different average sizes, porosities and swelling properties in contact with aqueous media.

Microspheres containing known % by weight of model drugs have been employed for the analysis of the associated time/pH-dependent "drug" release features.

Similarly, microspheres with encapsulated prednisolone have also been studied.

b) - A few samples of methylprednisolone (MP) or prednisolone (P) esters of gellan have been prepared and their drug-release properties studied using directly the powdery derivatives suspended in dilute aqueous phosphate buffer (pH 7.4, 25 C).

In other analogous experiments, thin layers of mixed aqueous gels of gellan and MP-gellan were left in contact with aqueous phosphate buffer (pH 7.4) and 0.2 M NaCl at 25 C (overall ionic strength : 0.25 M).

In essence, the results of approaches a) and b) and of preclinical tests show that products based on gellan esters can perform advantageously in controlled drug-release applications, in particular for nasal and ocular uses.

Acknowledgements : this research has been partially supported by the Italian National Research Council, CNR, Progetto Finalizzato Chimica II, Rome, Italy.

References

Crescenzi, V. ; Dentini, M.; Segatori, M.; Tiblandi, C.; Callegaro, L.; Benedetti, L. Carbohydr. Res., 1992 (special issue dedicated to microbial polysaccharides).

Crescenzi, V.; Dentini, M.; Coviello, T. In Frontiers in Carbohydrate Research-2, Chandrasekaran, R., Ed; Elsevier, London, 1991, pp 100-114.

Microbiology and Physiology Symposium III

J. G. Zeikus and G. Gottschalk: *Co-Chairs*

Primary Metabolites: Session A
J. G. Zeikus and G. Gottschalk: *Co-Chairs*

Secondary Metabolites: Session B
A. L. Demain: *Chair*

Production of Polyhydroxyalkanoic Acids by Bacteria and Their Application

Hans G. Schlegel* and **Alexander Steinbüchel,**
Institut für Mikrobiologie der Georg-August-Universität, 3400 Göttingen, Federal Republic of Germany
(FAX:49 551 393793)

Polyhydroxyalkanoic acids (PHAs) are the most widely distributed carbon storage products of prokaryotes. Although "bacterial bodies" were known (Meyer, 1899) and described as sudanophilic, lipid-like granules about hundred years ago (Meyer, 1901), the chemical nature was recognized only in 1927 when Lemoigne (1926, 1927) identified 3-hydroxybutyric acid as the hydrolysis product of the granules in *Bacillus megaterium.* Due to studies of several *Bacillus* strains (Macrae and Wilkinson, 1958) and phototrophic bacteria (Doudoroff and Stanier, 1959) poly (3-hyroxybutyric acid), PHB, became more widely known. In *Alcaligenes eutrophus* H16, which is presently used for industrial production of PHB, the formation of abundant amounts of the polymer was observed during growth studies (Schlegel et al., 1961; Schlegel and Gottschalk, 1962; Wilde, 1962). PHB turned out to be only one representative among a whole class of bacterial storage polyesters.

Hydroxyalkanoate Units in Bacterial Copolyesters

Beside 3-hydroxybutyric acid almost 40 different fatty acids became known as constituents of the PHAs. The first copolymer consisting of 3-hydroxybutyric acid (3HB) and 3-hydroxyvaleric acid (3HV) was isolated from flocs of activated sludge (Wallen and Rohwedder, 1974). The method to produce a copolyester of 3HB and 3HV, poly(3HB-*co*-3HV), from *A. eutrophus* was filed as a patent by ICI in 1982. 3-Hydroxyoctanoate (3HO) was detected in 1983 in the laboratory of Witholt in Groningen as the main constituent of a polyester accumulated by octane-grown cells of *Pseudomonas oleovorans*, (De Smet et al., 1983). Further studies revealed in various different bacteria the occurrence of other straight-chain saturated 3-hydroxyalkanoic acids with chain-lengths ranging from 3 to 12 carbon atoms, as well as 3 hydroxyalkenoic acids (5 to 14 carbon atoms) with one or two double bonds. Furthermore, hydroxyalkanoic acids with an aromatic side chain, or hydroxyalkanoic acids with the hydroxygroup in the 4- or 5-position, and even branched or brominated 3-hydroxyalkanoic acids were identified (see Steinbüchel 1991a, b for recent reviews).

It may be mentioned that the bacterial PHAs are related to synthetic poly-(2-hydroxyalkanoic acids) such as polylactic acid or polyglycolic acid which are used in surgery and drug-delivery.

Related to these bacterial polymers is the homopolyester of malic acid detected in a *Penicillium* and a *Physarum* species. This polymalic acid is water-soluble.

Properties of PHA

The physical properties of the PHAs are determined by the chemical composition as well as by the degree of polymerization. In the case of copolymers the properties are assumably influenced by the distribution of the monomers with the copolymer strand, too. The homopolyester PHB and the copolyester poly(3HB-*co*-3HV) are so far the most intensively studied PHAs.

The molecular masses of PHB vary between M_r $0.01 \cdot 10^6$ and M_r $3 \cdot 10^6$, corresponding to 100 and 30 000 monomers. The variation of the reported values is mainly due to the biological systems, i.e., the bacterial strain, substrate, growth rate and temperature. The molecular mass of the PHB from *A. eutrophus* has been described to be between M_r $0.6 \cdot 10^6$ and $3 \cdot 10^6$. PHB from cells harvested in the early phase of PHB accumulation has the highest degree of polymerization; the latter decreased with prolonged fermentation to about 30 % of this value. The molecular mass may be the result of the rates of PHB synthesis and PHB degradation due to the concomitant function of PHB synthase and the intracellular depolymerase system (Pries et al., 1991). Other properties of PHB are the melting point 173-180°C; decomposition temperature ≥ 205°C; specific densitiy 1.25 g/cm^3 (see review A. Steinbüchel, 1991a for further data). PHB with its high degree of crystallinity has a briddle consistency. With increasing lengths of the side chains the crystallinity and the melting point decrease, and the flexibility increases.

Thus a copolyester like poly(3HB-*co*-3HV) has a much lower melting point than PHB and can be processed like thermoplastic synthetic polymers, e.g. polypropylene. Polyhydroxyoctanoic acid (PHO) is a flexible, elastomeric material. The physical properties of the large variety of homo- and copolyesters and their mixtures remain to be determined.

Biosynthesis and Precursors

The nature of the polyester depends on the specific bacterium, the biosynthetic enzyme, the genetic background, the substrate(s) and the growth conditions. Four biosynthetic pathways of PHA synthesis can so far be differentiated. Among about 270 bacterial species known to accumulate one or the other PHA (listed in Steinbüchel, 1991b) the majority is able to form PHB. In *A. eutrophus* the sequential function of three enzymes results in the conversion of acetyl-CoA to PHB: biosynthetic β-ketothiolase, NADPH-dependent acetoacetyl-CoA reductase and PHB synthase. *A. eutrophus* forms PHB if hexoses or organic acids such as lactate, pyruvate, succinate, crotonate, 3-hydroxybutyrate or carbon dioxide plus hydrogen are provided as substrates. If propionic acid, valeric acid, 3-hydroxyvaleric acid or other substrates degraded via propionyl-CoA are used as substrates, poly(3HB-*co*-3HV) is synthesized. The composition of the copolyesters can be varied by the substrate mixing ratio, i.e., for example the concentrations of hexose and propionate in the medium, and can result in copolyesters containing besides 3HB 10 to 90 mol % 3HV (Doi et al., 1988). *A. eutrophus* does not accumulate homo- or copolyesters containing higher chain-length monomers than C5. Probably the majority of these bacteria which are able to produce PHB, belong to the type represented by *A. eutrophus*. Experimental evidence for this assumption has been obtained for other strains of *Alcaligenes* and for phototrophic bacteria (Brandl et al. 1989; Liebergesell et al., 1991). We refer to the described route of PHA synthesis as the *Alcaligenes eutrophus* PHA-biosynthetic pathway. A modification of this pathway is functioning in *Rhodospirillum rubrum* (Moskowitz and Merrick, 1969).

Pseudomonas oleovorans and other pseudomonads belonging to the rRNA homology group I accumulate PHAs consisting of 3-hydroxyalkanoic acids of medium-chain-length if the cells are cultivated on alkanes, alkanols or alkanoic acids (Haywood et al., 1989; Huisman et al., 1989; Timm and Steinbüchel, 1990). If the cells are cultivated on octane, a polyester is formed which contains 3-hydroxyoctanoic acid (3HO) as the main constituent (about 90 mol %). Cultivation on hexoses does not result in poly(3HO) synthesis; polymer synthesis requires specific precursors. We call this route the *P. oleovorans* PHA-biosynthetic pathway.

Almost all pseudomonads belonging to the rRNA homology group I, except *P. oleovorans*, synthesize PHAs consisting of medium-chain-length 3-hydroxyalkanoic acids from hexoses, gluconate or ethanol. We call this novel pathway the *Pseudomonas aeruginosa* PHA-biosynthetic pathway. In *P. aeruginosa* the main constituent of the polyester is 3-hydroxydecanoic acid (3HD); 3-hydroxyhexanoic acid (3HH), 3HO and 3-hydroxydodecanoic acid (3HDD) were minor constituents of the polymer (Haywood et al., 1990; Timm and Steinbüchel, 1990). In some strains of *P. aeruginosa* the polymer amounted to 70 % of the dry weight, in others from 10 to 55 %.

The existence of at least two different types of PHA synthases is in accordance with the behaviour of recombinant bacteria. When a transconjugant of *P. oleovorans*, which in addition to its own synthase genes contained the PHB-synthetic genes of *A. eutrophus* H16, was cultivated on octanoate, a blend of the homopolymeric PHB and of the poly(3HO) or poly(3HO-*co*-3HHx) was accumulated. Cultivation on gluconate resulted in the accumulation of only PHB. This experiment and experiments with recombinant strains of other pseudomonads indicate that there are at least two different types of PHA synthases: one dependent on short-chain-length hydroxyacyl-CoA thioesters and the other dependent on medium-chain-length hydroxyalkanoyl-CoA thioesters. A PHA synthase active with both or dependent on long-chain-length hydroxyalkanoic acids (15 or more carbon atoms) has not yet been detected.

The PHB-Biosynthetic Operon

In *A. eutrophus* the synthesis of PHB from acetyl-CoA is accompanied by the function of only three enzymes (genes in bracketts): biosynthetic β-ketothiolase (*phbA*), NADPH-dependent acetoacetyl-CoA reductase (*phbB*) and PHB synthase (*phbC*). The genes are organized in a single operon (*phbCAB*), and *phbC* is preceded by a σ^{70}-like promoter sequence. The *A. eutrophus* operon can be transferred to other bacteria as a single unit; it is expressed in *E. coli* and in most pseudomonads tested. It was functionally active resulting in the accumulation of PHB in the recombinant strains (Slater et al., 1988; Schubert et al., 1988; Steinbüchel and Schubert, 1989; Timm and Steinbüchel, 1990). Recently, *phb*B and *phb*C from *A. eutrophus* have been expressed in *Arabidopsis thaliana*, and production of PHB has been achieved in the leaves of the transgenic plant (Poirier et al., 1992).

However, in *A. eutrophus* the functionally active products of *phbA* and *phbB* are not essential for PHB synthesis; both functions can be fulfilled by other enzymes. This was shown by the restoration of PHB synthesis in a mutant carrying Tn5 in *phbC* achieved by complementing by a DNA fragment from *Thiocystis violacea* which encodes only PHB synthase and β-ketothiolase but not acetoacetyl-CoA reductase (M. Liebergesell and A. Steinbüchel, unpublished results). Isoenzymes for β-ketothiolase and acetoacetyl-CoA reductase (Haywood et al., 1988a, b) or a route of PHA synthesis as encountered in *Rhodospirillum rubrum* may be alternatively involved in PHB synthesis in *A. eutrophus* (Steinbüchel and Schlegel, 1991). The only other bacterium, in which the structural genes for the PHB-biosynthetic enzymes have so far been shown to be clustered, is *Chromatium vinosum* (Liebergesell and Steinbüchel, 1992).

In *Zoogloea ramigera*, which is the only other PHB-accumulating bacterium genetically analyzed, only *phbA* and *phbB* seem to be organized in a single operon. The PHB synthase gene has not been identified (Peoples et al., 1987; Peoples and Sinskey, 1989).

In *P. oleovorans* the structural genes for two PHA synthases and one PHA depolymerase are clustered (Huisman et al., 1991).

Regulation of PHA Formation

In *A. eutrophus* PHA is accumulated when the cells are cultivated in a nutrient medium containing abundant carbon substrate, but limiting concentrations of nitrogen, phosphorus, sulfur or magnesium. Conditions of restricted oxygen supply have a similar effect. The three PHB biosynthetic enzymes are synthesized constitutively (Oeding, 1972). Regulation occurs on the enzyme level. The condensation of two acetyl-CoA moieties by the biosynthetic β-ketothiolase is competitively inhibited by free coenzyme A (Oeding and Schlegel, 1973; Haywood et al., 1988a). Under unrestricted growth conditions the concentration of coenzyme A is probably high. The dominant role of biosynthetic β-ketothiolase in the regulation of PHB synthesis is in agreement with results of studies on PHB-negative mutants. When the latter are cultivated on an organic substrate or CO_2 plus H_2 under conditions promoting PHB synthesis in the wild type large amounts of pyruvate are excreted (Cook and Schlegel, 1978; Steinbüchel and Schlegel, 1989). These observations support the assumption that the cells dispose of a strong substrate degradative apparatus, and that the control of pyruvate dehydrogenase by NADH and of β-ketothiolase by free coenzyme A are the main regulatory valves governing the flow of metabolites.

Another aspect deserves attention. The rate of storage product accumulation results from both the synthetic and degradative processes in the cell. Studies on intracellular degradation of PHB are scarce and considered only the physiological level (Hippe and Schlegel, 1967; Doi et al., 1990). The rates of synthesis and degradation in different bacteria are apparently very different (Merrick and Doudoroff, 1964). The control of these rates will probably affect the efficiency of PHB synthesis and the product quality such as the molecular mass (Steinbüchel and Schlegel, 1991; Steinbüchel, 1991b).

Application of PHA

Due to a variety of favorable properties the PHAs lend themselves to become major comodity chemicals. So far only PHB (Seebach and Zueger, 1982; Seebach, 1988) and the copolyester poly-(3HB-co-3HV), known as "Biopol", were considered for or introduced into, respectively, industrial application (Byrom, 1987). The essential properties calling for attention, concern (i) the thermoplasticity, (ii) chirality of the building blocks, (iii) production from renewable carbon sources or by photosynthesis directly, (iv) biodegradability, (v) biological compatibility. Possibilities of actual applications have been discussed by Byrom (1987) and Steinbüchel (1991a). The continuation of research on PHA will be profitable under theoretical as well as practical aspects.

References

Brandl, H.; Gross, R.A.; Knee, E.J.; Lenz, R.W.; Fuller, R.C. *Int. J. Biol. Macromol.* **1989**, *11*, 49-56.

Byrom, D. *TIBTECH* **1987**, *5*, 246-250.

Cook, A.M.; Schlegel, H.G. *Arch. Microbiol.* **1978**, *119*, 231-235.

De Smet, M.J.; Eggink, G.; Witholt, B.; Kingma, J.; Wynberg, H. *J. Bacteriol.* **1983**, *154*, 870-878.

Doi, Y.; Segawa, A.; Kawaguchi; Kunioka, M. *FEMS Microbiol. Lett.* **1990**, *67*, 165-170.

Doi, Y.; Tamaki, A.; Kunioka, M.; Soga, K. *Appl. Microbiol. Biotechnol.* **1988**, *28*, 330-334.

Doudoroff, M.; Stanier, R.Y.; *Nature* **1959**, *183*, 1440-1442.

Haywood, G.W.; Anderson, A.J.; Chu, L.; Dawes, E.A. *FEMS Microbiol. Lett.* **1988a**, *52*, 91-96.

Haywood, G.W.; Anderson, A.J.; Chu, L.; Dawes, E.A. *FEMS Microbiol. Lett.* **1988b**, *52*, 259-264.

Haywood, G.W.; Anderson, A.J.; Dawes, E.A. *Biotechnol. Lett.* **1989**, *11*, 471-476.

Haywood, G.W.; Anderson, A.J.; Ewing, D.F.; Dawes, E.A. *Appl. Environ. Microbiol.* **1990**, *56*, 3354-3359.

Hippe, H., Schlegel, H.G. *Arch. Mikrobiol.* **1967**, *56*, 278-299.
Huisman, G.W.; Leeuw, O. de; Eggink, G.; Witholt, B. *Appl. Environ. Microbiol.* **1989**, *55*, 1949-1954.
Huisman, G.W.; Wonink, E.; Meima, R.; Kazemier, B.; Terpstra, P.; Witholt, B.; *J. Biol. Chem.* **1991**, *266*, 2191-2198.
Lemoigne, M. *Bull. Soc. Chim. Biol.* **1926**, *8*, 770-782.
Lemoigne, M. *Ann. Inst. Pasteur* **1927**, *41*, 148-165.
Liebergesell, M.; Hustede, E.; Timm, A.; Steinbüchel, A.; Fuller, R.C.; Lenz, R.W.; Schlegel, H.G. *Arch. Microbiol.* **1991**, *155*, 415-421.
Liebergesell, M.; Steinbüchel, A. *J. Bacteriol.* **1992**, submitted for publication.
Macrae, R.M.; Wilkinson, J.F. *Proc. R. Phys. Soc. Edinburgh* **1958**, *27*, 73-78.
Merrick, J.M.; Doudoroff, M. *J. Bacteriol.* **1964**, *88*, 60-71.
Meyer, A. *Flora* **1899**, *86*, 428-469.
Meyer, A. *Zentralbl. Bakteriologie I* **1901**, *29*, 809-810.
Moskowitz, G.J.; Merrick, J.M. *Biochem.* **1969**, *8*, 2748-2755.
Oeding, V. Dissertation, Universität Göttingen, **1972**.
Oeding, V.; Schlegel, H.G. *J. Biochem.* **1973**, *134*, 239-248.
Peoples, O.P.; Masamune, S.; Walsh, C.T.; Sinskey, A.J.; *J. Biol. Chem.* **1987**, *262*, 97-102.
Peoples, O.P.; Sinskey, A.J. *Mol. Microbiol.* **1989**, *3*, 359-367
Poirier, Y.; Dennis, D.E.; Klomparens, K.; Somerville, C. *Science* **1992**, accepted for publication.
Pries, A.; Priefert, H.; Krüger, N.; Steinbüchel, A. *J. Bacteriol.* **1991**, *173*, 5843-5853.
Schlegel, H.G.; Gottschalk, G. *Angewandte Chemie* **1962**, *74*, 342-346.
Schlegel, H.G.; Gottschalk, G.; von Bartha, R. *Nature* **1961**, *191*, 463-465.
Schubert, P.; Steinbüchel, A.; Schlegel, H.G. *J. Bacteriol.* **1988**, *170*, 5837-5847.
Seebach, D. *Polym. Prepr.* **1988**, **29**, 173-174.
Seebach, D.; Zueger, M. *Helv. Chim. Acta* **1982**, *65*, 495-203.
Slater, S.C.; Voige, W.H.; Dennis, D.E. *J. Bacteriol.* **1988**, *170*, 4431-4436.
Steinbüchel, A.; Schlegel, H.G. *Mol. Microbiol.* **1991**, *5*, 535-542.
Steinbüchel, A. *Nachr. Chem. Tech. Lab.* **1991a**, *39*, 1112-1124.
Steinbüchel, A. In *Biomaterials*; Byrom, D., Ed; MacMillan Publishers, Basingstoke, **1991b**, pp 123-213.
Steinbüchel, A.; Schlegel, H.G. *Appl. Microbiol. Biotechnol.* **1989**, *31*, 168-175.
Steinbüchel, A.; Schubert, P. *Arch. Microbiol.* **1989**, *153*, 101-104.
Timm, A.; Steinbüchel, A. *Appl. Environ. Microbiol.* **1990**, *56*, 3360-3367.
Wallen, L.L.; Rohwedder, W.K. *Environ. Sci. Technol.* **1974**, *8*, 576-579.
Wilde, E. *Arch. Mikrobiol.* **1962**, *43*, 109-137.

Bioenergetics and Application of Metabolite Excretion

Wil N. Konings*, **Bert Poolman** and **Arnold J.M. Driessen**, Department of Microbiology, University of Groningen, P.O. Box 14, 9750 AA Haren, The Netherlands (FAX 31 (50) 635205)

Bacteria can release metabolites into the environment. The excretion of several of these metabolites is of commercial interest and large scale fermentation processes are used for the production of such metabolites. This commercial interest has led to attempts to enhance the rate and yield of metabolite production. In some cases novel bacterial species or strains have been successfully applied for the enhanced production of a metabolite. The conversion of a substrate to a product involves several metabolic activities (Fig. 1). First, the substrate has to be translocated across the cytoplasmic membrane from the medium to the cytoplasm. Subsequently, the substrate has to be converted to the product by one or more enzymatic reactions after which the product has to be translocated from the cytoplasm to the external medium. Up to now manipulation of metabolite production has focused mainly on the metabolic pathway leading to product formation. This includes the use of specific growth conditions, selection of mutants (e.g. alteration of the kinetic/allosteric properties of key enzymes), and the control of enzyme levels by genetic techniques. Crucial steps in product formation are the translocation processes of substrate and products across the cytoplasmic membrane. Especially the mechanisms by which products can be excreted will be discussed and suggestions for manipulating the excretion process will be given.

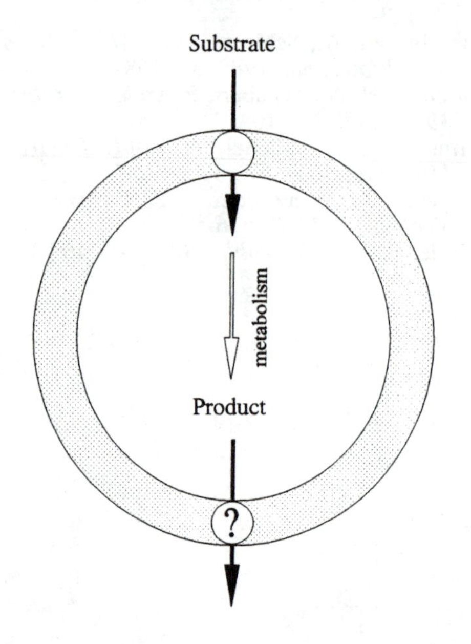

Figure 1. Schematic representation of the conversion of a substrate to a product by bacteria.

Mechanisms of product excretion

The excretion of products of metabolism from the cytoplasm to the environment can occur by similar mechanisms as are involved in the uptake of substrates (Konings et al., 1991). The most simple mechanism is **passive diffusion** of a product across the cytoplasmic membrane (Fig. 2, model 1). Hydrophobic compounds can dissolve rapidly into the inner side of the cytoplasmic membrane, subsequently diffuse across the membrane to the outer side, and then dissolve again into the external medium. The rate of passive diffusion depends on the partitioning of the product into the membrane (K), the thickness (ℓ in cm) and the surface area (A in cm^2) of the membrane, the diffusion coefficient (D in $cm^2 \cdot s$) of product in the membrane, and the concentration gradient of product in the cytoplasm and the external medium ($[c_{in}-c_{out}]$ in M) according to the equation:

$$J_{mem} = \frac{K \cdot D}{\ell} \cdot A \cdot (c_{in} - c_{out}) \qquad (M \cdot cm^3 \cdot \sec^{-1})$$

Passive fluxes are determined by the physical properties of the membrane which can be influenced by treatment of the cells with surfactants, antibiotics, detergents and others. By keeping the external concentration of product low, the product concentration gradient and thus the rate of passive diffusion will be maximized.

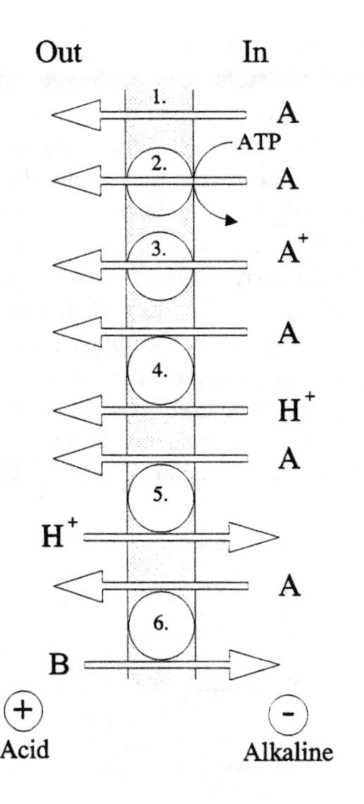

Figure 2. Excretion mechanisms in bacteria.

Passive diffusion of products can be counteracted by the action of active uptake systems for the product. Elimination of these uptake systems is needed to achieve optimal product excretion rates (Driessen et al., 1987a).

Excretion of products can also be mediated by active excretion mechanisms such as primary and secondary transport systems in which specific membrane proteins are involved in the translocation process across the cytoplasmic membrane. In **primary excretion systems** the energy for translocation is supplied by ATP or an analogous energy-rich phosphate-bond intermediate. "ATP-driven" excretion systems (Fig. 2, model 2) have been reported which mediate the excretion of compounds such as macromolecules (polysaccharides, ß-1,2-glucans, antibiotics and toxins) and xenobiotics (daunorubicin, doxorubicin, and carboxyfluorescein-analogues). Several of these systems belong to the multidrug resistance (Mdr) P-glycoprotein family (Hyde et al., 1990). Research on ATP-driven excretion systems in bacteria has been initiated recently, and the discovery of more systems can be expected in the near future. Excretion of products can also be facilitated by **secondary excretion systems**. In these protein-mediated translocation processes charges and/or ions (protons or sodium ions) are cotransported with the product, and the electrical gradient or electrochemical gradient of the co-excreted ion(s) contribute to the driving force for excretion. The most simple secondary excretion system catalyses the efflux of a product only (**uniport**) (Fig. 2, model 3). If this solute is negatively charged the electrical gradient (usually positive outside versus inside) will pull the product to the outside. In that case excretion can even occur against a product concentration gradient ($[product]_{out} > [product]_{in}$). Uniport excretion of neutral or positively charged products has to be driven by the product concentration gradient only and high product concentration gradients are needed to obtain high rates of efflux. Uniport efflux systems are not very common in bacteria as can be understood from the energetic point of view.

A second group of secondary efflux systems are the **symport** systems (Fig. 2, model 4) in which a product is excreted in a coupled movement with cation(s), usually proton(s). In this process the proton(s) moves against its electrochemical gradient (positive outside versus inside and acid outside versus inside). The driving force on the protons is therefore directed to the inside. Excretion of product can only occur when the product concentration gradient is directed to the outside and exceeds the force on the protons. Actually, in this excretion process the movement of the product drives the excretion of protons thereby generating an electrochemical proton gradient (proton motive force). Since the proton motive force can drive various energy requiring processes (e.g. ATP-synthesis, nutrient transport), the efflux process contributes to metabolic energy generation (conservation). This mechanism of product excretion has been termed "energy recycling". End-product efflux in symport with protons has been found for lactate in *E. coli* and the anaerobic organism *Lactococcus lactis* (Ten Brink & Konings, 1980; Otto et al., 1980). The high glucose and lactose fermentation rates can lead to lactate concentrations (above 300 mM) in these organisms. The excretion of lactate is mediated by a specific lactate transport system that catalyses the coupled movement of lactate together with two protons. The metabolic energy which is gained from the homolactic fermentation of lactose is 4 ATP by substrate level phosphorylation and one to two ATP equivalents by proton excretion coupled to lactate efflux. Similar energy-recycling mechanisms have been found for other products such as succinate in *Selenomonas ruminantium* (Michel & Macy, 1990). It is evident that high rates of product efflux require high product gradients (low $[product]_{out}$) and low electrochemical proton gradient.

The most attractive way of product excretion is by **antiport** (Fig. 2, models 5 and 6) mechanisms. A number of mechanisms have been reported in which a product is excreted in exchange for a proton(s) or another solute. In the exchange of product with proton(s) the product is excreted while proton(s) are taken up. This system is driven by the product concentration gradient and the electrochemical proton gradient (outside positive and acid versus inside). High rates of efflux can thus be obtained and even excretion of product against its concentration gradient can be achieved. Well-known proton antiport systems in bacteria are the excretion systems for Na^+ and Ca^{2+}-ions (Ambudkar & Rosen, 1990). Other transport proteins that belong to this class of excretion systems are the lysine efflux system of *Corynebacterium glutamicum* (Bröer and Krämer, 1991) the bacterial multidrug resistance (Bmr) carrier of *Bacillus subtilis* (Neyfakh et al., 1991), the phosphonium ion and ethidium efflux systems of *E. coli* (Purewal et al., 1990), the tetracyclin efflux system of *E. coli* (Yamaguchi et al., 1990).

The other class of antiport systems catalyses the excretion of a product in a coupled exchange with a substrate (Fig. 2, model 6), (Poolman, 1990). The exchanged solutes can be anions such as sugar-phosphates/phosphate, oxalate/formate, malate/lactate, or cations such as arginine/ornithine or neutral solutes such as lactose/galactose. In all these exchange processes the substrate concentration gradient is directed inwards while the product concentration is directed outwards. Both forces thus work together which allows a high rate of exchange. For the organisms, especially the fermentative ones, these exchange processes are very attractive since no metabolic energy is needed for the translocation of substrate or product. An example is the arginine deiminase pathway in which the conversion of arginine to ornithine only yields one ATP (Driessen et al., 1987b). Some exchange processes actually contribute to the production of metabolic energy. Examples are oxalate/formate exchange (oxalate fermentation) in *Oxalobacter formigenes* (Anantharam et al., 1989) and malate/lactate exchange (malolactic fermentation) in *L. lactis* (Poolman et al., 1991) (Fig. 3). In these processes the exchange reaction results in the net inward movement of a negative charge, leading to the generation of an electrical potential ($\Delta\Psi$). Furthermore, one proton is consumed in the conversion of substrate to product and this leads to an increase of the internal pH thus generating (or maintaining) a pH-gradient, inside alkaline (ΔpH). The resulting proton motive force can be used to drive ATP-synthesis as has been demonstrated for malolactic fermentation (Poolman et al., 1991) (Fig. 3).

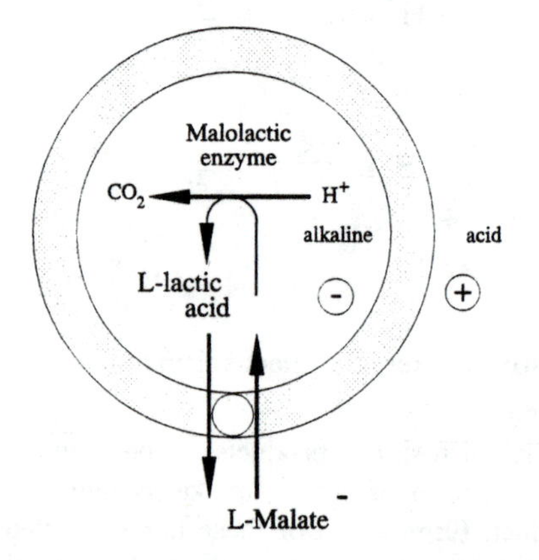

Figure 3. Malate/Lactic acid exchange during malolactic fermentation.

Manipulation of excretion processes

The rate of product excretion can often be maximized by increasing the product concentration gradient. This can be done by (i) increasing the rate of product formation internally, which leads to higher concentrations of product inside, (ii) decreasing the concentration of product externally, and (iii) decreasing the electrochemical proton gradient when the proton motive force counteracts product efflux or (iv) increasing the proton motive force in case of product/proton antiport. In the passive processes the rate of efflux can be further improved by increasing the permeability of the cytoplasmic membrane for instance by using surfactans or choosing growth conditions which lead to increased membrane permeability. In active efflux process, manipulation of the excretion processes is, in principle, possible by engineering of the transport proteins. For instance it is, in principle, possible to change the coupled movement of product and proton (ion). So has it already been shown that a solute/proton symport system can be converted into a solute uniport system. To change a solute/proton symport system into a solute/proton antiport system will obviously be more difficult. Alternatively, one might change the substrate specificity of already existing antiport systems. Improved excretion can also be achieved by expressing the genes of attractive excretion systems into the production organism. Also, the broad specificity multidrug resistance proteins, that are ATP-driven, might be used to enhance product excretion.

References

Ambudkar, S.V.; Rosen, B. In *The Bacteria*, Vol. XII Krulwich, T.A. Ed; Academic Press, New York 1990 pp. 247-271.

Anantharam, V.; Allison, M.J.; Maloney, P.C. *J. Biol. Chem.* **1989,** *264,* 7244-7250.

Bröer, S; Krämer, R. *Eur. J. Biochem.* **1991,** *202,* 137-143.

Driessen, A.J.M.; Hellingwerf, K.J.; Konings, W.N. *J. Biol. Chem.* **1987a,** *262,* 12438-12443.

Driessen, A.J.M.; Poolman, B.; Kiewiet, R.; Konings, W.N. *Proc. Natl. Acad. Sci. USA* **1987b,** *84,* 6093-6097.

Hyde, S.C.; Emsley, P.; Hartshorn, M.J.; Mimmack, M.M.; Gileadi, U.; Pearce, S.R.; Gallagher, M.P.; Gill, D.R.; Hubbard, R.E.; Higgins, C.F. *Nature* **1990,** *346,* 362-365.

Konings, W.N.; Poolman B; Driessen, A.J.M. *FEMS Microbiol. Rev.* **1991** 88/2.

Michel, T.A.; Macy, J.M. *J. Bacteriol.* **1990** *172,* 1430-1435.

Neyfakh, A.A.; Bidnenko, V.E.; Chen, L.B. *Proc. Nat. Acad. Sci. USA* **1991,** *88,* 4781-4785.

Otto, R.; Sonnenberg, A.S.M.; Veldkamp, H.; Konings, W.N. *Proc. Natl. Acad. Sci. USA* **1980,** *77,* 5502-5506.

Poolman, B. *Mol. Microbiol.* **1990,** *4,* 1629-1636.

Poolman, B.; Molenaar, D.; Smid, E.J.; Ubbink, T.; Abee, T.; Renault, P.P.; Konings, W.N. *J. Bacteriol.* **1991,** *173,* 6030-6037.

Purewal, A.S.; Jones, I.G.; Midley, M. *FEMS Microbiol. Lett.* **1990,** *68,* 73-76.

Ten Brink, B.; Konings, W.N. *Eur. J. Biochem.* **1980,** *111,* 59-66.

Yamaguchi, A.; Udagawa, T.; Sawai, T. *J. Biol. Chem.* **1990,** *265,* 4809-4813.

Physiological Improvements in Acetone Butanol Fermentation

Gerhard Gottschalk and Helga Grupe, Institut für Mikrobiologie der Georg-August-Universität, Grisebachstr. 8, W-3400 Göttingen, FRG

The acetone butanol fermentation as carried out by *Clostridium acetobutylicum* was applied on an industrial scale for about 40 years but was discontinued around 1950. During the last 15 years a renaissance of the research on this very interesting process has taken place and details are known now on the fermentative pathway and the optimal conditions for solvent production in batch as well as in continuous cultures. A still unsolved problem is the regulation of solvent formation. Evidence for a role of the ATP level and the NADH level will be presented.

Following the isolation of *Clostridium acetobutylicum* by Weizmann and the application of the acetone butanol fermentation on an industrial scale by the company Strange & Graham Ltd. the biotechnological production of acetone and n-butanol had its boom for about 40 years (see Jones and Woods, 1986; Bahl and Gottschalk, 1988). Dozens of fermentors with a working volume of 190.000 litres were in operation and solvents in the range of 100 tons were produced per day. After the Second World War the fermentation process was soon outcompeted by the petrochemical industry and the industrial process now is history. The oil-crisis in the 70ies triggered a revival of the interest in biotechnological processes for the production of chemicals; this included the acetone butanol fermentation and since about 1980 research on this process has boomed and considerable progress has been made in understanding this fermentation; but this progress has not been significant enough to allow the development of a fermentation technology which can compete with the existing petroleum-based technology.

Limitation of Product Formation

When *C. acetobutylicum* is grown in batch culture on glucose or maltose, acids are produced first; then the culture undergoes the so-called shift and solvents are produced. The maximal solvent concentration which can be reached is in the order of 2 %. It will be very difficult to gain more because of the toxicity of n-butanol. This solvent interferes with the proper function of the cytoplasmic membrane which has to be in an energized state in living organisms, meaning that a membrane potential and a certain pH has to exist across the membrane. Butanol is a very good drug to dissipate the proton-motive force at the membrane (Gottwald and Gottschalk, 1985). The composition of the cytoplasmic membrane of *C. acetobutylicum* is somewhat adapted to solvent stress, but as already mentioned a considerable increase of the final product concentration is most unlikely, and this is a severe drawback of this fermentation because it leads to high product recovery costs and large volumes of fermentation

1054–7487/92/0102$06.00/0

broth which has to be recycled or disposed. Therefore, product recovery has been in the centre of many research activities and it should be so in the future. The solvents may be recovered by destillation, by reverse osmosis or pervaporization using membranes, by adsorbance or liquid extaction (Ennis et al., 1986).

The acetone butanol fermentation can be carried out in continuous culture, preferentially at low pH and under phosphate limitation (Bahl et al., 1982). Other conditions allowing continuous solvent formation include limitation by the nitrogen source, sulfate or magnesium and growth in a turbidostat (Gottschal and Morris, 1982; Bahl and Gottschalk, 1988).

Shift from Acids to Solvents

The mechanism underlying the shift from acid to solvent production is of special interest. We know that enzyme induction is involved. This is true for the two enzymes involved in acetone formation, a CoA transferase and acetoacetate decarboxylase, and it is true for the two enzymes involved in butanol formation, NAD^+-specific butyraldehyde dehydrogenase and NAD^+-specific butanol dehydrogenase (Fig. 1).

What do we know about the triggering mechanism? Prior to a discussion of possible signals for the shift experimental tricks leading to an induction of the shift will be summarized.

The decrease of the pH from about 6.0 to 4.5 can be used to induce solventogenesis (Bahl et al., 1982). Because of the fact that *C. acetobutylicum* keeps the pH constant this means an increase of the internal proton concentration and simultaneously of the internal concentration of acids (Gottwald and Gottschalk, 1985). A similar effect in terms of inducing the shift can be observed by the addition acids, such as butyrate or propionate and valerate (Jewell et al., 1986). A combination of aceto-acetate and butyrate with uncouplers such as FCCP and CCCP has also been found to induce solvent formation (Huesemann and Papoutsakis, 1986). A specific induction of butanol production can be obtained by CO gassing or the addition of methylviologen (Kim et al., 1984; Kim and Zeikus, 1985; Rao and Mutharasan, 1986).

So there are two sets of conditions under which the metabolism of *Clostridium acetobutylicum* may be shifted to solventogenesis: first, the decrease of the pH by the production of acids or by the addition of acids and uncouplers which leads to an increase of the internal concentration of protons and acids and an elevated requirement for ATP; second, the inhibition of the evolution of molecular hydrogen by CO, H_2 pressure or methylviologen addition which should result in an increased NAD(P)H/NAD(P) ratio.

It is assumed that the generation of the two signals (low ATP, high NAD(P)H) leads to a complete shift, the formation of acetone and butanol. If only NAD(P)H is high (e.g. addition of methylviologen) acetone is not produced but butanol (Fig. 2). This notion is supported by the observed decrease of the ATP level during acid production and corresponding changes of the redox potential and the NAD(P)H level.

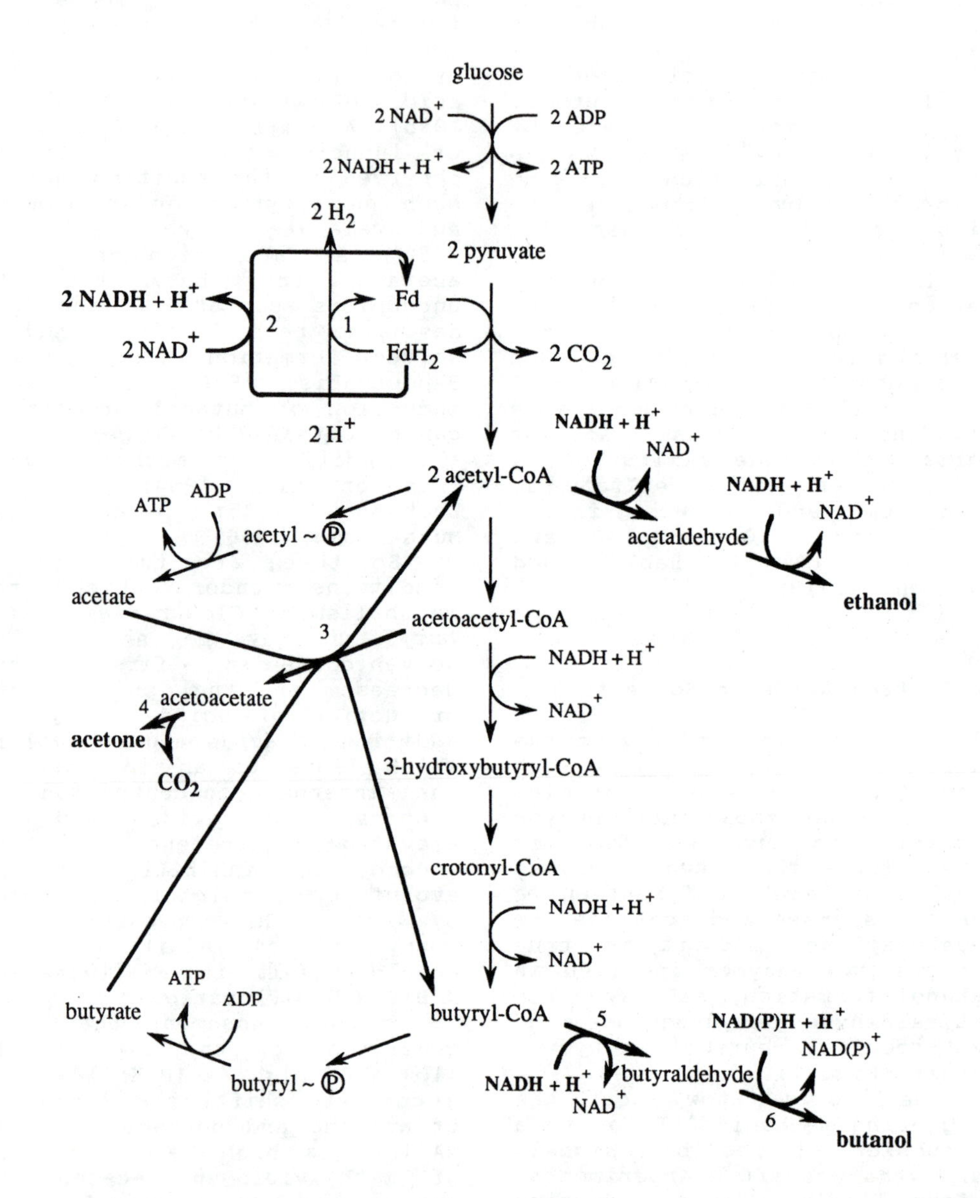

Fig. 1 Pathway of Acidogenesis and Solventogenesis. 1, hydrogenase; 2, NADH:Fd oxidoreductase; 3, CoA transferase; 4, acetoacetate decarboxylase; 5, butyraldehyde dehydrogenase; 6, butanol dehydrogenase; --->, acidogenic reactions; --->, additional solventogenic reactions.

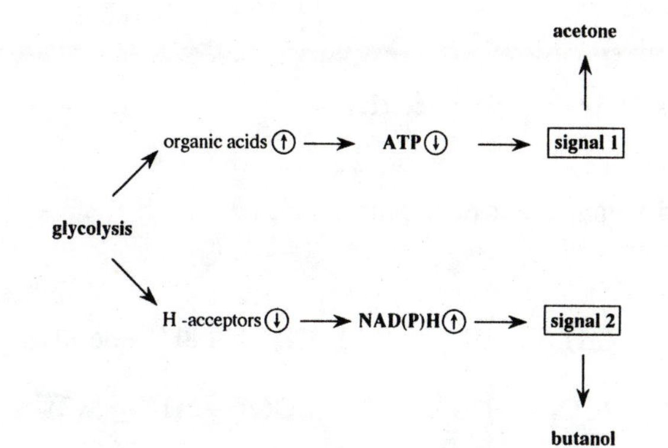

Fig.2 A two-signal model for the shift from acidogenesis to solventogenesis. Arrows within circles indicate increased or decreased levels of metabolites.

Acknowledgement

This work was supported by the Bundesministerium für Forschung und Technologie, Forschungsschwerpunkt "Grundlagen der Bioprozeßtechnik".

References

Bahl, H.; Andersch, W.; Braun, K.; Gottschalk, G. *Eur. J. Appl. Microbiol. Biotechnol.* **1982**, *14*, 17-20.
Bahl, H; Andersch, W.; Gottschalk, G. *Eur. J. Appl. Microbiol. Biotechnol.* **1982**, *15*, 201-205.
Bahl, H.; Gottschalk, G. In *Biotechnology*; Rehm, H.J.; Reed, G. Eds.;1988, Vol. 6b; pp. 1-30.
Ennis, B.M.; Gutierrez, N.A.; Maddox, I.S. *Process Biochem.* **1986**, *21*, 131-147.
Gottschal, J.C.; Morris, J.G. *Biotechnol. Lett.* **1982**, *4*, 477-482.
Gottwald, M.; Gottschalk, G. *Arch. Microbiol.* **1985**, *143*, 42-46.
Huesemann, M.; Papoutsakis, E.T. *Biotechnol. Lett.* **1986**, *8*, 37-42.
Jewell, J.B.; Coutinho, J.B.; Kropinski, A.M. *Curr. Microbiol.* **1986**, *13*, 215-219.
Jones, D.T.; Woods, D.R. *Microbiol. Rev.* **1986**, *50*, 484-524.
Kim, B.H.; Bellows, P.; Datta, R.; Zeikus, J.G. *Appl. Environ. Microbiol.* **1984**, *48*, 764-770.
Kim, B.H.; Zeikus, J.G. *Dev. Ind. Microbiol.* **1985**, *26*, 1-14.
Rao, G.; Mutharasan, R. *Appl. Microbiol. Biotechnol.* **1989**, *30*, 59-66.

Physiological, molecular, and Applied Aspects of Acetogenic Bacteria.

Lars G. Ljungdahl
Center for Biological Resource Recovery, and Department of Biochemistry, University of Georgia, Athens, GA 30602, USA

Acetogenic bacteria are unique in that they ferment a variety of substrates to acetate and have also the ability to synthesize acetate from CO_2 or other one-carbon compounds. Many of them grow autotropically and heterotrophically. It is well documented that they are important in anaerobic habitats. They are found in soils, muds, sediments in fresh and sea water, in intestinal tracts of man, animals and insects, and in man-made environments like sewage sludge digestors and composts.

The first acetogen isolated was *Clostridium aceticum* (Wieringa, 1940). It was not much investigated until Braun et al (1981) revived it from a spore preparation. Fontaine et al (1942) isolated *Clostridium thermoaceticum* from horse manure. It was the only acetogen available until 1967 and, consequently, it is the most studied acetogen. Presently, about 40 different acetogens are known. Most of them have been discovered within the last eight years. They are now subjects of many studies that include biochemical, physiological, medical, and ecological aspects. Recent reviews of acetogens include those by Fuchs (1986), Ljungdahl (1986), Wood (1989), and Wood and Ljungdahl (1991).

The acetyl-CoA Pathway

The interesting feature of acetogens is their ability to synthesize acetate from CO_2. This occurs via the recently discovered acetyl-CoA pathway, so named because acetyl-CoA is the first 2-carbon compound formed. The pathway is often referred to as the Wood pathway in honor of H.G. Wood, who much contributed to its elucidation.

C. thermoaceticum ferments fructose, glucose, and xylose almost exclusively to acetate. The fermentation can be considered to involve two steps as follows:

$$C_6H_{12}O_6 + 2H_2O \rightarrow 2CH_3COOH + 8H^+ + 8e^- \quad (1)$$

$$2CO_2 + 8H^+ + 8e^- \rightarrow CH_3COOH + 2H_2O \quad (2)$$

$$\text{Sum: } C_6H_{12}O_6 \rightarrow 3CH_3COOH \quad (3)$$

Reaction (1) represents the fermentation of glucose via the glycolytic pathway to pyruvate, which is further metabolized to acetate and CO_2. Reaction (2) represents the acetyl-CoA pathway by which the synthesis of acetate from CO_2 occurs. It involves reduction of CO_2 and as shown the reducing equivalents were generated in reaction (1). However, many other reductants serve as sources of electrons including H_2 and CO. It has been shown that *C. thermoaceticum* grows autotrophically on a gas mixture of H_2 and CO_2 or only CO (Kerby and Zeikus, 1983). That autotrophic growth occurs using the acetyl-CoA pathway means that this pathway is coupled to electron transport phosphorylation or chemosmosis (Hugenholtz and Ljungdahl, 1990b).

The acetyl-CoA pathway has been elucidated in studies of *C. thermoaceticum*. Carbon dioxide is fixed in two reactions catalyzed by formate dehydrogenase (FD) and carbon monoxide dehydrogenase/acetyl-CoA synthase (CODH). The latter enzyme catalyzes the reduction of CO_2 to CO and the synthesis of acetyl-CoA (Ragsdale and Wood, 1985). It is a nickel-iron-zinc protein (Ragsdale et al., 1983).

Formate formed in the FD-reaction is the precursor of the methyl group of acetyl-CoA. It is reduced as one-carbon intermediates of tetrahydrofolate (THF) to 5-methyl-THF. Enzymes involved are formyl-THF synthetase, methenyl-THF cyclohydrolase, methylene-THF dehydrogenase, and methylene-THF reductase. The methyl group of methyl-THF is transferred to the cobalt of a protein bound corrinoid. This protein is designated the corrinoid/Fe-S protein reflecting its content of 5-methoxybenzimidazolylcobamide and a [4Fe-4S] cluster (Ragsdale et al., 1987). Carbon dioxide destined for the carbonyl group of acetyl-CoA is reduced to the level of CO which binds to a nickel-iron cluster forming a NiFeC species part of CODH (Gorst and Ragsdale, 1991). CODH then catalyzes the formation of acetyl-CoA by condensing the Co-methyl group of the corrinoid/Fe-S protein, CoA, and the CO of the NiFeC species.

All enzymes of the acetyl-CoA pathway have been purified from *C. thermoaceticum*. Many of them have unusual properties. As mentioned CODH is a nickel-iron-zinc protein (Ragsdale et al., 1983). Formate dehydrogenase contains tungsten, selenium, and iron (Yamamoto et al., 1983). The discovery of tungsten in FD was the first demonstration of a biological role for tungsten. Two other W-proteins are present in *C. thermoaceticum*; a 60 kDa protein with unknown function (Ljungdahl and Andreesen, 1975), and an aldehyde reductase (White et al., 1989).

Interesting aspects of the THF series of enzymes can also be noted. The methenyl-THF cyclohydrolase and methylene-THF dehydrogenase reactions are catalyzed by a single bifunctional enzyme (Ljungdahl et al., 1980). The methylene-THF reductase is a flavoprotein containing an unusual iron-sulfur cluster (Park et al., 1991). This enzyme is associated with the cytoplasmic membrane and may function in the coupling of the acetyl-CoA pathway with the generation of ATP by electron transport (Hugenholtz and Ljungdahl, 1990b). Properties of the enzymes of the acetyl-CoA pathway are discussed in the review by Wood and Ljungdahl (1991).

Energy Generation in Acetogenic Bacteria

Andreesen et al. (1973) noticed that cell yields of *C. thermoaceticum* growing on sugars were higher than expected considering only substrate generation of ATP. They suggested that acetate synthesis from CO_2 was coupled to the generation of ATP. The discovery of cytochromes and menaquinones in *C. thermoaceticum* and *Clostridium formicoaceticum* (Gottwald et al., 1975) strengthened this idea, as did the fact that many acetogens were found to grow autotrophically (Fuchs, 1986; Wood and Ljungdahl, 1991).

Two b-type cytochromes, the menaquinone, MK-7, and a flavoprotein have been isolated from cell membranes of *C. thermoaceticum* and *Clostridium thermoautotrophicum* (Ivey, 1987; Das et al., 1989). An important observation was that membranes of *C. thermoautotrophicum* contain CODH and methylene-THF reductase and that these membranes carry out energy requiring reactions such as amino acid transport using energy generated by the oxidation of CO (Hugenholtz and Ljungdahl, 1990a).

C. thermoaceticum and *C. thermoautotrophicum* have H^+-ATPase (Ivey, 1987; Mayer et al., 1986) and their growth is independent of sodium (Yang and Drake, 1990). Other acetogens are strongly dependent on Na^+ for growth and apparently have Na^+-ATPase systems (Geerlings et al., 1989; Heise et al., 1989; Yang and Drake, 1990). Cytochromes have not been demonstrated in these bacteria. Thus, it appears as energy generation in anaerobic autotrophic acetogens occurs by different mechanisms. The elucidation of these is a new and challenging endeavor within the realm of anaerobic microbiology.

Molecular Biology Aspects of *C thermoaceticum*

The application of the powerful tools of molecular biology in studies of acetogens has just started. The main aims are to obtain the primary sequences of the enzymes of the acetyl-CoA pathway by cloning and DNA sequencing. All genes coding for proteins of *C. thermoaceticum* so far cloned into *E. coli* are expressed to varying degrees. A methyltransferase transferring the methyl group of methyl-THF to the corrinoid/Fe-S protein, and the subunits of the corrinoid/Fe-S protein are expressed at about 1% of the soluble *E. coli* protein (Roberts et al., 1989), formyl-THF synthetase to 30% (Lovell et al., 1988), and a leucine dehydrogenase 8.3% (Shimoi et al., 1987). Although the enzymes are expressed they are not all active. The formation of active enzyme seems to be correlated with the absence of a metal cluster in the enzyme. Thus formyl-THF synthetase, methyltransferase, and leucine dehydrogenase are active, whereas CODH, and methylene-THF reductase are inactive. These latter enzymes contain metal clusters.

E. coli seems to use the transcription and translation elements present in the *C. thermoaceticum* DNA inserts. Potential candidates for these elements have been identified by homology to *E. coli* consensus sequences (Morton et al., 1992). The sequences have been reported for formyl-THF synthetase (Lovell et al., 1990), CODH (Morton et al., 1991), and methylene-THF reductase (Chou, 1990).

Formyl-THF synthetase (Lovell et al., 1988), and leucine dehydrogenase (Shimoi et al., 1987) cloned in *E. coli* have been purified. They are identical with the enzymes isolated from *C. thermoaceticum*. Like most enzymes from thermophilic bacteria these enzymes have high thermostability, a property which simplified the purification, which involved a heating step to denaturate the *E. coli* proteins.

Metabolic Diversity of *C. thermoaceticum*

It has been mentioned that *C. thermoaceticum* ferments glucose, fructose, and xylose, and that it grows autotrophically on CO or a mixture of CO_2 and H_2. Bache and Pfennig (1981) discovered a significant metabolic potential of acetogens when they found *Acetobacterium woodii* to use in the presence of CO_2 the methoxy group of methoxylated aromatic acids such as vanillic, syringic and sinapic acids. They found further that the acrylic acid side chain of compounds such as ferulic, caffeic, and sinapic acids serves as electron acceptor and is reduced to a propionic acid side chain (Tscheck and Pfennig, 1984). The ability to use the o-methyl group of phenylmethylethers has now been found for several acetogens including *C. thermoaceticum* (Wu et al., 1988), which has the capacity also to decarboxylate benzoic and similar acids (Hsu et al., 1990). That acetogens metabolize phenylmethylethers indicates they play a role in the degradation of lignin breakdown products in anaerobic environments (Young and Frazer, 1987).

Applied Aspects of Acetogens

The ability of *C. thermoaceticum* and other acetogens to ferment sugars and many other substrates to acetate as the sole product has prompted investigations to use them for industrial production of acetate. An example is the production of calcium magnesium acetate (CMA) from glucose and dolime (Wiegel et al., 1991). CMA is considered an environmentally safe and effective deicer (Chollar, 1984).

References

Andreesen, J.R.; Schaupp, A.; Neurauter, L.; Brown, A.; Ljungdahl, L.G. *J. Bacteriol.* **1973**, 114, 743-751.

Bache, M.; Mayer, F.; Gottschalk, G. *Arch. Microbiol.* **1981**, 130, 255-261.

Braun, M.; Mayer, F.; Gottschalk, G. *Arch. Microbiol.* **1981**, 128, 288-293.

Chollar, B.H. *Public Roads*, **1984**, 47, 113.

Chou, C.-F.; *Cloning, Expression and Sequencing of the Gene Encoding the Methylenetetrahydrofolate Reductase from Clostridium thermoaceticum*. Dissertation, Univ. of Georgia, Athens, GA, **1990**.

Das, A.; Hugenholtz, J.; van Halbeek, H.; Ljungdahl, L.G. *J. Bacteriol.* **1989**, 171, 5823-5829.

Fontaine, F.E.; Peterson, W.H.; McCoy, E.; Johnson, M.J.; Ritter, G.J. *J. Bacteriol.* **1942** 43 701-715.

Fuchs, G. *FEMS Microbial. Rev.* **1986**, 39, 181-213.

Geerlings, G; Schonheit, P.; Diekert, G. *FEMS Microbiol. Lett.* **1989**, 57, 253-258.

Gorst, C.M.; Ragsdale, S.W. *J. Biol. Chem.* **1991**, 260, 20687-20693.

Gottwald, M.; Andreesen, J.R.; LeGall, J.;

Ljungdahl, L.G. *J. Bacteriol.* **1975**, 122, 325-328.

Heise, R.; Muller, V.; Gottschalk, G. *J. Bacteriol.* **1989**, 171, 5473-5478.

Hsu, T.; Daniel, S.L.; Lux, M.F.; Drake, H.L. *J. Bacteriol.* **1990**, 172, 212-217.

Hugenholtz, J; Ljungdahl, L.G. *FEMS Microbiol. Lett.* **1990a**, 69, 117-122.

Hugenholtz, J.; Ljungdahl, L.G. *FEMS Microbiol. Rev.* **1990b**, 87, 383-390.

Ivey, D.M.; *Generation of Energy During CO_2 Fixation in Acetogenic Bacteria*; Dissertation, Univ. of Georgia, Athens, GA, **1987**.

Kerby, R.; Zeikus, J.G. *Curr. Microbiol.* **1983**, 8, 27-30.

Ljungdahl, L.G. *Annu. Rev. Microbiol.* **1986**, 40, 415-450.

Ljungdahl, L.G.; Andreesen, J.R. *FEBS Lett.* **1975**, 54, 279-282.

Ljungdahl, L.G.; O'Brien, W.E.; Moore, M.R.; Liu, M.-T. *Methods Enzymol.* **1980**, 66, 599-609.

Lovell, C.R.; Przybyla, A.; Ljungdahl, L.G. *Arch. Microbiol.* **1988**, 149, 280-285.

Lovell, C.R.; Przybyla, A.; Ljungdahl, L.G. *Biochemistry* **1990**, 29, 5687-5694.

Morton, T.A.; Runquist, J.A.; Ragsdale, S.W.; Shanmugasundaram, T.; Wood, H.G.; Ljungdahl, L.G. *J. Biol. Chem.* **1991**, 266, 23824-23828.

Morton, T.A.; Chou, C.-F.; Ljungdahl, L.G. In *Genetics of Anaerobic Bacteria*; Sebald, M., Ed.; Springer Verlag, Berlin, Germany, **1992**, In Press.

Park, E.Y.; Clark, J.E.; DerVartanian, D.V.; Ljungdahl, L.G. In *Chemistry and Biochemistry of Flavoproteins*; Muller, F., Ed.; CRC Press, Boca Raton, FL, **1991**, *Vol.* 1; pp 389-400.

Ragsdale, S.W.; Wood, H.G. *J. Biol. Chem.* **1985**, 260, 3970-3977.

Ragsdale, S.W.; Clark, J.E.; Ljungdahl, L.G.; Lundie, L.L.; Drake, H.L. *J. Biol. Chem.* **1983**, 258, 2364-2369.

Ragsdale, S.W.; Lindahl, P.A.; Munck, E. *J. Biol. Chem.* **1987**, 262, 14289-14297.

Roberts, D.L.; James-Hagstrom, J.E.; Garvin, D.K.; Gorst, C.M.; Runquist, J.A.; Bauer, J.R.; Haase, F.L.; Ragsdale, S.W. *Proc. Natl. Acad. Sci. USA* **1989**, 86-32-36.

Tscheck, A.; Pfennig, N. *Arch. Microbiol.* **1984**, 137, 163-167.

Wiegel, J.; Carreira, L.H.; Garrison, R.J.; Rabek, N.E.; Ljungdahl, L.G. In *Calcium Magnesium Acetate: An Emerging Bulk Chemical for Environmental Applications*; Wise, D.L.; Lerendis, Y.A.; Matgahlchi, M., Eds.; Elsevier, New York, NY, **1991**, pp 359-418.

White, H.; Strobl, G.; Feicht, R.; Simon, H. *Eur. J. Biochem.* **1989**, 184, 89-96.

Wieringa, K.T. *Antonie van Leeuwenhoek J. Microbiol Serol.* **1940**, 6, 251-262.

Wood, H.G. In *Autotrophic Bacteria*; Schlegel, Schlegel, H.G.; Bowien, B., Eds.; Science Tech. Madison, WI and Springer Verlag, Berlin, Germany, **1989**, 33-52.

Wood, H.G.; Ljungdahl, L.G. In *Variations in Autotrophic Life*; Shively, J.M.; Barton, L.L., Eds.; Academic Press, New York, NY, **1991**, 201-250.

Wu, Z.; Daniel, S.L.; Drake, H.L. *J. Bacteriol.* **188**, 170, 5747-5750.

Yamamoto, I.; Saiki, T.; Liu, S.-M.; Ljungdahl, L.G. *J. Biol. Chem.* **1983**, 258, 1826-1832.

Yang, H.; Drake, H.L *Appl. Environ. Microbiol.* **1990**, 56, 81-86.

Young, L.Y.; Frazer, A.C. *Geomicrobial J.* **1987**, 5, 261-293.

Thermophilic Enzymes: New Sources, Uses and Biodesigns

J. Gregory Zeikus, Saroj Mathupala, Yong-Eok Lee, Sergey Podkovyrov, Badal C. Saha, Menghsiao Meng, and Michael Bagdasarian.
Departments of Biochemistry and Microbiology, Michigan State University, East Lansing, Michigan 48824
(FAX: 517-337-2122)

Enzymes from thermophilic bacteria are used in biotechnology because of their high activity and stability under harsh processing conditions. The molecular physiological and biochemical features of starch and xylan degrading enzymes from thermoanaerobic bacteria are described in relation to their activities, cellular organization, genetic composition, catalytic and thermostability domains, mechanism of catalysis, and process utility of native versus engineered proteins. Structure-function studies on amylopullulanase, cyclodextrinase and α-glucosidase of *Clostridium thermohydrosulfuricum* 39E; endoxylanase, ß-xylosidase, and acetylesterase of *Thermoanaerobacter* B6A-RI; and, xylose isomerase of *C. thermosulfurogenes* 4B are reported. The amylases and xylanases of thermoanaerobes are comprised of highly thermostable and active enzymes that are organized into hydrolysomes that efficiently couple extracellular polymer degradation with intracellular metabolism of soluble saccharides.

Recently, enzymes from thermophilic bacteria have been used in biotechnology including: the DNA polymerase of *Thermus aquaticus* for the PCR technique; the α-amylase of *Bacillus stearothermophilus* for starch liquefaction; and, the alcohol dehydrogenase of *Thermoanaerobium brockii* for synthesis of certain pheremones and fragrances. The known diversity of eubacterial moderate thermophiles and archaebacterial hyperthermophiles has grown enormously and attention has now focused on the biotechnological analysis of their thermostable enzymes.

Thermoanaerobic bacteria grow readily on biopolymers such as cellulose, xylan and starch. The cellulases of *Clostridium thermocellum* are organized into a cellulosome and they serve as the current model for molecular biological understanding of cellulose degradation (Lamed et al., 1988). Less is known about starch and xylan degradation by thermoanaerobic bacteria (Zeikus et al., 1991). Amylases have widespread use in starch bioprocessing for saccharide, sweetener and alcohol production. Xylanases are used in bread making and animal feeds and are being explored for biobleaching of paper and modification of fibers.

Thermoanaerobic bacteria evolved early on earth and stress was placed on evolution of enzymes with high stability and catalytic efficiency because of the limited chemical free energy available from their catabolism. Thermoanaerobic bacteria contain unique genes that encode for very active and stable enzymes (Lamed et al., 1988; Zeikus et al., 1991).

The purpose of this report is to review recent progress (S. Mathupala, Y.-E. Lee, M. Meng, and S. Podkovyrov, manuscripts in preparation) in our laboratory on the molecular physiology and biochemistry of saccharidases from thermoanaerobes that are involved in active conversion of starch or xylans. At present, genus and species assignments of saccharolytic thermoanaerobes that produce ethanol, lactate, acetate, H_2 and CO_2 as end products are based on spore formation ability, gram staining and substrate utilization. In order to define the genetic relationships among xylanolytic thermoanaerobes and to facilitate the establishment of new species names we have compared DNA homologies of these saccharolytic thermoanaerobes by DNA-DNA hybridization. This kind of analyses has revealed the existence of three generic groups among the amylolytic and xylanolytic thermo-

1054–7487/92/0110$06.00/0

anaerobes known to date: Group I includes one species of *C. thermocellum* which exhibited less than 14% homology to other species examined; Group II comprised of *Thermoanaerobacter* strains B6A-RI and LXII and *C. thermosulfurogenes*; and, Group III comprised of *C. thermohydrosulfuricum* 39E, *Thermoanaerobium brockii* HTD4 and *Thermoanaerobacter ethanolicus* JW200. These results suggest taxonomic re-assignments with sporeforming and non-sporeforming strains in the same genus and species.

Amylases

When *Clostridium thermohydrosulfuricum* 39E grows on starch it produces a starch binding receptor and an active amylopullulanase activity which are localized in a putative amylosome on the cell surface. *C. thermohydrosulfuricum* produces at least three different novel, thermostable amylolytic enzymes: amylopullulanase (Mathupala et al., 1990) which cleaves both α-1,4 and α-1,6 linkages in starch, α-glucosidase which cleaves α-1,6 and α-1,4 linkages in oligosaccharides (Saha and Zeikus, 1991); and, a cyclodextrinase (Saha and Zeikus, 1990) which readily degrades smaller oligosaccharides. These three enzymes are active and stable at pH 5.5 and 70°C. The α-glycosidase has a molecular weight of 160,000 and displays an optimum temperature of 75°C.

Amylopullulanase is a glycoprotein which displays higher affinity for pullulan than starch. It produces high conversion syrup from starch hydrolysis. The gene encoding for amylopullulanase was cloned from *C. thermohydrosulfuricum* 39E and expressed in both *E. coli* and *B. subtilis*. Lac Z fusion proteins and nested deletion mutants constructed from the gene allowed identification of the gene regions encoding for excretion, catalysis and thermostability of the expressed enzyme. A 4.8 Kb *Apu* gene region was identified for expression of a 140,000 MW amylopullulanase protein. Nested deletion mutants with a size of below 2.9 Kb lost activity or stability. A computer analysis comparison of the amino acid sequence of amylopullulanase of *C. thermohydrosulfuricum* 39E to *Aspergillus oryzae* α-amylase was used to identify the putative catalytic residues. By site-directed mutagenesis of the amylopullulanase (*Apu*) gene, Asp 628 or Asp 734 were changed to Asn 628 or Asn 734, respectively by single point base mutations. Glu 657 was changed to Gln 657. Complete loss of pullulan and starch hydrolysis was detected in all three mutations. These results indicate that a single active site in amylopullulanase analogous to α-amylase is responsible for cleavage of α-1,4 and α-1,6 glycosidic linkages.

The cyclodextrinase gene from *C. thermohydrosulfuricum* was cloned and expressed in *E. coli*. The nucleotide sequence of the CDase gene encoded a protein with a molecular weight of 68,028. The purified cyclodextrinase displayed Km values for α-, ß-, and γ CDs of 2.5, 2.1 and 1.3 mm, respectively. The products of cyclodextrin and oligodextrin hydrolysis by CDase were maltose and glucose. The deduced amino acid sequence of CDase was comprised of three regions homologous with those conserved in the sequences of α-amylases and pullulanases.

Xylanases

Thermoanaerobacter strain B6A and B6A-RI produces amylases, xylanases, xylose isomerase and ß-galactosidase that display activity in the same general pH and temperature range. A single step process for conversion of starch or lactose in milk to fructose sweetener was developed using xylose grown cells (Lee et al., 1990).

When *Thermoanaerobacter* strain B6A-RI is grown on xylan, it produces a xylan binding receptor and active endoxylanase activity which are localized in a putative xylanosome on the cell surface. *Thermoanaerobacter* strain B6A-RI produces multiple enzymes for xylan degradation including endoxylanases, ß-xylosidase, acetyl esterase and arabinofuranosidase. These enzymes are active at pH 5.5 and 70°C.

A cosmid clone encoding for endoxylanase, ß-xylosidase and acetylesterase was isolated from a genomic library of *Thermoanaerobacter* strain

B6A-RI. Genes encoding for an endoxylanase and ß-xylosidase were cloned and expressed in *E. coli.*

A 3.5 Kb *xyn*A gene region was identified for expression of a 125 kDa MW, monomeric, endoxylanase protein. The cloned enzyme differed from native endoxylanase which was glycosylated and displayed higher thermostability. Computer analysis comparison of the amino acid sequence of *Thermoanaerobacter* endoxylanase with other endoxylanases indicated that it was in the family "F" of ß-glycanases (Gilkes et al., 1991). The gene regions encoding for endoxylanase excretion and glycosylation were identified.

A 1.5 Kb *xyn*B gene region was identified for expression of a 55 kDa MW monomeric ß-xylosidase protein. The amino acid sequence deduced from the ß-xylosidase gene was compared to other ß-xylosidases. *Thermoanaerobacter* ß-xylosidase showed high homology (i.e., 60%) to ß-xylosidase from *Bacillus pumillus* (Xu et al., 1991).

The ß-xylosidase and endoxylanase genes of *Thermoanaerobacter* B6A-RI were part of a contiguous fragment in an *E. coli* cosmid clone. Studies in progress are examining the hypothesis that these xylanase genes are part of an operon.

Xylose Isomerases

Xylose isomerase activity is responsible for the conversion of xylose to xylulose prior to xylolysis and fermentation by bacteria. Xylose isomerase (i.e., glucose isomerase) is also used in biotechnology to convert glucose into high fructose corn syrup. Commercial sources of this enzyme have relatively low activity due to poor substrate affinity and low thermostability. Development of more active and thermostable glucose isomerases would be of value in production of higher concentration of fructose due to favorable chemical equilibrium at temperatures > 60°C.

Saccharolytic thermophiles of the genus *Thermoanaerobacter* and *Clostridium* produce thermostable xylose isomerase (Lee and Zeikus, 1991). The xylose isomerase was purified to homogeneity from *C. thermosulfurogenes* (Lee and Zeikus, 1991). The enzyme was very thermostable and displayed an optimal activity at 80°C.

The gene encoding the thermophilic xylose isomerase of *C. thermosulfurogenes* has been cloned and its nucleotide sequence was determined (Lee et al., 1990). The amino acid sequenced deduced from the coding sequence of the gene exhibited considerable homology to the sequences of other xylose isomerases studied to date. Surprisingly, the sequence of the thermophilic enzyme from *C. thermosulfurogenes* had a much higher degree of homology to the sequences of thermolabile enzymes of *E. coli* and *B. subtilis* than to the moderately thermostable enzymes of *Streptomyces* and *Arthrobacter*.

Amino acids that were predicted, by x-ray diffraction studies on *Arthrobacter* xylose isomerase, to be part of the active center were found to be highly conserved. An analysis of the putative catalytic domain indicated that only one amino acid difference occurred in the substrate binding site of *Thermoanaerobacter* versus *Arthrobacter* (i.e., tryptophan replaced a methionine in the thermophile enzyme).

Factitious enzymes developed by site-directed mutagenesis of the thermophilic xylose isomerase were used to demonstrate that Hist 101 was essential for catalysis by providing hydrogen binding to the substrate; and, that Trp 139 affects the size of the substrate binding site and influences the affinity of the enzyme to its substrate. A Gln 101 mutant enabled glucose isomerase activity to function below pH 6.0. A double mutant enzyme with phenylalanine 139 and serine 186 resulted in development of an enzyme with higher catalytic efficiency towards glucose than xylose (Lee et al., 1990; Meng et al., 1991).

Direction for Future Study

We are continuing molecular biology studies on thermophilic amylases and xylanases with the aim to 1) determine if juxtaposition between different amylases or xylanases enables enhanced rate of polymer degradation and limits release of soluble oligomers into the environment; 2) determine the molecular and biochemical properties of starch and xylan binding receptors and their juxtaposition to amylopullulanase and endoxylanase, respectively; 3) demonstrate the molecular basis for thermostability and thermophilicity of amylopullulanase and endoxylanase; and, 4) evaluate the industrial utility of these thermophilic amylases and xylanases.

References

Gilkes, N. R.; Henrissat, B.; Kilburn, D. G.; Miller, R. C.; Warren, R. A. J. *Microbiol. Rev.* **1991**, 55:303-315.

Lamed, R. E.; Bayer, E.; Saha, B. C.; Zeikus, J. G. *Proc. 8th Int. Biotechnol. Symp.*, Paris, **1988**; pp 371-383.

Lee, C.; Bagdasarian, M.; Meng, M.; Zeikus, J. G. *J. Biol. Chem.* **1990**, 265:19082-19090.

Lee, C.; Bhatnagar, L.; Saha, B. C.; Lee, Y.-E.; Takagi, M.; Imanaka, T.; Bagdasarian, M.; Zeikus, J. G. *Appl. Environ. Microbiol.* **1990**, 56:2638-2643.

Lee, C.; Saha, B. C.; Zeikus, J. G. *Appl. Environ. Microbiol.* **1990**, 56:2895-2901.

Lee, C.; Zeikus, J. G. *Biochem. J.* **1991**, 273:565-571.

Mathupala, S.; Saha, B.C.; Zeikus, J.G. *Biochem. Biophys. Res. Comm.* **1990**, 166:126-132.

Meng, M.; Lee, C.; Bagdasarian, M.; Zeikus, J. G. *Proc. Natl. Acad. Sci.* **1991**, 88:4015-4019.

Saha, B. C.; Zeikus, J. G. *Appl. Environ. Microbiol.* **1990**, 56:2941-2943.

Saha, B. C.; Zeikus, J. G. *Appl. Microbiol. Biotechnol.* **1991**, 35:568-571.

Xu, W.-Z.; Saima, Y.; Negoro, S.; Urabe, I. *Eur. J. Biochem.* **1991**, 202:1197-1203.

Zeikus, J. G.; Lee, C.; Lee, Y.-E.; Saha, B. C. *Enzymes in Biomass Conversion*, ACS Symposium Series 460, **1991**; pp 36-51.

Gramicidin S and its Producer, *Bacillus brevis*

A.L. Demain*, D. Kuhnt, T. Azuma, and R. Prakash
Massachusetts Institute of Technology, Cambridge, MA 02139
(FAX: 617-253-8550)

There is no doubt that secondary metabolites are natural products. Over 40% of filamentous fungi and actinomycetes produce antibiotics when they are freshly isolated from nature. Soil, straw, and agricultural products often contain antibacterial and antifungal substances. We may call these "mycotoxins," but they are nevertheless antibiotics. Indeed, one of our major public health problems is the natural production of such toxic metabolites in the field and during storage of crops. The natural production of ergot alkaloids by the sclerotial (dormant overwintering) form of *Claviceps* sp. on the seed heads of grasses and cereals has led to widespread and fatal poisoning ever since the Middle Ages. Natural soil and wheat straw contain patulin and aflatoxin is known to be produced on corn in the field. Trichothecin is found in anise fruits, apples, pears, and wheat. Microbially produced siderophores have been found in soil and microcins, enterobacterial antibiotics, have been isolated from human fecal extracts. The microcins are thought to be important in colonization of the human intestinal tract early in life.

Antibiotics have also been shown to be produced in unsterilized, unsupplemented soil, in unsterilized soil supplemented with clover and wheat straws, in mustard, pea, and maize seeds, and in unsterilized fruits. A further indication of natural antibiotic production is the possession of antibiotic resistance plasmids by most soil bacteria.

The widespread nature of secondary metabolite production and the preservation of the multigenic biosynthetic pathways in nature indicate that secondary metabolites serve survival functions in organisms that produce them. There are a multiplicity of such functions, some dependent on antibiotic activity and others independent of such activity. Indeed, in the latter case, the molecule may possess antibiotic activity but may be used for an entirely different purpose.

Secondary metabolites appear to serve (i) as competitive weapons used against other bacteria and fungi, amoebae, plants, insects, and large animals; (ii) as metal-transporting agents; as agents of (iii) plant-microbe relations, (iv) nematode-microbe symbiosis, and (v) insect-microbe symbiosis; (vi) as sexual hormones; and (vii) as differentiation effectors. Although antibiotics do not appear to be obligatory for sporulation, some secondary metabolites (including antibiotics) stimulate spore formation and inhibit or stimulate germination. Formation of secondary metabolites and spores appear to be regulated in a similar way. This could serve to insure secondary metabolite production during sporulation for the following possible reasons: to slow down germination of spores until a less competitive environment and more favorable growth conditions appear, to protect the dormant or initiated spore from consumption by protozoa, or to cleanse the immediate environment of competing microorganisms during germination.

Antibiotic production and differentiation

Development is composed of two phenomena, growth and differentiation. Differentiation encompasses both morphological differentiation (morphogenesis) and chemical differentiation (secondary metabolism). Secondary metabolites are made by chemical differentiation processes but function in both morphological and chemical differentiation.

Of the various functions postulated for secondary metabolites, the one which has received the most attention in recent years is the view that these compounds, especially antibiotics, are important compounds in the transition from vegetative cells to spores.

The close relationship between sporulation and antibiotic formation suggests that certain secondary metabolites involved in germination might be produced

1054–7487/92/0114$06.00/0

during sporulation and that the formation of these compounds and spores could be regulated by a common mechanism or by similar mechanisms.

Gramicidin S

In the last 15 years, considerable evidence has been obtained indicating that gramicidin S (GS) is an inhibitor of the phase of spore germination known as "outgrowth" in *B. brevis*. The cumulative observations are as follows.

(i) Initiation of germination (i.e., darkening of spores under phase microscopy) is similar in the parent and GS-negative mutants.

(ii) GS-negative mutant spores outgrow in 1 to 2 h whereas parental spores require 6 to 10 h. The delay in the parent is dependent on the concentration of spores and hence the concentration of GS.

(iii) Addition of GS to mutant spores delays their outgrowth so that they now behave like parental spores; the extent of the delay is concentration dependent and time dependent.

(iv) Preparation of parental spores on media supporting poor GS production results in spores which outgrow as rapidly as mutant spores.

(v) Removal of GS from parental spores by extraction allows them to outgrow rapidly.

(vi) Addition of the extract to mutant spores delays their outgrowth.

(vii) Exogenous GS hydrolyzed by a protease does not delay outgrowth of mutant spores. Parental spores treated with protease outgrow rapidly.

(viii) Exponential growth is not very sensitive to GS.

(ix) A mixture of parental spores and mutant spores show parental behavior, i.e., the mixture is delayed in outgrowth. This indicates that some of the GS externally bound to parental spores is released into the medium. This release could act as a method of communication by which a spore detects crowded conditions.

(x) Uptake of alanine and uridine into spores and respiration are inhibited by GS.

What is the value to the producing organism of inhibiting germination in the outgrowth stage, during which spores of bacilli are thought to have lost their resistance to factors such as heat? Wouldn't inhibition at this stage make spores more susceptible to attack by other organisms, and wouldn't rapid outgrowers (e.g., nonproducing mutants) be selected for? It turns out that initiated (i.e., phase-dark) spores of the GS-producing *B. brevis* are still resistant to heat, starvation, solvents, and even sonication (Daher *et al.*, 1985).

It appears that the produced GS is the basis of the hydrophobicity of dormant or initiated *B. brevis* spores. After outgrowth ceases, the resulting vegetative cells are hydrophilic (Rosenberg *et al.*, 1985). Since water-insoluble organic matter constitutes the chief source of soil nutrients, it is quite possible that the hydrophobicity of *B. brevis* spores and initiated spores aids in their search for nutrients to insure vegetative growth after germination.

A second question involves the mechanism by which the outgrowing spores recover from GS inhibition and finally develop into vegetative cells. One possibility is destruction of GS towards the end of the outgrowth stage. *B. brevis* ATCC 9999 produces an intracellular serine protease (Piret *et al.*, 1983). This type of enzyme is generally considered to be necessary for sporulation of bacilli. The *B. brevis* enzyme has the ability to cleave GS between valine and ornithine residues (Kurotsu *et al.*, 1982). Although it would appear that the intracellular enzyme might function to destroy GS and allow vegetative growth from outgrown spores, our data indicate that GS is not destroyed as the outgrowing spores develop into vegetative cells (Bentzen *et al.*, 1990). Furthermore, the recovery is not due to selection of spores whose outgrowth is resistant to GS. Another possibility is that GS kills outgrowing spores and the delay in outgrowth is merely the time required by a small population of surviving spores to germinate and become vegetative cells. Although we have confirmed the findings (Murray *et al.*, 1985) that GS kills a large proportion of outgrowing spores, the same residual fraction of survivors is seen despite our increasing of the GS concentration (Bentzen and Demain, 1990). This lack of effect of increased concentration of GS on killing is in contrast to the increasing delay in outgrowth caused by the increased GS concentration and makes unlikely a connection between the degree of killing and the length of the outgrowth stage. At this point, it appears that GS (because of its inhibition of oxidative phosphorylation, transport, and/or transcription) slows down, but does not totally inhibit, the macromolecular processes of outgrowth until a point is reached where all the outgrown spores have the proper machinery to differentiate into vegetative cells. During this process, GS is excreted into the extracellular medium.

It is thus probable that GS serves the initiated spore as a means of sensing a high population density and preventing vegetative growth until there is a lower density of *B. brevis* spores with which to compete for nutrients. However, proof of such a hypothesis will require experimentation of an ecological nature. Alternative hypotheses might be that GS in and on the dormant and initiated spores protects them from consumption by amoebae or that GS excretion during germination initiation and outgrowth eliminates microbial competitors in the environment and that the delay in outgrowth and death of a part of the outgrowing spore population is merely "the price the strain must pay" for such protection.

Recent studies in our laboratory have shown that the GS-producing parent has no survival advantage over its GS-nonproducing mutant in mixed culture (Kuhnt and Demain, unpublished). However, we have obtained data showing that the parent is better able to survive in competition with antibiotic-producing fungi than is the mutant in such mixtures. The roles of germination inhibition and antagonism in the natural setting may or may not be related; this is a question for future studies.

Other recent studies (Azuma, Prakash and Demain, unpublished) have shown that GS can inhibit vegetative growth, sporulation and GS production but only at much higher concentrations than required to inhibit germination. Thus the outgrowth stage of germination remains the most important target of GS in the producer organism.

References

Bentzen, G.; Piret, J.M.; Daher, E.; Demain, A.L. *Appl. Microbiol. Biotechnol.* 1990, *32*, 708-710.

Bentzen, G.; Demain, A.L. *Curr. Microbiol.* 1990, *20*, 165-169.

Daher, E.; Rosenberg, E.; Demain, A.L. *J. Bacteriol.* 1985, *161*, 47-50.

Kurotsu, T.; Marahiel, M.A.; Muller, K.D.; Kleinkauf, H. *J. Bacteriol.* 1982, *151*, 1466-1472.

Murray, T.; Lazaridis, I.; Seddon, B. *Lett. Appl. Microbiol.* 1985, *1*, 63-65.

Piret, J.M.; Millet, J.; Demain, A.L. *Eur. J. Appl. Microbiol. Biotechnol.* 1983, *17*, 227-230.

Rosenberg, E.; Brown, D.R.; Demain, A.L. *Arch. Microbiol.* 1985, *142*, 51-54.

Microbial Secondary Metabolites Affecting Lipid Metabolism

Hiroshi Tomoda* and Satoshi Ōmura, Research Center for Biological Function, The Kitasato Institute, Minato-ku, Tokyo 108, Japan (FAX: +81-3-3444-6637)

Four kinds of microbial inhibitors affecting different sites of lipid metabolism have been discovered. Cerulenin produced by *Cephalosporium* sp. and thiotetromycin produced by *Streptomyces* sp. inhibit fatty acid synthase. Triacsins produced by *Streptomyces* sp. are specific inhibitors of acyl-CoA synthetase and a fungal β-lactone 1233A was rediscovered as a potent and specific inhibitor of HMG-CoA synthase. Purpactins and glisoprenins produced by fungal strains were found to inhibit acyl-CoA: cholesterol acyltransferase.

Lipid metabolism is involved in various diseases such as atherosclerosis, hypertension, obesity diabetes and so on. Control of lipid metabolism by drugs would be important for treatment or prevention of these diseases. Additionally, such drugs will be used as a useful biological tool especially in the field of lipid research since the mechanism of lipid metabolism and the involved enzymes have not been well understood. Our research group has focussed on microbial inhibitors of lipid metabolism and discovered four kinds of enzyme inhibitors affecting fatty acid synthase, acyl-CoA synthetase, HMG-CoA synthase or acyl-CoA: cholesterol acyltransferase (ACAT). Some were discovered fortunately as antibiotics, and others were discovered by theoretically purposed screening systems for enzyme inhibitors using intact animal cells or intact microorganisms, or by conventional enzyme assay systems. In this paper, the discovery and recent achievement of these inhibitors are described.

Inhibitor of Fatty Acid Synthase

Cerulenin produced by *Cephalosporium caerulens* and thiotetromycin by *Streptomyces* sp. were originally isolated as an antigfungal (Ōmura, 1976) and antianaerobic antibiotic (Ōmura *et al.*, 1983), respectively. Later studies on their mechanism of action revealed that the two drugs are inhibitors of fatty acid synthase. Cerulenin inhibits all known types of fatty acid synthases, both the multifunctional enzyme complex (type I) and nonaggregated enzyme system (type II) except for the synthase from the cerulenin-producing fungus (Kawaguchi *et al.*, 1979). Cerulenin inhibits the enzyme by selectively binding to the cysteine residue in the active center of the condensing reaction domain (Funabashi *et al.*, 1989). Since cerulenin is the first antibiotic having such a unique mode of action, it has been widely used as a biochemical tool in the field of lipid research (Ōmura, 1976). Thiotetro-

1054–7487/92/0117$06.00/0

mycin is a specific inhibitor of type II synthase. The mode of action of thiolactomycin structurally related to thiotetromycin has been studied extensively. According to the results, acetoacetyl-ACP synthase involved in bacterial fatty acid biosynthesis was regarded as a target for thiolactomycin inhibition (Jackowski *et al.*, 1989).

Inhnbitor of Acyl-CoA Synthetase

Fatty acid metabolism in *Candida lipolytica* was studied extensively (Mishina *et al.*, 1978). This yeast possesses two functionally distinct acyl-CoA synthetases, designated ACS I and II. ACS I localized in microsomes and mitochondria is responsible for the synthesis of acyl-CoA solely for cellular lipids, whereas ACS II localized in peroxisomes provides acyl-CoA which is exclusively degraded via β–oxidation to yield acetyl-CoA. Acyl-CoA for cellular lipid synthesis is also provided via fatty acid synthase (FAS). To screen acyl-CoA synthetase inhibitors, two kinds of mutant strains L-7 (defective in ACS I) and A-1 (defective in FAS) were used as test organisms (Tomoda and Ōmura, 1990).

Four structurally related triacsins were isolated from *Streptomyces* sp. (Ōmura *et al.*, 1986). Triacsins C and D are identical to WS-1228A and B, respectively, previously reported as vasodilaters (Tanaka *et al.*, 1982).

Triacsins were found to inhibit acyl-CoA synthetase activity from *Pseudomonas* spp., rat liver and Raji cells (Tomoda *et al.*, 1987a). Triacsin C is the most potent with IC_{50} values of 3.6-8.7 μM, followed by triacsin A with 12-18 μM. Triacsins B and D are much less potent than triacsins A and C. ACS II from *C. lipolytica* was less sensitive to triacsins than ACS I.

Acyl-CoA plays important roles in eukaryotic and prokaryotic cells as 1) a metabolic intermediate in β-oxidation and elongation of fatty acids, 2) an acyl-donor in the biosyntehsis of complex lipids and in protein acylation, 3) a stimulator in intracellular protein transport, and 4) a regulator in many metabolic pathways. Triacsins are expected to be a useful tool to demonstrate these roles of acyl-CoA. In fact, we showed that long chain acyl-CoA synthetase is essential for animal cell proliferation by utilizing the drugs (Tomoda *et al.*, 1991a).

Inhibitor of HMG-CoA Synthase

Analogs of compactin and mevinolin, fungal HMG-CoA reductase inhibitors, have been developed and marketed as hypocholesterolemic agents. To screen out microbial enzyme inhibitors of mevalonate biosyntehsis, Vero cells were used as a test organism (Tomoda and Ōmura, 1990).

A fungal β-lactone 1233A (F-244, L-659,699), originally isolated as an antibiotic (Aldridge *et al.*, 1971), was found to inhibit mevalonate biosynthesis. Several lines of experiments demonstrated that 1233A specifically inhibits HMG-CoA synthase with an IC_{50} value of 0.2 μM (Ōmura *et al.*, 1987, Tomoda *et al.*, 1987b). When HMG-CoA synthase from rat liver was incubated with [^{14}C]1233A, only the band corresponding to the subunit of the enzyme was radiolabeled. A linear relationship between the amount of $[^{14}C]$1233A bound to HMG-CoA synthase and the inhibition of the enzyme activity was observed. These findings indicate that 1233A inhibits HMG-CoA synthase by covalently binding to the enzyme.

Inhibitor of ACAT

Acyl-CoA: cholesterol acyltransferase (ACAT) plays an important role in cholesterol ester accumulation in atherogenesis and in cholesterol absorption from the intestines. Inhibitors of ACAT activity are expected to be effective for treatment or prevention of atherosclerosis and hypercholesterolemia. ACAT inhibitors of natural origin have been rarely reported. A conventional enzyme assay using rat liver microsomes was used to screen ACAT inhibitors.

Two kinds of novel ACAT inhibitors, purpactins produced by *Penicillium* sp. (Tomoda *et al.*, 1991b) and glisoprenins by *Gliocladium* sp., were discovered. The inhibitory activity was not so potent with IC_{50} values of 121-126 µM for purpactins and 46-61 µM for glisoprenins. Beauvericin, a known cyclodepsipeptide, was found to be a very potent ACAT inhibitor with a IC_{50} value of 3.0 µM. The ACAT inhibitory activity of these compounds was evaluated in an intact cell assay system using J774 macrophages. The effects on cholesteryl ester formation as ACAT activity and cell viability were examined. Among the compounds, beauvericin showed the highest specificity i.e. cytotoxicity vs. ACAT inhibition in this cell assay system.

References

Aldridge, D.C., Gil, D., Turner, W. B., *J. Chem. Soc., (C)* **1971,** 3888-3891.

Funabashi, H., Kawaguchi, A., Tomoda, H., Ōmura, S., Okuda, S., Iwasaki, S., *J. Biochem.,* **1989,** *105*, 751-755.

Jackowski, S., Murphy, C. M., Cronan, Jr., J. M., Rock, C. O., *J. Biol. Chem.,* **1989,** *264,* 7624-7629.

Kawaguchi, A., Tomoda, H., Awaya, J., Ōmura, S., Okuda, S., *Arch. Biochem.. Biophys.* **1979,** *197,* 30-35.

Mishina, M., Kamiryo, T., Tashiro, S., Numa, S., Eur. *J. Biochem.,* **1978,** *82,* 347-354.

Ōmura, S., *Bacteriol. Rev.* **1976,** *40,* 681-697.

Ōmura, S., Iwai, Y., Nakagawa, A., Iwata, R., Takahashi, Y., Shimizu, H., Tanaka, H., *J. Antibiot.* **1983,** *6,* 109-114.

Ōmura, S., Tomoda, H., Xu, Q.M., Takahashi, Y., Iwai, Y., *J. Antibiot.* **1986,** *39,* 1211-1218.

Ōmura, S., Tomoda, H., Kumagai, H., Greenspan, M.D., Yudkovitz, J. B., Chen, J. S., Alberts, A. W., Martin, I., Mochales, S. Monaghan, R. L. Chabala, J. C., Schwartz, R. R., Patchett, *A. A..,* J. Antibiot., **1987,** *40,* 1356-1357.

Tanaka, H., Yoshida, K., Itoh, Y., Imanaka, H., *J. Antibiot.,* **1982,** *35*, 157-163.

Tomoda, H., Igarashi, K., Ōmura, S., *Biochim. Biophys. Acta,* **1987a,** *921,* 595-598.

Tomoda, H., Kumagai, H., Tanaka, H., Ōmura, S., *Biochim. Biophys. Acta,* **1987b,** *922*, 351-356.

Tomoda, H., Ōmura, S., *J. Antibiot.*, **1990,** *43*, 1209-1222.
Tomoda, H., Igarashi, K., Cyong, J. C., Ōmura, S., *J. Biol. Chem.*, **1991a,** *266*, 4214-4219.
Tomoda, H., Nishida, H., Masuma, R., Cao, J., Okuda, S., Ōmura, S., *J. Antibiot.* **1991b,** *44*, 136-143.

Secondary Metabolites from Fusarium moniliforme and Related Genera

Bettina Brueckner
Friedrich Schiller University Jena, Department of Microbiology
Neugasse 24, O-6900 Jena, Germany (FAX: 003778 8222345)

The fungus Fusarium moniliforme (perfect stage Gibberella fujikuroi) is well known for its ability to produce a group of phytohormons, the gibberellins. Beside gibberellins, this widespread fungus produces some other biological active secondary metabolites. The trichothecines (T-2-toxin, HT-2-toxin, diacetoxyscirpenol, deoxynivalenol) are a closely related group of sesquiterpene mycotoxins produced by certain species of Fusarium. Many other toxins have been isolated from Fusarium moniliforme cultures, e.g. moniliformin, fusarins and fumonisin B1. Fusarin C and 8Z-fusarin C have been shown to be mutagenetic. Fumonisin B1, a C20 aminopolyol, produced symptoms of leukoencephalomalacia and also seems to be a cancer-promoting factor in rats. Moreover, Fusarium moniliforme and related strains produce other active metabolites, e.g. pigments, phytotoxins, antibiotics and oestrogens. Many Fusarium strains are able to produce two pigment groups - the carotenoids and a family of polyketides (bikaverin, 8-O-methyljavanacin) which can act as antibiotics or phytotoxins. Therefore, Fusarium moniliforme and related genera are sources of different useful but also harmful bioactive metabolites. The formation is dependent on the conditions and strain specifity.

There is little doubt that the genus Fusarium currently constitutes one of the most important groups of toxigenic fungi. A number of facts explain their role as producers of a broad spectrum of harmful mycotoxins: 1. Fusarium species are extremely common in all geographical regions of the world. They can grow over a wide temperature range. 2. Fusarium species are active plant pathogens attacking roots, stems, and fruits and, therefore, causing a number of diseases. 3. Fusaria are also saprophytes and are found everywhere in decaying vegetation. 4. They have a good system of distribution producing microconidia and macroconidia as well. 5. Many strains are able to produce several different toxins at the same or at a different time depending on substrate, temperature and moisture. Beside mycotoxins the genus Fusarium is fomous because of the ability to produce a number of different secondary metabolites. It is not surprising that the scientists of many countries have turned to research on these biologically

1054–7487/92/0121$06.00/0

active natural products as a source of new agricultural or pharmaceutical compounds for use against target pests. Among the secondary metabolites from Fusarium species are plant growth hormones such as the gibberellins, antibiotics, phytotoxins, pigments andestrogenic compounds, the zearaleuones (Fig. 1).It can be predict that there are many secondary metabolites of Fusaria yet to be discovered and characterized. In some laboratories have been investigated the antibiotic-producing potential of a number of Fusarium strains. Many patterns of inhibition of microbial growth were found, most of which could not be attributes to any known compound (Hesseltine, 1977).

Interestingly, some mycotoxins and plant hormones in the same time can act as phytotoxins, pigments can have phytotoxic and antibiotic activities and zearalenone appears to be the only mycotoxin whose effect is primarily an estrogenic one. It is obviously from Fig. 2 that mevalonic acid is the structural unit of many secondary metabolites, including gibberellins, carotenoids, trichothecene mycotoxins and sterols. Another group of secondary metabolites is synthesized from acetyl-CoA or malonyl-CoA via polyketide pathway. The enniatin antibiotics are derived from amino acids. The enzymes involved in secondary metabolism are apparently of low specifity, which usually results in the production of particular chemical families of structurally related metabolites. In this paper, the secondary metabolites of the genus Fusarium are classified into three groups with respect to the biosynthetic pathways. The chemical structures of some important metabolites are given in Fig. 3.

Terpenoid Pathway

The terpenoid intermediate mevalonic acid, which is incidentally derived from acetyl-CoA is used in biosynthesis of sterols and may, when the growth rate of a culture is retarded, be used in synthesis of secondary metabolites including gibberellins, trichothecene toxins and carotenoid pigments.The first example and in the same time the most important secondary products of one member of the genus, Fusarium moniliforme (perfect stage Gibberella fujikuroi) are the gibberellins. The gibberellins are one of five classes of hormones that control the growth and the differentiation in plants. It was a Japanes who first discovered the GA3 as natural product from the fungus Gibberella fujikuroi. This discovery ultimately led to the isolation and identification of gibberellins in plants (Phinney, 1983). The economic importance of the gibberellins as plant hormones led to an extensive search for these compounds and to the biotechnological production of gibberellic acid and GA4/GA7 mixture (Martin, 1983). Up to now are known 79 different gibberellins, from which 26 can be produced by the fungus (Takahashi et al., 1986).In most plants the outstanding effect of the gibberellins is to elongate the primary stalk, to increase the rate of cell division, to increase the size of many fruits, especially grapes and to induce flowering. Their greatest commercial uses have been in increasing the size of fruits and the yield of sugarcane (Nickell, 1982). Development of biennial plants can be made so rapid that seed crops may be obtained from lettuce and sugar beets in one year instead of the usual two years (Demain, 1983). Furthermore, the gibberellins are used to reduce the time needed for malting of barley, to increase the yield of vegetables as well as to allow their earlier marketing. Another direction of the mevalonate metabolism in the mycelium of Fusarium species is the synthesis of carotenoids, especial-

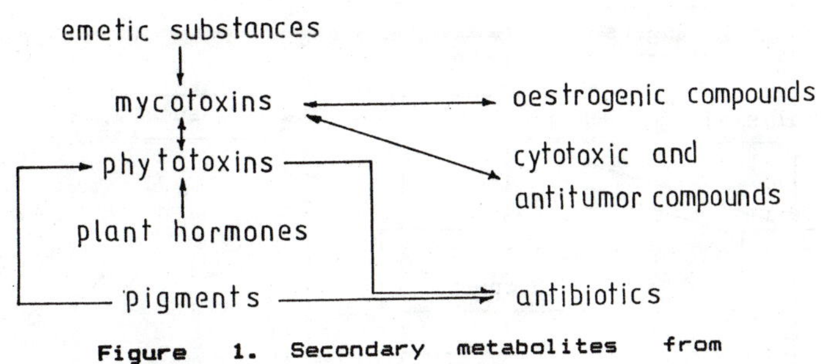

Figure 1. Secondary metabolites from Fusarium species

ly of the orange pigment neurosporaxanthin when the fungus is grown in the light (Avalos et al., 1985; Avalos & Cerda-Olmedo, 1986; 1987; Avalos et al., 1988). The basal level of this pigment depends on the strain and is particularly low in highly gibberellin producing strains because of the competition for mevalonic acid. On the other hand the findings of pigment mutants, which produce high amounts of carotenoids without photoinduction led to the conclusion that this organism is of interest in the industrial production of carotenoids (Avalos et al., 1985). The last examples of secondary metabolites with terpene structure are some trichothecene mycotoxins. These toxins are a group of biologically active secondary metabolites predominantly associated with species of Fusarium, but also produced by many different fungi. Much of the interest in these compounds has risen from their probable involvement in mycotoxicoses of farm animals. Furthermore, these compounds became well known after their implication in a disease of humans called alimentary toxic aleukia, described by Joffe (1971). So far in Fusarium moniliforme T-2 toxin (Betina, 1984), diacetoxyscirpenol (Betina, 1984; Bamburg, 1983) and trichothecolon (Chakrabarti & Ghosal, 1986) were found. The toxicological effects of T-2 toxin and diacetoxyscirpenol to animals are described in some publications (Hoerr & Carlton, 1981; Wermter, 1980; Betina, 1984). Furthermore, exocellular production of ergosterol and its esters may be obtained by culturing some fungi of the genus Fusarium on media based on n-paraffins (Nakao et al., 1975). With Fus. sp. S-19-5 (ATCC 20192) in stirred and aerated culture 0.65 g/l ergosteryl palmitate and 0.2 g/l free ergosterol were produced.

Polyketide Pathway

The next group of secondary metabolites from Fus. moniliforme includes substances of the polyketide pathway of biosynthesis. Beside carotenoids many strains also synthesize some red pigments with naphtoquinone structure such as bikaverins, 8-O-methyljavanicin and fusarubin (Fig. 3). Some of these pigments have biological activities. Bikaverin was found by Tchechoslovakian workers searching for antiprotozoal substances. The pigment extracts have a high antibiotic activity against Leishmania brasiliensis (Balan et al., 1970). In the course of the screening for plant growth regulators among Fusarium metabolites Steyn et al. (1979) found a growth inhibitor of lettuce seedlings in the culture filtrate of Fus. moniliforme isolated from rice. This red pigment called 8-O-methyljavanicin promotes the second leaf sheath of Oryza sativa 75% with 100 ppm solutions. Fusarium solani was shown to produce a group of antibiotic pigments of the

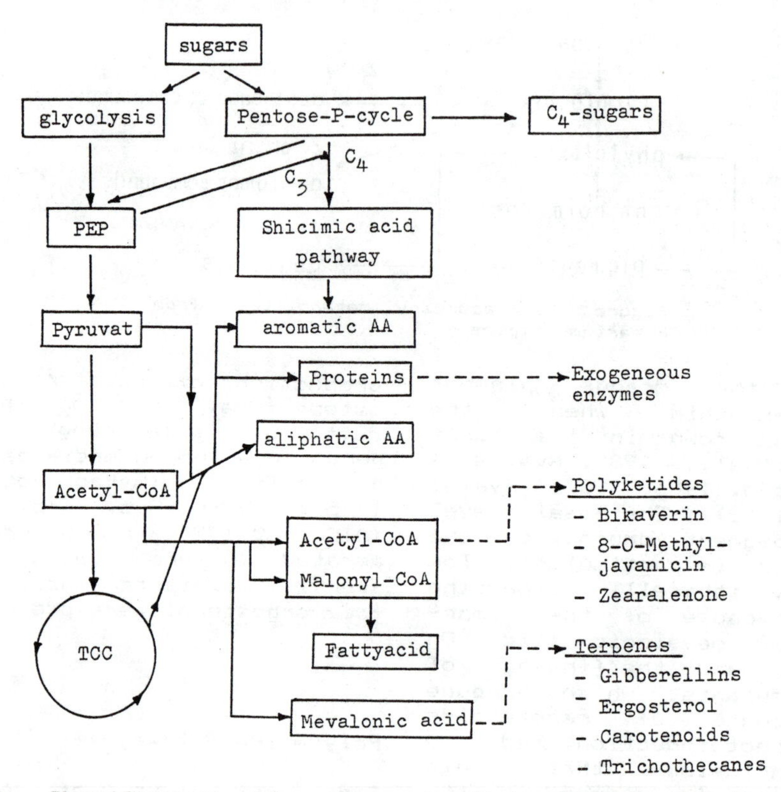

Figure 2. Classification of some secondary metabolites from Fusarium moniliforme and related species according to their linkage with the intermediate metabolism

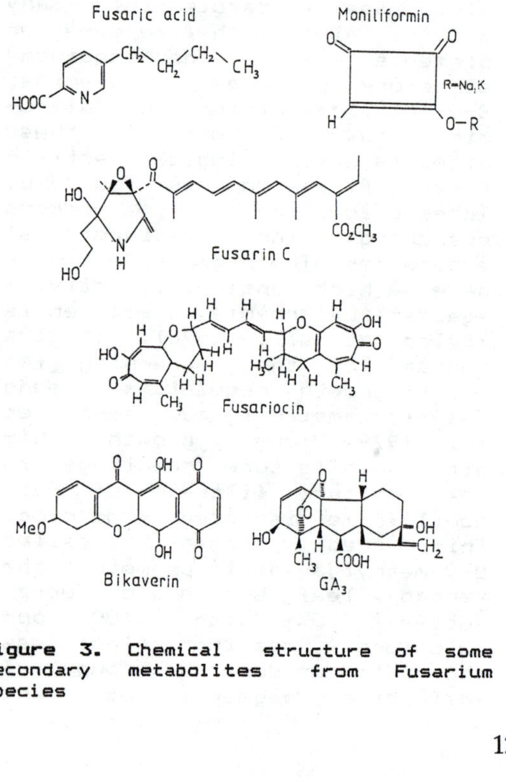

Figure 3. Chemical structure of some secondary metabolites from Fusarium species

fusarubin type: Fusarubin, O-ethyl-fusarubin, hydroxy-dihydrofusarubin and O-ethyl-hydroxydihydrofusarubin (Gerber & Ammar, 1979). These compounds are of interest because of their antitumor activity. The antibiotic bostrycoidin, another member of the fusarubin group, was first isolated from the fungus Fus. bostrycoides and later found in a wide range of Fusarium species (Parisot, 1989). Bostrycoidin revealed interesting antibiotic properties in particular against Mycobacterium tuberculosis. Fusarin C and 8Z-Fusarin C are other polyketide secondary metabolites which were isolated originally from Fus. moniliforme and later from strains of various Fusaria (Barrero et al., 1991). Fusarins are pigments with highly mutagenic and toxic properties. Fusarin C has been associated with the incidence of human esophageal cancer (Farber

& Sanders, 1986). The conditions leading to fusarin and gibberellin production in Gibb. fujikuroi do not always coincide. Knowledge of other secondary metabolites is relevant in the industrial production of gibberellins. Another example of secondary metabolites produced from acetyl-CoA via polyketide pathway is zearalenone. Extreme cases of hyperestrogenism in swine led to the discovery of this substance and its production by Fus. roseum but also by Fus. moniliforme on maize (Mirocha et al., 1979). Zearalenone is the only mycotoxin which was found useful commercially. The LD50 of this substance is about 16 g/kg in rats. Because of its anabolic property to increase the growth rate of animals and its estrogenic activities zearalenone has found economic application as a growth promotant in cattle (Leslie, 1983). In the fungus Gibb. zeae zearalenone enhances perithecia formation. Therefore, the function of a fungal sex hormone has been attributed to zearalenone.

Cyclic Peptides

Some Fusarium species were found to produce cyclic peptides with antibiotic or immunosuppressive properties. Of all isolated depsipeptide antibiotics valinomycin is the best investigated, followed by the enniatins. These compounds were determined in the mycelium of different Fusarium species as Fus. sambucinum, Fus. orthoceas, Gibb. baccata, Fus. avenaceum and Fus. moniliforme (Vesonder & Hesseltine, 1981). Valinomycin and enniatins are powerful bioactive, apparently by interaction with membrane components. These substances often have also phytotoxic properties. The newest data indicated that the phytotoxic substances produced by Fus. roseum are very similar but not identical to the cyclodepsipeptides enniatins, valinomycin antibiotics and beauvericin produced by other fungi, including Fus. roseum (Blain et al., 1991). From culture fluid of Fus. solani different cyclosporins could be isolated (Anke, 1986). The cyclosporins are cyclic peptides representing novel and highly potent immunosuppressive metabolites. In the same time, they exhibit strong antifungal activities, especially against Aspergillus and Curvularia.

Other Pathways

Fusaric acid, a further secondary metabolite with diverse biological activities, is an aspartie acid derivate. The substance was first isolated by Yabuta et al. (1934) and was subsequently implicated in wilt diseases of plants (Phinney, 1983). The presence of fusaric acid in plants impairs water permeability resulting in a loss of turgor and ionic unbalance. The substance has also antibiotic, pharmacological and insecticidal properties and is moderately toxic to animals (LD50 intraperitoneal about 100 mg/kg in mice; Burmeister et al., 1985). Beside fusaric acid and trichothecenes different strains of Fus. moniliforme can produce some further mycotoxins defining cronic toxicological effects in animals, e.g. moniliformin, fumonisins and fusariocin C, all produced by Fus. moniliforme (Scott, 1991). The chemical structures of these mycotoxins are well-known today, but it was not found any information on the biosynthetic pathway. All these compounds occur under natural conditions and are considered to be important with respect to human and veterinary health. Therefore, the species of genus Fusarium are characterized by the occurence of many secondary metabolites sometimes with extremely complicated structures and different biological activities.

References

Anke, K. In Biotechnology; Rehm, H.J.; Reed, G. Ed.; Verlag Chemie Weinheim; 1986; Vol. 4.
Avalos, J.; Casadesus, J.; Cerda-Olmedo, E. Appl. Envir. Microbiol. 1985, 49, 187-191.
Avalos, J.; Mackenzie, A.; Nelki, D.S.; Bramley, P.M. Biochimica et Biophysica Acta 1988, 966, 257-265.
Balan, J.; Fuska, J.; Kuhr, I.; Kuhrova, V. Folia Microb. 1970, 15, 479-484.
Bamburg, J.R. In Biological and biochemical actions of trichothecene mycotoxins; Hahn, F.E., Ed.; Springer-Verlag: Berlin-Heidelberg-New York-Tokyo, 1983; Vol. 8.
Barrero, A.F.; Sanchez, J.F.; Oltra, J.E.; Tamayo, N.; Cerda-Olmedo, E.; Candau, R.; Avalos, J. Phytochem. 1991, 30, 2259-2263.
Betina, V. Mycotoxins. Production, isolation, separation and purification; Elsevier: Amsterdam-Oxford-New York-Tokyo, 1984.
Blain, F.; Bernstein, M.; Khanizadeh, S.; Sparace, S.A. Phys. & Biochem. 1991, 81, 105-108.
Burmeister, H.R.; Grove, M.D.; Peterson, R.E.; Weisleder, D.; Plattner, R.D. Appl. Envir. Microbiol. 1985, 50, 311-314.
Chakrabarti, D.K.; Ghosai, S. Appl. Envir. Microbiol. 1986, 37, 217-219.
Demain, A. Science 1983, 219, 709-714.
Faber, J.M.; Sanders, G.W. Appl. Envir. Microbiol. 1986, 51, 381-384.
Gerber, N.N.; Ammar, M.S. J. Antibiotics 1979, 32, 685-688.
Hesseltine, C.W. In Mycotoxins in human and animal health; Rodricks, J.V.; Hesseltine, C.W.; Mehlman, M.A. Eds.; Pathotox Publ., Inc.; pp 341-344.
Hoerr, F.J.; Charlton, W.W. Fd. Cosmet. Toxicol. 1981, 19, 185-188.
Joffe, A.Z. In Microbiological toxins; Kandis, S.; Ciegler, A.; Ajl, J.S. Ed; Acad. Press: New York, 1971, Vol. 7; pp 139-189.
Leslie, J.F. Phytopathol. 1983, 73, 1005-1008.
Martin, G.C. In The biochemistry and physiology of gibberellins; Crozier, A. Ed; Praeger Publ.: New York, 1983, Vol 2; pp 395-444.
Mirocha, C.J.; Pathre, S.V.; Christensen, C.M. In Secondary products of metabolism; Rose, A.H. Ed; Acad. Press: London-New York-San Francisco, 1979, Vol. 3; pp 467-522.
Nakao, Y.; Kuno, M.; Suzuki, M. US Patent 3.884.759; 1975.
Nickell, L.G. In Plant growth regulators; Springer-Verlag: Berlin-Heidelberg-New York, 1982; pp 32-44.
Phinney, B.O. In The biochemistry and physiology of gibberellins; Crozier, A. Ed; Praeger Publ.: New York, 1983; Vol. 1, pp 19-52.
Parisot, D. J. Antibiotics 1989, 42, 1189-1199.
Scott, P.M. J. Assoc. Off. Anal. Chem. 1991, 74, 120-128.
Steyn, P.S.; Wessels, P.L.; Marasas, W.F.O. Tetrahedron 1979, 35, 1551-1553.
Takahashi, N.; Yamaguchi, I.; Yamane, H. In Chemistry of plant hormones; Takahashi, N. Ed; CRC Press, Boka Radon, Florida; pp 57-151.
Wermter, R. Dissertation Univ. M nchen, 1980.
Yabuta, T.; Kambe, K.; Hayashi, T. J. Agric. Chem. Soc. Japan 1934, 10, 1059-1068.

Molecular Biology of Antibiotic Production in *Streptomyces glaucescens*

C. Richard Hutchinson*[1,2], Heinrich Decker[1], Patrick G. Guilfoile[1,2], Evelyn Wendt-Pienkowski[1], Ben Shen[1] and Richard G. Summers[1].

[1]School of Pharmacy and [2]Department of Bacteriology, University of Wisconsin, Madison, WI 53706
(FAX:608 262 3134)

Streptomyces glaucescens uses a cluster of 12 genes and their protein products to produce the anthracycline antibiotic, tetracenomycin (tcm) C, and provide resistance to it. The function of each *tcm* gene has been established by mutation and sequence analysis plus its expression in a streptomycete host. Replacement of specific *tcm* genes with their homologs from other bacteria has the potential to create novel metabolites.

Microorganisms produce a large variety of biologically active substances by the so-called secondary metabolic pathways. This capacity is presumed to be non-essential for growth and yet to provide the organism with some advantage in the ecosystem. Fortunately, such processes produce many medically useful antibiotics with potent antiinfective, antitumor, and immunosuppressive properties (Crandall and Hamill, 1986). *Streptomyces glaucescens*, a representative of the most prolific bacterial genus among the antibiotic-producers, makes an anthracycline metabolite, tetracenomycin (tcm) C (Fig. 1B), which is a potent inhibitor of the growth of other streptomycetes and exhibits moderate cytotoxicity towards some tumor cells (Weber et al., 1979). We are studying the biochemistry and genetics of Tcm C formation as a model of bacterial secondary metabolism, with the intention of developing ways to manufacture new drugs by genetic engineering.

Organization of the Tcm C Gene Cluster

Tcm C production is determined by a cluster of 12 genes contained within an approx. 13 kilobase (kb) region of the *S. glaucescens* genome (Fig. 1A). We have been able to show that other stretomycetes can produce Tcm C upon transformation with this region cloned on a high copy number vector. Ten structural genes organized into at least two transcriptional units, *tcmGHIJKLMNO* and *tcmP*, encode the pathway enzymes, and two other genes provide the necessary resistance to Tcm C: *tcmA* encodes a transport protein for Tcm C export and *tcmR* encodes a repressor that regulates *tcmA* expression to coincide with Tcm C production (Guilfoile and Hutchinson, 1992a,b).

Functions of the Tcm C Genes

Mutations affecting Tcm C production have characteristic phenotypes that often reflect the accumulation of intermediates of Tcm C biosynthesis (Motamedi et al., 1986; Yue et al., 1986). For example, a *S. glaucescens* TcmVII mutant produces Tcm E but not Tcm A2 or Tcm C (Fig. 1B). Since only the *tcmP* gene restores Tcm C production to this mutant, we can deduce that *tcmP* encodes an *O*-methyltransferase that catalyzes the addition of a methyl group to the COOH of Tcm E. TcmIa mutations, in contrast, do not cause diffusible metabolites to accumulate, yet permit Tcm C production when fed Tcm F2 or any of the other pathway intermediates shown after it in Fig. 1B.

The *tcmIa* locus contains four genes, *tcmKLMN*, that encode the components of a type II polyketide synthase (PKS): *tcmK* provides a β-ketoacyl:ACP synthase that assembles the decaketide (Fig. 1B) from acetyl and malonyl Coenzyme A (CoA) and *tcmL* may be a subunit of this enzyme (Bibb et al., 1989); *tcmM* produces an acyl carrier protein (ACP) to which the malonylCoA substrate and growing poly-β-ketoacyl chain are attached (Bibb et al., 1989); and *tcmN* makes a trifunctional enzyme whose N-terminus is believed to catalyze partial cyclization and dehydration of the decaketide product of the TcmKLM proteins, but whose C-terminus provides the means for addition of a methyl group to the C-3 hydroxyl of Tcm D3 (Fig. 1B) (Summers et al., 1992). The exact manner in which a type II PKS operates has not been elucidated (nevertheless, its mechanism must parallel that of a fatty acid synthase (Hopwood and Sherman, 1990)), but we now can investigate this matter with a cell-free system we have developed that makes Tcm F2 from acetylCoA and malonylCoA in vitro.

1054–7487/92/0127$06.00/0

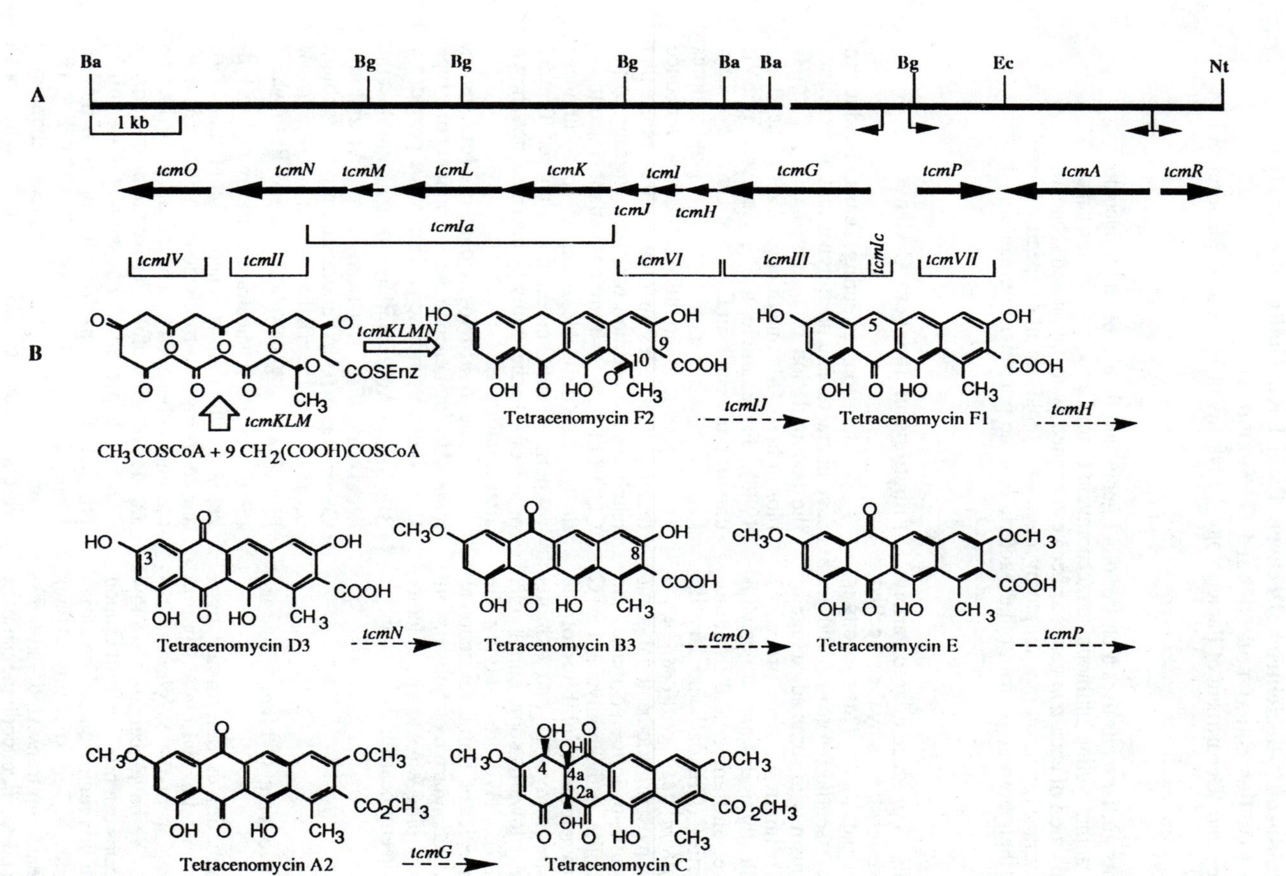

Fig. 1. (A) A restriction map of the Tcm C gene cluster, showing representative restriction sites [Ba = *Bam*HI, Bg = *Bgl*II, Ec = *Eco*RI and Nt = *Not*I] and the locations of the *tcm* genes (thick arrows in the direction of gene transcription) and mutations (open brackets). The locations of four promoters are shown by angled arrows just underneath the restriction map. (B) The biosynthetic pathway for Tcm C, illustrating the chemical structures of known intermediates. The genes governing each step are shown above the dashed arrows.

After Tcm F2 is made, it is cyclized at positions 9 and 10, then oxidized at C-5 to form Tcm F1 and Tcm D3 in sequo (Fig. 1B). The *tcmHIJ* genes govern these two steps. The behavior, in *S. glaucescens* wild-type and TcmVI mutant strains, of plasmids constructed from various combinations of these genes along with *tcmKLMN* lead to the belief that the product of *tcmH*, possibly together with that of *tcmI*, directs the hydroxylation, leaving *tcmJ* as another cyclase gene.

The Tcm C pathway is continued with the assistance of three *O*-methyltransferase genes, *tcmN* (C-terminus), *tcmO* and *tcmP*, to add the C-3, C-8 and COOH methyls in that order, then is completed by a second, novel hydroxylase encoded by *tcmG* that adds the C-4, 4a and 12a hydroxyls to Tcm A2 (Fig. 1B). How each of these hydroxyls is added is unsettled; we know only that the ones at C-4 and C-12a come from molecular oxygen and that at C-4a presumably from water (Anderson et al., 1989), and that the Tcm G protein is sufficient to introduce all three of them.

In the absence of Tcm C, the TcmR protein is bound to an operator region just upstream of the *tcmA* promoter to repress this gene until its product is required. Tcm C derepresses expression of the *tcmA* gene by binding to the TcmR repressor. The TcmA protein is thought to export Tcm C from the cell, by means of proton motive force, and thereby to protect the cell from the deleterious effects of this potent antibiotic. Expression of the *tcmGHIJK-LMNO* operon and *tcmP* is also likely to be regulated, in this case by the two promoters indicated in Fig. 1A, which should transduce the signals that trigger secondary metabolism in response to as yet undetermined environmental changes.

Enhanced Production of Tcm C Metabolites

Operating on the assumption that increased gene dosage can achieve a higher titer of pathway enzymes, we asked what would happen if we introduced the *tcmKLM* genes into *S. glaucescens* under the control of a strong, constitutive promoter (*ermE**) and cloned on a high copy number plasmid (Gramajo et al., 1991). To our delight, Tcm D3 was overproduced about 10-fold in the wild-type background and about 40-fold in the TcmIV mutant. A similar effect was obtained with just the *tcmM* gene, suggesting that the ACP was normally present in a limiting amount. That the production of Tcm C was not also raised may mean that another step, such as Tcm A2 --> Tcm C, becomes rate-limiting under these conditions, or less likely, that the level of resistance to Tcm C is a limiting factor.

Production of New Metabolites

Using recombinant DNA methods to construct novel combinations of antibiotic production genes in vivo has the potential of producing new metabolites. This was in fact first demonstrated by Hopwood et al. (1985), but the full potential of this approach to drug discovery has only just been tapped (Hutchinson, 1992).

In the case of Tcm C, we are examining the roles of the *tcmKLMN* PKS genes by substituting each one of them with the homologous genes from other organisms and determining the effect of this on the formation of Tcm F2 and Tcm C. Precise replacement of the *tcmL* gene, cloned together with *tcmK* and *tcmM* in a plasmid vector, with its *Streptomyces coelicolor* homolog, *actI orf2* (Malpartida et al., 1992), resulted in the partial restoration of Tcm C production in a Δ*tcmL* mutant. The same construction caused the formation of two new substances in small amounts, whose chemical structures are being determined. These results suggest that further interchanges of the *tcm* genes with their homologous cousins will produce additional new substances, some of which might have an interesting antibiotic activity.

References

Anderson, M. G., Khoo, C. L.-Y. and Rickards, R. W. *J. Antibiotics* **1989**, *42*, 640-642.

Bibb, M. J., Biro, S., Motamedi, H., Collins, J. F., and Hutchinson, C. R. *EMBO J.* **1989**, *8*, 2727-2736.

Crandall, L. W. and Hamill, R. L. In *Antibiotic-Producing Bacteria*; Queener, S.W. and Day, L.E., Ed.; The Bacteria. A Treatise on Structure and Function; Academic Press, N.Y., 1986, Vol. 9; pp 355-402.

Gramajo, H. C., White, J., Hutchinson, C. R. and Bibb, M. J. *J. Bacteriol.* **1991**, *173*, 6475-6483.

Guilfoile, P. G. and Hutchinson, C. R. *J. Bacteriol.*, **1992a**, *174*, in press.

Guilfoile, P. G. and Hutchinson, C. R. *J. Bacteriol.*, **1992b**, *174*, in press.

Hopwood, D. A., Malpartida, F., Kieser, H. M., Ikeda, H., Duncan, J., Fujii, I.,

Rudd, B. A. M., Floss, H. G. and Omura, S. *Nature* **1985**, *314*, 642-644.
Hopwood, D. A. and Sherman, D. H. *Annu. Rev. Genet.* **1990**, *24*, 37-66.
Hutchinson, C.R. *Pharm. Technol.*, **1992**, *16*, 22-31.
Malpartida, F., Cabellero, J. L., Martinez, E. and Hopwood, D. H. **1992**, unpublished data.
Motamedi, H., E. Wendt-Pienkowski, and Hutchinson, C. R. *J. Bacteriol.* **1986**, *167*, 575-580.
Summers, R. G., Wendt-Pienkowski, E. and Hutchinson, C. R. *J. Bacteriol.* **1992**, 174, 000-000.
Weber, W., Zähner, H., Siebers, J., et al. *Arch. Microbiol.* **1979**, *121*, 111-116.
Yue, S., Motamedi, H., Wendt-Pienkowski, E., and Hutchinson, C. R. *J. Bacteriol.* **1986**, 167, 581-586.

Genes, Enzymes, and Control of the Production of β-Lactam Antibiotics

Juan F. Martín*, Santiago Gutiérrez, Eduardo Montenegro, Juan José R. Coque, Francisco J. Fernández, Javier Velasco, Santiago Gil, Francisco Fierro, Javier G. Calzada, Rosa E. Cardoza, and Paloma Liras
University of León, Faculty of Biology, Section of Microbiology, 24071 León, Spain (FAX:34 87 291506)

Several enzymes involved in penicillin, cephalosporin and cephamycin biosynthesis have been purified to homogeneity. Gene-enzyme relationships have been studied. The three genes of the penicillin biosynthetic pathway pcbAB, pcbC and penDE are located in a single cluster. The cephalosporin biosynthetic pathway in C. acremonium is encoded by at least two separate clusters of genes: pcbAB-pcbC and cefEF-cefG. Twelve genes related to cephamycin biosynthesis are located in a single cluster in N. lactamdurans. Different arrangements of genes, and regulatory mechanisms, were found in the fungal and bacterial cephalosporin producers.

Penicillin, Cephalosporin and Cephamycin Biosynthesis

Penicillins, cephalosporins and cephamycins are β-lactam antibiotics formed by condensation of L-α-aminoadipic acid, L-cysteine and L-valine. The three amino acids are linked together to form δ-(L-α-aminoadipyl)-L-cysteinyl-D-valine (ACV). Formation of the tripeptide ACV is carried out by the enzyme ACV synthetase (ACVS). In the second step, ACV is oxidatively cyclized by removal of four hydrogen atoms to form the penam nucleus (a β-lactam fused to a thiazolidine ring) of isopenicillin N (IPN) which is present in all penicillins. From IPN the pathway diverges to hydrophobic penicillins in P. chrysogenum and A. nidulans and to cephalosporins and cephamycins in various molds and actinomycetes (Martín and Liras, 1989). Cephalosporins and cephamycins (7-α-methoxycephalosporins) contain the cephem ring system (a β-lactam fused to a dihydrothiazine ring).

The Penicillin Biosynthetic Genes are in a Single Cluster

Three genes pcbAB, pcbC and penDE encode all enzyme activities required for penicillin biosynthesis (Martín et al., 1991). The gene pcbAB encoding the ACVS of P. chrysogenum was cloned using two different strategies: i) complementation of mutants npe5 and npe10 of P. chrysogenum blocked in penicillin biosynthesis, and ii) transcriptional mapping of the regions around the previously cloned (see below) pcbC-penDE cluster (Díez et al., 1990). P. chrysogenum DNA fragments, cloned in λEMBL3 or cosmid vectors, from the upstream region of the pcbC-penDE cluster carry a gene (pcbAB) that complemented the deficiency of ACVS of mutants npe5 and npe10, and restored penicillin production to mutant npe5. A protein of about 400 kDa was observed in SDS-PAGE of cell-free extracts of complemented strains that was absent in the npe5 and npe10 mutants. Transcriptional mapping studies showed the presence of one long transcript of about 11.5 kb that hybridized with several probes internal to the pcbAB gene, and two small transcripts of 1.15 kb that hybridized with the pcbC or the penDE gene, respectively (Fig. 1). The gene has been completely sequenced. It includes an open reading frame of 11,376 nt that encodes a protein with a deduced Mr of 425,971. No introns appear to occur in the pcbAB gene. Three repeated domains were found in the α-aminoadipyl-cysteinyl-valine synthetase that have high homology with the amino acid sequence of the gramicidin S synthetase I and tyrocidine synthetase I. The pcbAB is linked to the pcbC and penDE genes and is transcribed in the opposite orientation to them (Fig. 1).

1054–7487/92/0131$06.00/0

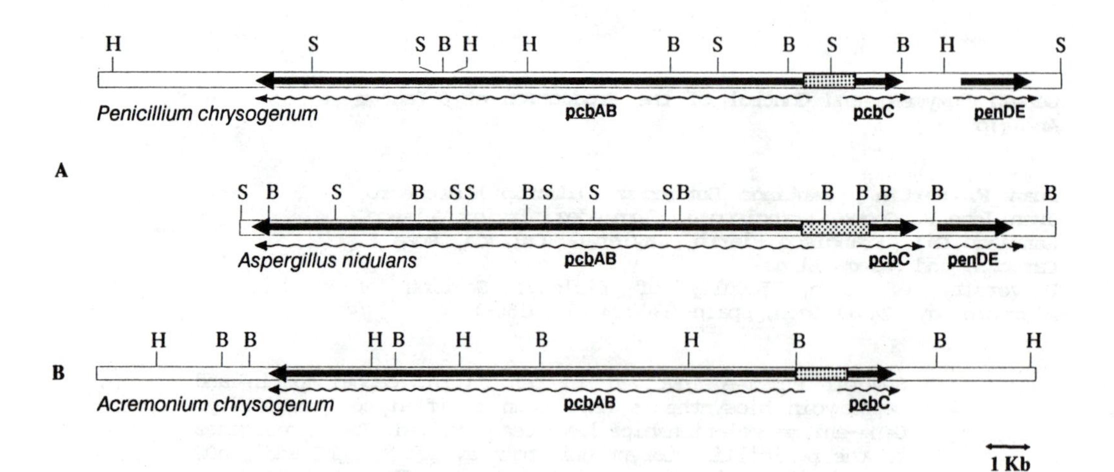

Fig. 1. Clusters of genes involved in (A) penicillin biosynthesis in P. chrysogenum and A. nidulans, and (B) early steps of cephalosporin biosynthesis in C. acremonium. Note the opposite orientation of the pcbAB and pcbC genes, which are expressed from a bidirectional promoter (stippled boxes). Wavy lines indicate the transcripts known to be formed from the cloned genes. (Adapted from Martín, 1992).

The pcbC gene of P. chrysogenum AS-P-78 was cloned using a probe corresponding to the amino terminal end of the enzyme. It contained an ORF of 996 nt encoding a polypeptide with an Mr of about 38,000. The pcbC gene does not contain introns and is expressed in E. coli minicells (Barredo et al., 1989a). The deduced amino acid sequence of the IPN synthase of P. chrysogenum is very similar to that of Streptomyces griseus (García-Domínguez et al., 1991), N. lactamdurans (Coque et al., 1991a) and other actinomycetes and filamentous fungi (Martín et al., 1991; Aharonowitz et al., 1992).

In the last step of penicillin biosynthesis the α-aminoadipyl side-chain of IPN is exchanged for phenylacetic acid. This reaction, which is catalyzed by the IPN:acyl-CoA acyltransferase (IAT) occurs only in penicillin-producing strains. This enzyme does not exist in C. acremonium and other cephalosporin producers due to the absence of the gene (Gutiérrez et al., 1991b). The IAT of P. chrysogenum has been purified to homogeneity. The purified preparation that catalyzed the formation of benzylpenicillin from phenylacetyl-CoA and 6-APA or IPN was shown to contain three proteins of 40, 29 and 11 kDa. The N-terminal sequence of the 29 kDa protein was used to isolate a fragment of P. chrysogenum DNA that contained an open reading frame with three introns (Barredo et al., 1989a). The deduced amino acid sequence of the open reading frame encodes a 40 kDa protein. The 11- and 29-subunits are probably formed by proteolytic cleavage of the 40 kDa protein encoded in the cloned gene.

The A. nidulans penDE gene is very similar to the previously cloned penDE gene of P. chrysogenum (Montenegro et al., 1990). Both genes contain three introns in similar positions

The Cephalosporin Biosynthetic Pathway is Encoded by Two Clusters of Genes

A 24 kb region of C. acremonium C10 DNA was cloned by hybridization with the pcbAB gene of P. chrysogenum (Gutiérrez et al., 1991a). The pcbAB was found to be closely linked to the pcbC gene forming a cluster of early cephalosporin-biosynthetic genes. A functional ACVS was encoded by a 15.6 kb EcoRI-BamHI DNA fragment, as shown by complementation of an ACVS-deficient mutant of P. chrysogenum. Two transcripts of 1.15 and 11.4 kb were found by Northern hybridization of C. acremonium RNA with probes internal to the pcbC and pcbAB genes, respectively (Fig. 1). An open reading frame of 11,136 bp was located upstream of the pcbC gene that matched the 11.4 kb transcript initiation and termination regions. It encoded a protein of 3712 amino acids with a deduced Mr of 414,791. The nt sequence of the gene showed 62.9% similarity to the pcbAB gene encoding the ACVS of P. chrysogenum; 54.9% of the amino acids were identical in both ACV synthetases. Three highly repetitive regions occur in the amino acid sequence of C. acremonium ACVS similar to the three repetitive domains in the sequence of P. chrysogenum ACVS. These regions probably correspond to amino acid-activating domains in the ACVS protein. In addition, a thioesterase domain was present in the ACVS's of both fungi.

A bifunctional enzyme, deacetoxycephalosporin C synthase/ hydroxylase (DAOCS/DACS) converts penicillin N to deacetylcephalosporin C (DAC). The last step of the cephalosporin biosynthetic pathway involves the conversion of DAC to cephalosporin C by the enzyme acetyl-CoA:DAC acetyltransferase (Félix et al., 1980). The DAC-acetyltransferase has been purified to homogeneity and shown to have a molecular weight of 52,000 Da by gel filtration and SDS-PAGE (J. Velasco and J.F. Martín, unpublished).

The gene (cefG) encoding the DAC acetyltransferase of C. acremonium C10 has been cloned (Gutiérrez et al., 1992). It contains two introns and encodes a protein of 444 amino acids with a relative mass of 49,269 that correlates well with the molecular weight deduced by gel filtration. The cefG gene is linked to the cefEF gene (encoding the bifunctional DAOC synthase/ hydroxylase) but it is expressed in opposite orientation to the cefEF gene (Fig. 2). Two transcripts of 1.2 and 1.4 kb were found in C. acremonium that correspond to the cefEF and cefG genes, respectively. The cloned cefG complemented the deficiency of DAC acetyltransferase in the non-producer mutant C. acremonium ATCC 20371 and restored cephalosporin biosynthesis in this strain. Heterologous expression of the cefG genes took place in Penicillium chrysogenum.

Most (if not all) of the cephalosporin biosynthetic genes are linked in two clusters. The pcbAB gene of C. acremonium is linked to the pcbC gene (Gutiérrez et al., 1991a) and they are located in chromosome VI (Skatrud and Queener, 1989), whereas the cluster of late genes, cefEF and cefG, is located on chromosome II. Only the location of the IPN epimerase gene, cefD, remains unknown.

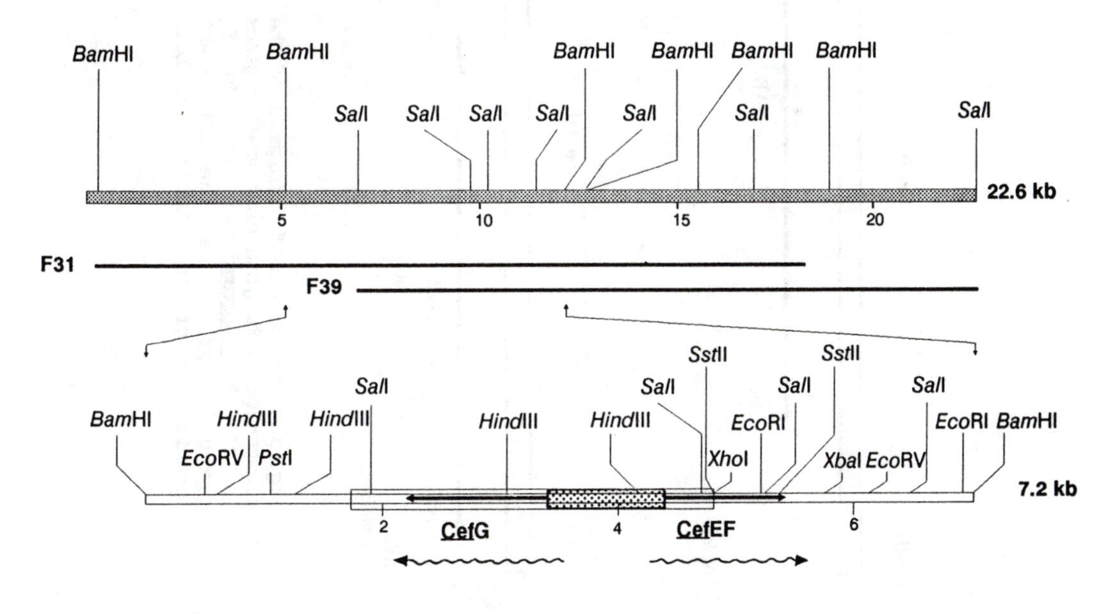

Fig. 2. Region of a C. acremonium C10 DNA region containing the cluster of late cephalosporin biosynthetic genes. F31 and F39 correspond to fragments cloned in two different phages. Note the opposite orientation of the cefEF and cefG genes, which are also expressed from a bidirectional promoter (stippled box). Wavy lines indicate the transcripts known to be formed from the cloned genes. (Adapted from Gutiérrez et al., 1992).

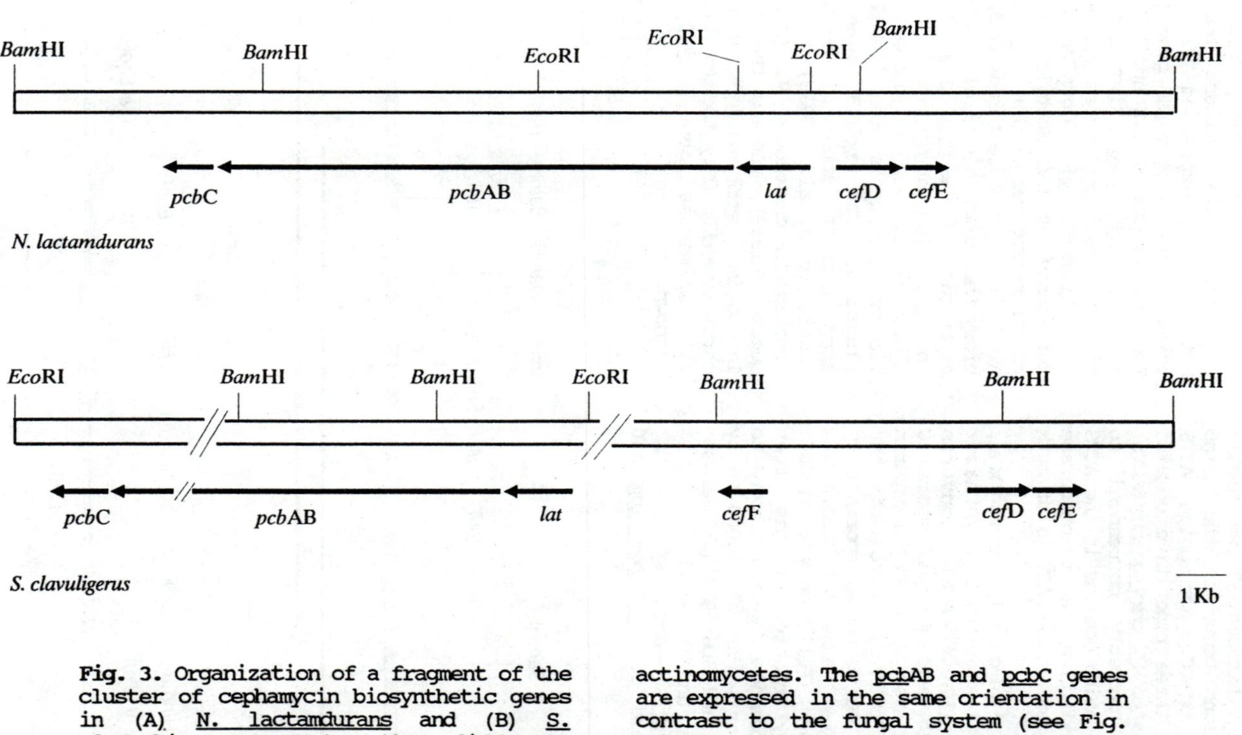

Fig. 3. Organization of a fragment of the cluster of cephamycin biosynthetic genes in (A) N. lactamdurans and (B) S. clavuligerus. Note the different organization of the cluster in both actinomycetes. The pcbAB and pcbC genes are expressed in the same orientation in contrast to the fungal system (see Fig. 1). (Adapted from Coque et al., 1992).

The Cephamycin Gene Cluster is Present in a Single DNA Fragment

A 34 kb fragment of the Nocardia lactamdurans DNA carrying the cluster of cephamycin biosynthetic genes was cloned in λEMBL3 by hybridation with probes internal to P. chrysogenum pcbAB and S. griseus pcbC genes (Coque et al., 1991a). At least 12 genes related to cephamycin C biosynthesis are located in this cluster. The pcbAB and pcbC genes were found to be closely linked together and expressed in the same orientation, in contrast to the divergent expression of the pcbAB-pcbC genes of fungi. The pcbAB encodes a large (3649 amino acids) multidomain ACVS with a deduced Mr of 404,134. This enzyme contains three repeated domains and a consensus thioesterase active-site sequence similar to those of the fungal ACVS's. The pcbC gene encodes an IPNS of 328 amino acids with a deduced Mr of 37469. This protein lacks one of the two cysteine residues conserved in the 8 previously sequenced cyclases (Coque et al., 1991a) (Fig. 3). A gene (lat) encoding lysine-6-aminotransferase was found upstream of the pcbAB gene, separated by a small intergenic region of 64 bp (Coque et al., 1991b). The lat gene contains an ORF of 1353 nt encoding a protein of 450 aminoacids (Mr 48811).

The cefD and cefE genes of Nocardia lactamdurans encoding IPN epimerase and DAOC synthase have been located 0.63 kb upstream from the lysine-6-aminotransferase (Coque et al., 1992). cefD contains an ORF of 1197 nt encoding a protein of 398 amino acids, with a deduced Mr of 43622 and contains a consensus pyridoxal phosphate binding sequence. cefE is located at the 3' end of cefD, with the ATG starting codon overlapping the final TGA codon of cefD. It encodes a protein of 314 amino acids (Mr 34532) which lacks deacetoxycephalosporin C hydroxylase activity (Coque et al., 1992). The cephamycin gene cluster has a different arrangement in N. lactamdurans and S. clavuligerus, what is probably important in the control of gene expression in the different β-lactam producers.

References

Barredo, J.L.; Cantoral, J.M.; Alvarez, E.; Díez, B.; Martín, J.F. Molec. Gen. Genet. **1989a**, 216, 91-98.

Barredo, J.L.; van Solingen, P.; Díez, B.; Alvarez, E.; Cantoral, J.M.; Kattevilder, A.; Smaal, E.B.; Groenen, M.A.M.; Veenstra, A.E.; Martín, J.F. Gene **1989b**, 83, 291-300.

Coque, J.J.R.; Martín, J.F.; Calzada, J.G.; Liras, P. Molec. Microbiol. **1991a**, 5, 1125-1133.

Coque, J.J.R.; Liras, P.; Láiz, L.; Martín, J.F. J. Bacteriol. **1991b**, 173, 6258-6264.

Coque, J.J.R.; Martín, J.F.; Liras, P. Molec. Gen. Genet. **1992**, submitted.

Díez, B.; Gutiérrez, S.; Barredo, J.L.; van Solingen, P.; van der Voort, L.H.M.; Martín, J.F. J. Biol. Chem. **1990**, 265, 16358-16365.

Félix, H.R.; Nüesch, J.; Wherli, W. FEMS Microbiol. Lett. **1980**, 8, 55-58.

García-Domínguez, M.; Liras, P.; Martín, J.F. Antimicrob. Agents Chemother. **1991**, 35, 44-52.

Gutiérrez, S.; Díez, B.; Montenegro, E.; Martín, J.F. J. Bacteriol., **1991a**, 173, 2354-2365.

Gutiérrez, S.; Díez, B.; Alvarez, E.; Barredo, J.L.; Martín, J.F. Molec. Gen. Genet., **1991b**, 225, 56-64.

Gutiérrez, S.; Velasco, J.; Fernández, F.J.; Martín, J.F. J. Bacteriol., **1992**, in press.

Martín, J.F.; Liras, P. In Advances in Biochemical Engineering/Biotechnology; Fiechter, A. Ed; Springer-Verlag, Berlín, Heidelberg, **1989**, Vol. 39, pp 153-187.

Martín, J.F.; Ingolia, T.D.; Queener, S.W. In Molecular Industrial Mycology: Systems and Applications for Filamentous Fungi; Leong, S.A.; Berka, R.M. Ed; Marcel Dekker, Inc, New York, **1991**, pp 149-196.

Martín, J.F. J. Ind. Microbiol., **1992**, in press.

Montenegro, E.; Barredo, J.L.; Gutiérrez, S.; Díez, B.; Alvarez, E.; Martín, J.F. Molec. Gen. Genet., **1990**, 221, 322-330.

Ramos, F.R.; López-Nieto, M.J.; Martín, J.F. Antimicrob. Agents Chemother., **1985**, 27, 380-387.

Revilla, G.; Ramos, F.R.; López-Nieto, M.J.; Alvarez, E.; Martín, J.F. J. Bacteriol., **1986**, 168:947-952.

BIOCATALYSIS AND BIOTRANSFORMATION SYMPOSIUM IV

D. S. Clark and H. Yamada: *Co-Chairs*

Immobilized and Two-Phase Systems: Session A
J. B. Jones and H. Yamada: *Co-Chairs*

Designed Biocatalysts: Session B
S. J. Benkovic and T. Beppu: *Co-Chairs*

Biocatalysis in Synthesis and Biotransformation: Session C
D. S. Clark and J. Tramper: *Co-Chairs*

Progress Towards the Rational Design of Immobilized Protein Systems: Preliminary Results with Immobilized Antibodies

Thomas Spitznagel and Douglas Clark*, Department of Chemical Engineering, University of California, Berkeley, CA 94720. (FAX:510 642-4778)

Binding parameters for both randomly immobilized whole antibody MOPC-315 and site-specifically immobilized Fab' fragments are reported. As the loading of protein increases, both immunosorbents lose partial specific activity towards a small hapten (molecular weight = 341). A further decrease in specific activity is observed when the immobilized whole antibody is assayed toward a large hapten (molecular weight = 50 kilodaltons). This additional decrease is not seen for the immobilized Fab' fragments. These results suggest that the activity loss of immobilized Fab' fragments is due to distortion or crowding of some (40%) combining sites, and that the further decrease observed for whole antibodies is due to improper orientation(s) of immobilized antibody molecules. Electron paramagnetic spectroscopy indicates that the conformation of the active immobilized antibody binding site is unaltered in both systems.

Due to their remarkably high specificity, immobilized monoclonal antibodies are playing an important role in biotechnology. Much attention has focused on the application of affinity chromatography, in which monoclonal antibodies are immobilized onto a solid support in order to purify proteins from complex mixtures. Immobilized antibodies also have many proven and potential applications in immunotherapy and immunodiagnostics (Rodwell et al., 1985; Klausner, 1987). Yet another application is in immunosensors, in which monoclonal antibodies or antibody fragments are immobilized onto electrodes or silicon chips for the detection of dilute substances (North, 1985). The use of antibodies will no doubt expand even further as catalytic antibodies (Jacobs, 1991) and techniques for producing antibodies in bacteria come of age (Skerra et al., 1988; Wood et al., 1985). However, a frequent drawback of immobilized antibodies, as well as any immobilized protein, is the relatively low specific activity of the immunoconjugate.

The primary factors considered in determining the activity of an immunosorbent are the total amount of protein immobilized (loading) and the number of active binding sites per antibody molecule (n value). Many studies have shown that the activity of an immobilized antibody is generally less than that of its soluble counterpart. Typical values of n for immobilized systems range from 0.2 to 1.9 (Eveleigh et al., 1977; Olson et al., 1989; Chase, 1984), indicating that a significant fraction of activity can be lost upon immobilization. However, although there is general recognition of this loss in activity, there is little consensus on its exact causes.

Upon immobilization, one can envision several different situations that could lead to reduced antibody activity. For example, an orientation effect is encountered when the binding sites are immobilized adjacent to the support and rendered inaccessible to antigen. Another possibility is molecular crowding. In this case, antibody binding sites interfere with each other due to a high local concentration of antibody. A third possibility arises if the binding sites are still active, but the antibody is inaccessible to the antigen because the hapten cannot penetrate the matrix. Finally, it is possible that upon immobilization, interactions, e.g., covalent bonds, between the support and the antibody may induce a conformational change in the binding pocket. Each of these effects, as well as many possible combinations, have been suggested to explain reduced binding efficiencies of immobilized antibodies. However, the majority of these explanations are speculative and have not been verified by direct observation.

In this report we describe the immobilization of the monoclonal antibody MOPC-315 and its

1054–7487/92/0138$06.00/0

Fab' fragments. By varying the immobilization chemistry, hapten size, and antibody loading, we have investigated possible causes of antibody inactivation upon immobilization.

Antibody System

MOPC-315 monoclonal antibody is specific for the dinitrophenyl (DNP) epitope. In this study, we compare the activity of MOPC-315 toward 3-(5-fluoro-2,4-dinitroanilino)-proxyl (FDNP-SL), a small hapten of molecular weight 341, and toward a large hapten (FAB-DNP; molecular weight = 50 kilodaltons) prepared by covalently coupling a DNP moiety to the terminal thiol group of a Fab' fragment from an antibody not specific for DNP groups (Figure 1). Binding assays were performed by measuring concentrations of free and bound FDNP-SL using electron paramagnetic resonance (EPR) spectroscopy (Figure 2). For binding assays of the large hapten, the concentration of FAB-DNP in solution was determined from its absorbance at 343 nm.

Randomly Immobilized MOPC-315

Whole MOPC-315 antibody was immobilized at varying concentrations onto aminopropylated controlled pore glass activated with glutaraldehyde. Typical coupling efficiencies (the percent antibody immobilized from solution) ranged from 50 to 75 %. Loadings and equilibrium binding parameters determined for each of the immunosorbents in the presence of FDNP-SL are presented in Table 1. The n values range from the soluble limit of n = 2.0 for the lowest loading to n = 1.0 for the highest loading. Thus, at the highest loading, which corresponds to the theoretical monolayer coverage, approximately 50 % of the original activity remains. Equilibrium association constants, K, also appeared to decrease somewhat, but this change is probably not significant.

Table 1 also lists the association constants and n values for the large hapten, FAB-DNP. Once again, in the limit of the lowest loading, there is no significant decrease in the n value (n = 2.0). However, at a higher loading, the n value is approximately 32 % of the original

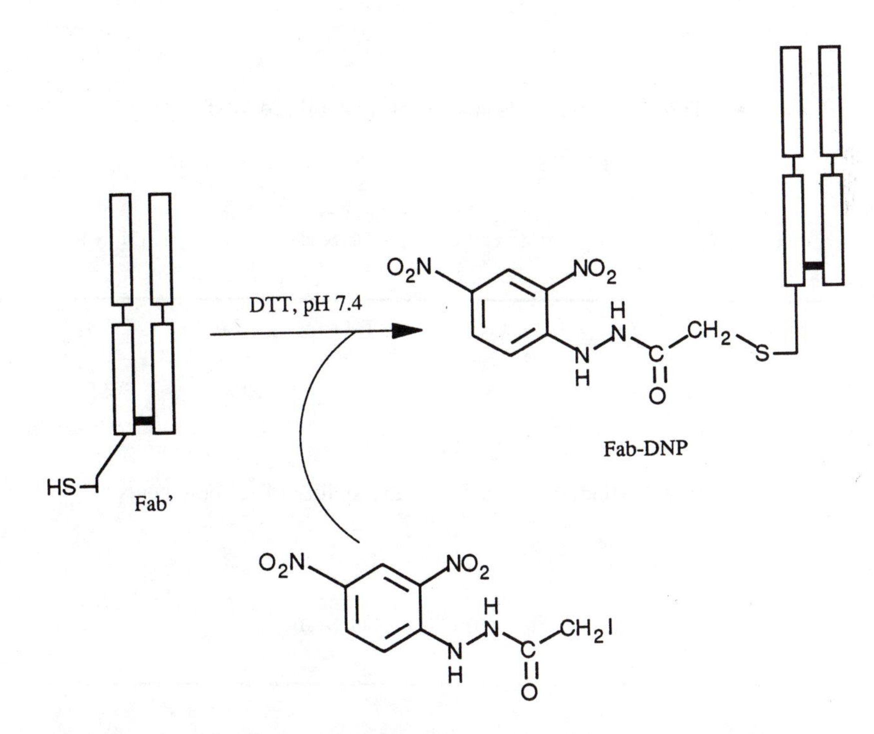

Figure 1. Preparation of the large hapten, Fab-DNP. Fab' fragments (non-reactive to DNP) were reacted with N-(2,4dinitroanilino)-2--iodoacetamide in the presence of dithiothreitol. The thickened lines represent interchain disulfide bonds.

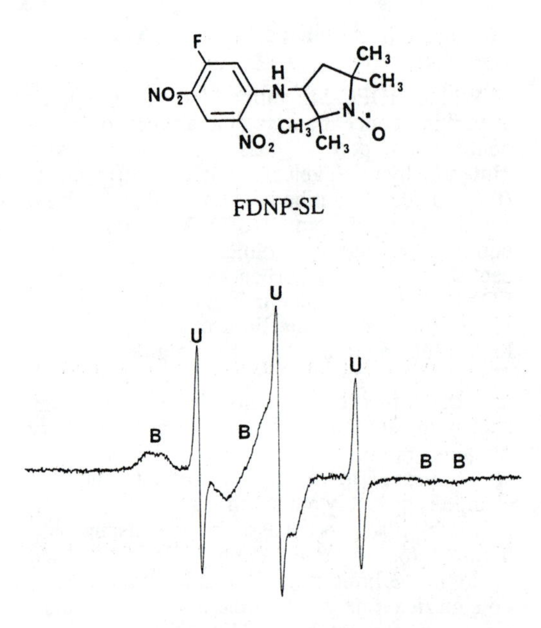

Figure 2. Top: The spin-labeled hapten, FDNP-SL. Bottom: EPR spectrum of FDNP-SL in presence of MOPC-315 showing signal of bound (B) and unbound (U) hapten.

soluble-antibody value. This decline represents a greater decrease in activity than was seen for the higher-loaded immunosorbent toward the small hapten. Hence, the size of the hapten appears to make a greater difference as the antibody loading increases. In addition, there was no significant decrease in the value of the equilibrium binding constant. These results suggest that at least some of the immobilized antibody molecules are inactive toward the large hapten due to crowding or orientation effects.

Site-Specifically Immobilized Fab' Fragments

In order to investigate the importance of binding-site orientation, similar experiments have addressed the influence of loading on immobilized Fab' fragments. The immobilization chemistry site-specifically attaches the Fab' fragment via a single sulfhydryl group at the tail of the fragment far from the active binding site. Table 2 shows the loadings, n values, and equilibrium binding constants for the different preparations. As before, the n value for a relatively low loading was the same as that of the soluble Fab' fragment (n = 1.0). However, as the loading increases, at least two observations can be made: 1) n values

Table 1. Binding Parameters of Immobilized MOPC-315

FDNP-SL			Fab-DNP		
Loading (µmol/L bead)	n	K (M^{-1} x 10^5)	Loading (µmol/L bead)	n	K (M^{-1} x 10^5)
75.1	2.1	3.1	114.8	2.0	1.8
251.4	1.0	1.5	178.6	0.69	3.5

Table 2. Binding Parameters of Immobilized Fab' Fragments

FDNP-SL			Fab-DNP		
Loading (µmol/L bead)	n	K (M^{-1} x 10^5)	Loading (µmol/L bead)	n	K (M^{-1} x 10^5)
85.2	1.0	1.9	85.2	0.93	2.6
394.9	0.61	3.6	394.4	0.60	4.0

for high Fab' loadings (ca. 61%) are slightly higher than the values of comparable preparations of immobilized whole antibody, and 2) higher loadings are much more readily obtained with the Fab' system.

As shown in Table 2, no significant differences in n values were seen between the small and large haptens for both loadings of Fab' fragments. Hence, immobilized whole antibodies and Fab' fragments both exhibit lower n values as the protein density increases. However, the n value depends on the hapten size for the immobilized whole antibody but not for the immobilized Fab' fragments, suggesting that different mechanisms are at least partly responsible for the reduced activities of the two systems.

EPR Studies of the Active Site

The possibility of conformational changes upon immobilization was also investigated by measuring EPR spectra of the soluble and immobilized antibody in the presence of FDNP-SL. The spectrum of the spin label in the antibody combining site is governed by the overall configuration and environment of the binding pocket. As the hapten is more tightly bound, and hence less mobile, the lineshapes of the spectrum broaden. The EPR spectra (not shown) indicate that there are no significant differences in the conformation of the *active* antibody binding site for either immobilized system. This conclusion is also supported by the relatively constant value of the equilibrium binding constant. Hence, immobilization does not appear to induce gross conformational changes in the active antibody binding site. However, the binding-site conformation(s) of inactive immbobilized antibody molecules, which are not probed by the DNP spin label, could differ considerably.

Conclusions

The binding parameters of randomly immobilized MOPC-315 and site-specifically immobilized Fab' fragments have been investigated. At low loadings both systems are essentially fully active and at high loadings both systems lose significant activity. Further, randomly immobilized antibody exhibits a further decrease in specific activity when assayed in the presence of a large hapten. In all cases studied, the conformation of the active combining sites was apparently unaltered.

Although these results do not clearly elucidate the mechanism(s) of inactivation, a preliminary interpretation is possible. Assuming that all of the immobilized Fab' fragments are uniformly oriented with their combining sites facing out from the surface, the 40% loss of activity at the high loading is presumably due to severe distortion or crowding of a subpopulation of combining sites. This deactivation mechanism is independent of hapten size. The additional decrease in activity exhibited by the immobilized whole antibody (a total loss of 50% for the small hapten and 60% for the large hapten) appears to reflect a difference between the two types of immunosorbent. A likely cause is thus an orientation effect, i.e., close proximity between the antibody combining site and the support surface. This effect would be unique to immobilized whole antibodies, and could account for the observed dependence on hapten size.

References

Chase, H.A., *Chem. Eng. Sci.*, **1984**, *39*, pp 1099-1125.

Eveleigh, J.W. and Levy, D.E., *J. of Solid-Phase Biochem.*, **1977**, *2*, pp 45-78.

Jacobs, J.W., *Bio/Technology*, **1991**, *9*, pp 258-262.

Klausner, A. *Bio/Technology*, **1987**, *5*, pp 551-556.

North, J.R., *Trends in Biotechnology*, **1985**, *3*, pp 180-186.

Olson, W.C., Spitznagel, T.M. and Yarmush, M.L., *Mol. Immun.*, **1989**, *26*, pp 129-136.

Rodwell, J.D., and McKearn, T.J., *Bio/Technology*, **1985**, *3*, pp 889-894.

Skerra, A. and Pluckthun. A., *Science*, **1988**, *240*, pp 1038-1041.

Wood, C.R., Boss, M.A., Kenten, J.H., Calvert, J.E., Roberts, N.A., and Emtage, J.S., *Nature*, **1985**, *314*, pp 446-449.

The Influence of the Support Material on Enzymatic Synthesis in Organic Media

Patrick Adlercreutz*, Ernst Wehtje, Mats Reslow and Bo Mattiasson,
Dept. of Biotechnology, Chemical Center, Lund University, P.O.Box 124,
S-221 00 Lund, Sweden
(FAX: +46 46 104713)

Enzymes adsorbed or deposited on porous support materials have been succesfully used as catalysts in organic media. However, the support must be chosen with great care. During the immobilization procedure, partial inactivation of the enzyme may occur, especially when a small amount of enzyme is immobilized on a support with a large surface area. The enzyme can be protected by the addition of proteins or polyethylene glycol. The partitioning of water to the components of the reaction mixture influences the catalytic activity of the enzyme. Normally, high reaction rates are obtained with supports of low aquaphilicity (water absorbing capacity). Even in experiments carried out at fixed enzyme hydration (fixed water activity) the support influences both the total activity of the enzyme and the relative rates of different reactions catalyzed by the same enzyme.

It is now a well established fact that enzymes can be catalytically active and stable in organic solvents. Several different ways of using enzymes in organic solvents have been developed (Laane, et al., 1987; Dordick, 1989). One of the best methods is to immobilize the enzyme on a solid support before adding it to the reaction medium. Since enzymes are normally not soluble in organic media, there is no need for covalent linkages between the support and the enzyme. Consequently, simple immobilization methods like adsorption can be employed. Many different kinds of supports have been used; it is often advantageous to use a porous support so that the enzyme is spread on a large surface area. It has been noted that the choice of the support makes a large impact on the catalytic performance of the enzyme. One can distinguish between direct effects of the support on the enzyme and indirect effects caused by the influence of the support on the partition of water, substrates and products between the bulk solvent and the micro-environment of the enzyme.

Water Partition Effects

Water plays a crucial role for biocatalysis in organic media; it partitions between the enzyme, the support and the solvent. The thermodynamic water activity is a good way to characterize the degree of hydration in the system because it is equal in all phases at equilibrium. Useful information can be obtained from measurements of water uptake under different conditions by the components of the system. The solubility of water in the reaction medium at different water activities can be easily determined (Adlercreutz, 1991). It is also possible to determine the water uptake by the enzyme and the support at different water activities so that adsorption isotherms are obtained (Adlercreutz, 1991). This can easily be done in air; adsorption isotherms of a few supports are shown in Fig. 1. Measurements of water adsorption isotherms of supports in organic solvents are scarce, but a few observations show that these are similar to the adsorption isotherms in air, at least at low water activities (Halling, 1990).

Aquaphilicity of Supports

A simple method to measure water uptake of supports in solvents was developed (Reslow, et al., 1988). The dry support is suspended in water-saturated solvent and the amount of water taken up by the support is measured after equilibration. The ratio of the amount of water on the support to the amount of water in the solvent under standard conditions is called the

1054–7487/92/0142$06.00/0

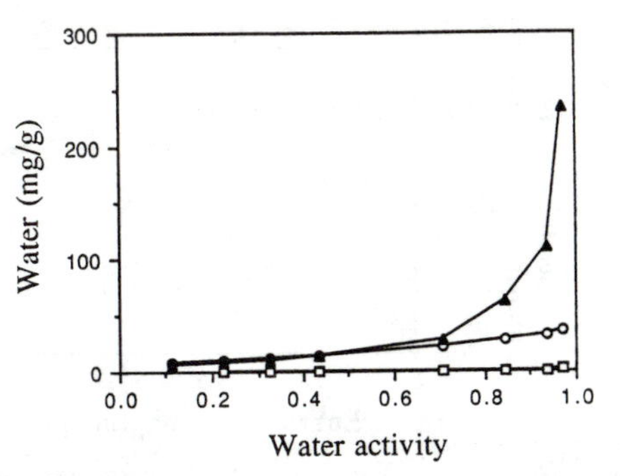

Fig. 1. Water adsorption isotherms at 25 °C in air for Accurel PA-6 (▲), hexyl-CPG (O) and Celite (❑). Data from (Adlercreutz, 1991).

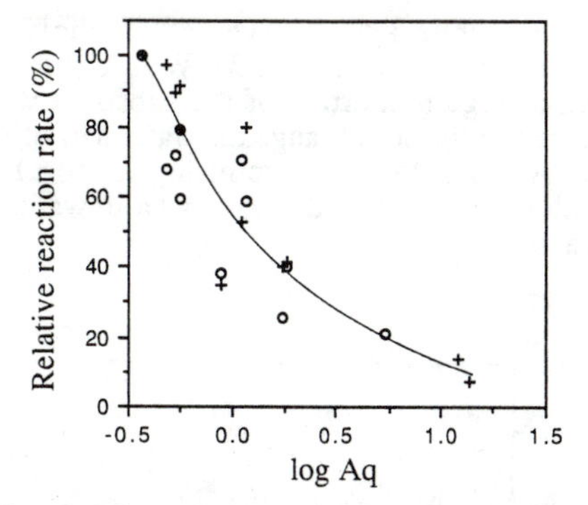

Fig. 2. The relative reaction rate obtained with α-chymotrypsin (O) and with horse liver alcohol dehydrogenase (+) when deposited on support materials with different aquaphilicity (Aq). The reaction catalyzed by α-chymotrypsin was the esterification of N-acetyl-L-phenylalanine with ethanol. Alcohol dehydrogenase catalyzed the reduction of cyclohexanone with the concomitant oxidation of ethanol. Diisopropyl ether was used as reaction medium. The highest reaction rates obtained in each reaction were set at 100 %. Data from (Reslow, et al., 1988).

aquaphilicity of the support. This parameter does not give as much information as a complete adsorption isotherm, but it is much easier to measure and it is useful for comparisons between supports.

When reactions are carried out using the same enzyme on different supports with a fixed amount of water present, supports of high aquaphilicity absorb large quantities of water, yielding an enzyme which is less hydrated than an enzyme immobilized on a low-aquaphilicity support. The reduced hydration of the enzyme normally reduces its catalytic activity.

The aquaphilicity of several supports was determined and two different enzymes (α-chymotrypsin and horse liver alcohol dehydrogenase, HLADH) were immobilized on the supports. A fixed amount of water was added to each reaction mixture. Among the supports there were several derivatives of controlled pore glass (CPG) with the same particle size and pore diameter, so that differences due to the morphology of the particles were minimized. The catalytic activity decreased with increasing aquaphilicity of the support (Fig. 2). Of course, it could be argued that the supports with high aquaphilicity could also provide high activity, provided some extra water is added. However, the competition for water, as described above, is only part of the cause of the activity differences observed. When different amounts of water were added to α-chymotrypsin on hexyl- or glucose-CPG, the esterification activity was higher on the hexyl-CPG, regardless of the amount of water added. Obviously, there is also some other positive effect of the low-aquaphilicity supports.

Reactions at Controlled Water Activity

In order to study the effects of the support in more detail, a technique for carrying out reactions at fixed water activity was developed. Using this technique, catalytic activities of enzymes immobilized on different supports can be measured at a fixed level of enzyme hydration, so that the indirect effects of the supports due to their influence on the partitioning of water are minimized. In this technique the enzyme preparation and the substrate solution are equilibrated separately in an atmosphere of controlled relative humidity (water activity) before the reaction is started. Using this technique, it was observed that, at fixed water activity, the support caused large effects on the activity of the enzyme (Adlercreutz, 1991). For HLADH, the catalytic activity increased with increasing water activity and with a few supports a maximum was observed in the range A_w = 0.9-1.0. Celite was the best support and supports of higher aquaphilicity gave lower reaction rates.

Interesting results on the relative rates of different reactions catalyzed by the same enzyme were obtained with α-chymotrypsin, catalyzing an alcoholysis reaction with hydrolysis of the same substrate as a competing reaction. On most supports, the alcoholysis activity increased with increasing water activity, but with the polyamide support Accurel PA-6, the highest rates were observed at low water activity. The rates of the hydrolysis reaction increased with increasing water activity with all supports. At the lowest water activity used (0.33), no hydrolysis was detected. Consequently, by using the polyamide support at a low water activity, hydrolysis was suppressed while a high level of alcoholysis activity was maintained (Fig. 3). With Celite as support, a large proportion of the substrate was hydrolyzed in the entire range of water activities studied. Accurel PA-6 is a promising support for applications in which hydrolysis is an unwanted side-reaction.

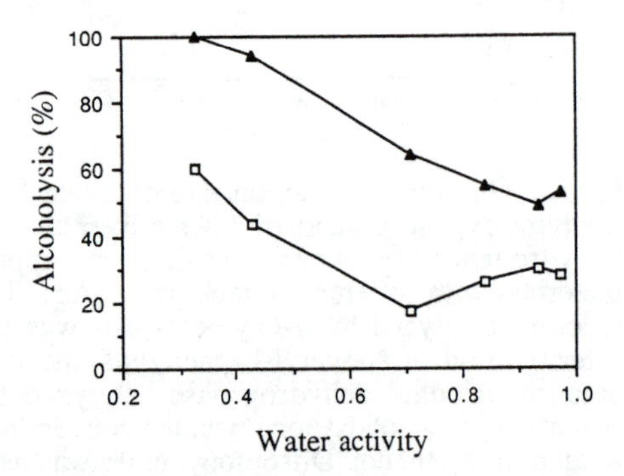

Fig. 3. Reaction specificity of α-chymotrypsin at different water activities with Accurel PA-6 (▲) or Celite (❑) as support. The reactions catalyzed were the alcoholysis of N-acetyl-L-phenylalanine ethyl ester with 1-propanol and the competing hydrolysis of the same substrate. The ratio of the rate of the alcoholysis reaction and the total reaction rate is plotted versus the water activity. Diisopropyl ether was used as reaction medium. Data from (Adlercreutz, 1991).

Enzyme Inactivation by the Support. Stabilization by Additives

One direct effect of the support on the enzyme is partial inactivation of enzyme during adsorption on the support. This effect can be quite pronounced when low amounts of enzyme are immobilized on highly porous supports. When a lipase was immobilized on Celite, full catalytic activity was expressed only when the enzyme loading was about 20 mg of enzyme per g of Celite or higher (Wehtje, et al., submitted)(Fig. 4). At lower enzyme loadings, the enzyme was partially inactivated during immobilization. Inactivation was avoided by addition of substances like proteins or poly(ethylene glycol). These additives should be present during the immobilization procedure; additives added after enzyme immobilization had no effect. With an addition of 20 mg poly(ethylene glycol) 6000 per g of Celite, considerably higher specific activities were obtained at low enzyme loadings and full catalytic activity was expressed at an enzyme loading as low as 5 mg per g of Celite. At high

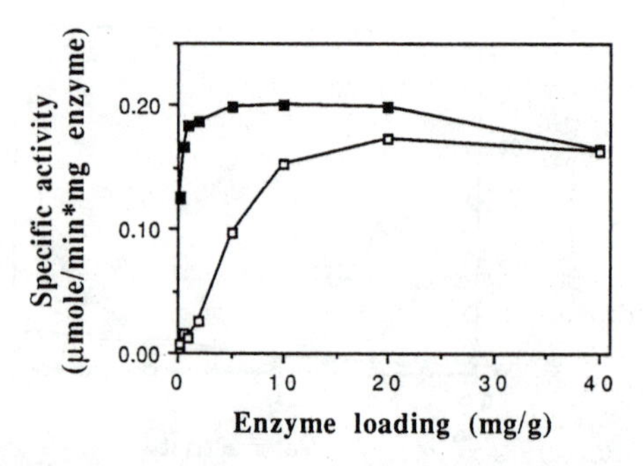

Fig. 4. Specific activity of lipase P as a function of the enzyme loading on Celite without (❑) or with (■) the addition of 20 mg poly(ethylene glycol) 6000 per g of Celite. The reaction studied was the esterification of oleic acid and ethanol in hexane. Data from (Wehtje, et al., submitted).

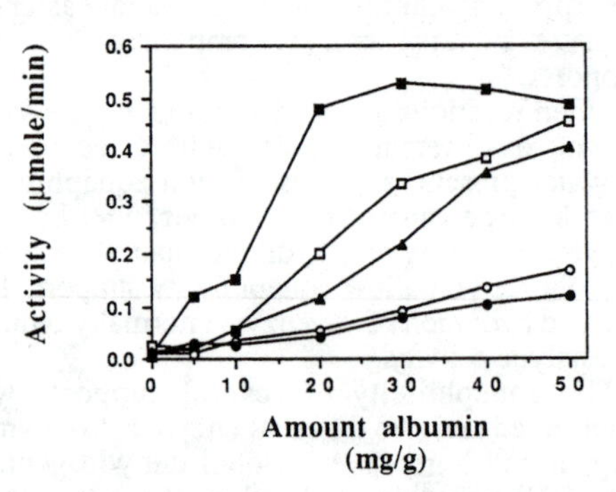

Fig. 5. Activity of mandelonitrile lyase, immobilized on different controlled pore glass supports, as a function of the amount of albumin. The enzyme (0.25 mg/g CPG) was immobilized after the addition of albumin. The supports had different mean pore diameters: 2917 Å (■), 1273 Å (❑), 729 Å (▲), 240 Å (○), and 77 Å (●). Data from (Wehtje, et al., submitted).

enzyme loadings, the addition of poly(ethylene glycol) had a negligible effect.

It was observed that the inactivation process was dependent on the surface area of the support. On highly porous supports like controlled pore glass with large surface areas, quite large amounts of albumin were needed to protect the enzyme from inactivation (Fig. 5). The amount of protein needed for protection roughly corresponded to monolayer coverage of the surface of the support, provided the pores were large enough for enzyme and additive to penetrate.

Experimental

Enzyme immobilization. The enzyme was dissolved in an aqueous buffer solution. This solution was mixed with the support (1-2 ml/g) and water was removed at reduced pressure.

Adjustment of water activity. Chambers containing saturated salt solutions were used for adjustment of water activity. The immobilized enzyme preparation and the substrate solution were incubated separately in these chambers and equilibration was achieved via the gas phase. After equilibration, usually overnight, the enzyme and the substrate were mixed and the reaction started.

Enzymatic conversions. Reactions were carried out in small (usually 4-10 ml) glass reactors. The enzyme preparation was suspended in the organic solvent containing the substrate. The reactors were shaken at constant temperature (normally 25 °C). Samples were taken regularly and were analyzed by HPLC or GLC.

Measurement of the aquaphilicity of a support. (Reslow, et al., 1988). To 50 mg of dry support, 250 mg of water-saturated diisopropyl ether was added and the container was closed. After 16 h the water content in the solvent (amount H_2O, solvent) was measured by gas chromatography. The amount of adsorbed water was calculated (amount H_2O, support). The aquaphilicity of the support, Aq, was defined as (amount H_2O, support)/(amount H_2O, solvent).

References

Adlercreutz, P. *Eur. J. Biochem.*, **1991**, *199,* 609-614.

Biocatalysis in organic media; Laane, C.; Tramper, J.; Lilly, M. D., Eds.; Elsevier: Amsterdam, **1987**.

Dordick, J. *Enzyme Microb. Technol.*, **1989**, *11,* 194-211.

Halling, P. *Biochim. Biophys. Acta*, **1990**, *1040,* 225-228.

Reslow, M.; Adlercreutz, P.; Mattiasson, B., *Eur. J. Biochem.*, **1988**, *172,* 573-578.

Wehtje, E.; Adlercreutz, P.; Mattiasson, B., *(submitted).*

Reactor Engineering for Multiphasic Biotransformations

J M Woodley, Department of Chemical and Biochemical Engineering,
University College London, London WC1E 7JE, UK
(Fax: (0)71 388 0808)

Multiphasic biotransformations are now being used in the synthesis of a variety of fine chemicals. The heterogeneous nature of the reaction media arises from the range of physical states and relative solubilities of substrate(s) and product(s) as well as the variety of biocatalyst forms. However, although such biotransformations are finding application they are poorly understood. Here, some of the implications of heterogeneity are described and future research identified.

Interest in biotransformations is increasing as their potential for novel chemistry and environmentally acceptable processes and products is recognised. Consequently biocatalysts are now finding use in different forms (immobilized cell or enzyme, dissolved enzyme or intact cell) and in different reaction media dependent upon the physical states and relative solubilities of substrate(s) and product(s) (Lilly and Woodley, 1985). This has led to multiphasic biotransformations. In this paper some of the implications of heterogeneity for biotransformation reactor and process engineering are discussed.

Unlike multiphasic chemical reactions not all permutations of phase contacting and subsequent separation are likely for biotransformations. Biocatalysts require a certain amount of water to function and consequently they require association, either with an aqueous phase or a solid which itself carries the necessary water. Aside from catalyst requirements the substrates and products may be predominantly soluble in an aqueous, organic or gaseous phase and this leads to the likely multiphasic biotransformation scenarios, some examples of which are given in Table 1.

Features of multiphasic biotransformations

The key advantage of introducing organic or gaseous carrier phases for non-aqueous substrate(s) or product(s), to a solid or aqueous catalyst-rich phase is the high concentrations of products obtainable whether the reactor is operated in either batch or continuous mode (Lilly et al., 1987). In addition the adverse effects of inhibitory substrate(s) and/or products(s) can be eliminated or reduced by preferential partitioning of the inhibitory compound away from the reaction zone. Potentially easier product recovery may be effected by phase separation exploiting the relative solubilities of the reaction components. However, the presence of more than one phase in the reactor also introduces two further considerations: first the implication for reactor and process engineering of the need to contact and then separate the phases effectively and secondly, understanding the interaction between the biocatalyst and interfacial phenomena. These will be addressed in the subsequent sections.

1054–7487/92/0146$06.00/0

Table 1: Some examples of multiphasic biotransformations

Process phases present	Reaction	Catalyst
L-L+	Ester hydrolysis	Pig pancreatic lipase
S-L+	Fat interesterification	Microbial lipase (I)
L-S	Benzyl penicillin deacylation	Penicillin acylase (I)
G-S	Ethanol oxidation	Alcohol oxidase (I)
L-S-L+	Ester hydrolysis	Microbial lipase (I)
G-L-S	Phenol oxidation	*Mycobacterium*
G-L-S-L+	Benzene hydroxylation	*P. Putida*

Where G,L,L+ and S represent gas, aqueous liquid, organic liquid and solid catalyst phases respectively and (I) represents an immobilized biocatalyst

Multiphasic bioprocessing

Catalyst and substrate, whether dissolved in aqueous phase, dissolved in a carrier organic solvent or forming the non-aqueous phase itself, need to be contacted in the reactor. However, characteristic of multiphasic biotransformations is that there are several options available to achieve effective biocatalysis. This is well illustrated by the hydroxylation of toluene to its cis-glycol catalysed by *Pseudomonas putida*. In this biotransformation the poorly water-soluble organic substrate, toluene, may presaturate the aqueous phase prior to catalyst contact, be fed directly to the aqueous phase reactor, or be introduced via a carrier organic solvent. The conclusion drawn from our evaluation was that toxic effects of the toluene, beneath aqueous phase saturation concentration, were such that introduction by a second liquid phase looks the most attractive option, although research is continuing into the concomitant problems of phase separation (Hack et al, In press). In the region of liquid or gas phase contact presence or absence of solids is important. Solids may be contacted with liquid phase in a packed bed at high concentration. However, if additional phases are present, i.e. a second liquid phase or a gas phase, and are also to be efficiently contacted, dispersion is required and hence the stirred tank is the preferred configuration. This has important implications because it means that two-liquid phase biotranformations cannot be carried out within a packed bed, for example. Channelling leads to very poor interphase transfer. In a stirred tank attrition may occur at too high a concentration (>10% reactor volume) and too high a degree of agitation. Although operation in a stirred tank is flexible (many different types of biotransformation can be run in the same multi-product vessel) there are limitations to phase hold-up dictated by the mixing (eg. Lilly et al, 1990). The general findings for introduction of a gas, organic liquid or solid to an aqueous phase is illustrated in Fig. 1 where the limitations upon hold-up (defined specifically as phase ratio for organic phase) are outlined. Where solid or organic liquid is added to the aqueous phase hold-up is constant in a batch or

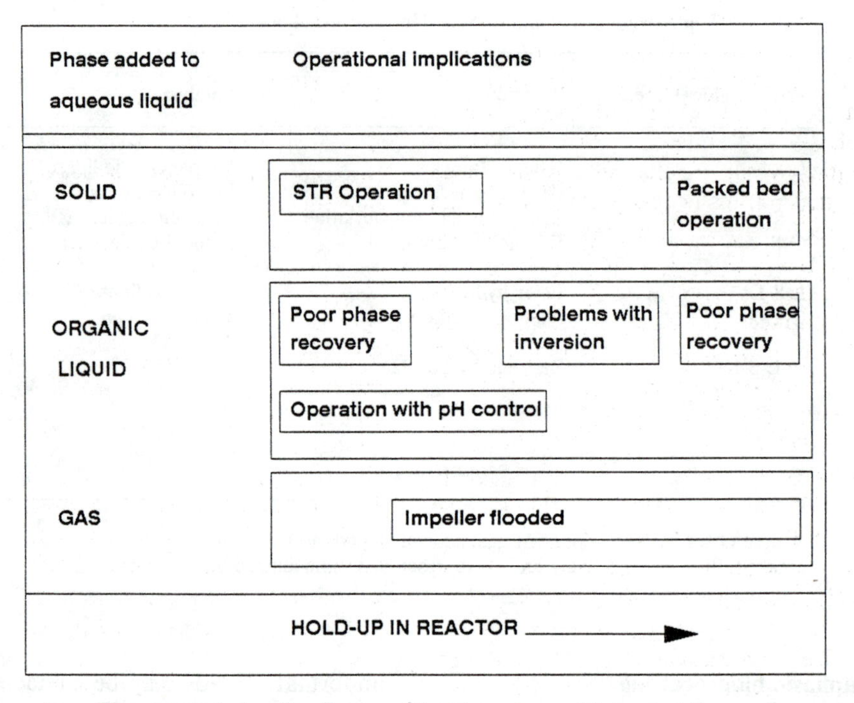

Figure 1: Operational implications of hold-up in multiphasic biotransformations

continuously operated stirred vessel, but gas needs to be continuously supplied to maintain hold-up. This may have implications for stripping organic solvent for example. Equally crucial after phase contacting is the ability downstream to separate the phases. Studies of the pig liver esterase catalysed benzyl acetate hydrolysis clearly reveal the difficulty of aqueous-organic liquid phase separation following agitation at optimum levels for biocatalytic activity (Dhillon et al, In press). Cells too may be damaged by direct aqueous-organic liquid interfacial contact in a mutiphasic medium leading to weakening and subsequent solid-liquid phase separation problems downstream.

Interfacial Effects

If contact of phases is carried out in the presence of the catalyst then interfacial effects will be implicated. The necessity or otherwise of catalyst-interface contact has been the subject of much study. For microbial catalysts studies of the *Arthrobacter simplex* catalysed hydrocortisone Δ^1-dehydrogenation have revealed damaging effects of the aqueous-organic solvent interface upon biocatalyst activity and stability for a range of solvents . Immobilization of the Gram-positive *A.simplex* within calcium alginate beads affords significant protection from direct interfacial contact (Hocknull and Lilly, 1990). However, such improved protection is not so clearly revealed for gram-negative *Pseudomonas putida* used to catalyse the hydroxylation of naphthalene to its corresponding dihydrodiol (Harrop et al, In press).

While most enzymes are severely damaged by both gas-liquid and aqueous-organic liquid interfacial contact (Williams et al, 1987), lipases require the presence of an interface to function efficiently (Kierkels et al, 1990). Understanding interfacial effects upon biocatalysts is crucial to the further development of these reactors and processes. The use of a Lewis cell (defined interface apparatus) as a tool to clarify the rôle of the interface has proved powerful; in particular it can be used as part of a series of experiments to elucidate reaction location (interfacial or within the bulk of a phase) (Woodley, 1990). Where reactions occur in the bulk of a phase then mass transfer becomes key to the observed kinetics. However, whether the reaction occurs at an interface or in the bulk phase, partitioning of components between phases is important. This can be used to advantage for example in partitioning a toxic poorly water-soluble organic substrate away from the biocatalytic reaction zone. Where toxic or inhibitory products are to be removed, mass transfer as well as partitioning will play a rôle.

Future Directions

The technology of multiphasic biotransformation is an expanding area with some important fine chemical targets. The biochemical engineering principles underlying these processes have been outlined in this paper. Despite the development of a set of bioprocess research tools to assist understanding there are still two key issues to be resolved:

* the precise rôle of the interface in both enzymic and microbial catalysis including evaluation of the rôle of the amount and nature of the interface for a given catalyst and the frequency and residence time of the catalyst-interface contact.

* the basis for scale-up of multiphasic biotransformations including knowledge of inhibitory or toxic concentrations to be controlled and rate-limiting interphase transfer steps.

There is an exciting programme of research underway to answer these questions.

References

Dhillon, P.S.; Tyler, D.W.; Woodley, J.M. *Enzyme Microb. Technol.* In Press

Hack, C.J.; Woodley, J.M.; Lilly, M.D.; Liddell, J.M. *Biocatalysis* In press.

Harrop, A.J.; Woodley, J.M.; Lilly, M.D.; *Enzyme Microb. Technol.* In press

Hocknull, M.D.; Lilly, M.D. *Appl. Mirobiol. Biotechnol.* **1990**, 33, 148-153

Lilly, M.D.; Woodley, J.M. In *Biocatalysts in Organic Syntheses*; Tramper, J.; van der Plas, H.C.; Linko, P. Eds.; Elsevier, Amsterdam, The Netherlands, **1985**, pp 179-192

Lilly, M.D.; Dervakos, G.A.; Woodley, J.M. In *Opportunities in Biotransformation*; Copping, L.G.; Martin, R.E.; Pickett, J.A.; Bucke, C.; Bunch, A.W. Eds.; Elsevier, London, UK. **1990**, pp5-16

Lilly, M.D.; Brazier, A.J.; Hocknull, M.D.; Williams, A.C.; Woodley, J.M. In *Biocatalysis in Organic Media*; Laane, C.; Tramper, J.; Lilly, M.D. Eds; Elsevier, Amsterdam, The Netherlands, **1987** pp 3-17

Williams, A.C.; Woodley, J.M.; Ellis, P.A.; Lilly, M.D. In *Biocatalysis in Organic Media*; Laane, C.; Tramper, J.; Lilly, M.D. Eds.; Elsevier, Amsterdam, The Netherlands, **1987**, pp 399-404

Woodley, J.M. In *Biocatalysis*; Abramowicz, D.A. Eds.; Van Nostrand Reinhold, New York, NY, **1990**, pp 337-355

Opportunity for Chiral Synthesis with Redesigned Lactate Dehydrogenases.

Helen M. Wilks, Kathleen M. Moreton, Keith W. Hart, Josep L. Gelpi, Guy M. Casy, Richard B. Sessions, Christine L. Willis, Anthony R. Clarke & J. John Holbrook*, Molecular Recognition Centre, University of Bristol, Bristol BS8 1TD, U.K. (FAX +44 272 303497).

The ability to design enzymes with new substrate specificities enables their use in the synthesis of single enantiomers of intermediates for the production of chiral effects chemicals. The design and construction and use is described of a *'chemist's enzyme'*: that is an enzyme which, together with the wild type enzyme yields 100% ee of a wide range of S-2-hydroxyacids with β- and τ- branched side chains and functionalized allylic alcohols.

It has long been anticipated that the specificity of enzyme catalysis will be exploited in the simple synthesis of enantiomerically pure chemicals needed for modern drugs and other effects chemicals to be used in the environment (De Camp, 1989). Natural enzymes are under evolutionary pressure to transform compounds found in the same cell as the enzyme and in their natural state will only fortuitously be found useful for the chemoenzymic conversion of 'unnatural' intermediates required for chiral pharmaceuticals synthesis. For some years, and in the wake of some failed early attempts at enzyme redesign it was held (reviewed by Knowles, 1985) that enzyme structure was too complex to possibly obtain new catalysis or substrate specificity rationally.

Enzyme Redesign.

That pessimistic view was countered by a 10^7-fold switch in substrate specificity achieved by a single amino acid substitution (Gln102Arg) which transformed a lactate dehydrogenase (LDH) framework into a malate dehydrogenase which was twice as active as that found in the same *B. stearothermophilus* bacterium in nature (Wilks *et al.*, 1988). This is still the largest switch in enzyme specificity (with good catalysis) achieved. Since that time there is an accelerating rate of reporting the successful redesign of enzymes with synthetically-useful properties. From our own laboratory we mention a bacterial enzyme which is active without the addition of the expensive allosteric activator fructose-1,6-bisphosphate (Clarke, *et al.*, (1987), an L-2-hydroxyacid dehydrogenase (Wilks *et al.*, 1990), a thermostable enzyme which can interconvert NAD and NADP for chirals production (Feeney, *et al.*, 1990). Most recently we have made a semi-synthetic specific phenyl-lactate dehydrogenase for monitoring phenylpyruvate in the urine of phenylketonurics on a nominally phenylalanine-free diet (Holbrook *et al.*, 1992). An important commercial redesign recently reported (Meng *et al.*, 1991) is the conversion of a xylose-fructose isomerase into a glucose-fructose isomerase for the final step of the bulk process in making high fructose syrup from enzyme-hydrolysed corn (maize) starch .

R- and S-Hydroxyacids

Chiral 2-hydroxyacids are versatile intermediates for the synthesis of chiral effects chemicals and are easily available from the LDH-catalysed of reduction of 2-ketoacids - themselves available from many simple synthetic routes (Cooper *et al.*, 1983). It would be ideal to have redesigned and overexpressed catalysts available for both the **R**- and S-2-hydroxyacids and although we have a low resolution x-ray structure of the D-LDH from a thermotolerant *Lactobacillus* (Dunn, C.R.; Rawas, A; Holbrook, J.J. unpublished results) the structural detail is not yet sufficient for rational redesign and this present paper is limited to the high resolution structure of the L-LDH from *B. stearothermophilus* (Wigley *et al.*, 1992) with the D-hydroxyacids being obtained in 80% yield and >97% ee in an inversion process developed by C.L. Willis (unpublished) based upon the Mitsunobu reaction.

The Need for a Broad Specificity LDH

In theory at least once the principles of catalyst design are mastered on a particular protein framework it should be possible to produce a catalyst for the reduction of any 2-ketoacid to its S-2-hydroxyacid (Clarke *et al.*, 1991). The thermostable LDH is robust and because it is expressed at 40% of *E. coli* protein is very cheap. Yet it still expensive to have to design a new version of the enzyme for each new target pharmaceutical. We thus conceived of an alternate strategy: the design of a redox catalyst which while retaining absolute S-specificity would have its catalytic vacuole redesigned so that it would accommodate a very wide range of 2-ketoacids - a *'Chemist's Enzyme'*. Thus there would only be a need to design and synthesize the catalyst once and it would be available to reduce many ketoacids to S-2-hydroxyacids.

Design of a *' Chemist's LDH '*.

Substrate specificity and catalysis in L-LDH is best understood from the presence within the enzyme of a catalytic vacuole (Holbrook *et al.*, 1975; Dunn *et al.*, 1991). Only small and correctly charged substrates allow the protein to engulf the substrate in an internal vacuole that is isolated from solvent protons, in which water is frozen and hydride transfer is rapid. Solvent friction limits the rate of forming the vacuole and, with good substrates, the maximum rate of catalysis. Because maximum rate is limited by physics, and not appreciably by amino acid sequence (see Wilks *et al.*, 1992) it has been possible to alter the sequence of the surface of this vacuole to obtain varied substrate specificities without too great a risk that the physical, viscosity-limited, maximum rate of catalysis will be altered.

Table 1. Reductions to S-2-hydroxyacids.

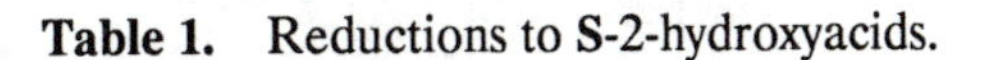

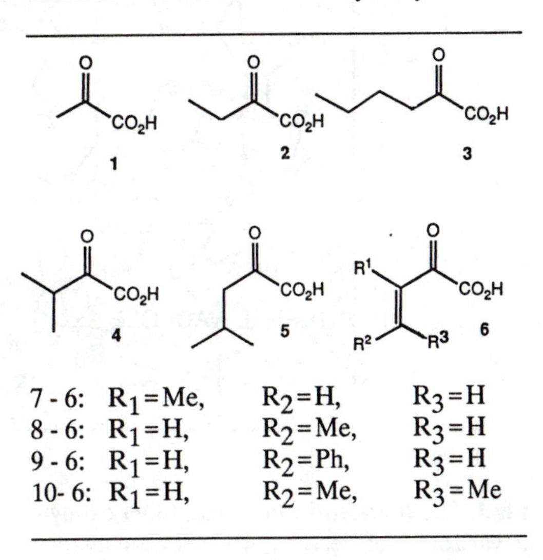

7 - 6:	R_1=Me,	R_2=H,	R_3=H
8 - 6:	R_1=H,	R_2=Me,	R_3=H
9 - 6:	R_1=H,	R_2=Ph,	R_3=H
10- 6:	R_1=H,	R_2=Me,	R_3=Me

Fig. 1 shows a cross section through the *B. stearothermophilus* L-LDH catalytic vacuole. The size of substituted pyruvates (Table 2) which can be accommodated in the natural vacuole is limited. Most enzymes show very low activity with long (**3**) and branched ketoacids (**4, 5, 6-10** e.g. Table 1). To obtain a broad specificity enzyme it was necessary to generally enlarge the size of the internal catalytic vacuole and in such a way that there was sufficient flexibility to enable the protein to tightly close around a wide range of side chain sizes. Because of the extreme sensitivity of the closed vacuole to charge imbalance we have made no attempt in this initial design to make a vacuole which

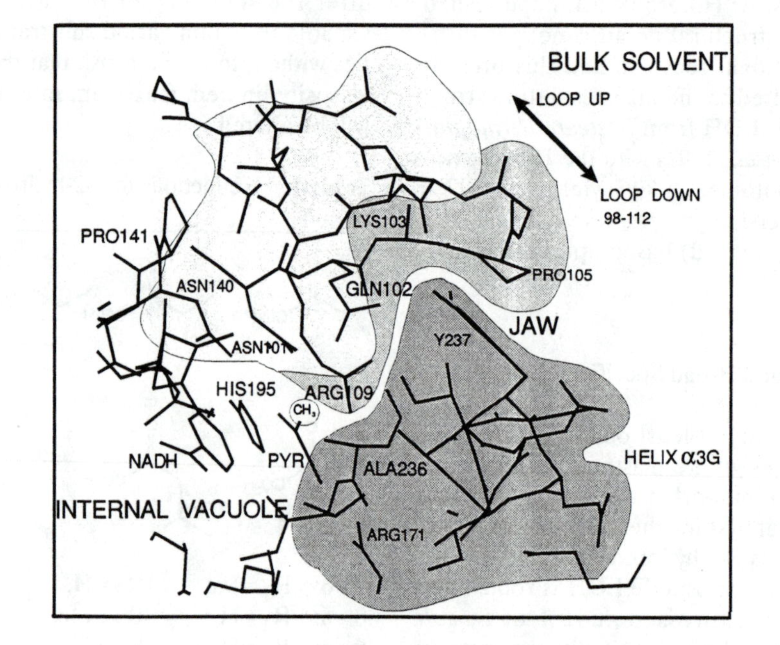

Fig.1. The *B. stearothermophilus* LDH catalytic vacuole. For design purposes we consider the catalytic vacuole to contain a *catalytic pathway*: that is the amino acids Arg109 (sub strate >C=O polarization), Arg171 (main binding of the α-carboxyl), Asn141 and Thr246 (H-bonds to substrate >C=O and α-carboxyl), His195-Asp168 (the couple which acts as the proton source/sink) which are almost invariant in all LDHs and a *substrate specificity region*: consisting of the underside of the βD αD loop (residues 99-108 *n.b.* no residue is named 104) particularly Gln102 and Asn101, and the surface of helices α2G/α-3G CH_3 of Thr246, Ala235 and Ala236. Specificity residues vary widely in different LDHs.

would, without protein redesign, accommodate both neutral and charged substrate side chains.

The design process involved building the *ciscoid* isomer of 4-methyl-2-oxo-3-pentenoate into the x-ray coordinates of the crystal complex of *B. stearothermophilus* L-LDH with NADH, oxamate and fructose-1,6-bisphosphate (FBP). The catalytic pathway (NADH C4: Arg171: Arg109: Asn140: Asp168: Thr246 and the glyoxylate moiety of the substrate) was held at the same position as in the complex with oxamate. The *ciscoid* rotamer of **10** was calculated *in vacuuo* to be more stable by 2.4 kCal than other rotamers. Using the Biosym Associates display program INSIGHT running on a Silicon Graphics SG35 computer the protein side chains which collided with the new substrate were noted. Chemical intuition was then used to replace these with other natural amino acid side chains to avoid the side chain clashes with the new substrate. Both the original native enzyme with pyruvate and the mutant ^{102,103,105}Gln,Lys,Pro^{235,236}Ala,Ala ---> ^{102,103,105}Met,Val,Ser^{235,236}Gly,Gly were then energy minimized using Biosym Associates program DISCOVER v. 2.5. Only a sphere of atoms within 14Å of the carbonyl oxygen of the substrate were allowed to move. To the frozen waters observed in the original crystal structure (Wigley *et al.*, 1992) were added random waters to fill all non-protein volume within 22Å of the substrate carbonyl. Minimization used an all atom consistent force field with a cutoff of 10Å. The relaxation process enabled the clash with the terminal CH_3- of the substrate to be relieved in the Met102 mutant, but not with the native sequence.

Table 2 Catalytic Properties of Enzymes

Substrate	Enzyme	$k_{cat}(s^{-1})$	K_M(mM)
1[a]	Native	250	0.06
1[a]	MVSGG	32	4
2[a]	Native	186	2.1
2[a]	MVSGG	1.1	3.3
3[a]	Native	29	3.4
3[a]	MVSGG	185	7.1
4[a]	Native	0.26	22
4[a]	MVSGG	11.3	22
5[a]	Native	0.33	6.7
5[a]	MVSGG	18.5	14.3
7[a]	Native	2	7
8[b]	Native	6.1	4.0
10	Native	<0.03	large
10	MVSGG	32	4

From [a]Wilks *et al.*, 1990; [b]Casy *et al.*, 1992.

Table 3. Chiral Alyllic Alcohol Preparations.

Cmpnd 6 R_1, R_2, R_3	Enz.	Conc$^{n.}$ (mM)	Days	Yield %	ee %S
Me, H, H	W.T.	15mM	2.8	60	>99[a]
H, Me, H	W.T.	50mM	1.2	98	>97[a]
H, Ph, H	W.T.	20mM	2.3	85	>98[a]
H, Me,Me	MVS	15mM	5	91	>99[b]

[a]Casy *et al.*, 1992. [b]Reaction conditions: 6 units MVSGG LDH (Wilks *et al.*, 1990), 5 units yeast FDH, 1 mmol **6**(R_1=H, R_2=Me, R_3=Me), NADH (0.02 mmol), Na formate (3.1 mmol), Fru-1,6-BP (0.4 mmol) in 80ml 5 mM Tris-HCl pH 6 under N_2 in a pH stat. At 5 days acidify to pH 2, and extract with ethyl acetate.

The ^{235,236}Ala,Ala ---> ^{235,236}Gly,Gly mutations were inserted to allow helix α2G to become more flexible in an attempt to obtain mutations which would facilitate the binding not only of *ciscoid* isomer of 4-methyl-2-oxo-3-pentenoate but to a wide range of γ-branched substrates. The results of Table 1 (**3,4,5,10**) indicate that this objective was broadly attained. The present state of modelling is that when known active mutant enzyme:novel substrate complexes are subjected to 50 ps of molecular dynamics in water then the substrate remains in the catalytic vacuole with the vacuole (Dunn *et al.*, 1991) intact. For known inactive mutant enzyme:

novel substrate pairs the dynamics results in a collapse of the active centre vacuole (Clarke *et al.*, 1991).

Validity of the design was tested by a preparative scale chiral reductions of 4-methyl-2-oxo-3-pentenoate and other branched chain ketoacids in experiments where reducing equivalents were provided by formate. The results (Table 3) demonstrate the efficient syntheses of a range of novel allylic 2-hydroxyacids which are of considerable potential value for chiral synthesis of pharmaceuticals and effects chemicals.

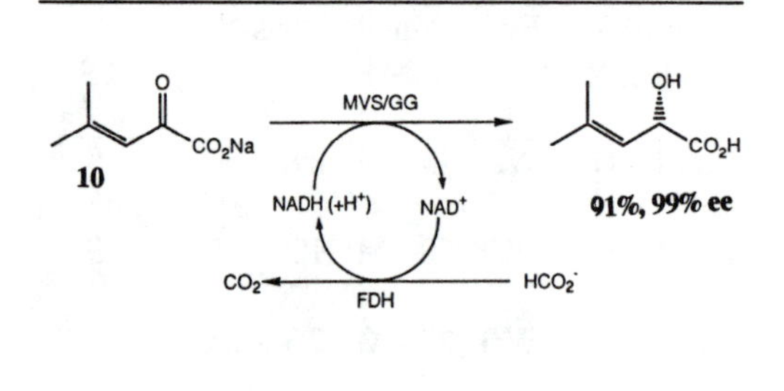

We thank The SERC and The British Council (Acciones Integradas) for support.

References.

Bur, D.; Luyten, M. A.; Wynn, H.; Provencher, L.R.; Jones, J. B.; Gold, M.; Friesen, J.D.; Clarke, A.R.; Holbrook, J. J. *Can. J. Chem.* **1989**, *67*, 1065-70.

Casy, G.; Lee, T.V.; Lovell, H. *Tetrahedron Lett.* **1992**, *33*, 817-20

Clarke, A.R.; Colebrook, S.; Cortes, A.; Emery, D.C.; Halsall, D.J.; Hart, K.W.;Jackson, R.M.; Wilks H.M.; Holbrook, J.J. *Biochem. Soc. Transact.* **1991**, *19*, 576-581

Clarke, A.R.; Wigley, D.B.; Barstow, D.A.; Chia, W.N.; Atkinson, T.; Holbrook, J.J. *Biochim. et Biophys. Acta* **1987**, *913*, 72-80

Clarke, A.R.; Wigley, D.B.; Chia, W.N.; Barstow, D.; Atkinson, T.; Holbrook, J.J. *Nature* **1986**, *324*, 699-702.

Cooper, A.J.L.; Ginos, J.Z.; Meister, A. *Chem. Rev.* **1983**, *83*, 321-358

De Camp, W.H. *Chirality* **1989**, *1*, 2-6.

Dunn, C.R.; Wilks, H.M.; Halsall, D.J.; Atkinson, T.; Clarke, A.R.; Muirhead, H.; Holbrook, J.J. *Phil. Trans Roy. Soc. (London)* **1991**, *332*, 177-185

Hart, K.W.; Clarke, A.R.; Wigley, D.B.; Waldman, A.D.B.; Chia, W.N.; Barstow, D.A.; Atkinson, T.; Jones, J.B.; Holbrook, J.J. *Biochim. et Biophys. Acta* **1987**, *914*, 294-298

Feeney, R.; Clarke, A.R.; Holbrook, J.J. *Biochem. Biophys. Research Commun.* **1989**, *166*, 667-672

Holbrook, J.J.; Wilks, H.M.; Clarke, A.R. *U.K. Patent Application* **1992** No. GB9202033.8.

Holbrook, J.J.; Liljas, A;. Steindel, S.J.; Rossmann, M.G. *The Enzymes* **1975**, *11a*, 191-293.

Knowles, J.R. *Science*, **1985**, *236*, 1252-1255

Meng, M.; Lee, C.; Bagdasarian, M.; Zeikus, J.G. *Proc Natl Acad Sci U S A*, **1991**, *88*, 4015-4019

Wigley, D.B.; Gamblin, S. J.; Turkenburg, J. P.; Dodson, E. J.; Piontek, K.; Muirhead, H.; Holbrook, J. J. *J. Molec. Biol.* **1992**, *223*, 317-335

Wilks, H.M.; Halsall, D.J.; Atkinson, T.; Chia, W.N.; Clarke, A.R.; Holbrook, J.J. *Biochemistry* **1990**, *89*, 8587-8591.

Wilks, H.M; Cortes, A.; Emery, D.C.; Halsall, D.J.; Clarke, A.R.; Holbrook, J.J *Annals N. Y. Acad. Sci.* **1992** Accepted November 1991

Wilks, H.M.; Hart, K.W.; Feeney, R.; Dunn, C.R.; Muirhead. H., Chia, W.N.; Barstow, D.A.; Atkinson, T.; Clarke, A.R.; Holbrook, J.J. *Science* **1988**, *242*, 1541-4.

Mechanism of the Reaction Catalyzed by Mandelate Racemase: Lessons from Protein Engineering, Chemical Modification, and X-Ray Crystallography

John A. Gerlt and **John W. Kozarich**, Department of Chemistry and Biochemistry, University of Maryland, College Park, MD 20742
George L. Kenyon, Department of Pharmaceutical Chemistry, School of Pharmacy, University of California, San Francisco, CA 94143
David J. Neidhart, Abbott Laboratories, Abbott Park, IL
Gregory A. Petsko, Rosenstiel Basic Medical Sciences Research Center, Brandeis University, Waltham, MA 02254

Many enzyme-catalyzed reactions involve abstraction of a weakly acidic proton from a carbon adjacent to a carbonyl/carboxylic acid group by a weakly basic functional group in the active site. Using mandelate racemase as a prototype for these reactions, we have investigated the mechanism by which a substrate α-proton having a pK_a of 29 in solution can be rapidly abstracted (k_{cat} = 700 sec^{-1}) by a base having a apparent pK_a of ~6. We believe that the rapid abstraction of the α-proton by either the amino group of a Lys residue or the imidazole group of a His residues is critically dependent upon interaction of the carboxylate group with an essential Mg^{2+}, the ε-ammonium group of a Lys residue, and the protonated carboxylic acid group of a Glu residue.

Pseudomonads can utilize a wide variety of organic compounds as sole carbon and energy sources. The utilization of **R**-mandelic acid by *Pseudomonas putida* is dependent upon its conversion to benzoic acid by the enzymes of the mandelate pathway (Fig. 1); benzoic acid is converted to acetyl CoA and succinyl CoA by enzymes of the β-ketoadipate pathway (Fig. 2).

We have cloned and sequenced a restriction fragment encoding the genes of the mandelate pathway (Tsou et al., 1990). The enzymes of mandelate metabolism are homologous to other enzymes: **S**-mandelate dehydrogenase is homologous to glycolate oxidase (spinach), ferricytochrome b_2 (yeast), and lactate monooxygenase; benzoylformate decarboxylase is homologous to pyruvate decarboxylase, pyruvate oxidase, and acetolactate synthase; and the NAD^+/$NADP^+$-dependent benzaldehyde dehydrogenase is homologous to a large family of aldehyde dehydrogenases. We did not expect to find an open reading frame encoding a mandelamide hydrolase, whose identity was deduced by its homology to other amide hydrolases. Nor did we anticipate the surprising discovery that mandelate racemase would be homologous to muconate lactonizing enzyme (in the β-ketoadipate pathway) (Neidhart et al., 1990). Given these homologies, our studies of the enzymology of mandelate metabolism in *P. putida* is relevant to many problems in mechanistic enzymology. Although we are studying mechanistic questions about each of these enzymes, this article will focus on our studies on the mechanism of the reaction catalyzed by mandelate racemase. In particular, we are concerned with understanding how a weakly basic catalyst in the active site (pK_a ~ 6) can abstract the very weakly acid α-proton (pK_a ~ 29) at a very rapid rate (k_{cat} = 700 sec^{-1}).

The Active Site of Mandelate Racemase

Large amounts of mandelate racemase are readily available since the gene can be expressed at high levels in *P. aeruginosa*, a Pseudomonad whose genome does not encode a mandelate racemase. The homogeneous enzyme has been crystallized, and, at present, the structure has been refined at 1.8 Å resolution (Neidhart et al., 1991). While diffusion of mandelate into the crystals has been difficult (since the enzyme is crystallized from $(NH_4)_2SO_4$ and sulfate is a competitive inhibitor), it has nevertheless been possible to solve the structure of a complex of the enzyme with **R**-atrolactate and of the covalent adduct that is formed by irreversible alkylation of

1054–7487/92/0155$06.00/0

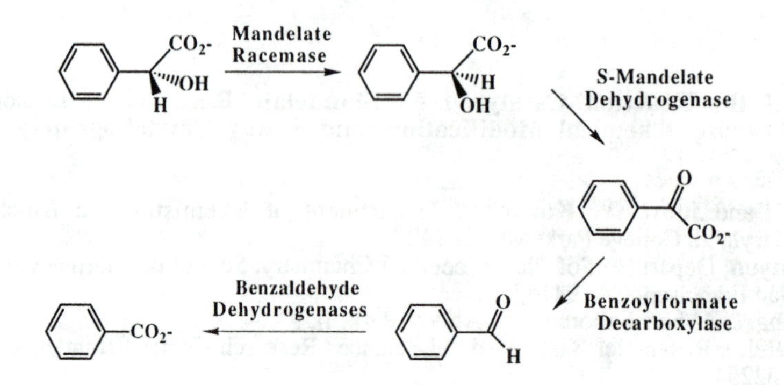

Fig. 1. The mandelate pathway.

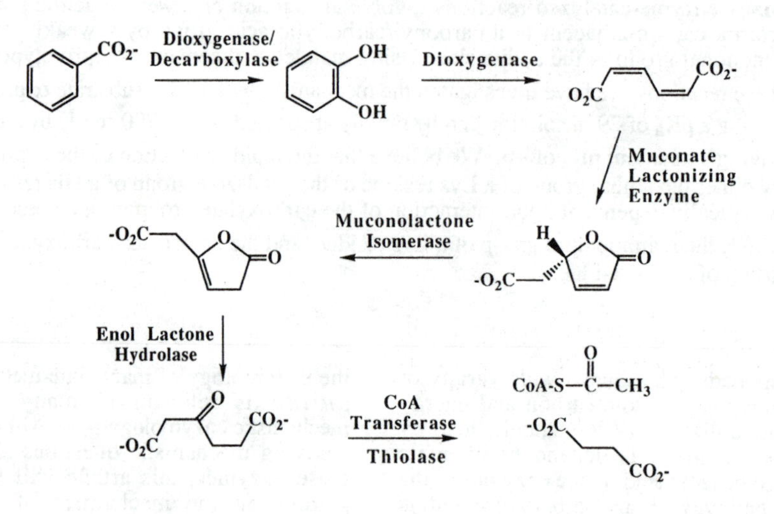

Fig. 2. The β-ketoadipate pathway.

the enzyme with **R**-α-phenylglycidate (*vide infra*) (Landro et al., 1992). An schematic diagram of the active site is shown in Fig. 3; crucial active site residues are identified, along with assignments for their functions in catalysis. The data supporting these assignments will be summarized in the remainder of this article

Two Basic Catalysts for C-H Bond Cleavage

The high resolution structure reveals that the ε-amino group of Lys 166 and the imidazole group of His 297 are appropriately located to act as the acid/base catalysts in C-H bond making and breaking. Lys 166 is the **S**-specific base, and His 297 is the **R**-specific base. These assignments are consistent with chemical data which are consistent with the conjugate acid of the **S**-specific base being polyprotic and the conjugate acid of the **R**-specific base being monoprotic (Powers et al., 1991).

We have generated and characterized the mutant in which His 297 is replaced with Asn (H297N) (Landro et al., 1991). The structure of H297N has been determined to 2.2 Å resolution; no distortion of the active site by the substitution was detectable. H297N is catalytically inactive with respect to racemization of either enantiomer of mandelate. However, it catalyzes the elimination of bromide ion from **S**- but not **R**-(*p*-bromomethyl) mandelate at a rate comparable to that of wild type enzyme with both enantiomers. More importantly, it catalyzes the exchange of the α-proton from **S**- but not **R**-mandelate at a rate comparable to that of wild type enzyme with either enantiomer. The observation of exchange in the absence of racemization is persuasive evidence that a transiently stable intermediate is generated in the active site. This implies that the racemization reaction catalyzed by wild type enzyme also involves the formation of a stabilized intermediate; kinetic isotope effect studies support a stepwise rather than concerted mechanism for the racemization reaction catalyzed by wild type enzyme. [The identity of this intermediate will be discussed in a later section.] The partial reactions catalyzed by H297N also support the proposal that two acid/base catalysts (Lys 166 and His 297) are required for racemization of mandelate.

We have not yet been successful in the comparable inactivation of Lys 166 by mutagenesis. However, we have obtained evidence for its participation in catalysis by characterizing the nature of the inactivation of wild type enzyme by α-phenylglycidate (Landro et al., 1992). The enantiomers of α-phenylglycidate have been separately prepared in enriched forms; note that **R**-α-phenylglycidate has the same relative configuration as **S**-mandelate (Fig. 4). Only the **R**-enantiomer of α-phenylglycidate irreversibly alkylates and inactivates mandelate racemase; the **S**-enantiomer of is not an covalent inactivator. The structure of the inhibited enzyme has been determined at 2.2 Å resolution. This structure clearly reveals that the ε-amino group of Lys 166 has been alkylated, with opening of the epoxide ring (Fig. 5). Thus, the product of this reaction is effectively **S**-mandelate covalently attached to Lys 166 through a methylene bridge. This demonstrates that the ε-amino group of Lys 166 is unusually reactive, as might be expected for an essential acid/base catalyst.

An Acid Catalyst for C-H Bond Cleavage

The structure of the enzyme inhibited with **R**-α-phenylglycidate has been particularly informative since the sulfate anion present in the structure of wild type racemase is displaced by the mandelate covalently attached to Lys 166; thus, the interactions of mandelate with active site residues can be visualized rather than modeled. [A remarkably similar structure was obtained for **R**-atrolactate bound in the site, so we believe that the structural features that are observed and the mechanistic hypotheses they suggest are quite pertinent to the mechanism of the reaction.]

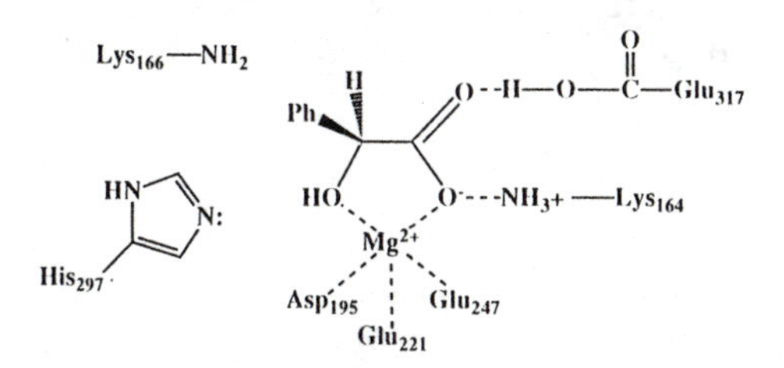

Fig. 3. A schematic diagram of the active site of mandelate racemase.

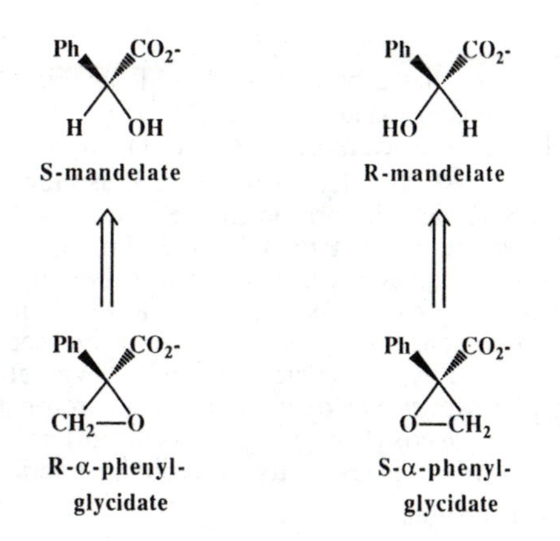

Fig. 4. A comparison of the structures of the enantiomers of mandelate and α-phenylglycidate.

As indicated in the schematic diagram shown in Fig. 3, the α-hydroxyl group and one of the carboxylate oxygens of mandelate are coordinated to the essential Mg2+ (which is coordinated (monodentate) to the carboxylate groups of Asp 195, Glu 221, and Glu 247). Thus, that the α-hydroxyl group of mandelate is absolutely essential for reaction can be understood. The metal-coordinated carboxylate oxygen is also hydrogen bonded to the ε-ammonium group of Lys 164.

Curiously, the second carboxylate oxygen of mandelate is within hydrogen bonding distance

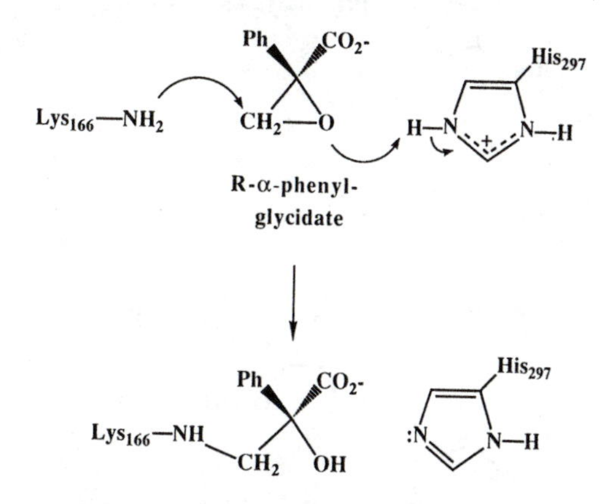

Fig. 5. The mechanism of inactivation of mandelate racemase by α-phenylglycidate.

of the γ-carboxylate group of Glu 317. This requires that a proton be shared by the metal coordinated mandelate and Glu 317. Since the charge on the carboxylate group is neutralized by the Mg^{2+} and Lys 164, it is reasonable to propose that the proton in the hydrogen bond would be located on Glu 317, i.e., Glu 317 exists as the acid when substrate is bound. The crucial question is "Is there a mechanistic requirement for a proton source in the reaction catalyzed by mandelate racemase?" We believe the answer to this question is "Yes, protonation of the carboxylate group to form an enol(ate) intermediate is necessary for rapid abstraction of the α-proton."

Proposed Mechanism for the Reaction Catalyzed by Mandelate Racemase

We have pointed out that thermodynamics demands that the pK_a of the α-proton of an aldehyde, ketone, thioester, or carboxylic acid is reduced by ~15 pK_a units by concerted protonation of the carbonyl/carboxylic acid group (Gerlt et al., 1991). Since the pK_a of the α-proton of mandelic acid is 22.0 (Chiang et al., 1990), general acid catalysis would reduce the effective pK_a of this proton to ~7. From the pH dependence of both kcat for wild type enzyme and the rate of exchange of the α-proton of **S**-mandelate catalyzed by H297N, we have estimated that the pK_as of the lysine and histidine bases in the active site are ~6. Thus, acid catalysis reduces the pK_a of the basic catalysts involved in C-H bond breaking/making. The corollary to the involvement of concerted abstraction of the α-proton and protonation of the Mg^{2+}-coordinated carboxylate group is that the reaction proceeds *via* a Mg^{2+}-coordinated enol/enolate intermediate (the Mg^{2+}-coordinated carboxylate oxygen is unprotonated; the carboxylate oxygen which was originally hydrogen bonded to the carboxylic acid functional group of Glu 317 is protonated in the intermediate).

Thus, we propose that mandelate racemase catalyzes facile abstraction of the α-proton of mandelate by a concerted general acid-general base catalyzed mechanism that leads to the formation of an enol/enolate intermediate. By microscopic reversibility, the ketonization of the enol/enolate intermediate will also be general acid-general base catalyzed, with the carboxylate group of Glu 317 serving as the general basic catalyst and the conjugate acid of either Lys 166 or His 297 serving as the general acidic catalyst (Fig. 6). Note that when the enol/enolate intermediate is present in the active site, both Lys 166 and His 297 would be protonated, so the ketonization reaction occurs in a stereorandom manner.

We are investigating various aspects of this mechanism, including seeking evidence for the participation of Glu 317 as an acidic catalyst and providing an explanation for the unexpectedly low pK_a for the ε-ammonium group of Lys 166

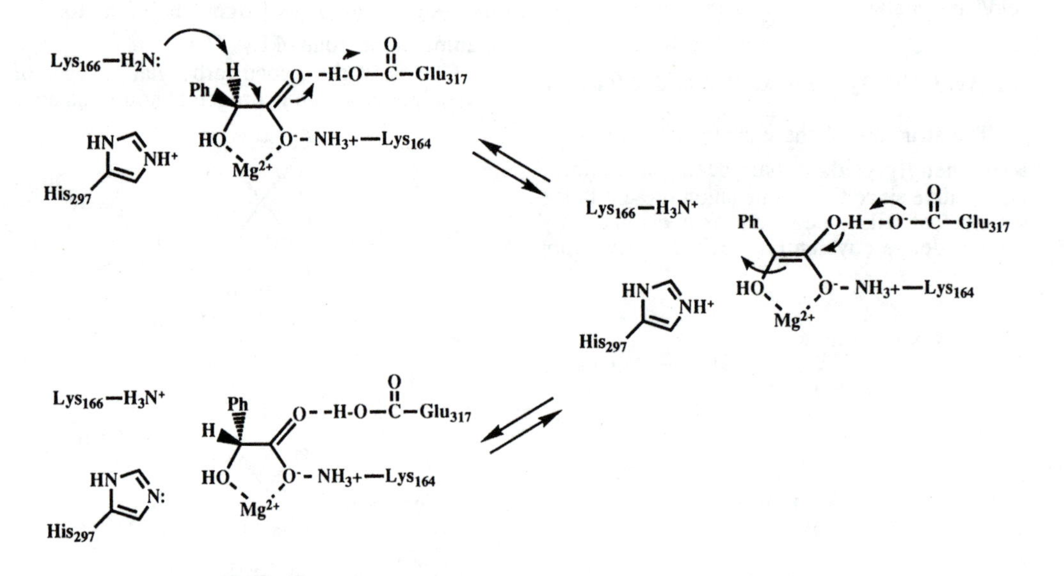

Fig. 6. Proposed mechanism of the reaction catalyzed by mandelate racemase.

(~ 6), or, alternatively, the "normal" pK_a for the imidazolium group of His 297 (also ~ 6).

Acknowledgment

This research was supported by NIH AR-17323, GM-34572, and GM-40570.

References

Chiang, Y.; Kresge, A. J.; Pruszynski, P.; Schepp, N. P.; Wirz, J. *Angew. Chem. Int. Ed. Engl.* **1990**, *29*, 792.

Gerlt, J. A.; Kozarich, J. W.; Kenyon, G. L.; Gassman, P. G. *J. Am. Chem. Soc.* **1991**, *113*, 9667.

Landro, J. A.; Kallarakal, A.; Ransom, S. C.; Gerlt, J. A.; Kozarich, J. W.; Neidhart, D. J.; Kenyon, G. L. *Biochemistry* **1991**, *30*, 9274.

Landro, J. C.; Neidhart, D. J.; Kenyon, G. L.; Kozarich, J. W.; Gerlt, J. A. *Biochemistry*, submitted for publication.

Neidhart, D. J.; Kenyon, G. L.; Gerlt, J. A.; Petsko, G. A. *Nature* **1990**, *347*, 692.

Neidhart, D. J.; Howell, P. L.; Petsko, G. A.; Powers, V. M.; Li, R.; Kenyon, G. L.; Gerlt, J. A. *Biochemistry* **1991**, *30*, 9264.

Powers, V. M.; Koo, C. W.; Kenyon, G. L.; Gerlt, J. A.; Kozarich, J. W. *Biochemistry* **1991**, *30*, 9255.

Tsou, A. Y.; Ransom, S. C.; Gerlt, J. A.; Buechter, D.; Babbit, P.; Kenyon, G. L. *Biochemistry* **1990**, *29*, 9856.

Optical Resolution of Pantoyl Lactone by a New Fungal Lactonase

Hideaki Yamada*[1,2], Sakayu Shimizu[1], Michihiko Kataoka[1], Keiji Sakamoto[1], Kentaro Shimizu[1], and Masao Hirakata[1],
[1]Department of Agricultural Chemistry, Kyoto University, Sakyo-ku, Kyoto 606 (FAX: 8175 753 6128), and [2]Department of Biotechnology, Kansai University, Suita, Osaka 564, Japan

Kinetic resolution of DL-pantoyl lactone can be carried out by a specific fungal hydrolase. We found that many mold strains belonging to the genera *Fusarium*, *Gibberella* and *Cylindrocarpon* specifically hydrolyze D-(-)-pantoyl lactone. When *Fusarium oxysporum* cells were incubated with DL-pantoyl lactone under automatic pH control (pH 6.8-7.2), the D-(-)-isomer was efficiently hydrolyzed. The resultant D-(-)-pantoic acid in the reaction mixture showed a high optical purity, and the coexisting L-(+)-isomer remained without any modification.

Commercial production of pantothenate depends exclusively on chemical synthesis. The conventional chemical process involves reactions yielding racemic pantoyl lactone from isobutyraldehyde, formaldehyde and cyanide, optical resolution of the racemic pantoyl lactone to D-(-)-pantoyl lactone with quinine, quinidine, cinchonidine, brucine and so on, and condensation of D-(-)-pantoyl lactone with β-alanine. A problem of this chemical process is the troublesome resolution of the racemic pantoyl lactone and the racemization of the remaining L-(+)-isomer. Therefore, most of recent studies in this area have been concentrated to development of efficient method to obtain D-(-)-pantoyl lactone. We have reported that there are several microbial reactions useful for the synthesis of chiral intermediates for D-pantothenate production (Fig. 1; Hata *et al.*, 1987, 1989, 1990; Kataoka *et al.*, 1990a, b, c, 1991, 1992a, b, c; Shimizu & Yamada, 1986, 1989, 1990; Shimizu *et al.*, 1984, 1987a, b, c, 1988a, b; Yamada & Shimizu, 1985a, b, 1988, 1992). In the present paper, a stereospecific hydrolysis of DL-pantoyl lactone to D-(-)-pantoic acid using a microbial lactonase is proposed.

Kinetic resolution of DL-pantoyl lactone can be carried out by specific fungal hydrolases. We found that many mold strains belonging to the genera *Fusarium*, *Gibberella* and *Cylindrocarpon* specifically hydrolyze D-(-)-pantoyl lactone to D-(-)-pantoic acid (Fig. 1, reaction ⑦). On the other hand, several yeast strains hydrolyzed only the L-(+)-isomer (Fig. 1, reaction ⑧). For practical purposes, the former reaction is more advantageous than the latter, because, in the latter case, optical purity of the remaining D-(-)-pantoyl lactone is low unless the hydrolysis of L-(+)-pantoyl lactone is complete. Among various *Fusarium* strains tested, *Fusarium oxysporum* AKU 3702 showed the highest hydrolysis activity and gave D-(-)-pantoic acid of high optical purity (>95% *ee*). When *Fusarium oxysporum* cells were incubated in 70% (w/v) aqueous solution of DL-pantoyl lactone for 24 hours at 30°C with automatic pH control (pH 6.8-7.2), about 90% of the D-(-)-isomer was hydrolyzed. The resultant D-(-)-pantoic acid in the reaction mixture showed a high optical purity (96% *ee*) and the coexisting L-(+)-isomer remained without any modification (Shimizu & Yamada, 1990; Yamada & Shimizu, 1992).

The enzyme responsible for this hydrolysis was isolated from *Fusarium oxysporum* cells and crystallized. It is a kind of aldonolactonase with a molecular mass of 125,000. The enzyme is composed of two identical subunits, each of which contains 1-2 mol of Ca^{2+} and about 15% carbohydrate. Ca^{2+} is necessary for the enzyme activity. Ca^{2+} also plays an important role as a

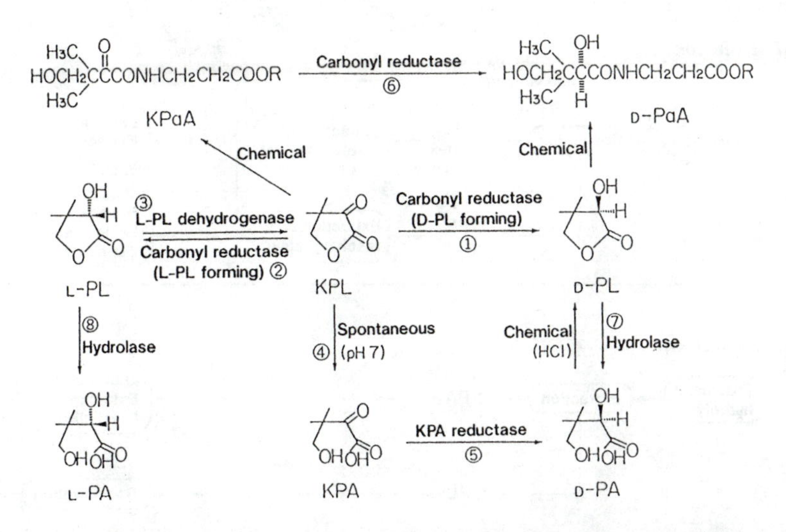

Fig. 1. Reactions involved in the enzymatic transformation to D-(-)-pantoyl lactone or D-(+)-pantothenate. D-PL, D-(-)-pantoyl lactone; L-PL, L-(+)-pantoyl lactone; KPL, ketopantoyl lactone; KPA, ketopantoic acid; D-PA, D-(-)-pantoic acid; L-PA, L-(+)-pantoic acid; KPaA, 2'-ketopantothenate; D-PaA, D-(+)-pantothenate.

stabilizer of the enzyme. Mannose is a major component of the carbohydrate. The enzyme catalyzes reversible hydrolysis of several sugar lactones, such as D-galactonolactone, L-mannonolactone, D-gulonolactone and D-gluconolactone (Fig. 2). The enzyme greatly favors the hydrolytic direction under the neutral or mild alkaline conditions. Reaction equilibrium at pH 6.0 is about 50% when D-(-)-pantoyl lactone is the substrate. Several aromatic lactones, *i.e.*, dihydrocoumarin, homogentisic acid lactone and 2-coumaranone, are also good substrates of the enzyme. All the sugar lactones which serve as substrate have a downward hydroxyl group at the 2-position, when the lactone rings are drawn according to the Haworth system. The corresponding enantiomers are competitive inhibitors.

Practical hydrolysis of the D-(-)-isomer in a racemic mixture is carried out using immobilized mycelia of *Fusarium oxysporum* as the catalyst. Stable catalyst with high hydrolytic activity can be prepared by entrapping the fungal mycelia into calcium alginate gels. When the immobilized mycelia were incubated in a reaction mixture containing 400 g/l DL-pantoyl lactone for 16 hours at 30°C under the conditions of automatic pH control, 90-95% of the D-(-)-isomer was hydrolyzed (optical purity, 90-95% *ee*). After repeated reactions for 100 times (*i.e.*, 100 days), the immobilized mycelia retained more than 90% of their initail activity (Yamada & Shimizu, 1992).

Comparison of the enzymatic process proposed by the present study and the conventional chemical process for the resolution of DL-pantoyl lactone is shown in Fig. 3.

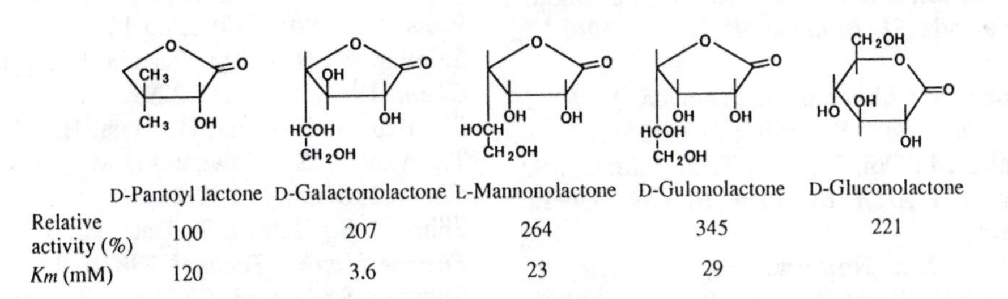

	D-Pantoyl lactone	D-Galactonolactone	L-Mannonolactone	D-Gulonolactone	D-Gluconolactone
Relative activity (%)	100	207	264	345	221
Km (mM)	120	3.6	23	29	

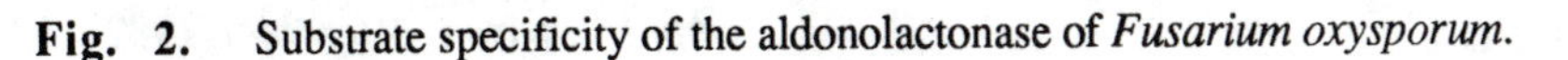

Fig. 2. Substrate specificity of the aldonolactonase of *Fusarium oxysporum*.

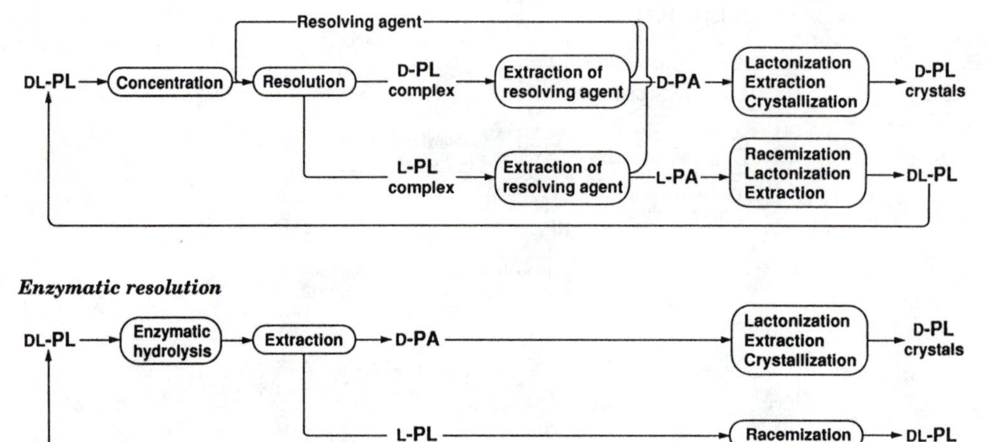

Fig. 3. Comparison of enzymatic and conventional chemical resolution processes for DL-pantoyl lactone. DL-PL, DL-pantoyl lactone; D-PL, D-(-)-pantoyl lactone; L-PL, L-(+)-pantoyl lactone; D-PA, D-(-)-pantoic acid; L-PA, L-(+)-pantoic acid.

References

Hata, H.; Shimizu, S.; Yamada, H. *Agric. Biol. Chem.* **1987,** *51,* 3011-3016.

Hata, H.; Shimizu, S.; Hattori, S.; Yamada, H. *Biochim. Biophys. Acta* **1989,** *990,* 175-181.

Hata, H.; Shimizu, S.; Hattori, S.; Yamada, H. *J. Org. Chem.* **1990,** *55,* 4377-4380.

Kataoka, M.; Shimizu, S.; Yamada, H. *Agric. Biol. Chem.* **1990a,** *54,* 177-182.

Kataoka, M.; Shimizu, S.; Doi, Y.; Yamada, H. *Appl. Environ. Microbiol.* **1990b,** *56,* 3595-3597.

Kataoka, M.; Shimizu, S.; Doi, Y.; Sakamoto, K.; Yamada, H. *Biotechnol. Lett.* **1990c,** *12,* 357-360.

Kataoka, M.; Shimizu, S.; Yamada, H. *Recl. Trav. Chim. Pays-Bas* **1991,** *110,* 155-157.

Kataoka, M.; Doi, Y.; Sim, T.-S.; Shimizu, S.; Yamada, H. *Arch. Biochem. Biophys.* **1992a,** in press.

Kataoka, M.; Nomura, Y., Shimizu, S.; Yamada, H. *Biosci. Biotech. Biochem.* **1992b,** in press.

Kataoka, M.; Shimizu, S.; Yamada, H. *Eur. J. Biochem.* **1992c,** in press.

Shimizu, S.; Yamada, H. In *Biotechnology;* Rehm, H.-J.; Reed, G., Eds.; VCH Verlagsgesellschaft: Weinheim, **1986,** Vol. 4, pp 159-184.

Shimizu, S.; Yamada, H. In *Biotechnology of Vitamins, Pigments and Growth Factors;* Vandamme, E.J., Ed.; Elsevier Applied Science: London, **1989,** pp 199-219.

Shimizu, S.; Yamada, H. In *Chemical Aspects of Enzyme Biotechnology: Fundamentals;* Baldwin, T.O.; Raushel, F.M.; Scott, A.I., Eds.; Plenum Press, New York, **1990,** pp 151-163.

Shimizu, S.; Hata, H.; Yamada, H. *Agric. Biol. Chem.* **1984,** *48,* 2285-2291.

Shimizu, S.; Yamada, H.; Hata, H.; Morishita, T.; Akutsu, S.; Kawamura, M. *Agric. Biol. Chem.* **1987a,** *51,* 289-290.

Shimizu, S.; Hattori, S.; Hata, H.; Yamada, H. *Enzyme Microb. Technol.* **1987b,** *9,* 411-416.

Shimizu, S.; Hattori, S.; Hata, H.; Yamada, H. *Appl. Environ. Microbiol.* **1987c,** *53,* 519-522.

Shimizu, S.; Hattori, S.; Hata, H.; Yamada, H. *Eur. J. Biochem.* **1988a,** *174,* 37-44.

Shimizu, S.; Kataoka, M.; Chung, M.C.-M.; Yamada, H. *J. Biol. Chem.* **1988b,** *263,* 12077-12084.

Yamada, H.; Shimizu, S. *Ullman's Encyclopedia of Industrial Chemistry;* VCH Verlagsgesellschaft: Weinheim, **1985a,** Vol. A4, pp 150-170.

Yamada, H.; Shimizu, S. In *Biocatalysts in Organic Synthesis;* Tramper, J.; van der Pals, H.C.; Linko, P., Eds.; Elsevier: Amsterdam, **1985b,** pp 19-37.

Yamada, H.; Shimizu, S. *Angew. Chem. Int. Ed. Eng.* **1988,** *27,* 622-642.

Yamada, H.; Shimizu, S. In *Enzyme Engineering 11;* Clark, D.S.; Estell, D.; Dordick, J., Eds.; The New York Academy of Sciences: New York, **1992,** in press.

Synthesis of Novel Materials Using Nonaqueous-Based Biocatalysis

Jonathan S. Dordick*
Department of Chemical and Biochemical Engineering, University of Iowa, Iowa City, IA 52242

A variety of novel polymeric materials based on sugars have been prepared using either wholly enzymatic catalysis or chemoenzymatic catalysis in nonaqueous media. These polymers include poly(esters, acrylates, amides, urethanes, and acetylenes). Most of these materials are highly hydrophilic and hygroscopic, and are potentially useful for water-absorbent applications. Moreover, the sugar comprises the bulk of the material's weight and is biodegradable. The high selectivity of enzymatic catalysis coupled with the high productivity and reaction breadth of chemical catalysis makes the chemoenzymatic approach particularly attractive.

The application of enzymes in organic solvents is now well-established (Dordick, 1991; Klibanov, 1986; Zaks and Russell, 1988). The exquisite selectivities and fast catalytic rates of enzymes under ambient conditions provide commercially competitive alternatives to traditional organic synthesis to be developed in many diverse areas including oxidoreductions, condensations, among others (Dordick, 1989). Nowhere is this more evident than in the use of nonaqueous biocatalysis for the synthesis and modification of polymers.

The incorporation of sugars into polymers such as poly(esters, amides, or acrylates) is a difficult task in traditional chemistry due to the large number of equally reactive hydroxyl groups on the sugar molecule. For example, sucrose contains 8 hydroxyl groups; 3 primary and 5 secondary. Regioselective acylation of sucrose via chemical means (e.g., through the use of acid chlorides and anhydrides) is somewhat selective for primary groups, and can be shifted to secondary groups if the primary hydroxyls are blocked with bulky reagents (e.g., trityl, tosyl, or sulfuryl chlorides (Otake, 1972; Ballared et al., 1980; Hough et al., 1975)). However, little or no differentiation between the three primary hydroxyls are observed.

As opposed to chemical catalysts, the use of enzymes in organic solvents provided us with an opportunity to develop a synthetic strategy for sugar-based polyester synthesis. Specifically, enzyme-catalyzed regioselective acylation of sugars offers an alternative to the poor selectivity of chemical synthesis. Our first synthetic strategy was to incorporate sucrose (as a diol) into a polyester backbone by condensing the sugar with a diacid derivative (Patil et al., 1991a). A screen of 45 commercially available hydrolases (e.g., lipases, esterases, proteases, and carbohydrases) resulted in the identification of proleather (an alkaline protease from *Bacillus* sp., sold by Amano) as a suitable enzyme for this synthesis. The activity of this enzyme in pyridine was found to be highly dependent upon the pH from which the enzyme was lyophilized from with an optimal pH of 9.5.

The proleather-catalyzed polyester synthesis was performed by suspending 15 mg/mL of the pH-adjusted enzyme in 25 mL pyridine containing 0.1 M sucrose and 0.1 M di-(2,2,2-trifluoroethyl)adipate. This adipate derivative provided for a highly reactive leaving group (trifluoroethanol) which is known to improve the rate of transesterification Koskikallio, 1969). After 25 days, the polymeric material had an M_w of 2100 and an M_n of 1600 for a polydispersity of 1.3. Figure 1 depicts the structure of the poly(sucrose adipate).

The poly(sucrose adipate) product was also biodegradable. To test this, 10 mg/mL polyester (prepared from the di-trifluoroethyladipate reaction) was dissolved in 0.1 M phosphate buffer, pH 7, and 2.0 mg/mL Proleather was added and the reaction shaken at 100 rpm at 25°C. Aliquots of the reaction were periodically removed and analyzed by GPC. Within 48 h, 2.65 mg/mL free sucrose was liberated from the polymer, representing a biodegradative yield of 31%. The formation of free sucrose indicates that the Proleather can hydrolyze <u>both</u> the 6 and 1' ester linkages during biodegradation. Furthermore, the resulting biodegraded material had a M_w = 1,300 as compared to 2,100 for the fractionated sucrose polyester starting material. Continued biodegradation occured with values of M_w ca. 900 after 9 days. Therefore, significant depolymerization of the sucrose polyester takes

1054–7487/92/0164$06.00/0

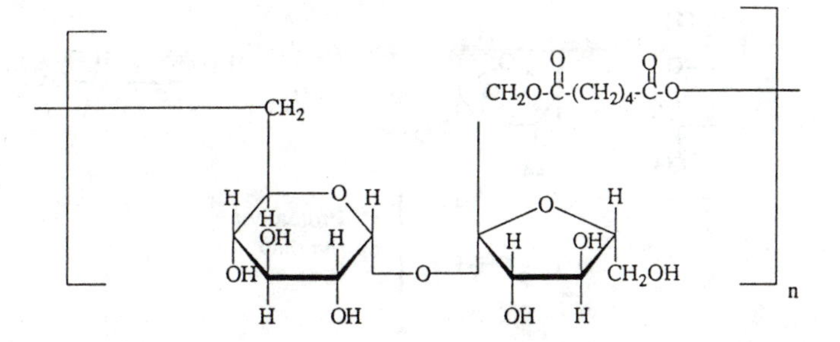

Figure 1. Synthesis of poly(sucrose adipate) by proleather catalysis in pyridine (Patil et al., 1991a).

place enzymatically in aqueous solution; a prerequisite for successful biodegradability. Interestingly, the same enzyme that catalyzed the synthesis of the polymer in pyridine, catalyzed its breakdown in aqueous buffer.

The power of selective enzymic synthesis of sugar-based polymers can be enhanced by taking advantage of chemoenzymatic synthesis. In this manner, the high regioselectivity of enzymatic catalysis can be coupled with the high reaction productivity and synthetic versatility of chemical catalyst. In this regard, we have synthesized two sucrose-containing polymers (Patil et al., 1991b). The first is a poly(sucrose adipamide) with sucrose remaining as a component of the polymer backbone; and the second is a poly(sucrose acrylate) wherein the sucrose exists as a polymer side chain. In both instances, the high degree of regioselectivity afforded by enzymatic catalysis is used to modify the sucrose before chemical polymerization is performed.

The synthesis of each polymer involved an initial enzymatic reaction. For the poly(sucrose adipamide) proleather was employed to catalyze the formation of sucrose 6,1'-ditrifluoroethyladipate followed by polycondensation with ethylenediamine. Figure 2 depicts the synthetic strategy employed in this synthesis and Table I outlines the resulting polymer properties. The enzymatic portion of this synthesis is complete (using 100 mg/mL enzyme) in 24 h, while the chemical polymerization is complete in an additional 24 h.

The synthesis of poly(sucrose acrylate) was also performed by an initial acylation of sucrose using proleather. In this case, vinyl acrylate (0.3 M) was dissolved along with 0.1 M sucrose, and 15 mg/mL proleather was added. The suspension was stirred at 45^{0}C for 24 h and the sucrose 1'-acrylate product isolated by silica gel chromatography. The sucrose monoacrylate was then polymerized in DMF via hydrogen peroxide and potassium persulfate. Figure 3 depicts the reaction scheme and the properties of the polymer are described in Table I. Clearly, the molecular weights obtained, specifically with the poly(sucrose acrylate) are substantially higher than from that obtained by the purely enzymic synthetic route.

Monosaccharide derivatives could be used for polyacrylate synthesis in addition to sucrose. The range of acrylate products include galactose, glucose, and mannose O-glycosides. Acylation is consistently, and solely, at the 6-position. Polymerization with the initiator ABIN yields polymers with molecular weights > 100,000. Chemoenzymatic synthesis represents a significant advantage over either purely chemical approaches (enhanced regioselectivity) and purely enzymatic routes (greater speed and, in the case of poly(sucrose acrylate), substantially higher molecular weights).

Table 1. Properties of Polymers Prepared Via Chemoenzymatic Methods.

Poly(sucrose adipamide)	Poly(sucrose acrylate)
Mw = 8,100	Mw = 91,000
Mn = 4,800	Mn = 57,000
Mw/Mn = 1.7	Mw/Mn = 1.6
Water-insoluble	Water-soluble
Soluble in polar organics	Soluble in polar organics

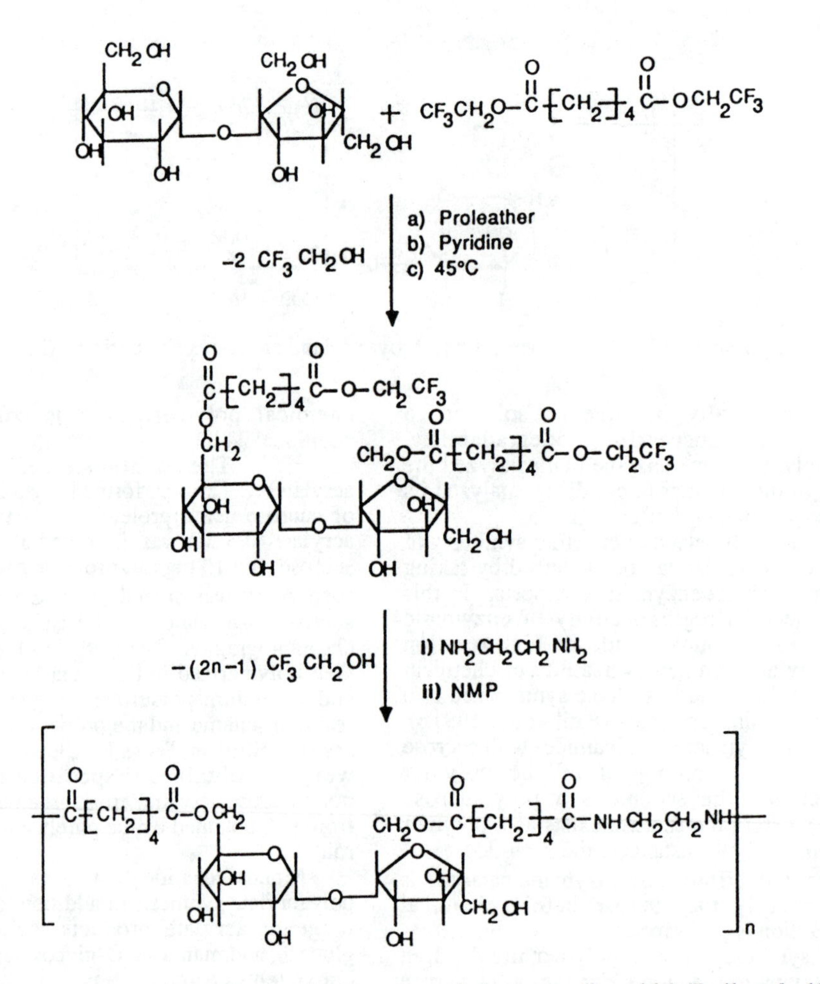

Figure 2. Chemoenzymatic synthesis of poly(sucrose adipamide) (Patil et al., 1991b).

A chemoenzymatic approach has also been used to prepare two novel sugar-based polyacetylenes. In one case, propargyl alcohol has been used as an aglycon acceptor in the β-galactosidase-catalyzed transglycosylation of lactose to give *O*-propargyl-β-D-galactopyranoside. Chemical polymerization with $AlBr_3$ resulted in the formation of poly(propargyl galactopyranoside) with M_w = 1,300 and M_n = 1,100 and yields of 95%. Polymerization catalyzed by $[(C_6H_5)_3P]_2Ni(CO)_2$ resulted in polymers with M_w = 37,400 and M_n = 36, 800, however yields were 3% and the polymerization took 2 weeks. The polymer was both water-soluble and highly hygroscopic and picked up its weight in water by standing open to air in 5 minutes. In the second case, lipase-catalyzed transesterification of *O*-methyl-β-D-glucopyranoside with ethyl propiolate resulted in the 6-propiolate ester which upon reaction with $[(C_6H_5)_3P]_2Ni(CO)_2$ resulted in polymers with M_w = 38,400 and M_n = 26,100, however, yields were < 4%. These materials may have application as hydrophilic polyacetylenes for use as organic conductors or as water-absorbents with uses as hydrogels.

In conclusion, the high specificity of enzymatic catalysis has now been extended to the synthesis and modification of polymers. Polyfunctional monomers or polymers can now be used to prepare novel materials via wholly enzymatic or combined chemoenzymatic syntheses. These materials have a variety of potential uses ranging from biodegradable plastics to drug delivery matrices to phenolic resins.

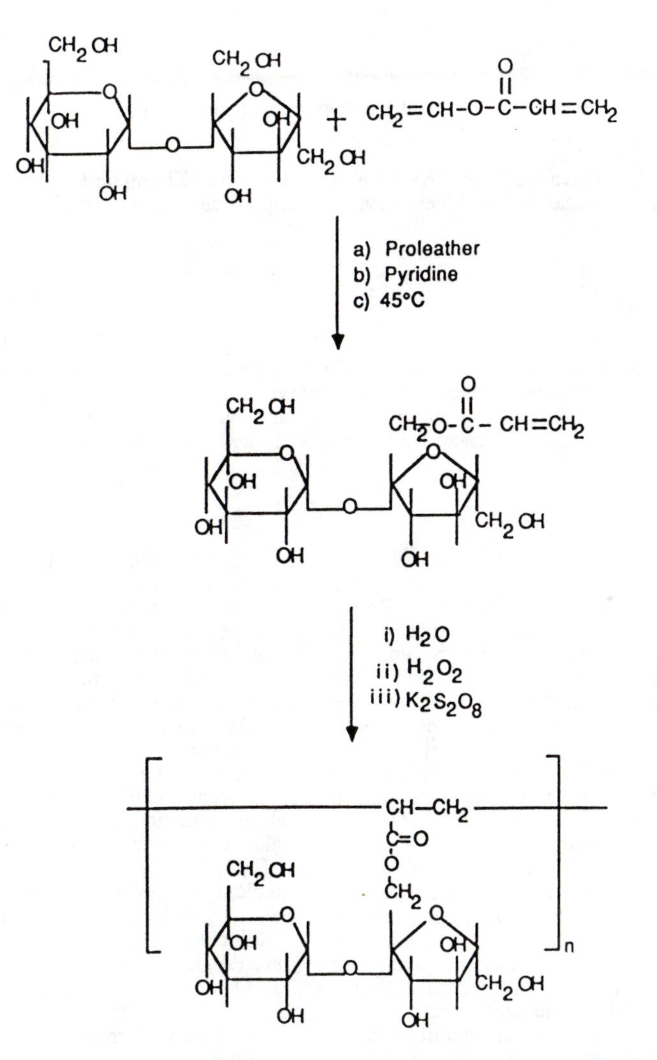

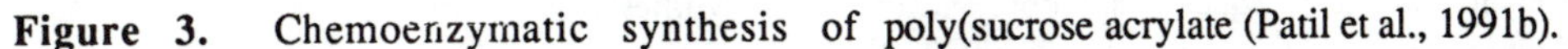

Figure 3. Chemoenzymatic synthesis of poly(sucrose acrylate (Patil et al., 1991b).

References

Ballard, J. M., L. Hough & A. C. Richardson. *Carbohydr. Res.* **1980**, *83*, 138.

Dordick, J. S. Principles and Selected Applications of Nonaquoeus Enzymology. In *Applied Biocatalysis* ; Blanch, H. W.; Clark, D. S. Eds., Marcel Dekker. New York, **1991**; pp 1-51.

Dordick, J. S. *Enz. Microb. Technol.* **1989**, *11*, 194.

Hough, L., S. D. Phadnis & E. Tarelli. *Carbohydr. Res.* **1975**, *44*: 37.

Klibanov, A. M. *Chemtech*, **1986**, *16*, 354.

Koskikollio, J. In *The Chemistry of Carboxylic Acids and Esters*, Patai, S., Ed. Interscience, London, **1969**, Chapter 3.

Otake, T. *Bull. Chem. Soc. Jpn.* **1972**, *43*, 3199.

Patil, D. R.; Rethwisch, D. G.; Dordick, J. S. *Biotechnol. Bioeng.* **1991a**, *37*, 639.

Patil, D. R.; Dordick, J. S.; Rethwisch, D. G. *Macromolecules*, **1991b**, *24*, 3462.

Zaks, A. & A. J. Russell. *J. Biotechnol.* **1988**, *8*, 259.

Engineering Enzymes for the Synthesis of Carbohydrates and Peptides

Chi-Huey Wong*, Tetsuya Kajimoto, Kevin K.-C. Liu, Ziyang Zhong and Yoshitaka Ichikawa, Department of Chemistry, The Scripps Research Institute, La Jolla, CA 92037
(FAX:619 554 6731)

The techniques of recombinant DNA and protein engineering are now readily available for the development of effective and stable enzymes for large-scale synthesis of chiral intermediates and bioactive molecules, especially carbohydrates, oligosaccharides, peptides, and their conjugates and related substances.

Many enzymes have been developed for the stereocontrolled synthesis of chiral synthons and bioactive molecules (Davies et al., 1989). Recent focus in this field has been extended to the development of more effective and stable enzymes for the synthesis of molecules with increasing complexity. One class of such complex molecules are carbohydrates and their lipid or peptide conjugates, especially those exist on cell surfaces (Ciba Found. Symp., 1989). These glycoconjugates are important recognition and communication elements (Ciba Found. Symp., 1989; Geisow, 1991); however, most of their structures and precise functions at the molecular level have not been determined and demonstrated. Part of the reasons is the difficulty encountered in the isolation, characterization and synthesis of these molecules. From synthetic point of view, enzyme-based technology seems to be well suited for the preparation of glycoconjugates and related substances, as these molecules are multifunctional and highly soluble in aqueous solution, and many enzymes are now available for the transformation of these compounds (Toone et al., 1989). The following describe new practical enzyme-based technologies recently developed in our laboratories for the synthesis of sugar- and peptide-related substances.

Aldolases for carbon-carbon bond forming reactions

Catalytic asymmetric aldol condensation is one of the most effective methods for carbon-carbon bond forming reactions (Heathcock, 1981). Enzyme-catalyzed aldol addition reactions have been shown to have great potential in this regard (Toone et al., 1989; Kajimoto et al., 1991). More than 20 aldolases are known and several of them have been overexpressed and explored for synthesis (Kajimoto et al., 1991; Kajimoto et al., 1991; Liu et al., 1991; Gautheron-Le Narvor et al., 1991; Ziegler et al., 1988; Hung et al., 1991). All the aldolases explored so far possess two common features: first, they are highly specific for the donor substrates (i.e. dihydroxyacetone phosphate, pyruvate, or phosphoenolpyruvate) but flexible for the acceptor components; second, the stereoselectivity in aldol condensations is often controlled by the enzyme not by the substrate, with some exceptions observed in the sialic aldolase reactions (Gautheron Le-Narvor et al., 1991). Although enantiomerically aldehydes are often used as substrates in enzymatic aldol reactions, both thermodynamic and kinetic approaches can be utilized to prepare a single diastereomeric aldol product starting with a racemic aldehyde substrate. A particular useful application of aldolases is the synthesis of aza sugars as key building blocks for the development of glycosidase and glycosyltransferase inhibitors (Fig. 1) (Kajimoto et al., 1991; Kajimoto et al., 1991; Liu et al., 1991).

Glycosyltransferases for oligosaccharide synthesis

Enzymatic synthesis of oligo-saccharides based on glycosidases (Nilsson, 1988; Stangier et al., 1991) and glycosyltransferases (Toone et al., 1989; Wong et al., 1982; Hindsgaul et al., 1991) has been well documented. Although glycosidases are less expensive and readily available, the sugar nucleotide-dependent glycosyltransferases seem to be more suitable for the synthesis of complex oligosaccharides as the enzymatic reactions are stereo- and regioselective for various complex acceptor structures. The major problems are that glycosyltransferases are not readily available and that sugar nucleotides are too expensive to be used as stoichiometric reagents for large-scale synthesis. Furthermore, the reactions often exhibit product inhibition caused by the released nucleoside phosphates. A simple solution to these problems is to regenerate the sugar nucleotide from the released nucleoside phosphate after glycosidic bond formation (Fig. 2). Of eight sugar nucleotides commonly found in mammalian systems as

1054–7487/92/0168$06.00/0

Retrosynthesis of Azasugars

Enzyme-catalyzed aldol condensation + Pd-mediated reductive amination

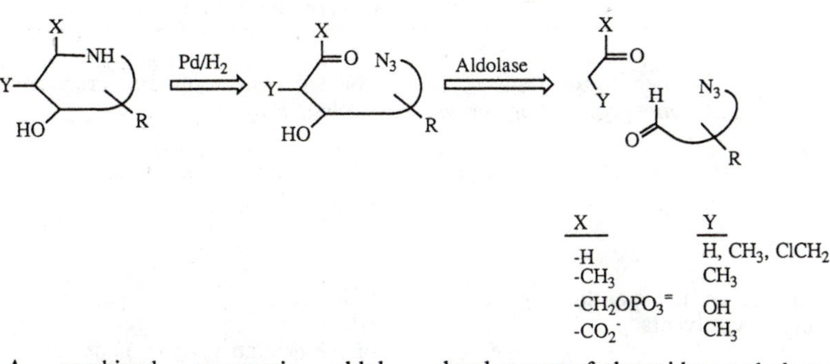

Figure 1. A combined enzymatic aldol condensation and Pd-mediated reductive amination provides a variety of aza sugars useful for the development of glycosidase and glycosyltransferase inhibitors.

substrates for glycosyl-transferases, six of which now can be regenerated *in situ* for use in oligosaccharide synthesis (Table 1). These regeneration systems are available for the synthesis of oligosaccharides on >10 g scales. The enzymes can be coimmobilized in a polymor support such as cross-linked polyacrylamide and recovered for reuse. No product inhibition was observed. The regeneration systems can also be combined in one pot to form more than two glycosidic bonds with the use of two or more glycosyltransferases (Ichikawa et al., 1991). The remaining issues to be solved in enzymatic oligosaccharide synthesis will be to develop procedures for the regeneration of UDP-N-acetylglucosamine and UDP-N-acetylgalactosamine, and to make glycosyltransferases readily available. Although several cloning systems based on CHO cells have been developed (Paulson et al., 1989), it is still desirable to develop more practical systems to express the catalytic domains of glycosyltransferases in micro-organisms (Aoki et al., 1990; Wong et al., 1991). The regeneration systems now available should be useful for the large-scale synthesis of bioactive oligosaccharides for therapeutic evaluation. Sialyl Le^X, the ligand of endothelial leukocyte adhesion molecule (Phillips et al., 1990), for example, perhaps can be prepared on 0.1-1 kilogram scales for use to test its clinical value as antiinflammatory agent.

With regard to substrate specificity, glycosyltransferases seem to accept a variety of unnatural substrates *in vitro* as indicated in the studies of galactosyl (Wong et al., 1991) and fucosyltransferases (Dumas et al., 1991). Some substrates are even better accepted than natural substrates. These studies provide a new direction to the synthesis of novel oligosaccharides.

Table 1. Sugar nucleotides used as donor substrates for glycosyltransferases and their regeneration

Glycosyltransferase	Sugar Nucleotide Regeneration	Reference
Galactosyltransferase	UDP-Galactose	1
Glucosyltransferase	UDP-Glucose	1
Sialyltransferase	CMP-Sialic Acid	2
Glucuronyltransferase	UDP-Glucuronic Acid	3
Mannosyltransferase	GDP-Mannose	4
Fucosyltransferase	GDP-Fucose	4
N-Acetylglucosaminyltransferase	UDP-N-acetylglucosamine	-
N-Acetylgalactosaminyltransferase	UDP-N-acetylgalactosamine	-

1. Wong, C.-H.; Haynie, S.L.; Whitesides, G.M. J. Org. Chem. 1982, 47. 5416.
2. Ichikawa, Y.; Shen, G.-J.; Wong, C.-H. J. Am. Chem. Soc. 1991, 113, 4698.
3. Gygax, D.; Spies, P.; Winkler, T.; Pfaar, V. Tetrahedron 1991, 47, 5119.
4. Wong, C.-H.; Liu, K.K.-.C.; Kajimoto, T.; Chen, L.; Zhong, Z.; Dumas, D.P.; Liu, J.L.-C.; Ichikawa, Y.; Shen, G.-J. Pure & Appl. Chem., in press.

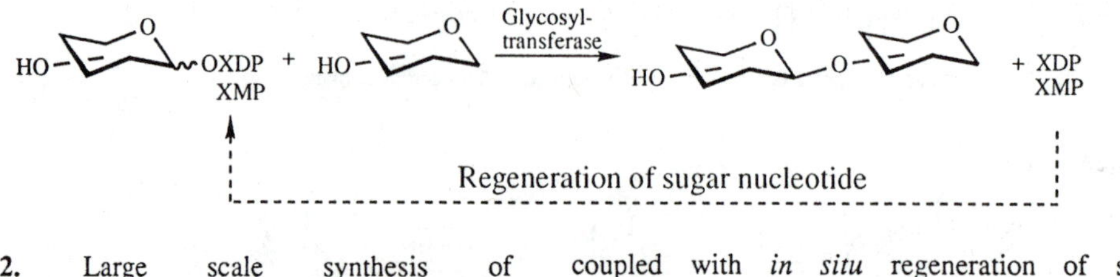

Figure 2. Large scale synthesis of oligosaccharides based on glycosyltransferases coupled with *in situ* regeneration of sugar nucleotides.

Engineering proteases for peptide ligation in aqueous and organic solvents

A major problem in synthetic peptide chemistry is the lack of effective methods for the coupling of large peptide segments. Synthesis of peptides with more than 100 amino acids, particularly those containing unnatural amino acids, still represents a significant problem. Although the biological method based on *in vitro* site-directed mutagenesis can be used to introduce unnatural amino acids into large peptides, only small amounts (~ μg) of peptides or proteins can be prepared (Noren et al., 1989). An alternative method is protease-catalyzed peptide segment coupling (Schellenberger et al., 1991). The method, however, has the disadvantages that proteases are generally unstable under the reaction conditions (e.g. polar organic solvents) and that the reaction is often contaminated with undesired hydrolytic reactions. The half life of subtilisin BPN' in anhydrous dimethylformamide (DMF), for example, is about 20 minutes, which is not long enough for large scale processes. An improvement of the enzyme stability in DMF has been accomplished with the use of site-directed mutagenesis to increase the half life to about one day at room temperature (Wong et al., 1990). A second variant of subtilisin BPN' with a half life of about 14 days in dry DMF has recently been prepared. This variant contains the following changes: Met50phe, Asn76Asp, Gly169Ala, Gln206Cys, Asn218Ser (Zhong et al., 1991). The variant is also more stable than the wild type enzyme in aqueous solution. This work has provided a substantial amount of experimental information useful for the design of enzymes to be used in polar organic solvents. Based on the subtilisin work, several changes must be carried out simultaneously to make enzymes stable and active in polar organic solvents. These include the minimization of surface charges, the enhancement of internal polar interactions (e.g. H-bonding, ionic, and metal-coordinating forces), the optimization of internal hydrophobic interactions and conformational restrictions. These rules are in general consistent with those predicted previously for the protein crambin (Arnold, 1988). The ultimate desired mutant enzymes often come from the study of several individual site-specific changes followed by the investigation of their cumulative effects. Random mutagenesis followed by selection of the desired variants under the desired condition is another effective approach. In the case of subtilisin BPN', surface charge minimization (e.g. Lys217→Tyr) and conformational restriction (e.g. Gly169→Ala) have proven to be the most effective. Each of the changes improves the enzyme stability in DMF by ~3 kcal/mol as measured by its irreversible inactivation. Surface charge minimization has also proven to be effective for the stabilization of α-lytic protease (Martinez et al. 1991) and subtilisin E (Chen et al., 1991) in aqueous DMF.

Kinetically controlled peptide segment coupling catalyzed by serine proteases in aqueous solution or organic cosolvents is a common method. Coupling of large peptide segments in aqueous solution is particularly desirable as many large unprotected peptides are only soluble in aqueous solution. To enhance the ratio of aminolysis to hydrolysis in aqueous solution, one can change the active-site of the enzyme to alter the nature of catalysis. The ratio of aminolysis to hydrolysis, for example, increases substantially when the active-site His of chymotrypsin is methylated at the ε-2N, or when the active-site Ser of subtilisin BPN' is converted to Cys (Wong et al., 1992). The reasons for the enhancement of aminolysis to hydrolysis are mainly due to the increase of affinity and reactivity of the acyl intermediate for the amine nucleophile (West et al., 1990). These active-site modified serine proteases will be useful for the coupling of peptide segments containing natural or unnatural amino acids (e.g. D-, phosphorylated, or glycosylamino acids) in aqueous solution.

Conclusion

The practicality of enzyme-based synthetic technology can be improved via protein engineering to alter the performance of enzymes. As more enzymes become available via advanced recombinant techniques, enzyme-catalyzed synthesis of complex biomolecules such as glycoconjugates and their intermediates will become a very effective and practical way for large scale processes.

This research was supported by NIH, NSF, Cytel, Office of Naval Research and DOE through the Materials Science Lawrence Berkeley Laboratory.

References

Aoki, D.; Appert, H.E.; Johnson, D.; Wong, S.S.; Fukuda, M.N. *EMBO* **1990**, *9*, 3171.

Arnold, F.H. *Protein Engineering* **1988**, *2*, 21.

Carbohydrate Recognition in Cellular Function. Ciba Foundation Symposium 145, John Wiley & Sons, Chichester, U.K. **1989**.

Chen, K.; Robinson, A.C.; van Pam, M.E.; Martinez, P.; Economou, C.; Arnold, F.H. *Biotechnol. Prog.* **1991**, *7*, 125.

Davies, H.G.; Green, R.H.; Kelly, D.R.; Roberts, S.M. *Academic Press*, London, U.K. **1989**.

Dumas, D.P.; Ichikawa, Y.; Wong, C.-H.; Lowe, J.B.; Nair, R.P. *BioMed. Chem. Lett.* **1991**, *1*, 551.

Dumas, D.P.; Ichikawa, Y.; Wong, C.-H.; Lowe, J.B.; Nair, R.P. *BioMed. Chem. Lett.* **1991**, 1425.

Gautheron-Le Narvor, C.; Ichikawa, Y.; Wong, C.-H. *J. Am. Chem. Soc.* **1991**, *113*, 7816.

Geisow, M.J. *TIBTECH* **1991**, *9*, 221.

Heathcock, C.H. *Science* **1981**, *214*, 395.

Hindsgaul, O.; Kaur, K.J.; Gokhale, U.B.; Srivastava, G.; Alton, G.; Palcic, M.M. ACS Symposium Series 466, American Chemical Society, Washington, D.C. **1991**, pp 38-50.

Hung, R.R.; Straub, J.A.; Whitesides, G.M. *J. Org. Chem.* **1991**, *56*, 3849.

Ichikawa, Y.; Liu, J.L.-C.; Shen, G.-J.; Wong, C.-H. *J. Am. Chem. Soc.* **1991**, *113*, 6300.

Kajimoto, T.; Liu, K.K.-C.; Pederson, R.L.; Zhong, Z.; Ichikawa, Y.; Porco, J.A.; Wong, C.-H. *J. Am. Chem. Soc.* **1991**, *113*, 6187.

Kajimoto, T.; Chen, L.; Liu, K.K.-C.; Wong, C.-H. *J. Am. Chem. Soc.* **1991**, *113*, 6678.

Liu, K.K.-C.; Kajimoto, T.; Chen, L.; Zhong, Z.; Ichikawa, Y.; Wong, C.-H. *J. Org. Chem.* **1991**, *56*, 6280.

Martinez, P.; Arnold, F.H. *J. Am. Chem. Soc.* **1991**, *113*, 6336.

Nair, R.P. *BioMed. Chem. Lett.* **1991**, *1*, 551.

Nilsson, K.G. *TIBTECH* **1988**, *6*, 256.

Noren, C.J.; Anthony-Cahill, S.J.; Griffith, M.C.; Schultz, P.G. *Science* **1989**, *244*, 182.

Paulson, J.; Colley, K.J. *J. Biol. Chem.* **1989**, *264*, 17615.

Phillips, M.L.; Nudleman, E.; Gaeta, F.C.A.; Perez, M.; Singhal, A.L.; Kahomori, S.-I.; Paulson, J.S. *Science* **1990**, *250*, 5365.

Schellenberger, V.; Jakubke, H-D. *Angew Chem. Int. Ed. Engl.* **1991**, *30*, 1437.

Stangier, P.; Thiem, J. ACS Symposium Series No. 466, American Chemical Society, Washington, D.C. **1991**, pp 64-78.

Toone, E.J.; Simon, E.S.; Bednarski, M.D.; Whitesides, G.M. *Tetrahedron* **1989**, *45*, 5365.

West, J.B.; Hennen, W.J.; Lalonde, J.L.; Bibbs, J.A.; Zhong, Z.; Meyer, E.F.; Wong, C.-H. *J. Am. Chem. Soc.* **1990**, *112*, 5313.

Wong, C.-H.; Haynie, S.L.; Whitesides, G.M. *J. Org. Chem.* **1982**, *47*, 5416.

Wong, C.-H.; Chen, S.-T.; Hennen, W.J.; Bibbs, J.A.; Wang, Y.-F.; Liu, J.L.-C.; Pantoliano, M.W.; Whitlow, M.; Bryan, P.N. *J. Am. Chem. Soc.* **1990**, *112*, 945.

Wong, C.-H.; Ichikawa, Y.; Krach, T.; Gautheron, C.; Dumas, D.P.; Look, G. *J. Am. Chem. Soc.* **1991**, *113*, 8137.

Zhong, Z.; Liu, J.L.-C.; Dinterman, L.M.; Finkelman, M.A.J.; Mueller, W.T.; Rollence, M.L.; Whitlow, M.; Wong, C.-H. *J. Am. Chem. Soc.* **1991**, *113*, 683.

Ziegler, T.; Straub, A.; Effenberger, F. *Angew. Chem. Int. Ed. Engl.* **1988**, *29*, 716.

BIOREACTOR ENGINEERING SYMPOSIUM V

B. C. Buckland and M. D. Lilly: *Co-Chairs*

Transport Phenomena in Bioreactor Systems: Session A
M. Moo-Young and K. van't Riet: *Co-Chairs*

Animal Cell Culture: Session B
W.-S. Hu and P. Gray: *Co-Chairs*

Microbial Fermentations and Economics: Session C
B. C. Buckland and M. D. Lilly: *Co-Chairs*

Transport Phenomena in Novel Bioreactors: Design of Airlift-Based Devices

Murray Moo-Young* and **Yusuf Chisti**
Department of Chemical Engineering, University of Waterloo, Ontario, Canada N2L 3G1
(FAX:519 746 4979)

Of the many newly proposed non-mechanically agitated bioreactor designs, very few have gained commercial acceptance. Notable among these is the airlift system. Here we examine the principal aspects of transport phenomena in airlift-based reactor designs.

Relative fragility of the newer biocatalysts (animal cells, insect cells, recombinant microorganisms, plant cells) frequently accompanied by instability has added a new dimension to bioreactor design. Airlift type bioreactors are well suited to the new applications where they achieve the requisite heat, mass and momentum transfer in a gentle hydrodynamic environment.

Reliable design and scale-up of an airlift bioreactor requires the ability to predict the magnitude of the induced liquid circulation, gas holdup, gas-liquid and solid-liquid mass transfer, heat transfer and mixing among other variables. A knowledge of how those parameters are affected by the geometry of the reactor (e.g., aspect ratio; riser-downcomer cross-sectional area ratio) and the operational conditions (e.g., gas injection rate; fluid properties) is essential. Major advancements in the engineering know-how of the airlift systems have enhanced the reactor design and scale-up capability sufficiently that dependable design is possible with reduced pilot scale development. Advances in the prediction of hydrodynamic and transport characteristics are discussed below.

Hydrodynamics. The liquid circulation - up-flow in the riser, downflow in the downcomer - is a most important property of airlift reactors. Liquid circulation has a controlling influence on mixing, turbulence, bubble size, gas-liquid and solid-liquid mass transfer, heat transport and the magnitude of shear forces [1]. The circulation arises due to the gas holdup difference between the riser and the downcomer; however, the circulation in turn affects the holdup. The gas holdup and liquid circulation are inseparable and correlations which attempt to do so are not particularly successful at predicting either. A notable equation for prediction of the circulation velocity (U_{Lr}) in water-like media was obtained by energy balance over the airlift loop [2, 3]; the equation is

$$U_{Lr} = \left[\frac{2gh_D(\varepsilon_r - \varepsilon_d)}{\dfrac{K_T}{(1-\varepsilon_r)^2} + K_B\left(\dfrac{A_r}{A_d}\right)^2 \dfrac{1}{(1-\varepsilon_d)^2}}\right]^{0.5} \quad (1)$$

The validity of equation (1) has been demonstrated over a broad range of reactor geometries (external- and internal-loop airlifts) and scales [2]. Other investigators [4] have provided evidence in support of equation (1). The equation relates reactor geometry (h_D, A_r, A_d, K_T, K_B) to liquid circulation velocity; however, the equation does not explicitly show the effect of gas flow rate (U_{Gr}) on liquid circulation. Equation (1) must be used with other vertical two-phase flow correlations [5] of the form

$$\varepsilon_r = \frac{U_{Gr}}{\alpha + \beta\,(U_{Gr} + U_{Lr})^y} \quad (2)$$

Note that equation (2) is obtained by independent variation of U_{Gr} and U_{Lr} in pipes [5]; the equation cannot by itself predict the liquid velocity or gas holdup in airlift systems. When α, β and y are not known for a particular fluid, those parameters must be

1054–7487/92/0174$06.00/0

experimentally obtained in small-scale developmental work as described elsewhere [3]. The variety and complexity of biofluids does not at present allow for generalizations on the constants in equation (2). The procedure for simultaneous calculation of gas holdups and liquid circulation velocity using equations (1) and (2) has been detailed previously [2, 3].

Although equation (1) applies to low viscosity newtonian flows, it can be easily extended to more viscous fluids. The use of energy balance method for prediction of circulation in non-newtonian pseudoplastic media is also effective [6]; however, the technique is relatively complex and has not been tested as exhaustively as equation (1).

Shear effects. The magnitude of liquid circulation velocity determines the fluid-wall shear stress as well as the dimensions of the turbulent microeddies in the bulk fluid. Perhaps the most turbulence-sensitive biocatalysts are the plant cells, which due to their large size, may be susceptible to damage by relatively larger fluid eddies (hence to damage at lower power inputs) than other cells. Numerous reports of successful culture of plant cells in airlift type systems [7] are testimony to the low-shear capability of the airlift devices. Of course, the airlift bioreactors are already well proven in commercial animal cell culture.

Gas-liquid mass transfer. The influence of liquid circulation on gas-liquid mass transfer is primarily through gas holdup. The procedure for the estimation of liquid circulation also provides the gas holdup at any operating riser gas velocity [3, 6]. The holdup can be correlated [3, 8, 9] to the overall volumetric gas-liquid mass transfer coefficient, k_La_L:

$$k_La_L = \psi \frac{A_r}{(A_r + A_d)} \frac{\varepsilon_r}{(1 - \varepsilon_r)} \quad (3)$$

Equation (3) is based in theoretical fundamentals [3]. The coefficient ψ is a constant for a given fluid. In physical terms, ψ amounts to $6k_L/d_B$ which must be experimentally determined on a small scale [3]. Because the k_L and d_B behaviour in different fluids is hard to infer [8], the experimentally determined ψ ensures a more confident scale-up.

Other approaches to k_La_L prediction have also been proposed [10].

Solid-liquid mass transfer. Very little data are available on solid-liquid mass transfer in airlift bioreactors. Although not necessary for freely suspended microorganisms, such data are required for slurries of larger particles, for example for suspensions of immobilized enzymes. Limited investigations in an external-loop reactor [11] have provided the equation

$$\frac{k_s d_p}{D} = 2 + 0.48 \left(\frac{\rho_L E^{1/3} d_p^{4/3}}{\mu_L} \right)^{0.72} \left(\frac{\mu_L}{\rho_L D} \right)^{1/3} \quad (4)$$

Equation (4) has the general form $Sh = 2 + x\, Re^c\, Sc^{1/3}$ which has previously been applied to slurries in stirred tanks [12, 13], fluidized beds [14], and bubble columns [13], and also to fixed bed reactors [15]. The solid-liquid mass transfer attainable in airlift systems is comparable to that in other devices mentioned. Hence the airlift reactors may not only perform capably as immobilized enzyme reactors, but may prove superior to the more common devices because of the other advantages of airlifts [8].

Split-cylinder type internal loop airlifts incorporating packed beds of inert matrices (vermiculite, polyurethane foam) in the downcomer have been developed for hybridoma culture [16]. In such systems the medium is oxygenated in the riser and gas-free medium percolates through the packing. Although high viable cell densities have been reported [16] for the small units tested, the question of oxygen depletion along the depth of deep beds remains to be experimentally addressed for practical scaleup of the reactor concept.

Heat transfer. Theoretical developments, confirmed by experiment, have shown [17] that the film heat transfer coefficient depends on the superficial gas velocity raised to 1/4. Data on air-water in some small concentric draft tube airlifts [17] have produced the correlation

$$h_T = 13340\left(1+\frac{A_d}{A_r}\right)^{-0.7} U_{sg}^{0.275} \qquad (5)$$

where h_T is an average heat transfer coefficient. Wider applicability of equation (5) has not been checked.

The heat transfer performance of airlift devices exceeds that of bubble columns [17].

Hybrid airlift reactors. Although highly viscous fermentations have been attempted with some success in airlift reactors, the gas-liquid mass transfer can still be a limitation in those applications due to large gas bubbles. Placement of static mixers in the riser or the downcomer of the airlift can yield major improvements in k_La_L through breakage and redispersion of gas bubbles [18]. This hybrid scheme of operation has been demonstrated for non-newtonian pseudoplastic media (K = 10^{-3} - 10 Pa•s^n; n = 0.5 - 1.0) in an external-loop airlift [18]. For comparison, the k_La_L was correlated with the riser gas velocity by the equation $k_La_L = \lambda\ (U_{Gr})^z$ where the exponent z was not sensitive to K-value of the fluid or to the presence of static mixers. However, λ was significantly affected by the static mixers: λ_M (i.e., λ in presence of static mixers) was always higher than λ. The improvement in λ due to static mixers depended mainly on the consistency index (K) of the fluid. The relationship among λ, λ_M and K was found to be

$$\frac{\lambda_M}{\lambda} = 4.43\ K^{0.12} \qquad (6)$$

Thus, the higher the K-value, the greater was the k_La_L intensification due to static mixers. For otherwise identical conditions, the presence of static mixers improved k_La_L by 30-500 %, depending on the fluid.

Preliminary theoretical evaluations of the airlift-trickle bed hybrid system (discussed earlier) have shown that enough liquid circulation can be generated even in deep beds (> 3 m) of 0.003 m diameter glass spheres as may be used for immobilization of enzymes or anchorage dependent animal cells. On a small scale, anchorage-dependent animal cells (as opposed to hybridomas) have been grown in this type of device [19].

In this short overview we limited the scope of discussion to our work; other perspectives on airlift technology are due to Merchuk [20], Joshi et al. [21] and Scragg [22].

Nomenclature

- A_d Downcomer cross-sectional area (m^2)
- A_r Riser cross-sectional area (m^2)
- a_L Gas-liquid interfacial area per unit liquid volume (m^{-1})
- c Exponent (-)
- D Diffusivity (m^{2}s^{-1})
- d_B Diameter of bubble (m)
- d_p Diameter of particle (m)
- E Energy input per unit mass (Wkg^{-1})
- g Gravitational acceleration (ms^{-2})
- h_D Height of gas-liquid dispersion (m)
- h_L Height of gas-free liquid (m)
- h_T Mean heat transfer coefficient (Wm^{-2}K^{-1})
- K Consistency index of fluid (Pa•s^n)
- K_B Pressure loss coefficient for flow turnaround at the bottom of airlift (-)
- K_T Pressure loss coefficient for flow turnaround at the top of airlift (-)
- k_L Overall gas-liquid mass transfer coefficient based on liquid film (ms^{-1})
- k_s Solid-liquid mass transfer coefficient (ms^{-1})
- n Flow behaviour index (-)
- Re Reynolds number (-)
- Sc Schmidt number (-)
- Sh Sherwood number (-)
- U_{Gr} Superficial gas velocity in riser (ms^{-1})
- U_{Lr} Superficial liquid velocity in riser (ms^{-1})
- U_{sg} Superficial gas velocity based on outer tube diameter (ms^{-1})
- x Coefficient (-)
- y Exponent in equation (2) (-)
- z Exponent (-)
- α Parameter in equation (2) (ms^{-1})
- β Coefficient in equation (2) (-)
- ε_d Downcomer gas holdup (-)
- ε_r Riser gas holdup (-)
- λ Coefficient (s^{z-1}m^z)
- λ_M λ in presence of static mixers (s^{z-1}m^z)
- μ_L Viscosity of liquid (Pa•n)
- ρ_L Density of liquid (kgm^{-3})
- ψ Constant (= $6k_L/d_B$) (s^{-1})

References

1. Moo-Young, M. and Chisti, Y., *Biotechnology*. **1988,** *6*(11), 1291-1296.
2. Chisti, M. Y., Halard, B. and Moo-Young, M., *Chem. Eng. Sci.* **1988,** *43*, 451-457.
3. Chisti, M. Y., *Airlift Bioreactors*, Elsevier: London, 1989; pp 203-229; 266-272; 251-261.
4. Wachi, S., Jones, A. J. and Elson, T. P., *Chem. Eng. Sci.* **1990,** *46*, 657-663.
5. Hills, J. H., *Chem. Eng. J.* **1976,** *12*, 89-99.
6. Chisti, M. Y. and Moo-Young, M., *J. Chem. Technol. Biotechnol.* **1988,** *42*, 211-219.
7. Taticek, R. A., Moo-Young, M. and Legge, R. L., *Plant Cell, Tissue and Organ Culture* **1991,** *24*, 139-158.
8. Chisti, M. Y. and Moo-Young, M., *Chem. Eng. Commun.* **1987,** *60*, 195-242.
9. Chisti, M. Y. and Moo-Young, M., *Biotechnol. Bioeng.* **1988,** *31*, 487-494.
10. Bello, R. A., Robinson, C. W. and Moo-Young, M., *Chem. Eng. Sci.* **1985,** *40*, 53-58.
11. Mao, H. H., Chisti, Y. and Moo-Young, M., *Chem. Eng. Commun.* **1992,** in press.
12. Calderbank, P. H. and Moo-Young, M. B., *Chem. Eng. Sci.* **1961,** *16*, 39-54.
13. Sano, Y., Yamaguchi, N. and Adachi, T., *J. Chem. Eng. Japan.* **1974,** *7*, 255-261.
14. Arters, D. C. and Fan, L. -S., *Chem. Eng. Sci.* **1986,** *41*, 107-115.
15. Ohashi, H., Sugawara, T. and Kikuchi, K. and Konno, H., *J. Chem. Eng. Japan.* **1981,** *14*, 433-438.
16. Phillips, H. A., Scharer, J. M. and Moo-Young, M., Poster presented at the *Cell Culture Engineering III* conference, Palm Coast, FL, February 2-7, 1992.
17. Ouyoung, P. K., Chisti, M. Y. and Moo-Young, M., *Chem. Eng. Res. Des.* **1989,** *67*, 451-456.
18. Chisti, Y., Kasper, M. and Moo-Young, M., *Canad. J. Chem. Eng.* **1990,** *68*, 45-50.
19. Whiteside, J. P. and Spier, R. E., *Biotechnol. Bioeng.*, **1981,** *23*, 551-556.
20. Merchuk, J. C. In *Encyclopedia of Fluid Mechanics*; Cheremisinoff, N. P., Ed.; Gulf Publishing: Houston, 1986, Vol. 3; pp 1485-1511.
21. Joshi, J. B., Ranade, V. V., Gharat, S. D. and Lele, S. S., *Canad. J. Chem. Eng.* **1990,** *68*, 705-741.
22. *Bioreactors in Biotechnology - A Practical Approach*; Scragg, A. H., Ed.; Ellis Horwood: New York, 1991; pp 126-157.

Advanced Measuring Techniques for Mixing and Mass Transfer in Bioreactors

Andreas Lübbert
Institut für Technische Chemie, Universität Hannover
Callinstr.3, D-3000 Hannover, Germany (FAX 49 511 762 3004)

Recent developments in measuring techniques to characterize the fluiddynamical behavior of bioreactors are reviewed. Only those techniques are discussed which can be used during cultivation processes in reactors of pilot or production scale.

Bioreactors are built to provide microorganisms in production systems that environment necessary to optimally convert substrates into the desired product. This logistic problem is solved by inducing complex multiphase flows. In order to achieve a more thorough understanding of the hydrodynamics in bioreactors, the availability of detailed data is a necessary prerequisite. Unfortunately, powerful measuring techniques which work in bioreactors during their real operation are sparce. Some of the available techniques are reviewed in this paper.

Integral Measuring Techniques

A first approach to characterize the fluid-dynamical behaviour of bioreactors is usually performed by means of integral measurement data.

Motion of the Gas Phase

Integrally the gas phase, which is continuously fed to aerobic cultures, is characterized by gas residence time distributions (RTD). They are determined by means of tracer measurements using a gaseous tracer, usually helium. In the classical implementation of this technique, the tracer is added in form of Dirac pulses. The helium signal, then measured in the gas leaving the dispersion, is essentially proportional to the gas-RTD.

The signal-to-noise ratio of gas-RTD measurements has been enhanced by more than an order of magnitude using pseudostochastic test signals instead of single Dirac pulses (Lübbert et al., 1987). Hence, either the total measuring time or the amount of tracer can be reduced considerably.

One difficult problem of RTD-measurements is the determination of the helium concentration just as the gas is leaving the dispersion and before it becomes mixed with the gas in the headspace of the reactor. This problem was solved (Bodemeier et al., 1992) with a membrane probe which is capable of measuring the tracer gas concentration within the dispersion. As depicted in Figure 1, even in viscous media a high signal-to-noise ratio of the RTDs becomes possible.

In airlift bioreactors, the signal-to-noise ratio is high enough to detect structures in the time-of-flow distributions. As has been shown (Lübbert et al., 1990), the amount of gas recirculated as well as the characteristic circulation times can be detected.

Since the membrane probes allow to measure the tracer concentration at appropriate positions within the reactor, the fluid flow within individual parts of the flow system can be investigated separately. Interesting information can be obtained on the coalescence/redispersion behaviour of these reactors (Lübbert et al., 1990) if the tracer is added at different places to the multiphase flow.

This extension of classical RTD-measurements elucidate a general trend in integral measurement techniques, namely to bridge the gap to local techniques.

Integral Motion of the Liquid Phase

From the quantities characterizing the motion of the continuous liquid phase, induced by the agitation systems, the circulation and mixing times are of main importance to reactor performance. They can also be

1054–7487/92/0178$06.00/0

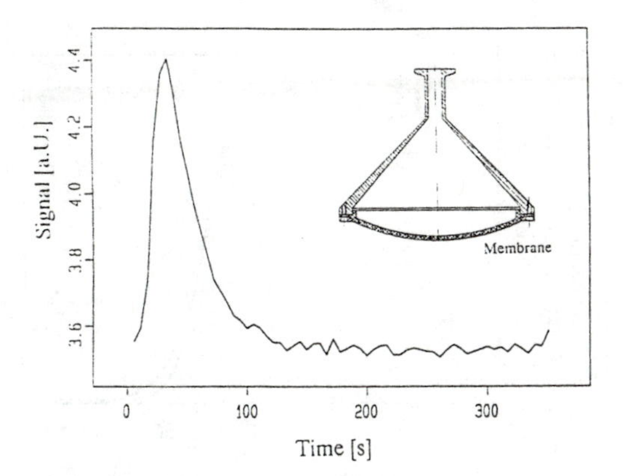

Figure 1. Membrane probe for gas residence time distribution measurements in bioreactors. The probe is attached to the vacuum system of a mass spectrometer. The example of a RTD was obtained during a cultivation of *Penicillium crysogenum*.

measured by stimulus response techniques. However, in liquids, the choice of appropriate tracers/detector systems is much more difficult, because most material tracers - when applied at sufficiently high concentrations - either affect the biological system or the rheology of the cultivation media. The use of fluorescent tracers and fluorimeters is presently preferred.

Since it often is extremely difficult to find an adequate tracer/detector system, flow-follower techniques (e.g. Bryant, 1977) have continuously been improved. Flow-followers are constructed as inert and sterilizable macroscopic solid particles of several millimeters in diameter. They are adjusted to neutral buoyancy with respect to the multiphase fluid, such that they follow the mean flow of the dispersion through the reactor. The flow followers are registrated whenever they pass a control area within the reactor. In this way, the circulation time distribution of the dispersion can be estimated. The group around Dudukovic (Devanathan et al., 1990) improved this technique by using γ-radiating pills, which can easily be detected with scintillation counters. With an array consisting of several counters posed around the reactor, it is possible to keep track of the single flow follower on its random path around in the reactor. From such measurements, the structure of the liquid phase motion, e.g., in bubble column reactors, can be investigated in much more detail. Besides the main circulation structure, dead-space regions, secondary circulations structures, etc. can be identified.

Flow followers, unfortunately, necessitate long total measuring time to obtain statistically significant characteristic functions. In order to more efficiently use the measuring time, Bodemeier et al. (Bodemeier et al., 1992) developed flow followers which transmit an identification code via radiofrequency transmission to the detectors in the reactor. In this way, several flow followers can be used simultaneously in the reactor.

Local Measuring Techniques

Gas Phase Properties

On the millimeter scale, the rising bubbles dominate the flow of both phases. In airlift reactors, they also act as primary agitators and, hence, are responsible for local mixing. This is why the bubbles are so important in bioreactors.

The primary bubble properties determining the oxygen transfer into the liquid phase, are the local gas holdup and the specific interfacial area. Both are more or less strongly coupled via the mean bubble diameter.

The simplest way of measuring local gas holdups is to use the old one-point conductivity probe technique. At the measuring point, it is able to distiguish between bubble gas and continuous liquid phase. Hence, it can determine the local gas holdup, which is simply the relative time the probe detects a gas phase.

The underlying principle of the conductivity probes is simple, however, in practice, some problems appear with electrochemical effects at the electrodes. Also, small bubbles may follow the liquid phase motion around the probe tip. Moreover, deformations of bubbles may occur when the wire tips enter a bubble.

The electrochemical effects can be removed by using optical probes. The selective detection of larger bubbles can only be reduced by using very thin wires. However, it is better to avoid any contact between a probe and the bubbles within the measuring volume. This can be achieved by using ultrasound transmission (Stravs and von

Stockar, 1987) or reflection techniques (Bröring et al., 1991, 1992).

Ultrasound radiation can be transmitted through the liquid phase which, in microbial cultures, often is opaque. However, the ultrasound is reflected almost perfectly at gas-liquid interfaces. Hence, rising bubbles passing the ultrasound beam stochastically attenuate or interrupt the transmission signal measured at the receiver. The mean attenuation is mainly a function of the specific gas-liquid interfacial area, anlogeous to Calderbank's light transmission technique (Calderbank, 1958). Unfortunately, the latter cannot be used in most bioreactors, since it requires optical transparency.

Since the transmittance of ultrasound in real cultivations is restricted to a few centimeters, both, the ultrasound probe as well as the detector must be placed into the dispersion. The resulting inconvenience, however, can be drastically reduced with ultrasound reflection techniques, in which only a single probe is used to transmit ultrasound pulses and to receive the signal reflected by bubbles in the beam path back onto probe. If the amplitude of the reflections is analysed, the specific interfacial area, the bubble number density, the Sauter diameter as well as the local gas holdup can be estimated (Bröring et al., 1992). If the frequency shift of the received ultrasound pulses as compared to the pulses transmitted (Doppler shift) is analysed, the velocity components of the bubbles directed along the beam path can be determined (Bröring et al., 1991). Figure 2 presents a typical bubble velocity distribution measured in a pilot-scale stirred tank bioreactor during a cultivation.

Bubble size distributions can be measured by withdrawing a small representative sample of the dispersion through a glass capillary. There the bubbles appear as elongated slugs. By means of a light barrier system the volumes of the slugs are easily measured and converted into the diameters of the equivalent spherical bubbles. If the gas composition of the slugs is analysed by a mass spectrometer, the technique can be used to obtain valuable information on coalescence processes (Prince and Blanch, 1990). This, however, requires additional tracer bubbles to be released into the reactor at the gas distributor.

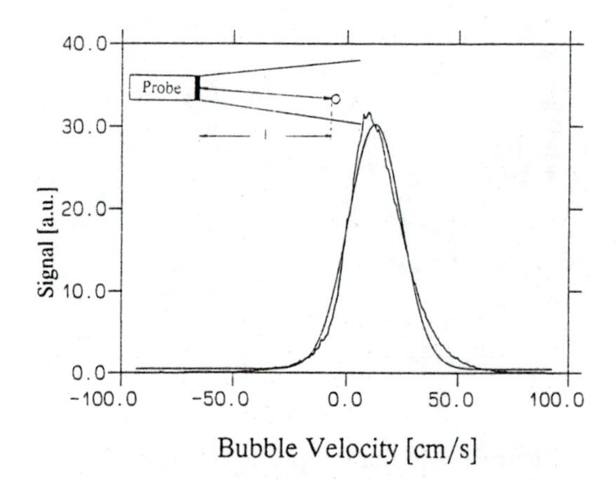

Figure 2. Typical bubble velocity distribution measured during a cultivation of *Penicillium crysogenum* in a 1 m^3 stirred tank bioreactor

<u>*Liquid-Phase Properties*</u>

Traditionally, fluid-dynamical information is most often represented by flow velocity vector fields. In multiphase flows within bioreactors, the flow velocities of the continuous liquid phase is of utmost interest, since the transport of the nutrients is primarily performed by liquid phase motions.

Locally highly resolved fluid velocity fields are difficult to measure during real cultivations, since the influence of the bubbles on the measuring signals of hot-film or laser Doppler anemometers are extremely difficult to correct. As an alternative, the heat pulse technique was developed, which essentially is a local stimulus-response technique using heat to mark the liquid-phase fluid elements. As depicted in Figure 3, the technique requires two probes to be immersed into the flow some centimeters apart from each other, a tiny heat source and an appropriate fast thermometer probe. The arrangement primarily measures the time-of-flow distribution of those flow elements which are marked at the heater and happen to arrive later on at the detector. From the known distance between both probes, the mean velocity of the particles along the line, defined by them, can be obtained. Since the flow elements are marked pseudostatistically, the signal-t-noise ratio of the results is so high that measurements are possible under nearly all rheological conditions which are met in bioreactors. Figure 4 shows a result measured

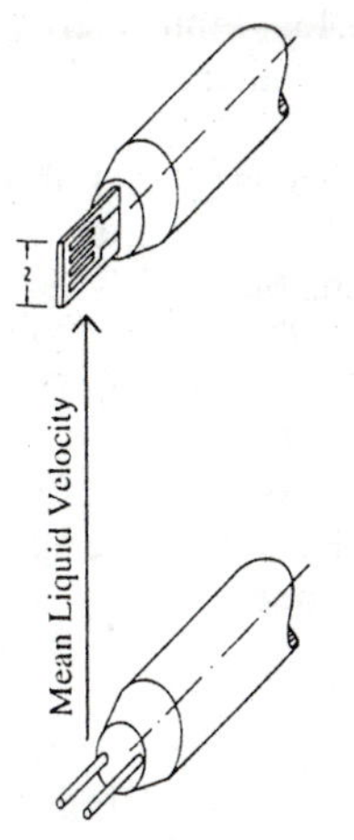

Figure 3. Arrangement of the heat pulse technique to measure the time-of-flow distribution between two points in the flow. From this distribution, the mean velocity and the mixing mechanism can be determined.

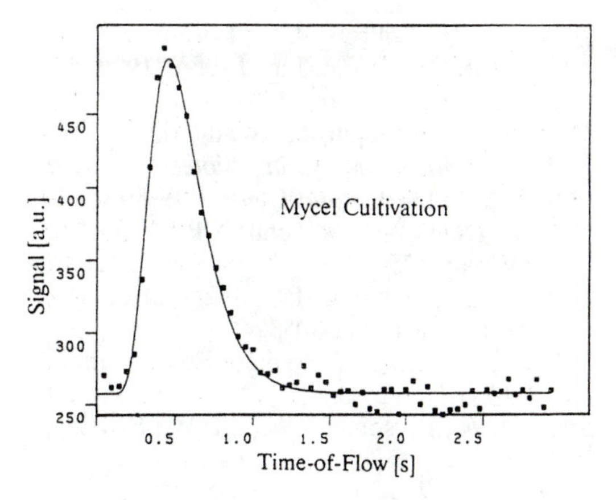

Figure 4. Characteristic result of a time-of-flow distribution as measured during a cultivation of *Cephalosporicum acremonium* in a bioreactor, where peanut meal of about 20 vol. % was used as substrate.

during a viscous cultivation whith high solid holdups.

The heat pulse technique leads to estimates of time-of-flow distributions averaged over a total measuring time of several minutes, hence, with respect to the time resolution, it cannot compete with the anemometers used in single phase flows. However, the method is capable to additionally provide information about the local mixing behavior in the liquid phase (Lübbert and Larson, 1990). With the heat pulse technique, it was possible to clarify the role of rising bubbles in the mixing of the liquid phase.

Concluding Remarks

By means of advanced local measuring techniques, many details of the fluiddynamics in bioreactors can be investigated. However, the expense necessary to acquire all the data necessary for a comprehensive modelling of a given reactor during real cultivations, is very high. The general strategy must, therefore, be to investigate the fluiddynamics in a model medium which is rheologically similar to the original production system. In a next step, the models must then be varified by measurements in the original system at a few critical places in the flow systems.

Most measuring devices discussed are not commercially available. The reason is that the developments are too expensive for the small market. Hence, it is recommended that bioreactor developers cooperate with the developers of the advanced measuring devices.

References

Bodemeier, S.; Claas, W.; Lübbert, A., AIChE Symp. Series, **1991**, submitted

Brentrup, L.; Weiland, P.; Onken, U., Chem.-Ing.-Techn., **1980**, *52*, pp 72-73, Synopse 758

Bröring, S.; Fischer, J.; Korte, T.; Sollinger, S.; Lübbert, A., Can.J.Chem.Eng., **1991**, *69*, pp 1247-1256

Bryant, J., Adv.Biochem.Eng., **1977**, *5*, pp 101-123

Calderbank, P.H., Trans.Inst.Chem.Eng.,' **958**, *36*, pp 443-463

Devanathan, N.; Moslemian, D.; Duducovic, M.P., Chem.Eng.Sci., **1990**, *45*, pp 2285-2291

Fischer, J.; Bröring, S.; Lübbert, A., Chem. Eng.Technol., **1991**, in press

Fischer, J.; Bröring, S.; Lübbert, A., AIChE Symp. Series, **1991**, submitted

Genenger, B.; Lohrengel, B.; Lorenz, M.; Vo gelpohl, A., Chem.-Ing.-Techn., **1990**, *62*, pp 862-863, Synopse 1894

Lübbert, A.; Fröhlich, S.; Schügerl, K., In: *Mass Spectrometry in Biotechnological Process Analysis and Control*, E. Heinzle, M. Reuss, eds., Plenum Publ. Corp., **1987**, pp 125-142

Lübbert, A.; Larson, B., Chem.Eng.Sci., **1990**, *45*, pp 3047-3053

Lübbert, A.; Larson, B., Chem.Eng.Technol., **1987**, *10*, pp 27-32

Lübbert, A.; Larson, B.; Wan, L.W.; Bröring, S., I.Chem.E. Symposium Series, **1990**, *121*, pp 203-213, Hemisphere

Prince, M.J.; Blanch, H.W., AIChE J., **1990**, *36*, pp 1485-1499

Schmidt, J.; Nassar, R.; Lübbert, A., Colum bus, Ohio, Paper submittet to Chem.Eng.Sci., **1992**

Stravs, A.A.; Pittet, A.; Stockar, U. von; Reilly, P.J., Biotechn.Bioeng., **1986**, *28*, pp 1302-1309

Stravs, A.A.; Wahl, J.; Stockar, U. von; Reilly, P.J., Chem.Eng.Sci., **1987**, *42*, pp 1677-1688

Oxygen Transfer in Industrial Air Agitated Fermenters

Roy Carrington, Keith Dixon* and **Anthony J. Harrop,** Pfizer Ltd., Ramsgate Road, Sandwich, Kent. CT13 9NJ, England. (Fax: (44) 304 616760)
Graeme Macaloney, Alberta Research Council, P.O.Box 8330, Station F, Edmonton, Alberta, Canada, T6H 5X2.

Investigations into the oxygen transfer and mixing characteristics of a $20M^3$ "bubble column" fermenter operating an industrial, highly viscous, non-Newtonian, antibiotic fermentation indicated the following:
1. The internal helical cooling coil acted as a "leaky" draft tube, allowing some radial mixing between the riser and downcomer.
2. Mixing times of between 14 and 18 seconds were observed with viscosities of up to ~900 cp. Dissolved oxygen tension (D.O.T.) was measured throughout the vessel, no poorly oxygenated zones were observed.
3. K_La was found to increase almost linearly with increasing power input and decrease exponentially with increasing viscosity.
4. Installation of a solid draft tube resulted in decreased oxygen transfer efficiency and increased mixing times compared to the "leaky draft tube" provided by the cooling coil.
5. The oxygen content of the micro-bubbles was found to vary between 8% and 15%, micro-bubbles contribute up to 15% of overall oxygen transfer.

Pfizer operate, commercially, a number of different fermenter designs. The present program was undertaken to enable detailed comparisons of mixing and mass transfer characteristics of different configurations of air agitated fermenters to be made under controlled, commercial process conditions.

Materials and Methods

Experiments were carried out in a $20M^3$ pilot scale bubble column fermenter fitted with an internal helical cooling coil (Fig.1).

The fermentation studied was a commercial streptomyces antibiotic fermentation which used a complex medium and produced a viscous non-Newtonian broth. For simplicity the apparent viscosity was measured at a single shear rate (10 sec^{-1}) using a Brookfield viscometer (model LVF) and a number 3 spindle at 60 r.p.m.

Investigation of Circulation Time and Pattern

Tracer studies were performed at various apparent broth viscosities up to 900 cp by injecting 50% w/w NaOH solution at two different points, A and B, at the top of the fermenter (Fig 1). The progress of the shots around the vessel was monitored using the six pH probes also indicated in Fig.1. When the NaOH was added at point A the sequence and times of response of the six probes were as recorded in Table 1. When the NaOH was added at point B on the opposite side of the vessel from the pH probes the size of the responses was much smaller and in general about 2 seconds later than those seen when additions were made at point A. However, the probe response sequence was essentially the same.

The results indicate that the vessel behaves like an airlift fermenter. It has a region of good mixing in the zone above the cooling coils, while the coils themselves act as a "leaky" draft tube and promote flow down the annulus with backmixing taking place between the coils into the riser section.

The draft tube like behaviour of the cooling coils gives liquid velocities of about 1m/sec which result in circulation times of 9-12 seconds and mixing times of 14 to 18 seconds.

Measurement of the Dissolved Oxygen Profile

The fermenter was fitted with one mobile and four static D.O.T. electrodes (Fig 1). All D.O.T. electrodes were spanned to read 98% saturation 2 hours after inoculation. The static electrodes gave the same D.O.T. profile as

1054–7487/92/0183$06.00/0

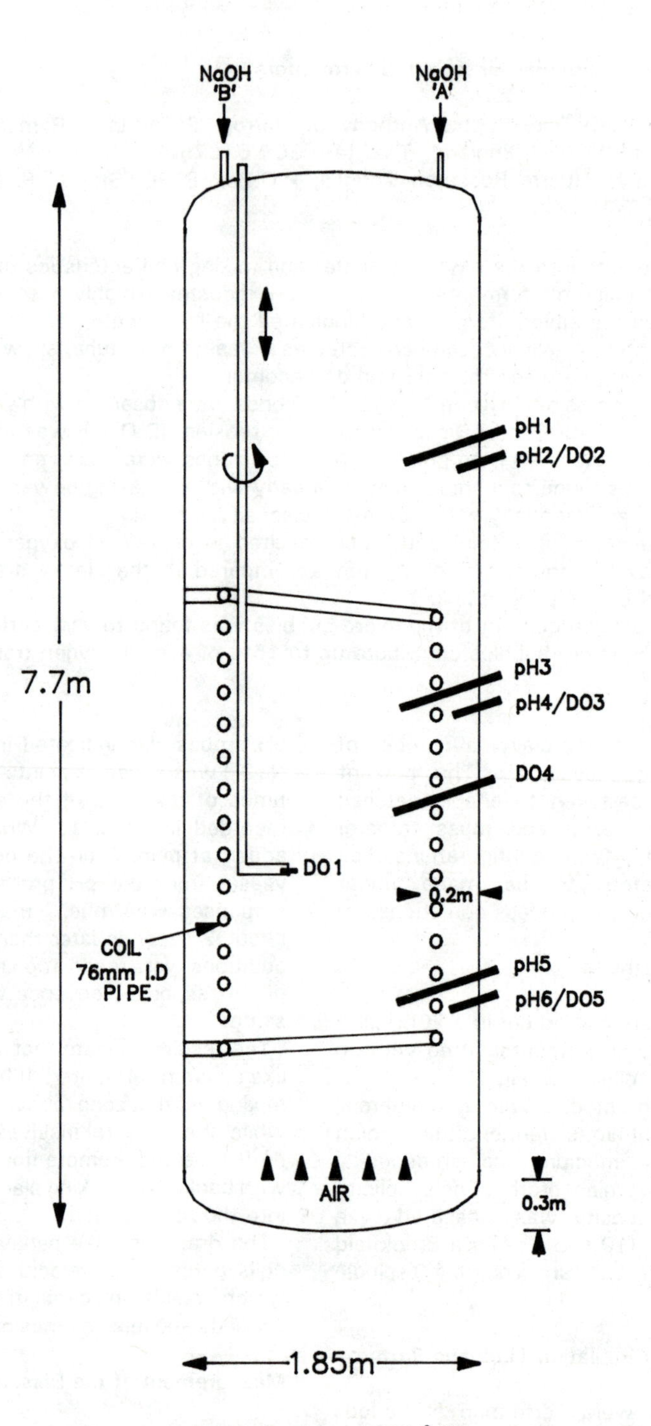

Fig. 1. Idealised diagram of $20M^3$ fermenter.

each other throughout the fermentation cycle. No reproducible D.O.T. gradient was observed between individual static electrodes over a large number of fermentation trials. The mobile D.O.T. electrode was used to measure D.O.T. at different heights and radial positions in the vessel and at different viscosities throughout the fermentation timecourse. The results showed that at any given height above the sparger, the D.O.T. at different radial positions was constant.

Mobile D.O.T. electrode readings at different heights were compared with those of the static electrodes. Static electrode readings varied with time due to changes in respiration through the course of the experiment. Results (Table 2) showed that at low viscosity the D.O.T. was constant throughout the vessel at any height and any point in time. As viscosity increased there was some evidence of partial oxygen depletion at the bottom of the vessel. The higher reading given by the mobile electrode at 880 and 1350 cp is thought to be a calibration difference rather than an actual difference in D.O.T.

Air Lift Configuration

The $20M^3$ fermenter was modified to a true air lift design by installing a continuous draft tube within the helical coil. Measurements of K_La at different power inputs and viscosities were made, for comparison with the "bubble column" configuration. Also D.O.T. profiles, circulation rates and mixing times were measured. The overall circulation time was found to be 9-12 seconds, similar to that obtained for the "bubble column" configuration. The mixing time was found to be 18-24 seconds, compared to a mixing time of 14-18 seconds for the "bubble column."

Mixing time is thought to increase due to the presence of the continuous draft-tube, preventing back mixing, and radial mixing between the riser and downcomer streams, thus resulting in poorer overall mixing.

Once again, the results of the mobile D.O.T. electrode indicated the D.O.T. to be remarkably constant throughout the fermenter. There was some evidence of slightly lower D.O.T. near the sparger and measurements indicated that oxygen concentrations in the broth leaving the downcomer were slightly lower than those in the bulk of the vessel.

K_La was measured for the draft tube configuration at various levels of power input and viscosity. Results are compared to those of the "bubble column" configuration in Fig 2. These show that the introduction of a continuous draft tube resulted in a 5 to 25% reduction in K_La, the greatest reduction being observed at high viscosity. Thus the "leaky" draft tube design gives better oxygen transfer than the continuous draft tube.

Effect of Micro-bubbles

At high viscosity the bubble size distribution was essentially bi-modal, consisting of a few large, highly coalesced bubbles and numerous micro-bubbles. The micro-bubbles were "trapped" in the broth, their low rising velocity unable to overcome the broth yield stress. In order to determine the contribution to oxygen transfer of these micro-bubbles, experiments were conducted to determine their percentage hold-up, average oxygen content and diameter. Results indicated that the micro-bubbles ranged from < 0.1 mm, to 2 mm diameter, and typically constituted 25% of the total gas hold-up. Considering the small size and large number of these bubbles, they have a very large total area for oxygen transfer compared to the larger bubbles. Also, the micro-bubbles were shown to have an oxygen content between 8% and 15%, indicating considerable potential for oxygen transfer.

In order to quantify the effect of micro-bubbles on K_La it was necessary to measure their rate of generation at different aeration rates. The method used was first described by J. Philip in 1987(1).

A glass column (16 cm dia and 2 m high) was filled with fermentation broth which had been de-aerated by holding under a vacuum for 1 hour to remove all micro-bubbles. Air was then sparged into the broth, at a constant rate. Periodically, aeration was stopped and the height of the broth in the column was measured after allowing rising air to escape. From this it was possible to determine the rate of increase in hold-up due to the accumulation of micro-bubbles. A curve was plotted of increase in hold-up against time. (Fig 3).

During fermentation at steady state, the rate of production of micro-bubbles is equal to their rate of coalescence and escape. When de-aerated broth is used, micro-bubbles are generated at a constant rate but their initial concentration is so low that the rate of coalescence is negligible. Therefore, the initial increase in trapped air is equal to the rate of

TABLE 1 **pH RESPONSE TIMES FROM TRACER STUDY ADDITION POINT A**

Probe No	Response Time after addition of Sodium Hydroxide shot (seconds)	Order of Response
1	2.50	2
2	1.25	1
3	5.25	5
4	3.50	3
5	5.75	6
6	5.00	4

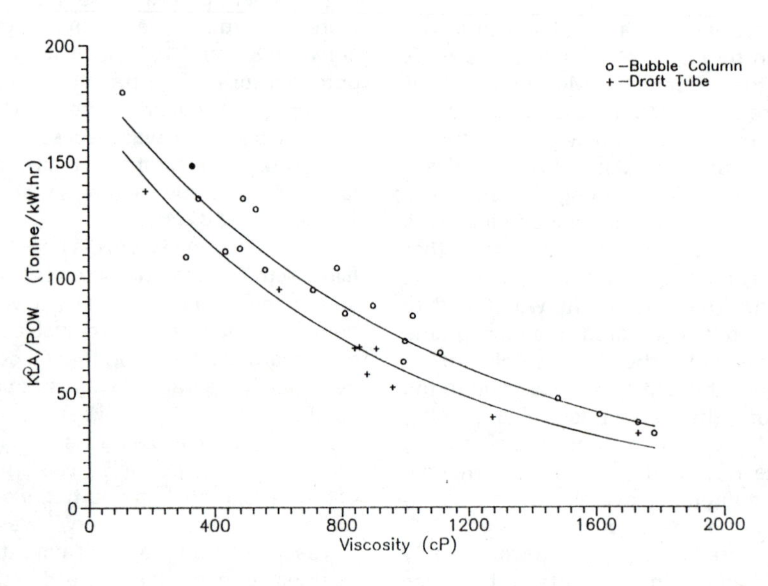

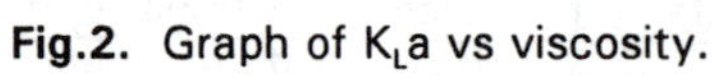

Fig.2. Graph of K_La vs viscosity.

TABLE 2 DISSOLVED OXYGEN PROFILE

Apparent Viscosity (cp)	Fermentation Time (Normalised Hours)	Mobile D.O.T. (% Sat^n)	Height above Sparger (m)	Average static D.O.T. (% Sat^n)	ΔD.O.T. Mobile - Static (% Sat^n)
420	35.8	36.1	5.5	39.2	3.1
420	35.4	38.4	4.7	39.7	1.3
420	35.0	37.7	3.3	39.1	1.4
420	34.6	37.4	1.9	40.7	3.3
420	34.3	35.4	0.5	38.0	2.6
880	36.9	47.8	5.4	39.4	8.3
880	36.6	46.4	4.7	37.7	9.5
880	36.3	42.7	3.3	33.5	9.3
880	36.0	37.0	1.9	30.8	6.6
880	35.7	36.3	0.5	31.0	5.3
1350	82.5	44.3	5.9	36.5	7.8
1350	83.3	34.5	4.0	30.5	4.0
1350	83.7	32.0	2.6	28.0	4.0
1350	84.2	26.1	1.2	23.3	3.2
1350	84.6	16.4	0.5	17.0	(0.4)

formation of micro-bubbles. This can be obtained from the initial slope of the curves of hold-up against time (Fig 3). The fraction of sparged air going to trapped air can also be calculated.

Experiments were carried out under a range of conditions varying from the same volumetric air flow rate to the same superficial gas velocity as those employed in the 20 M^3 fermenter. A range of broth viscosities were used, varying from 614 cp to 1420 cp.

Results (Table 3) show that over a wide range of operating conditions between 0.4% and 0.7% of the sparged air forms micro-bubbles. This agrees well with the work carried out by J. Philip (1). From a knowledge of micro-bubble formation rate and oxygen content, it is possible by mass balance to calculate the average oxygen transfer rate from the micro-bubbles to the broth. This gives an average micro-bubble contribution of 3.7 m.mol/L/hr which represents between 7% and 15% of the overall oxygen transfer rate. Also, given the micro-bubble hold-up and generation rate it is possible to calculate the mean residence time, this approximates to 6.4 minutes.

Thus, overall conclusions are that micro-bubbles build up very rapidly, steady state is reached within about 10 minutes and the mean residence time is approximately 6 minutes.

TABLE 3 **MICRO-BUBBLE FORMATION**

Air Flow Rate (l/min)	Superficial Gas Velocity (m/min)	Viscosity (cp)	Rate of Formation of Micro-Bubbles (l/min)	Percentage of Sparged air Forming Micro-Bubbles (%)
180	8.9	1185	0.40	0.4
82	4.1	1420	0.19	0.5
82	4.1	614	0.19	0.5
15	0.7	980	0.05	0.7

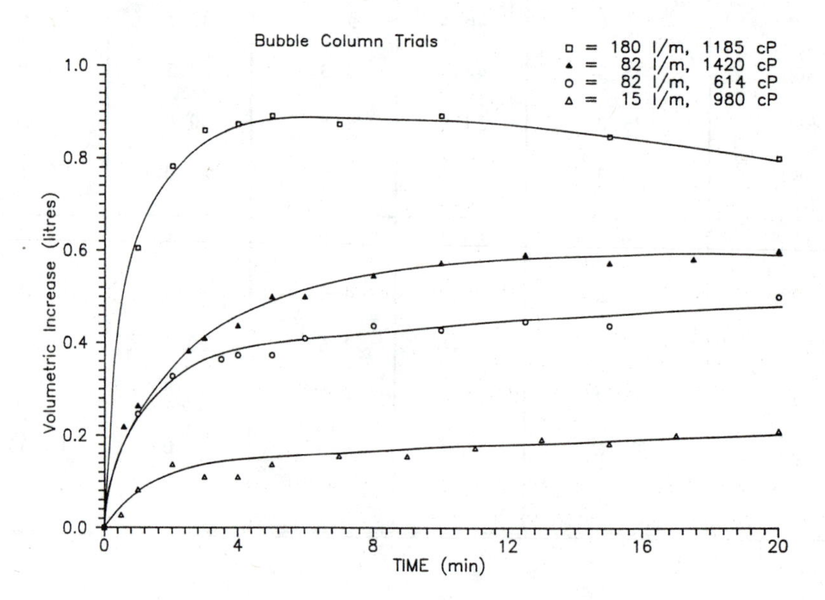

Fig.3. Graph of volume increase vs time.

Their contribution to overall oxygen transfer is of the order of 15% or less. Hence, though micro-bubbles provide the bulk of the air/liquid interfacial area they are of minor importance in terms of overall oxygen transfer.

Considering their long residence time and the diffusion gradient generated, these micro-bubbles must suffer diffusion limitation to be left with 10% oxygen after 6 minutes. Micro-bubbles behave as rigid spheres with little motion relative to the surrounding broth, thus, surfactants probably migrate to the gas/liquid interface and present a barrier to diffusion, and hence, oxygen transfer.

Acknowledgement

We would like to thank Professor J. F. Davidson of Cambridge University for his kind assistance throughout the period of this work.

References

1) J. Philip - *Viscous Liquids in Bubble Columns,* phD thesis, Christ's College, Cambridge, 1987.

FOAMING, MASS TRANSFER, AND MIXING: INTERRELATIONS IN LARGE SCALE FERMENTORS

K. Van't Riet and H. M. Van Sonsbeek, Wageningen Agricultural University, Department of Food Science, Food and Bioprocess Engineering Group, P.O. Box 8129, 6700 EV Wageningen, The Netherlands (FAX: 31–8370–82237)

Abstract

Mass transfer relations predict the oxygen concentration in the stirrer region for a 'standard' broth. When the broth is viscous a standard apparent viscosity can be introduced and the oxygen concentration can be calculated again. However the broth is not 'standard' and in particular the characteristics as dependent on foaming conditions are hardly known. Coalescence properties and foaming properties are related. This leads to changes in oxygen concentration at addition of chemical antifoam.

Another complicating aspect is the relation between mass transfer and mixing. Mass transfer mainly takes place in the stirrer region. For large scales this leads to considerable oxygen concentration profiles, dependent on the position in the vessel and a minimum required oxygen concentration in the stirrer region.

Foaming

In general liquids will foam only when impurities are present. In fermentations these are usually proteins that are always present: either extracellular proteins or cell lytic products. Foaming, can be advantageous for mass transfer, but causes very serious operating problems, and therefore has to be nullified. Mechanically foam destruction is often proposed but consumes too much energy and is not fully reliable. A chemical antifoam system will always be necessary.

Antifoam agents

Most antifoam agents are strong surface-tension-lowering substances. When they are added to a culture liquid (or protein solution), foaminess is strongly suppressed. Antifoam liquids are usually composed of oils, fatty acids, esters, polyglycols and siloxanes and destabilize protein films by 1),hydrophobic bridges between two surfaces, 2) displacement of the absorbed protein, 3) rapid spreading on the surface of the film. Especially the second mechanism is often relevant. Low molecular weight surfactants, like fatty acids, monoglycerides and phospholipids, give a lower surface tension than proteins. Therefore the absorbed protein will be displaced and for instance the repulsion between two film surfaces is reduced.[1]

Effects of antifoam agents on mass transfer

Mass-transfer rates (k_l) are dependent on the mobility of the surface of bubbles. In tap water bubbles with a radius less than 1 mm have a rigid interface, and bubbles with a radius greater than 2 mm have a mobile interface, which is advantageous for the mass-transfer rate. Proteins will stabilize bubble surfaces to some extent, yet, and influence on k_l due to this effect is not expected in fermentations, because the diameter of these bubbles is usually greater than 3 mm.

Proteins can shift the fermentation broth from coalescing to non-coalescing, which enables much smaller bubbles. Yet, this effect is minimised because both foam stability and oxygen transfer from the gaseous to the liquid phase depend on surface phenomena. Conditions which cause the collapse of bubbles in foam also favour the coalescence of bubbles within the liquid phase, resulting in larger bubbles with reduced surface to volume ratios and hence a reduced rate of oxygen transfer. Therefore foam control in a production fermenter should be a compromise between minimizing the foam height and maximizing mass transfer.[1]

Foam control

Foam control is not simply a matter of preventing foam formation, but has far-reaching effects, up to the productivity of a fermenter. In figure 1 three assumed k_la profiles can be seen. For increasing liquid heights the k_la decreases for two reasons: 1) The volume increases at equal energy input; 2) Antifoam addition is necessary to keep the foam level below overflow condition. The assumed sharp decrease of k_la above a liquid height of 9.7 m is due to abundant antifoam addition. From these k_la profiles figure 2 can be calculated. It shows that the optimum total oxygen transfer, and therewith the optimal productivity of the fermentor, is strongly dependent on the k_la characteristics. Only in case of a very limited dependance of k_la with liquid height should the fermentor be filled up to the top. In other cases the optimum is at lower broth heights.

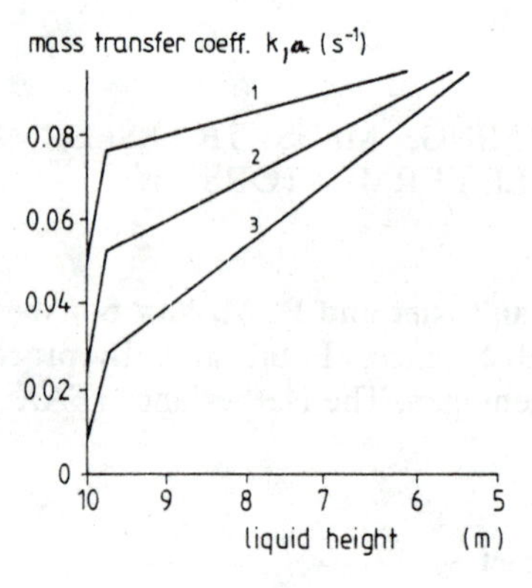

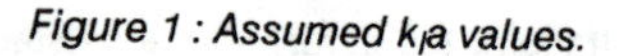

Figure 1 : Assumed k_la values.

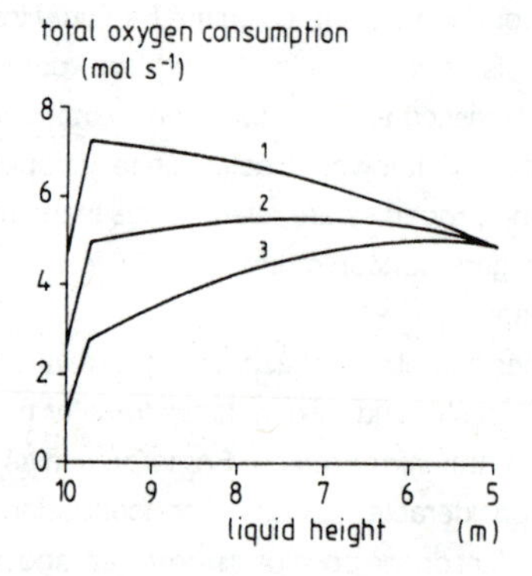

Figure 2 : Calculated total oxygen-consumption values.

Mixing

Another aspect that is of great importance for mass transfer in bioreactors is mixing. For aerated stirred-tank reactors, correlations are available to estimate mixing times for the liquid phase under turbulent conditions.[2] For bubble columns, mixing of the liquid phase is dependent on the flow regime in the vessel. The homogeneous regime can only occur at low superficial velocities together with an evenly distributed sparger. In all other cases the heterogeneous regime occurs. For both regimes equations that describe mixing in the liquid phase are available.[2] Also for air-lift loop reactors it is possible to estimate mixing times, although an iteration procedure is necessary.[2]

Scale-up

Especially with respect to mass transfer and mixing scale up of bioreactors is very complicated. The available tools for scale up (or scale down) are:[3]

1 dimensional analysis
2 rules of thumb
3 mechanistic equations
4 time constants

Dimensional analysis is a powerful scale-up technique but collecting the parameters and variables is a rather arbitrary process and keeping all dimensionless groups constant at scale up is impossible for fermentations.[3] Rules of thumb for bioreactor design are numerous and originate often from a selection of equations derived from literature. Rules of thumb for scale-up purposes are usually based on constant P/V, k_la, v_{tip} or p_{O2}. Problems can show up when applying these rules of thumb. For instance by keeping P/V constant, k_la will be constant, but the mixing time increases at scale scale up.

True mechanistic equations do not exist, but empirical equations with a good record and well tested in a well defined range are known. Therewith the rules of thumb are reliable.[3]

Critical times for mass transfer

Time constants are defined as capacity divided by rate. The advantage of working with time constants is that they are a quick manner to estimate the importance of the different mechanisms involved in biotechnological processes. Limitations of processes can be quantified by calculating their critical time constant. In aerated stirred-tank reactors mass transfer is assumed to take place only in the stirrer region.

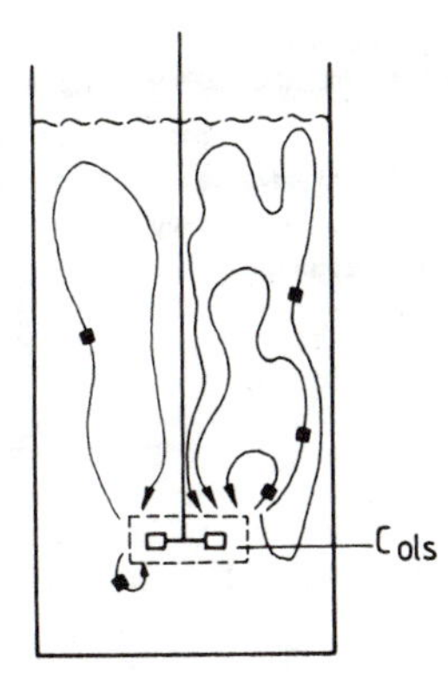

Figure 3 : Schematic representation of trajectories of (segregated) fluid elements. Fluid element and its trajectory: -■-; Region where oxygen mass transfer occurs: ---------.

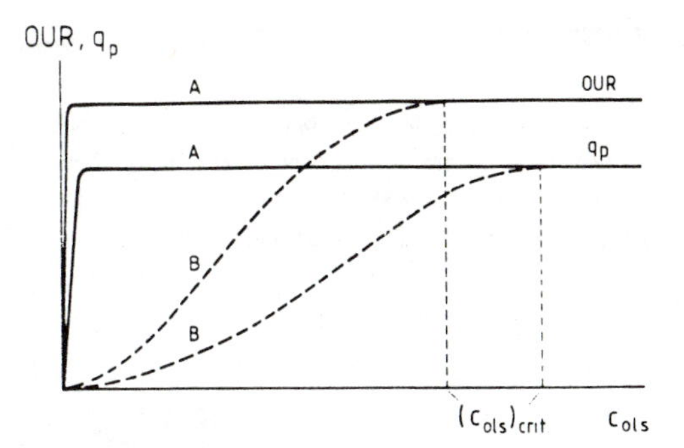

Figure 5 : Example of the influence of C_{ol} on OUR and q_p. Line A: A very small-scale, ideally mixed fermenter. Line B : A large-scale, stirred fermenter.

The critical time value for oxygen is then :

$$t_{cro} = \frac{C_{ols}}{OUR}$$

where C_{ols} is the concentration in the stirrer region and OUR is the oxygen uptake rate, which is considered to be constant in time. This means that all fluid elements with a residence time value that is larger than t_{cro} will be depleted of oxygen before returning to the stirrer region (figure 4).

If the time constant for mixing exceeds t_{cro} than local oxygen depletion in the vessel occurs. To prevent this the C_{ols} value should be high. However, for mass transfer C_{ols} should be as low as possible to obtain a maximal driving force. These two requirements contradict each other and a compromise should be found. A method to determine the C_{ols} under critical conditions is schematically given in figure 5. For small-scale, ideally mixed reactors, the critical oxygen concentration at which biological production (q_p) and OUR starts to decrease is very small. For large scale vessels this value is much higher because of local depletion of oxygen. In case such measurements cannot be made a first estimate of t_{cro} is the circulation time t_c being $1/4$ t_m (t_m is the mixing time).

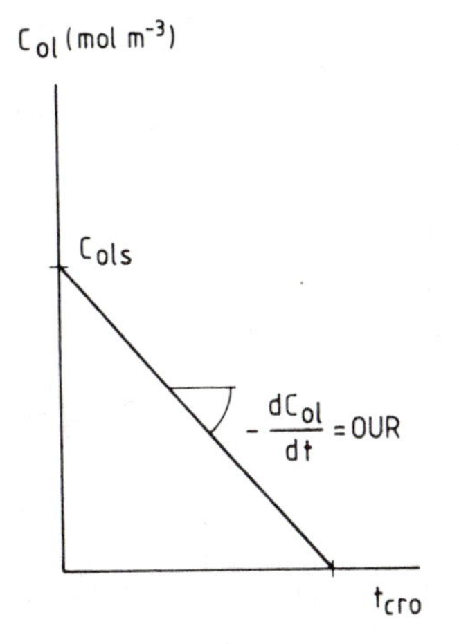

Figure 4 : Critical time for oxygen depletion.

The example of the critical times concept does not apply to the bubble column because mass transfer takes place in the whole column. Also in the air-lift loop reactor the single point addition of oxygen does not hold at all. Mass transfer takes place in the whole riser part of the air lift. Oxygen depletion can occur by liquid passing through the downcomer, but this usually only occurs for columns exceeding a height of about 10 m. Mass-transfer limitation can also occur in the upper part of very tall bubble columns or air-lift loop reactors.

Conclusion

It is made clear that the influence of both foam control and mixing, have complex consequences for mass transfer. Considering design or scale-up these aspects require careful optimization with respect to mass transfer. The critical time concept can be a good tool for this purpose.

References

1 Prins A., Van 't Riet K. (1987). Proteins and surface effects in fermentation: foam, antifoam and mass transfer. Trends in Biotechnology, 5, 296-301.

2 Van 't Riet K., Tramper J. (1991). Basic Bioreactor Design. Marcel Dekker, Inc., New York.

3 Kossen N.W.F. (1991). Scale-up in biotechnology. Proceedings: "Recent advances in industrial applications of biotechnology", Kusadasi, Turkey, September 16-27, Nato Advanced Study Institute.

New Agitators v Rushton Turbines: A Critical Comparison of Transport Phenomena

A W Nienow, Rhône-Poulenc Professor of Biochemical Engineering, SERC Centre for Biochemical Engineering, The University of Birmingham, U.K. (FAX: 021-414-5324)

Multiple Rushton turbines of about ⅓ of the vessel diameter have been the agitation system of choice for stirred bioreactors since the late 1950's. However, basic transport phenomena studies in the 1970's, highlighted their deficiencies. Since the mid-1980's, similar studies have shown that many of these weaknesses can be overcome (or at least improved upon). On the basis of studies mainly conducted in the SERC Centre, Rushton turbines are compared with new agitators including the Scaba 6SRGT and other hollow-blade agitators, the Lightnin' A315, the Prochem Maxflo T and the Ekato Intermigs. Weaknesses and new problems are also outlined.

Agitators are required to achieve a set of physical mixing objectives, e.g., bulk fluid and gas phase mixing, air dispersion, oxygen transfer, heat transfer, solid suspension (for microcarrier culture, for example), etc. It should be possible to design for these conditions and this requires knowledge of the best type of agitator and air sparger, the optimum feed position for antifoam, for chemicals for pH control (and for nutrients in fed batch operations). The final specification must also cover the agitators size, speed and power input. The biological performance depends on the interaction of these physical aspects of mixing with the sensitivity of the particular bioprocess to them, e.g., level of dissolved oxygen, pH variation, mechanical stress from turbulence or bubble bursts, etc. This interaction is biological system specific. In addition, there may also be feedback effects, e.g., agitation conditions affect mycelial morphology which may lead to rheological complexity (high viscosity) which in turn will require more intense agitation to satisfy the physical demands of mixing.

However whilst recognising that the final biological performance is system specific (Nienow, 1990a), this review considers only generic physical aspects.

Strengths and Weaknesses of the Rushton Turbines and other Radial Flow Agitators

Power Draw

The flat blade and disc of a Rushton turbine leads to the formation of two low pressure trailing vortices at the rear of each of the six blades (Nienow and Wisdom, 1974; Van't Riet and Smith, 1975). This in turn leads to a high power number under unaerated conditions which leads to high torque for a given power input and a high initial cost (Muskett and Nienow, 1987). When aerated, air is concentrated in these low pressure regions (ventilated cavities) (Fig. 1), leading to a local increase in pressure, and a ~50% fall in power (Nienow et al, 1977). This leads to a severe loss of oxygen mass transfer potential (Van't Riet, 1979) though it can be reduced by modifying the sparger (Nienow et al, 1988). The fall is a very complex function of agitator speed, aeration rate, number of impellers (Hudcova et al, 1989), scale and fluid properties and even now is not capable of being accurately predicted (Nienow, 1990b).

By changing the shape of the blade from a flat plate to a plate of parabolic cross section, the trailing vortex and the associated ventilated

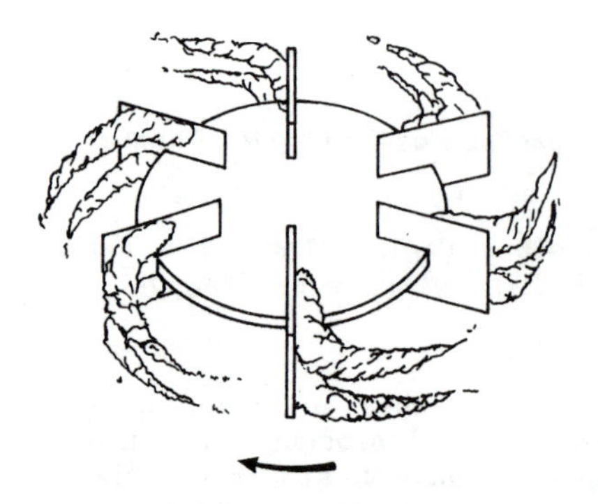

Fig. 1. A Rushton turbine dispersing air (adapted from Nienow and Wisdom, 1974)

cavities are essentially eliminated. Thus, a low power number is obtained; anand the power remains constant (or nearly so, even in high viscosity broths) on aeration. Such an impeller is the Scaba 6SRGT (Saito et al, 1990) (Fig.2). The I.C.I. Gasfoil is very similar (Cooke et al, 1988) and the Chemineer CD6 is intermediate between the Rushton and the Scaba (Dawson, 1992).

Flooding

An impeller is flooded when the air flow into it is too high for it to handle. This condition is indicated when the bulk flow pattern in the vessel normally associated with the particular agitator (radial in the case of the Rushton turbine) is lost to be replaced by a centrally flowing air-broth plume up the middle with a liquid-flow as an annulus (Nienow et al, 1985). A good impeller is one which for a given power input can handle a large air flow before flooding and the 6SRGT excels the

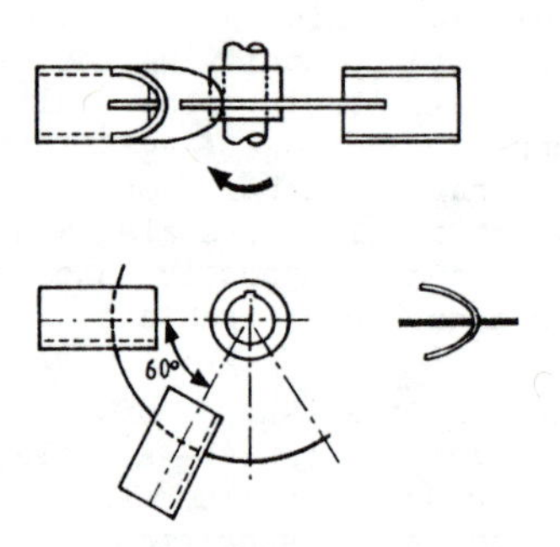

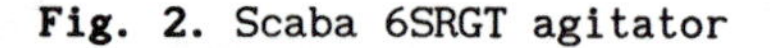

Fig. 2. Scaba 6SRGT agitator

Rushton in this regard too (Saito et al, 1990).

Bulk Blending (Homogenisation)

Low power number, large diameter impellers are better for bulk blending than large power number, small diameter ones, especially at equal torque, power and speed (Nienow, 1990b). Thus, the 6SRGT is better than the Rushton. However, all radial flow agitators give very poor top-to-bottom blending in multiple agitator configurations as is often found in large scale fermenters (Cronin and Nienow, 1989). Good bulk blending (homogenisation) is the feature that inevitably deteriorates markedly on the large scale leading to lower concentrations of oxygen in the broth away from the agitators; and a high concentration of chemicals close to feed points, especially if addition is onto or near the top surface. Ideally, all such additions should be made into the impeller zone (Cronin and Nienow, 1989).

Bulk blending is probably the critical feature in high viscosity fermentations such as Xanthan (Xhao Xueming et al, 1991) and mycelial broths (Nienow, 1990c).

Heat Transfer

Heat transfer only becomes a critical process requirement on the largest scale, especially in very shear thinning broths, prone to cavern formation (Nienow and Elson, 1988). Large impeller-to-vessel diameter ratio impellers (Xhao Xueming et al, 1991), that do not lose pumping capacity on aeration (Cooke et al, 1988) and good bulk blending help in this respect too.

Strengths and Weaknesses of Axial Flow Agitators

To improve top-to-bottom bulk blending, low power number downward pumping axial flow agitators work well unaerated. However, once aerated, such impellers are prone to flooding accompanied by very big falls in power draw since they lose pumping capacity (Chapman et al, 1983). Also, there are inherent instabilities in flow pattern because the downward flow produced by the agitator is opposed by the upward flow of the sparged air. The problem of flooding and fall in power has been partially overcome by the development of hydrofoil agitators with large solidity ratios ((the horizontally projected area of the blades)/(the horizontal area swept out by the blades)). Impellers such as the Prochem Maxflo T (Fig.3) (Buckland et al, 1988) and the Lightnin' A315 (Fig.4) (Nienow and Buckland, 1989) have have solidity ratios greater than 0.9. However, the problem of flow and associated torque instabilities leading to possible excessive vessel vibrations (Balmer et al, 1987) remains.

Extensive work at Birmingham has shown that with upward-pumping, pitched blade turbines, flow and torque instabilities are eliminated (Bujalski et al, 1990). The fall in power on aeration is less than with Rushtons and more air can be handled before flooding occurs. Preliminary work (Hass, 1987) has shown similar improved performance with the Prochem Maxflo T.

Ekato Intermig agitators are always used in at least pairs (Fig.5). They cause flows which are a mixture of upward and downward pumping when unaerated. On aeration, they become radial flow agitators because large cavities form, especially in viscous broths (Dawson, 1992). Thus, they are prone to compartmentalisation; and also large mechanical vibration instabilities can develop (Kipke, 1982).

Mass Transfer Rate ($k_L a$) and Hold-Up

Our extensive studies using a variety of techniques for measuring $k_L a$ have shown that equal power per unit volume and superficial air velocity in the same coalescing fluids (and in ones with modest coalescence inhibition) leads to the same $k_L a$ or hold-up, regardless of the impeller type (e.g., see Chapman et al, 1983; Balmer et al, 1987; Bujalski et al, 1990) in low viscosity fluids.

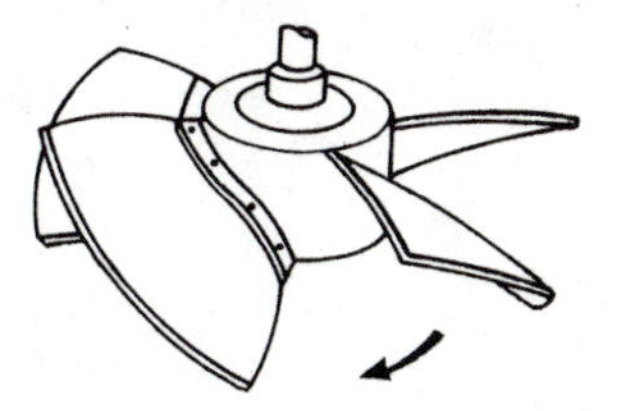

Fig. 3. Prochem Maxflo T agitator

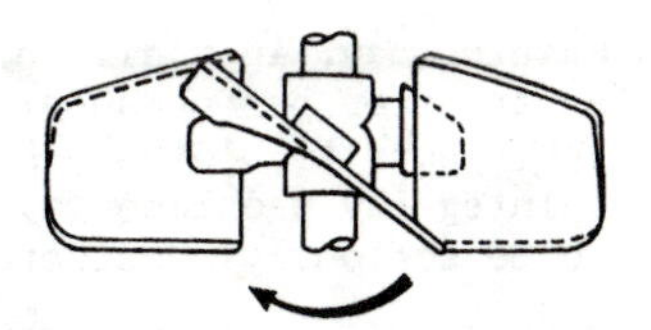

Fig. 4. Lightnin' A315 agitator

This conclusion was also drawn by Van't Riet (1979) in his exhaustive survey of literature $k_L a$ values, resulting in a generalised correlation for $k_L a$, independent of impeller type, for water. Recent work has confirmed this for modern impellers both in our laboratory, e.g., Intermig, A315 (Dawson, 1992) and elsewhere, e.g., I.C.I. Gasfoil (Cook et al, 1988) and Scaba 6SRGT (Shell, 1991).

As the fluid becomes more viscous and shear thinning, low power number agitators enable higher agitation speeds to be used. This leads to a lower viscosity and higher $k_L a$ (Cooke et al, 1988; Dawson, 1992). Overall $k_L a \propto (\text{viscosity})^{-0.5}$ (Hickman and Nienow, 1986).

Some very recent results also suggest that impellers which give large amounts of air recirculation at equivalent powers, e.g., Prochem's, give enhanced hold-ups in fluids where coalescence is severely inhibited (Machon et al, 1991).

Conclusion

Replacing standard Rushton turbines by larger, low power number agitators which do not lose so much power when aerated; which are able to handle more air without flooding; and which give better bulk blending and superior heat transfer characteristics is certainly possible. Designing new systems with higher speeds and larger diameters compared to the

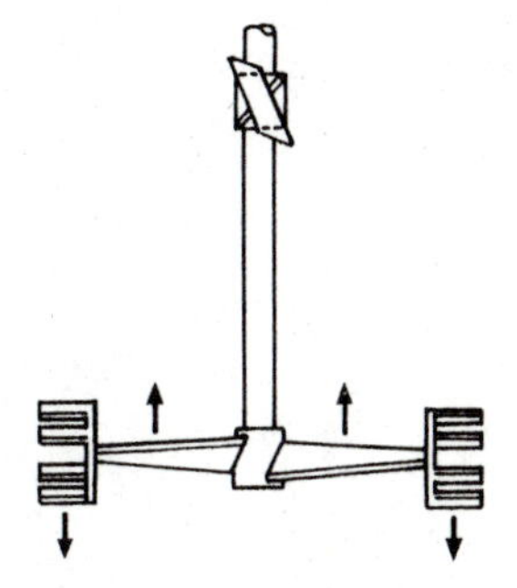

Fig. 5. Ekato Intermig agitators

standard Rushton may, in addition, give better mass transfer performance in both low and high viscosity broths. However, mechanical integrity problems may arise and need to be seriously considered.

References

Balmer, G.J.; Moore, I.P.T.; Nienow, A.W. In *Biotechnology Processes; Scale-up and Mixing*; Ho, C.S.; Oldshue, J.Y. Ed., A.I.Ch.E., N.Y., **1987**, pp.116-127.

Buckland, B.C.; Gbewonyo, K.; DiMasi, D.; Hunt, G.; Westerfield, G.; Nienow, A.W. *Biotechnol.Bioeng.*, **1988**, *31*, 737-742.

Bujalski, W.; Nienow, A.W.; Liu Huoxing; *Chem.Eng.Sci.*, **1990**, *45*, 415-421.

Chapman, C.M.; Nienow, A.W.; Cooke, M.; Middleton, J.C. *Chem.Eng.Res. Des.* **1983**, *61*, 82-95 and 182-185.

Cooke, M.; Middleton, J.C.; Bush, J. *Proc. 2nd Int.Conf. on Bioreactor Fluid Dynamics*, BHRA, Cranfield, U.K., **1988**, pp 70-100.

Cronin, D.G.; Nienow, A.W.; *In Proc. 3rd Bioreactor Project Res.Symp.*, (Moody, G.W. Ed.), Nat.Eng.Lab., East Kilbride, U.K., **1989**, pp 17-34.

Dawson, M.K. Ph.D. Thesis, University of Birmingham, U.K., **1992**.

Hass, V.C. *M.Sc. Project Report*, University of Birmingham, **1987**.

Hickman, A.D.; Nienow, A.W.; *Proc.Int.Conf. on Bioreactor Fluid Dynamics*, BHRA, Cranfield, U.K.**1986**, pp 301-316.

Hudcova, V.; Machon, V.; Nienow, A.W.; *Biotechnol.Bioeng.*, **1989**, *34*, 617-628.

Kipke, K.; *Proc. 4th European Mixing Conf.*, BHRA, Cranfield, U.K., **1982**, pp 389-398.

Machon, V.; McFarlane, C.M.; Nienow, A.W. *Proc. 7th European Mixing Conf;* KVIV, Belgium, **1991**, pp 243-249.

Muskett, M.J.; Nienow, A.W. *Fluid Mixing III*, I.Chem.E.Symp.Ser. **1987**, No.108, pp.33-48.

Nienow, A.W.; Wisdom, D.J.; *Chem.Eng. Sci.* **1974**, *29*, 1994-7.

Nienow, A.W.; Wisdom, D.J.; Middleton, J.C. *Proc. 2nd European Mixing Conf.*, BHRA, Cranfield, U.K., **1978**, pp F1-1-F1-16 and discussion, pp. X53-X55.

Nienow, A.W.; Warmoeskerken, M.M.C.G.; Smith, J.M.; Konno, M. *Proc. 5th European Mixing Conf.*, BHRA, Cranfield, U.K., **1985**, pp.143-154.

Nienow, A.W.; Liu Huoxing; Wang Haozhung; Allsford, K.V.; Cronin D.; Hudcova, V. *Proc. 2nd Conf. on Bioreactor Fluid Dynamics*, BHRA Cranfield, U.K., **1988**, pp.159-177.

Nienow, A.W.; Elson, T.P. *Chem.Eng. Res.Des.*, **1988**, *66*, 5-15.

Nienow, A.W.; Buckland, B.C.; *A.I.Chem.E. Annual Meeting*, Chicago, **1990**, Paper 75a.

Nienow, A.W. *Proc. 5th European Cong. Biotechnology*; Christiansen, C.; Munck, L.; Villadsen, J. Ed., Munksgaard, Denmark, **1990a**, Vol.II, pp 791-796.

Nienow, A.W. *Chem.Eng.Prog.*, *86*, **1990b**, 61-71.

Nienow, A.W. *Trends in Biotech.* **1990c**, *8*, 224-233.

Saito, F.; Nienow, A.W.; Chatwin, S.; Moore, I.P.T. *CHISA 90*, Prague, Czechoslovakia, Aug.**1990**, Paper J.3.5.

Shell KSLA Laboratory, *Private communication*, **1991**.

Van't Riet, K. *Ind.Eng.Chem. (P.D.D.)*, **1979**, *18*, 357.

Van't Riet K.; Smith, J.M. *Chem.Eng. Sci.* **1975**, *30*, 1083.

Xhao Xueming; Nienow, A.W.; Kent, C.A.; Chatwin, S.; Galindo, E. *Proc. 7th European Mixing Conf.*, KVIV, Belgium, **1991**, pp 277-283.

Maximising Mammalian Cell Expression of Heterologous Proteins in Continuous Flow Bioreactors.

Peter P Gray*, K Jirasripongpun, C Gebert
Bioengineering Centre, Department of Biotechnology, University of New South Wales, Kensington 2033, Australia
(FAX: 612 313 6710)

The production of r-hGH and r-hFSH by recombinant CHO cell lines growing on microcarriers in perfusion suspension culture is described. Continuous medium feeds coupled with subsequent additions of microcarriers have been used to obtain high cell densities at the end of the growth phase, prior to switching to a production phase on protein free medium. Cell densities of 5 x 10^7 cells/ml have been obtained using non-porous microcarriers (Cytodex 2), which are greater than the maximum cell densities which could be obtained using a porous microcarrier (8.4 x 10^6 cells/ml for Cultisphere G). At the high cell densities (> 10^7 cells/ml), it was possible to maintain the same specific productivities as observed at the low cell density fermentations, providing the perfusion rate of the culture was maintained at 0.3 vols /10^6 cells/day or above. At these rates, volumetric productivities of r-hGH of 0.78 g/l of reactor volume/day were observed.

The use of mammalian cells as hosts for the production of recombinant DNA derived heterologous proteins is now widespread.

Much of the early mystique surrounding the large scale growth of animal cells is now disappearing, and it is now realised that it is quite possible to routinely operate/analyse/optimise such fermentations using methods commonly applied to other microbial biocatalysts, providing the strict requirements for asepsis are observed.

Accordingly mammalian cell fermentations are now routinely performed using most of the common modes of bioreactor operation, viz. batch, fed-batch, repeated batch and continuous operation. Continuous culture with an internal recycle of cells, often referred to as perfusion culture, is widely used. For such continuous flow systems, whether the cells are in suspension cultures with an internal retention, in fluidized beds or in membrane-based bioreactors, the main criteria will be to produce the heterologous protein in the correct form, at high productivities and at the highest purity possible. The overall productivity from these continuous flow systems will be controlled by three main factors irrespective of the reactor configuration employed, viz: the number of viable cells; the specific productivity of the cells (q_p); and the stability of the cell line, which will control the time for which the bioreactor can be operated.

In this paper factors important in obtaining high productivities of heterologous protein from recombinant CHO cells growing in suspension cultures will be discussed.

Materials and Methods

The CHO cell lines grown in the study were constructed so as to express high levels of recombinant hGH and hFSH without the need for high levels of gene amplification (Friedman *et al*, 1989).

The cells were grown on microcarriers in an airlift reactor with an operating volume of approximately 2 litres. A schematic diagram of the bioreactor is shown in Figure 1.

Microcarriers and their attached cells were retained in the reactor by stainless steel screens of 100 µm pore size. Cells were grown to confluence on a DMEM:COONS F12 medium containing 10% FCS. The medium was then replaced with a protein free production medium (DMEM:COONS F12) containing up to 80 µM zinc sulphate to induce transcription.

All other procedures and assays were as previously reported (Gray *et al*, 1990).

Results

High cell density fermentations:

It is possible to grow CHO cells either as attached cultures, in suspension, or in the form of microflocs. We have found that the cells have a preference for growing as attached cultures, and for this reason growth on microcarriers has proved to be simple and effective. Cells on microcarriers are relatively easy to retain in bioreactors during perfusion culture using the type of stainless steel screens described in Fig.1.

1054–7487/92/0197$06.00/0

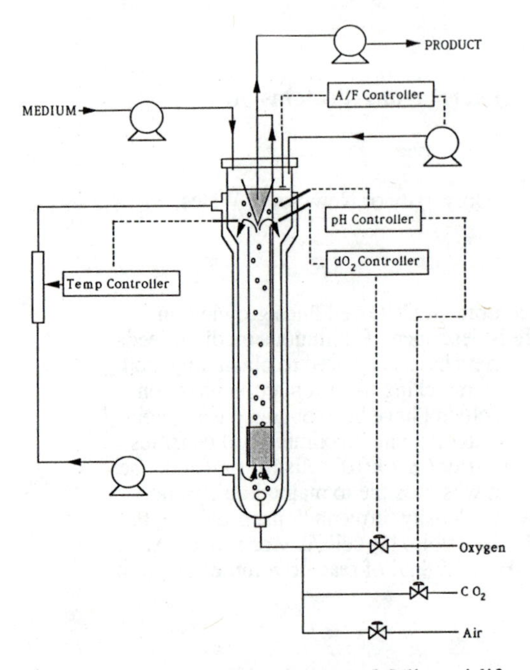

Figure 1. Schematic diagram of 2 litre airlift reactor used for growing the cells showing upper conical 100 μm stainless steel mesh used for taking off media during growth and production phases and lower 100 μm stainless steel mesh positioned at the bottom end of the draft tube and used for media changeover at the end of the growth phase.

Such systems make cell recycle less of a problem than for cells growing as suspension cultures. Also, microcarriers usually have less diffusional problems resulting from multicell layering which can be experienced with microflocs.

A standard batch fermentation of CHO cells growing on DMEM:COONS F12/10% FCS medium and the manufacturer's recommended level of Cytodex (3g/l), results in final cell densities of 3-5 x 10^6 cells/ml, with more than 80% of the microcarriers confluent. It was found that it was possible to extend the period of growth and reach higher final cell concentrations if a continuous feed of a medium containing 2.5% FCS (in DMEM: COONS F12) was commenced at a cell density of 3 x 10^6 and more microcarriers added to the culture to a final concentration of 15 g/l. The combined effect of a supply of additional nutrients, plus the additional surface area for growth resulted in a 10 fold increase in final cell density to a level of 5 x 10^7 cells/ml (**Fig 2**). The concentration of r-hGH in the reactor once the medium was changed to a production medium and expression induced, is also shown in **Fig 2**. In **Fig 3** it can be seen that the percentage of microcarriers which were confluent reached 80-90 %, and that as the cell density increased, the feed rate was increased to the final value of 12 reactor volumes per day which was maintained during the production phase. Operational considerations resulted in the run being completed after 14 days; in other experiments the production phase was maintained for greater than 28 days with no decreases in productivity. At these high cell densities it was found that the specific productivity of hGH q_{hGH} was the same as at the lower cell densities, providing the feed rate the culture was maintained at 0.3 volumes/10^6 cells/day or greater **(Fig 4).** At lower feed rates, the supply of several key amino acids became limiting (Crowley et al, 1991). Using solid microcarriers such as Cytodex, cell growth is only possible on the surface of the beads. Intuitively it would seem that by using a porous microcarrier, where cell growth can occur throughout the bead, that it would be possible to obtain cell densities which were higher than those obtained with the solid microcarrier. Accordingly, similar feeding regimes were used with the cells growing on porous gelatine microcarriers, Cultisphere-G. The manufacturers recommend usage of these microcarriers at a level of 1 g/l. Perfusion experiments using the protocols described for **Figs 2 & 3** were used with Cultisphere concentration up to 7 g/l. The max. cell concentrations obtained ranged from 8.24 to 8.46 x 10^6 cells/ml; i.e. it was not possible with the porous carrier to reach the final cell densities obtained with the non-porous carriers. Growth rates observed for the porous carriers (0.025h^{-1}) were less than those observed for the solid carriers (0.04h^{-1}).

It is interesting to compare these results with those quoted for other cell culture systems aimed at process intensification. Such comparisons are often difficult owing to the lack of all the data required for accurate comparisons. For example, in fluidized bed/packed bed bioreactor systems the total liquid volume of the bioreactor and all external loops is required in order to allow comparisons of productivities. This is necessary in order to make a valid comparison with data such as that shown in **Fig 1** for a suspension bioreactor, where concentration terms are based on the total liquid volume. At 15 g/l of Cytodex 2, it is estimated that the settled volume of the beads is less than 20% of the total reactor volume. A cell concentration of 5 x 10^7 cells/ml as shown in **Fig 2** translates then to a cell concentration of at least 2.5 x 10^8 cells/ml of packed microcarriers. This figure compares favourably with the figure of 2 x 10^8 cells/ml matrix quoted for one of the better known porous microcarrier systems (Runstadler & Young, 1991).

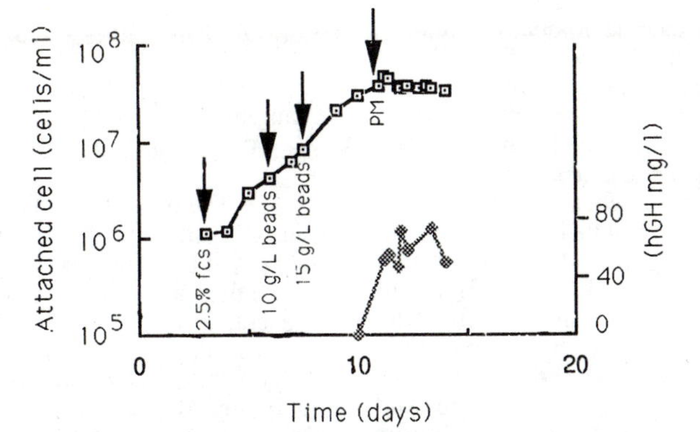

Figure 2. Growth of CB515 in 2 litre airlift reactor on Cytodex 2 microcarriers. Arrows indicate the time of commencing the 2.5% FCS containing growth medium feed; spent media was removed at the same rate as the feed. Arrows also indicate times of addition of extra microcarriers and the final microcarrier concentration after addition. PM indicates time of complete media change and the commencement of feeding of production media.

▣ attached cell concentrations
◈ concentration of hGH.

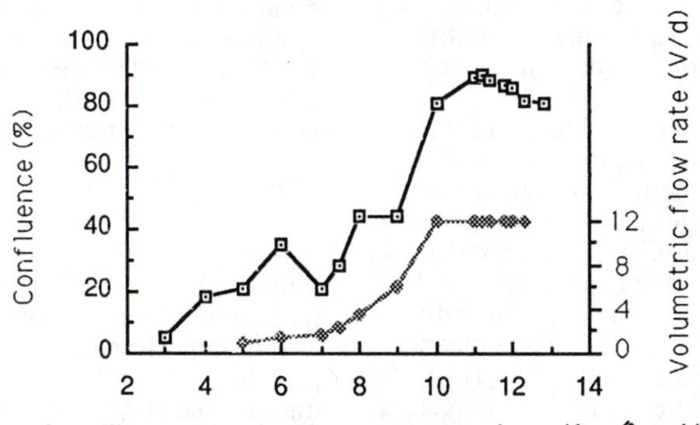

Figure 3. Data from Figure 1 showing the volumetric flow rate: reactor volumes/day ◈ and bead confluence ▣

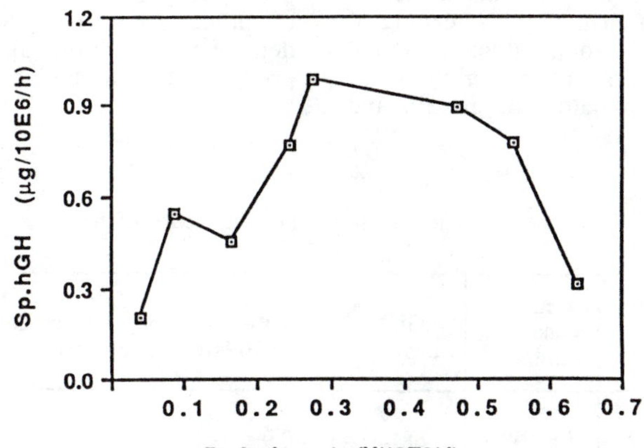

Figure 4. Specific productivity of hGH (q_{hGH} µg/ 10^6 cells/hr) vs. perfusion rate of production media (reactor volumes/10^6 cells/day.

Maintenance of high specific productivity (q_p):

The specific productivity of a cell line expressing a heterologous protein will be a function of both the genetic construction used to develop the recombinant cell line and the environment. The genetic construction used for the cell lines in this work makes use of techniques which allow high level expression (Friedman *et al*, 1989) without the need for high levels of amplification with their inherent stability problems. We have studied the expression of a simple single chain polypeptide (hGH) and a two chain glycoprotein (hFSH). The work indicates that as the size and complexity of the protein increases, factors which influence intracellular protein processing steps become more important in maximising q_p. For example, for hGH q_{hGH} values of 0.5-1.0 μ gms hGH synthesized/10^6 cells/hr are observed in both low (~10^6 cells/ml) and high (~10^7 cells/ml) cell density perfusion cultures, providing the feed rate of medium is maintained at approximately 0.3 reactor volumes/ 10^6 cells/day. Extensive environmental optimization has shown that q_{hGH} can be increased up to approx. 3 μg/10^6 cells/hr.

For the larger, more complex glycoprotein FSH, specific productivities of the order of q_{FSH} of 0.01 μg/ 10^6 cells/hr were observed for a large number of separate transfections using varying genetic constructs. Manipulation of promoters and constructs had minimal influence on the overall productivity of these more complex molecules. It was therefore thought that co-translational or early post-translational events were the rate-limiting steps of protein production. Optimis-ation of environmental factors which influenced either these steps directly or through general cellular metabolism was achieved using fractional factorial experimental designs. This method of experimental design enabled the screening of large numbers of factors, which included metabolites, precursors, stimulators of metabolism and stimulators of protein and oligosaccharide processing steps. The results of such a design are shown in **Fig 5** It was found that activation of co-translational events by manipulation of lipid metabolism could increase specific productivities to 0.1-0.5 μg/ 10^6 cells/ hr.

The overall volumetric productivity of the type of continuous perfusion reactor used in this work will be a function of the cell concentration in the bioreactor and the specific productivity of the cells. Representative figures for the data shown above are shown in **Table 1**.and demonstrate that volumetric productivies as high as 0.78g/l/day can be obtained in the high cell density fermentations

As already discussed, optimisation carried out at low cell densities have shown that it is possible to increase q_{hGH} up to 3 μg/10^6 cells/ hr. Such specific productivities, when combined with a high cell density fermentation would result in volumentric productivities of 2.9 g hGH/l reactor volume/day. Such high productivities are obtained under closely controlled environmental conditions, with cell viabilities approaching 100% and where there are minimal problems with respect to transferring nutrients/waste products to and from cells.

Conclusions

It is now possible to operate mammalian cell fermentations producing heterologous proteins at volumetric productivities which are respectable even when compared with more 'traditional' fermentation products. As the protein becomes larger and more complex, co-translational and/or early post-translational events become increasingly important as the rate-limiting steps in protein production. The use of good experimental design allows the identification of factors which can maximise the specific productivity of such systems. The combination of optimised specific productivities and high cell density fermentations allows high volumetric productivities to be obtained in suspension bioreactors.

Table 1 Volumetric reactor productivities as a function of the cell concentration.

Feed rate reactor vols/day	Cell conc^n/ ml	[hGH] mg/l	q_{hGH} μg/10^6/h	Volumetric productivities g/l/day reactor volume
1.0	3 x 10^6	68	0.9	0.068
12.0	4 x 10^6	65	0.8	0.780

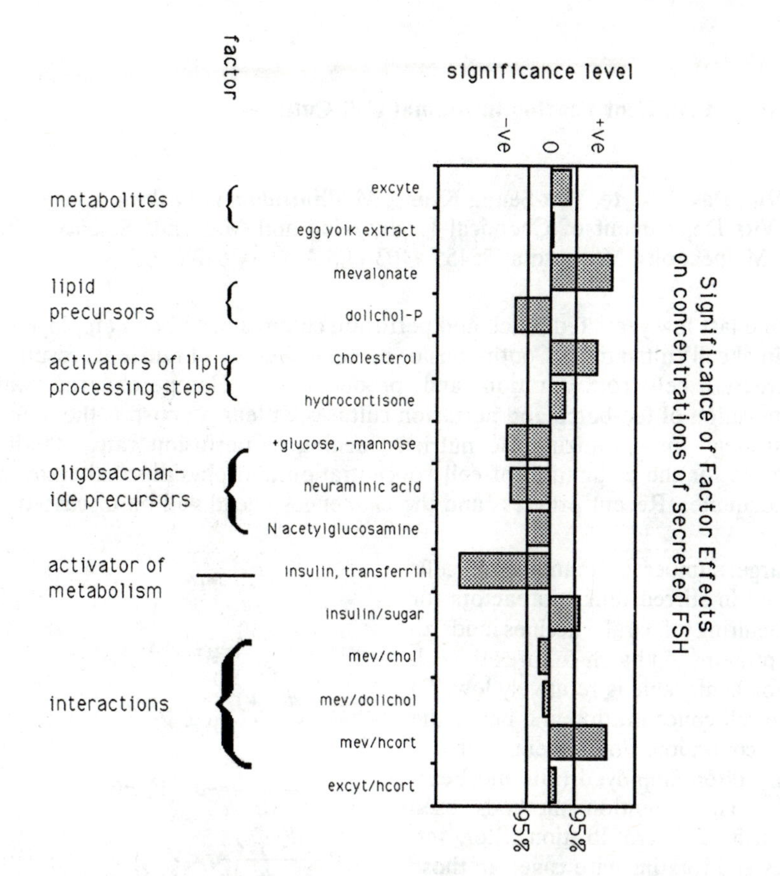

Figure 5 Example of results from a screening experiment, which compared the effectiveness of various factors to stimulate FSH secretion in CHO cells. A positive result indicates a stimulation, a negative result indicates a suppression. and the zero line indicates the control level. The 95% confidence limits are also shown.

References

Crowley, J.M: Marsden, W.L; and Gray, P.P. Effect of dilution rate on the metabolism and product formation of a recombinant mammalian cell line growing in a chemostat with internal recycle of cells. Animal Cell Culture and Production of Biologicals, Sasaki, R. and Ikura, K., Eds., Kluwer Academic Publishers, 1991, pp.275-281.

Friedman, J.S; Cofer, C.L; Anderson, C.L; Kushner, J.A; Gray, P.P; Chapman, G.E; Stuart, M.C; Lazarus, L; Shine, J; and Kushner, P.J. High expression in mammalian cells without amplification. *Bio/Technology* 7, 1989, pp.359-362.

Gray, P.P; Crowley, J.M; and Marsden, W.L. Growth of a recombinant Chinese Hamster Ovary cell line and high level expression in protein free medium. In *Trends in Animal Cell Technology*, Murukami, H., Ed., VCH I Kodansha, Tokyo, 1990, pp.265-270.

Runstadler, P.W; Young, M.W. In *Animal Cell Culture and Production of Biologicals*, Sasaki, R. and Ikura, K., Eds., Kluwer Academic Publishers, 1991, pp.103-119.

Toward Advanced Nutrient Feeding in Animal Cell Culture

Wei-Shou Hu*, David Gryte, Yun-Seung Kyung, Madhusudan V. Peshwa, Cheryl M. Perusich and Hugo Vits Department of Chemical Engineering and Materials Science, University of Minnesota, Minneapolis, Minnesota, 55455-1103 U.S.A. (Fax 612 626 7246)

In the last few years fed-batch and perfusion cultures have been employed in animal cell cultivation. Continuous or dynamic feeding of nutrients results in increased cell concentration and productivity. Hampering the wide application of fed-batch and perfusion cultures, at least in part, is the lack of a strategy for optimizing the nutrient feeding or perfusion rate. On-line sensors for the estimation of cell concentration and physiological state are inadequate. Recent progress and the challenges ahead will be discussed.

A large number of mammalian cells are cultivated in stirred tank bioreactors for the manufacturing of viral vaccines and recombinant proteins. In batch cultures the cell concentration achievable is relatively low. To increase the cell concentration and hence the productivity, continuous flow systems with cell retention are often employed (Hu and Peshwa, 1991). The retention methods used include centrifuge, microfiltration filter, settling devices and rotating wire cage. In those perfusion systems, cell concentration reaches a higher value and a longer production period is often achieved. Besides continuous flow bioreactors with cell retention, fed-batch cultures have also been employed in which limiting nutrients were fed continuously and resulted in increased productivity (Büntemeyer et al., 1991). In another case, glucose was fed continuously to reduce the lactic acid production (Hu et al. 1987). The execution of a program for perfusion or dynamic feeding requires on-line estimation of cell concentration and other variables pertinent to the growth or physiological state of cells. The lack of adequate real-time information hampers the control of these processes and the development of more advanced strategies (Hu and Piret, 1992).

Fed-batch culture

In a fed-batch culture the dynamics of the system can be described by the following equations.

$$\frac{dV}{dt}=F(t)$$

$$\frac{d(x_v V)}{dt}=\mu x_v V-\mu_d x_v V$$

$$\frac{d(x_d V)}{dt}=\mu_d x_v V$$

$$\frac{d(s_i V)}{dt}=-q_i x_v V+F(t)s_{oi}$$

$$\frac{d(m_j V)}{dt}=q_j x_v V$$

$$\frac{d(pV)}{dt}=q_p x_v V$$

$$\begin{aligned}\frac{d(OS.V)}{dt} &\approx -\sum_i \zeta_i q_i x_v V+\sum_j \zeta_j q_j x_v V \\ &\quad +2HV\frac{d[CO_2]}{dt}+\zeta_B V\frac{dB}{dt} \\ &= -\sum_i \zeta_i \frac{ds_i}{dt}+\sum_j \zeta_j \frac{dm_j}{dt} \\ &\quad +2HV\frac{d[CO_2]}{dt}+\zeta_B V\frac{dB}{dt}\end{aligned}$$

A Monod type equation relating μ to a limiting substrate concentration, s, is frequently used. Animal cell growth is also affected by osmolality. The rate of change in osmolality is affected by the rates of substrate consumption, metabolite production, the corresponding number of dissociable ionic species and the activity coefficients for those compounds. In addition, osmolality is also influenced by the carbon dioxide partial pressure in the gas phase and base addition for pH control.

1054–7487/92/0202$06.00/0

In a sodium bicarbonate buffered medium, sodium and bicarbonate ions contribute up to 75 mM of ionic species. $NaHCO_3$-free medium can be adjusted to isotonic osmolality by increasing the concentrations of key components. As cell growth proceeds, nutrients can be fed to sustain the growth. This should allow a higher cell concentration to be achieved provided that metabolite accumulation does not become inhibitory. A similar strategy can be used in continuous culture.

Continuous Culture with Cell Retention

For continuous cultures with cell retention, the flow terms are added to the balance equations. A cell retention device can be either internal or external to the reactor. In order to attain a steady state, a purge stream containing cells is necessary. Considering a simple case in which a microfiltration membrane is used; one of the exhaust streams passes through microfiltration membrane and is cell-free, while the other carries cells at a concentration equal to that in the bioreactor. The balance equations for viable and dead cells can be written as

$$V\frac{dx_v}{dt}=V\mu x_v-F(1-\sigma)x_v-\mu_d x_v V$$

$$V\frac{dx_d}{dt}=V\mu_d x_v-F(1-\sigma)x_d$$

The retention factor, σ, is the ratio of volume flow rate of the cell-free stream to that of the cell-containing stream. The other equations for substrates, metabolites, product and osmolality are identical to those of simple continuous culture.

$$V\frac{ds_i}{dt}=-q_i x_v V+F(s_{oi}-s_i)$$

$$V\frac{dm_j}{dt}=q_j x_v V-Fm_j$$

$$V\frac{dp}{dt}=q_p x_v V-Fp$$

$$V\frac{dOS}{dt}\cong-\sum_i \zeta_i q_i x_v V+\sum_j \zeta_j q_j x_v V$$

$$+2HV\frac{d[CO_2]}{dt}+\zeta_B V\frac{dB}{dt}-F(OS_o-OS)$$

$$=-\sum_i \zeta_i\frac{ds_i}{dt}+\sum_j \zeta_j\frac{dm_j}{dt}$$

$$+2HV\frac{d[CO_2]}{dt}+\zeta_B V\frac{dB}{dt}$$

Here the volume of the culture is assumed to be constant.

Status of Control of Fed-batch and Perfusion Cultures

From the equations shown above it can be seen that a large number of variables are needed to characterize the state of the culture, including V, x_v, x_d, μ, μ_d, q, and possibly other variables not yet identified representing the physiological state of cells. The number of variables affecting the state of the culture is also large, including OS, s, and m. Currently the strategy of nutrient feeding is largely based on repetitive experimentation with off-line chemical analysis to determine the "optimal" nutrient concentration and specific consumption rates. Much work has been done on kinetic models in the past few years, however, the applicability of these models for any predictive value in a fed-batch or perfusion culture is uncertain. A rational strategy for the selection of nutrient feeding rate in a fed-batch culture, or the flow rate and the retention factor, or the time point at which perfusion is initiated in a perfusion culture, has yet to be developed. In one reported study the perfusion rate was varied according to the residual glucose concentration measured off-line in an attempt to sustain a constant glucose concentration at a level sufficient to support a maximum growth rate (Seamans and Hu, 1990). However, the growth rate gradually decreased even though the glucose concentration was maintained at a relatively high level. Eventually cell concentration reached a plateau; however, the growth limiting factor which caused the cell concentration to level off has not been identified. Without a kinetic model simple feeding can still be carried out, as illustrated in the last example, provided that real-time information on culture conditions can be attained.

On-line Sensing in Animal Cell Bioreactors

A number of on-line measurement methods are available. Among the simplest are the measurement of base addition to estimate the lactic acid production, the estimation of oxygen uptake rate, and the measurement of light scattering or turbidity to estimate biomass.

Base addition

A large fraction of glucose consumed by mammalian cells is converted to lactic acid. To maintain constant pH in the culture, a stoichiometric amount of base is used to neutralize the lactic acid. This is the basis for lactic acid production rate measurement. The amount of base added at any time can be measured by a load cell. One source of error of such an estimation is the bicarbonic acid whose concentration is strongly affected by CO_2 partial pressure.

For on-line process control, one would like to examine not only the amount of lactic acid produced, but also the rate of production. A small error can cause a relatively large deviation in the estimated rate estimation. Furthermore, the stoichiometric ratio between glucose and lactate and the relationship between glucose metabolism and physiological state may vary during the cultivation. Other means of estimating cell growth and the physiological state are necessary.

Dissolved oxygen concentration

Oxygen uptake rate can be determined by inducing a transient wherein a gas of known composition is circulated through the fermenter for a certain period of time during which the dissolved oxygen level of the culture is monitored. A mass balance for oxygen in a culture gives

$$\frac{dC}{dt} = k_L a(C^* - C) - OUR$$

The oxygen uptake rate (OUR) of culture can be solved as:

$$OUR = \frac{(C_o - C_f)}{(t_f - t_o)} + \frac{\int_{t_o}^{t_f} k_L a[C^* - C(t)]dt}{(t_f - t_o)}$$

where C_o is the initial dissolved oxygen concentration, C_f the final dissolved oxygen concentration, t_f the final time. Numerical integration can be used to evaluate the above integral using an on-line microcomputer.

Turbidity

A number of on-line optical sensors for light scattering or optical density measurements are commercially available. Many perfusion cultures suffer from extensive fluctuations, possibly as a result of variations in flow rate or retention ratio. Such fluctuations can be minimized if a turbidity sensor is employed. We examined a laser turbidity probe recently and a very good correlation between total cell concentration and the sensor output signal was observed. By employing a filter, specific growth rate was estimated (Konstantinov et al.).

Concluding Remarks

Fed-batch and perfusion cultures allow for a high cell concentration and productivity to be achieved in animal cell bioreactors. To date, a rational way of selecting the operating conditions for feeding or perfusion is not available. A number of simple techniques can be used to estimate some variables related to the physiological state of cells (lactic acid production rate, oxygen consumption rate, specific growth rate). However, more sophisticated methodology, possibly employing expert system or neural network, is highly desirable for system identification and control.

Acknowledgements

This work was supported in part by grants from National Science Foundation, U.S.A. (BCS8552670) and from Boehringer Mannheim, Germany.

Nomenclature

B	The amount of base added
H	Henry's law constant
m	concentration of metabolite
OS	osmolality
p	concentration of product
q	specific rate of substrate consumption, metabolite or product formation
s	concentration of substrate
V	volume of the culture
x	cell concentration
ζ	number of dissociable ionic species
μ, μ_d	specific growth rate or death rate
σ	retention factor

subscripts

d	dead cells
i	substrate species i
j	metabolite species j
o	concentration in the feed
v	viable cells

REFERENCES

Büntemeyer, H., Lütkemeyer, D., Lehmann, J. *Cytotechnol.*, **1991** 5, 57-67

Hu, W-S., Dodge, T.C., Frame, K.K. and Himes, V.B. *Dev. Biol. Standard.* **1987** 66, 279-290.

Hu, W-S. and Peshwa, M.V. *Can. J. Chem. Eng.* **1991** 69, 409-420.

Hu, W-S. and Piret, J.M. Large scale mammalian cell culture: methods, applications and products. *Curr. Opinion Biotechnol.* **1992** (in press)

Konstantinov, K., Pambayun, R., Matanguihan, R., Yoshida, T., Perusich, C.M. and Hu, W.-S. On-line monitoring of hybridoma growth using a laser turbidity sensor. (Submitted for publication).

Seamans, T.C. and Hu, W-S. *J. Ferm. Bioeng.* **1992,** 70, 241-245.

Physiological State and Heterologous Protein Expression in Chinese Hamster Ovary Cells

Alan T. Bull, Anthony J. Baines, Paul M. Hayter and Ian Salmon, Biological Laboratory, University of Kent, Canterbury, Kent, CT2 7NJ, United Kingdom.
(FAX: +44 (0)227 463482).

The growth of a recombinant human interferon-gamma producing CHO cell line has been analyzed in batch and glucose-limited chemostat cultures on a partially optimized, serum-free growth medium. The proportion of different IFN-γ glycoforms varied markedly as a function of batch process time and of dilution rate, while the cell population showed heterogeneity with respect to aggregate formation and the capacity to synthesize and process IFN-γ. These physiological effects are discussed in the context of process optimization and control.

Optimization studies of recombinant therapeutic protein production in animal cells in our laboratory have focussed on protein yield, protein authenticity and heterogeneity within the cell population. These parameters are related to the physiological state of the host cell, consequently we have sought to obtain an understanding of the effects of culture conditions on cell physiology and production kinetics as a route to rational optimization. Recombinant human interferon-γ (IFN-γ) production has been selected for these studies. The core polypeptide of IFN-γ contains 146 amino acids and is N-glycosylated at Asn-28 and Asn-100. All experiments have been made with a mutant CHO-K1 cell line DHFR$^-$ co-amplified for IFN-γ expression by methotrexate selection. This communication considers the results of experiments made in batch and chemostat cultures.

Culture Conditions

The cell line was grown initially in RPMI 1640 containing 7% adult bovine serum. Adaptation to a serum-free medium was made by sequential passage on reduced concentrations of serum using specific growth rate as the criterion for successful adaptation (Hayter et al., 1991). Subsequently a Plackett-Burman statistical design (Plackett and Burman, 1946) was used to identify a group of positive medium components which, in combination, stimulated the specific growth rate, cell and IFN-γ concentrations by greater than 40% (Castro et al., 1992). Recombinant protein synthesis was completely stable over periods in excess of 2000h (>50 generations) in chemostat culture.

Batch Kinetics and Product Profile

Growth in stirred fermenter cultures produced ca. 10^6 cells ml^{-1}, a maximum specific growth rate of 0.03h^{-1}, and 4000-6000 IU IFN-γ ml^{-1} (Hayter et al., 1991). The specific rates of growth and IFN-γ production were positively correlated and declined in parallel as glucose was exhausted. The most efficient glycosylation of IFN-γ also was associated with exponential growth under which conditions 60-70% of the protein was fully glycosylated (2N). The proportion of the 2N glycoform fell to 30-40% at the end of growth at which stage the non-glycosylated (ON) IFN-γ was 20-30% of the total (Curling et al., 1990).

Under certain conditions cell growth and IFN-γ can be dissociated (e.g. the absence of glutamine from the medium; Hayter et al., 1991). Given the difficulty of defining the exact relationship between growth and IFN-γ expression in batch cultures, subsequent investigations were made in chemostats.

Chemostat Studies

Steady state glucose-limited chemostat cultures have been established with glucose feed concentration (S_R) of 2.75mM (Hayter et al., 1992). During prolonged culture at a dilution rate (D) of 0.015h^{-1} the volumetric productivity

1054–7487/92/0206$06.00/0

of IFN-γ remained constant but the specific rate of IFN-γ production was reduced when S_R was increased to 4.25mM. The CHO cells showed adaptation to continuous culture by increasing their efficiency of glucose and glutamine utilization and the specific rates of amino acid consumption. Because of these time-dependant changes in cell physiology the analysis of different steady state conditions was made on separately established cultures.

To date we have studied IFN-γ production at D value ranging from 0.01 to $0.02h^{-1}$. Cell viability fell significantly (to 79%) only at the lowest D, thus specific growth rate and D are numerically equivalent over most of this range. The change in residual glucose concentration as a function of D was atypical for a growth-limiting substrate and suggests that some other nutrient (e.g. glutamine) was growth-limiting at low D, or, that dual substrate limitation of the type described by Egli and Schmidt (1990) was occurring. However, the specific range of glucose utilization was a linear function of D and the data imply that little or none of the glucose was consumed for maintenance energy purposes. Interferon titres were highest at the lowest D examined. Moreover, the proportions of 2N, 1N and ON IFN-γ at $0.01h^{-1}$ were 70, 24 and 6 respectively and similar to those of early exponential batch culture and to those of the 'natural' glycosylation pattern. At all other dilution rates the glycosylation profile was 56 (2N), 33 (1N) and 11 (ON).

Heterogeneity of the Cell Population

The cell line used in these studies produces aggregates in suspension culture. The aggregate or clump size distribution remained constant under steady state growth conditions but the proportion of large clumps increases as D is lowered. The physiology of cells in the centres of clumps will be affected by impaired mass transfer of nutrients and O_2, while secretion of IFN-γ also may be restricted. Thus, product heterogeneity (glycosylation, protein processing) may arise in response to the imposed culture conditions but also may reflect heterogeneity of cells within the population. We have found (Newsam et al., 1992) that specific cell-cell attachments occur under all culture conditions and represent vinculin-rich adherens junctions. Such junctions give rise to the large cell clumps referred to above. A second level of cell heterogeneity has been revealed using anti-IFN-γ immunofluorescent antibodies. All growing cells have IFN-γ detectable within the endoplasmic reticulum but only a proportion have IFN-γ also associated with the Golgi apparatus. This observation implies that there may be variable residence times for IFN-γ and hence variable exposure to processing enzymes in the Golgi.

Control of IFN-γ Heterogeneity

The relationship between CHO cell physiology and the state of IFN-γ heterogeneity is complex. Glycosylation is affected by glucose availability and site occupancy is maximized when glucose is in growth excess. However, not all our data are consistent with the hypothesis that high glycolytic flux rates are conducive for maximum glycosylation. Thus, the changes observed in amino acid metabolism at high glucose concentrations (Hayter et al., 1992) may be indicative of a smaller pool of biosynthetic precursors and explain the decreased rate of IFN-γ production and IFN-γ glycosylation. Therefore, our current research is focussed on the behaviour of amino acid-limited cultures.

Finally, the finding that IFN-γ titres, the proportion of the 2N glycoform and cell aggregate size all have their highest values in populations growing at low dilution rates, have implications for process optimization. We are working towards developing biomass recycle reactors for animal cell processes and such systems will permit operation at low D values while providing high volumetric productivity.

Acknowledgements

We thank Lucy Gettins and Malcolm Gould for excellent technical assistance; Wellcome Biotechnology for the CHO cell line; Celltech for monoclonal antibodies; and the SERC, British Biotechnology. Glaxo, ICI, Smith-Kline Beecham and Wellcome Research for support of the programme of which the work described here is part.

References

Castro, P.M.L.; Hayter, P.M.; Ison, A.P.; Bull, A.T. Appl Microb. Biotechnol., submitted, **1992**.

Curling, E.M.A.; Hayter, P.M.; Baines, A.J.; Bull, A.T.; Gull, K.; Strange, P.G.; Jenkins, N. Biochem. J. **1990**, *272*, 333-337.

Egli, T.; Schmidt, Ch. R. In *Mixed and Multiple Substrates and Feedstocks*, Hamer, G.; Egli, T.; Snozzi, M. Ed; Hartung-Gorre-Verlag, Konstanz, **1990**, pp. 45-53.

Hayter, P.M.; Curling, E.M.A.; Baines, A.J.; Jenkins, N.; Salmon, I.; Strange, P.G.; Bull, A.T. Appl. Microb. Biotechnol. **1991**, *34*, 559-564.

Hayter, P.M.; Curling, E.M.A.; Baines, A.J.; Jenkins, N.; Salmon, I.; Strange, P.G.; Tong, J.M.; Bull, A.T. Biotech. Bioeng. **1992**, *32*, in press.

Newsam, R.; Coppen, S.R.; Hayter, P.M.; Bull, A.T.; Baines, A.J. Proc. ESACT Meeting, Manchester April **1992**, in press.

Plackett, R.L.; Burman, J.P. Biometrika **1946**, *33*, 305-325.

Industrial Large Scale Operations Using Animal Cells Technology

Michael C. Comer* and Karl-Heinz Sellinger, Boehringer Mannheim GmbH, P.O. Box 1152, Penzberg, D-8122, Germany (Fax 49-88 56 60 30 28)

The developments in recombinant DNA and hybridoma technologies have made it possible to express many natural or modified molecules in in-vitro systems. These biological products offer tremendous potential as therapeutic and diagnostic agents for human and animal health care. A variety of new bioreactors have been developed to meet the needs of manufacturing. These systems often support higher cell concentrations with a concomitant increase in unit volume productivity. However, from a manufacturing perspective, the process encompasses not only the bioreactor but also all the auxiliary and downstream processing. The importance of an integrated approach for process design and bioreactor evaluation will be discussed.

Animal cell culture processes have gone through a transformation in the 1980's. Previously, animal cell processes were limited to the production of viral vaccines and a few biologics for which the demand is relatively small in terms of mass quantity. With the therapeutic use of antibodies and recombinant proteins, the demands can often be greater than tens of kilogram a year. To meet this new need in manufacturing processes employing animal cells, a large number of bioreactors were developed. Many systems, such as hollow fiber systems (for review, Piret and Cooney, 1990), ceramic modules and automated roller bottles with extended surface, can be easily adopted for processes requiring relatively small quantity of products. Others, such as collagen (Runstadler et al. 1990) or synthetic polymer (Reiter et al. 1991) based on macroporous beads in fluidized bed, or stirred tanks with cell retention devices (Comer et al., 1990; Kitano et al., 1986; for review Hu and Peshwa, 1991) are likely to be more amenable to scale-up.

Despite the efforts in development not all those bioreactors have met equal success. Some have found inroad into manufacturing, others remained largely a research and development tool. Many factors affected these different outcomes, but may not be related to technology per se. One element of consideration which is sometimes not emphasized, is the flexibility and easiness that a new bioreactor can be integrated in a production process.

As has already been mentioned, the new developments in the fermentation processes may lead to higher cell densities and increased product concentration. As a consequence also, to longer process times in the larger dimensions and thereby variations in cell viability, cell death and eventually, cell lysis. Therefore it is to be expected that batch to batch variations in concentrations of extraneous proteins or indeed "immature" formation of product molecules, may be released into the culture supernatant.

It is therefore extremely important for the downstream process parts of the operation that fluctuation ranges for cell density, viability, duration of growth, product concentration and possible extraneous protein formation etc., be well defined, agreed and adhered to by those concerned in the complete operations. This is not only a requirement for good quality control and good manufacturing practice but also it would be economically absurd to modify the downstream process on a "run to run" basis to accommodate broad variations in the fermentation process. The downstream

1054–7487/92/0209$06.00/0

process therefore, should be so designed to compensate for minor limits of variation within the acceptable range of fluctuation mentioned above.

In order to achieve this modern and precise techniques for "In Process Control" have to be established, such as ELISA (for product and major contaminating proteins), HPLC, sugar analysis, electrophoresis, isoelectric focusing etc.. This, in order to accomplish a monitoring and control function, to ensure the quality and consistency of the desired product.

The integration of the fermentation technology in large scale production processes requires therefore a communal approach to the process development, the downstream processing and the availability of the necessary analytical tools, in order to achieve a reproducible, high quality and indeed, last but not least, an economically manufactured product.

REFERENCES

Comer, M.J., Kearns, M.J., Munster, M., Lorenz, T., Szperalski, B., Koch, S., Behrendt, U., and Brunner, H., *Cytotechnol.* **(1991)** 3, 295-299.

Piret, J.M. and Cooney, C.L., *Biotech. Adv.* **(1990)** 8,763-783.

Hu, W.-S. and Peshwa, M.V., *Can. J. Chem. Eng.* **(1991)** 69, 409-420.

Kitano, K., Shintani, Y., Ichimori, Y., Tsukamoto, K., Sasai, S., and Kida, M., *Appl. Microbiol. Biotechnol.* **(1986)** 24, 282-286.

Reiter, M., Blüml, G., Caida, T., Zach, N., Unicluggauer, F., Doblhoff-Dier, O., Noe, M,, Plail, R., Huss, S., Katinger, H., *Bio/Technology*, **(1991)** 9:1110-1102.

Runstadler Jr., P.W., Tung, A.S., Hayman, E.G., Ray, N.G., Sample, J.v.G., and DeLucia, D.E. In *Large Scale Mammalian Cell Culture Technology*, **(1990)** A.S. Lubiniecki Ed., Marcel Dekker Inc., pp. 363-391.

Systems for Large-Scale Continuous Perfusion of Mammalian Cells Grown in Suspension

David C. Cohen, Mokhtar Mered, Robert Simmons, Alfred C. Dadson, Jr., Carlos Figueroa, and Craig W. Rice*
Miles, Inc., Berkeley, California 94701

Specific perfusion rate is a parameter which normalizes the flow of medium through a perfused culture to the unit cell density (*eg*., nl/cell/day), rather than the volume of the reactor. Because of its cell-specific perspective, the term is independent of both the volume of the vessel and the cell density. Analytical application of specific perfusion rate shows that it is a primary driver of cellular metabolic rates (nutrient utilization, protein production). Predictive application allows calculation of large scale performance based on data derived from smaller scale and lower cell density systems. Application to operation control, by holding constant the rate of specific perfusion, results in a volumetric productivity that is directly proportional to cell density, as well as a titer that is constant, and independent of cell density.

Work reported here comes from an experimental fermentation program that is directed toward optimizing processes that are based in large-scale continuous perfusion reactors. Process optimization attends both to maximizing yield and taking best advantage of the opportunity afforded by continuous perfusion to approach constancy of cell-based and volumetric measures of performance over time. In this context, process development effort includes the investigation of cell retention/recycle systems and development of operation control strategies.

From the experience with a hybridoma that secretes a therapeutic monoclonal antibody, as well as with other cells and products, there has emerged an appreciation of the value of the specific perfusion rate parameter. Data utilizing this parameter during industrial scale process development exercises have shown a consistency across a considerable range of volume and cell density scale, as might be expected of a single cell-based normalization perspective. The relationships among process parameters that are discussed below follow as a consequence of the driving force that specific perfusion exerts on cell metabolism.

The utility of the specific perfusion rate parameter is ultimately realized as a means of controlling the fermentation process, by holding *constant* the rate of cell specific perfusion. When culture systems are held at a constant specific perfusion rate, cellular specific rates of metabolism and production become constant. When the specific production rate is thus held constant, volumetric productivity becomes a direct function of cell density. Further, the product titer becomes constant, and independent of the cell density.

Definition and Context of Specific Perfusion Rate

Perfusion rates are commonly used in a volumetric sense, in terms either absolute (liters/day) or relative (fold-volume change/day). The specific perfusion rate is one in which the rate of perfusion in absolute volume terms is normalized to the single cell or other unit of cell mass, thus yielding, *eg*., nl/cell/day.

Another conceptual approach is to consider the reciprocal of cell density (cells/medium volume); this reciprocal value would describe the volume of medium available for a single cell to call its own. If this value is multiplied by a relative fold-change rate/day, then again, a nl/cell/day rate term is yielded.

Units of the rates of substrate utilization and protein production also can be normalized to cell-specific rates, in contrast to normalization to medium volume. Such specific metabolic rate parameters lend themselves well to comparison with the specific rates of perfusion and production.

*Current Address: Genentech Corporation, South San Francisco, California 94080

1054–7487/92/0211$06.00/0

The Specific Productivity Rate and Other Metabolic Rates are Functions of the Specific Perfusion Rate

The data reported in the figures to follow were derived from a data pool of approximately 800 daily measurements of a hybridoma fermentation process that was run in vessels spanning a 20-fold range of volume, with cell densities ranging between 0.2 and 40 x 10^6 cells/ml, volumetric perfusion rates of 0.1 to 10 volume changes/day, and with several types of cell retention/recycle system. The basic relationships described for this particular cell have also been observed with other cell types.

Fig. 1 shows specific productivity (pg/cell/-day) as a function of specific perfusion. The slope of the regression line, and the breakpoint where the specific productivity levels off are characteristic of both the cell line and the medium. The terms of the slope of the linear portion of the curve reduce to the product titer (pg/nl or mg/l). Thus, titer is seen as a function of the specific perfusion rate (see below).

Other cell-specific rates of metabolism are also found to be a function of the specific perfusion rate. Fig. 2 shows specific glucose utilization to increase with increasing specific perfusion. Similar plots emerge if the utilization of other substrates (such as glutamine), or the production of by-products (such as lactate) are tracked.

The plateau at which the specific productivity levels off (Fig. 1) represents the maximal rate that the cell is capable of sustaining. The rate of specific perfusion at which this occurs can be seen as the level at which the medium, or more specifically, the nutrient components of the medium are fully utilized. Increasing the specific perfusion rate beyond this point is inefficient with regard to medium utilization, and can have the effect of reducing product titer (Fig. 3). At high rates of specific perfusion, non-utilized medium is, in effect, diluting the product.

Despite the variation in conditions from which data were derived (with regard to culture volume, volumetric perfusion rate, and cell recycle system), the data show a remarkable unity. The relationship between specific perfusion rate and specific metabolic rates thus transcends these variables. The picture which emerges is simple and understandable, one in which cellular metabolism is being driven by the availability of nutrients conveyed by the medium.

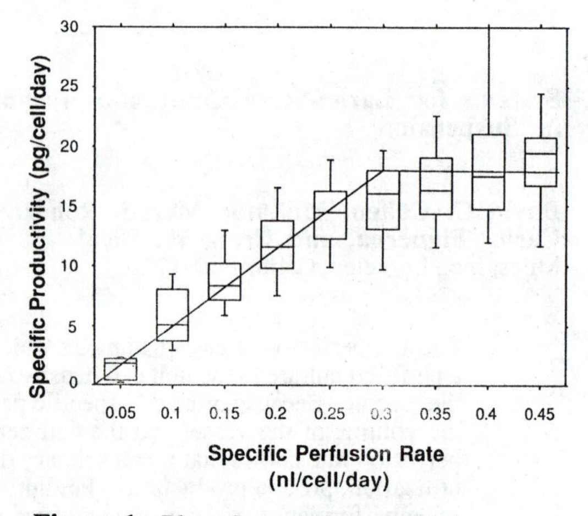

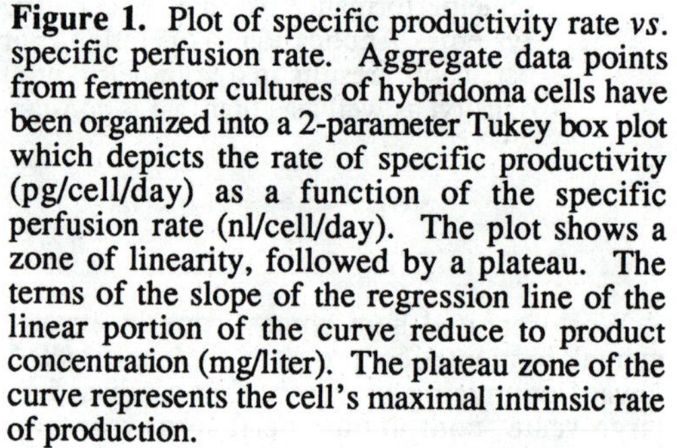

Figure 1. Plot of specific productivity rate *vs.* specific perfusion rate. Aggregate data points from fermentor cultures of hybridoma cells have been organized into a 2-parameter Tukey box plot which depicts the rate of specific productivity (pg/cell/day) as a function of the specific perfusion rate (nl/cell/day). The plot shows a zone of linearity, followed by a plateau. The terms of the slope of the regression line of the linear portion of the curve reduce to product concentration (mg/liter). The plateau zone of the curve represents the cell's maximal intrinsic rate of production.

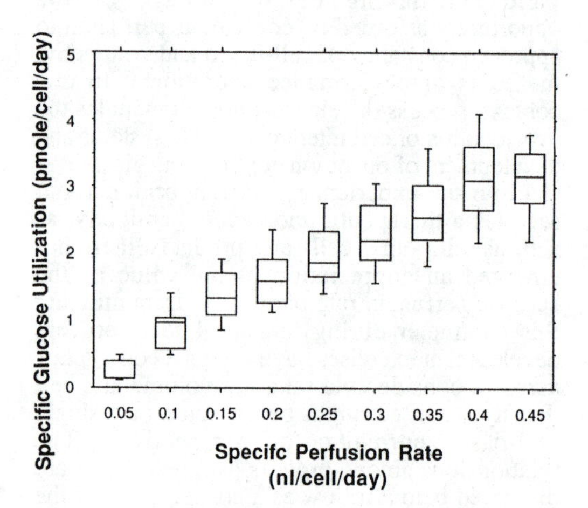

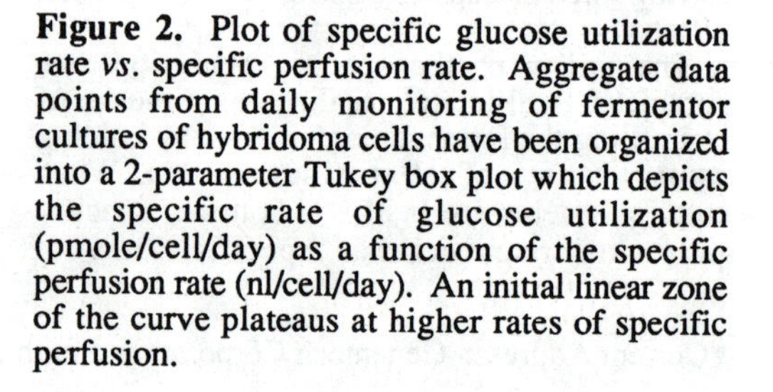

Figure 2. Plot of specific glucose utilization rate *vs.* specific perfusion rate. Aggregate data points from daily monitoring of fermentor cultures of hybridoma cells have been organized into a 2-parameter Tukey box plot which depicts the specific rate of glucose utilization (pmole/cell/day) as a function of the specific perfusion rate (nl/cell/day). An initial linear zone of the curve plateaus at higher rates of specific perfusion.

Specific Perfusion Rate as a Process Control

If the specific perfusion rate is varied purposefully, or allowed to vary as a consequence of other process variables, then the metabolic

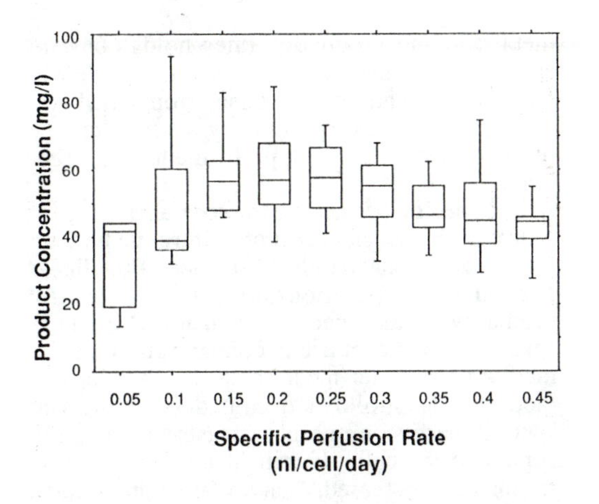

Figure 3. Plot of product concentration *vs.* specific perfusion rate. Aggregate data points from daily monitoring of fermentor cultures of hybridoma cells have been organized into a 2-parameter Tukey box plot which depicts the product concentration (mg/liter) as a function of the specific perfusion rate (nl/cell/day). The product concentration is constant between 0.1- and 0.3 nl/cell/cay, but shows a downward trend at higher rates of specific perfusion.

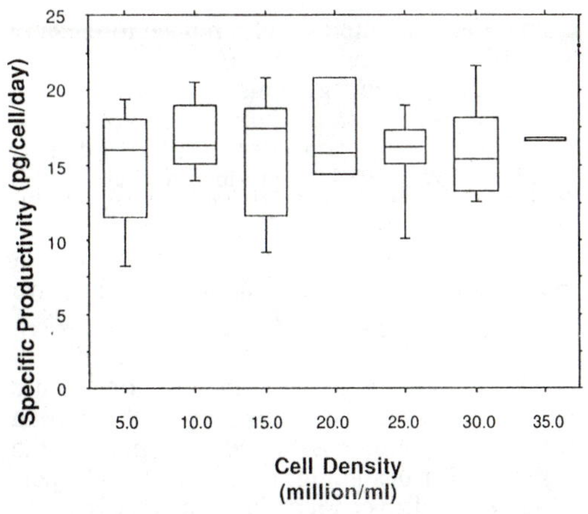

Figure 4. Plot of specific productivity rate *vs.* cell density. Aggregate data points from daily monitoring of fermentor cultures of hybridoma cells have been organized into a 2-parameter Tukey box plot which depicts the rate of specific productivity (pg/cell/day) as a function of the cell density (millions/ml). The specific productivity is independent of cell density.

consequences and relationships such as those above can be described. If, on the other hand, the rate of specific perfusion is held constant, then the consequence of varying cell density can be examined. From the study of this hybridoma, cultured in its particular medium, it was clear that a specific perfusion rate of 0.3 nl/cell/day was optimal in terms of driving cell metabolism maximally, without diluting the product with excess medium.

The consequences of operating fermentors at a constant specific perfusion rate were thus examined. Once again, the data depicted in figures 4 and 5 were derived from fermentors of varying volume, and varying cell recycle system. During an extended fermentation campaign that was conducted with a specific perfusion rate of 0.3 nl/cell/day, cell density varied as a result of variation during innoculum build up, and through manipulations of the cell density controls.

Fig. 4 shows that the specific productivity rate was constant over a wide range of cell density when the specific perfusion rate was held constant. This verifies, therefore, that the rate of specific productivity is independent of cell density.

Fig. 5 shows that the rate of volumetric productivity (mg/liter reactor volume/day) is a linear function of cell density, between 1 and 40 x 10^6 cells/ml when the specific perfusion rate is held constant. The terms of the slope of the

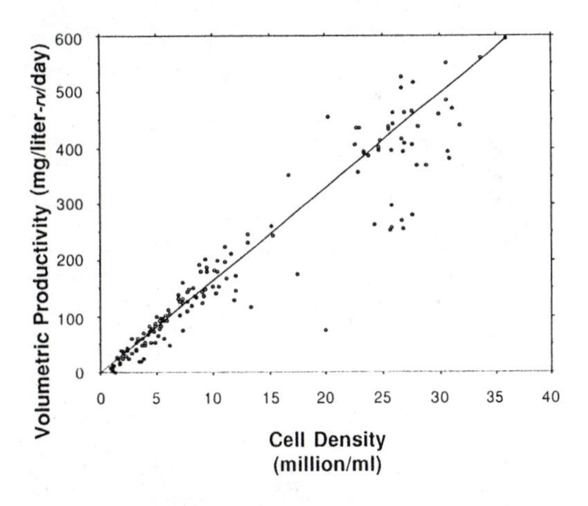

Figure 5. Plot of volumetric productivity rate *vs.* cell density. Aggregate data points from daily monitoring of fermentor cultures of hybridoma cells have been organized into a 2-parameter scatter plot that depicts volumetric productivity (mg/liter reactor volume/day) as a function of cell density (millions/ml). The volumetric productivity is seen to be a linear function of cell density over a range that includes cultures with a density of less than 1 million/ml to greater than 35 million/ml. The data points fit the regression line well, with an r^2 value of 0.89.

regression line, interestingly, reduce to those of specific productivity (mg/10^6 cells/day, or pg/cell/day). The figure also graphically illustrates the relationship whereby the volumetric productivity rate is equivalent to the cell density multiplied by the specific productivity rate.

Conclusions and Broader Applications of the Specific Perfusion Rate Parameter

The relationship between the rates of specific perfusion and specific productivity, that has been introduced here, establishes a quantifiable approach for designing, optimizing, and operating mammalian cell fermentation processes. As the terms of the parameters suggest, and experimental data have shown, the relationship between the rate of specific perfusion and cellular metabolic and production rates holds constant across a wide range of culture volume scale and cell density. Thus, data obtained from small volume scale and low cell density systems can be used to predict process performance at larger scale.

The specific perfusion parameter also has great utility in the development of culture media with greater nutritional depth. The slope of the linear portion of the relationship between specific productivity and specific perfusion (Fig. 1) is sensitive to the nutrient concentration in the medium. A richer medium causes the slope to increase. This follows from understanding that specific perfusion rate is a measure of medium supply to the cell. If critical nutrient concentrations are increased, then less medium volume needs to be supplied to each cell; the optimal rate of specific perfusion is lowered, and the product titer increases.

REDUCTION TO PRACTICE **Barry C. Buckland**
Merck Research Laboratories, P.O. Box 2000, Rahway, NJ 07065 USA
(Fax: 908 594-5468)

A description will be given of important new fermentation products introduced by the Pharmaceutical Industry during the 1980's. These new products fall into the broad categories of vaccines, therapeutic proteins, and secondary metabolites: the special process issue for each type will be discussed. Emphasis will be on recent developments in fermentation technology especially in the area of process monitoring and control. An example will be given (Lovastatin production) which will provide a framework for describing an interdisciplinary approach to process development.

New Fermentation Products

During the past decade, a number of important new pharmaceutical products made by fermentation were introduced. These fall into 3 broad classifications: natural products (secondary metabolites), therapeutic proteins (using recombinant DNA) and vaccines (Table 1). From a process technology viewpoint, the secondary metabolites need to be considered separately because of the much larger scale of operation (Table 1) required for production.

Proteins made by recombinant DNA technology represent a new group to the field. In terms of scale of operation and degree of product definition, they lie between vaccines and secondary metabolites.

Vaccines made by microbial fermentations

Historically, vaccines have been relatively poorly defined in respect to characterization of the final product. This has now started to change both because of spectacular improvements in analytical capabilities as well as a revolution in the way that vaccines are developed.

Two examples will be given which illustrate this revolution: one for a vaccine to prevent Hepatitis B and the second for a vaccine to prevent meningitis. Recombivax® represented the first introduction of the application of genetic engineering to express a protein which can be used as a vaccine. A surface antigen protein from the Hepatitis B virus was successfully expressed in yeast and purified to result in an effective replacement for the traditional vaccine made from the infected blood serum.

The conjugate vaccine used for meningitis owes nothing to recombinant DNA. Instead, this vaccine was made by purifying the outer membrane protein (OMPC) from _Neisseria meningititis_ and chemically conjugating this with the membrane polysaccharide from _Haemophilus influenzae_. The main challenge in fermenter design for this type of process to ensure adequate containment of these pathogenic cultures.

Therapeutic proteins

Therapeutic proteins have been introduced using _E. coli_, yeast or various animal cells as host. This area provides many examples of fruitful interdisciplinary collaborations.

The first key dialog occurs between the Bioprocess Engineer and the molecular biologist. Several key decisions MADE BY THE MOLECULAR BIOLOGIST will have a direct impact on process economics and the ease of scale-up. Questions which need to be resolved include the choice of host, expression system and promoter. The right choice depends on the characteristics of the final product (is it glycosylated? extent of cross-linking?).

For a glycosylated protein, it is usually desirable to use an animal cell culture host. The optimum cell line for process scale-up is one that can be readily adapted to suspension culture. Then, the scale-up approach can be

1054–7487/92/0215$06.00/0

TABLE 1 SCALE OF OPERATIONS FOR THERAPEUTIC DRUGS

TYPE	DOSE	ANNUAL PRODUCTION 1,000,000 DOSES
IMMUNOMODULATOR OR VACCINE	0.0001 - 0.1 mg	0.1 - 100 grams
HORMONE	1 - 10 mg	1 - 10 kg
ENZYME	100 mg	100 kg
ANTIBIOTIC	1 gram	1000 kg

remarkably similar to that for microbial fermentations. In fact, we have designed fermenters up to the 1,000 liter scale which can be used for either microbial cells or suspension culture mammalian cells. Superimposed on this choice is the fact that different cell lines glycosylate proteins differently: the resulting product often has different properties in vivo eg. half-life can vary over a wide range.

The second dialog occurs between the fermentation and purification groups. Many purification issues can be solved by changes in the fermentation. For example, changes in media components can greatly influence filtration properties, choice of defoamer will affect chromatography, and choice of components will influence the degree of process validation (if you do not put something in, then you do not have to prove that it has been removed). The most difficult problems result from a small change in the final protein. Obviously, it is a formidable challenge to separate a mixture which contains the final product of 180 amino acids mixed with 10% of an impurity which contains 179 amino acids. This is a purification challenge best resolved at the fermentation step by determining conditions under which the amino peptidases are not induced. A similar problem occurs when, under certain culture conditions, one amino acid can be substituted for another (eg. norleucine substitution for methionine).

There are no difficult problems related to fermenter design for recombinant hosts such as E. coli or Saccharomyces. However, facility design much more closely resembles that used historically for Biologics such as vaccines (Bader, 1992). Vaccines developed in the 1950's to 1970's were not well defined as product entities: the process was defined instead. Therefore, great focus was put on procedures.

Because a number of the new recombinant protein products are high value added and are run at a small scale (human growth hormone, for example) there has been less incentive to question whether some of these very expensive procedures make sense for products which, with modern analytical capabilities, are often well defined. The net result has been an astronomical cost in facility design for fermentation products. It now costs as much to build a 3,000 liter scale facility for Biologics as for a 200,000 liter scale facility for an antibiotic. Also, even though titers are now reasonable for a recombinant protein (1g/ℓ), the cost of manufacture/kg of bulk drug is about two orders of magnitude higher than that of an antibiotic at 10 g/l titer.

At some point, some of the assumptions that have led to such expensive facilities and processes will have to be questioned. Do all of these refinements result in an improvement in quality?

Secondary metabolites

Four major new secondary metabolites were introduced to the marketplace in the 1980's, each of which can be considered "miracle drugs". Cyclosporin (an immunoregulant) has helped to revolutionize organ transplant procedures by greatly reducing rejection. Imipenem (derived from Thienamycin) is a modified carbapenem which has the broadest spectrum of any known antibiotic. Lovastatin is a secondary metabolite used for reducing cholesterol levels. Finally, Ivermectin has been used to prevent "African River Blindness" as well as being a superb antiparasitic drug for veterinarian use.

One of the best kept secrets (unintentionally kept as a secret) in the 1980's in Biochemical Engineering was that working on secondary metabolites was a fascinating, important and rewarding experience. Furthermore, as well as demonstrated therapeutic applications, the four products listed added together have higher sales than all of the recombinant products added together. The focus of attention in the 1980's was on a narrow spectrum of the total pharmaceutical industry, ie. proteins made as a result of genetic engineering.

The Lovastatin story illustrates many important points and will be described here. After many years of research, HMG Co A reductase was identified as the key rate limiting enzyme for the biosynthesis of cholesterol. An intensive search for inhibitors included the screening of fermentation broths from soil isolates and this resulted in the discovery of an exceptionally good inhibitor (Lovastatin) produced by a fungus Aspergillus terreus. (Alberts et al., 1980).

In contrast to a simple bacterial fermentation such as E. coli, the main process challenge in this type of fermentation is the adequate supply of oxygen to the culture at a very large scale of operation (up to 250,000 liters). As the cell density of Aspergillus terreus increases, the broth becomes very viscous and oxygen supply becomes more difficult. Qualification, obtained experimentally showed that the oxygen transfer coefficient is inversely proportional to broth viscosity.

Two approaches were taken to improve the oxygen transfer capabilities. The first was to install a new type of hydrofoil axial impeller which provided improved bulk mixing for equivalent power/volume while still giving good oxygen transfer. (Buckland et al, 1989). The second approach was microbiological. Procedures were developed which resulted in culture growth in the form of pellets. If the pellets were small enough, the apparent viscosity could be reduced by a factor of five and oxygen transfer adequate even into the interior of the pellet (Gbewonyo et al, 1992). As the oxygen transfer coefficient is improved in these ways, it follows that the mass transfer coefficients for the supply of other key nutrients will also be improved. Thereby, in a very large fermenter, a better control of the individual cell environment will result and will inevitably give more reproducible performance. Not only to obtain higher yields, but also to result in better control of the amount of undesirable by products.

Because of the complexity of secondary metabolite biosynthesis, genetic engineering has yet to make a major impact in this area. The biggest recent changes have come from improved characterization of the fermentation, primarily obtained off-line. Recent attempts to convert this information on-line have included the application of robotics. An example of this approach will be given during this talk.

Conclusion

Recent advantages in genetic engineering have made it possible to manufacture a great variety of therapeutic proteins using fermentation. Advances in vaccines have led to a new class of conjugate vaccines.

Regardless of the type of fermentation, the trend in all of these processes is toward improved process monitoring and control. Enormous advances in analytical capabilities have greatly enhanced our ability to monitor fermentation processes off-line including improved characterization of the final product.

References

Alberts et al. Proc. Nat'l. Acad. Sci. 1980. 77: 3957-3961.

Buckland, B.C.; Gbewonyo, K.; Hallada, T.; Kaplan, L.; Masurekar, M. In Novel Microbial Products for Medicine and Agriculture. Ed. Demain. 1989. Society for Industrial Microbiology. 161-169.

Gbewonyo, K.; Hunt, G.; Buckland, B.C. 1992. Bioprocess Engineering. In Press.

Bader, 1992. IBS9. Symposium. This volume.

Garnick, R.L.; Solli, N.J.; Papa, P.A. Analytical Chemistry, 1988. 60. 2546-2557. The Role of Quality Control in Biotechnology: An Analytical Perspective.

The Influence of the Physical Environment in Fermenters on Antibiotic Production by Microorganisms

Malcolm D. Lilly, Andrew Ison, and **Parvis Ayazi Shamlou,** Department of Chemical and Biochemical Engineering, University College London, Torrington Place, London WC1E 7JE, UK (FAX:071 388 0808)

The interaction between the physical environment in a fermenter and the behaviour of three different antibiotic producing microorganisms are described. Both the synthesis of difficidin by *Bacillus subtilis* and of penicillin by *Penicillium chrysogenum* were reduced at low dissolved oxygen values. Cultures of *Saccharopolyspora erythraea* displayed non-Newtonian rheological properties which only partly correlated with changes in morphology during fermentations. There was a good relationship between morphological changes and the rate of penicillin synthesis by *Penicillium chrysogenum*.

Introduction

The objective of fermentation scale-up is to maintain, where possible, the optimum chemical and physical environments in the fermenter at all scales. In an aerated stirred tank both the agitation and aeration affect the distribution of nutrients, including oxygen, and shear forces in the fermentation broth (Lilly, 1983). These impact on culture morphology, broth rheology and product formation and it is important, therefore, to understand the interactions between them. Some results for three antibiotic producers, *Bacillus subtilis, Saccharopolyspora erythraea* and *Penicillium chrysogenum* are reported here.

Influence of DOT on antibiotic synthesis

The mixing times (t_m) in aerated stirred fermenters increase with fermenter volume (V) according to

$$t_m = \text{constant} \times V^{0.3}$$

reaching about 100 s in a 100 m^3 vessel. If the time constant for the rate of oxygen transfer is shorter then there will be a vertical dissolved oxygen concentration gradient as observed, for example, by Manfredini et al. (1983) for chlorotetracycline and tetracycline production by strains of *Streptomyces aureofaciens* in a 112 m^3 fermenter. It is also possible in viscous fermentations to have radial DOT gradients in fermenters, especially in the stirrer region (Oosterhuis and Kossen, 1984). Thus microorganisms in large fermenters are subjected to fluctuating dissolved oxygen concentrations. Previously we reported that the specific penicillin production rate (q_{pen}) of *Penicillium chrysogenum* strain P1 fell at DOT values below 35% air saturation and that the q_{pen} decreased substantially compared to the value observed at a constant DOT of 30% air saturation when the cultures were subjected to cycling DOT (23 - 37% air saturation) with a cycle time of 2 min (Vardar and Lilly, 1982).

The syntheses of many other antibiotics by various organisms are known to be influenced by DOT. With mycelial organisms any change in the morphology can affect the response to DOT (Wang and Fewkes, 1977) so we have selected for further study of the effects of dissolved oxygen the synthesis of difficidin and its hydroxylated derivative,

1054–7487/92/0219$06.00/0

oxydifficidin, by *Bacillus subtilis* (Zimmerman et al., 1987). The bacterium has been grown in batch culture at 28°C on a soluble starch/lactate/Pharmamedia medium at various constant DOT values by regulating the agitation and aeration conditions. Whereas oxydifficidin is produced for over 50 h at a rate essentially independent of DOT, the rate of synthesis of difficidin is reduced below 20% air saturation and only reaches a maximum value above that DOT (Suphantharika et al., to be published). Studies on the effect of cycling of DOT on this antibiotic fermentation are in progress.

Morphological measurements by image analysis

The development of image analysis techniques for measurement of the morphology of filamentous cultures over the last few years (Adams and Thomas, 1988; Packer and Thomas, 1990) now allows the determination of the influence of the fermentation conditions on morphology to be determined more precisely. It is also possible to measure the biomass concentration and the extent of cell vacuolisation during fermentations of fungi, such as *Penicillium chrysogenum*, (Packer et al., in press).

Morphology and rheology of *S.erythraea* fermentations

Image analysis has been used to study the relationship between broth morphology and rheology for three different Actinomycete fermentations (Warren et al., to be published). For *Saccharopolyspora erythraea*, an erythromycin producer, the rheological behaviour fitted a power law. During the initial growth phase, lasting 50 - 70 h, the consistency index rose with biomass concentration. Within 10 h of the start of the fermentation the flow index had fallen from 1 to 0.2 and then was constant at that value until late in the fermentation. The culture consisted of short hyphae with a few branches. The mean main hyphal length of the culture remained in the range, 15 - 25 μm throughout the fermentation. There appeared to be little correlation between morphology and rheology of the culture.

Relationship between culture morphology and penicillin synthesis

The relationships between the agitation conditions, morphology and product synthesis have been examined with *Penicillium chrysogenum* strain P1 which was grown in batch culture on a medium containing sucrose, lactose, peptone, phenoxyacetate and mineral salts. The absence of undissolved solids allowed biomass measurements to be made. Experiments were done in 5, 100 and 1000 litre working volume fermenters each fitted with three turbine impellers. Morphological measurements on dispersed mycelia showed that the mean main hyphal length reached a maximum at the end of the rapid growth phase and then declined over the next 100 h at a rate dependent on the fermenter size and agitation speed.

At 5 L scale the peak oxygen uptake and carbon dioxide evolution rates were delayed by several hours at high agitation speeds (up to 1300 rpm). The mean main hyphal length at that time was much shorter (about 130 μm) than the value observed at lower agitation speeds (about 300 μm). The onset of penicillin synthesis was also delayed and the rate of synthesis over the next 75 h was lower. The mean main hyphal lengths and specific production rates for different agitation speeds at the three fermentation scales were correlated with various parameters (Makagiansar, 1992).

We observed previously with the same strain, growing on a complex solids-containing medium with glucose feeding at 10 and 100 litre scales, that impeller tip speed was a poor scale-up parameter but that there was a good correlation between penicillin production rate and $P/D_i^3t_c$ (where P is the agitator power,

D_i is the impeller diameter and t_c is the calculated circulation time) and also, to a lesser extent, with calculated liquid circulation frequency (Smith et al., 1990). The results for the recent studies at three scales of fermentation gave a good correlation between q_{pen} with $P/D_i^3t_c$ (Fig. 1) but less well with the calculated circulation frequency. The mean main hyphal length during the penicillin production phase also correlated well with $P/D_i^3t_c$ (Makagiansar et al., submitted for publication). This term is a measure of the maximum shear stress due to agitator power dissipation and the frequency with which mycelia pass through the region of high shear. It was interesting that in the 1000 L fermenter it was not possible to reach a value of the term sufficient to reduce the specific penicillin production rate even when full power was being drawn (9.4 KW). Other published data and correlations are also being examined for shear sensitive microorganisms with a view to developing scale-up relationships between fermenter hydrodynamics, culture morphology and penicillin production. Our initial analysis suggests that cell interaction with turbulent eddies plays a key role in determining the extent of mechanical damage to cells.

The authors wish to thank the Science and Engineering Research Council for support of this work, and Dr C. R. Thomas, Birmingham University (UK) and Dr K. Dixon, Pfizer (UK) for their important contributions to these studies.

References

Adams, H.L.; Thomas, C. R. *Biotechnol. Bioeng.* **32**, 707-12.

Lilly, M. D. In *Bioactive Microbial*

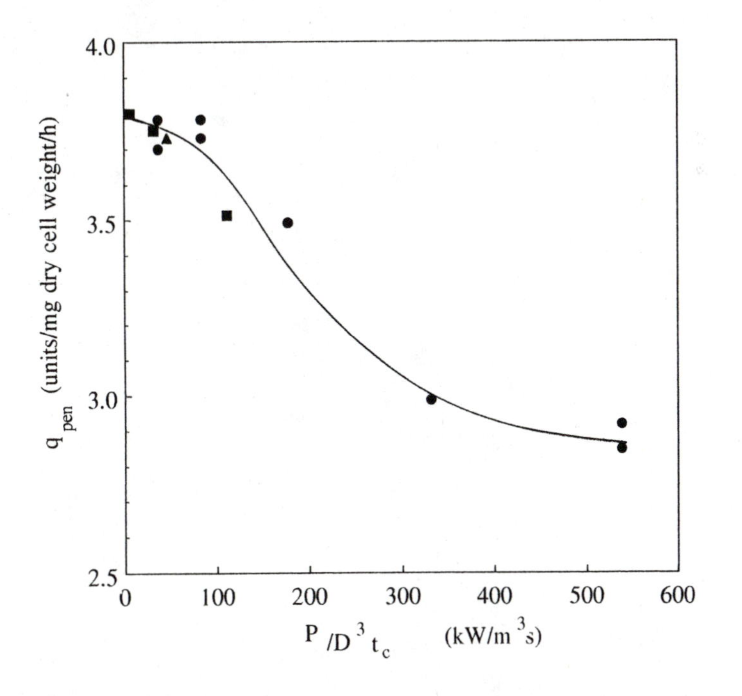

Fig. 1. Correlation of specific penicillin production rate (q_{pen}) with $P/D_i^3t_c$ for fermentations at 5 (●), 100 (■) and 1000 (▲) litre scale (Makagiansar et al., submitted for publication).

Products 2; Winstanley, D. J.; Nisbet, L. J. Ed; Academic Press: London, 1983, pp 79-89.

Makagiansar, H. Y. *PhD thesis,* **1992**, Univ. of London.

Makagiansar, H. Y.; Ayazi Shamlou, P.; Thomas, C. R.; Lilly, M. D., submitted for publication.

Manfredini, R.; Cavallera, V.; Marini, L.; Donati, G. *Biotechnol. Bioeng.* **1983**, 25, 3115-3132.

Oosterhuis, N. M. G.; Kossen, N. W. F. *Biotechnol. Bioeng.* **1984**, 26, 546-50.

Packer, H.; Thomas, C. R. *Biotechnol. Bioeng.* **1990**, 35, 870-81.

Packer, H.; Keshavarz-Moore, E.; Lilly, M. D.; Thomas, C. R. *Biotechnol. Bioeng.*, in press.

Smith, J. J.; Lilly, M. D.; Fox, R. I. *Biotechnol. Bioeng.* **1990**, 35, 1011-23.

Suphantharika, M.; Ison, A.; Lilly, M. D., submitted for publication.

Vardar, F.; Lilly, M. D. *Europ. J. Appl. Microbiol. Biotechnol.* **1982**, 14, 203-211.

Wang, D. I. C.; Fewkes, R. C. J. *Dev. Ind. Microbiol.* **1977**, 18, 39-56.

Zimmerman, S. B.; Schwartz, C. D.; Monaghan, R. L.; Pelak, B. A.; Weissberger, B.; Gilfillan, E. C.; Mochales, S.; Hernandez, S.; Currie, S. A.; Tejera, E.; Stapley, E. O. *J. Antibiotics* **1987**, 40, 1677-81.

Warren, S.; Lilly, M. D.; Thomas, C. R.; Dixon, K., to be published.

"Lights Out" Production of Cephamycins in Automated Fermentation Facilities

Hideo Eiki*, Isao Kishi, Tomohisa Gomi and Masao Ogawa,
Institute of Manufacturing Technology, Yamanouchi Pharmaceutical Co., Ltd.
Takahagi, Ibaraki 318, Japan (FAX:293 24 2707)

Some fermentation methods for preparing cephamycin derivatives have been attempted using *Streptomyces oganonensis,* from which oganomycin G, one of the 3-thiol-substituted derivatives, has been selected as the key material for cefotetan synthesis. In order to accelerate the fermentation studies as well as to establish an economical manufacturing system, some automated fermentation facilities were constructed. These have been operated through networks of programmable controllers and computers. Since 1983, the mass production of oganomycin GG derived from oganomycin G has been conducted in such highly automated facilities without a night shift system.

7-Methoxycephalosporins, namely, cephamycins, are produced by *Streptomyces* species (Higgens and Kastner, 1971; Nagarajan *et al.*, 1971). Cephamycins show strong resistance against gram-negative bacteria producing beta-lactamases. One of the limitations of early cephalosporins was their poor activity against these bacteria. Therefore, it was expected that cephamycins would be promising materials in order to synthesize new beta-lactams having strong and broad antibacterial activities.

Our company also isolated one of the cephamycin producers, designated *Streptomyces oganonensis*, and attempted some fermentation methods for preparing new cephamycin derivatives. Oganomycin G and its derivative, oganomycin GG (Osono *et al.*, 1980) were promising materials for cefotetan synthesis (YM09330, Iwanami *et al.*, 1980).

In order to accelerate the fermentation studies of oganomycin G, as well as to achieve industrial production, it seemed essential to economize in the operations of the fermentation facilities. Industrial fermentation has been mostly conducted by a night shift system with skilled workers, though working conditions need to be improved to eliminate the night shift system. Our company has neither such a night shift system nor skilled workers in the fermentation field. These problems may be resolved if the fermentation facilities are automatically operated without the night shift system, that is, as a "lights out" production system.

Since 1975, the authors have been working to establish some automated fermentation facilities.

In this paper, we briefly review oganomycin fermentation and outline the automated fermentation processes.

Attempts at Obtaining Cephamycin Derivatives

Some fermentation methods employing *S. oganonensis* were attempted to prepare cephamycin derivatives (Chart 1). The strain produces cephamycin A and B. Under mild conditions similar to those of fermentation, cephamycin A and B decompose at rates of 85 % and 70 % a day, respectively.

When p-hydroxy-cinnamate is added to the medium, the main products become stable oganomycin A and B. (Gushima *et al.*, 1981). Evidently, the instability of cephamycin A and B results from the alpha-methoxy group at the 3-acyloxymethyl groups.

In the mixed culture of *S. oganonensis* and *Torulopsis sp*. producing the 3-esterase, a great amount of oganomycin E, that is, 3-hydroxy-methyl-cephamycin, accumulates in the broth (Abe and Eiki, 1989). Cephamycins are

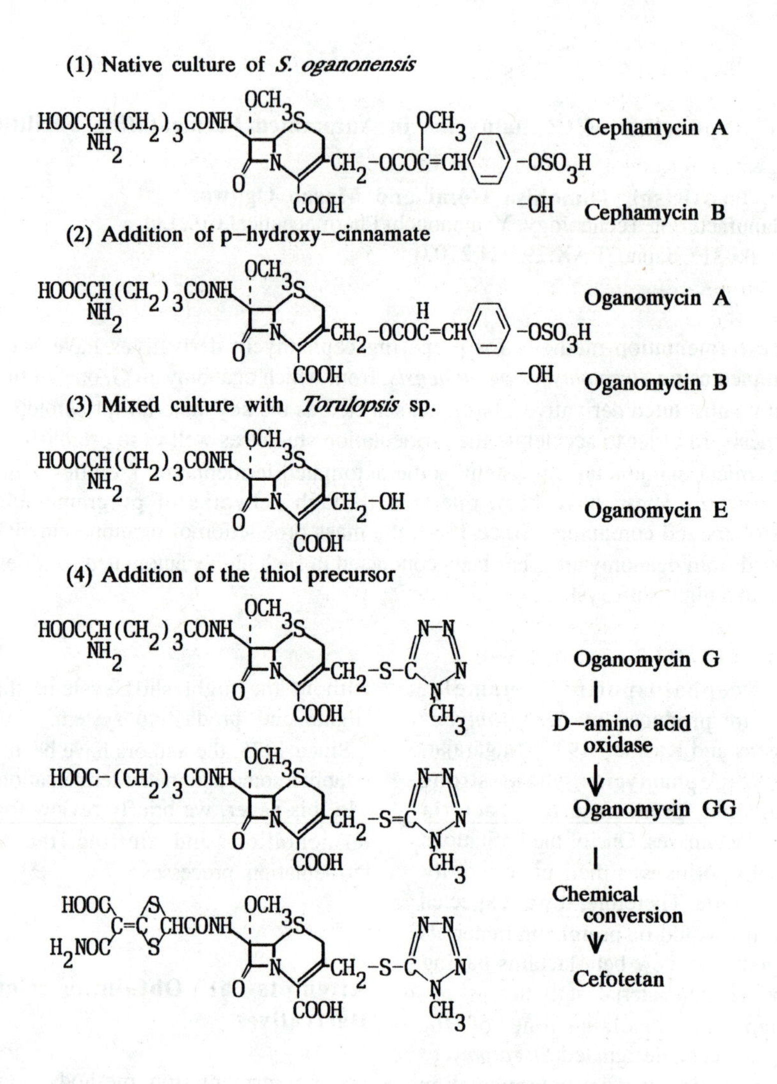

Chart 1. Preparation of cephamycin derivatives using *S. oganonensis*.

quantitatively converted to stable oganomycin E by the esterase. However, oganomycin A, B and E are considered to be unsuitable fermentation products. They are too stable to use in chemical synthesis of new beta-lactams.

Numerous sulfur compounds have been observed to displace the acetoxy group from cephalosporins (Cocker *et al.*, 1965). When 5-mercapto-1-methyl-tetrazole is added to the medium, the 3-thiol-substituted cephamycin, oganomycin G accumulates directly. Moreover, oganomycin G is converted to oganomycin GG using D-amino acid oxidase from *Trigonopsis variabilis* (Shanmuganathan and Nickerson, 1962). Consequently, oganomycin GG has been selected as the key material for cefotetan synthesis.

Automated Fermentation Facilities for Experimental and Pilot Use

Three automated fermentation facilities for experimental and pilot uses were constructed. The design concepts were mainly to eliminate skilled work from the daily operation of the fermentors, and to ensure flexibility of various fermentation studies. Moreover, we had to achieve safety in operations, extension to special functions, and reduction in investment.

These facilities are composed of fermentors, operation panels, programmable controllers, and personal computers, *etc.* (Chart 2). Two laboratory systems consist of 5 and 14 fermentors respectively, ranging from 30 l to 500 l, and some laboratory analyzers, such as HPLCs and autoanalyzers. The pilot plant system consists of 9 fermentors including 3 kl and 20 kl vessels. These automated facilities have been controlled by the networks of some programmable controllers (Memocon-SC series, Yasukawa Electric Mfg.) which can conduct logic and sequence controls along with computer communication. The controllers deal with discrete inputs and outputs, analog inputs and outputs, and other functions including data handling, calculation, counting, and timing. The analog outputs of the facilities, except exit air line valves of the 3 and 20 kl fermentors, were omitted to save expenses. However, each analog control of temperature, agitation speed, concentration of dissolved oxygen, and feeding rate, *etc.* has been properly controlled by the discrete outputs of time comparative pulses depending on the deviation between the present and set values.

Operation of each fermentor is freely selected from the main tasks consisting of empty sterilization, main sterilization, fermentation, discharge, washing, or reset. Regarding other control modes, such as inoculation, feeding, agitation, and pH control, *etc*., the operators can select from individual special functions programmed previously.

Present and set values or the status of the process control can be watched via digital indicators and/or computers. The set values for

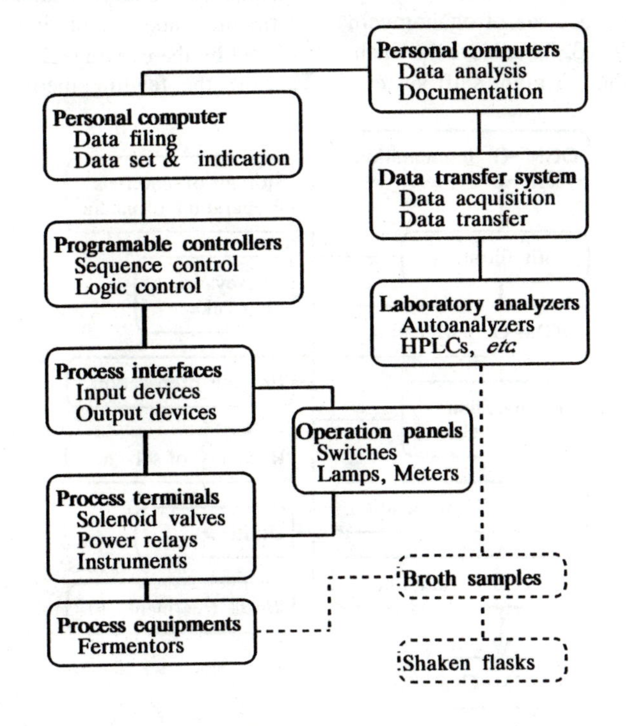

Chart 2. The network of the laboratory fermentation system.

the process control are changeable using digital switches and/or computers. These data as well as the data from the laboratory analyzers are regularly accumulated in computer files. After every batch, the batch records are printed out with figures and tables. This documentation system has also been applied to laboratory experiments using shaken flask cultures.

In case of emergency at night or on a holiday, the control system warns the security center to call the operators.

The facilities allow us to apply some sophisticated approaches, such as automatic estimation of the average concentration of dissolved oxygen in a fermentor (Eiki and Osono, 1990). The operators and staff have also been trained in the operations of the fermentation systems. Consequently, we have shown that such automated facilities promise the feasibility of the oganomycin manufacturing system without a night shift system.

Manufacturing Processes of Oganomycin GG

The manufacturing processes of oganomycin GG is illustrated in Chart 3. To achieve the batch-wise production without a night shift system, ordinary administration and information exchanges among the processes as well as the operations and controls have to be economically conducted. Achieving the weekly scheduled operations was one of the main concepts behind the manufacturing processes. Therefore, all of these facilities have been automated using the network of programmable controllers and computers just like for the fermentors mentioned above.

The present status of the process operations can be seen as follows. Two main fermentors are each operated at a batch a week. Raw medium materials stocked in the outdoor silos are automatically weighed and transferred into the indoor material hoppers before the medium preparation. In the early morning of the day on which the start of the next batch is scheduled, the main fermentation is finished. After transferring the broth to the reservoir, the fermentor is washed with alkaline hot water. By the time the operators come to the office, these jobs are almost complete. Several operators then perform their tasks which include empty sterilization, medium preparation, medium sterilization in the fermentor, and inoculation. The new batch is started by the evening. During the fermentation period, the feeding media are automatically

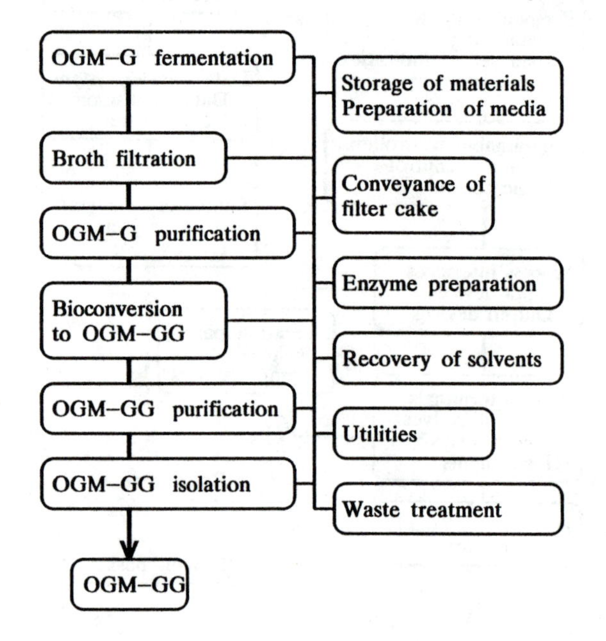

Chart 3. The manufacturing processes of oganomycin GG. Abbreviation: OGM–G, oganomycin G; OGM–GG, oganomycin GG.

prepared, sterilized, and added to the main fermentors. Of course, other control items including agitation speed, aeration, inner pressure, temperature, concentration of dissolved oxygen, and data filing, *etc*. have also been conducted in each automatic mode.

Additionally, the filtration process of the broth is controlled along with the adsorption of oganomycin G on resin columns, from which oganomycin G is transferred to the enzyme reactor for conversion to oganomycin GG. The product is isolated by centrifugal separators via the purification processes consisting of resin treatment, solvent extraction and crystallization. Moreover, the preparation of D-amino acid oxidase and recovery of solvents have been automatically performed. All of these processes which continue for 24 hours a day are conducted by daytime workers only.

Since 1983, the mass production of oganomycin GG has been conducted using these automated facilities on a weekly schedule, without a night shift system or Sunday workers. Investment in the automated facilities seemed more expensive than in ordinary facilities. However, the workers have benefitted from the automated facilities which have eliminated the night shift system. Additionally, these facilities are profitable in saving manpower.

References

Abe, M.; Eiki, H. *Japan kokai*, **1989**, 89-25,785.

Cocker, J.D.; Cowley, B.R.; Cox, J.S.G.; Eardley, S.;Gregory, G.I.; Lazenby, J.K.; Long, A.G.; Sly, J.C.P.; Somerfield, G.A. *J. Chem. Soc.* **1965**, 5015-5031.

Eiki, H.; Osono, T. *J. Ferment. Bioeng.* **1990**, 69, 313-316.

Gushima, H.; Watanabe, S.; Saito, T.; Sasaki, T.; Eiki, H.; Oka, Y.; Osono, T. *J. Antibiotics* **1981**, 34, 1507-1512.

Higgens, C.E.; Kastner, R.E. *Int. J. Syst. Bacteriol.* **1971**, 21, 326-331.

Iwanami, M.; Maeda, T.; Fujimoto, M.; Nagano, Y.; Nagano, N.; Yamazaki, A.; Shibanuma, T.; Tamazawa, K.; Yano, K. *Chem. Pharm. Bull.* **1980**, 28, 2629-2636.

Nagarajan, R.; Boeck, L.D.; Gorman, M.; Hamill, R.L.; Higgens, C.E.; Hoehn, M.M.; Stark, W.M.; Whitney, J.G. *J. Amer. Chem. Soc.* **1971**, 93, 2308-2310.

Osono, T.; Watanabe, S.; Saito, T.; Gushima, H.; Murakami, K.; Susaki, K.; Takamura, S.; Miyamoto, T.; Oka, Y. *J. Antibiotics* **1980**, 33, 1074-1078.

Evolution in Fermentation Facility Design from Antibiotics to Recombinant Proteins

Fredric G. Bader, Genetics Institute, 1 Burtt Rd., Andover, MA 01810 (Fax: 508-470-2070).

The fermentation process has evolved at a steady pace in supporting the manufacture of drug products. However, bulk drug fermentation facilities have changed little in the past 30 years. The recent introduction of recombinant DNA proteins as human pharmaceuticals, has created a new challenge to fermentation specialists. These products are being treated as biologicals, requiring processing facilities that closely resemble aseptic pharmaceutical manufacturing facilities. The modern fermentation specialist needs to become familiar with the impact of biological facility and processing requirements to be productive in biopharmaceuticals.

The fermentation process has evolved substantially from the early days of the uncontrolled batch fermentation. Implementation of sterilizable sensors, sterile addition technology, improved strain development and reliable computer control systems have led to the development of sophisticated fed batch and continuous processing. These achievements have reduced the cost of producing an antibiotic like penicillin by a factor of thousands; enabling its economic availability throughout the world.

The basis for discussing antibiotic fermentation processes is shown in Table 1. Bulk antibiotic fermentation facilities most closely resemble large chemical plant installations. These facilities are frequently out door installations consisting of large scale equipment. They have a heavy reliance on solvents and solvent recovery systems and demand hefty utility systems for support.

By contrast, Table 2 shows the basis for discussing recombinant proteins. Here, the fermentation systems are substantially smaller and are exclusively indoor facilities. Recombinant fermentation systems require smaller utility support systems, but all utility systems, process equipment and raw materials tend to be of the highest quality possible.

The two major differences between antibiotic and recombinant fermentations are the scale of operation and the regulatory requirements for approving the product as a human pharmaceutical. Antibiotics are considered drugs and are covered by The Center of Drug Evaluation and Research of the FDA. Recombinant proteins are more frequently considered to be biologicals and are covered by the Center for Biological Evaluation and Research of the FDA. This distinction may appear insignificant to those untrained in the regulatory requirements of these two agencies. Infact, the distinction is quite significant.

Table 3 provides a comparison between requirements for Drugs and Biologicals. In both cases, the products must be licensed by the FDA and must conform to a pre-established set of release specifications. They differ in that biological product release requires that the product be produced by a tightly specified process in a facility that has been filed, inspected and licensed by CBER. This is not the case for drugs. The 'product by process' and 'product by facility' approach of biologicals requires that all process or facility changes be submitted to CBER as formal amendments and approved prior to implementation. In the case of drugs, only

TABLE 1: **BASIS FOR ANTIBIOTICS**

DRUGS	-	PRODUCT BY SPECS
SCALE	-	50,000 TO 250,000 LITER
RAW MATERIALS	-	AGRICULTURAL WASTE PRODUCTS
BATCH SIZE	-	THOUSAND KILOGRAM
PRODUCT COST	-	TENS OF DOLLARS PER KILOGRAM
ORIENTATION	-	BULK CHEMICAL PROCESS

TABLE 2: **BASIS FOR rDNA PROTEINS**

BIOLOGICALS	-	PRODUCT BY SPECS PRODUCT BY PROCESS PRODUCT BY FACILITY
SCALE	-	100 TO 10,000 LITERS
RAW MATERIALS	-	HIGH QUALITY REQUIREMENTS
BATCH SIZE	-	GRAMS
PRODUCT COST	-	TENS OF DOLLARS PER MILLIGRAM
FACILITY ORIENTATION	-	ASEPTIC PHARMACEUTICAL

TABLE 3: **COMPARISON - DRUGS VERSUS BIOLOGICALS**

AREA	DRUGS	BIOLOGICALS
LICENSE	NDA	PLA
RELEASE SPECS	YES	YES
PROCESS SPECS	GENERAL	SPECIFIC
FACILITY LICENSE	NO (DMF)	YES (ELA)
FOCUS	DRUG PRODUCT	DRUG SUBSTANCE
CHANGE CONTROL	FLEXIBLE	RIGID

major changes require approval prior to implementation. As a result biological process development and improvement strategies require greater regulatory input and will occur at a slower pace.

Table 4 shows a summary list of the impact of biological regulations when compared to those of drugs. Typically, the licensed process and facility must have been used to produce Phase III clinical material. This requires that one get to full scale in the commercial facility much earlier than with drugs. It also requires that the process be 'locked in' much earlier in the product development cycle.

Certain characteristics of proteins affect the nature of the processes that must be used to produce them. Proteins are administered as parenterals, are susceptible to biological degradation and tend to retain endotoxins. To protect both the patient and the proteins, processing must occur in a controlled atmosphere using high grade raw materials, utilities and validated equipment. Proteins also have a high potency and can be immunologically active at very low levels. This increases concerns about cross contamination when more than one product is produced in the same facility. To deal with these issues, the fermentation engineer must be prepared to address facility design practices that are closely related to those used in aseptic pharmaceutical processing facilities. Table 5 lists some of the issues that must be dealt with in designing a modern recombinant protein production facility.

Table 6 shows a generic timeline to design, build and start-up a fermentation producing facility. In the case of a drug product like an antibiotic, production of commercial material can typically occur during the fourth year of the project. In the case of a biological protein, facility and process approval requirements will delay commercial production until the beginning of year seven. As a rough estimate, biological requirements increase the carrying cost of the facility and the cost of money to a point where a biological facility costs over twice as much to bring on stream as an equivalent drug facility.

The biological regulatory requirements provide us with certain benefits and concerns as outlined in Table 7. The net effect is higher consumer safety at a higher cost of product. The balance between these two variabilities is a matter of public policy. The role of the fermentation engineer is to adapt to the biological rules that govern recombinant protein production to produce the highest quality product at the lowest cost possible.

TABLE 4: **IMPACT OF BIOLOGICAL DIFFERENCES**

CLINICAL PRODUCTION IN COMMERCIAL FACILITY

EARLY PROCESS LOCK-IN

ASEPTIC PHARMACEUTICAL STYLE PROCESSING

HEAVY CONCERN ON CROSS CONTAMINATION

FDA HAS TO APPROVE PROCESS AND FACILITY

FDA HAS TO APPROVE ALL CHANGES

TABLE 5: INCREASE FACILITY EMPHASIS IN BIOLOGICALS

1. FACILITY LAYOUT
2. ENGINEERING CONTROLS OF CONTAMINATION
3. EQUIPMENT DESIGN TO ASEPTIC STANDARDS
4. STYLE OF CONSTRUCTION
5. ENVIRONMENTAL MONITORING
6. VALIDATED CLEANING/CHANGEOVER OR DEDICATED
7. UTILITIES: WFI
   HIGH QUALITY PROCESS WATER
   CLEAN STEAM
   HIGH PURITY GASES
   CLEAN-IN-PLACE
   HVAC

TABLE 6: FACILITY TIME-LINE COMPARISON

YEAR	DRUG ACTIVITY	BIOLOGICAL ACTIVITY
1	DESIGN	DESIGN
2	CONSTRUCTION	CONSTRUCTION
3	CONSTRUCTION	CONSTRUCTION
4	VALIDATE - PRODUCE	VALIDATE
5	PRODUCE	CLINICALS & FILE
6	PRODUCE	INSPECTION/APPROVAL
7	PRODUCE	PRODUCE

TABLE 7: BIOLOGICAL ORIENTATION

BENEFITS:

MORE CAREFUL PLANNING
HIGHER CONSISTENCY/RELIABILITY
LOWER CONSUMER RISK

CONCERNS:

HIGH PRODUCT COST
HIGH REGULATORY RISK
IMPEDED COST REDUCTION
HIGH CAPITAL RISK
POTENTIAL MARKET SHORTAGE
IMPEDIMENTS TO ORPHAN DRUGS

Comparison of Different Reactor Designs and Performances

Karl Schügerl
Institut für Technische Chemie, Universität Hannover
Callinstr. 3, D-3000 Hannover, Germany (FAX: (49) 511 762 3004)

In the chemical industry, the production costs are mainly influenced by chemical and other running expenditures. The same holds true for the manufacturing of biotechnological bulk products. In contrast to the high-added-value products, the costs of product separation and purification are low compared to the product formation costs.

In the case of production with aerobic microorganisms, the key factors are the costs for chemicals (mainly substrate) and energy (including aeration and cooling). The reactor costs are relatively unimportant. Therefore, for the selection of suitable reactors, not only their volumetric performance (e.g., volumetric productivity), but their specific performance with respect to the key parameters (chemicals, energy) are decisive.

According to industrial practice, the medium composition has a much larger effect on the process performance than on the reactor type, upon which, however, the optimal medium composition depends, at least for filamentous molds.

Comparison of reactors and their performances can be carried out on different levels, i.e., with regard to

- the oxygen transfer rate and efficiency,
- the cell mass productivity and efficiency,
- the metabolite or enzyme productivity.

The efficiency can be related to the substrate or energy consumption.

Comparison of Reactors Based on the Oxygen Transfer Rate and Efficiency and Cell Mass Production Rate and Efficiency

It can be shown that the volumetric and specific performance data depend on the medium properties, specific power input and specific aeration rate.
Therefore, the prerequisites of the reactor comparison are the use of the same medium and specific operation variables.

In low viscosity cultivation medium and at specific power inputs below 2.5 kWm^{-3}, the oxygen transfer rates Q_{O2} (kg O_2 m^{-3} h^{-1}) are with air: 1.0-3.6 (stirred tank), 1.9-2.2 (stirred loop), 0.3-4.8 (bubble column), 5-8 (airlift tower internal loop), 1.6-3.2 (airlift tower external loop) and the efficiencies of the oxygen transfer E_{O2} (kg O_2/kWh) are: 0.8-2.6 (stirred tank), 0.9-1.1 (stirred loop), 2.2.-12 (bubble column), 2-4 (airlift internal loop), 3-10 (airlift external loop).

The cell mass productivity R_X (kg m^{-3} h^{-1}) is given by

$$R_X = Q_{O2} \, Y_{X/O2} \tag{1}$$

and the efficiency of cell mass productivity E_X (kg cell mass /kWh) by

$$E_X = E_{O2} \, Y_{X/O2} \tag{2}$$

At constant yield coefficient of cell production with respect to oxygen consumption $Y_{X/O2}$ (e.g., *Trichosporon cutaneum*), R_X and E_X can be calculated from Q_{O2} and E_{O2} (Adler and Fiechter, 1986). At oxygen transfer limitation

$$Q_{O2} = k_L a \, C_{O2}^{*}, \tag{3}$$

where C_{O2}^{*} is the dissolved oxygen saturation value. Since C_{O2}^{*} is approximately constant, R_X and E_X can be calculated by means of $k_L a$ and $Y_{X/O}$.

According to Schlüter et al. (1992) for the volumetric mass transfer coefficient $k_L a$ in *Trichosporon cutaneum* medium system, the following relationship holds true:

$$k_L a \left(\frac{v}{g^2}\right)^{1/3} = C_4 \left[\frac{P/V_L}{R \, \{v g^4\}^{1/3}}\right]^a \left[\frac{q_G}{V_L}\left(\frac{v}{g^2}\right)^{1/3}\right]^b \tag{4}$$

1054–7487/92/0232$06.00/0

where v = kinematic viscosity of the medium
g = acceleration of gravity
R = density of the medium
$C_4 = 7.94*10^{-4}$, a = 0.62 and b = 0.23 for flat bladed disc turbine DT and
$C_4 = 5.89*10^{-4}$, a = 0.62 and b = 0.19 for INTERMIG stirrer.

The validity range of Eq (4) is:
$0.5 < P/V_L < 16$ kW/m^3
$0.0038 < q_G/V_L < 0.027$ s^{-1}.

The difference between the performances of DT and INTERMIG stirrers is slight.

In highly viscous non-Newtonian media, a more complex relationship is valid. According to Herbst et al. (1992), for Xantan production, which is controlled also by the oxygen transfer rate, the following relationship recommended by Kawase and Moo-Young (1988) can be applied:

$$k_La = 0.675\, D_M^{0.5} \frac{R^{3/5}\{P/V_L R\}^{[9+4n]/10[1+n]}}{\{k/R\}^{[1+n]/2}\, \sigma^{3/4}} \left(\frac{u_G^{0.5}}{u_B}\right)\left(\frac{\mu_{eff}}{\mu_W}\right)^{0.25} \qquad (5)$$

where u_G = the superficial gas velocity (m/s),
u_B = 0.265 m/s the bubble rise velocity (m/s),
D_M = the molecular diffusivity of oxygen in the liquid phase(m^2/s),
n = flow behavior index (1),
k = fluid consistency factor (Pa s^n),
μ_{eff} = effective viscosity (Pa s),
μ_W = viscosity of water (Pa s),
σ = surface tension of the medium (N/m),

Eq(5) holds true for DT, but for INTERMIG, the k_La calculated by this equation had to be modified by a factor of 3. This indicates that, in contrast to low viscosity media, in highly viscous media the stirrer performances are different. Eqs(1) and (2) hold true for 70-, 300-, 1500-, and 3000-l bioreactors.

If catabolite repression occurs at higher substrate concentrations (e.g., in the case of *Saccharomyces cerevisiae*), substrate limitation is imperative. The prerequisite for substrate limitation, however, is avoiding oxygen transfer limitation. Therefore, Eqs.(1) and (2) give the upper limit for R_X and E_X. Under industrial baker's yeast production conditions, the production of ethanol is avoided by suitable operating conditions (low substrate concentration, highly dissolved oxygen concentration), thus, no change of oxidative to oxidative-fermentative metabolism occurs. Under these conditions, oxygen balance along the reactor combined with the Monod model and constant yield coefficients $Y_{X/S}$ and $Y_{X/O2}$ give satisfactory results.

If oxygen transfer limitation prevails, as is evident from the calculated oxygen concentration distribution along the tower or tower loop reactor (especially in the downcomer of the reactors with an outer loop), *scale-down* measurements are necessary to evaluate the dynamics of the metabolism's change, and investigations must be carried out in the pilot plant or in the production reactor in order to determine the liquid circulation rate through the riser and downcomer and the mean contact time of the cells in these oxygen-limiting regimes.

According to our *scale-down* measurements, the biological parameter change depends on the duration and frequency of the oxygen-limiting phase.

In a 10 m high and 5 m^3 volume pilot plant airlift tower loop reactor, at u_G = 17.5 cm s^{-1}, the liquid circulation time was determined to be 15.5 s. (Fröhlich et al., 1991). Therefore, in this reactor, no correction of biological data was necessary. In a 200 m^3 reactor, however, this correction of the biological data was necessary.

Comparison of Reactors with Respect to the Enzyme and Metabolite Productivity

As long as the formation is a primary metabolite closely connected to the cell growth, the product formation rate R_P can be calculated from the cell growth rate R_X by means of the relationship

$$R_P = R_X Y_{P/X} \qquad (6)$$

if the coefficient $Y_{P/X}$ is constant during the production. A typical example is the homofermentative lactic acid production with *Lactobacilli*.

A most difficult task is a comparison is

based on the product formation rate of enzymes or secondary metabolites.

Since it is not possible to make general conclusions, two examples are presented to show the problems with these systems.

The first example is the production of an intracellular fusion protein of the enzyme EcoRI (restriction endonuclease) and SpA protein with a recombinant *E. coli* K12 JM103 (Brandes, 1991; Wu, 1992), which carried expression plasmids pMTC48 (50-110 copies), repression plasmids pRK248 (2-8 copies) and protection plasmids EcoR4 (40-60 copies).

pMTC48 provides the cell with an ampicillin resistence and two independent promotors P_{LacUV5} and P_R. This allows the chemical induction with IPTG through promotor P_{LacUV5} as well as by the temperature induction through promotor P_R. The resulting protein is a fusion protein from EcoRI and SpA (*Staphilococcus aureus*). pEcoR4 methylase plasmid carries the resistance gene against chloramphenicol and reduces the cell toxicity of EcoRI.

The performances of a small 1.5-l stirred tank reactor and a 60-l airlift tower loop reactor with a draft tube are considered and compared.

The influence of the induction time on the enzyme activity was investigated in the batch-culture of *E. coli* with system C.

In the 1.5-l reactor, the optimal IPTG concentration was 0.5 mmol l^{-1}, and in the 60-l reactor 4 mmol l^{-1}. On account of the high costs of IPTG and the relatively high concentration required for an efficient induction, the temperature shift induction would be preferred.

The optimal induction temperature was 38 ^oC.

A comparison of the chemical (IPTG) induction with the temperature shift induction in the 1.5-l reactor showed that the latter provided enzyme activities five times higher with the temperature shift, which is in good agreement with the ratio of the efficiencies of the Lambda promoter to the Lac promoter, which is equal to five, determined in *E. coli* (Vidal-Ingigliardi and Raibaud, 1985). In the 60-l AL, however, the enzyme activity with temperature induction was lower than that with chemical induction. This is partly caused by the reduction of the CFU during the temperature induction and partly by the uneven temperature and lower temperature increase rate (Brandes, 1991).

These effects become more serious in larger reactors, since they depend on the heat transfer rate into the medium (heat transfer coefficient, the heater/cooler surface area and mixing intensity).

The second example is the production of penicillin V with the fungus *Penicillium chrysogenum* in a 30-l stirred tank and in a 100-l airlift tower reactor with an outer loop.

This is a typical secondary metabolite production with well-distinguished growth and production phases. The aim is the production of large amounts of active cell mass during a reasonably short growth phase and of high amounts of antibiotics of a very low specific growth rate during the production phase. On account of the high viscosity of the filamentous mycel, however, high specific power input is necessary to supply the cells with a satisfactory amount of oxygen.

Investigations employing different stirrer types and impeller diameters showed that the attained volumetric mass transfer coefficients were by a factor of two larger with INTERMIG 066 impellers than with those in a flat-bladed disc turbine 033 (DT 03) at about the same specific power input. Similar result were reported on PROCHEM HYDROFOIL impellers employed in antibiotic production by Buckland et al. (1988).

By changing the fungus morphology from filamentous mycel to pellet, the viscosity of the mold was reduced by a factor of 3-4, and the volumetric oxygen mass transfer coefficient into the medium was enhanced by the same factor. However, the same productivity can only be maintained with pellet suspension if the pellet diameter is smaller than 0.4 mm. Increasing the pellet diameter from 0.6 to 1.2 diminishes the productivity by a factor of two (Möller et al. 1992). On account of the more uniform distribution of the energy dissipation rate (EDR) in bubble column and airlift tower loop reactors, maintaining a uniform pellet size distribution there is easier than in a stirred tank, in which the EDR varies from the impeller edge to the reactor wall by a factor of hundred.

It was possible to obtain this pellet size with a low productive strain. The productivity was 30% lower than in the stirred tank reactor due to the lower cell mass concentration in the reactor; but the specific productivity with respect to the power input was 2.5 times higher, with regard to the substrate and oxygen consumption 20 and 30% higher, respectively, than in the stirred tank reactor.

However, it was not possible to maintain the optimal pellet size with a highly productive strain; the product concentration attained in an airlift tower loop reactor was only half (10 g l^{-1}) of that in the stirred tank reactor (20 g l^{-1}). This is partly caused by the lower cell mass concentration (23 g l^{-1} instead of 35-40 g l^{-1}) and the larger pellet size (at an average of about 0.7 mm up to 140 h and 0.4 mm after that) (Möller, 1987).

Systematic investigations are imperative to optimize the pellet morphology with different strains. In general, suitable precultures (appropriate spore concentration in the first preculture, filamentous mycel, or small pellets with diameters < 0.2 mm in the second preculture) and diluted medium in the main culture are essential for achieving the optimal pellet size and density.

References

Adler, I.; Fiechter, A., *Bioprocess Engng. 1*, **1986**, pp 51-59.

Brandes, L. *Dissertation*, University Hanno ver, **1991**.

Buckland, B.C., Gbewonyo, K., Jain, D., Glazomitzky, K., Hunt, G., Drew, S.W., ' in *2nd Internat. Conference on Bioreactor Fluid dynamics*, Cambridge, UK: 21-23 Sept., King, R. Ed; Elsevier Applied Science Publisher, London, **1988**, pp 1-15.

Fröhlich, S., Lotz, M., Korte, T., Lübbert, A., Schügerl, K., Seekamp, M., In *Biotechnol. Bioeng. 38*, **1991**, pp 43-55.

Herbst, H., Schumpe, A., Deckwer, W.D. In *Chem. Eng. Technol.*, **1992**, submitted.

Kawase, Y., Moo-Young, M.(1988), In *Chem. Eng. Res. Dev. 66*, **1988**, pp 284-288.

Möller, J. *Dissertation*, University of Hanno ver, **1987**.

Rüffer, M. (1991) personal communication.

Schlüter, V., Yonsel, S., Deckwer, W.D., In *Chem.-Ing.-Techn.*, **1992**, submitted.

Vidal-Ingigliardi, D., Raibaud, O., In *Nucleids Acids Research 13*, **1985**, pp 5919-5926.

Wu Xiaoan, *Dissertation*, University of Han nover, **1992**.

DOWNSTREAM PROCESSING SYMPOSIUM VI

C. L. Cooney, A. Hershman, and M. Hoare: *Co-Chairs*

Biological Principles in Aid of Purification: Session A
F. H. Arnold and M. Uhlen: *Co-Chairs*

Membrane Processes: Session B
C. L. Cooney and K. H. Kroner: *Co-Chairs*

Chromatographic Processes: Session C
N. B. Afeyan and P. O. Hedman: *Co-Chairs*

Extractive Processes: Session D
H. W. Blanch and J. Cabral: *Co-Chairs*

Protein Folding and Refolding: Session E
A. Hershman and M. Uhlen: *Co-Chairs*

Stability, Formulation, and Quality Control: Session F
R. Middaugh and C. R. Hill: *Co-Chairs*

Protein Engineering to Facilitate Purification and Folding of Recombinant Proteins

Tomas Moks, Elisabet Samuelsson and Mathias Uhlén

Department of Biochemistry and Biotechnology, Royal Institute of Technology
S-10044 Stockholm, Sweden (Fax: int-46-824 5452)

Protein engineering where genes or gene fragments are spliced together to form gene fusions is an attractive tool in the aid of purification of recombinant proteins. Using affinity handles, various biospecific or biochemical interactions can be used to facilitate the down stream processing. In addition, the fusion protein technique can also be used to keep misfolded or aggregated proteins in solution during in vitro refolding. Using Staphylococcal protein A as a solubilizing fusion partner, it was possible to refold human peptide hormones at relatively high protein concentrations in the absence of any denaturing agents, in a process well suited for large scale production.

Recombinant DNA techniques have allowed in vitro fusions of genes or gene fragments to improve the properties of recombinant proteins in production and down stream processing. In this review we will discuss the use of such protein engineering to facilitate the purification of proteins and the use of solubilizing fusion partners to enhance the in vitro refolding of recombinant disulphide containing proteins. Cleavage methods to separate the desired protein from the fusion partners are also discussed.

Gene Fusion to Facilitate Purification

In traditional downstream processing after fermentation a number of different unit operations are used to recover the protein from other components. During this procedure, properties like electric charge, solubility, size, hydrophobicity, specific affinity etc. of the different components in the complex mixture, form the basis of the separation and each purification step has to be optimized for the specific separation situation. With the protein fusion technique it is now possible to invert the issue and modify the product to fit already available optimized separation operations. The usually unique properties of the purification handle that is fused to the product are used in the separation. If necessary, this handle is later removed to obtain a native product.

Many protein or peptide ligands with useful properties for separation have been presented and they have their inherent advantages and disadvantages. Table 1 lists some of the systems developed. The protein ligands, like protein A and monoclonal antibodies give very high selectivity, but the column material is more sensitive to proteolytic degradation and the scale-up is relatively difficult since the sterilization using alkali or heat is not possible. The use of mono clonal antibodies might also be relatively expensive. However, for many products these systems based on protein-protein interactions are still of great interest, since an essentially pure product can be obtained in a one-step procedure with almost 100% yield (Moks et al., 1987). An advantage with the flag system (Hopp et al., 1988) is that a mild elution can be accomplished, which is important for many products which are sensitive to extreme pH or high salt conditions.

Table 1. Principles used for gene fusion systems to facilitate the recovery of proteins

Interaction	Examples
Protein-protein	Protein A
Enzyme-substrate	ß-galactosidase
Antibody-antigen	Flag peptide
Peptide-ion exchange	Poly (Arg)
Peptide-IMAC	Poly (His)
Peptide-HIC	Poly (Phe)
2-phase system	Poly (Trp)
Carbohydrate-receptor	Maltose-binding protein
DNA-binding	Lac operator
Vitamin-receptor	Streptavidin

Systems based on ion-exchange, metal-affinity (Arnold, 1991) or aqueous two-phase

1054–7487/92/0238$06.00/0

extraction (Köhler et al., 1991) where the ligands are less sensitive to proteolysis and regeneration, makes these approaches easier to incorporate into large scale processes. However, the resolution obtained is usually much poorer than what is obtained in the systems based on protein-protein interactions. This affects both the yield and the purity and additional purification steps must be introduced to get acceptable quality of the final product. An interesting alternative is to produce histidine containing fusion proteins as inclusion bodies and the recover these by specific Ni^{2+} affinity chromatography at denaturing conditions, such as 4 M urea. For a review, see Arnold (1991).

In conclusion, several efficient alternative gene fusion systems exist. Various considerations, such as localization, solubility, yield, purity and the final use of the gene fusion product influence the choice of the optimal systems for a particular protein.

Enhanced Solubility During In Vitro Refolding

One of the major problems associated with the expression of foreign genes in bacteria is related to the folding of the gene product into its native conformation. In fact, many expression systems in E.coli lead to the formation of aggregates of denatured proteins, often referred to as inclusion bodies, with only a small fraction maturing into the desired native protein. This has lead to the development of complicated refolding schemes where the insoluble aggregates are dissolved in strong denaturants such as urea or guanidine-HCl. However, the low solubility of unfolded protein necessitates very dilute protein solutions to avoid reaggregation and precipitation during the refolding, which has made the systems complicated and expensive.

Recently, a new approach to facilitate the in vitro refolding of recombinant proteins with low solubility was described (Samuelsson et al., 1991). By using the highly soluble protein A domain as a solubilizer of misfolded and multimeric forms of the human peptide hormone IGF-I, high yields of native protein could be recovered without the problems of aggregation and precipitation of incorrect structures. Since the refolding easily can be performed at protein concentrations as high as 1-2 mg/ml, without the need for any strong denaturants or expensive redox chemicals, the approach is well suited for large scale production of recombinant proteins. The approach to solubilize proteins by gene fusion techniques has recently been extended for in vitro refolding of IGF-II (Forsberg et al., 1991).

Cleavage Methods

Depending on the final use of the product it may be necessary to remove the purification handle during the downstream processing. For many technical applications the purification handle may be harmless or may even make the product more easy to assay. However, in case of pharmaceuticals it is likely that the handle should be removed. This is in many instances a bottle-neck in the current use of fusion protein.

Two main principle methods are available for cleavage of the fusion protein: Hydrolysis with chemicals or hydrolysis using specific enzymes. A number of chemicals can be used to hydrolyse specific peptide bonds, with hydroxyl-amine (Asn-Gly), formic acid (Asp-Pro) and cyanogen bromide (C-terminus of Met) as the most frequently applied methods. The chemical methods have the advantage of being relatively cost-effective and easy to scale-up, but they provide harsh conditions that may be detrimental for the product. In addition, larger proteins are likely to contain several sensitive sites and a correct N-terminal is not always possible to achieve. This disadvantage may be reduced by using enzymatic hydrolysis of the cleavage site. Proteases are available that are highly specificity for more complex amino acid sequences. However, the enzyme must work efficiently in immobilized state to avoid that otherwise unnecessary further protein separation steps must be involved.

Enzymatic cleavages of fusion proteins have recently been the focus of an extensive review (Carter, 1990). In a large survey of Forsberg (personal communication), different methods for cleavage of peptide hormones, such as insulin-like growth factor I and II, from a fusion partner based on protein A was compared. Interestingly, a mutant variant of subtilisin called Ala64-subtilisin has been replaced by an alanine residue was found to give high yields of correct cleavage products. More than 80% of the products were normally recovered after cleavage of several fusion proteins. The specificity of this protease is very high, recognizing the sequence Ala-Ala-His-Tyr, and cleaving C- terminal of the Tyr residue. Although site-specific cleavages of fusion proteins still can be considered as a bottle-neck in the preparation of recombinant proteins using gene fusion techniques, the Ala64-subtilisin demonstrates that efficient and specific proteases can and probably will be engineered in the future.

Concluding Remarks

Protein engineering has proven to be an attractive technique to facilitate the downstream processing of recombinant proteins expressed in bacteria. So far the technique has mainly been used as a research tool for production of immunogens and various proteins for structural and/or functional studies. A few examples of large scale production of pharmaceutical proteins also exist.

However, the use of these techniques for production of low cost industrial enzymes is still limited and not many cases of eucaryotic expression systems using gene fusions have been reported. It is likely that new recombinant proteases with high specificity will be developed in the future and thus make fusion proteins to an attractive alternative in wider applications.

References

Arnold, F.H.,Haymore, B.L., *Science*, **1991**, 252, 1796-1797

Carter, P., *Protein purification: from Molecular Mechanism to Large-Scale Process*; Ladish, M.R., Willson, R.C., Painton, C.C. and Builder, S.E., Ed.; ACS Symposium Series No. 427., American Chemical Society, **1990**, 181-193.

Forsberg, G., Samuelsson, E., Wadensten, H., Moks, T. and Hartmanis, M., In *Techniques in protein chemistry*; Angeletti, R.H., Ed., Academic press, San Diego, California, **1992**.

Hopp, T.P., Pricklet, K.S., Price, V.L., Libby, R.T., March, C.J., Cerretti, D.P., Urdal, D.L. and Conlon, P.J., *Bio/Technology*, **1988**, 6, 1204-1210.

Moks, T., Abrahmsén, L., Österlöf, B., Josephsson, S., Östling, M., Enfors, S.-O., Persson, I., Nilsson, B. and Uhlén, M., *Bio/Technology*, **1987**, 5, 379-382.

Samuelsson, E., Wadensten, H., Hartmanis, M., Moks, T. and Uhlén, M., *Bio/Technology*, **1991**, 9, 363-366.

Uhlén, M., Moks, T., *Methods in Enz.*, **1990**, 185, 129-143

Köhler, K., Ljungkvist, C., Kondo, A., Veide, A. and Nilsson, B., *Bio/Technology*, **1991**, 9, 642-646.

Photolithography, Chemistry and Biological Recognition

Christopher P. Holmes and **Stephen P.A. Fodor**
Affymax Research Institute, 4001 Miranda Ave.
Palo Alto, California 94304

New technologies for generating and screening large chemical libraries have recently been described. For example, recombinant techniques allow the creation of of very large numbers of randomly generated peptides (Cwirla et al., 1990; Devlin et al., 1990; Scott and Smith, 1990, Cull et al., 1992). Using biological selection criteria these libraries can be rapidly screened to identify molecules of interest. Alternatively, new approaches to the generation of synthetic chemical libraries have also been introduced (Fodor et al., 1991, Houghten et al., 1991, Lam et al., 1991). Among these is a technique which is capable of producing a spatially addressable array of chemical compounds. In the case of peptides, this is accomplished by combining solid-phase peptide synthesis and photolithography. The high coupling yields of Merrifield solid-phase chemistry allow efficient peptide synthesis, and the spatial resolution of photolithography affords miniaturization. The merging of these two technologies is accomplished through the substitution of a photolabile amine protecting group for the traditional Fmoc or Boc employed in the Merrifield synthetic procedure. Recent advances in this technology also allow the synthesis of oligonucleotide arrays.

The key points of this technology are illustrated in Figure 1. A synthesis substrate is prepared for amino acid coupling through the covalent attachment of photolabile nitroveratryl-oxycarbonyl (NVOC) protected amino linkers (Patchornik et al., 1970). Since the amines are unreactive until the photolabile protecting groups are removed, selective deprotection by light dictates where chemical coupling can occur. Following illumination, the first of a set of activated amino acids, each bearing a photolabile protecting group on the amino terminus, is exposed to the entire surface. Coupling only occurs at those sites addressed by light in the preceding step. The solution of amino acid is removed, and the substrate is again illuminated through a second mask, rendering a different region available for reaction with a second protected building block.

In a light-directed synthesis, the pattern of masks and the order of reactants define the products and their respective locations on the support. The position of each compound is thus precisely known and hence its interactions with other molecules can be directly probed. Incubation of the array with a biological receptor and subsequent evaluation of binding generate a powerful structure-activity-relationship (SAR) database. One convenient method to measure binding interactions is to incubate the array with a fluorescently tagged receptor (or receptor reporter molecule, e.g., a fluorescently tagged antibody) and to detect bound fluorescence with a confocal microscope. The fluorescence intensity at each synthesis site will depend on the affinity of the receptor for the compound, the concentration of receptor, and the number and density of interacting sites on the array. The format of the assay can also be configured to allow for real-time measurements of binding interactions. Other methods of reading the array are also possible. Radiolabeled or chemiluminescent receptors could be used, followed by appropriate detection. Characterizations other than binding might also be addressed. For example, catalytic cleavage of a peptide at a site between a fluorescent donor or acceptor pair would lead to increased fluorescence.

We have developed general formalisms describing the relationship between the desired set of products and the appropriate photolithographic masks. The process is conveniently expressed in matrix notation (Fodor et al., 1991). If a particular position is to receive a new monomer group, it is addressed by a 1 (light on condition), if not, it receives a 0. For example, to form ACD from the ordered set of reactants {A,B,C,D}, the light switch consists of a vector (switch vector) [1,0,1,1]. The collection of switch vectors specifies the photolithographic mask for each chemical step and provides a map of the location of each product in the array.

There are many potential strategies for ligand construction or optimization that can be used in a light-directed synthesis. One may choose to form the largest set of compounds for a given number of chemical steps. Using single amino acids as building blocks, 2^n compounds can be formed in n chemical steps. Thus for a synthesis employing 16 amino acid couplings, a maximum of 65,536 independent compounds can be generated. The resulting peptides range in length from 1 to 16 residues, with a mean length of 8. Alternatively, one could form all combinations of

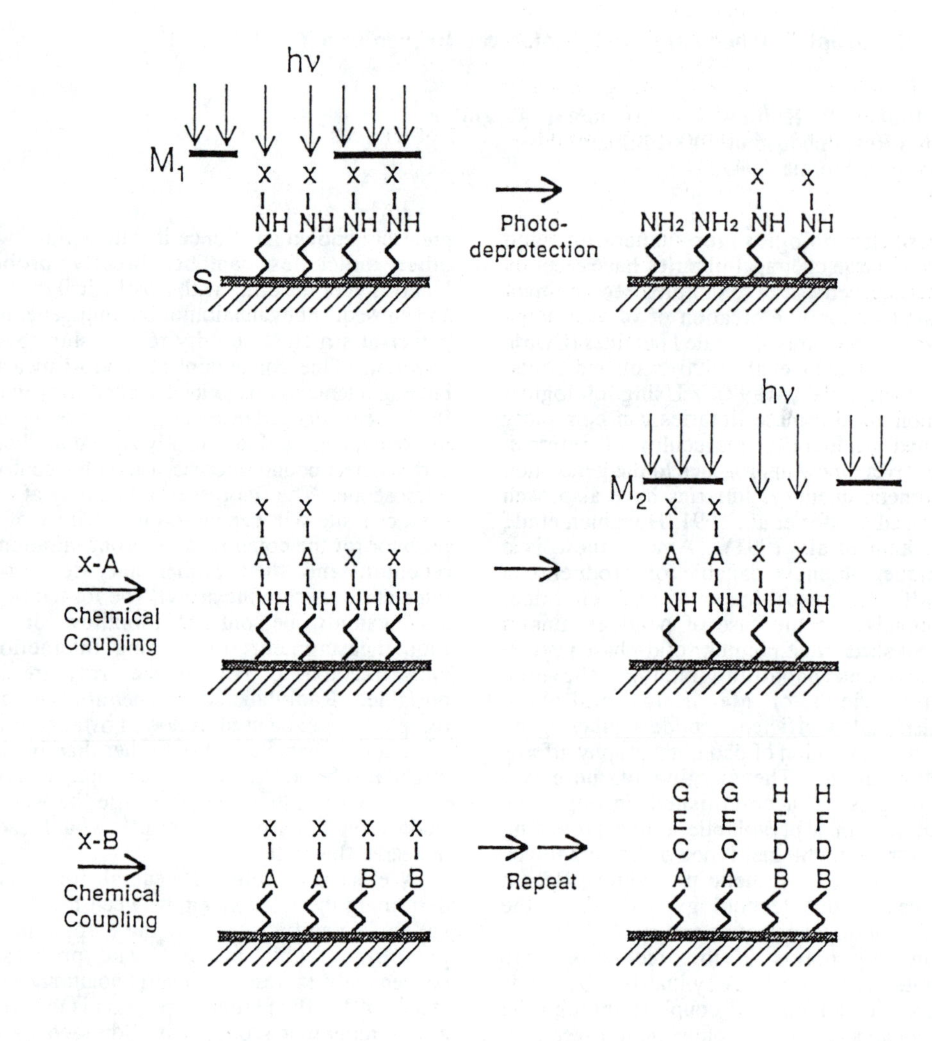

Figure 1. Light-directed spatially addressable parallel chemical synthesis. A substrate containing amino groups blocked by photochemical protecting groups is selectively illuminated through a mask (M_1). The first of a set of building blocks (each bearing a photochemical protecting group X) is exposed to the entire substrate. Chemical coupling occurs only at those regions which were exposed to light in the preceding step. A different mask (M_2) is then used to photodeprotect selectively a different region of the substrate. A second photoprotected group X-B is added and reacts at those sites addressed by M_2. The process is repeated to obtain the desired set of products.

a set of chemical building block units for a given length. A straightforward synthesis format is X^l, where X is the number of chemical building blocks, and l is the length of the peptides. For example, when X = 20 (e.g., the genetically encoded amino acids) and l = 2, then 400 dipeptides or dipeptidyl substitutions would be formed.

The number of compounds which can be conveniently formed in the array depends largely on the photolithographic resolution. Currently, arrays of tens of thousands of peptides can be routinely generated and screened, and the present limit for high contrast photodeprotection is about 20 µm. Arrays with this resolution can yield densities of 250,000 compounds contained within an area of 1 cm by 1 cm.

The light-directed method is well suited for rapid analoging of known ligands. For example, the peptides in the array might be made resistant to proteolysis by introducing D or other unnatural amino acids. The introduction of a variety of chemical buildings blocks can be used to confer desirable physical properties into the peptides (e.g., hydrophobicity, charge character, conformational restriction, etc.). The evaluation of the subtle influences conferred by novel building blocks, in conjunction with the simultaneous screening for biological recognition of the peptides, offers the potential for the rapid optimization of a peptide lead in the area of drug discovery.

References

Cull, M.G., Miller, J.F., and Schatz, P. J. *Proc. Natl. Acad. Sci. USA* **89**, 1865 (1992).

Cwirla, S.E., Peters, E. A., Barrett, R. W., and Dower, W. J. *Proc. Natl. Acad. Sci. USA* **87**, 6378 (1990).

Devlin, J. J., Panganiban, L. C., and Devlin, P. E. *Science* **249,** 404 (1990).

Fodor, S.P.A., Read, J. L., Pirrung, M. C., Stryer, L., Lu, A. T., and Solas, D. *Science* **251**, 767 (1991).

Houghten, R. A., Pinilla, C., Blondelle, S. E., Appel, J. R., Dooley, C. T., and Cuervo, J. H. *Nature* **354**, 84 (1991).

Lam, K. S., Salmon, S. E., Hersh, E. M., Hruby, V. J., Kazmierski, W. M., and Knapp, R. J. *Nature* **354**, 82 (1991).

Patchornik, C. A., Amit, B., and Woodward, R. B. *J. Amer. Chem. Soc.* **92**, 6333 (1970).

Scott, J. K., and Smith, G. P. *Science* **249,** 386 (1990).

Novel Metal-Affinity Adsorbents Prepared by Template Polymerization

Sean Plunkett, Pradeep K. Dhal and Frances H. Arnold*
Division of Chemistry and Chemical Engineering 210-41, California Institute of Technology, Pasadena, CA 91125, USA (FAX: 818-568-8743)

The development of cost-effective affinity separations will eventually require that expensive and labile affinity ligands be replaced by small chemical mimics. We have begun to develop a new class of metal-complexing polymers that exhibit selectivity for individual molecules. Designed for metal-affinity or ligand-exchange separations, these polymer contain arrays of metal ions (Cu(II)) that bind complementary arrays of functional groups (imidazole) on the target molecule. We achieve this by preparing the polymers using the target molecule as the template to arrange the metal complexes in the correct three-dimensional orientation for recognition. This template polymerization scheme using metal complexes as the recognition elements may have applications in the preparation of polymers for selective recognition of biological molecules such as proteins.

Template polymerization offers a powerful route to preparing functional polymeric matrices that selectively bind specific molecules (the template). The process involves interaction of an appropriate functional monomer with the desired template molecule to form a template-monomer assembly, followed by its polymerization in the presence of crosslinking agents. During polymerization, the geometry of the template-monomer moiety is captured in the growing polymer matrix, and the crosslinking stabilizes this structure. After polymerization, the template molecule can be removed by breaking up the template-polymer interaction. A polymer produced in this manner exhibits selectivity for the template molecule over structural analogs. This selectivity has been attributed to appropriate orientation of the functional groups on the polymers towards the complementary sites on the template molecules (Wulff, 1986).

The interactions previously used in template polymerizations have included covalent bonding, hydrogen bonding, hydrophobic interactions, and electrostatic interactions (Shea et. al., 1990; Ekberg and Mosbach, 1989). We have recently added metal-ligand coordination to this list (Dhal and Arnold, 1991). Although polymers that rely on hydrogen bonding, hydrophobic and electrostatic interactions for molecular recognition exhibit faster rebinding kinetics compared to systems based on reversible covalent bonds, the inherently weak nature of these noncovalent interactions limit their recognition capabilities. Metal-ligand coordination overcomes this limitation by providing a strong interaction with rapid rebinding kinetics. Furthermore, due to their high functional group specificity and mild formation conditions, metal-ligand coordination is well-suited to molecular recognition of biomolecules and may open the way to a template polymerization-based system for molecular recognition of biological macromolecules.

Metal-Ligand Coordination and Molecular Recognition

Bioseparations based on metal-ligand coordination are well known and extensively studied (Sulkowski, 1985; Porath, 1988; Arnold, 1991). In immobilized metal-affinity chromatography (IMAC), a metal ion is trapped by an immobilized chelating group such as iminodiacetic acid (IDA). The metal ion thus affixed to the support matrix still retains at least one coordination site for complexation with substrate molecules. Typically, a first row transition metal is used as the coordinating ion; these ions are generally exchange-labile and provide rapid kinetics which are compatible with chromatographic separations (Sundberg and Martin, 1974). Copper(II) iminodiacetate (Cu(II)IDA) immobilized on a hydrophilic support, for example, can resolve proteins that differ in their surface-accessible histidine contents (Sulkowski, 1985; Porath, 1988; Arnold, 1991).

We have studied metal-affinity two-phase aqueous extraction of proteins as an alternative bioseparation based on metal-ligand coordination and also as a technique for measuring the Cu(II)IDA-surface histidine binding constant. Cu(II)IDA was attached to polyethylene glycol (PEG) and used in aqueous PEG/dextran. The average association constant between a single surface histidine and Cu(II)IDA was determined to be $2.2 \times 10^3\ M^{-1}$ in the PEG-rich phase (Suh and Arnold, 1990). In terms of bond energy, this translates to nearly 5 kcal/mol--significantly stronger than hydrogen bonding in aqueous solution. Based on these observations and the inherent attractive features of metal-affinity separations (Arnold, 1991), we felt that the coordination of Cu(II)IDA with imidazole would provide an excellent route to the preparation of selective polymeric matrices by template polymerization. To demonstrate this concept, we set out to produce a rigid polymer using an imidazole-containing template molecule bound to a polymerizable Cu(II)IDA derivative. This would result in a chelating matrix with a strategic distribution of binding sites that favor subsequent preferential rebinding of template molecules.

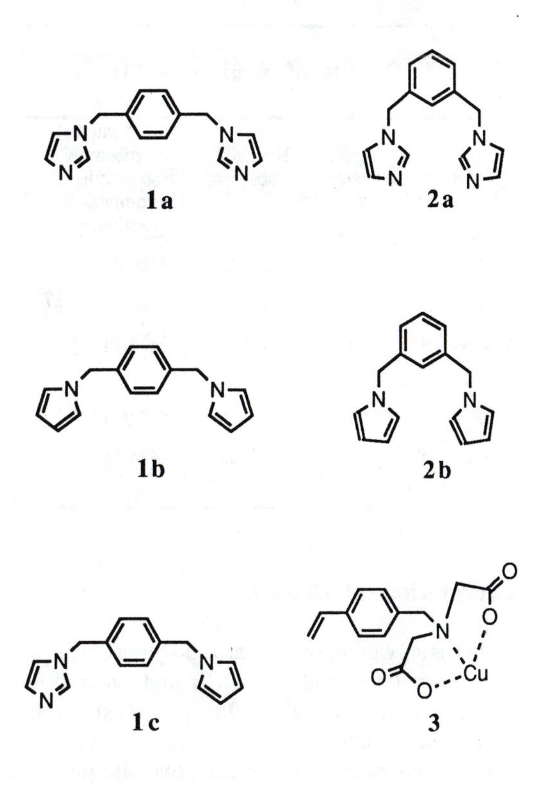

Template Polymerization of Protein Analogs

For the purposes of template polymerization using metal coordination, a protein can be thought of as a set of imidazole rings attached to a molecular scaffolding provided by the protein backbone. In an effort to construct a simple model of a protein for our preliminary template polymerization studies, we created a series of structurally similar bis-imidazole compounds (**1a**, **2a**). These "protein analogs" were allowed to interact with two equivalents of a copper-chelating monomer (**3**) to produce the soluble monomer-template assembly. Formation of nitrogen-Cu(II) coordination bonds was evident from the blue shift of λ_{max} in the visible region upon addition of the templates (Dhal and Arnold, 1991) and characteristic changes in the copper ion esr spectral patterns (Dhal and Arnold, 1992). Following preorganization of the monomer-template assemblies in solution, polymerization was allowed to proceed, using ethylene glycol dimethacrylate as the crosslinker, azobisisobutyronitrile as a free radical initiator and methanol as a porogenic agent. The polymers obtained by this method are macroporous, with very high surface areas (> 200 m^2/g), and the functional groups are highly accessible to solvents and reagents (Dhal and Arnold, 1992).

Substrate recognition and selective binding characteristics of the resulting templated polymers were determined in competitive rebinding experiments using an excess of the two bis-imidazole substrates (Table 1) (Dhal and Arnold, 1991). The copper-containing templated polymers exhibited selectivity for their templates (P-2, $\alpha_{1a/2a} = 1.15$; P-3, $\alpha_{2a/1a} = 1.17$), while the copper-free templated polymers and the copper-containing polymer prepared with no template (P-1) did not.

Table 1. Results of Rebinding Studies

Polymer	Original template	Rebinding Substrates	Relative substrate selectivity in competitive rebinding
P-1	----	1a + 2a	$\alpha_{1a/2a} = 1.02$
P-2	1a	1a + 2a	$\alpha_{1a/2a} = 1.17$
P-2, no Cu	1a	1a + 2a	$\alpha_{1a/2a} = 1.04$
P-3	2a	1a + 2a	$\alpha_{2a/1a} = 1.15$
P-3, no Cu	2a	1a + 2a	$\alpha_{2a/1a} = 1.04$

Recognition Mechanism

In analogy to other templated polymer systems that have been studied (Wulff and Shauhoff, 1991; Shea and Sasaki, 1989), at least three possible mechanisms for the observed selectivity can be envisioned: cavity fitting, two-site metal-ligand coordination (two-site binding), or some combination of cavity fitting and metal-ligand coordination (single-site binding). The cavities are certainly heterogeneous, and selectivity most likely results from a combination of all three binding possibilities. In order to elucidate some aspects of the recognition mechanism, two types of pseudo-templates were synthesized: one having both imidazole rings replaced by pyrrole (**1b**, **2b**), and the second having a single imidazole replaced by a pyrrole ring (**1c**). The pyrrole-for-imidazole substitution provided us with molecules that were very similar in geometry and steric bulk to **1a** and **2a**, but lacking one or both of the copper-coordinating nitrogens. The bis-pyrrole pseudo-templates both bound very weakly to the copper-containing polymers. Additionally, no binding of the bis-imidazole templates was seen in rebinding studies performed on copper-free templated polymer (Dhal and Arnold, 1991). These observations underscore the critical role for nitrogen-Cu(II) coordination in rebinding and suggest that cavity fitting alone cannot be responsible for binding or selectivity.

To further examine the nature of the polymer binding sites, saturation binding studies were performed using the single-imidazole pseudo-template **1c** and N-benzyl imidazole. The maximum binding capacities of the polymer P-2 for its bis-imidazole template and for the corresponding single-imidazole pseudo-template were found to be very similar. The capacity of the same polymer for N-benzyl imidazole was nearly twice that found for the two larger molecules (Dhal and Arnold, 1992). This observation is consistent with the view that the templating procedure has created a distribution of pairs of closely-spaced copper ions.

Discussion

The results of the rebinding experiments using the bis-imidazole substrates demonstrate that this polymerization technique can produce selective polymeric matrices. The selectivities seen (α = 1.15-1.17) are significant when the similarity of the templates is considered--we were unable to separate **1a** and **2a** by reverse phase HPLC or traditional ligand-exchange chromatography on commercially available matrices. Higher selectivities are observed for separations of less similar substrates (Dhal and Arnold, 1992). Furthermore, there has been little attempt thus far to improve or optimize the preparation of these templated polymers.

The mechanism of selectivity remains unclear. While the experiments described here have eliminated the possibility of cavity fitting as the sole mechanism for substrate recognition, there is no direct evidence yet for two-site binding. Even with evidence of two-site binding of the bis-imidazole templates to these matrices, some contribution to selectivity from cavity fitting would not necessarily be precluded. The polymers most likely contain a heterogeneous population of sites. Further work with different pseudo-templates, as well as nmr and esr spectroscopy, may provide more information on the nature of the binding sites.

Outlook

The ability to produce metal-complexing polymers that recognize protein analogs may eventually provide the basis for a new class of materials for recognition of biological molecules--in essence, synthetic antibodies. With the relatively high bond strength of metal-ligand coordination interactions, as few as two interactions could lead to high specificity in protein recognition, based solely on distribution of surface histidines. The spectrum of applicability could be widened by the use of other metal ions such as Fe(III), Zn(II) or Ca(II), alone or in combination.

In order to apply this technique to biological molecules, however, there are still many obstacles to overcome. Most biomolecules will require an aqueous-based polymerization system, operating under relatively restrictive pH and temperature ranges. Water-soluble reagents are required, and these must not be denaturing to proteins or react with the myriad functional groups found on protein surfaces. With larger substrates such as proteins, diffusion into and out of the templated binding sites may become a critical factor. Materials must be produced that are suitable for chromatography, with good mechanical strength and durability. Despite these considerations and difficulties, the potential utility of these robust metal-complexing polymers makes their further investigation and development worthwhile.

References

Arnold, F. H. *Bio/Technology* **1991**, *9*, 151.

Dhal, P. K.; Arnold, F. H., submitted **1992**.

Dhal, P. K.; Arnold, F. H. *J. Am. Chem. Soc.* **1991**, *113*, 7417.

Ekberg, B.; Mosbach, K. *Trends Biotechnol.* **1989**, *7*, 92.

Porath, J. *Trends Anal. Chem.* **1988**, *7*, 254.

Shea, K. J.; Stoddard, G. J.; Shavelle, D. M.; Wakui, F.; Choate, R. M. *Macromolecules* **1990**, *23*, 4497.

Shea, K. J.; Sasaki, D. Y. *J. Am. Chem. Soc.* **1989**, *111*, 3442.

Suh, S. S.; Arnold, F. H. *Biotech. Bioeng.* **1990**, *35*, 682.

Sulkowski, E. *Trends Biotech.* **1985**, *3*, 1.

Sundberg, R. J.; Martin, R. B. *Chem. Rev.* **1974**, *74*, 471.

Wulff, G.; Shauhoff, S. *J. Org. Chem.* **1991**, *56*, 395.

Wulff, G. "Polymeric Reagents and Catalysts," *ACS Symposium Series* **1986**, *308*, 186.

The Role of Immunoaffinity Chromatography in Isolation of Proteins of Therapeutic Interest

G. Mitra*, P. Ng and **M. Wong**
Pharmaceutical Division, Miles Inc., Berkeley, California 94701 (FAX:510-420-5478)

Recombinant factor VIII (rFVIII) expressed in hamster cells has been subjected to preparative immunoaffinity chromatography utilizing a monoclonal antibody directed to the light chain of factor VIII. Experience with repeated use of the column is discussed. Contaminant protein and DNA clearance across the chromatographic step is estimated to be $>10^3$ fold. Antibody release in the product can be controlled and subsequently separated from rFVIII via anion exchange chromatography.

Therapeutic proteins expressed in recombinant fermentor harvests need to be extensively purified from media proteins, cell substrate proteins and DNA before they are suitable for human administration. Allowable levels of cell substrate DNA in therapeutic preparations vary between <100 pg/dosage to <10 pg/dosage depending upon the regulatory agency in question. Residual levels of cell substrate proteins can be antigenic and to allow an adequate safety margin an acceptable residual level might be <0.2 μg/dosage. To arrive at these levels in the final product the clearance factor of cellular protein and DNA through the purification cascade needs to be at least $>10^5$ fold. Immunoaffinity chromatography, due to its high specificity, allows for significant removal of protein and DNA contaminants in a single step. For scale-up of immunoaffinity chromatography it is critical to demonstrate multiple reuse without pyrogenicity or fouling of the column.

We report here our experience with preparative scale immunopurification of rFVIII expressed in hamster cells. The antibody utilized is directed towards the light chain (80,000 daltons) of the factor VIII molecule which is thought to be associated with the 90,000-210,000 dalton portions via Ca^{+2} bridges (Eaton et al., 1986). The effect of coupling chemistry, antibody release rate in the product and its subsequent separation and the extent of contaminants removal are discussed.

Materials and Methods

For random orientation of the antibody polyhydroxy silica support (Controlled Pore Glass, 500 Å porosity) was activated with periodate, essentially as described by Ohlson et al. (1978), reacted with the monoclonal antibody directed towards the light chain of factor VIII and the Schiff base stabilized with sodium cyanoborohydride. For directed orientation the hydroxyl groups of the monoclonal antibody, mostly in the Fc domain, were activated with periodate according to O'Shannessy (1990) at pH 5.0 and reacted with Avid gel F (Bioprobe International, Tustin, California), the support backbone being a copolymer of oligoethylene glycol, glycidylmethacrylate and pentaerythroldimethacrylate.

Procoagulant activity of factor VIII was assayed by a one-stage Activated Partial Thromboplastin Time Test (APTT) modified from the methods of Langdell et al. (1953) and Proctor et al. (1961).

Mouse IgG was detected by a standard sandwich ELISA technique with rabbit anti-mouse IgG as coat and biotin labeled anti-mouse IgG for detection. Assay sensitivity was approximately 1 ng/ml. Cell substrate protein was also estimated by sandwich ELISA assay, the detection antibody being labeled with biotin. Detection sensitivity was approximately 1 ng/ml. A Mono Q column was obtained from Pharmacia LKB Biotechnology (Piscataway, New Jersey).

1054–7487/92/0248$06.00/0

DNA clearance across the immunoaffinity step was demonstrated via cell substrate DNA assay by DNA hybridization (Kafatos et al., 1979). Sample DNA was denatured and immobilized on membrane filters. Probe DNA was labeled with ^{32}P and incubated on the membrane filters for hybridization. Excess unhybridized probe was rinsed followed by autoradiography; DNA standards of known concentration allowed quantitation. Detection sensitivity was approximately 10 pg of cell substrate DNA. DNA clearance was also verified via extraneously added ^{32}P labeled cell substrate DNA to the protein solution immediately prior to the immunoaffinity step.

Results

A periodate activated silica support bound antibody column (random orientation) has been repeatedly utilized for adsorption/elution of rFVIII in immidazole buffer (0.02 M) at pH 6.9, 1 M calcium chloride being utilized for elution. As indicated in Fig. 1, the antibody column provided essentially quantitative recovery for approximately 180 runs. The leaching of the antibody in the eluted product was within the 160-325 ng/ml range for the first four runs and subsequently decreased to ≤5 ng/ml for the rest of the study.

Separation of immunopurified rFVIII from mouse immunoglobulin was obtained on an anion exchange column, as indicated in Fig. 2. On Mono Q column with 0.02 M imidazole buffer at pH 6.9 rFVIII bound tighter, allowing for a clean separation from the antibody.

Clearance of cell substrate DNA and cell substrate protein across the immunoaffinity step for three consecutive runs is shown in Table 1. The data has been normalized for 1000 u of rFVIII; cellular protein clearance was 4670-6003 fold and cellular DNA clearance, assayed by dot blot hybridization, was 1064-5710 fold. Table 2 shows the clearance of ^{32}P labeled cell substrate DNA,

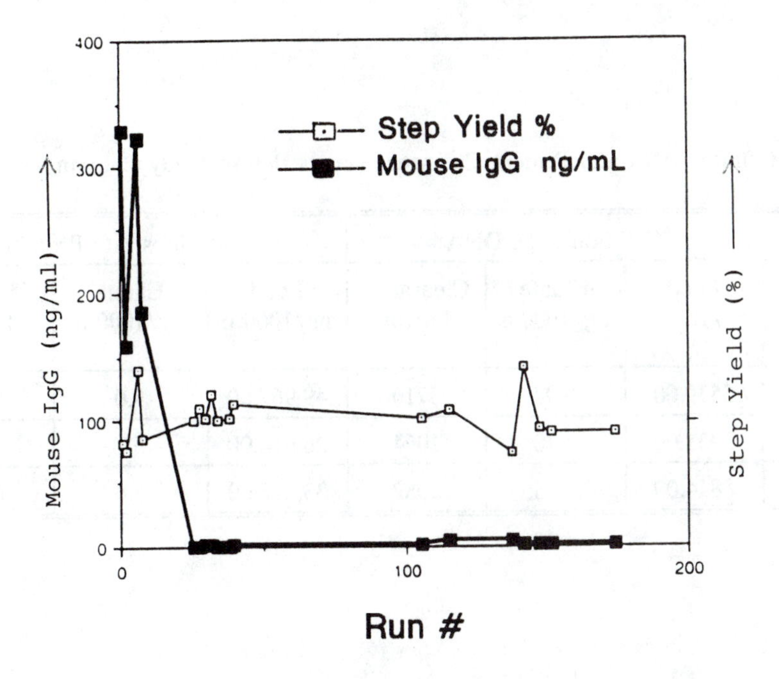

Fig. 1. Step yield of rFVIII and mouse IgG concentration in column eluate as a function of repeated antibody column runs.

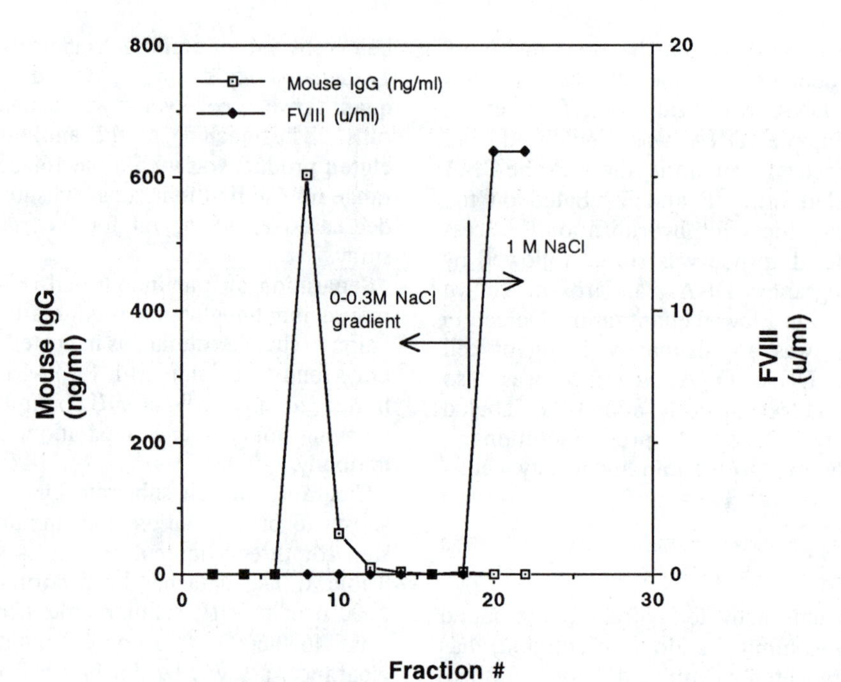

Fig. 2. Separation of mouse IgG and rFVIII on Mono Q column.

Table 1. Cellular DNA and Protein Clearance across the Antibody Column

	Cell Substrate DNA			Cell Substrate Protein		
Run #	Feed ng/1000 u	Eluate ng/1000 u	Clearance Factor	Feed ng/1000 u	Eluate ng/1000 u	Clearance Factor
1	5539.00	0.97	5710	49,967.00	9.0	5552
2	3935.00	3.70	1064	26,618.00	5.7	4670
3	8814.00	2.86	3082	33,616.00	5.6	6003

Table 2. Clearance of ^{32}P-labeled Cell Substrate DNA across the Antibody Column

Run #	Antibody Column Load cpm	Antibody Column Flow-through and Wash cpm
1	3.75×10^7	4.12×10^7
2	1.66×10^7	1.66×10^7
3	2.10×10^7	1.95×10^7

the flow-through and the wash fraction essentially accounting for almost all of the extraneously added radioactivity.

Fig. 3 shows the comparison of rFVIII binding capacity for the randomly immobilized column and the site oriented (coupled via carbohydrate linkage to the Fc domain) column with equivalent antibody loading (2.5 mg/ml). For the former, >10% of applied rFVIII was detected in the flow-through at an rFVIII loading condition of ≥1000 units/ml support, for the latter an rFVIII loading of up to 4000 units/ml support resulted in equivalent amounts of rFVIII in the flow-through fractions.

Discussion

Repeated use of the immunoaffinity column is mandated in preparative scale and the experience to date supports reuse for at least 180 times (Fig. 1) without any significant loss of column capacity. Unlike other column support materials (ion exchange, size exclusion, etc.) the antibody column cannot be regenerated via 0.1 N NaOH for depyrogenation between runs, which results in non-reversible denaturation of the bound antibody. Pyrogen free operation was ensured via operation at +5°C under a clean room condition. All buffers and protein solutions were sterile filtered through a 0.2μ absolute filter prior to use. No bacteriostatic agents or protease inhibitors were introduced as these are unlikely to be acceptable to the regulatory authorities.

Silica support was found to be fairly inert to microbiological attack and its rigidity allowed space velocities in excess of 25 cm/min. Similar flow properties from polyhydroxy silica support based antibody columns had been previously reported by Roy et al. (1984) for purification of recombinant leucocyte A interferon. The initial higher release rate of the bound antibody is believed to be due to a small fraction which is loosely bound during immobilization; following the first few runs the IgG release rate was consistenly near detection limit of the mouse IgG ELISA (<5 ng/ml). The very low levels of mouse IgG in eluates together with consistent undiminished antigen binding capacity argue for stability of this type of linkage in immunoaffinity chromatography.

Factor VIII has been reported to have an isoelectric point (pI) around pH 5.2-5.7 (Hamer et al., 1986) and pH 4.8-5.3 (Gelsema et al., 1979). The light chain specific monoclonal antibody of subclass IgG1 has a PI of 6.7-6.8. The observed order of separation on the Mono Q column (Fig. 2) is hence consistent and the use of the anion exchange chromatography ensured a residual mouse IgG value of <10 ng/1000 u of FVIII.

As indicated in Table 1 and Table 2, the immunoaffinity step ensures at least 3 orders of magnitude reduction of cell substrate DNA and protein from rFVIII. The high degree of purity of recombinant products mandated by the regulatory authorities is thus facilitated by the specificity of the antigen-antibody reaction.

As indicated in Fig. 3, an approximate four fold increase in column capacity resulted from the randomly coupled to the site oriented column. Efficiency as defined by (actual Ag bound)/(Ag bound if 100% of bound Ab is active) increased from 15% to 60%. Apart from reduction in manufacturing costs, increased antigen capacity also has the potential of reducing the non-specific interactions, thereby lowering the level of carryover cell substrate protein and DNA.

Data reported here demonstrate the technical feasibility of preparative scale immunoaffinity chromatography for rFVIII expressed in hamster cells. Prolonged maintenance of antigen binding capacity of the column together with low antibody release rates are two contributory factors towards a practical industrial process. Immunopurified rFVIII has been safely utilized in human clinical trials (Schwartz et al., 1990) for a period of over 3 years.

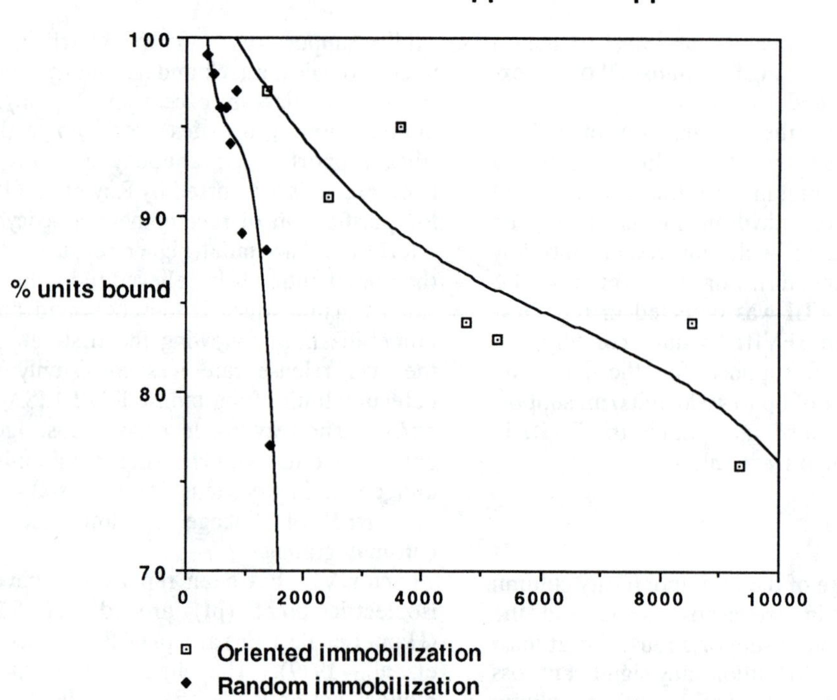

Fig. 3. Factor VIII binding capacity of the antibody column for random immobilization as well as directed immobilization.

References

Eaton, D.L.; Vehar, G.A. *Prog. Hemostas.* **1986**, *8*, 47-70.

Gelsema W.J.; DeLigny, C.L.; Van der Veen, N.G. *J. Chromat*. **1979**, *171*, 171-181.

Hamer, R.J.; Koedam, J.A.; Beeser-Visser, N.H.; Sixma, J.J. *Biochim. Biophys. Acta* **1986**, *873*, 356-366.

Kafatos, F.C.; Jones, C.W.; Efstratiadis, A. *Nucleic Acids Res*. **1979**, 7, 1541-1552.

Langdell, R.D.; Wagner, R.H.; Brinkhouse, K.M. *J. Lab Clin. Med.* **1953**, *41*, 637-647.

Ohlson, S.; Hansson, L.; Larsson, P.O.; Mosbach, K. *FEBS Lett*. **1978**, *93*, 5-9.

O'Shannessy, D.J.; Wilchek, M. *Anal. Biochem.* **1990**, *191*, 1-8.

Proctor, R.R.; Rapaport, S.I. *Am. J. Clin. Path*. **1961**, *36*, 212-219.

Roy, S.K.; Weber, D.V.; McGregor, W.C. *J. Chromatogr*. **1984**, *303*, 225-228.

Schwartz, R.S.; Abildgaard, C.F.; Aledort, L.M.; Arkin, S. et al. *New Eng. J. of Med*. **1990**, *323*, 1800-1805.

Electrically Enhanced Cross-Flow Filtration of Biosuspensions

A.Brors and K.H.Kroner*,
GBF-Gesellschaft für Biotechnologische Forschung mbH, Bioprocess Division,
W-3300 Braunschweig, Germany (FAX: 0049-531-6181-515)

Studies about the effects of different types of electric fields on the cross-flow filtration of microbial suspensions are presented. In the case of an *E. coli* fermentation broth a 3-4 fold flux increase and a reduction of power consumption in comparison to conventional cross-flow was found, by applying instationary fields. The technical feasibility of the method is discussed.

Cross-flow microfiltration is increasingly used in the bioindustries as an alternative tool for the separation of biosuspensions. However, the method is limited in its performance due to the formation of highly resistent sublayers and membrane fouling. Currently numerous groups are investigating methods to overcome the limitations. Beside other means, such as the optimization of the hydrodynamics, the application of an electric field may be a suitable aid to enhance performance, because particles suspended in solution are generally charged and thus are susceptible to electro-kinetic forces. But so far little has been reported with respect to biotechnical applications.

General Aspects and Objectives

The theory of electrofiltration distinguishes three clearly separated regimes of filtration behaviour defined by the main filtration resistances $R_{membrane}$, R_{cake}, and R_{film} ($J=\Delta P/R_m+R_c+R_f\Delta P$). If the electric field, defined by the field strength E [V/m], overcomes the limiting resistances R_c and R_f due to electrophoretic movement of the charged particles, a substantial increase of the filtration rate should be achieved. At a field strength $E>E_{cr}$, where E_{cr} is defined as the field strength at which the filtration velocity is equal to the electrophoretic velocity, a clear liquid boundary layer is assumed, where flux is related only to R_m. Generally, a linear relation between flux and field strength is assumed (Henry et al 1977).

The electrokinetic properties of particles in solution are characterized by the Zeta-Potential and the electrophoretic mobility μ [m^2/V s]. Biological cells in aqueous solution generally show a negative Zeta-Potential, but values depend strongly on the type of organism, the state of growth and the environmental conditions and thus are difficult to predict. For some microorganisms values of μ have been reported in the range of $0.11\text{-}1.63\cdot10^{-8}$ m^2/Vs (Brors et al 1990). This would lead to electrophoretic velocities of $1.1\text{-}16.3\cdot10^{-5}$ m/s, corresponding to flux values of about 40-580 L/hm^2 at a field strength of 10V/cm. For microorganism suspensions flux rates are typically around 50 L/hm^2 or $1.4\cdot10^{-5}$ m/s, respectively (Kroner et al 1984). A significant flux improvement can be assumed if the electrophoretic velocity exceeds this value and if the radial migration rate, defined by the hydrodynamics, is comparably small. In this respect, it was suggested that particle sizes should generally be smaller than 6-8 μm (Wakeman et al 1987), which would be the case for microorganism. On the other hand there might be limitations with respect to the conductivity of the aqueous biosuspensions in terms of electric power to be supplied and electrolysis effects. In order to test the feasibility of the method a case study for the separation of *E. coli* fermentation broth was performed.

Experimental Conditions

Experiments have been carried out with freshly fermented *E. coli* (DSM 498) in a standard salt medium and glucose as the main carbon source, harvested at the end of the exponential growth phase (pH 7.0; conductivity

6.5 mS; μ_{cells} 1.3-1.5·10^{-8} [m^2/Vs]; C_{cells} ca. 0.02 v/v). A selfconstructed flat-channel module (Channel: L/W/H: 27.5cm· 4cm· 0.1cm; Membrane area: 96 cm^2), fitted with titanium electrodes parallel to the membrane, on the retentate and filtrate side, respectively, was used. The electrodes were directly exposed to the medium and not separately flushed (see Fig.1). A microporous polyamide membrane with 0.2 μm cut-off was used (Ultipore N66, Pall Corp.) having a slightly negative Zeta-Potential at pH 7. The module was arranged in a test rig, supplied with a PC monitoring and control system, as earlier described (Brors et al 1991). Experiments were mainly carried out in a 'batch-recycle mode'.

Results and Discussion

During earlier studies with different microorganisms 2 to 7 fold higher fluxes have been achieved in comparison with conventional cross-flow filtration when applying a constant DC-field up to 100-120 V/cm (Brors et al 1990 and 1991), see Tab.1. No regular order of the measured flux increases with respect to the electrophoretic mobility of the organisms can be seen, although its influence is evident. In contrast to commonly held theories no linear relation was found between field strength and flux (see Fig.2, curve A). In principle one can clearly distinguish between two different regimes of operation, which may reflect the main resistances R_f and R_c. Though the flux enhancement was substantial, our results indicate that complete removal of the particles away from the membrane seems impossible even if $E > E_{cr}$, because the remaining resistances are obviously higher than R_m. One of the reasons for these findings may be the fact that a broad Zeta-Potential distribution of the microbial particles was generally found (Brors et al, 1991), with a non-negligible proportion of positively charged particles, which may be transported electrophoretically to the membrane. Further, if particles are in contact with the membrane they may be adsorbed and it becomes difficult to remove them by the electric forces. However, the most severe problem in applying constant DC-fields with a high field strength was the high conductivity of the broths, leading to high current consumption with associated Joule-heating and electrolysis.

In order to reduce these problems we have studied the application of pulsed fields, which are also proposed by other authors (Wakeman et al 1987 and Bowen et al 1989), and alternating current fields. In Fig. 2 our experiments with four different field-types applied are summarized. Flux increase with field strength was found to be higher for the instationary fields compared to the stationary DC-field; even at much lower field strengths the flux was significantly enhanced, following the electrophoretic velocity term up to 2 V/cm. Surprisingly, the AC-field behaviour is similiar to that of the pulsed DC-field. This may be attributed to the

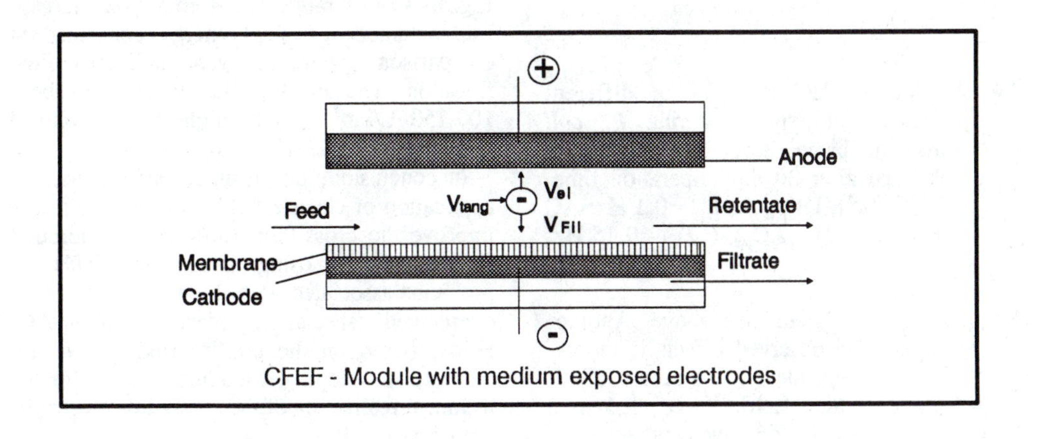

Fig. 1: Schematic drawing of the flow channel arrangement in the CFEF-module and f-orces acting on a charged particle.

Tab.1 Summarized Results for the Cross-Flow Electrofiltration of Biosuspensions - Relation between Performance and Electrokinetic Properties

Organism	μ [$m^2/Vs\cdot10^{-8}$]	$^*V_{el}$ [μm/s]	$^*V_f$ [μm/s]	$^*J_E$ [L/hm^2]	J_E/J_O [-]
E. coli / DSM 1328	0.12	12	12	43	2
Baker's Yeast	0.49	49	141	508	7
L. plantarum	0.64	64	12	43	3.8
E. coli / DSM 498	1.38	138	32	115	3.5
Methylomonas sp. M 15	1.44	144	36	130	5.1

*at E = 100V/cm constant DC-field
Conditions: ΔP 0.2-0.5 bar; 0.5-1.0 m/s; pH 6-7; cell conc. 2-4% v/v

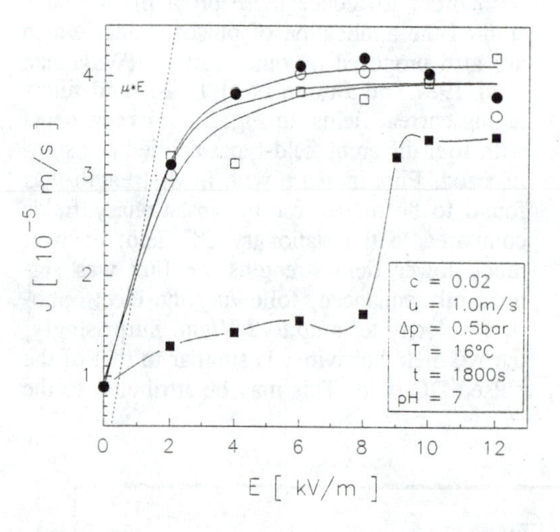

Fig. 2: Flux vs.field strength for different electric field-types, filtering *E. coli* broth in 'batch recycle mode'. Values obtained after 30 min. operation time. DC_{const} (■); DC_{puls} (□), f=0.1 Hz; AC (●) f=0.2 Hz; AC_{puls} (O) f=0.15 Hz.

charge distribution mentioned above. Another way to explain the observed effects is the assumption that, independent of the kind of field, the pulsation of the electric forces led to a disturbance of the boundary layer at the wall of the membrane and thus increased mass transfer. Therefore a frequency dependency can be assumed. From our current results we expect that the optimal frequency is in the order of 1 Hz, which is close to the values defined for the effects of pulsating flow onto mass transfer (e.g.Finnigan et al, 1990).

However, the main advantages of the application of the instationary fields are the drastic reduction in power consumption and suppression of electrolysis. Taking the total power consumption into account (pump and field) a minimum between 2 and 4 V/cm results for the pulsed fields, as can be seen from Fig.3. At this range a 3-4 fold flux increase was obtained with less energy consumed in comparison to the conventional cross-flow filtration. The absolute flux values are about 100-150 L/hm^2, which might be economical (Kroner, et al 1984).

In conclusion, our results confirm that the application of electric fields is a suitable aid to improve the cross-flow filtration of microbial suspensions. By using instationary fields the problems associated with this technique can be overcomed, especially by means of pulsed AC-fields. Based on the present findings we are currently developing a module concept for the technical realization of the cross-flow electrofiltration principle.

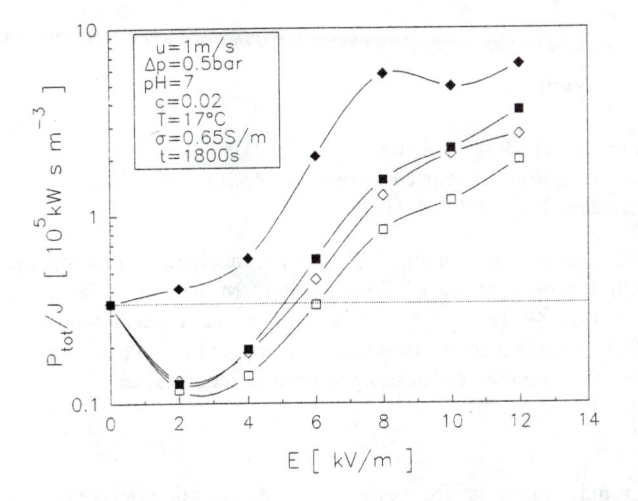

Fig. 3: Power consumption vs. field strength for the experiments shown in fig. 2. Dotted line indicates the limiting value without field. (♦) DC $_{const}$; (◇) DC_{puls}; (□) AC_{plus};(■) AC.

References

Bowen, W.R.; Kingdon, R.S.; Sabuni, H.A.M. *J.Membrane Sci.* **1989**, 40, 219-229.

Brors, A.; Kroner, K.H. In *Int.Congress on Membranes and Membrane Processes, ICOM'90*; Proceedings Vol.1, North American Membrane Society, Chicago IL, **1990**, pp. 616-618.

Brors, A.; Kroner, K.H.; Hustedt, H.; Deckwer, W.D. In *Biochemical Engineering Stuttgart*; Reuss, M.; Chmiel, H.; Gilles, E.D.; Knackmuss, H.J. Ed.; Gustav Fischer, Stuttgart D, **1991**, pp.287-292.

Finnigan, S.M.; Howell, J.A. *Desalination*, **1990**, 79, 181-202.

Henry, J.D.; Lawler, L.F.; Kuo, C.H.A. *AIChE J.*, **1977**, 23,(6), 851-859.

Kroner,K.H.; Schütte, H.; Hustedt, H.; Kula, M.R. *Process Biochemistry*, **1984**, 19, 67-74.

Wakeman, R.J.; Tarleton, E.S. *Chem.Eng.Sci.*, **1987**, 42, (4), 829-842.

Membrane Chromatography

Karl Günter Briefs and **Maria-Regina Kula**
Institute of Enzymetechnology, Heinrich-Heine-University Düsseldorf,
D-(w)-5170 Jülich, Germany, (FAX: 2461 61 2490)

We analyzed recently the application of a stack of modified microporous membranes for protein chromatography (K.G. Briefs and M.-R. Kula, *Chem. Eng. Sci.*, **(1992)**, 47, 141-149. Here we present a mathematical model to describe elution profiles from ionexchange membranes during gradient elution and compare model predictions with experimental data using albumin as model protein.

Common chromatographic supports are porous particles 10 - 100 µm in size, which may carry ligands for specific binding modes on the surface. There exist conflicting demands with regard to mechanical stability of the support, capacity, and productivity for a chromatographic process, especially for the separation of proteins. The dynamic capacity of 100 µm particles employed for preparative purposes deminishes with increasing flow rate due to the diffusive transport within the pores. We investigated the application of modified microporous membranes for adsorption (Champluvier and Kula, 1991), and in chromatography mode (Briefs and Kula, 1992) using a stack of 96 membranes (17 mm bed height). Porous membranes of 1 - 3 µm pore diameter provide sufficient binding capacity on their internal surface, and exhibit low pressure drop (≤ 0.8 MPa at 10 cm/min flow rate).

We have shown, that mass transfer influence is negligible up to flow velocities of 20 cm/min for ionexchange chromatography using bovine serum albumin (BSA) as a model. A break-through curve was recorded for BSA at 4.0 cm/min flow velocity and analyzed with regard to dispersion in the stack. From this experiment a Bodenstein number of 0.02 was determined. Since in process chromatography high loadings of the support are required the adsorption isotherm becomes highly nonlinear. The isotherm can be described by a Langmuir or better by a Freundlich equation.Under these conditions a balance over a single representative pore was established as illustrated in fig 1 and extended to the cross section.

Changes in time and space are described by equation 1

$$\frac{\partial c}{\partial t} + v\frac{\partial c}{\partial z} + \frac{2}{r_p}\frac{\partial q}{\partial t} = D\frac{\partial^2 c}{\partial z^2} \qquad (1)$$

Since kinetic effects can be neglected adsorption is well characterized by Freundlich isotherms

$$q = b(i)\ c^{o(i)} \qquad (2)$$

This description is still incomplete since protein chromatography is most often performed changing the pH or ionic strength in the eluent. In the following discussion a model and experimental data are presented, describing ionexchange chromatography for elution with a gradient in ionic strength over a membrane stack using BSA as a model protein.

Experimental

The membranes and the membrane holder employed are described in detail by Briefs and Kula (1992). BSA was measured by absorption at 280 nm. It was dissolved in 20 mm Tris/HCl buffer pH 8, applied to a DEAE-Immunodyne membrane stack, pore size 3 µm, with a cross section of 2.5 cm^2, and equilibrated against the same buffer. One ml sample was applied to the stack containing 1, 3.5 or 10 mg BSA, respectively. Elution was accomplished by a linear gradient from 0 - 1 M NaCl in the buffer. The flowrate was held constant at 3 cm/min.

Mathematical model

Inserting equations 2 into equation 1 leads to a differential equation, which is highly non-linear and difficult to solve. Therefore we adopted an approach developed in partition chromatography, the HETP concept (height equivalent to a theoretical plate, Martin

1054–7487/92/0258$06.00/0

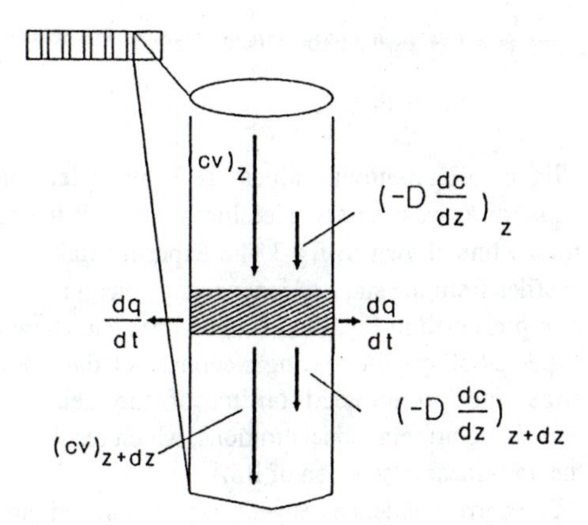

K; $q = Kc$	distribution coefficient for protein
M_s; $l = M_s i$	distribution coefficient for salt
l	adsorbed ion concentration of NaCl (mole/cm^2)
m	actual plate number for salt
M	total number of plates for salt
n	actual plate number for protein
N	total number of plates for protein
o	Freundlich equation parameter
q	adsorbed protein on pore surface (mg/cm^2)
Q_p	cross section of a pore (µm^2)
r_p	membrane pore radius (µm)
t	time (s)
v	liquid flow relocity (cm/s)
V_p	pore volume (ml)
V_G	gradient volume (ml)
z	axial position (cm)
z	length of membrane stack (cm)
$\zeta = A_p/V_p$	(1/cm)
$\theta = tQ_p/V_p$	dimensenless time
$\theta_{pr} = t_{pr} Q_p/V_p$	dimensenless time of sample addition

Fig. 1: Balance over a single pore in a modifiied microporous membrane

and Synge, 1941). The HETP concept has been extended by Nelson et al. (1978) to include non-linear adsorption and applied first to protein chromatography by Yamamoto et al. (1983). We transfered the approach to analyze membrane chromatography. The simplification in the mathematical treatment is reached substituting the space coordinate dz (equation 1) by a certain number of theoretical plates. This simplification is valid under the inherent assumptions that the number of plates are constant and the theoretical plates are equivalent throughout the stack. The number of theoretical plates is also a measure for dispersion, the relationship was given by Grubner et al (1967) as

$$N = \frac{v\ z_e}{2\ D_s} \qquad (3)$$

For the model describing ion exchamne chromatography with linear gradient elution we will first set up the balance equation for protein

$$\frac{V_p}{N}\frac{\partial c_n}{\partial t} = v\ Q_p\ c_{n-1} - v\ Q_p\ c_n - \frac{A_p}{N}\frac{\partial q_n}{\partial t} \qquad (4)$$

Together with the equation for the distribution coefficient of protein:

$$q_n = K(c_n, i)\ c_n \qquad (5)$$

a transformation will yield:

$$\frac{d\,c_n}{d\,\theta} = \frac{N[c_{n-1} - c_n] - c_n\ \zeta\ (\frac{\partial K}{\partial i})\ (\frac{\partial i}{\partial \theta})}{1 + \zeta\ K + c_n\ \zeta(\frac{\partial K}{\partial c_n})} \qquad (6)$$

A solution of equation 6 is possible under the following boundary conditions

$$\begin{array}{ll} c_1 = c_2 = \ldots = c_N = 0 & \text{for } \theta = -\theta_{pr} \\ c_0 = c_a & \text{for } -\theta_{pr} < \theta \le 0 \\ c_0 = 0 & \text{for } \theta > 0 \end{array}$$

Two terms in equation 6, the distribution coefficient K(i, cn) and the ionic strength and their derivations i, di/dt need to be determined. A balance equation for ionic strength after the plate model is given by

$$\frac{V_p}{M}\frac{\partial i_m}{\partial t} = v\ Q_p\ i_{m-1} - v\ Q_p\ i_m - \frac{A_p}{M}\frac{\partial l_m}{\partial t} \qquad (7)$$

For a linear gradient in ionic strength and under the assumption that the distribution coefficient Ks for salt is constant, and approximately zero, and dispersion effects for salt can be neglected, equation 8 results

$$\begin{array}{ll} i_m = i_a & \text{for} \quad \theta \le \frac{m}{M} \\ i_m = i_a + \frac{\Delta i\ V_p}{\Delta V_G}\{\theta - \frac{m}{M}\} & \text{for } \theta > \frac{m}{M} \end{array} \qquad (8)$$

The derivation is easy to obtain and must then be introduced in equation 6. If the adsorption isotherm is measured experimentally and in this way a distribution coefficient determined, then equation 6 is representing a non-linear first order differential equation, which can be solved numerically beginning from the first plate and ending at the last one, which will yield a description for the eluate of the membrane stack. The numerical solution presented below was carried out with a library program (NAG, 1988) using the method of Adams.

Results and Discussion

The adsorption isotherms for BSA were determined at 20 °C at various buffer compositions. The results are presented in Fig. 2

The linear portion of the adsorption isotherm decreases with increasing ionic strength.Saturation of the membrane is no longer observed at high concentrations of protein and salt . Therefore the experimental curves were modeled as Freundlich isotherms: $q = b(i)c^{o(i)}$. Transforming all the Freundlich isotherms in a mathematical equation will show that they fit very well the following equations, which will give the isotherms, the distribution coefficient and the derivations

$$q = b_1 \exp(-b_2 i)\ c^{(o_1 i + o_2)}$$

$$K(i, c) = b_1 \exp(-b_2 i)\ c^{(o_1 i + o_2 - 1)}$$

$$\frac{\partial K}{d c} = b_1 \exp(-b_2 i)\ (o_1 i + o_2 - 1)\ c^{(o_1 i + o_2 - 2)}$$

$$\frac{\partial K}{d i} = (-b_2 b_1) \exp(-b_2 i)\ c^{(o_1 i + o_2 - 1)} + b_1 \exp(-b_2 i)\ o_1 \ln c\ c^{(o_1 i + o_2 - 1)}$$

Since all unknown values are now determined, equation 6 was solved yielding as a result the continous line shown in fig 3.The experimental elution profiles from the stack of ionexchange membranes are also presented in Fig. 3 as broken lines and show the expected shape: increasing steepness of the leading edge and pronounced tailing of the peak with increasing protein concentrations, which results from the non-linear adsorption of BSA.

The correspondence between experiment and model is good for a first approximation. To improve the fit the value for N should be adjusted or redetermined

The prediction of elution profiles of a multicomponent mixture helps to find optimal separation conditions. For such a problem equation 6 must be solved in parallel for each component. Results will be valid only, if the different compounds move independently through the stack and concentrations are low enough to avoid displacement. Otherwise equation 5 must be modified to include further interaction terms.

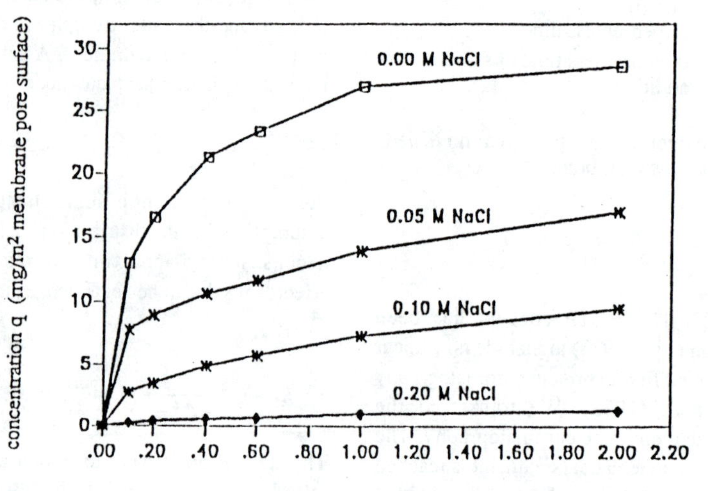

Fig. 2: Isotherms for BSA adsorption as a function of salt concentration (buffer : 0.02 M Tris/HCl, pH 8, 20 °C, DEAE-Immunodyne membrane, 3 µm pore size)

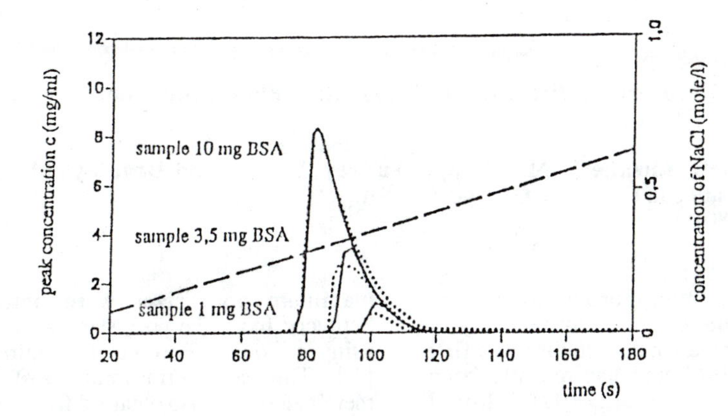

Fig. 3: Calculated (——) and experimental (...) elution diagrams for BSA from a DEAE-Immunodyne membrane stack (N = 136)

List of Symbols

A_p	surface area of a pore for adsorption (m^2)
b	Freundlich equation parameter
Bo	Bodenstein number = $\frac{2 r_p v}{D}$
C	protein concentration (mg /ml)
D	dispersion coefficient (cm^2/s)
i	ion concentration of NaCl (mole/ml)

Acknowledgments

K.G. Briefs was supported by a scholarship within the Program "Applied Biology and Biotechnology" from the BMFT.

References

Briefs, K.-G.; Kula, M.-R. *Chem. Eng. Sc.* **1992**, 47, 141-149

Champluvier, B.; Kula, M.-R. *J. Chromatogr.* **1991**, 539, 315-325

Grubner, O.; Zikanova, A.; Ralek, M. *J. Chromatogr.* **1967**, 28, 209-218

Martin, A.J.P.; Synge, R.L.M. *Biochem. J.* **1941**, 35, 1358-1368

Nelson, W.C.; Silarski, D.F.; Wankat, P.C. *Ind. Eng. Chem. Fundam.* **1978**, 17, 32-38

Yamamoto, S.; Nakanishi, K.; Matsuno, R.; Komikubo, T. *Biotechnol. Bioeng.* **1983**, 25, 1465-1483

Industrial Scale Recovery of Proteins by Tangential Flow Filtration.

Robert van Reis*, Elizabeth M. Clapp, Audrey W. Yu and **Bradley M. Wolk.**
Genentech Inc., South San Francisco, CA 94080, USA.
(FAX: 415 266-3880)

Industrial scale microporous membrane tangential flow filtration (TFF) for recovery of proteins from mammalian cell culture was first accomplished in 1985 and has recently been described (van Reis et al., 1991). Initial development efforts focused on the use of existing microporous TFF membranes, experimental capacity versus rate analysis, linear scale-up techniques and design of industrial scale equipment suitable for recovery of human recombinant DNA pharmaceuticals.

Progress has since been made in the area of fluid dynamic optimization of TFF for mammalian cell-protein separation (van Reis et al., 1990). Process performance was optimized as a function of wall shear rate, flux, and membrane pore size. Membranes previously available for dead-end filtration were investigated and subsequently commercialized in TFF systems with steam-in-place capability.

A systematic study of membrane design specifically aimed at TFF recovery of proteins from mammalian cell culture has now been undertaken. Process capacity has been determined by studying chemical fouling, particulate loading and protein retention for various experimental membranes.

Cell Culture

Cultures of Chinese Hamster Ovary cells transfected by recombinant DNA methods to express a 43 kD protein were grown in stainless steel fermenters at 37 °C to densities of (2-10) x 10^6 cells mL^{-1} in proprietary serum-free cell culture media.

Membranes

Experimental hollow fiber membranes were provided by Sepracor Inc. (Marlborough, MA) in linear scale down modules. All membranes are slightly anisotropic and consist of polyethersulfone hollow fibers which have been modified with a covalently bound cellulosic coating. Scanning electron micrographs of the membranes are shown in Fig. 1-3. Initial studies utilized three different membranes as shown in Table 1. The minimum, mean flow and maximum pore sizes were obtained from Sepracor Inc. The pore sizes were determined using a Coulter Porometer (Coulter, Hialeah, FL). The water permeability (Lp) data for the membranes was also obtained from Sepracor Inc. The performance of two experimental membranes (A and D) was compared with a standard mammalian cell harvest membrane (H).

Table 1. Membrane Characteristics

Membrane	Pore Size (min, mean, max.) [μm]	Permeability [$L\ m^{-2}\ h^{-1}\ kPa^{-1}$]
A	0.10, 0.12, 0.22	5.7
D	0.18, 0.26, 0.49	38
H	0.29, 0.36, 0.54	86

Fluid Dynamics

The hollow fiber modules were operated at a wall shear rate $\gamma_w = 4000\ s^{-1}$ in the laminar flow regime as defined by a Reynolds number N_{Re} = 185 with a constant flux $J_f = 32\ L\ m^{-2}\ h^{-1}$:

$$\gamma_w = 4Q/n\pi r^3 \quad (1)$$

$$N_{Re} = 2\rho Q/\mu n\pi r \quad (2)$$

where; Q = cell suspension feed rate [$m^3\ s^{-1}$], n = number of hollow fibers [-], r = hollow fiber inner radius [m], ρ = cell suspension density [kg m^{-3}], and μ = cell suspension dynamic viscosity [Ns m^{-2}]. Axial pressure drop (ΔP) and transmembrane pressure (TMP) were calculated as:

$$\Delta P = 8QL\mu/n\pi r^4 \quad (3)$$

$$TMP = (P_F + P_R)/2 - P_f \quad (4)$$

where; L = hollow fiber length [m], P_F = cell suspension feed pressure [Pa], PR = cell

1054–7487/92/0262$06.00/0

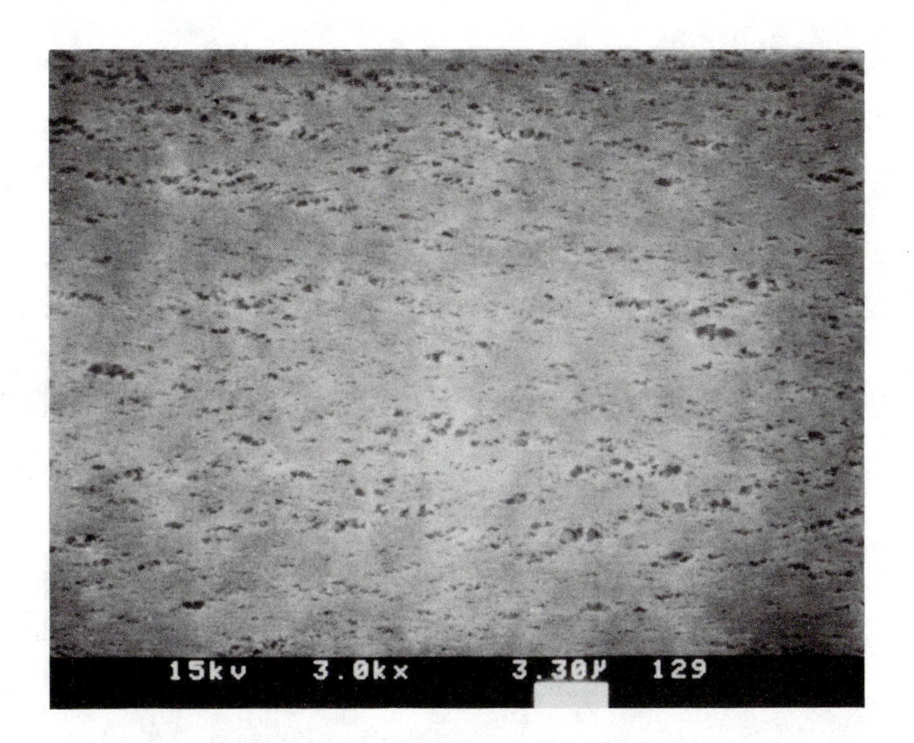

Fig.1. Scanning electron micrograph of internal surface of hollow fiber membrane A.

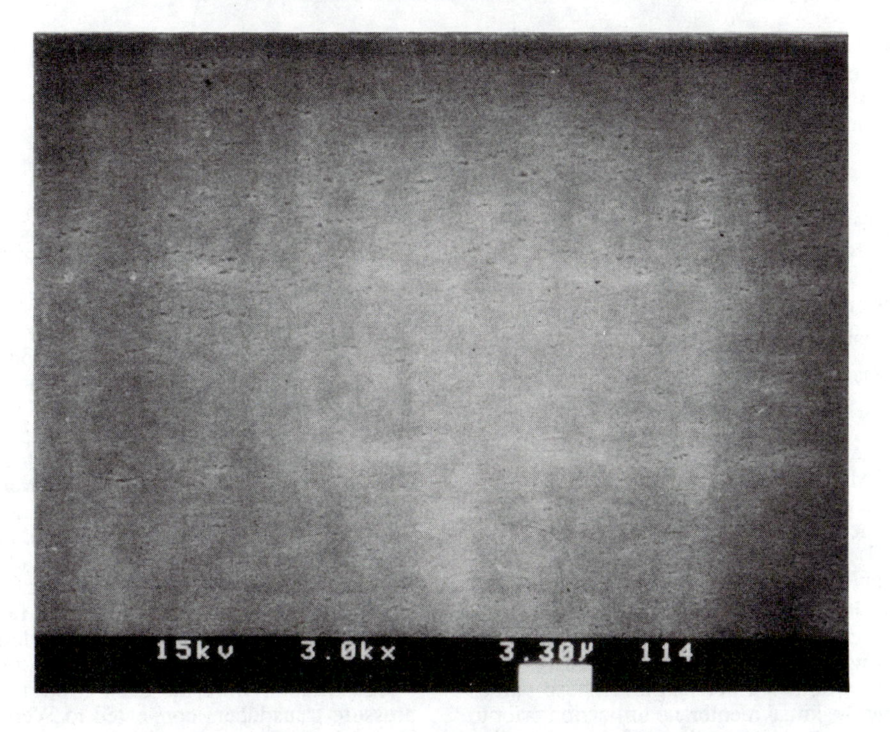

Fig.2. Scanning electron micrograph of internal surface of hollow fiber membrane D.

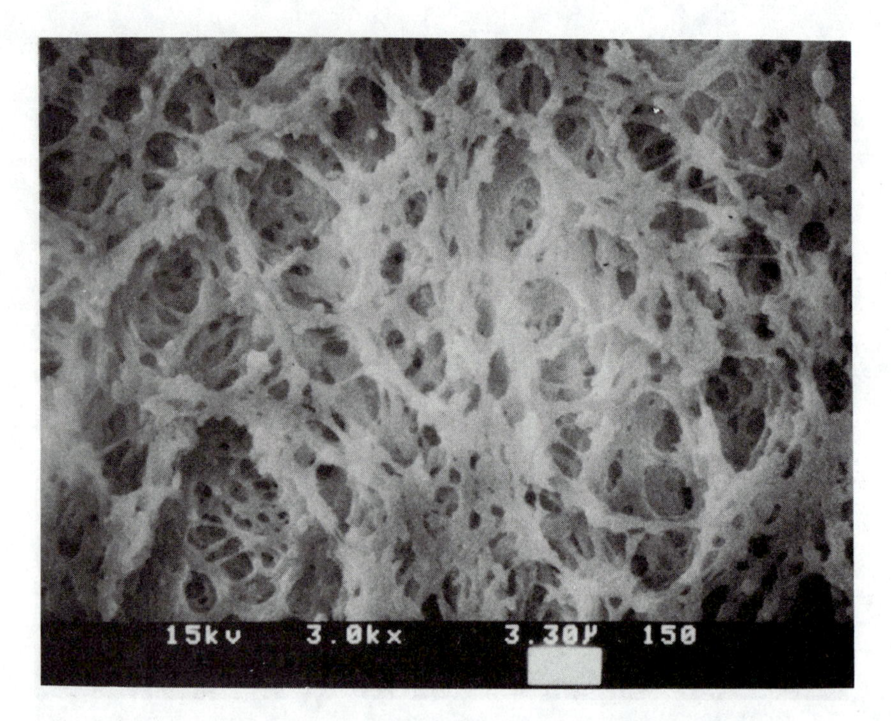

Fig.3. Scanning electron micrograph of internal surface of hollow fiber membrane H.

suspension retentate pressure [Pa], and Pf = filtrate pressure [Pa]. Fluid dynamic parameters for the hollow fiber modules and cell culture suspension are shown in Table 2.

Table 2. Fluid Dynamic Parameters

Hollow fiber length, L	0.56	m
Fiber inner diameter, D	$6x10^{-4}$	m
Feed density, ρ	1030	kg m^{-3}
Feed dynamic viscosity, μ	$1x10^{-3}$	Ns m^{-2}

High permeability membranes may yield an initial TMP of less than half the axial pressure drop, ΔP, resulting in Starling flow. The initial membrane utilization and inlet and outlet fluxes (Table 3) were calculated using finite element numerical analysis based on the initial water permeabilities (Table 1) and Eqs. 3 and 4 using an average flux of 32 $Lm^{-2}h^{-1}$. Numerical analysis was done using a step value $\Delta L = L/100$ where L = hollow fiber length. These values represent the initial membrane utilization prior to the onset of chemical fouling. Chemical fouling and particulate loading will, however, rapidly decrease the permeability.

Table 3. Membrane Utilization - Water

Membrane	A	D	H
Area [%]	93	60	53
Inlet Jf [L m^{-2} h^{-1}]	69	183	464
Outlet Jf [L m^{-2} h^{-1}]	-4	-111	-373

Experiments

All experiments were carried out on 100 L fermenters at 37 °C. Two to three different hollow fiber modules were connected in parallel and run simultaneously in a cell concentration mode. One or two Waukesha Model 6 (Waukesha, Wisconsin) positive displacement feed pumps were used and each individual filtrate rate was controlled with a separate Masterflex (Cole-Parmer, Chicago, IL) peristaltic pump. The concentration factor did not exceed threefold during any of the experiments. The combined feed flow rate was measured using a Sanitech (Andover, NJ) flow meter and filtrate flow rates were measured with graduated cylinders and a stopwatch. All pressures were measured with Spectramed (Oxnard, CA) Model TNF-R pressure transducers connected to Wedgewood Technology (San Carlos, CA) pressure monitors. A total of six experiments (including 14 module tests) were run.

Results and Discussion

The effects of chemical fouling and initial particulate loading were determined by comparing the experimental water permeabilities with process permeabilities as shown in Table 4.

Table 4. Membrane Permeability.

Membrane	Water Lp	Process Lp
	$[L\ m^{-2}\ h^{-1}\ kPa^{-1}]$	
A	5.0	2.4
D	22	5.2
H	68	17

The water permeabilities were somewhat lower than the manufacturer's values (Table 1), which would be expected when comparing data obtained on short fiber lengths in a dead-end mode to data obtained on full length fibers in a TFF mode. Based on the initial process permeabilities, a new set of data was calculated for membrane utilization and inlet and outlet filtrate fluxes as shown in Table 5.

Table 5. Membrane Utilization - Process

Membrane	A	D	H
Area [%]	100	98	65
Inlet Jf $[L\ m^{-2}\ h^{-1}]$	48	66	139
Outlet Jf $[L\ m^{-2}\ h^{-1}]$	17	-0.8	-70

The effects of particulate loading were studied by measuring TMP as a function of time and plotting the data as a function of filtrate volume processed per unit area of membrane, as shown in Fig. 4. A TMP of 70 kPa has previously been established as a practical endpoint (van Reis et al., 1991) for process capacity determination. The retention characteristics for these three membranes are, however, significantly different as shown in Fig. 5. If a process is designed based on a 15-fold concentration factor and a four diavolume diafiltration (maximum 20% dilution of the harvested cell culture fluid), then 98% yields can be obtained with a retention R=0.4. A conservative endpoint could therefore also be based on reaching this retention value. By combining the data in Fig. 4 and Fig. 5 one would arrive at process capacities for the three membranes as shown in Table 6.

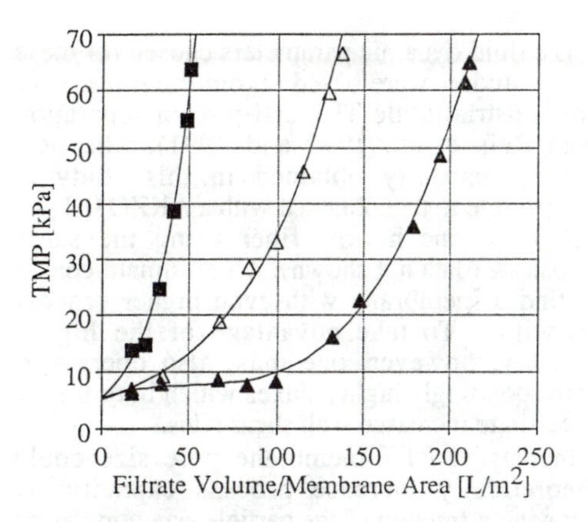

Fig.4. Particulate loading capacity for experimental membranes A (Δ), and D (▲), and standard membrane H (■). Particulate loading capacity was determined by measuring transmembrane pressure as a function of filtrate volume processed per square meter of membrane area at a constant flux of 32 $L\ m^{-2}\ h^{-1}$.

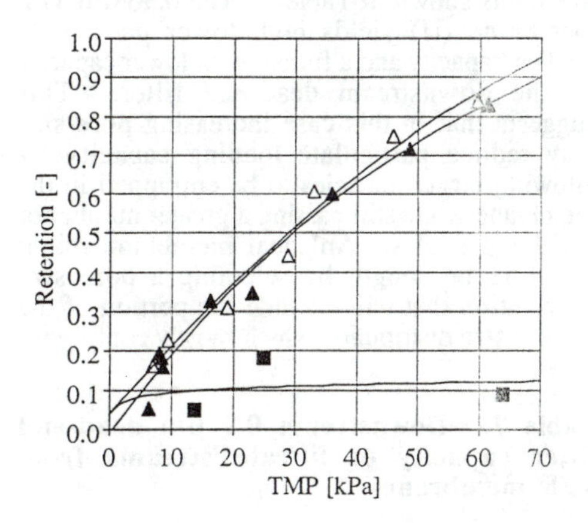

Fig.5. Retention of a 43 kD protein as a function of transmembrane pressure during particulate loading at a constant flux of 32 $L\ m^{-2}\ h^{-1}$ for experimental membranes A (Δ) and D (▲) and standard membrane H (■).

Table 6. Process Capacity

Membrane	Process Capacity
	$[L\ m^{-2}\ h^{-1}]$
A	76
D	150
H	54

The fluid dynamic parameters chosen for these initial studies were based on our previous work on industrial scale TFF cell-protein separation (van Reis et al., 1990 and 1991). The best process capacity obtained in this study is comparable to that obtained with a AKZO 0.2 µm polypropylene hollow fiber using the same feedstock (data not shown). Our ultimate goal is to find a membrane with even higher process capacity. To take advantage of the higher capacity, however, one must also operate at correspondingly higher fluxes which in turn may necessitate increased wall shear rates.

Increasing TFF membrane pore size could theoretically increase process capacity by allowing a fraction of the particle size population to pass through the membrane. It is therefore crucial to study the downstream dead-end 0.2 µm sterile filtration capacity on the TFF filtrate stream. Single use dead-end filters will typically contribute more to the cost of the operation than the cost of the TFF modules which may be re-used 100 times (van Reis et al. 1991). The downstream capacity of Millipore 0.2 µm Durapore® dead-end filters run at a flux of 765 L m^{-2} h^{-1} is shown in Table 7. The 0.36 µm TFF membrane (H) yields both lower particulate loading capacity and a filtrate with lower capacity on the downstream dead-end filter. This suggests that in this case increasing pore size may reduce particulate loading capacity by allowing larger particles to be entrapped in the membrane while still passing a greater number of smaller particles. An ideal membrane might therefore be sought by selecting a pore size distribution that will exclude the portion of the particle size distribution which mostly contributes to particulate loading while only allowing particles less than 0.2 µm to pass. In addition it would be desirable to have an effective pore size (after initial chemical fouling) which allows passage of the protein of interest without retention.

Table 7. Downstream 0.2 µm dead-end filter capacity of filtrate streams from TFF membranes.

Membrane	Downstream Capacity [L m^{-2}]
A	6.9×10^3
D	N.D.
H	3.9×10^3

Conclusions

A systematic approach to the design of TFF membranes for recovery of proteins from mammalian cell culture has been implemented. Initial studies have resulted in a threefold increase in process capacity. The effect of membrane characteristics on particulate loading and protein retention have also been demonstrated. A matrix of membrane characteristics including pore size, morphology, hydraulic resistance and surface chemistry has been devised for further optimization of TFF cell-protein separation. Correlation of particulate loading as a function of particle size and pore size distribution will also be explored.

Acknowledgments

The authors would like to acknowledge Mr. Sam Williams of Sepracor Inc., who was responsible for development of the membranes series used in these studies. All linear scale down modules used in these studies were provided by Sepracor Inc. The authors would also like to thank everyone in Cell Culture Process R&D, Cell Culture Operations, Fermentation Operations and Product Recovery Operations at Genentech who provided outstanding collaboration and support during this project.

References

van Reis, R.; Leonard, L.C.; Builder, S.E. Engineering Foundation Conference on Recovery of Biological Products V. **1990**.

van Reis, R.; Leonard, L.C.; Hsu, C.C.; Builder, S.E. *Biotechnol. Bioeng.* **1991**, *38*, 413-422.

Membrane Processing of Concentrated Biological Suspensions

Geoffrey Taylor, Ian Grant, Andrew Ison, Michael Hoare.
The Advanced Centre for Biochemical Engineering (SERC Interdisciplinary Research Centre), Department of Chemical & Biochemical Engineering, University College London, Torrington Place, London WC1E 7JE, UK
(Fax: 071 388 0808)

The batch concentration by ultrafiltration of biological suspensions follows two distinct phases. At low concentration the flux is pressure related and is often invariant with concentration. This is followed by a transition to a region where there is a steep decline in flux with concentration. Such behaviour is observed for protein precipitate suspensions where the requirement has been to concentrate to sufficient level before proceeding to a drying stage and also for fermentation broths where the objective is to maximize recovery of low molecular weight metabolites without recourse to excessive dilution by washing of the retentate. The hydrodynamic effects determining this transition point are discussed.

It is well recognized that the concentration of soluble protein using membranes is limited by so-called concentration polarization effects leading to relatively low flux rates at only moderately high protein concentration. The design of membrane systems to allow operation at high cross flow velocities to limit polarization effects leads to improvement in flux rates provided the inherently higher pressures involved do not lead to increased fouling of the membrane structure. Studies of the recovery of an intracellular enzyme (alcohol dehydrogenase from *Saccharomyces cerevisiae*) have shown that hollow fibre membranes can be operated at high cross flow velocities (up to 3 m s^{-1}, equivalent Reynolds number range of 1500 to 3000) for several hundred passes over the membrane surface without loss of enzyme activity other than an initial loss due to surface adsorption (Narendranathan and Dunnill, 1982).

One way to overcome polarization effects when handling protein solutions is to seek to operate with a substantial part of the protein in the precipitate phase. Fig. 1 examines the effect of membrane length on the flux versus concentration profile for soluble and isoelectric point precipitated protein from the same source (Devereux et al., 1986). The lower flux rates for precipitated compared with soluble protein at low concentration are due to a greater tendency to adsorb to and hence foul the membrane (Bentham et al., 1988). Both precipitate concentration curves are characterized by a region where the flux remains approximately constant with concentration and then by a transition to a sharp decline in flux with increased concentration. The pilot-scale rig was operated using a centrifugal pump with back pressure control on the retentate outlet such that the inlet pressure was held to a maximum value as specified by the membrane manufacturers. In the first region the flux is primarily controlled by the transmembrane pressure drop, this being greater for the shorter membrane and hence results in higher

1054–7487/92/0267$06.00/0

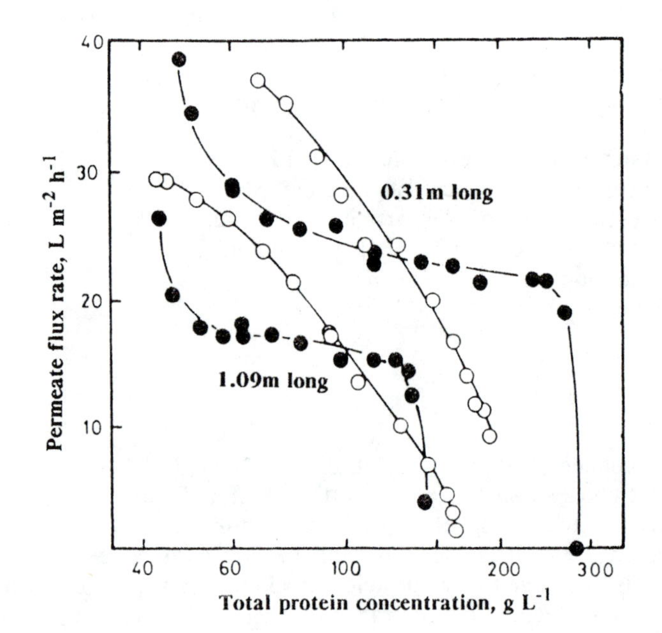

Fig. 1 Comparison of the pilot plant scale concentration of soluble and precipitated protein using a Romicon PM50 hollow fibre membrane cartridge. -○- soluble protein, pH9; -●- isoelectric point precipitated protein, pH4.8. Fibre lengths as given.

flux rates. The transition point relates to the phase of operation where the retentate outlet pressure has already reached zero and then, due to increased retentate apparent viscosity, where the flow conditions over the membrane surface have reached some critical condition such that the membrane behaves in a polarized fashion and a sharp decline in flux is observed.

This behaviour is more clearly seen in Fig. 2 where the replacement of the centrifugal pump with a positive displacement pump gives better control of the operation during the region of sharp flux decline (Taylor et al., to be published). Similar behaviour is observed for the concentration of fermentation broths where, as for the precipitated protein, there is a considerable proportion of soluble protein present (Grant et al., to be published).

One means of optimizing the engineering design of the membrane separation rig is to identify the means whereby the transition point to steep flux decline occurs at higher retentate concentrations. To examine this transition point a wide range of experiments have been carried out under total permeate and retentate recycle to examine the effect of flow conditions on the flux rate achieved. The variables examined include the concentration of the retentate, the hollow fibre membrane diameter, the membrane length and the ratio of soluble to precipitated protein. One such set of results is given in Fig. 3 and in this and other cases the transition point is determined by the membrane wall shear

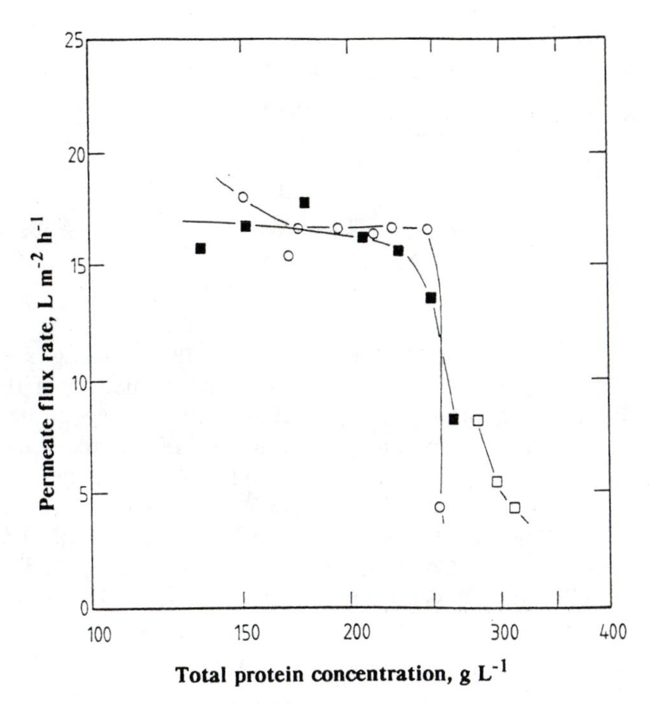

Fig. 2 Permeate flux rate during the concentration of precipitated soya protein suspensions: the effect of recirculation pump type on the flux rate profile. Membrane type, Romicon XM50; hollow-fibre length, 0.315 m, inner diameter, 1.52 mm; membrane area, 0.7 m^2. Experimental conditions: ○ , centrifugal pump; operating temperature, 42°C; inlet and outlet retentate pressures, 172 kPa, 103 kPa; initial fraction of total protein soluble, 0.19; ■ , □ , positive displacement pump; operating temperature, 27°C; inlet and outlet retentate pressures; ■ , 172 kPa, 103 kPa; □ , 172 kPa, 69 kPa; initial fraction of total protein soluble. 0.23.

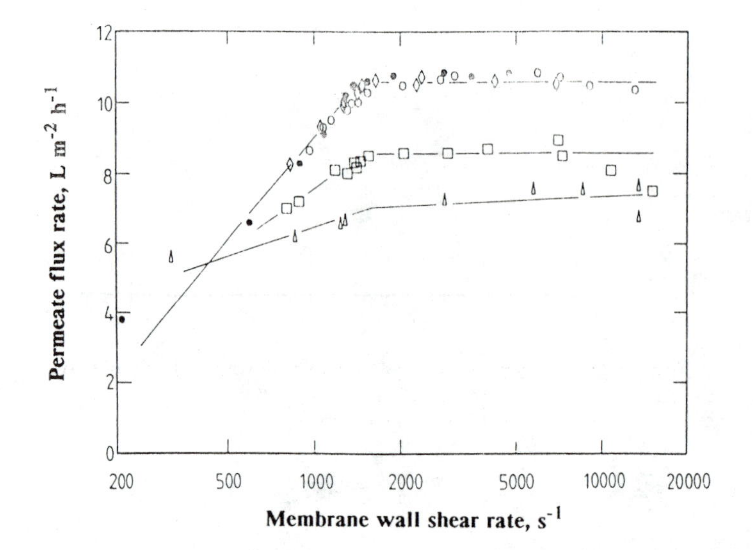

Fig. 3 Dependence of the permeate flux rate on the membrane shear rate at difference precipitated soya protein con-concentrations. Operating temperature, 25°C; membrane type, Romicon PM50; hollow fibre length 0.315 m, inner diameter, 1.1 mm; membrane area, 0.70 m^2. Length-averaged transmembrane pressure drop, 138 kPa. Membrane shear rate evaluated using rheological parameters determined from the corresponding flow curves. Total protein concentration, fraction of total protein soluble, dry weight: △ , 104 kg m^{-3}, 0.24, 0.140 kg kg^{-1}; □ , 144 kg m^{-3}, 0.22, 0.174 kg kg^{-1}; ○ , 180 kg m^{-3}, 0.20, 0.206 kg kg^{-1}; ◇ , 213 kg m^{-3}, 0.19, 0.235 kg kg^{-1}; ● , 231 kg m^{-3}, 0.19, 0.248 kg kg^{-1}

rate rather than velocity or wall shear stress or Reynolds number. Further analysis requires an understanding of the effect of flux rate itself on the onset of the transition point.

Acknowledgements

The financial support of the Science and Engineering Research Council, UK for two of the authors (Geoffrey Taylor and Ian Grant) is gratefully acknowledged.

References

Bentham, A.C.; Ireton, M.J.; Hoare, M.; Dunnill, P. *Biotechnol. Bioeng.* **(1988)** Vol. 31, 984-994

Devereux, N.; Hoare, M.; Dunnill, P. *Chem. Engng. Commun.* 45 **(1986)**, 255-276

Narendranathan, T.J.; Dunnill, P. *Biotechnol. Bioeng.* XXIV **(1982)** 2103-2107

Adapting Chromatography for Initial Large-Scale Protein Recovery

Per Hedman and **Ann-Kristin Barnfield Frej**, Pharmacia BioProcess Technology AB,
S-751 82, Uppsala, Sweden
(FAX: +46-18-129843)

The capture of a protein or peptide product from a fermentation or crude cell homogenate using adsorption techniques has required extensive clarification of the feed-stock. The preferred method for adsorption of proteins or peptides from unclarified feeds is adsorption in expanded or fluidized beds. Studies of a new adsorbent intended for use in expanded beds show axial dispersion and dynamic capacities comparable to those measured in packed beds.

Over the last couple of decades there has been a tremendous development of chromatography media for protein separations. For analytical purposes, the development has resulted in smaller particle sizes, which make higher column efficiencies possible. The higher back-pressures resulting from the use of smaller particle sizes are not a major concern in the analytical laboratory.
For process scale chromatography, the development of media has been divergent, resulting in a number of products. Each of these products and its proposed mode of operation meet a set of design criteria (Table 1). There is no single product or mode of operation which meets all such criteria.

Available alternatives include very large diameter media, small diameter media, low-priced media for single-use batch adsorption, filter and membrane based media used in different configurations and most recently, media for use in expanded or fluidized beds.

In order to know which alternatives to use for each separation problem, it is necessary to define the objectives for a particular separation. The relative importance of the different process design objectives differs from one process to another. Consequently, for each process one has to determine which is the most useful adsorbent and which is the most useful mode of operation.

Using large particle media in order to avoid high pressures and minimize equipment costs

For the determination of many components in a single sample, high selectivity is not as important as high efficiency. This is why small particle diameters and high pressure systems are dominant in the analytical laboratory. In process scale chromatography, most steps rely more on selectivity than on a high number of plates. Furthermore, high-pressure equipment is more expensive for larger scale operations. This is why low-pressure operations are preferred.

For initial process steps e.g. capturing a protein from a large volume, a high efficiency (large

Table 1 Process design compromises

Minimize proteolytic breakdown	Adsorb product rapidly from entire feedstock using a large column, single cycle adsorption, high flow-rate
Minimize media inventory	Run many cycles, pool collected fractions
Minimize validation/ QA workload	Avoid pooling subfractions from many cycles Use inexpensive disposable media
Reduce volume	Use slow flow-rate for elution, high loading of small column
Remove as much host protein as possible	Use highly selective ligand and/or use small particle size, slow flow-rate and low loading for high efficiency
Minimize equipment cost	Use large particles in order to keep pressures low

1054–7487/92/0271$06.00/0

number of plates) is not required since resolution is not of primary concern. However, a certain number of plates is required in order to achieve an acceptable yield. This number depends on the ratio between the product concentration and the product-ligand complex dissociation constant. Usually, such small disso-ciation constants are achievable that the number of plates required for a high yield is obtained in short beds containing media with 90 µm and larger particles.

Some feed-stocks are quite viscous (e.g. feed-stocks containing DNA or ethanol, especially when cold). These feeds will generate high back pressures with 90 µm packings even at normal flow velocities (100 - 300 cm/h). In order to handle such feeds a new medium, SP Sepharose® Big Beads, was developed. This cation exchanger, having a particle size range from 100 to 300 µm, yields much lower back pressures than 90 µm media (Fig. 1). At pressures below 300 kPa SP Sepharose Big Beads permits flow velocities of 2000 cm/h with 25 ^{o}C water and 500 cm/h with 65 % ethanol.

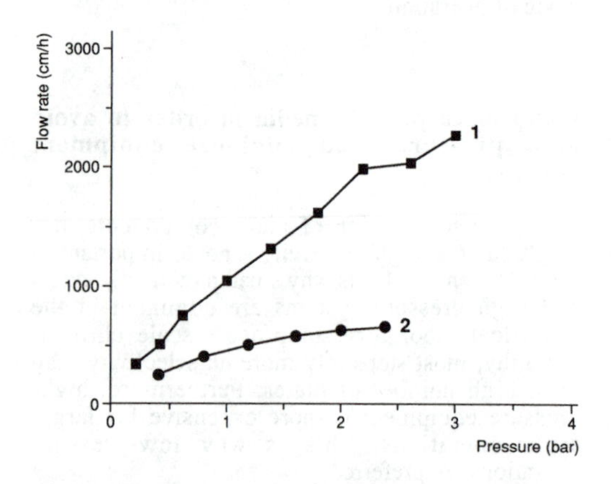

Fig. 1 Pressure drop as a function of flow-rate for SP Sepharose Big Beads.1. Bed height 20 cm in a BioProcess™ Glass Column (BPG 300) measured with water at 25 ^{o}C2. Bed height 30 cm in a Process Plastic Column (PP 113) measured with 65 % ethanol at 20 ^{o}C.

When reducing the back pressure by increasing the particle size, the mass transfer rate is reduced due to the longer diffusional path length. It should be noticed that this effect is stronger for larger proteins with low diffusivity and of much less importance for peptides with higher diffusivity (Fig. 2).

Feeding unclarified feeds into packed columns

Fermentation broths and crude cell homogenates contain a number of proteases which will reduce product recovery and produce degradation products that are difficult to remove. The significance of this problem depends on the susceptibility of the protein of interest to proteolytic cleavage (by the proteases present in the feed-stock). However, it is always desirable to minimize product exposure to proteases.

The "exposure" is a function of time and the likeli-hood that a susceptible peptide bond is moved close enough to the enzyme´s active site. This probability is reduced when the protein is adsorbed onto a chromatography gel. For this reason, it is desirable to bind the protein of interest rapidly to an adsor-bent, without spending time removing particulate matter such as host cells, cell debris, etc. However, this has not been possible with packed columns.

The pressure drop over a 26.5 cm bed of DEAE Sepharose Fast Flow rose to over 300 kPa after feeding one column volume of an *E. coli* homo-genate (25% wet weight) at 30 cm/h. With a yeast homogenate, the pressure drop was 220 kPa after 3.2 volumes were fed through the column. Similar results were observed with Phenyl Sepharose Fast Flow.

An *E. coli* homogenate (0.6 % dry weight after homogenization) was obtained after passing the cells (in the fermentation broth) 3 times through a homogenizer (APV Gaulin 30 CD, 60, 70 and 80 MPa). The homogenate was fed through a 2 cm high bed of larger size composite agarose beads at 300 cm/h linear flow-rate without causing an appreciable back pressure. The available homogenate volume was 95 times the volume of sedimented adsorbent. Even after recirculation of the homogenate, corresponding to a passage through the bed of 2100 column volumes homogenate, the pressure only reached 30 kPa and remained stable. However, feeding 53 column volumes of undisrupted *E. coli* culture (1.8% dry weight) through the column caused a rapid increase in back pressure to more than 400 kPa. Inspection showed the 60 µm gel retaining net was blocked. In a similar experiment, where a 100 µm net was used, the back pressure

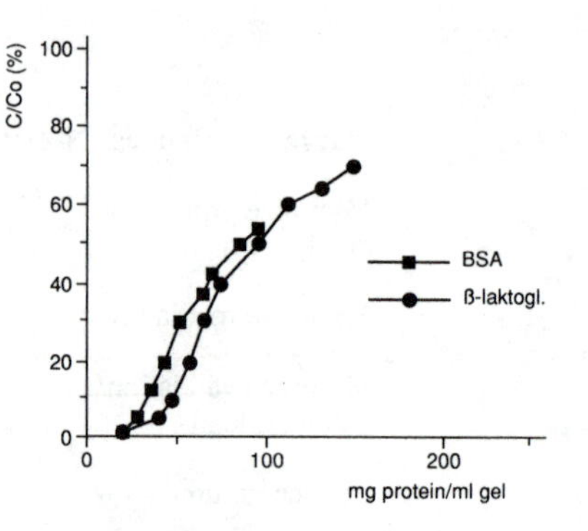

Fig. 2 Binding capacity of SP Sepharose Big Beads measured with frontal analysis (outlet concentration C, inlet concentration C_0) in acetate pH 5 (BSA) and formate pH 4.1 (ß-lactoglobulin) at 300 cm/h linear flow-rate.

did not exceed 30 kPa after recirculation of an *E. coli* suspension through the bed at 300 cm/h.

These experiments demonstrate that extensive testing with different feed-stocks and different equipment configurations is required before a new technique, such as adsorption from unclarified feed-stocks, can be adopted on a commercial scale using packed beds.

Adsorption in stirred tank reactors

Packed beds of commercially available media tend to act as depth filters for all feed-stocks that are normally used. The build-up of particulate matter in the bed with time will eventually block the flow. This problem is avoided by carrying out the adsorption in a stirred tank. This method has been used for many years on a commercial scale for the isolation of plasma coagulation Factor IX with DEAE Sephadex® A-50 (Brummelhuis, 1980). In this case there is no feed-stock subject to proteases left in a holding tank "waiting for reaching the adsorption reactor". However, the contact time may be long to achieve high recovery since this is a single step adsorption procedure as opposed to the multi-step (or multi-plate) process in a packed bed.

Adsorption in fluidized beds

The adsorption of proteins in a multi-stage fluidized bed reactor was described by A. Buijs and J. A. Wesselingh (Buijs et al., 1980). The higher efficiency of a multi-step process compared with a completely back-mixed well fluidized bed, was achieved by inserting a number of plates with suitably sized holes into the adsorption column. Another approach, the use of magnetically stabilized fluidized beds was described by M.A. Burns and D.J. Graves (Burns et al., 1985).

N.M. Draeger and H. A. Chase (Draeger et al. 1990) found that by limiting the expansion of a fluidized bed the back-mixing is reduced. This indicates that a "multi-plate" adsorption is possible in an expanded bed.The reduction of the dynamic adsorption capacity when expanding a packed bed by a factor of two was small.

The Q Sepharose Fast Flow used in these experiments was never intended for use in expanded beds. It is obvious from Draeger´s and Chase´s experiments that particles with higher sedimentation velocities are desirable. Commercially available composite particles based on porous silica have been tried in fluidized beds. We have found that such particles, do not resist cleaning with sodium hydroxide due to the large silica surface area (data not shown).

New media for use in expanded beds

A number of prototype ion exchange media with different particle sizes, particle size distributions and densities were synthesized. Resistance to sodium hydroxide washes and repeated autoclaving was specified when starting this project. For expanded beds, particles larger and denser than media such as Sepharose Fast Flow have been found suitable.

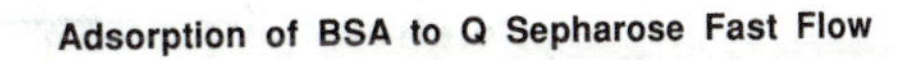

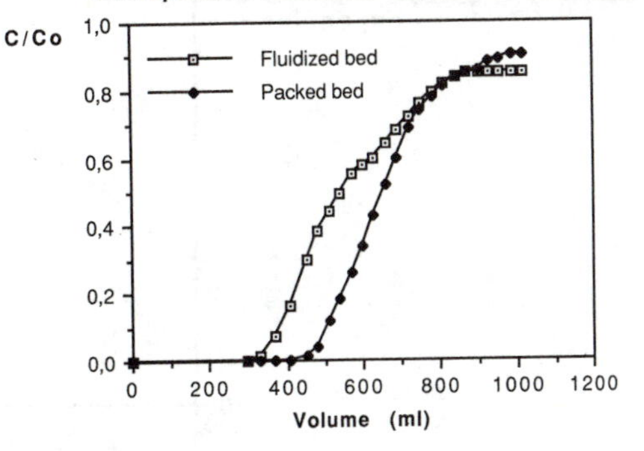

Fig. 3 Breakthrough curves obtained in a packed bed and in a bed expanded by a factor of 2. Data from Draeger and Chase.

These prototype media are compatible with flow-velocities of 300 cm/h (Fig. 4), although the density and viscosity of the feed may change this figure.

The binding capacity for these prototype media in expanded bed mode is higher at 300 cm/h flow velocity than that which Chase reported for Q Sepharose Fast Flow at much lower flow velocities. From the frontal analysis experiments, (Fig. 5) it appears that the axial dispersion (e.g. as judged by the slope at 50% saturation) is approximately the same in a packed bed as in an expanded bed. The dynamic capacity, though, is higher in the packed bed (Fig. 5). The first of these media prototypes to be evaluated did not lose its characteristics after several days storage in 1 M sodium hydroxide at elevated temperatures. Preliminary studies show retained binding and sedimentation properties after autoclaving.

For dilute feeds in a large volume scale, it is necessary to use large diameter columns. As the axial dispersion is not considerably higher than for a packed bed, it is desirable to use low height to diameter ratios to obtain high productivity. This results in an engineering problem, which we are currently addressing.

Conclusion

Composite agarose media show adsorption rates sufficiently high to allow for high recovery at flow velocities of 300 cm/h after a single passage through an expanded bed. Such particles should be useful in the "contaminated" fermentation area since they withstand autoclaving and prolonged cleaning with sodium hydroxide.

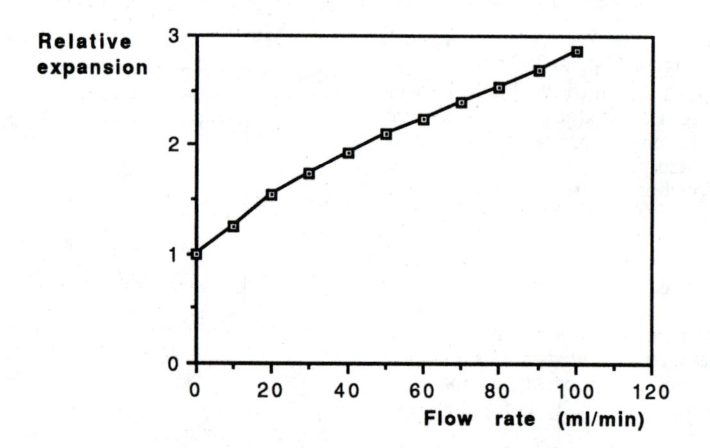

Fig. 4 Expanded bed height vs. liquid flow-rate measured in a 5 cm diameter column with dense prototype agarose beads.

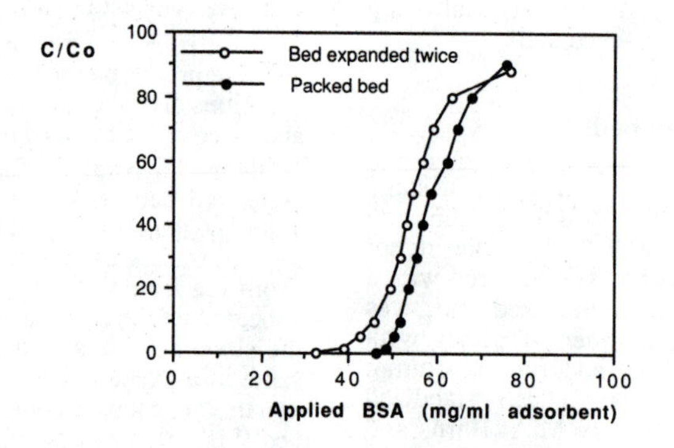

Fig. 5 Breakthrough curves obtained in a 5 cm diameter column using prototype dense DEAE-derivatized beads and a 0.6 mg/ml feed solution.

Acknowledgements

The authors wish to thank Dr. Howard Chase, University of Cambridge, UK, for adsorption data obtained with Q Sepharose Fast Flow and Ms. Susanne Kämpe, Pharmacia BioProcess Technology AB, for providing expansion and breakthrough data for the dense prototype beads described above.

References

Brummelhuis, H. G. J. In in *Methods in Plasma Protein Fractionation* ; Curling, J. M., Ed.; Academic Press: London, UK, **1980**, pp. 117-128.

Buijs, A.; Wesselingh, J. A. *J. Chromatog.* **1980**, 201, 319-327.

Burns, M.A.; Graves, D. J. *Biotech. Prog.* **1985**, 1(2), 95-103.

Draeger, N. M.; Chase, H. A. *I. Chem. Eng. Journ.* **1990**, Symp. Ser. 118, 161-172.

Chromatographic Tools for the Purification and Analysis of Biomolecules: The Next Generation

Noubar B. Afeyan, PerSeptive Biosystems, Inc., 38 Sidney Street, Cambridge, MA. 02139
(FAX: 617 621 2575)

Surface-mediated separations via chromatography are used for the purification and quantitative analysis of complex protein mixtures. Porous particles functionalized with the appropriate surface chemistries are used today. Speed, capacity and resolution together determine the overall separation performance and these are strongly related to the morphology of the porous particles.

Since the advent of Perfusion Chromatography™, that is, the use of flow-through particles to perform high throughput - high speed separations, the direct coupling between the three performance variables has been broken. New particles designed to perform at production scale, high performance and high speed separations using low pressure drop (peristallic pump) systems will be discussed. At 1000 cm/hr, a 10 cm bed of these particles generates less than 2 bar pressure drop while demonstrating chromatographic properties equivalent to operation at 50 - 10 cm/hr (10,000 plates/meter).

High speed separations also form the basis for a new technique, ImmunoDetection™, which combines antibody-based assays with a flow-through column format. The result is real time immunoassays performed in seconds or minutes. Application of Immunodetection for on-line monitoring of a product and contaminants in a bioprocess will be presented as a case study.

High throughput in downstream processing has allowed scaled-down operations, elimination of concentration (UF) steps, improved activity yield (due to residence time) and a dramatically different plant design and economics. These aspects will be explained with concrete case studies. Finally, prospects for automated multi-step chromatography with on-line monitoring and control, to purify a complex mixture to the requisite purity, will be presented.

1054–7487/92/0275$06.00/0

Industrial Perspectives on the Purification of Biological Molecules: Process Development and Scale-Up

Henry Helnsohn and Bryan Lawlis, Genentech Inc, 460 Point San Bruno Blvd., South San Francisco, CA 94080
(FAX: 415-266-1437)

In this presentation we will discuss process strategies for both industrial and pharmaceutical products.

There are fundamental economic differences in consumer and food industry products versus pharmaceutical products. These differences significantly influence process development strategies and ultimate implementation of processes in the plant. Typically, the former businesses, for economic reasons, require "spartan" approaches to processing, whereas the latter industries utilize a variety of highly resolving purification methodologies.

We will provide four examples of problems encountered in large scale processing of proteins produced by recombinant technology. Two of these examples are taken from commercial enzyme business and two will be from pharmaceutical processes. The strategy for solving problems encountered will vary significantly in approach and all will revolve around the choice of when and how to solve separation problems using chromatography.

Two examples drawn from the industrial enzyme business are the purification of subtilisin by a novel route and the processing of recombinant calf chymosin. In the production of subtilisin, because of the large quantities of material and low cost of final product, chromatography is difficult to justify economically. In this product a biopolymer which was introduced during fermentation could not be removed by conventionally inexpensive alternatives. Our approach to removal of this impurity employed one of the oldest existing technologies for purification - crystallization. Conditions and methods of processing by this unique approach at large scale will be discussed.

Calf chymosin is known to be one of the most valuable food enzymes on a $/G basis. Successful purification of this protein from recombinant sources could justify chromatography if used in a cost effective manner. Our original approach was to process material from recombinant fungi using hydrophobic interaction chromatography. Subsequently it was discovered that highly purified chymosin could be manufactured using liquid-liquid extraction. It was necessary to then separate product from added polyethylene glycol by conventional ion exchange chromatography.

Because of the value of the proteins produced, pharmaceutical recovery processes allow more latitude for the implementation of numerous and expensive chromatography steps in manufacture. In spite of this fact some problems are better solved by the simplest approach.

We provide one example of a proteolytic clip which was generated during fermentation. Methods were devised to resolve these highly similar species and subsequently through cooperation with fermentation development personnel, the proteolytic species was reduced significantly. This resulted in improved chromatography performance and yield.

Our last example of strategies to purification of pharmaceutical proteins involves studies directed toward optimizing resolving capacity in a chromatography unit operation. In this example the impact of backbone of the resin was investigated and it was discovered that strikingly different results could be obtained with a variety of resin backbones.

1054–7487/92/0276$06.00/0

Liquid-Liquid Extraction of a Recombinant Protein with a Reversed Micellar Phase

Mário J.L. Pires and **Joaquim M.S. Cabral** *
Laboratório de Engenharia Bioquímica
Instituto Superior Técnico, Av. Rovisco Pais Lisboa 1000 PORTUGAL
(FAX: 351-1-8480072)

Recombinant cytochrome b_5 was extracted to a reversed micellar phase of a cationic surfactant CTAB in cyclohexane / decanol. Depending on the experimental conditions - temperature, pH, and ionic strength, the migration to the organic phase is controlled by either hydrophobic (pH close to the pI and high ionic strength) or electrostatic (pH far from pI and low ionic strength) interactions.

The use of reversed micelles as a tool for separation of proteins is mainly based on trial and error experiments, despite some fundamental studies on the interaction of these molecules with several surfactants (Luisi, Giomini, Pileni and Robinson 1988). Usually a set of experimental conditions (pH, ionic strength and type of surfactant) is chosen that allows the solubilization of the protein in the water pool of the reversed micelle, via an electrostatic interaction with the charged heads of the surfactant molecule (Rahman, Chee, Cabral and Hatton 1988; Dekker Hilhorst and Laane 1989; Hatton 1989). However there are examples where the protein interacts with the hydrophobic alkyl chains of the surfactant, and therefore is extracted to the organic phase even at unfavourable pH (Aires-Barros and Cabral 1991; Camarinha Vicente, Aires-Barros and Cabral 1990). This work describes the extraction of a recombinant protein - cytochrome b_5 - to a reversed micellar phase of a cationic surfactant, and the influence of the operational conditions on the electrostatic or hydrophobic control of the interaction with the micellar aggregate.

Experimental Methods

1) Cytochrome b_5 Production and Purification

Cytochrome b_5 from rat was cloned into the plasmid pUC 13 and expressed in *Escherichia coli* TB1 (von Bodman, Schuller, Jollie and Sligar 1989). The protein is produced by fermentation in LB medium (Sigma) suplemented with ampicilin (100 μg/ml) (Boheringer Mannheim) in erlenmeyers of 2 l with one liter of culture, at 37 °C, pH 7.0 and agitation 250 rpm. After 16-20 hours of fermentation the cells were concentrated 10 times, up to 50-70 g/l, and sonicated with a Labsonic 2000 B.Braun at 200 W, 4°C, over 5 min. The cell debris were removed by centrifugation, and the cytochrome in the

1054–7487/92/0277$06.00/0

supernatant was purified in two chromatographic steps - an anionic exange step with a DE 52 (Whatman), and a gel filtration with Sephadex G 75 (Pharmacia), up to a ratio of A_{411}/A_{280} of 5.

2) Liquid-Liquid Extraction Experiments

Cytochrome b_5 was extracted to a reversed micellar phase of a cationic surfactant 100 mM CTAB (hexadecyltrimethylammonium) (Sigma) in an organic phase containing cyclohexane (p.a. Merck) and 10 % (v/v) decanol (p.a. Merck). Each extraction was performed in a jacketed vessel, at constant temperature, with agitation with a magnetic stirrer, for 5 min. Each phase had an initial volume of 2 ml. The phases were separated by centrifugation, and the protein determined with a spectrophotometer using the extinction coefficient at 411 nm (ε=130 mM^{-1} cm^{-1}). It was studied the effect of temperature, ionic strength (using different concentrations of KCl), and pH (using buffered protein solutions), on the extraction process.

Results and Discussion

Cytochrome b_5 is a small protein (13600 Da) with a low isoelectric point (pI 4.4). This protein can be extracted from an aqueous solution to a reversed micellar phase of 100 mM CTAB in cyclohexane/10% (v/v) decanol . The extraction depends on three main parameters - ionic strenght, pH and temperature. If this process is mainly due to an electrostatic interaction between the protein and the reversed micelle, it should be expected a strong dependence on the pH of the aqueous solution, as it controls the overall charge of the protein, and on the ionic strenght, since it controls the magnitude of this interaction. By controling these three parameters it was possible to observe a shift in the type of forces that are responsible for the solubilization of cytochrome b_5 in the micellar phase. Cytochrome b_5 was extracted at ranges of pH between 4.4 and 7.0, and ionic strenght between 0.1 and 1.0 M KCl. The results are shown in Fig.1. Since the protein precipitates at pH below 4.0, the extraction was only performed above the isoelectric point. It can be seen that at

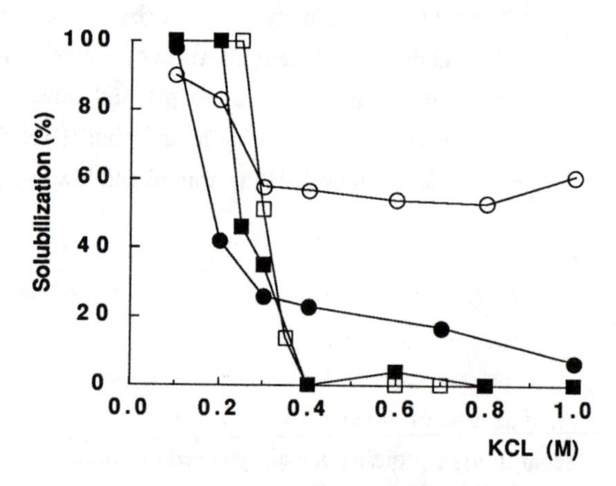

Fig. 1. Extraction of cytochrome b_5 to a micellar phase of 100 mM CTAB / cyclohexane/ 10 % (v/v) decanol at several pH and ionic strengths.(○-pH 4.4; ●-pH 5.0; ■-pH 6.0; □-pH 7.0)

pH far above the pI (6.0 and 7.0) the extraction is mainly governed by an electrostatic interaction with the positive surfactant. At high ionic strenght this interaction is damped and the extraction does not occur. On the other way, at pH close to pI the extraction seems to be the result of an hydrophobic interaction between the protein and the surfactant, as above 0.4 M KCl it is independent on the ionic strenght of the aqueous phase. The initial decrease in the extraction can be atributed to the dampening of electrostatic interactions that still exists at low ionic strength.

Since hydrophobic forces are strongly dependent on temperature (becoming less important at lower temperatures) the extraction of cytochrome b_5 was tested at several temperatures, between 10 and 40 °C. Two sets of experiments were designed at different ionic strengths, using 0.1 and 0.8 M KCl, and at pH values of 4.4 and 7.0. The results are shown in Fig. 2. As it can be seen, the extraction is only dependent on the temperature when the pH of the aqueous phase is close to the isoelectric point. As the temperature is lowered, the extraction of cytochrome b_5 markedly decreases. At higher ionic strenght (0.8 M KCl) the curve is shifted to the left, i.e., the extraction seems to be more afected by the temperature, which is in agreement with the results shown in Fig.1. The hydrophobic interaction between an acidic protein, as cytochrome b_5 and a cationic surfactant (CTAB) may be explained by the existence of an hydrophobic region at the cytochrome surface that can interact with the micellar agregate under conditions of pH close to pI and high ionic strength. This region was identified in calf cytochrome b_5, and was described as a hydrophobic patch which is surrounded by negatively charged residues with a total solvent exposed area of 350 Å^2 (Mathews and Czerwinsky 1984). The patch contains the hydrophobic groups Phe-35, Pro-40, Leu-70 and Phe-74. In the rat sequence Phe-74 is replaced Tyr-74 (Mathews and Czerwinsky 1984).

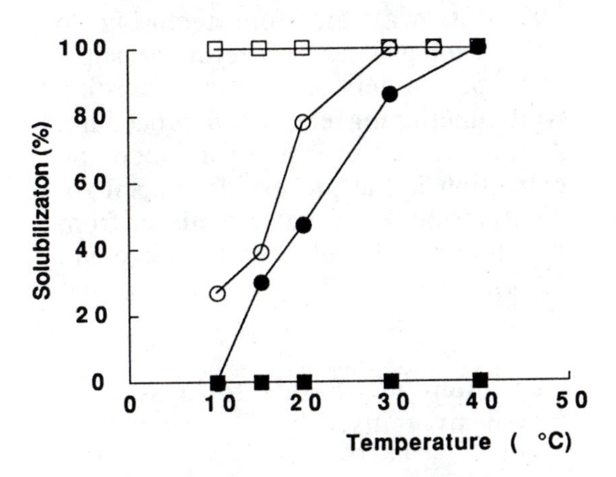

Fig. 2. Temperature effect on the extraction of cytochrome b_5 from an aqueous phase at pH 7.0 (❑,■) or pH 4.4 (O,●) with 0.1 (O,❑) or 0.8 (●,■) M KCl to a reversed micellar phase of 100 mM CTAB/cyclohexane/10 % (v/v) decanol.

Conclusions

The extraction and encapsulation of recombinant cytochrome b_5 with CTAB reversed micelles is controlled by two mechanisms, based on electrostatic or hydrophobic interactions, which depend on the operational conditions -temperature, pH and ionic strength.

References

Aires-Barros, M.R., Cabral, J.M.S. Biotech. Bioeng. **1991**, 38 1302-1307

von Bodman, S., Schuler, M., Jollie, D., Sligar, S. Proc. Natl. Acad.Sci. USA **1986**, 83, 9443-9447.

Camarinha Vicente, M.L., Aires-Barros, M.R., Cabral, J.M.S. Biotechnol. Tech. **1990**, 4, 137-142.

Dekker, M., Hilhorst, R., Laane, C. Anal. Biochem. **1989**, 178, 217-226.

Hatton, T.A. in Surfactant-Based Separations; Samehorn J.F. and Harwell, J.H. Eds. Marcel Dekker:New York. **1989** pp 55-59.

Luisi, P.L., Giomini, M., Pileni, M.P., Robinson, B.H., Biochim.Biophys.Acta **1988**, 947, 209-241.

Mathews, F.S., Czerwinski, E.W. in The Enzymes of Biological Membranes; Martonosi, A.N.,Ed; Plenum Press:New York and London **1984**, Vol 4; pp 235-300.

Rahman, R.S., Chee, J.Y., Cabral, J.M.S., Hatton, T.A. Biotechnol. Prog. **1988**, 4, 218-224.

Recovery of Fused Proteins by Liquid - Liquid Extraction

Sven-Olof Enfors*, **Kristina Köhler**, and **Andres Veide**
Department of Biochemistry and Biotechnology, Royal Institute of Technology, S-100 44 Stockholm (FAX: +46 8 7231890)

Genetic engineering can be used to modify protein products to control their partitioning in aqueous two-phase systems. Based on data from extraction of *E.coli* ß-galactosidase from cell homogenate, a continuous one-step extraction process was developed for ß-galactosidase fusion proteins in a PEG 4000/potassium phosphate system. Tryptophan is suggested to be the key amino acid in determining the high partition coefficient of the ß-galactosidase. Fusion of only six tryptophan residues to a 16 kD model protein (ZZ) increased the partition coefficient from 1.6 to 96.

Recovery and purification of proteins from cell suspensions or cell homogenates are mostly based on the combination of a large number of unit operations, which puts a large emphasis on the performance achieved in each step. The efficiency of each operation is traditionally optimised by varying different physico-chemical parameters to adjust the separation technique to fit the properties of the product and the impurities. However, genetic engineering techniques have now made it possible to go the other way around, and adjust the product to fit existing well-optimised separation techniques. This is achieved by modifying the amino acid content of a protein by addition, deletion or substitution of parts of the gene that codes for the target protein. So far, this principle has mainly been used to fuse proteins or amino acids to either terminal of the target protein. Examples are fusion of the target protein to IgG binding proteins for IgG affinity chromatography (Moks et.al., 1987), poly-his for immobilised metal (Ni) affinity chromatography (Hochuli et.al., 1988) and poly-Arg for cat-ion exchange chromatography (Sassenfeld and Brewer, 1984).

Aqueous two-phase systems have been well recognised in the scientific downstream processing literature, but relatively little industrial usage of this technique has so far been reported. There are several factors that may contribute to this, one of which is that this technique still depends much on trial and error methodology, even if systematic approaches now are gradually becoming available (Kuboi, et.al., 1991). It was therefore tempting to employ the genetic engineering principle to adopt a protein product to an existing well functioning extraction procedure. A candidate process for such an extraction is the process developed for purification of ß-galactosidase from disintegrated *E. coli* cells (Veide et.al., 1984).

Extraction of ß-galactosidase fusion proteins

Recovery and purification of ß-galactosidase from *E. coli* may be achieved by continuous one-stage extraction in a poly(ethylene)glycol 4000 - potassium phosphate system (Veide et.al., 1984). Since almost all impurities

1054–7487/92/0280$06.00/0

are partitioned to the salt-rich bottom phase, while the ß-galactosidase is recovered to the PEG phase with a partition coefficient around 20, any other proteins that could be partitioned to this PEG phase would be equally well processed in this system.

A number of *E.coli* ß-galactosidase fusion proteins has been investigated (Köhler et.al. 1991a). Fig 1 describes the principle composition and the partition coefficients of these proteins in a PEG 4000/potassium phosphate system. It is noteworthy that ß-galactosidase, which in this system had a partition coefficient of 17, still dominates over the model products SpA (part of the staphylococcal protein A) and SpA-StrpG (fusion between SpA and parts of streptococcal protein G) which both have partition coefficients of 0.7. According to Albertsson's theory (Albertsson,1982) the partition coefficient of a complex of two molecules should be equal to the product of the partition coefficients of the two proteins, provided the contact area between the two molecules is insignificant. Application of this model to the partitioning of the fusion proteins shown in Fig 1 would result in a partition coefficient of 4.1, since the ß-galacto-sidase is a tetramere and the gene fusion technique results in four units of fusion partner per unit of ß-galactosidase.The observed partition coefficients were slightly lower: 3.5 for SpA-ßgalacto-sidase and 2.8 for SpA-StrpG-ß-galacto-sidase, respectively, which may be explained if some of the ß-galactosidase surface is masked by the fusion partner.

However, even very small changes of the primary structure may have considerable effects on the partitioning. A change in the linker region between SpA and ß-galactosidase involving only a dozen amino acids of more than 1000 of the fusion protein was enough to change the partition coefficient (Köhler et.al., 1991).

Even larger sensitivity of ß-galactosidase fusion proteins to small changes in the primary structure has been reported (Andersson and Bailey,1987). Change of the five amino acids Arg-Ser-Arg-Ile-Pro to Leu-Gly-Ser-Pro-Ala in the N-terminal of an ompF-ß-

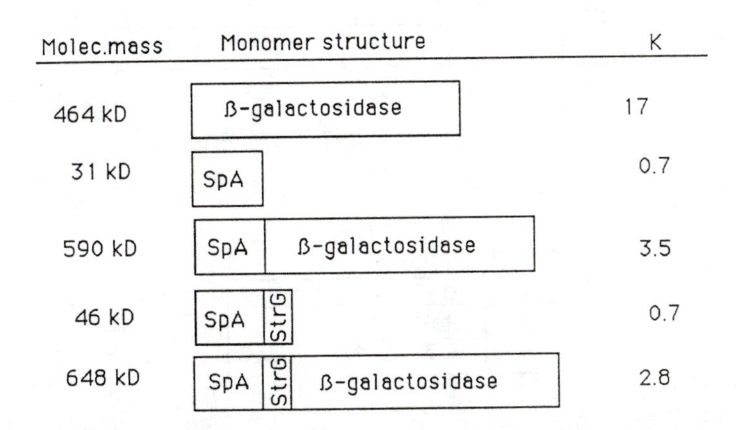

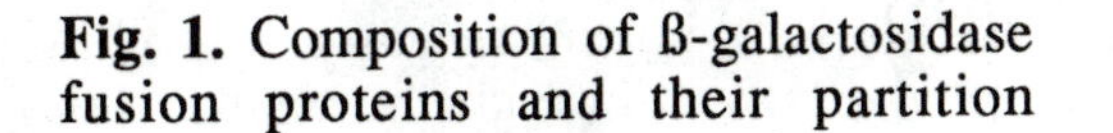
Fig. 1. Composition of ß-galactosidase fusion proteins and their partition coefficient in PEG 4000/ potassium phosphate.

galactosidase fusion protein with additional 1038 amino acids almost doubled the partition coefficient in a PEG/Dextran system from 5.7 to 10.4.

Tryptophan as partition ligand

The extreme partition coefficient of *E.coli* ß-galactosidase in the system applied should be attributed to some unique property of the protein. Unequal partitioning of a protein in an aqueous two-phase system is commonly ascribed a combination of several effects. Köhler et.al suggested that the behaviour of *E.coli* ß-galactosidase might be caused by relatively much tryptophan residues exposed at the protein surface (Köhler et.al. 1991b) . This was based on two main observations: Firstly, ß-galacto-sidase contains 3.45 times more tryptophan than the mean E.coli cytoplasmic protein (Huddleston et.al. 1991). Secondly, analysis of available literature data on partition coefficients of dipeptides (Diamond et.al., 1989) revealed that Trp tended to partition the peptides towards the PEG phase.

In order to test their hypothesis Köhler et. al. designed a series of fusion proteins containing increasing amounts of Trp-residues (Köhler et.al.,1991b). As model product was used an IgG binding derivative of staphylococcal protein A, the so called ZZ protein. The partition coefficient of this protein in 10.8% PEG 4000/ 13.3% potassium phosphate was 1.6. The tryptophan residues were inserted as multiples of the tetrapeptide Ala-Trp-Trp-Pro eight amino acids from the C-terminus of ZZ (Fig 2). The partition coefficient increased with increasing number of Trp-units up to 96, when three tetra-peptides (containing six Trp-residues) were employed.

As a control, a corresponding series of ZZ-fusion proteins were designed in which the unpolar amino acid isoleucine (Veide et.al., unpublished) was substituted for tryptophan (Fig 2). It is obvious from the partition coefficients shown in Fig 2 that the strong partitioning effect of the Trp-containing tetrapeptide is caused only by the two Trp- residues.

The mechanism of the strong Trp effect on protein partitioning is not

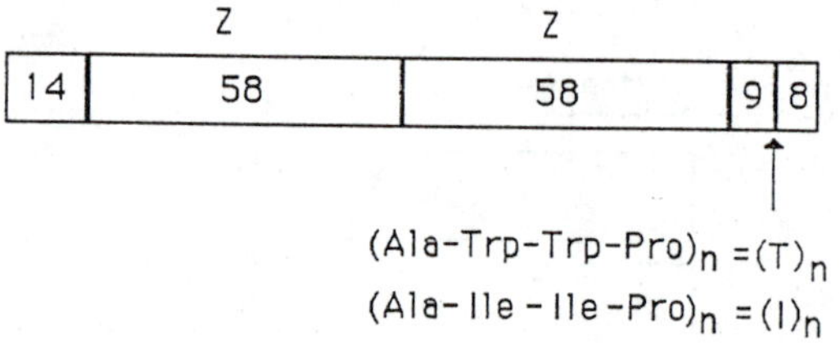

Z Z

14	58	58	9	8

$(Ala\text{-}Trp\text{-}Trp\text{-}Pro)_n = (T)_n$

$(Ala\text{-}Ile\text{-}Ile\text{-}Pro)_n = (I)_n$

Partition coefficients:

n=	0	1	3
$(T)_n$	1.6	12	96
$(I)_n$	2.3	4.8	4.7

Fig. 2. Composition of the ZZ-fusion proteins and their partition coefficient in 10.8% PEG 4000/13.3% potassium phosphate (Köhler, Veide and Enfors, 1991).

clear. Kuboi et.al (Kuboi et.al.,1991) have related the partition coefficients of several amino acids in a PEG/dextran system and found that the logarithm of the partition coefficient was directly proportional to the relative hydrophobicity (kJ/mol) of the amino acids. In this series of amino acids Trp was the most hydrophobic acid and it showed also the highest partition coefficient. However, the observed hydrophobicity dependence was not so dramatic as the effects caused by insertion of only six Trp-residues into the ZZ-protein. This fact indicates that other factors than pure hydrophobicity may be involved in the function of Trp as a partition ligand in PEG-salt systems (Huddleston et. al., 1991).

Concluding remarks

Fusion of only a few tryptophan units was sufficient to drastically increase the partition coefficient of a model protein in a phase system that contains very little of other *E. coli* cell constituents. If the phenomenon is general , fusion of a few tryptophan units to the product protein may be used for fast isolation of proteins from cell disintegrate for further purification, if required, by conventional chromatographic techniques. The extraction technique is then used mainly to circumvent the bottleneck of recovering the product from the viscous and particulate raw extract.The partition coefficient is, however, not the only parameter that may be altered by genetic manipulations of the protein structure. Effects on folding, proteolysis and, in case of secreted proteins, protein transport, must also be considered. In all cases positive as well as negative consequences may be observed when the protein structure is modified by genetic engineering techniques.

References

Albertsson, P.-Å. *Methods Biochem. Analysis* **1982**, *29*, 1-24.

Andersson, E.; Bailey, J. E. *Biotechnol. Techn.* **1987**, *1*, 207-212.

Diamond, A. H.; Xia, L.; Hsu, J. T. *Biotechnol. Technol.* **1989**, *3*, 271-274.

Hochuli, E.; Bannwarth, W.; Döbeli, H.; Gentz, R.; Stüber, D. *Bio/Technology* **1988**, *6*, 1321-1325.

Huddleston, J.; Veide, A.; Köhler, K.; Flanagan, J.; Enfors, S.-O.; Lyddiatt, A. *TIBTECH* **1991**, *9*, 381-388.

Kuboi, R.; Tanaka, H.; Komasawa, I. In *International Solvent Extraction Conference*; Elsevier: Kyoto, 1991; pp (in press).

Köhler, K.; Veide, A.; Enfors, S.-O. *Enzyme Microb. Technol.* **1991a**, *13*, 204-209.

Köhler, K.; Ljungquist, C.; Kondo, A.; Veide, A.; Nilsson, B. *Bio/Technology* **1991b**, *9*, 642-646.

Moks, T.; Abrahamse'n, L.; Österlöv, B.; Josephsson, S.; Östling, M.; Enfors, S.-O.; Persson, I.; Nilsson, B.; Uhle'n, M. *Bio/Technology,* **1987**, *5*, 379-382.

Sassenfeld, H. M.; Brewer, S. J. *Bio/Technology* **1984**, *2*, 77-81.

Veide, A.; Lindbäck, T.; Enfors, S.-O. *Enzyme Microb. Technol.* **1984**, *6*, 325-330.

Molecular Selectivity in Structured Fluids

T. Alan Hatton
Massachusetts Institute of Technology, Cambridge, MA 02139
(FAX: 617-253-8723)

Structured fluids, such as micelles, reversed micelles, and polymers in solution, can provide unusual solvating environments for biological molecules. This is illustrated using the example of protein and amino acid solubilisation in organic phases, hosted by reversed micellar aggregates in these phases.

Large-scale bioproduct separations using solvent extraction technology have been carried out effectively for antibiotics and carboxylic acids, using conventional organic solvents, and proteins, using aqueous two-phase polymer systems. More recently there has been interest in exploiting the aggregative properties of certain surfactants in solution for the selective extraction of proteins and amino acids. These "structured fluids" have unusual solvating properties in that they can accommodate both the hydrophobic and hydrophilic solvation requirements of amphiphilic species, such as the many biological molecules of current or potential commercial interest. Examples of structured fluids include reversed micelles, which are nanometer-scale water droplets stabilized by surfactants, and phase-separated micellar solutions. These latter systems are characterized by two aqueous phases, one rich in micelles, and the other substantially depleted of surfactant; they have many features in common with aqueous two-phase polymer systems, another example of structured fluids. This presentation covers reversed micellar systems only.

The factors governing molecular selectivity in these systems include not only the expected contributions from hydrophobic, dipolar and electrostatic interactions, but also, in the case of interfacially-associated solutes, steric effects. These latter effects can be manifested in thermodynamic terms through the interfacial bending energy of the curved surfactant interfaces.

Protein Solubilization

The solubilization of proteins in reversed micellar organic solutions has been demonstrated clearly to depend on electrostatic interactions between the charged proteins and the ionic surfactants used to stabilize the micellar structures. With anionic surfactants, for instance, the pH should be below the isoelectric point of the protein for there to be any significant solubilization, since only then does the protein have charge opposite that of the surfactant and do attractive interactions come into play. Size exclusion effects also play a role, as indicated both by the effect of salt type and concentration on the micelle size and protein solubilzation capacity, and by the correlation between the pH of optimal transfer, relative to the pI, and the protein molecular weight. The more asymmetric the charge distribution over the protein surface, the greater the propensity for solubilization.

Selectivity of reversed micelles for protein separations can be enhanced through the addition of suitable cosurfactants with affinity headgroups recognised by the proteins of interest. Examples include the enhancement of concanavalin-A transfer through the addition of alkyl glucosides to the reversed micellar solution, the interaction of myelin basic protein with lecithin (but not lysolecithin!), α-chymotrypsin with alkyl boronic acids, and avidin with biotin. The efficiency of this affinity extraction process is dependent on the strength ofthe association

1054–7487/92/0284$06.00/0

between the protein and the affinity group on the surfactant, as well as the accessibility of the headgroup when the cosurfactant is located within the reversed micelle wall.

Solubilization of Amino Acids and Other Amphiphilic Solutes

Comprehensive studies on the solubilization of amino acids and other low molecular weight amphiphiles point, not surprisingly, to the importance of the hydrophobic effect in determining the partitioning of the solute to the structured interface afforded by the surfactant monolayer. This has been demonstrated with over 40 amino acids, as well as with alkyl glucosides, carboxylic acids, alkyl boronic acids and alcohols. Charge effects can be accounted for through simple consideration of the Boltzmann factor.

It has been shown, however, that steric factors can also be important. As the size of the micelle is decreased, the interfacial partition coefficient changes. It has been argued successfully that this is due to steric effects as the surfactant interfacial curvature increases. Short amphiphiles are squeezed out of the interface because they sample only the headgroup region of the surfactants, which is more compressed as the curvature increases. With longer amphiphiles, the hydrophobic moiety samples increasingly the tail region of the surfactant interface, which opens up as the curvature increases, thereby providing additional configurational freedom for the solute tail. In this case, the partition coefficient increases with increasing interfacial curvature.

Protein Stability and Protein Folding

Rainer Jaenicke,
Institut für Biophysik und Physikalische Biochemie, Universität Regensburg, W-8400 Regensburg (Federal Republic of Germany)
(FAX: 0040 941 943 2813)

Proteins exhibit marginal *stability*, equivalent to only a few weak interactions. Expressed in ΔG_{stab} per residue, the free energy is below the level of thermal energy. Molecular adaptation to extreme conditions may be accomplished either by minute local structural changes or by "extrinsic factors" not encoded in the amino-acid sequence. No general strategies of stabilization have been established yet. *Protein folding* may be described as "hierarchical condensation", with the consecutive formation of secondary structure, domains and structured monomers. The rate-limiting steps (cysteine oxidation, proline isomerization, subunit association) are catalyzed or directed by enzymes or chaperones.

Fundamentals of Protein Stability

The structure and stability of proteins are determined by the amino acid sequence and the solvent environment. Both are still not well understood in their thermodynamics: the *structure*, because neither the mode nor the code of protein folding has yet been solved; the *stability*, because empirical energy functions are insufficient to make unequivocal predictions with regard to the effective potential energy surface of unknown proteins.

Principles determining the spatial structure of proteins in aqueous solution are (i) the minimization of hydrophobic surface area, and (ii) the distribution of non-polar residues in the interior, and polar residues on the surface of the molecule. Evidently, these principles have important implications for the subtle balance of the net attractive and repulsive interactions between the residues and the solvent that govern protein dynamics and stability (Jaenicke, 1987, 1991a).

Globular proteins in solution exhibit only marginal free energies of stabilization. The subtle balance of hydrogen bonds, hydrophobic interactions, ion pairs and van der Waals forces yields $\Delta G_{stab} \leq 60$ kJ/mol, independent of the mode of denaturation. Thus, on balance, the stability of globular proteins (even those from extremophilic organisms) is based on the equivalent of a few hydrogen bonds, hydrophobic interactions or ion pairs (Jaenicke, 1991b). In certain cases, disulfide bonds have been shown to enhance protein stability with ΔG_{stab} increments much larger than one would expect for the restricted increase in entropy of the denatured state. The formation of cystine bridges *in vivo* occurs cotranslationally. It is catalyzed by protein disulfide isomerase (PDI) which resides in the endoplasmic reticulum, where folding of secretory disulfide-bonded proteins is known to occur (Jaenicke, 1991a).

Structural Elements, Fragments, Domains and Subunits

Protein stability is accomplished by the cumulative effect of non-covalent interactions at many locations within a given molecule. With regard to the stability of fragments, it is well-known that protein domains exhibit intrinsic stability (Jaenicke, 1987). Reducing the chainlength below a certain limit, it becomes evident that proteins are cooperative structures showing mutual stabilization of structural elements. In order to find out at which fragment size domains lose their capacity to fold independently, thermolysin was used as a model. The N-terminal portion of the enzyme is found to stabilize the all-helical C-terminal domain. This may be shortened to the 62 residues three-helix structure, without losing much of the stability of the native enzyme (Vita et al., 1989). In considering the intrinsic stability of local structural elements, a minimum length of 15 residues may still sustain native-like structure (Scholtz & Baldwin, 1992).

Whether the N- and the C-terminal ends of the polypeptide chain are important for protein stability depends on the protein. Taking RNase and LDH as examples, it has been shown that the N-terminal ends of both proteins can be cleaved without altering the overall topology; however, the stability is greatly affected (Jaenicke, 1991c). In the case of lactate dehydrogenase, or cleaving off the N-terminal decapeptide allows the native tetramer to be dissociated to the dimer with drastically reduced stability; the monomer is only accessible as short-lived intermediate on the pathway of reconstitution. The separate ("nicked") domains are able to pair correctly in a topologically correct manner. Evidently, the "structured monomer" possesses a native-like tertiary structure, but strongly reduced stability (Opitz et al., 1987).

Physiological Stress and Protein Stability

Presently there is no way to design proteins with specific stability characteristics. Therefore, it seems useful to try to elucidate the long-term experiment that has been going on in nature, where adaptation to extremes of physical conditions has led to proteins with highly specified stabilities toward temperature, pressure, pH and low water activity. In comparing mesophiles and extremophiles, it has been shown that the range of viability is commonly shifted rather than broadened (Jaenicke, 1981). Focusing on homologous proteins from organisms taken from different extreme environments, it turns out that adaptation to extremes of physical conditions tends to maintain "corresponding states" regarding overall structure, flexibility and ligand affinity. Generally, evolution is geared to maintain optimum function in widely differing solvent environments. It does so with an amazingly high degree of conservatism with respect to both the "topology" of proteins and their constituent amino acids. Even proteins from hyperthermophiles do not contain other than the 20 natural amino acids as amide-linked polypeptides. The increase in overall stability is of the same order as mentioned before: $\Delta\Delta G_{stab}$ for a typical thermophilic protein is 50 kJ/mol, again only marginal compared with the inner energy which is of the order of 10^4 kJ/mol. Faced with this ratio, it is evident that there may be many ways to accumulate the $\Delta\Delta G_{stab}$ required to stabilize a protein; on the other hand, one would predict that it is highly improbable to uncover general rules of molecular adaptation (Jaenicke, 1991b).

Fundamentals of Protein Structure

In developing a general theory of the structure of biological systems, one has to describe how the intrinsic physical and chemical properties of the building blocks and their mutual interactions combine to determine (i) the energetics of the 3D array of matter, (ii) the kinetics of its formation, and (iii) the mechanism underlying the transition from disorder to order. In the case of biological systems, evolution has been optimizing the correlation of structure and function. J. Kepler, in contemplating the six-cornered snow flake, was the first to point to the fact that biological structure is caused by its mode of action. Structure formation of biomolecules is a spontaneous process which occurs autonomously, in close analogy to physical processes such as crystallization. In the case of proteins, the relationship between the amino acid sequence and the corresponding unique 3D structure is obvious from denaturation/renaturation experiments. Their "model" character for the situation *in vivo* seems questionable because renaturation refers to the complete polypeptide chain, whereas the nascent chain may be assumed to fold cotranslationally, i.e. from the N- to the C-terminal end . The fact that both the native protein and the product of renaturation are indistinguishable suggests that the renaturation indeed mimics *in vivo* folding. However, in this comparison, cell-biological factors such as local solvent conditions, cotranslational and posttranslational modification, transcriptional or translational control, chaperone action, etc. have been neglected. Except for the effects of protein concentration, viscosity and specific ligands, hardly any attempts have been made to investigate possible influences of the *in situ* conditions on the folding process. Codon bias might be significant in providing high yields of active expression; on the other hand, directionality, variation in chain elongation, and glycosylation cannot play a significant role (Jaenicke, 1988, 1991a; Kurland, 1991; Gething & Sambrook, 1992; Kern et al., 1992).

Hierarchical Condensation

Considering the hierarchy of protein structure, the kinetic mechanism of protein self-organization is not a strictly sequential reaction proceeding from secondary structural elements to the native tertiary and quaternary structure. The reason is that the tertiary contacts themselves select the correct local structures and stabilize them, such that local structures with low stability and high folding and unfolding rates are in rapid equilibrium early in folding. What seems clearly established is the observation that structure formation starts from next-neighbor interactions which, in the time range of milliseconds, generate stretches of secondary structure. Subsequently, non-local interactions lead to metastable supersecondary structures and subdomains which, still within fractions of a second, collapse into the native-like "molten globule state". Domains fold and unfold as independent entities. The corresponding shuffling processes occur by a limited number of pathways, with the rate-limiting steps as late events. If, at this level, there is still significant hydrophobic surface area exposed to the aqueous solvent, association will end up in the geometrically and stoichiometrically unique quaternary structure. Domain pairing and subunit recognition are highly specific, even in the presence of excess heterologous proteins (Jaenicke, 1987, 1991a; Opitz et al., 1987). At the highest level of the structural organization, intrinsic structure determination is restricted to simple regular multimeric systems. Complex assembly systems need multistep procedures to mimic the natural sequential process, which is obviously determined by the genome organization and/or helper proteins (Jaenicke, 1991a).

Folding Kinetics and Catalysis of Folding

The synthesis of a polypeptide chain of average length takes of the order of 1 min. The rate of in vitro folding varies in a wide range. In the case of small proteins without disulfide bridges reactivation after complete denaturation ranges from < 1 s (staphylococcal nuclease and aldolase) to 10 s (myoglobin). Thus, in these cases, translation is rate-determining, and folding and synthesis may be assumed to occur in a synchronized fashion. For small cross-linked proteins, such as (oxidized) ribonuclease, the same holds true. Reoxidation, however, takes >10 min. In this case, the common *in vitro* approach differs drastically from the *in vivo* process. This may be assumed to show parallel folding and cross-linking, with translation followed, (i) by (PDI catalyzed) crosslinking, (ii) translocation through the membrane, and (iii) secretion. *In vitro* kinetics have clearly proven that the characteristic features of the chain fold are formed early on the folding pathway (Staley & Kim, 1990; Weissman & Kim, 1991); proline cis-trans isomerization is the last (rate-limiting) step (Fischer & Schmid, 1990; Schmid et al., 1992).

In the case of large polypeptide chains, molecular chaperones seem to be involved in the inhibition of premature folding and/or association. The way how intermolecular protein-protein interactions promote correct intramolecular interactions is difficult to visualize, and presently, no mechanistic details regarding the mechanism of chaperones are known, except that GroE inhibits aggregation as the competitive side reaction of correct folding, and that proteins incapable of correct folding are trapped and finally subjected to degradation (Buchner et al., 1991).

Yield of Native Expression and Reconstitution

The yield of structure formation *in vivo* and *in vitro* is determined by the competition of folding and "wrong aggregation" according to

$$A \xleftarrow{k_2} U \xrightarrow{k_1} N$$

with N,U,A, as native, unfolded and aggregated states, and k_1 and k_2 as first- and second-order rate constants, respectively (Jaenicke, 1987; Kiefhaber et al., 1991). Generally, protein biosynthesis and *in vivo* structure formation is assumed to yield 100% native protein without requiring extrinsic factors or components. The following observations show that this assumption may not always be true: 1. Protein structure controls function as well as compartmentation and turnover. During secretion, misfolded and misassembled polypeptides are retained in the ER and specifically degraded. This "quality control" by the ER leads to the *apparent* yield of 100% (Hurtley & Helenius, 1989). 2. As shown for the phage P22 tailspike protein, even under optimum growth conditions of the bacterium, the yield of in vivo folding is $<50\%$; under unbalanced physiological conditions, wrong conformers are produced which are continuously removed by proteolysis (Seckler & Jaenicke, 1992). 3. Chaperones are involved in folding-association, regulating the yield in an ATP-dependent way (Ellis & van der Vies, 1991). 4. Overexpression of recombinant protein leads in most cases to wrong aggregation and deposition of the nascent protein in inclusion bodies (Kiefhaber et al., 1991).

Oligomers : Association

Early stages during the (re-)folding of oligomeric proteins are identical with those involved in the self-organization of single-chain proteins. They end up with native-like "structured monomers" (M) which in a subsequent reaction undergo association to form the native quaternary structure (M_n). Thus, the overall process may be written as a sequential uni-bimolecular folding-association reaction, with "wrong aggregation" ($n\mathcal{M} \rightarrow A$) as side reaction:

$$n\mathcal{M} \xrightarrow{k_1} nM \xrightarrow{k_2} M_n$$
$$\downarrow k_{>2}$$
$$A$$

Whether folding (k_1) or association (k_2) are rate-determining in the sequential uni-bimolecular reaction depends on the protein, as well as the state of denaturation (Jaenicke, 1987). As one would predict for the "hierarchical condensation", the formation of the native secondary structure is formed before domains or the complete tertiary and quaternary structure condense to minimize the solvent accessible surface area.

References

Scholtz, J.M.; Baldwin, R.L. *Annu. Rev. Biophys. Biomol. Struct.* **1992**, *21*, 95-118.

Buchner, J.; Schmidt, M.; Fuchs, M.; Jaenicke, R.; Rudolph, R.; Schmid, F.X.; Kiefhaber, T. *Biochemistry* **1991**, *30*, 1586-1591.

Ellis, R.J.; van der Vies, S.M. *Annu. Rev. Biochem.* **1991**, *60*, 321-347.

Fischer, G.; Schmid, F.X. *Biochemistry* **1990**, *29*, 2205-2212.

Gething, M.-J.; Sambrook, J. *Nature* **1992**, *355*, 33-45.

Hurtley, S.M.; Helenius, A. *Annu. Rev. Cell. Biol.* **1989**, *5*, 277-307.

Jaenicke, R. *Progr. Biophys. Mol. Biol.* **1987**, *49*, 117-237.

Jaenicke, R. In *Protein Structure and Protein Engineering*; Winnacker, E.L.; Huber, R. Ed; Colloq. Ges. Biol. Chem. Mosbach, **1988**, *39*, 16-36.

Jaenicke, R. *Biochemistry* **1991a**, *30*. 3147-3161.

Jaenicke, R. *Eur. J. Biochem.* **1991b**, *202*, 715-728.

Jaenicke, R. *Ciba Foundation Symp.* **1991c**, *161*, 206-221.

Kern, G.; Schülke, N.; Schmid, F.X.; Jaenicke, R. *Protein Science* **1992** in press.

Kiefhaber, T.; Rudolph, R.; Kohler, H.-H; Buchner, J. *Bio/Technology* **1991**, *9*, 825-829.

Kurland, C.G. *FEBS Lett.* **1991**, *285*, 165-169.

Opitz, U.; Rudolph, R.; Jaenicke, R.; Ericsson, L.; Neurath, H. *Biochemistry* **1987**, *26*, 1399-1406.

Schmid, F.X.; Mayr, L.; Mücke, M.; Schönbrunner, E.R. *Adv. Protein Chem.* **1992**, *43*, in press.

Seckler, R.; Jaenicke, R. *FASEB Journal* **1992**, in press.

Staley, J.P.; Kim, P.S. *Nature* **1990**, *344*, 685-688.

Vita, C.; Fontana, A.; Jaenicke, R. *Eur. J. Biochem* **1989**, *183*, 513-518.

Weissman, J.S.; Kim, P.S. *Science* **1991**, *253*, 1386-1393.

Effect of Polyanions on the Folding and Unfolding of Acidic Fibroblast Growth Factor

Henryk Mach, Carl J. Burke, David B. Volkin, Jonathan M. Dabora, Gautam Sanyal, and C. Russell Middaugh*, Merck Sharp & Dohme Laboratories, WP26–331, West Point, PA 19486 (FAX: 215–661–5299)

Acidic fibroblast growth factor (aFGF) depends critically on an interaction with polyanions for its structural integrity and activity. The stoichiometry of its interaction with heparin is found by light scattering and analytical ultracentrifugation studies to be one aFGF per 4-5 monosaccharide units. The growth factor is stabilized against urea-induced unfolding through interaction with heparin and the thermodynamics of this phenomenon are quantitatively described. In addition, the kinetics of the unfolding and refolding of aFGF have been characterized and the role of heparin in this process delineated.

Acidic fibroblast growth factor (aFGF) is unusual in its dramatic dependence on sulfated polyanions such as heparin for its stability and biological activity (Burgess and Maciag, 1989). It has been suggested as a consequence of extensive circumstantial evidence that interaction with extracellular matrix polyanions actually controls the activity of this growth factor *in vivo* (Vlodavsky et al., 1991). The nature of this interaction, however, is incompletely understood. We have therefore conducted a series of studies to quantitatively define the nature of aFGF/heparin complexes. The stoichiometry of aFGF binding to heparin is determined and the kinetics of the urea induced unfolding and refolding of the growth factor characterized. Urea was selected as the structural perturbant because it is non-ionic and will therefore not directly perturb the salt bridges responsible for the protein/polysaccharide interaction.

Stoichiometry of the aFGF/Heparin Interaction

The average number of molecules of aFGF that bind to a single heparin polymer (avg. $M_r \sim 16$kDa) was estimated by a combination of static and dynamic light scattering measurements. As the molar ratio of aFGF to heparin is increased, the ratio of the experimental scattering intensity to the intensity observed in the absence of any aFGF/heparin interaction rises until a saturating value of 15 is reached (Fig. 1). This value corresponds to a maximum stoichiometry of 15 aFGF/heparin chain since values are proportional to the molecular weight of the complexes. The scattering intensity then begins to decline as free aFGF molecules begin to appear in solution. A similar trend is seen when the mean hydrodynamic radius of the aFGF/heparin complexes are measured by dynamic light scattering (Fig. 1). Both types of measurement suggest a maximum binding stoichiometry of 14-15 aFGF molecules per heparin polymer. Similar experiments were conducted with a lower molecular weight form of heparin (~4.8kDa) and a stoichiometry of 4 aFGF per heparin polymer was found (not illustrated). Fitting these data to a simple stoichiometric model

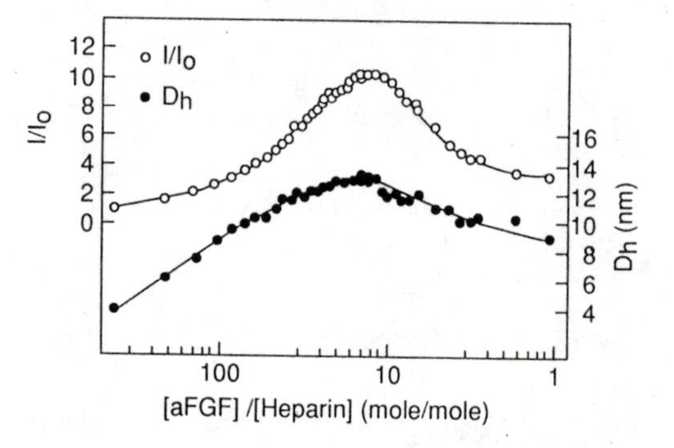

Fig. 1. Effect of the ratio of aFGF to heparin on the size of aFGF heparin complexes as measured by the ratio of the experiment scattering intensity (I) to that if no complex formation occurred (I_o) or by dynamic light scattering in the form of the hydrodynamic radius (D_h) determined from cumulant analyses of autocorrelation functions. All measurements were performed with 1 mg/ml aFGF in PBS at 6°C and 90° scattering angle with a Malvern 4700 spectrometer equipped with a Spectra Physics argon laser.

of aFGF/heparin binding yields average values of 14 and 4 moles of aFGF per high and low molecular weight heparin respectively under saturating conditions. Further analysis of these data suggests that approximately ten of these sites exhibit somewhat stronger interaction than the other 4. This corresponds to an aFGF binding site on heparin of 4-5 monosaccharide units per growth factor molecule, a surprisingly high density of aFGF molecules. Interestingly, tetrameric heparin fragments are the minimum size necessary to stabilize aFGF to thermal unfolding (Volkin et al., 1992). Furthermore, enzymatic digestion of heparin in the presence of aFGF also finds a tetrameric fragment as the smallest polysaccharide protected by the growth factor (manuscript submitted).

Partial quantitative confirmation of the light scattering results was obtained by equilibrium sedimentation experiments in which the distribution of the sedimenting species was monitored at 280 nm so that only the aFGF was detected. At weight (and ~molar) ratios of 0.3-1.0 aFGF to 16kDa heparin, molecular weights of 30-35kDa were observed, corresponding to 1:1 heparin/aFGF complexes. In contrast, at aFGF to heparin ratios of 3 and 10 to 1, a range of species containing 2-4 aFGF and 3-9 aFGF molecules respectively were observed. These values correspond well with those obtained by the light scattering measurements. Finally, measurement of the initial rate of reaction of the lysine residues of aFGF with o-phthalaldehyde reveals that maximum protection of lysine modification occurs at a 10:1 ratio of aFGF to heparin.

Equilibrium Unfolding of aFGF/Heparin Complexes

The unfolding free energy changes of aFGF at 4°C ($\Delta G°_{4°C}$) were examined as a function of heparin and NaCl concentration using urea as a structural perturbant and tryptophan fluorescence intensity (F) to detect the unfolding transition (Copeland et al., 1991). The results were analyzed by the two state linear extrapolation method employing a nonlinear least-squares fit to plots of $F_{350\ nm}/F_{300\ nm}$ vs. urea concentration as described by Santoro and Bolen (1988). The results of these experiments are summarized in Table 1 below. The NaCl experiments were conducted in the presence of a 3X weight excess of 16kDa heparin. It should be noted that aFGF begins to dissociate from heparin at NaCl concentrations between 0.2-0.5M.

Table 1. Unfolding Free Energy Changes for aFGF in the Presence of Varying Levels of Heparin (0.1M NaCl) and Salt (3X Heparin)at 4°C

aFGF +	$\Delta G°_{4°C}$ (kcal mol^{-1})
0X heparin	6.5 ± 0.1
0.1X heparin	4.5 ± 0.1
0.3X heparin	7.1 ± 0.2
1X heparin	8.6 ± 0.2
3X heparin	9.2 ± 0.5
10X heparin	8.9 ± 0.8
0M NaCl	7.7 ± 0.3
0.1M NaCl	9.2 ± 0.5
0.25M NaCl	9.8 ± 0.3
0.5M NaCl	8.4 ± 0.1
1M NaCl	8.9 ± 0.2

The $\Delta G°_{4°C}$ values of aFGF unfolding show a minimum at high aFGF/heparin ratios and independence from the effect of heparin as fewer aFGF molecules are bound per polysaccharide polymer. This suggests that aFGF initially binds to sites that strongly stabilize it and then to less stabilizing locations as the aFGF molecules begin to saturate the heparin polymer. The effect of sodium chloride on the unfolding of the growth factor as it is bound to these strong sites is relatively minor, although a small amount of destabilization is apparent at lower ionic strengths, presumably due to the lack of Debye-Hückel type charge shielding.

Kinetics of Unfolding

The kinetics of the urea-induced unfolding of aFGF were also investigated by monitoring changes in the intrinsic tryptophan fluorescence of the protein. It was assumed that aFGF can in principle be unfolded through one or both of two pathways involving aFGF in either a liganded or free state. This simple model predicts that if the dissociated form of the growth factor is unfolded, rates of unfolding should be heparin concentration dependent. Conversely, if just heparin associated aFGF unfolds, rates should be independent of heparin concentration as long as there is sufficient polysaccharide to complex all of the aFGF present. To attempt to differentiate between these two possibilities, the rate of urea induced unfolding of aFGF was measured as a function of heparin concentration. Data were analyzed in the form of plots of the reciprocal of the unfolding rate constant versus the ratio of heparin to aFGF (Fig. 2). At low heparin/aFGF ratios (0-5:1), the rate of refolding is strongly dependent on the amount of heparin present. Additional heparin above a 20:1 ratio, however, does not affect unfolding. Thus both pathways of unfolding contribute to the denaturation of aFGF in the presence of heparin.

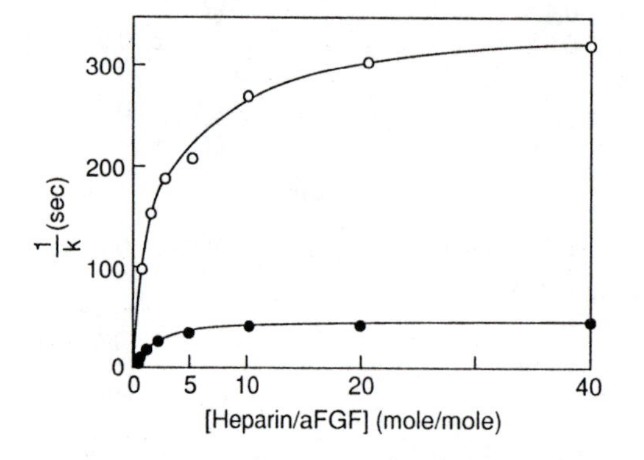

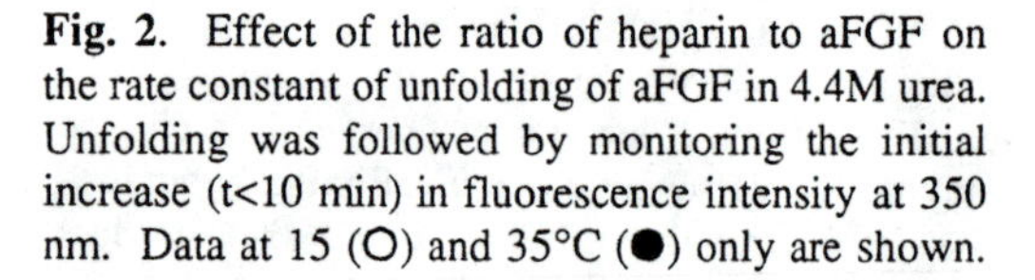
Fig. 2. Effect of the ratio of heparin to aFGF on the rate constant of unfolding of aFGF in 4.4M urea. Unfolding was followed by monitoring the initial increase (t<10 min) in fluorescence intensity at 350 nm. Data at 15 (O) and 35°C (●) only are shown.

Without bound heparin, the unfolding follows first order kinetics with a half-life of few seconds at ambient temperature and an apparent activation energy of 22.3 kcal/mole. Unfolding of heparin-liganded aFGF is approximately 150-fold slower with a decreased activation energy of 16.2 kcal/mole. The rate-limiting step of the unfolding process of the heparin-bound aFGF appears to have a smaller enthalpy of formation (15.6 vs. 21.7 kcal/mole) but larger entropy (TdS of -19.7 vs. -16.7 kcal/mole). The overall Gibbs free energy difference is only slightly smaller for heparin-bound aFGF (35.3 vs. 38.3 kcal/mole). Since the dramatic stabilization of aFGF by heparin does not appear to originate from an increased free energy of unfolding, the actual cause remains obscure. A model quantitatively linking affinity and kinetic criteria can be constructed and will be reported elsewhere. The dissociation constant found from this analysis is approximately 20nM throughout the 15 to 35°C temperature range examined. This suggests that the free energy change due to heparin binding is predominantly entropic and supports the electrostatic nature of this interaction. This was further supported by examining the effect of ionic strength on the aFGF/heparin interaction. The model of Pitzer (1977) was used to obtain binding constants in the absence of ions and the product of the number of charges involved in the protein/polysaccharide complex. An excellent fit to this model was obtained and a value for the K_d at zero ionic strength of 11pM was obtained. The apparent product of the charges of the ions involved in the complex was found to be 22. If we assume monovalent charges and a one to one correspondence in charge pairing, this suggests the existence of approximately 5 ion pairs in the complex. This can be compared to the 8 potential charge pairs that could be formed by the minimum tetrameric heparin unit necessary to completely stabilize aFGF.

Refolding

The same fluorescence changes employed above were used to follow the refolding of aFGF. Protein in 2M guanidine-HCl solutions was diluted to 0.1M Gdn·HCl and the single exponential kinetics measured (Dabora et al., 1991). In the absence of heparin, the protein was found to renature at temperatures up to 30°C with an activation energy of ~14 kcal/mole. No refolding was observed at higher temperatures. In the presence of heparin, however, the protein now refolds between 30 and 50°C with an activation energy of only 1.4 kcal/mole. The heparin has only a slight diminishing effect on the activation energy in the 0-30°C region (E_a ~ 10.3 kcal/mole). Thus, heparin or some other polyanion appears to be absolutely necessary for aFGF to fold into its native conformation at physiological temperatures.

Summary

One aFGF molecule appears to bind every four or five monosaccharides on a heparin polymer. Most bind with relatively high affinity, although evidence for some weaker binding sites is also found. These results correlate with the stabilization of aFGF to urea induced unfolding. The weaker binding sites (and correspondingly less stable protein) may also reflect the presence of protein-protein interactions between the bound aFGF molecules. The kinetics of unfolding suggest that the free protein unfolds in a few seconds but that the rate of denaturation of the heparin complexed protein is approximately 150-fold slower. A simple model predicts that ~5 ion pairs are formed by aFGF as it binds to heparin. Folding of the protein absolutely requires the presence of a polyanion such as heparin above 30°C. Thus, the folding and unfolding of aFGF and consequently the activity of the growth factor is potentially controlled through its interaction with polyanions.

References

Burgess, W. H.; Maciag, T. *Ann. Rev. Biochem.* **1989**, *58*, 575-606.

Copeland, R. A.; Ji, H.; Halfpenny, A. J.; Williams, R. W.; Thompson, K. C.; Herber, W. K.; Thomas, K. A.; Bruner, M. W.; Ryan, J. A.; Marquis-Omer, D.; Sanyal, G.; Sitrin, R. D.; Yamazaki, S.; Middaugh, C. R. *Arch. Biochem. Biophys.* **1991**, *289*, 53-61.

Dabora, J. M.; Sanyal, G.; Middaugh, C. R. *J. Biol. Chem.* **1991**, *266*, 23637-23640.

Pitzer, K. S. *Acc. Chem. Res.* **1977**, *10*, 371-377.

Santoro, M. M.; Bolen, D. W. *Biochemistry* **1988**, *27*, 8063-8068.

Vlodavsky, K.; Fuks, Z.; Ishai-Michaeli, R.; Bashkin, P.; Levi, E.; Korner, G.; Bar-Shavit, R.; Klagsbrun, M. *J. Cell Biochem.* **1991**, *45*, 167-176.

Volkin, D. B.; Tsai, P. K.; Dabora, J. M.; Middaugh, C. R. **1992**, *this volume*.

Folding of Pancreatic Trypsin Inhibitor

Peter S. Kim, Howard Hughes Medical Institute, Whitehead Institute for Biomedical Research, Department of Biology, M.I.T., Cambridge, MA 02142
(FAX: 617 258 5737)

Bovine pancreatic trypsin inhibitor (BPTI) is a small protein that has been used for extensive studies of protein folding (Creighton, 1978). We have designed, synthesized and characterized by two-dimensional nuclear magnetic resonance spectroscopy (2D-NMR), peptide models corresponding to two intermediates that have a crucial role early in the folding of BPTI (Oas et al., 1988; Staley et al., 1990). In both cases, we find that the peptide models fold as a stable subdomain, containing native-like elements of secondary and tertiary structure (Oas et al., 1988; Staley et al., 1990). These results suggest that subdomains are key determinants in protein folding, and that the predominant source of cooperativity is native tertiary interactions.

On the other hand (Creighton, 1978), earlier studies of the folding of BPTI revealed two features indicating that nonnative interactions have an important and specific informational role in protein folding. First, there were specific nonnative species (that is, molecules containing disulfide bonds not found in the native protein) that were populated at high levels in kinetic folding experiments. Second, two of the well-populated nonnative species had a critical role in guiding BPTI to the final native state.

We have reexamined the folding of BPTI, taking advantage of developments in separation technologies that have occurred since the original BPTI folding studies were performed. We find that only intermediates with native disulfide bonds are well populated during folding (Weissman et al., 1991). In addition, our kinetic studies indicate that the salient features of BPTI folding, including the rate-limiting disulfide rearrangement step, are determined in large part by native structure in these well-populated folding intermediates (Weissman et al., 1991). Thus, native interactions play a predominant role in the folding of BPTI.

This conclusion is supported by the demonstration that the native fold of BPTI can be obtained without nonnative disulfide species. BPTI containing only the 5-55 disulfide bond (with all other Cys residues replaced by Ala, so that nonnative disulfide bonds cannot form) is folded completely into a stable (melting temperature of ~40 °C) native conformation as determined by 2D-NMR (Staley et al., 1992). This molecule, denoted $[5\text{-}55]_{Ala}$, also is a functional trypsin inhibitor (Staley et al., 1992). The finding that $[5\text{-}55]_{Ala}$ is an essentially completely folded protein provides an explanation for many of the thermodynamic and kinetic properties of the [5-55] intermediate in the folding of BPTI.

References

Creighton, T. E. *Prog. Biophys. Mol. Biol.* **1978**, *33*, 231.
Oas, T. G. ; Kim, P. S. *Nature* **1988**, *336*, 42.
Staley, J. P.; Kim, P. S. *Nature* **1990**, *344*, 685.
Staley, J. P.; Kim, P. S. *Proc. Natl. Acad. Sci. USA* **1992**, *89*, 1519.
Weissman, J. S.; Kim, P. S. *Science* **1991**, *253*, 1386.

Folding for Profit: Renaturation of Recombinant Proteins from Inclusion Bodies as a New Downstream Process

Rainer Rudolph, Boehringer Mannheim GmbH, Nonnenwald 2, D-8122 Penzberg, Germany
(FAX 49-8856-602673)

With the advent of recombinant DNA technology, a practical application of protein folding emerged in the downstream processing of recombinant proteins. Upon over-expression of heterologous proteins in microorganisms or lower eucaryotes the natural folding pathway is frequently perturbed. Instead of the native, soluble protein large, inactive aggregates accumulate within the host cell. These inclusion bodies can be converted to active product by including *in vitro* renaturation in the downstreaming protocol.

Introduction

The cloning of foreign genes in host cells and the expression of the desired polypeptides in large quantities seemed to guarantee an unlimited supply of proteins of low natural availability. Host cells such as the Gram-negative microorganism *Escherichia coli* are easy to manipulate and easy to grow in large fermenters using inexpensive culture media. However, it soon became clear that expression of a polypeptide did not necessarily include the formation of the native, active protein (Itakura et al., 1977). For a large number of proteins the correct folding pathway is perturbed upon heterologous expression resulting in the formation of inactive, insoluble aggregates. However, aggregation is not restricted to proteins which are heterologous to the transformed cell (for review see Marston, 1986).

In order to obtain native, active proteins from inclusion bodies, *in vitro* folding processes must be included in the downstreaming protocol. Due to this industrial application, protein folding which had been an area of "pure " research, became a major topic in biotechnology. Both aspects of "Protein folding for pleasure and for profit" have been summarized by Pain (1987).

Over the last decades a large amount of information on the *in vitro* folding for pleasure has accumulated (for review see Jaenicke, 1991). Although a detailed description of folding pathways is still missing, it has been established that folding occurs through a succession of a definite number of intermediate conformational states. Parts of the secondary structure are formed rapidly during refolding. Subsequent slow steps which involve the reorganization of tertiary contacts are often limited by reshuffling of incorrectly formed disulfide bonds and/or by proline isomerization. Additional rate-limiting steps may be domain pairing and, in the case of oligomeric proteins, slow association of subunits (Teschner et al., 1987, Jaenicke & Rudolph, 1986).

Despite this wealth of information very little is known about the mechanisms responsible for the low yield of renaturation often encountered upon *in vitro* folding. Furthermore, it is a common observation that the yield of renatured protein decreases (with a concomitant increase of inactive aggregates) upon increasing the concentration of the protein to be renatured (Zettlmeissl et al., 1979). For industrial purposes, renaturation at low protein concentrations would necessitate large reaction volumes. Therefore, new methods had to be designed for the renaturation of recombinant proteins from inclusion bodies in order to ascertain a high yield of recovery at high protein concentrations (Rudolph, 1990).

1054–7487/92/0295$06.00/0

Recovery of Active Proteins From Inclusion Bodies

Cell lysis and isolation of the inclusion bodies

Inclusion bodies are generally compact particles, which can be easily collected by centrifugation of the cell lysate.

Inclusion body formation, which is usually considered to be a nuisance, has certain advantages: The desired product is enriched by simple liquid/solid separation at an early stage of downstream processing. In addition, inclusion body formation may help to prevent degradation of the expressed protein by intracellular proteinases.

Solubilization of the inclusion bodies

Inclusion bodies are not a paracrystalline array of correctly structured protein subunits. Therefore, solubilization requires the addition of rather drastic denaturants which were previously applied for the disintegration of "irreversibly" aggregated protein produced by heat or acid precipitation (Rudolph et al., 1979). Usually 6 M guanidinium chloride (GdmCl) or 8 M urea are used as denaturants. GdmCl is preferable to urea, which may contain cyanate leading to carbamylation of free amino groups of the polypeptide, especially upon long term incubation at alkaline pH values (Hagel et al., 1971). Since inclusion body isolates usually contain a certain amount of interchain disulfide bonds, a reducing agent should be included in the solubilization buffer.

Renaturation: Formation of correct disulfide bonds

Folding proteins with concomitant disulfide bond formation includes both the regeneration of the native, non-covalent interactions and the formation of covalent chemical bonds. The number of possible disulfide bridges increases rapidly with the number of cysteine residues.

Reoxidation of protein disulfide bonds is preferably performed by dilution of the solubilized, reduced inclusion body protein in "oxido-shuffling" buffer systems (Wetlaufer, 1984). Reduced and oxidized glutathion (GSH, GSSG) are commonly used as "oxido-shuffling" reagents, although other low molecular weight thiols may also be used. These reagents increase both the rate and the yield of protein renaturation/reoxidation by facilitating the reshuffling of incorrect disulfide bonds. However, regeneration of protein disulfide bonds *in vitro* is often hampered by practical difficulties such as instability or insolubility of the reduced polypeptide chains. These difficulties may often be eliminated by reversible chemical modification of protein thiols with concomitant introduction of charged residues. As an example reversible chemical modification may be achieved by the formation of mixed disulfides with glutathion (Rudolph, 1990).

Renaturation: Folding vs. aggregation

The pathway of correct folding (including disulfide bond formation) has been found to be in kinetic competition with the formation of inactive aggregates (Zettlmeißl et al., 1979; Kiefhaber et al., 1991). Aggregation may originate from non-specific (hydrophobic) intractions at the level of the unfolded polypeptide chain, as well as from incorrect domain interactions of partially structured folding intermediates (Goldberg et al., 1991). In both cases second (or higher) order aggregation reactions compete with first order folding steps. Consequently, aggregation predominates during renaturation above a limiting concentration of denatured protein.

However, high yields of renaturation per volume can be obtained provided that the actual concentration of denatured polypeptides is kept below a critical level (Rudolph & Fischer, 1990). This may be achieved by slow and continuous, as well as by discontinuous addition of denatured protein into the renaturation solution.

In certain cases, the yield of renaturation may be improved by the presence of non-denaturing concentrations of chaotrophic agents. Screening of a variety of compounds revealed that other additives such as alkylureas or carbonic acid amides can also be used to improve folding. Upon renaturation of human tissue-type plasminogen activator obtained from *E.coli* inclusion bodies, L-arginine, which has not been previously described as a "folding enhancer", induced a dramatic increase of the yield of renaturation (Rudolph,

1990). A similar effect of this additive was observed upon *in vitro* folding of a recombinant antibody Fab fragment (Buchner & Rudolph, 1991).

Outlook

Recently, helper proteins termed molecular chaperones, have been discovered which are involved in the correct folding and association of polypeptide chains both *in vitro* and *in vivo* (for review see Ellis & van der Vies, 1991). Coexpression of these molecular chaperones to obtain active, soluble protein from transformed microorganisms may become a major topic in the future (Gatenby et al., 1990).

At present, low molecular weight chemical reagents are used to facilitate *in vitro* folding for industrial purposes. These additives may be used to "chaperone" the correct folding pathway by reducing the formation of incorrect aggregates.

References

Buchner, J.; Rudolph,R. *Bio/Technology* **1991,** *9,* 157-162.

Ellis, R.J.; van der Vies, S.M. *Annu. Rev. Biochem.* **1991**, *60,* 321-347.

Gatenby, A.A.; Viitanen, P.V.; Lorimer, G.H. *Trends Biotechnol.* **1990**, *8*, 354-358.

Goldberg, M.E.; Rudolph, R.; Jaenicke, R. *Biochemistry* **1991**, *30,* 2790-2797.

Hagel, Pl; Gerding, J.J.T.; Fieggen, W.; Bloemendal, H. *Biochim. Biophys. Acta* **1971**, *243,* 366-373.

Itakura, K.; Hirose, T.; Crea, R.; Riggs, A.D.; Heyneker, H.L.; Bolivar, F.; Boyer, H.W. *Science* **1977**, *198*, 1056-1063.

Jaenicke, R. *Biochemistry* **1991**, *30*, 3147-3161.

Jaenicke, R.; Rudolph R. *Meth. Enzymol.* **1986**, *131*, 218-250.

Kiefhaber, T.; Rudolph, R.; Kohler, H.-H; Buchner, J. *Bio/Technology* **1991**, *9*, 825-829.

Marston, F.A.O. *Biochem. J.* **1986;** *240*, 1-12.

Pain, R. *TIBS* **1987**, *12,* 309-312.

Rudolph, R. In *Modern Methods in Protein and Nucleic Acid Research*; Tschesche H. Ed., Walter de Gruyter; Berlin, New York **1990**, pp 149-171.

Rudolph, R.; Zettlmeissl G.; Jaenicke, R. *Biochemistry* **1979**, *18*, 5572-5575.

Rudolph, R.; Fischer, S. **1990**, *US-Patent* 4 933 434.

Teschner, W.; Rudolph, R.; Garel, J.-R. *Biochemistry* **1987**, *26*, 2791-2796.

Wetlaufer, D.B. *Meth. in Enzymology* **1984**, *107,* 301-304.

Zettlmeissl, G.; Rudolph, R.; Jaenicke, R. *Biochemistry* **1979**, *18,* 5567-5571.

The Effect of Polyanions on the Stabilization of Acidic Fibroblast Growth Factor

David B. Volkin,* P.K. Tsai, Jonathan M. Dabora and **C. Russell Middaugh**, Merck Sharp & Dohme Research Laboratories, WP26-331, West Point, PA 194896 USA

The successful formulation of acidic fibroblast growth factor (aFGF) as a drug substance requires the stabilization of a markedly labile protein that partially unfolds under physiological conditions. Heparin is known to stabilize this polypeptide growth factor. Through the biophysical characterization of the thermal denaturation of aFGF in the presence of unmodified, fragmented and chemically derivatized heparins, the structural elements of this heterogeneous sulfated polysaccharide (size and sulfation level) required to stabilize aFGF are defined. A novel competitive binding assay is described as a convenient technique to examine the interaction of polyanions with aFGF. The specificity of the aFGF polyanion binding site is shown to be surprisingly weak with a wide variety of well defined, small sulfated and phosphorylated compounds also able to induce stability.

Acidic fibroblast growth factor (aFGF) is a potent mitogen for a wide variety of cell-types (Burgess and Maciag, 1989). The receptor-mediated biological activity of aFGF has generated considerable excitement and interest for potential clinical applications such as the acceleration of wound healing (ten Dijke and Iwata, 1989). Acidic FGF is a 15.9 kDa protein that is usually isolated by its affinity for heparin. This sulfated polysaccharide also stimulates aFGF mitogenic activity and enhances its stability against pH, temperature and proteolytic stress (Burgess and Maciag, 1989). In the absence of heparin, aFGF is extremely unstable and in the purified form is partially unfolded under physiological conditions (Copeland et al., 1991).

The successful formulation of aFGF into a drug substance requires defining conditions (shelf-life, storage temperature and dosage form) for the maintenance of the therapeutic protein without physical degradation and subsequent loss of biological activity. The nature of the stabilization of aFGF by ligands such as heparin is an important first step in understanding the potential mechanisms of aFGF inactivation. In particular, the specificity of the interaction of aFGF with polyanions is a crucial parameter in the design of either aqueous or lyophilized formulations.

Stabilization of aFGF by Heparin

Fluorescence spectroscopy is a convenient method to monitor the structural integrity of aFGF in solution as a function of temperature. Upon thermal denaturation of the protein, the wavelength emission maximum shifts from 309 to 350 nm (excitation 280 nm) due to unquenching of the molecule's single tryptophan residue (Copeland et al., 1991). As shown in Table 1-A, aFGF unfolds at 28°C in a PBS buffer, pH 7.2. As the weight amount of heparin ($M_r \sim 16{,}000$) is increased from 1/100X to 1/3X, aFGF is dramatically stabilized against thermal denaturation with T_m values increasing to 59°C. Heparin levels above this amount produce only small further increases in aFGF stability (e.g. 3X heparin, T_m = 61°C). The amount of heparin required to reach one-half the T_m maximum for aFGF is 0.1X (by weight). The stoichiometry and thermodynamic parameters of the interaction of aFGF with heparin are discussed further in this volume (Mach et al., 1992).

Heparin is an inherently heterogenous substance. Therefore, to better understand the ligand induced stabilization of aFGF, a series of well defined, enzymatically prepared fragments of heparin and chemically modified heparins were examined by fluorescence spectroscopy. As shown in Table 1-B, a tetrasaccharide is the smallest unit of heparin capable of stabilizing aFGF against thermal denaturation. Similar experiments with chemically modified heparins show that as the sulfation level of the derivatized heparin decreases, the thermal stability (T_m) of aFGF is diminished. For example, as shown in Table 1-C, heparin itself (13.5%S) significantly stabilizes aFGF (T_m = 61°C) while completely desulfated, N-acetylated heparin (1.5%S)

Table 1. The Effect of Polyanions on the Thermal Denaturation Temperature (T_m) of aFGF as Measured by Fluorescence Spectroscopy. (A) Heparin Concentration; (B) Well Defined Heparin Fragments; (C) Chemically Modified Heparins; (D) Various Polyanions. Experiemnts were Performed at 100 µg/ml Protein (6.3 µM) in PBS Buffer at pH 7.2.

	Ligand (Ligand:Protein Ratio (wt))	Tm (°C)
(A)	Heparin (M_r ~ 16,000)	
	Buffer alone	28
	1:100	31
	1:33	34
	1:10	43
	1:6	56
	1:3	59
	3:1	61
(B)	Heparin Fragments (1:1)	
	Disaccharide	30
	Tetrasaccharide	54
	Hexasaccharide	57
	Octasaccharide	50
	Decasaccharide	56
	Low MW heparin (M_r ~ 5000)	60
(C)	Chemically Modified Heparin (3:1)	
	Unmodified heparin (13.5%S)	61
	Desulfated heparin (1.5%S)	28
	Partially desulfated heparins (4.5-8%S)	49
	Epoxy-heparin (N.P.)	57
(D)	Polyanions (3:1)	
	Sulfated β-cyclodextrin	57
	Sucrose octasulfate	58
	Inositol hexasulfate	50
	Phytic acid	61
	Tetrapolyphosphate	50
(E)	Phosphorylated Inositols (3:1)	
	P_1	28
	P_2	31
	P_3	42
	P_4	51
	P_6	61

has no effect on aFGF stability (T_m = 28°C). Thus both the size (tetrameric fragment or larger) and sulfation level define heparin's ability to enhance the stability of aFGF.

Interaction of Polyanions with aFGF

In order to further investigate the specificity of the interaction of aFGF with unmodified and chemically derivatized heparins, a fluorescence based competitive binding assay was developed. When the antitryptanosomal agent suramin (a symmetrical polysulfonated naphthylurea) is incubated with aFGF, the drug induces reversible microaggregation of the growth factor and an increase in suramin fluorescence is observed. This enhancement is due to the direct binding of the drug to aFGF (manuscript submitted). Incremental addition of ligands such as heparin to this complex results in the loss of fluorescence enhancement as the heparin competes with suramin for the polyanion binding site. For example, as increasing amounts of heparin are added to a suramin/aFGF solution (see Figure 1), the enhancement of suramin fluorescence is lost with an IC50 value of 9 µg/ml heparin. (IC50 = concentration of ligand that results in the loss of one-half of the response.) This IC50 value corresponds to a molar ratio of 1:10 (ligand:protein) which is similar to the ratio of aFGF to heparin observed for one-half the T_m maximum during thermal denaturation experiments (see above). Similar experiments were then carried out with chemically modified heparins. As shown in Table 2, IC50 values increased as the percent sulfation decreased. These results directly correlate with the thermal denaturation experiments, i.e. increasing heparin sulfation results in enhanced interaction and stabilization of aFGF.

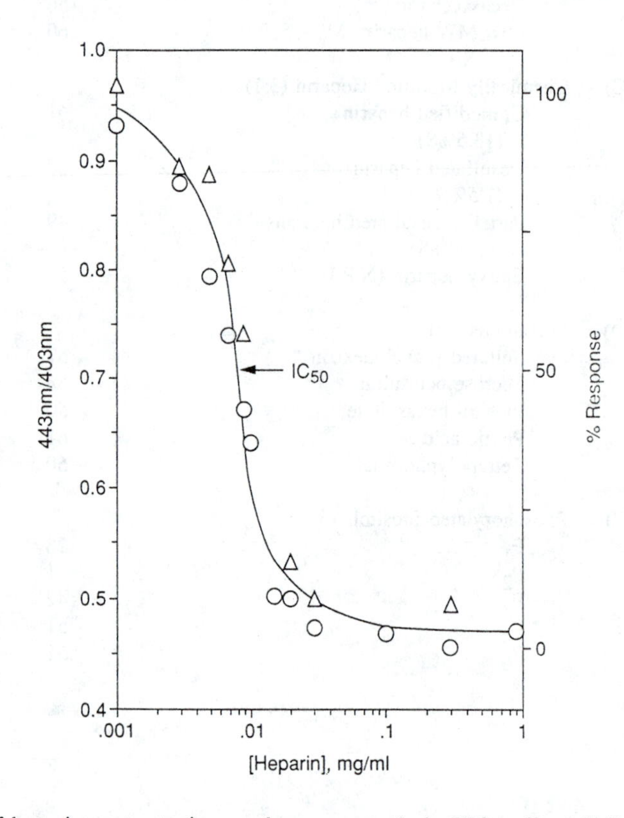

Fig. 1. The effect of heparin concentration on the fluorescence emission spectrum of suramin-aFGF complexes. Two different sources of heparin were examined; bovine (Δ) and porcine (O). Experiments were carried out with 6.3 µM aFGF and 10 µM suramin in PBS buffer (pH 7.2) at 10°C. Suramin-aFGF complex was excited at 315 nm on a Hitachi F-2000 fluorometer. The IC50 value is the concentration of heparin that results in the loss of 50% of the suramin fluorescence enhancement.

Table 2. Effect of Various Polyanions on the Fluorescence Emission Spectrum of the Suramin-aFGF Complex. IC50 Value is the Concentration of Ligand Required to Produce One-Half of the Response. For Details See Figure 1.

Ligand	IC50 (μg/ml)
Heparin (13.5%S, $M_r \sim 16{,}000$)	9
Desulfated heparin (1.5%S)	>10,000
Partially desulfated heparins (4.5-8%S)	500-1,000
Epoxy-heparin (N.P.)	20
Low MW heparin ($M_r \sim 5000$)	10
Inositol hexasulfate	33
Sulfated β-cyclodextrin	15
Phytic acid	7

This competitive binding assay provided a simple technique to screen the ability of other polyanions to interact with aFGF. We found that small, well defined sulfated ligands such as inositol hexasulfate and sulfated β-cyclodextrin can interact with aFGF (i.e. compete with suramin for the polyanion binding site) with IC50 values of 33 and 15 μg/ml, respectively. In addition, a highly phosphorylated, small molecular weight compound, inositol hexaphosphate, was as effective as heparin itself with an IC50 value of 7 μg/ml.

Stabilization of aFGF by Polyanions

The ability of a variety of well-defined low molecular weight polyanions such as inositol hexasulfate, sucrose octasulfate, sulfated β-cyclodextrin, phytic acid (inositol hexaphosphate) and tetrapolyphosphate, to stabilize aFGF against thermal denaturation was then measured by fluorescence spectroscopy. It was found that each of these ligands stabilize aFGF with T_m values from 50 to 61°C (Table 1-D). Moreover, simple inorganic salts such as sodium sulfate can also induce stabilization of aFGF although at higher concentrations. In order to determine the effect of the increasing charge of a ligand on aFGF stabilization, a series of phosphorylated inositol compounds were examined. As shown in Table 1-E, as the number of phosphate groups on inositol was increased from zero to six, the T_m values for aFGF increased dramatically from 28 to 61°C.

Conclusions

A wide variety of polyanions, including small sulfated and phosphorylated compounds, have been shown to interact and stabilize aFGF. In addition, a series of well defined heparin fragments and differentially phosphorylated inositol compounds were used to probe the extent and electrostatic requirements of the aFGF polyanion binding site. The surprisingly weak specificity of the aFGF polyanion binding site can be contrasted with the highly specific interaction of antithrombin III with a heparin pentamer sequence. Small changes in the pentamer such as removal of the 3-O-sulfate group in the third saccharide unit result in a dramatic loss of binding affinity and antifactor Xa activity (Petitou et al., 1991).

The recently published X-ray crystallographic structure of aFGF suggests that the basic amino acid residues involved in polyanion binding (Arg 122, Lys 118, Lys 112) form a cluster of positively charged residues on the protein surface (Zhu et al., 1991). The compactness and ready accessibility of the aFGF polyanion binding site probably accounts for the weak specificity of ligand binding described above. No changes in the fluorescence or CD spectrum of aFGF have been observed upon ligand binding implying no major conformational change in the protein. Thus it appears that the observed ligand induced stabilization of aFGF is primarily due to the favored interaction of ligands with the native versus unfolded form of the protein.

The addition of polyanions to formulations of aFGF significantly enhances the stability of the protein both in terms of storage temperature and shelf-life (not illustrated). Clearly, ligand induced stabilization of aFGF points to the major cause of aFGF inactivation - conformational instability. However, aFGF contains three free thiol groups, one of which (Cys 117) is found near or within the polyanion binding site. Defining conditions to protect these thiol groups from oxidation is another important consideration in producing a stable environment for aFGF. Thus, there may be multiple causes of aFGF inactivation. Since the potential mechanisms of protein degradation include a wide variety of conformational and covalent processes (Volkin and Klibanov, 1989), the design of a protein formulation for an unstable protein such as aFGF is an ongoing challenge.

References

Burgess, W. H.; Maciag, T. *Ann. Rev. Biochem.* **1989**, *58*, 575-606.

Copeland, R. A.; Ji, H.; Halfpenny, A. J.; Williams, R. W.; Thompson, K. C.; Herber, W. K.; Thomas, K. A.; Bruner, M. W.; Ryan, J. A.; Marquis-Omer, D.; Sanyal, G.; Sitrin, R. D.; Yamazaki, S.; Middaugh, C. R. *Arch. Biochem. Biophys.* **1991**, *289*, 53-61.

Mach, H.; Burke, C. J.; Volkin, D. B.; Dabora, J. M.; Sanyal, G.; Middaugh, C. R. **1992**, this volume.

Petitou, M.; Lormeau, J. C.; Choay, J. *Nature* **1991**, *350*, 30-33.

ten Dijke, P.; Iwata, K. K. *Biotechnology* **1989**, *7*, 793-798.

Volkin, D. B.; Klibanov, A. M. In *Protein Function, A Practical Approach*; Creighton, T. E., Ed.; IRL Press: Eynsham, Oxford, England, **1989**; pp 1-24.

Zhu, X.; Komiya, H.; Chirino, A.; Faham, S.; Fox, G. M.; Arakawa, T.; Hsu, B. T.; Rees, D. C. *Science* **1991**, *251*, 90-93.

Mass Spectrometry in Protein Characterization: A Rival for SDS-PAGE

Ronald Beavis*, Department of Physics, Memorial University of Newfoundland, St. John's, Nfld., Canada A1B 3X7
Brian Chait, Rockefeller University, 1230 York Avenue, New York, NY 10021

The recently developed technique of matrix-assisted laser desorption mass spectrometry is described, as it applies to protein analysis. The mass accuracy and resolution of this technique are a substantial improvement on SDS-PAGE, the current preferred method for assigning the molecular mass of a protein.

A frequently used technique for identifying proteins in mixtures is sodium dodecylsulfate - polyacrylamide gel electrophoresis, commonly called SDS-PAGE (Weber et al., 1969). SDS-PAGE will separate a complex mixture of proteins into a series of bands on a polyacrylamide gel that can be visualized by a variety of techniques. A useful feature of SDS-PAGE is that the position of a band can be correlated to the molecular mass of either a protein (or protein subunit). The relative migration distance of bands has been very useful in protein analysis, providing an estimate of the number of amino acid residues in an unknown molecule. The accuracy of the molecular mass measurement is assumed to be 5 - 10% for most proteins although certain classes of proteins (e.g., glycoproteins) are known to migrate anomalously.

A mass spectrometer with a matrix-assisted laser desorption ion source is now an alternate method of determining the molecular mass of proteins (or subunits) in mixtures (Hillenkamp et al., 1992; Beavis et al., 1990). Mass spectrometry measures the masses of isolated, gas phase protein ions directly by their behavior in static electric fields; it does not provide any separation of components in an analytical sense. The isolation of individual protein molecules in the gas phase removes effects caused by the solution behaviors of protein molecules, e.g., aggregation and conformation, which can disturb SDS-PAGE results.

Instrumental Requirements

The instrumentation that is normally available in analytical laboratories are either double-focussing magnet mass spectrometers or quadrupole mass spectrometers. Neither of these devices are well suited to using a matrix-assisted laser desorption ion source; there are currently no commercially available ion sources that can be used with these instruments. Matrix-assisted laser desorption is well-suited to time-of-flight mass spectrometers and several companies offer devices made exclusively for this purpose.

The mass spectrometer used in our laboratory was a linear time-of-flight analyzer built at Rockefeller University (Beavis et al., 1989). Ions are produced by illuminating the sample with laser light for approximately 10 nanoseconds. They are then accelerated by a static electric field to a kinetic energy of 30

1054–7487/92/0303$06.00/0

keV. The fast moving ions then pass through a field-free region and are detected. Because all the ions have the same kinetic energy, the lightest ions reach the detector first, followed by ions of increasing mass. The output of the detector has the appearance of a chromatogram, with the abscissa being the time-of-flight rather than a chromatographic retention time. A simple algorithm can then be used to convert the time-of-flight scale into mass.

Sample Preparation

The sample preparation method is used has been described in detail (Beavis et al., 1990). Briefly, 1 μl of a protein mixture (1 - 10 μM) was mixed with five microliters of a solution of the sinapinic acid "matrix" (50 mM) in 40% acetonitrile. Approximately 0.5 μl of this solution was placed on a metal probe tip and dried in room temperature air. The dried sample was dipped into 4 C distilled water for ten seconds to remove any highly soluble contaminants, chiefly inorganic salts, from the surface. Finally, the matrix/analyte deposit was inserted into the mass spectrometer and analyzed.

Sample deposits produced using caffeic or sinapic acid are fairly homogeneous: most spots on the surface will produce protonated protein molecule ions when irradiated by the laser. Deposits formed using other types of matrix, e.g., nicotinic acid or pyrazinoic acid, may have only small regions of the sample that will produce results.

An Example

Fig. 1 shows a portion of the mass spectrum obtained from a purified recombinant protein sample (HIV I protease). The main peak in the spectrum corresponds to the intact, protonated molecular ion of the expected protein. The calculated mass of the protein (MM_{calc} = 10,789.8 u) agrees very well with the measured mass (MM_{exp} = 10,789.5 u). The close agreement is a strong indication that the correct product has been made. The two, less intense, lower mass peaks do not correspond to any proteolysis product of the intact molecule: they are unidentified polypeptide species.

While Fig. 1 has a straightforward interpretation, Fig. 2 is the full mass spectrum of the same sample. Clearly, there are many unexpected peptides. The desired protein is peak at the right of the spectrum. The most intense of the lower mass peaks can be assigned by their molecular mass to sensible proteolytic cleavages of the protein. Some of the other peptides cannot be assigned. This level of detailed information about the contaminants of a presumed pure sample cannot be obtained using lower resolution techniques, such as SDS-PAGE.

It must be noted that these smaller peptides are not the result of fragmentation caused by the desorption process. The polypeptide chains (including disulphide bonds) are not damaged by the desorption process: the species in the spectrum correspond to intact molecules that were present in the original sample.

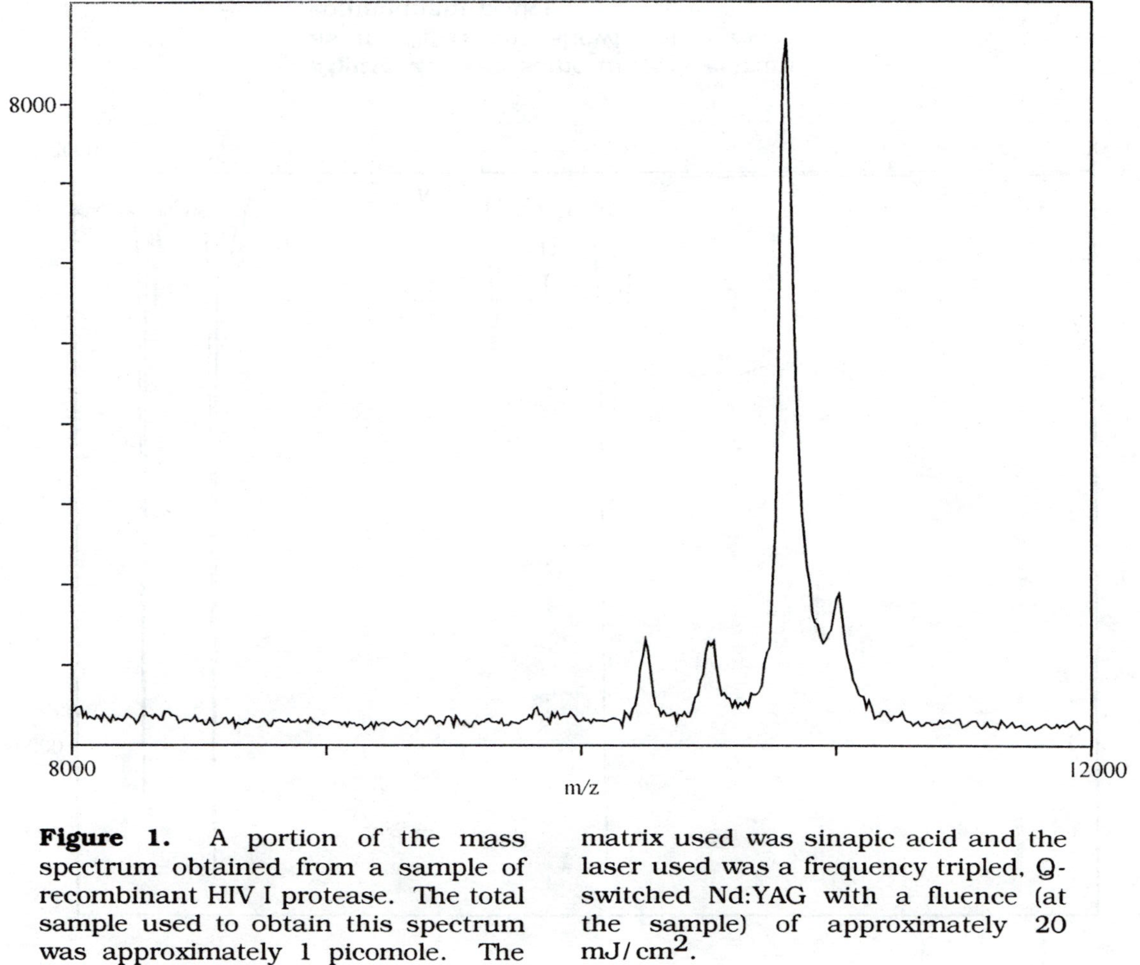

Figure 1. A portion of the mass spectrum obtained from a sample of recombinant HIV I protease. The total sample used to obtain this spectrum was approximately 1 picomole. The matrix used was sinapic acid and the laser used was a frequency tripled, Q-switched Nd:YAG with a fluence (at the sample) of approximately 20 mJ/cm^2.

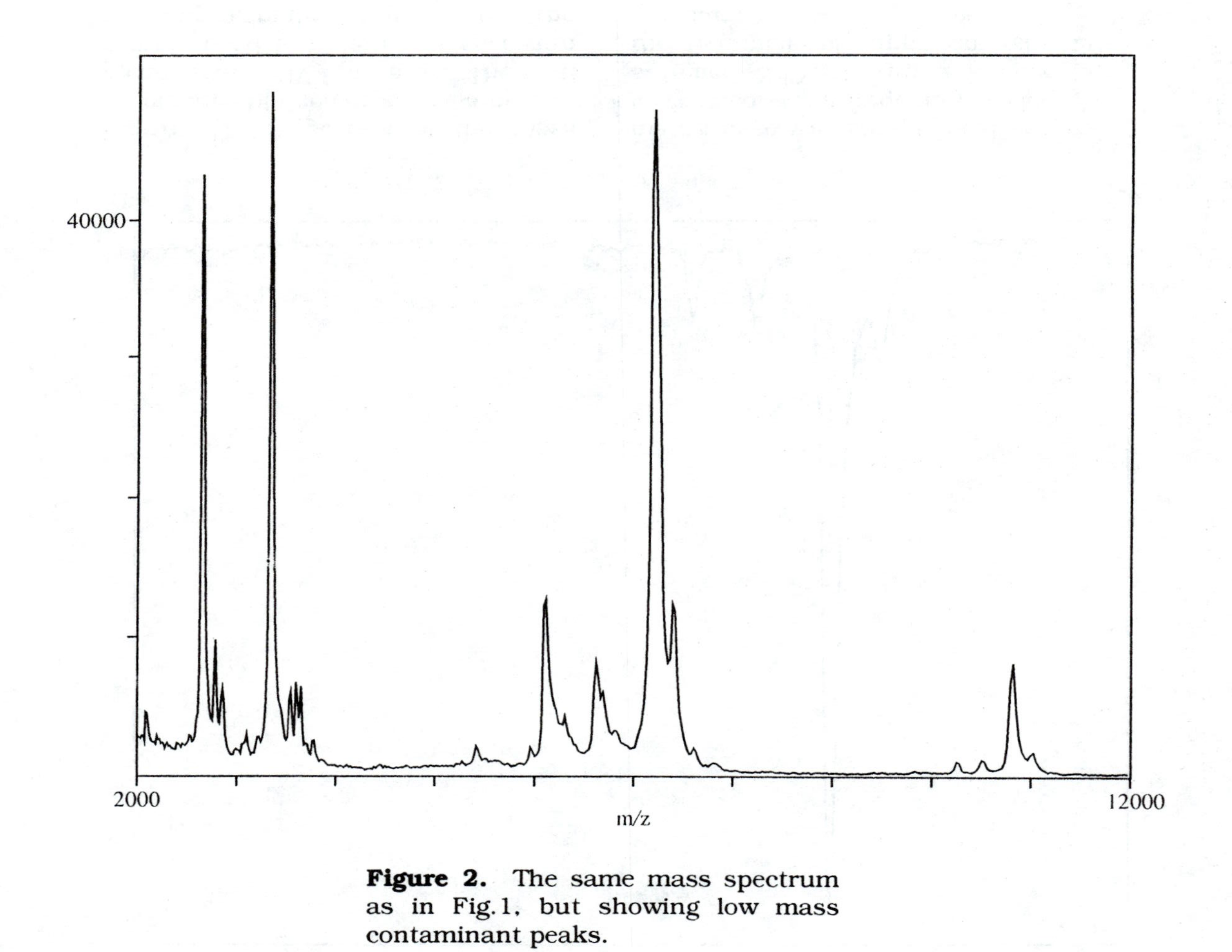

Figure 2. The same mass spectrum as in Fig.1, but showing low mass contaminant peaks.

References.

Beavis, R.C.; Chait, B.T. *Rapid Commun. Mass Spectrom.* **1989**, *3*, 233.

Beavis, R.C.; Chait, B.T. *Proc. Natl. Acad. Sci. (USA)* **1990**, 6873-6877.

Beavis, R.C.; Chait, B.T. *Anal. Chem.* **1990**, *62*, 1836.

Hillenkamp, F.; Karas, M.; Beavis, R.C.; Chait, B.T. *Anal. Chem.* **1991**, *63*, 1193A-1203A.

Weber, K.; Osborn, M. *J. Biol. Chem.* **1969**, *224*, 4406

Recombinant Human IFN-γ Produced by CHO cells: Effects of Culture Environment on Product Quality.

Elisabeth Curling*, Paul Hayter and Nigel Jenkins.
Biological Laboratory, University of Kent, Canterbury, Kent, CT2 7NJ, UK.
(FAX: 0227 763912)

Chinese hamster ovary (CHO) cell line secreting recombinant human IFN-gamma (Hu-IFN-γ) showed a marked change in the glycosylation profile of product secreted with time in stirred-batch culture. The relative proportion of non-glycosylated IFN-γ increased with time, coincident with a proportional decrease in the fully-glycosylated form. It was shown that individual glycosylation variants could be partially separated in a charge-dependent manner using cation-exchange chromatography to purify the IFN-γ from batch culture supernatant.

Human IFN-γ is a lymphokine normally secreted by T lymphocytes which has potent anti-viral activity (Johnson., 1985). The cDNA has been cloned and expressed in both *E.coli* (Gray et al., 1982) and CHO cells (Mutsaers et al., 1986). Normally, recombinant proteins expressed in CHO cells closely mimic the complex post-translational modifications seen in the natural form of the protein. The core 17.1kD polypeptide of human IFN-γ is normally N-glycosylated at Asn_{-100} and/or Asn_{-28} and the carboxyl-terminus of the glycoprotein is variably processed (Fig 1). Since a departure from a natural glycosylation pattern can lead to reduced activity and increased clearance rates *in vivo* (Ashwell and Hartford, 1982), a study of the effect of culture conditions on glycosylation fidelity was performed using spinner and fermenter batch cultures. Cation-exchange chromatography was then used to purify the IFN-γ secreted, and to discover if the glycosylation variants could be separated on the basis of charge.

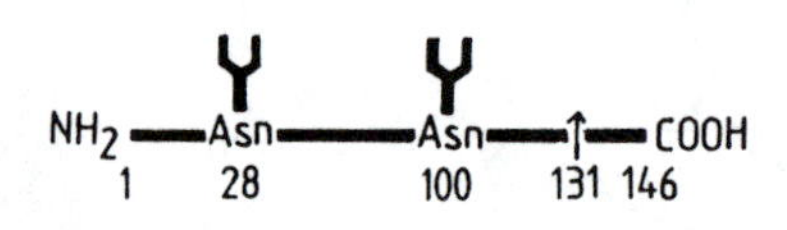

Figure 1. Summary of the positions of the glycosylation and proteolytic cleavage sites on human IFN-γ. Each of the two oligosaccharide chains is an N-linked bi-antennary structure with a sialic acid residue terminating each branch.

Cell culture and IFN-γ variant analysis.

A CHO-K1 mutant cell line deficient in the dihydrofolate reductase gene (DUK) was transfected with a plasmid containing both the Hu-IFN-γ and the DHFR gene. Amplification of DHFR-linked IFN-γ copy number was achieved by the addition of methotrexate to a final concentration of 0.1 μM. This cell line was obtained from Wellcome Research Ltd, UK. Cultures were performed in a 500ml spinner vessel or in a 2 litre stirred fermenter as described previously (Curling et al., 1990a, Hayter et al., 1991.). Cultures were seeded at 10^5 cells/ml in serum-free medium and daily samples were taken for metabolite, glucose and IFN-γ content. For glycosylation analysis, secreted IFN-γ was immunoprecipitated, resolved by SDS-PAGE electrophoresis and then silver-stained. Individual tracks were then scanned by densitometry as described (Curling et al., 1990a).

Fig. 2. shows the glycosylation profile of IFN-γ harvested daily from the spinner culture. A significant alteration in the pattern of glycosylation was noted with time in batch culture; the proportion of non-glycosylated IFN-γ increased while the fully-glycosylated IFN-γ decreased from 61% on day 2 to 29% on day 5. Similar results were seen when the cells were grown in 2 litre fermenters under pH and oxygen control in which either a 20mM or 10mM starting concentration of glucose was used (Curling, et al., 1990a). The change in the glucose concentration present at the start of batch culture had no effect on the shift towards the secretion of under-glycosylated IFN-γ (Results not shown).

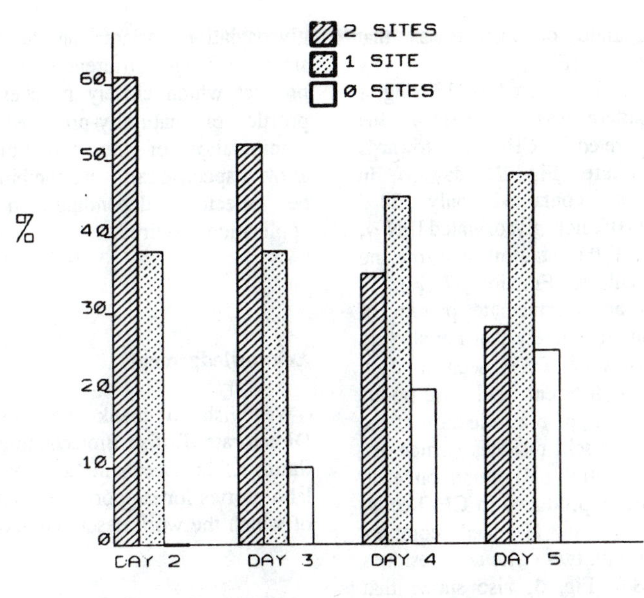

Figure 2. Shift towards under-glycosylated IFN-γ with time in batch culture. Daily samples were analysed by immunoprecipitation, SDS PAGE gel electrophoresis and silver staining, followed by scanning densitometry. (Reproduced with permission from Curling et al 1990b. Copyright 1990 Elsevier Science Publishers Ltd, UK).

Purification and partial separation of IFN-γ variants.

Secreted IFN-γ was purified using cation exchange chromatography. A 16ml column of P11 phosphocellulose (Whatman Biosystems, Kent UK) was pre-treated and equilibrated as described by Devos et al., (1984). The column was connected to an FPLC system (Pharmacia, UK). For IFN-γ purification, 100ml of CHO IFN-γ culture supernatant was loaded onto the column at 0.6ml/min. The column was washed and IFN-γ was eluted using a salt gradient, essentially as described by Devos et al., (1984). Four ml fractions were collected and those containing IFN-γ were detected using a sandwich ELISA (Hayter et al., 1991).

Fig. 3. shows the densitometry scan patterns of fractions 36, 37 and 38 (a, b and c respectively) which together contained the majority of the IFN-γ. These fractions eluted at salt concentrations of 0.26M (Fraction 36), 0.29M (Fraction 37) and 0.34M (Fraction 38) within a salt gradient of 0.1 to 1M NaCl. Analysis of these fractions showed that the fully-glycosylated form of IFN-γ (Mr of 24-26 kD) eluted slightly earlier than the singly-glycosylated form (Mr of 20-21kD), with the non-glycosylated form (Mr of 15.5-17kD) eluting last. Interestingly, Fraction 36 (Fig. 2. panel a)

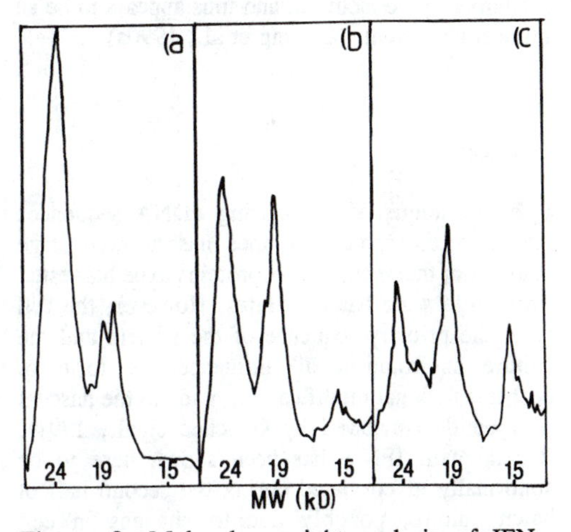

Figure 3. Molecular weight analysis of IFN-γ glycosylation variants after cation exchange chromatography and scanning densitometry of immunoprecipitated samples. Key: x axis = relative molecular weight; y axis = absorbance. Panel (a) = Fraction 36; (b) = Fraction 37 and (c) = Fraction 38, eluted at salt concentrations of 0.26M, 0.29M and 0.34M respectively. The three major peaks represent fully-glycosylated IFN-γ (24-26kD), IFN-γ with only one site conjugated with oligosaccharide (19-21kD) and non-glycosylated IFN-γ (15-17kD).

contained the same ratio of variants as that described for natural IFN-γ produced by lymphocytes; 67% fully-glycosylated to 33% singly-glycosylated. This pattern was also seen in the recombinant IFN-γ secreted by CHO cells towards the start of batch culture (Fig. 2., day 3). In contrast, Fraction 38 contained only 34% fully-glycosylated and 20% non-glycosylated IFN-γ, which closely matched the pattern seen on the 4th-5th day of batch culture. Fraction 37 (Fig. 2. panel (b)) contained an intermediate pattern of glycosylation variation. It is likely that the slightly different elution profiles of IFN-γ seen in each fraction were due to the different isoelectric points of each glycoform. This may be a reflection of the number of terminal sialic acid residues (which are negatively-charged at neutral pH) found on each IFN-γ variant. For IFN-γ produced in CHO cells, each oligosaccharide is an N-linked complex bi-antennary structure with two sialic acid residues (Mutsaers et al., 1986). Fig. 3. also shows that each major peak consists of minor variants which are likely to be the result of proteolytic cleavage of the polypeptide, at a pair of dibasic residues at amino acid position 131 (Rinderknecht et al., 1984). This added source of product variation could not be reduced by the addition of a range of protease inhibitors to the medium, and thus appears to be an intracellular event (Curling et al., 1990a).

Discussion

The technique of introducing cDNA sequences into culture-adapted cell lines has allowed large amounts of therapeutic glycoproteins to be harvested from large scale batch cultures. However, the cell type, the primary sequence of the protein and the culture environment all influence the form of post-translational modifications made to the nascent polypeptide (reviewed by Goochee et al., 1991). Recombinant IFN-γ has been shown here to be abnormally glycosylated during the second half of batch culture, possibly due to changes in cell metabolism influenced by the changing nutrient and metabolite balance in the culture medium. However, the pattern of glycosylation variants which yields the best biological activity for a given therapeutic can be selected by ion-exchange chromatography during down-stream processing (Goochee et al., 1991). The importance of vacant glycosylation sites on therapeutics has been studied using the drug erythropoetin, where the glycoconjugation of such sites affected both the biosynthesis and secretion rate (Dube et al., 1988). This paper provides some evidence that partial separation of (IFN-γ) glycosylation variants on the basis of relatively small charge differences can yield a purified product which closely matches the glycosylation profile of naturally-produced IFN-γ. Further manipulation of the purification conditions may allow a specific ratio and combination of variants to be selected, depending on the therapeutic application required.

Acknowledgements

We wish to thank the SERC Biotechnology Directorate, British Biotechnology Ltd, Glaxo, ICI, Smith-Kline Beecham and Wellcome Research laboratories for support of the research programme of which the work described above was a part.

References

Ashwell, G; Hartford, J. *Ann. Rev. Biochem.* **1982,** *51,* 531-554.

Curling, E.M.A.; Hayter, P.M.; Baines, A.J.; Bull, A.T.; Gull, K.; Strange, P.G.; Jenkins, N.J.; *Biochem. J.* **1990a**, *272*, 333-337.

Curling, E.M.A.; Hayter, P.M.; Baines, A.J.; Bull, A.T.; Strange, P.G.; Jenkins, N.J. In: *Separations for Biotechnology II,* Pyle, D.L. Ed; Elsevier Science Publishers Ltd, UK, **1990b**, pp481-490.

Devos, R; Opsomer, C; Scahill, S.J; Van Der Heyden, J; Fiers, W. J. *Interferon Research* **1984,** *4,* 461-468.

Dube, S.; Fisher, J.W.; Powell, J.S. *J. Biol. Chem.* **1988,** *263*, 17516-17521.

Goochee, C.F.; Gramer, M.J.; Andersen, D.C.; Bahr, J.B.; Rasmussen, J.R. *Biotechnology* **1991,** *9*, 1347-1355.

Gray, P.W.; Leung, D.W.; Pinnica, D.; Yelverton, E.; Najarian, R.; Simonsen, C.C.; Derynk, R.; Sherwood, P.J.; Wallace, D.M.; Berger, S.L.; Levinson, A.D.; Goeddel, D.V. *Nature* **1982,** *295,* 503-508.

Hayter, P.M.; Curling, E.M.A.; Baines, A.J.; Jenkins, N; Salmon, I.; Strange, P.J.; Bull, A.T. *Appl. Microbiol. Biotechnol.* **1991,** *34,* 559-564.

Johnson, H. *Lymphokines* **1985,** *11,* 33-46.

Mutsaers, J.H.; Kamerling, J.P.; Devos, R.; Guisez, Y.; Fiers, W.; Vliegenthart, J.F.G. *Eur. J. Biochem.* **1986,** *156,* 651-654.

Rinderknecht, E.; O'Connor, B.H.; Rodriguez, H. *J. Biol. Chem.* **1984**, *259,* 6790-6797.

Regulatory Strategy for Biotechnology Products in the United States and the European Economic Community

Kathy Lambert, Celltech Ltd., Slough, Berkshire, SL1 4EN, United Kingdom
(FAX: 44 753 536632)

The perspective for this paper is that of a manufacturer planning both USA and EEC applications for a medicinal product derived from recombinant DNA or hybridoma technology. Firstly, a comparison of registration systems and procedures for EEC and USA is made. Data requirements (with particular reference to downstream processing quality issues) for the FDA and the European authorities are then reviewed. Finally, recommendations for joint US/EEC applications are provided.

A key element of the FDA's regulatory approach to biotechnology products is that no new procedures or systems are needed for biotechnology since existing regulations are sufficiently broad in scope. FDA will assign review responsibilities on a product (rather than manufacturing process) basis. Thus US applications for biotech products may be NDAs (submitted to Center for Drug Evaluation and Research), PLAs (submitted to Center for Biologics Evaluation and Research) or pre-marketing authorisations (submitted to the Center for Devices and Radiological Health). In contrast, the European regulatory framework for biologicals includes national, multinational and supranational routes for registration and the route for an individual product is determined mainly by the methods used for development and manufacture. Biotechnology products, if marketing in more than one EEC country is planned, must go through the supranational high-tech/biotech (concertation) route defined under Council Directive 87/22/EEC. This means that an opinion from the Committee for Proprietary Medicinal Products (CPMP) must be obtained before any individual country can decide to authorise marketing of the product. The CPMP opinion is not binding, so the 12 member states (one of which, the rapporteur country, acts as contact between the applicant and the CPMP) are still able to make individual decisions on whether to grant a marketing authorisation.

Important differences between the USA and EEC registration procedures are as follows. For the EEC a specific format, as described in Volume II of the Rules Governing Medicinal Products in the European Community, is obligatory. An EEC draft Note for Guidance on the headings to be used for biotech products has recently been published. Expert reports, that is critical appraisals of each of the main parts of the dossier, in a specific format including summary tables, must be provided. Lastly, since the opportunity for interaction with the CPMP is relatively limited (CPMP meetings are held only every few months), the dossier must be as complete as possible at the start of the concertation procedure: this means early contact with the chosen rapporteur country prior to application is essential.

Both the FDA Center for Biologics Evaluation and Research and the CPMP

1054–7487/92/0311$06.00/0

Biotechnology/Pharmacy Working Party have produced guidelines on aspects of preparing and testing monoclonal antibody and recombinant-DNA product derived medicinal products. The CPMP Working Party guidelines on validation of virus removal/inactivation and on minimising the risks from bovine spongiform encephalopathy are also useful. Both the US and EEC guidelines cover the needs to be met for product license applications. For entry into initial clinical trials, some interpretation has to be made, concentrating on the key product safety issues, however, the scope of work to enter early clinical trials is broadly the same worldwide.

Inevitably some slight differences in emphasis appear at the PLA stage (for instance in the EEC, a separate biologicals establishment license application is not needed), however, the approach and key concerns for the FDA and European authorities are the same: lot to lot consistency, increasing emphasis on the risks from viruses and compliance with Good Manufacturing Practice standards for pharmaceuticals. These three principles should be applied when developing quality assurance programmes for specific products and applications from the guidelines.

For the consistency lots (ie. product to be used in pivotal Phase III clinical trials) both the FDA and CPMP are showing increasing concern that manufacturing process change be minimised, to ensure that both the product itself and the impurity profile of commercial lots will be equivalent to that of the Phase III investigational materials. Downstream processing (from primary recovery to purified unformulated bulk) is seen as the most critical part of the manufacturing process in terms of effect on product quality and consistency. To minimise effect on product quality, scale up during development of chromatographic processes should be accomplished by increasing the chromatographic bed width rather than height. Validation of such scale changes should include investigation of the consistency of impurity profiles throughout the steps of the purification process, using validated analytical procedures. The EEC have recently tightened their requirements on analytical validation. Good Manufacturing Practice issues for downstream processing include control of microbial contamination or degradation, together with full validation and a maintenance schedule for separation and purification equipment.

For chromatography, procedures for cleaning and sanitisation of gels, the number of cycles per gel batch permitted, and QC acceptance specifications for gels should be defined. Both EEC and FDA require criteria for reprocessing to be defined for each purification step. For validation of virus removal and inactivation, as required for processes based on culture of mammalian cell lines, it is possible to use exactly the same study design with a range of viruses for both US and EEC submissions. However, when relating such studies to the manufacturing situation, EEC have accepted infectious retrovirus assay results on fermenter samples at peak viable cell density as a basis for the retrovirus titre which has be to removed. However, the FDA require this virus loading to be estimated in terms of the total number of particles detected by electron microscopy at the bulk harvest stage.

Registration procedures in the EEC are changing rapidly and a current proposal to repeal Directive 87/22/EEC would take effect on 1st January 1993. The proposal allows for the establishment of a European Agency fór the Evaluation of Medicinal Products, for the centralised assessment of biotechnology products and for the CPMP opinion to be binding.

Recommendations for a manufacturer planning both USA and EEC applications for a biotech product are as follows:

i) define manufacturing method, scale and location as soon as possible (allowing for any FDA shared manufacturing licensing requirements);
ii) take into account the EEC Notice to Applicants dossier format;

iii) for EEC choose and involve the experts for expert reports as soon as possible;

iv) identify the EEC rapporteur country and discuss the project with them as soon as possible.

In conclusion, the overall concerns and scientific principles governing registration of biotechnology products in the EEC are very similar to those in the USA. This is a changing arena which will continue to provide challenges for the foreseable future.

References

CBER, FDA, Bethesda, MD 20892. *Points to Consider in the Production and Testing of New Drugs and Biologicals produced from Recombinant DNA Technology*, 1985.

CBER, FDA, Bethesda, MD 20892. *Points to Consider in the Manufacture of Monoclonal Antibody Products for Human Use*, 1987.

CBER, FDA, Bethsda, MD 20892. *Points to Consider in the Characterisation of Cell Lines used to produce Biologicals*, 1987.

Commission of the European Communities, Office for Official Publications of the European Communities L-2985 Luxembourg. *The Rules governing Medicinal Products in the European Communities*, 1989 (and 1990/1991 addenda), Vol I-IV.

CPMP, *Guidelines on the Production and Quality Control of Medicinal Products Derived by Recombinant DNA technology; Trends in Biotechnology*, 1987, 5, G1-G4.

CPMP, *Guidelines on the Production and Quality Control of Monoclonal Antibodies of Murine Origin intended for Use in Man; Trends in Biotechnology*, 1988, G5-G8.

CPMP, *Note for Guidance on the Validation of Virus Removal and Inactivation Procedures*, 1991, Doc III/8115/89-EN (Final).

CPMP, *Note for Guidance on Minimising the Risk of Transmission from Animals to Man of Agents Causing Spongiform Encephalopathies via Medicinal Products*, 1991, Doc III/3298/91-EN (Final).

CPMP, *Note for Guidance on Headings to be used in Part II of the Notice to Applicants for Biotechnological and Biological Products*, 1991, Doc III/3153/91-EN (Draft 7).

Bioinstrumentation and Bioprocess Control Symposium VII

G. N. Stephanopoulos and B. Sonnleitner: *Co-Chairs*

Sensors and Analytical Techniques: Session A
J. S. Schultz and I. Karube: *Co-Chairs*

Modeling of Bioprocesses: Intracellular and Extracellular Events: Session B
J. E. Bailey and G. Lyberatos: *Co-Chairs*

Control of Bioprocesses: Session C
G. N. Stephanopoulos and B. Sonnleitner: *Co-Chairs*

Bioinformatics: Session D
M. L. Mavrovouniotis and I. Endo: *Co-Chairs*

Development of Practical Electrochemical Biological Sensing Devices

I. John Higgins * and John Bolbot,
Cranfield Biotechnology Limited, Bedford, MK43 OAL, UK. (FAX: 44–908–217300)

Amperometric enzyme electrodes currently dominate the biosensor market. This is probably because they were the first type of biosensor to receive commercial attention and their developmental problems (although considerable) have not been as fundamental as experienced with other biosensor types. Enzyme electrodes can now be developed for a variety of analytes and screen-printing technology has facilitated mass production. They can be used with pocket-sized, portable devices or as part of more complex laboratory instrumentation. Electrochemical instruments have been developed for the rapid assessment of microbial numbers and for multiple amperometric immunoassay. The latter offers several advantages over its contemporary photometric competitors. Electrochemical-biological devices based on semiconductor technology will probably lead to a new generation of multifunctional 'smart' sensors.

Of all the possible biosensor options, amperometric enzyme electrodes currently hold some 90% of the technology market share, principally detecting simple clinical analytes. This domination (expected to continue for at least the next five years) is perhaps as much due to the problems that have been encountered in developing other types of biosensors as to the practical advantages (often fortuitous) which have spurred the development of enzyme electrodes:

- Suitability of many redox enzymes for detection of simple analytes.
- Availability of stable enzymes.
- Appropriate sensitivity.
- Broad linear dynamic range.
- Success with 'real' samples.
- Availability of simple low-cost instrumentation.
- Amenable to mass production.

Failure on one or more of these counts has made it likely that only rather specialised niche applications will accommodate optical, piezoelectric and BIOFET technology in the near future because considerable development is still required before low-cost devices become generally available. However, the next major breakthrough waiting to take biosensor technology forward will almost certainly be such 'affinity' sensors, most probably directed at disease diagnosis and monitoring (Ballantine and Swain, 1991). Amperometric biosensors have been comprehensively reviewed by Hendry *et al.* (1990).

The world's first successful commercial biosensor, a glucose analyser suitable for use with whole blood, is still produced by the Yellow Springs Instrument Company (Ohio, USA). In this instrument the oxidation of glucose by glucose oxidase produces hydrogen peroxide which is detected at a platinum electrode. This format has now become a standard for enzyme-electrode glucose analysers and has been adapted for a number of laboratory instruments.

In the last decade, the demand was realised for small portable instruments suitable for mass markets, and in 1987 MediSense Inc. (Cambridge, MA, USA and Abingdon, Oxon, UK) launched the 'ExacTech' range of pocket-sized devices for diabetics, which accurately measure blood glucose on a disposable enzyme electrode (**Fig. 1**). Based on inventions at Cranfield and at Oxford University, the ExacTech uses a ferrocene derivative as an artificial redox carrier (a mediator) to shuttle electrons from glucose oxidase to a carbon electrode. By circumventing the natural redox couple ($0_2/H_20_2$), the device is largely independent of oxygen concentrations and can also operate at lower potentials thus avoiding interference from other electroactive components in blood. Commercially, it highlights all of the advantages listed above, the development of screen-printing technology facilitating the mass production of enzyme electrodes both cheaply and consistently.

Similar enzyme-electrode formats can be used for monitoring other clinical analytes such as cholesterol, triglycerides, lactate, creatinine, urea and ammonia, as well as diversifying into other areas such as food freshness, fermentation monitoring and drink-drive testing. The

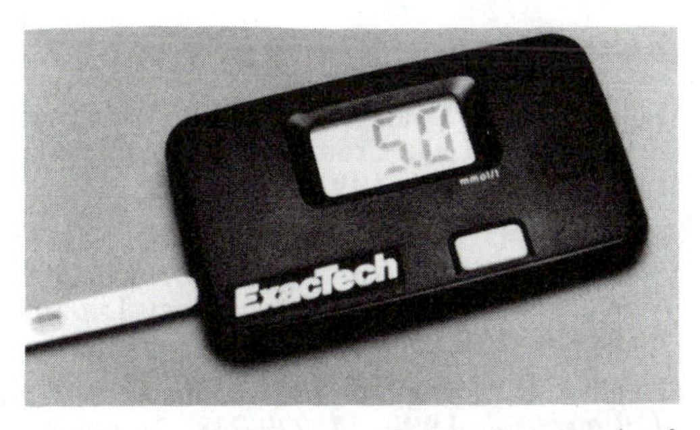

Figure 1. The 'ExacTech Companion' blood glucose meter manufactured by MediSense.

technology may also be incorporated into more complex instrumentation and some recent work at Cranfield Biotechnology Limited (CBL) has been directed towards increasing the analyte range of existing blood electrolyte analysers by adding one or more enzyme-electrode channels. Repeat-use electrodes (as opposed to one-use disposables) have also been developed which have immobilised mediators.

Other work at CBL has exploited fundamentally analogous electrochemistry for the rapid assessment of microbial numbers in 'real' food and environmental samples, a rapidly expanding market area (Ballantine *et al.*, 1992). The instrument (**Fig. 2**) uses a diffusible mediator which is reduced by metabolising microorganisms and produces a quantifiable current by carrying these 'metabolic electrons' to the anode. The instrument is quite exceptional in that proprietary developments have enabled low levels of microorganisms to be detected even in particulate fluids with very high levels of electrochemical interferants such as unpasteurised milk. Workers at Cranfield are also examining alternative possibilities for novel electrochemical microbial sensors based on electroactive compounds which partition into polarised cell membranes. This effect can be followed amperometrically to provide a measure of microbial population and results so far indicate that sensitivity may be in excess of 100-fold that of mediated microbial sensors.

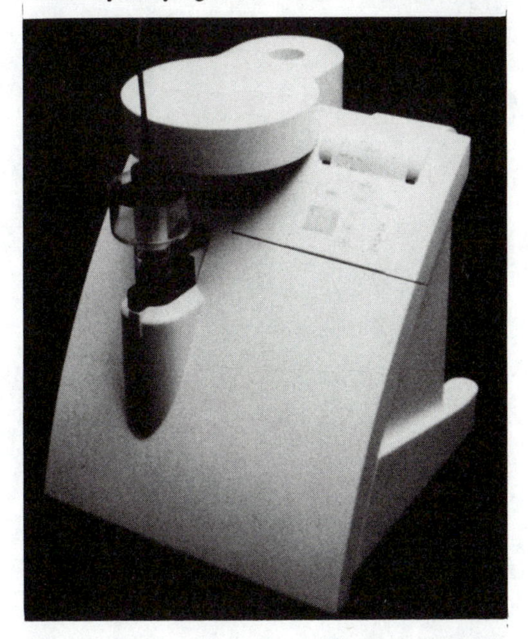

Figure 2. Instrument for rapid microbiological testing developed by CBL.

CBL has recently developed laptop and benchtop prototype instruments suitable for amperometric immunoassay based on enzyme amplification (**Fig. 3**). A sandwich immunoassay is performed in which the second antibody is conjugated to alkaline phosphatase. The alkaline phosphatase converts NADP to NAD which then catalyses a redox cycle between alcohol dehydrogenase and diaphorase, causing the reduction of ferricyanide which is detected amperometrically. This electrochemical detection method has been developed as an alternative to the widely used photometric system involving formazan dye and has the advantages of a much wider dynamic range, reduced instrumentation costs and (at least) similar or improved sensitivity.

Perhaps the most promising area of biosensor development is that based on semiconductor FET technology in which a biological interaction, either enzymatic or immunological, is used to sensitise the transistor gate (BIOFET). The technology offers the

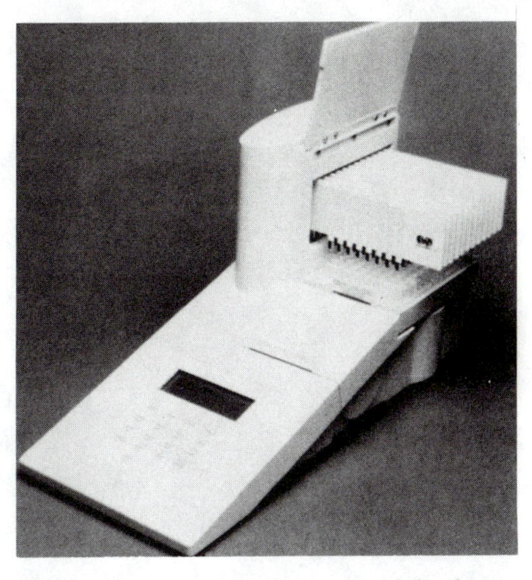

Figure 3. Instrument for multiple electrochemical immunoassay developed by CBL.

powerful integration of biosensors and microelectronics and has the concomitant opportunities of miniaturisation and mass production. CBL is currently a partner in an international project concerning development of the B-ASIC (Biosensitive-Advanced Silicon Integrated Circuit) in which a silicon chip is customised to several sensor applications with different biological components. The ultimate aim is miniature multifunctional devices capable of sophisticated signal processing - the so-called 'smart sensor'.

References

Ballantine, D.; Swain, A., Eds. In *Biosensors: A New Realism*; Cranfield Biotechnology Limited, Beds, UK. **1991**.

Ballantine, D.; Swain, A.; Griffiths, D. Eds. In *Rapid Microbiological Testing: Growth of a New Culture*; Cranfield Biotechnology Limited, Beds, UK. **1992**.

Hendry, S.P.; Higgins, I.J.; Bannister, J.V. In *J. Biotechnol.* **1990**, 15, 229-238.

Multi Enzyme Biosensors
- Increase of Sensitivity and Specificity by Coupling Enzyme Reactions

Frieder Scheller, Ulla Wollenberger, Dorothea Pfeiffer, Alexander Makower, and **Florian Schubert+**
Max-Delbrück-Center of Molecular Medicine, O-1115 Berlin, (FAX:00372/949 7008)
+Physikalisch-technische Bundesanstalt, W-1000 Berlin 10

Principles of coupling enzyme reactions for designing the analytical performance of biosensors are presented in this paper:

(i) signal augmentation by combining accumulation of an intermediate with analyte recycling in enzymatic reactions
(ii) elimination of interfering signals
(iii) quantification of allosteric enzyme effectors

Enzyme sequences

This line has been initiated in the late 70s by Rechnitz' group (Jensen and Rechnitz, 1979). who introduced enzyme sequences into enzyme electrodes. The scope of analytes has been expanded by sequential and parallel coupling of different enzymes with each other, within and with subcellular organelles, and with tissue slices. Examples include ensembles of sensors based on glucose oxidase, lactate monooxygenase and glucose -6-phosphate dehydrogenase (Scheller et al., 1988).

Sequences with accumulation

An increase of sensitivity is gained if an electrode inactive intermediate formed in the analyte conversion is accumulated in front of the polarized indicator electrode. Conversion of the accumulated intermediate upon addition of a cosubstrate results in an increased reaction rate.

The principle may be explained for the glycerol determination with a sensor using the sequence glycerol dehydrogenase (GlyDH)/ lactic dehydrogenase (LDH)/lactate monooxygenase (LMO).

The oxidation of glycerol in the presence of an excess of NAD^+ generates the intermediate NADH. The accumulated NADH is "stripped" by addition of an excess of pyruvate. Due to the high concentration and diffusivity of pyruvate the reaction rate considerably exceeds that for the steady state current which is determined by the glycerol influx from the bulk. Ten seconds after the start with pyruvate the current reaches a peak. Since the NADH is consumed in the LDH-LMO reaction the current decreases and approaches a steady state identical to that without accumulation of NADH. The peak current rises with accumulation time. A limit is set by the equilibrium of the reaction. Therefore, the peak current levels off at long accumulation time. At an accumulation time of 6 min the sensitivity in the kinetic mode (di/dt) is increased by a factor of 64.

1054–7487/92/0319$06.00/0

Amplification by analyte recycling

High signal amplification is obtained when the analyte is shuttled between two enzymes catalyzing the overall reaction between two cofactors. For glucose amplification GOD has been co-immobilized with GDH which reduces the gluconolactone in the presence of an excess of NADH. In this way an amplification of 10 is obtained. For the enzymatic recycling pair LOD/LDH the signal was amplified up to four orders of magnitude (Schubert et al., 1991).

In such sytems the analyte acts as a catalyst, being shuttled between both enzymes in the overall reaction of both cosubstrates. In this way, significantly more cosubstrate will be converted than the amount of analyte present in the enzyme membrane. Hence the change in the parameter indicated at the transducer will greatly exceed that obtained with one-way analyte conversion. On the other hand, the upper limit of linearity is decreased. The enzyme excess present in diffusion-controlled membranes is exploited in the analyte recycling. Therefore, the amplification factor decreases with progressive enzyme inactivation during operation of the sensor.

Combination of recycling and accumulation

ATP (adenosine-5'-triphosphate) was determined by recycling in an enzyme reactor with co-immobilized pyruvate kinase and hexokinase in the presence of glucose, NAD^+, and phosphoenolpyruvate. Recycling produces glucose-6-phosphate which is converted to an equivalent amount of NADH by glucose-6-phosphate dehydrogenase. Oxidation of NADH takes place at a graphite flow-through electrode modified with an adsorbed phenoxazine derivate.

The amplification factor is directly proportional to the residence time in the reactor. Thus, it is increased as the flow rate decreases; it becomes 350 at a flow rate of 0,07 ml/min. The amplification factor can be increased further by a controlled stop-

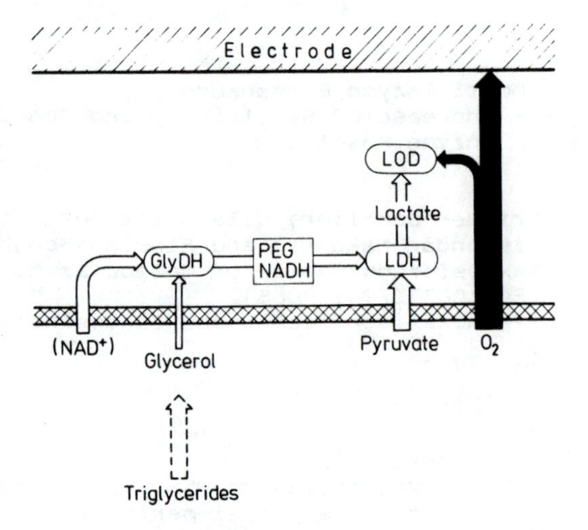

Fig. 1: Scheme of the sequential conversion of glycerol to give an oxygen depletion.

time recycling; it became 1200 at a stop-time of 12 min (Yang et al., 1991).

Elimination of interfering signals

Another aspect of the signal generation by adding the initiator substance is the elimination of electrochemically interfering signals of the sample. Whilst the basic current integrates all electrochemically active sample components, the current change reflects only the intermediate concentration. This advantage has been used in the determination of triglycerides. In the measuring procedure the triglycerides are hydrolyzed in alkaline methanol. After completion of this reaction the sample is injected into a buffered solution. The addition of the methanolic solution causes a change of the basic oxygen sensor output resulting from the different oxygen solubility in the sample and in the background solution. After establishing a stable baseline, pyruvate is added thus, avoiding this disturbance. In this way triglyceride concentrations have been determined in serum samples with good correlation to the established method (r = 0.993).

Measurements of AMP - the effector of glycogen phosphorylase b

Glycogen phosphorylase b-the key

enzyme of glycogenolysis-catalyzes the phosphorylation of the terminal glucose residue of glucogen to give free glucose-1-phosphate (G-1-P) according to the following equation:

$(\alpha\text{-D-glucose})_n$ + P ---> $(\text{glucose})_{n-1}$ + G-1-P

The rate of G-1-P formation represents the activity of the glucogen phosphorylase b. The G-1-P concentration is measured in the following three-enzymes sequence.

G-1-P alkaline Phosphatase → α-Glucose + P:

α-Glucose Mutarotase → ß-Glucose

ß-Glucose + O_2 Glucose Oxidase → Gluconolactone + H_2O_2

At the compromise pH of 7.5 the sensitivity is sufficient to measure 1 uM G-1-P with a serial CV > 5%. Thus 1 U/l glycogen phosphorylase b can be quantified from the slope of the current time curve.

Since the activity of glycogen phosphorylase b is controlled by the concentration of its effector AMP the combination with the above mentioned three-enzyme sequence allows the measurment of the AMP concentration. By fixing the glycogen phosphorylase together with the three-enzyme layer a new sensor for AMP has been developed. This multi-enzyme layer responds reproducibly to AMP in the lower micromolar range.

References

(1) Jensen,M.A., Rechnitz,G.A.; J.Membrane Sci. 1979, 5, 117

(2) Scheller,F., Renneberg,R., Schubert,F.; Methods Enzymology. 1988, 137, 29

(3) Schubert,F., Wollenberger,U. Pfeiffer,D., and Scheller,F.W.; Advances in Biosensors, 1991, 1, 77

(4) Yang,X., Pfeiffer,D., Johansson,G., Scheller,F.; Anal.Lett. 1991, 24, 1401

Improving biosensor and FIA performance via tailor-made enzymes and antibodies

Rolf D. Schmid and Ursula Bilitewski
Department of Enzyme Technology
GBF - Gesellschaft für Biotechnologische Forschung
D-3300 Braunschweig, Germany
(FAX: ++49-6181-302)

By screening of appropriate microorganisms, the following enzymes: hypoxanthine oxidase, L-amino acid oxidase, short-chain acyl-CoA oxidase and several cholinesterases were obtained. In cooperation with other groups at the GBF, the structure of glucose oxidase from Aspergillus niger was solved. The elucidation of glucose oxidase from Penicillium amagasakiense, of NADH oxidase from Thermus thermo-philus and of luciferase from Vibrio harveyi are underway. We are now starting to apply the techniques of protein design to obtain enzymes and antibodies with improved properties for biochemical analysis.

Biosensors and flow injection analysis systems (FIA) based on enzymes and antibodies depend not only on new transducer concepts, but also on the availability of enzymes and antibodies of appropriate selectivity, activity and stability. We have developed some new microbial enzymes with potential for biochemical analysis, and are in the process to modify several enzymes and antibodies, via protein engineering techniques, for improved performance in biosensors and FIA.

1. Hypoxanthine dehydrogenase and hypoxanthine oxidase

Hypoxanthine is an important analyte in the determination of freshness of fish flesh [1]. The xanthine oxidase generally used for analysis is isolated from milk and is low in activity. We have isolated a microbial hypoxanthine dehydrogenase from Pseudomonas putida F1 [2] and a microbial hypoxanthine oxidase from Cellulomonas cellulans [3]. Both enzymes show higher specific activity than the enzyme isolated from milk and higher specificity towards hypoxanthine as compared to xanthine. Thus they can advantageously be used for enzymatic assays requiring the sensitive detection of hypoxanthine.

2. L-Amino acid oxidase

The determination of selected or total L-amino acids is important in medical and food analysis and may also be applied to bioprocess control. Most commercial L-amino acid oxidases are isolated from snake venom; they are costly and exhibit a pronounced substrate specificity. We succeeded in isolating a microbial L-amino acid oxidase from Cellulomonas cellulans and purifying it to homogeneity [4]. The enzyme oxidizes all amino acids except L-proline and may be used, in combination with more specific L-amino acid oxidases, for the analysis of L-amino acid mixtures.

3. Acyl-CoA oxidase

Short-chain fatty acids constitute an important parameter for the quality assessment of raw milk samples and milk products. Fatty acids in serum or milk can be enzymatically determined by the coupled reaction of acyl-CoA synthetase/acyl-CoA oxidase, but the specificity of both enzymes for short-chain fatty acids is low. We screened a strain of Arthrobacter nicotinianae with high specificity for the oxidation of short-chain fatty acids. The strain can be used for the construction of a microbial sensor useful in the rapid detection of rancid raw milk or milk products [5,6] or mastitis [7]. Both the acyl-CoA synthetase and the acyl-CoA oxidase isolated from the strain show pronounced selectivity for short-chain fatty acids.

4. Cholinesterases

Cholinesterase inhibition is an established assay procedure for organophosphates and carbamates. Usually acetylcholine esterases from bovine erythrocytes of from fish organs are used along with butyrylcholine esterase from horse liver. We have purified bovine erythrocyte acetylcholine esterase to homogeneity, after papain digestion of a C-terminal oligopeptide [8]. In addition, we have screened, in collaboration with a group at Kyoto University [9], various cholinesterases from Pseudomonas strains. Though the microbial enzymes in general are less susceptible to inhibition by organophosphates and carbamates as compared to their mammalian counterparts [10], they might develop into valuable indicator enzymes for pattern analysis of inhibitor mixtures.

5. Antibodies specific for triazine herbicides

For the immunoassay of triazines in water samples, a range of highly specific poly- and monoclonal antibodies have become available in recent years [11]. We have produced several grams of a monoclonal antibody with high selectivity for simazine, a triazine herbicide, and purified, after papain digestion, the Fab fragment of this antibody nearly to homogeneity [12].

6. Glucose oxidase

Glucose oxidase is one of the major enzymes used in biochemical analysis. We were able to solve to structure of two varieties of this enzyme, from Aspergillus niger and from Penicillium amagasakiense, by x-ray analysis and molecular displacement methods [13-15]. The structure offers a host of information on the location of the active site of the enzyme. Since the genes coding for these enzymes have been cloned, protein design is the logical next step for tailoring the enzyme in view of biosensor applications.

7. NADH oxidase

Hydrogen peroxide-forming varieties of NADH oxidase are potentially useful for the combination of dehydrogenases with photometric, luminometric or amperometric assays. We have purified NADH oxidase from the thermophile Thermus thermophilus to homogeneity [16]. After crystallization of the enzyme, a native data set which allows x-ray analysis at about 2.3 A resolution was obtained [17]. The enzyme has been cloned and expressed in E. coli by the group of M. Sprinzl in Bayreuth, and the gene sequence has been established [18].

8. Bacterial Luciferase

Though bacterial luciferase has found less application in bioluminometric assays as compared to luciferase from insects, its independance from ATP and its dependency from long-chain aldehydes provide some interesting possibilities in the area of homogenous bioluminometric assay systems [19]. We have purified the enzyme from Vibrio harvey to homogeneity and obtained a native x-ray data set from its crystals. The enzyme has been cloned and expressed in E. coli by the group of J. Kuhn in Haifa, and the sequences of the heterodimer have been established [20].

9. Outlook

From the above examples it seems clear to us that there is considerable potential both in the screening of new microbial enzymes for biochemical analysis and in site-directed mutagenesis of important analytical enzymes. Pertinent applications are already under study

References

[1] Chemnitius, G. et al., *Anal. Chim. Acta,* in press

[2] Kim, J. M.; Schmid, R. D. in Biosensors, Schmid, R. D.; Scheller, F. Ed; *GBF Monographs, Vol. 13.* VCH Publishers, New York 1989, pp. 421 - 424

[3] Artolozaga, M.-J. et al., in preparation

[4] Braun, M. et al., *Appl. Microb. Biotechnol.*, submitted

[5] Ukeda, H. et al., *J. Agric. Chem.*, submitted

[6] Ukeda, H. et al., *Z. Lebensmitt. Untersuch. Forsch.* submitted
[7] Ukeda, H. et al., *Z. Lebensmitt. Uuntersuch. Forsch.* submitted
[8] Peters, C. et al., in preparation
[9] Yamada, H. et al., in preparation
[10] Safarik, I. et al, *Letters Appl. Microbiol. 13*, pp. 137-139 (1991)
[11] Hock, B. et al., *Anal. Lett 24(4)* pp. 529-549 (1991)
[12] Steingaß, H. et al., in preparation
[13] Kalisz, H. et al., *J. Mol. Biol.* 213, pp. 207-209, (1990)
[14] Hecht, H.-J. et al., *J. Mol. Biol.*, submitted
[15] Hendle, J. et al., *J. Mol. Biol.*, in press
[16] Reiser, C. O. A. et al., *Eur. J. Biochem.*, in press
[17] Erdmann, H. et al., in preparation
[18] Park, H.-T. et al., *Eur. J. Biochem.*, in press
[19] Lang, D. et al., *Enzyme Microb. Biotechn.*, in press
[20] Kuhn, J. et al., *Proc. Natl. Acad. Sci. USA 68*, pp. 2484-2487 (1972)

Near Infrared Spectroscopy - A Powerful Technique for At-line and On-line Analysis of Fermentations

Stephen V. Hammond and Ian K. Brookes* (Fax 212 573 1421)
Pfizer International Inc. New York. NY10017

ABSTRACT

This paper will describe the work carried out by Pfizer in the development of near infrared spectroscopy, for the analysis of product and nutrient concentrations, in commercial fermentation processes. The technique has been developed for the rapid monitoring of antibiotic and enzyme fermentations, as both an at-line and on-line method. In addition the use of the technique as a powerful development tool will also be described.

Key words: Near infrared, fermentation, analysis, fibre optics, robot.

The use of NIR for multicomponent analysis in the Agriculture and Food industries is well documented (Kaffa et al. 1982, Murray et at 1983). We believed that our fermentation processes could be analyzed in a similar way to those complex matrices, taking advantage of the unique ability of NIR to analyse neat, unprepared samples. The rapid analysis of samples would provide benefits in process monitoring and control. In particular we wished to measure fat, and product concentrations and growth level in our antibiotic fermentations, at-line initially, but progressing to on-line analysis as we gained experience. The applicability of the at-line measurement systems to enzyme production fermentations has also been investigated.

A robot based autosampler for non-attended NIR analysis of large batches of samples was also developed.

All concentration data has been removed from the figures, as the information is confidential.

EXPERIMENTAL

1) Safety

NIR analysis is in itself non-hazardous. Any hazard will be associated with the sample. Appropriate precautions must be taken when handling samples.

2) Apparatus

The work in this paper was performed using an NIRSystems 6500 scanning monochromator (NIRSystems, Silver Spring, Maryland 20904). The instrument was equipped with a sample transport module for all at-line work. A static sample cup module was used for flow cell work, and a fibre optic module for the on-line fibre optic work. The fibre used was standard NIRSystems "Interactance" probe (part No. NR-6640). The software used for this work was the NSAS package provided with the instrument.

3) Procedure for at-line calibration

Calibration was carried out by analysing in replicate 50-60 broth samples, using standard analytical methods, and collecting their reflectance spectra in a 10 mm pathlength cuvette. The samples were selected to be representative of all the variations, likely to be encountered; process changes, medium formulation variations and the natural variation in substrates. Several similar calibrations were produced for each parameter and subjected to a validation process resulting in selection of the most robust calibration for actual use.

DISCUSSION

1) At-line calibrations for antibiotic production

A statistical comparison of reference and NIR methods is shown in figure 1. The NIR methods are shown to give equivalent results to the reference assays. Figure 2 illustrates the correlation for antibiotic as a plot. Three parameters; antibiotic produced, cell growth and fat level are measured simultaneously from a single broth scan (fig 3). The spectroscopic details of these assays are given in figure 4.

2) Time dependant calibrations

For the measurement of produced antibiotic, the use of locally weighted regression (Naes etal.1991) has been effective in compensating

ALL NUMBERS HAVE BEEN NORMALISED TO PERCENTAGES

CONSTITUENT	REFERENCE ASSAY	S.E. REF METHOD	R	S.E.P. NIR METHOD
FAT	SOLVENT EXTRACTION	9.1	0.984	4.1
CELL GROWTH	DRY CELL WEIGHT	10.4	0.975	6.2
ANTIBIOTIC	HPLC	2.0	0.995	2.2

R = THE REGRESSION COEFFICIENT

S.E.P. = INSTRUMENT STANDARD ERROR OF PERFORMANCE =

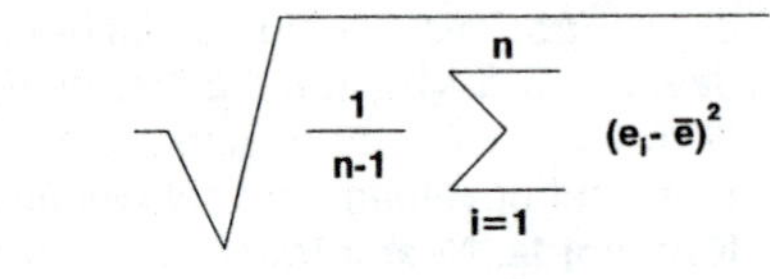

$$\sqrt{\frac{1}{n-1}\sum_{i=1}^{n}(e_i-\bar{e})^2}$$

WHERE $\bar{e} = \frac{1}{n}\sum_{i=1}^{n} e_i$ AND e_i = REFERENCE - NIR VALUE

S.E. = STANDARD ERROR OF THE REFERENCE ASSAY CALCULATED FROM THE POOLED STANDARD DEVIATION OF TRIPLICATE ASSAYS PERFORMED ON TWENTY DIFFERENT SAMPLES.

$$S = \sqrt{\frac{(S_1)^2 + (S_2)^2 + ... + (S_n^2)}{n}} \qquad S_n^2 = \frac{\sum (X_1 - \bar{X})^2}{r-1}$$

Fig.1. Statistical comparison of the NIR methods relative to the reference assays.

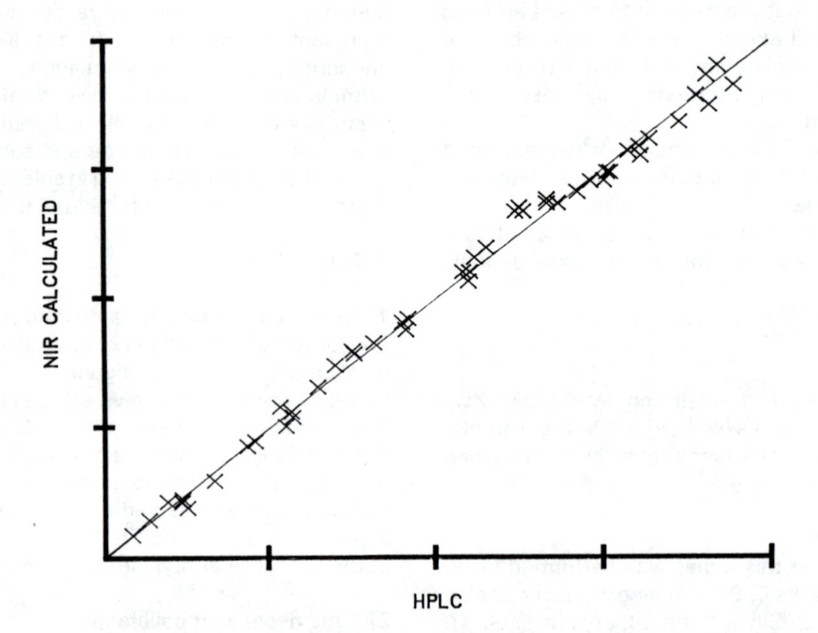

Fig.2. Correlation plot for at-line antibiotic measurement. HPLC data against NIR absorbance at 1660 nm, the C-H aromatic overtone. (Reproduced with permission from Hammond 1991. Copyright *NIR News*)

TIME	OIL	DRY WEIGHT	POTENCY
11:43:39	XXXX	XXXX	XXXX
11:45:22	XXXX	XXXX	XXXX
11:49:04	XXXX	XXXX	XXXX
11:50:58	XXXX	XXXX	XXXX
11:52:51	XXXX	XXXX	XXXX

Fig.3. Print out produced during routine analysis. Three constituents are measured simultaneously from one scan. Adapted from Hammond 1991.

for a changing background matrix as raw materials break down and cell mass increases, (fig 5). Up to three calibrations have been used in time sections. This improves the overall accuracy of the assays generated.

3) Applicability to Enzyme Production Fermentations

Figure 6 shows a correlation of laboratory measured and NIR calculated enzyme activity for an alkaline protease manufactured by Pfizer. The spectra used to produce the correlation are shown in figure 7. The Partial Least Squares Multivariate regression technique was used over the range 460-740 nm (lower order overtone absorptions of protein) to produce the correlation, modelling the very reproducible spectral changes associated with increased enzyme activity. The correlation shows the excellent potential for monitoring enzyme production by NIR spectroscopy.

4) Benefits of moving to NIR testing for antibiotic fermentations

The assay time for each antibiotic sample has been reduced from two hours down to two minutes. The shorter assay time has allowed increased assay frequency leading to very much improved process control and better process understanding. All of which has improved productivity.

CONSTITUENT	NIR ABSORBANCE
FAT	1716nm, THE CH_2 OVERTONE ABSORPTION OF FATTY ACIDS
CELL GROWTH	2394nm, THE CH_2 COMBINATION ABSORPTION OF TECHOIC ACIDS
ANTIBIOTIC	1660nm, THE C-H AROMATIC OVERTONE

Fig.4. NIR absorbances of functional groups used to construct calibrations.

5) On-line fat measurement

Having established robust systems for at-line analysis of antibiotic fermentations our effort has been directed to moving the system on-line for the measurement of fat concentration. Two different approaches have been successful at producing good real time spectra of broth. The first approach has been to pump broth out of the reactor through a custom-made flow cell and return the scanned broth back into the reactor (fig 8) using an aseptic circulation loop. The drawback to this approach is the risk of contaminating the fermenter.

The second approach was to scan the broth in situ through a glass window in the reactor wall using a remote fibre optic sensor. Thus avoiding the risk to the aseptic environment of the fermenter (fig 9). In the remote optics system the detectors are placed at the end of a fibre bundle thus avoiding the loss of intensity which occurs when the reflected signal has to be transmitted back via a second fibre bundle to detectors in the instrument itself.

Good spectra of the broth were obtained by both approaches, producing good calibrations for fat concentration (fig 10). Figure 11 shows the spectral region used to correlate laboratory fat data with the observed change in the spectral absorption. This is a well formed symmetrical peak at 1716 nm, characteristic of the CH_2 absorptions of fatty acids. The calibration for fat has been tested by monitoring fat level on-line, whilst sampling the bio-reactor and performing conventional solvent extraction analysis. The agreement between the two sets of analytical results is very good (fig 12). This on-line approach allows tight process control and improved process performance.

5. Automated Analysis.

The use of fibre optic probes has been extended to the automated analysis of small scale fermentation broths used in mutation screening programmes. Trays of 24 samples are placed in a reading rack by a robot arm, the arm then changes hands and manipulates a fibre optic probe scanning each sample in the tray in turn (fig 13). The robot arm can exchange and scan up to 100 trays on a fully automated basis. The probe consists of two

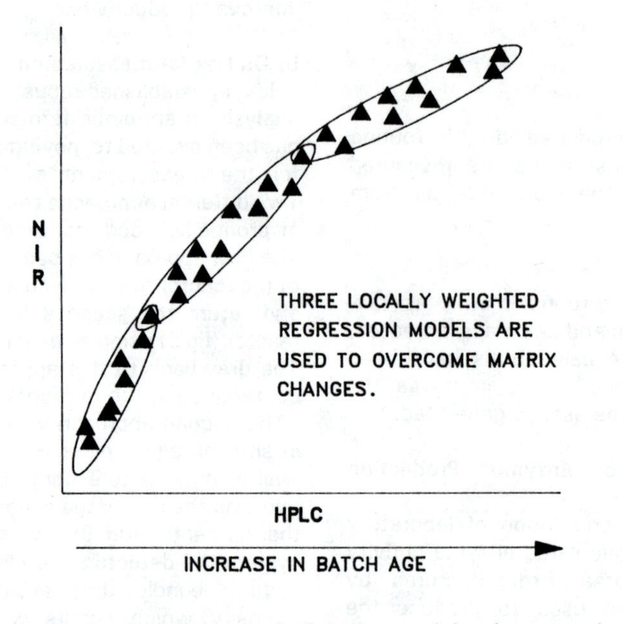

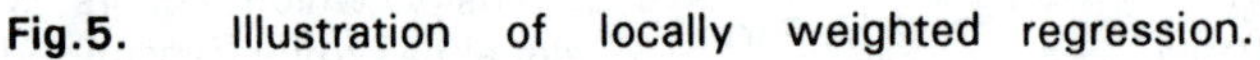

Fig.5. Illustration of locally weighted regression.

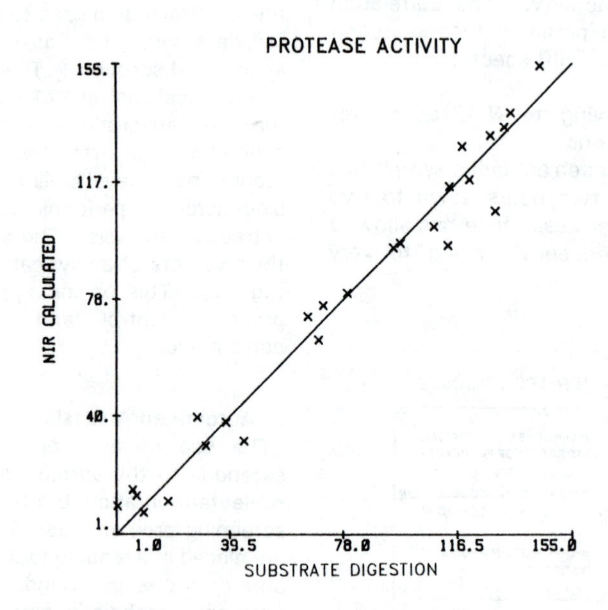

Fig.6. Correlation plot for at-line enzyme activity. Substrate digestion against NIR absorbance over the range 460-740 nm.

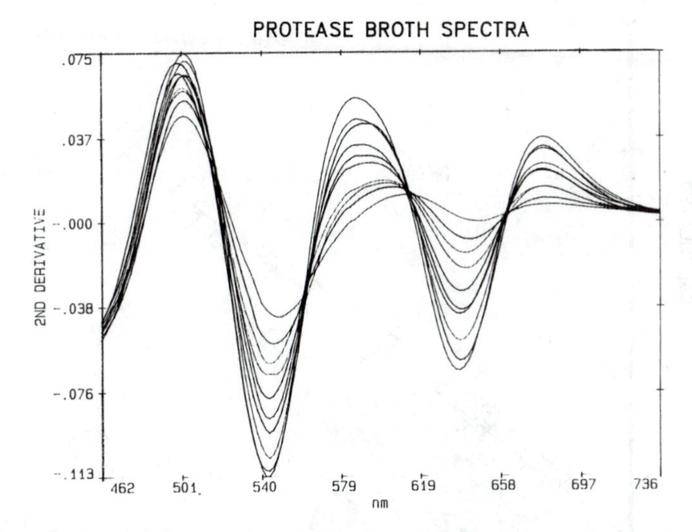

Fig.7. Spectral region used for PLS multivariate regression analysis of enzyme activity.

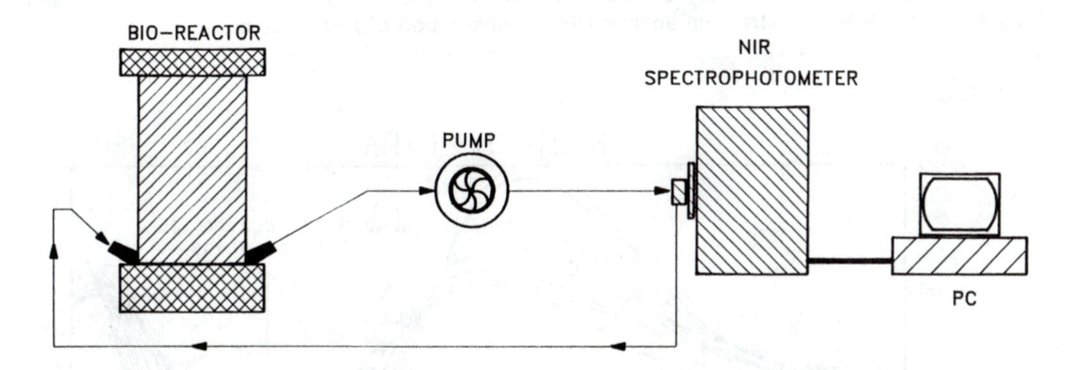

Fig.8. Illustration of the "flow cell loop" on-line system. (Adapted from Hammond, 1991).

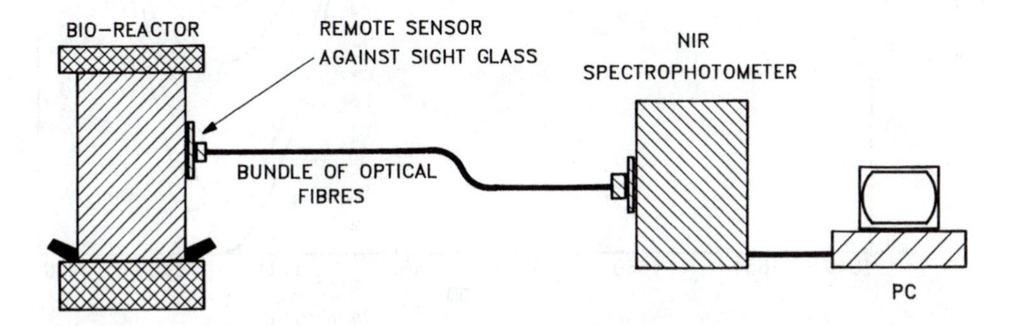

Fig.9. Illustration of the "remote fibre optic" on-line system. (Adapted from Hammond, 1991).

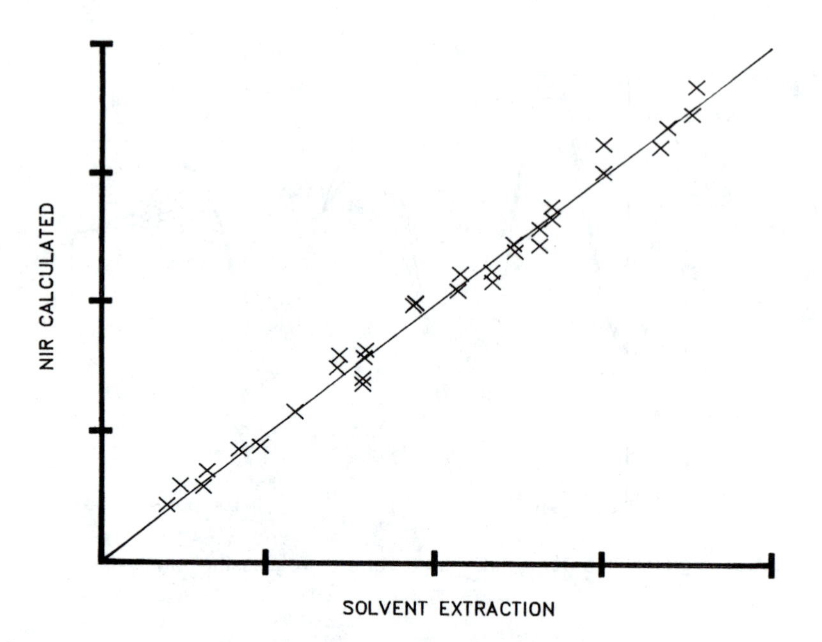

Fig.10. Correlation plot for on-line fat measurement. Solvent extraction against NIR absorbance at 1716 nm, the CH_2 overtone absorption of fatty acids.

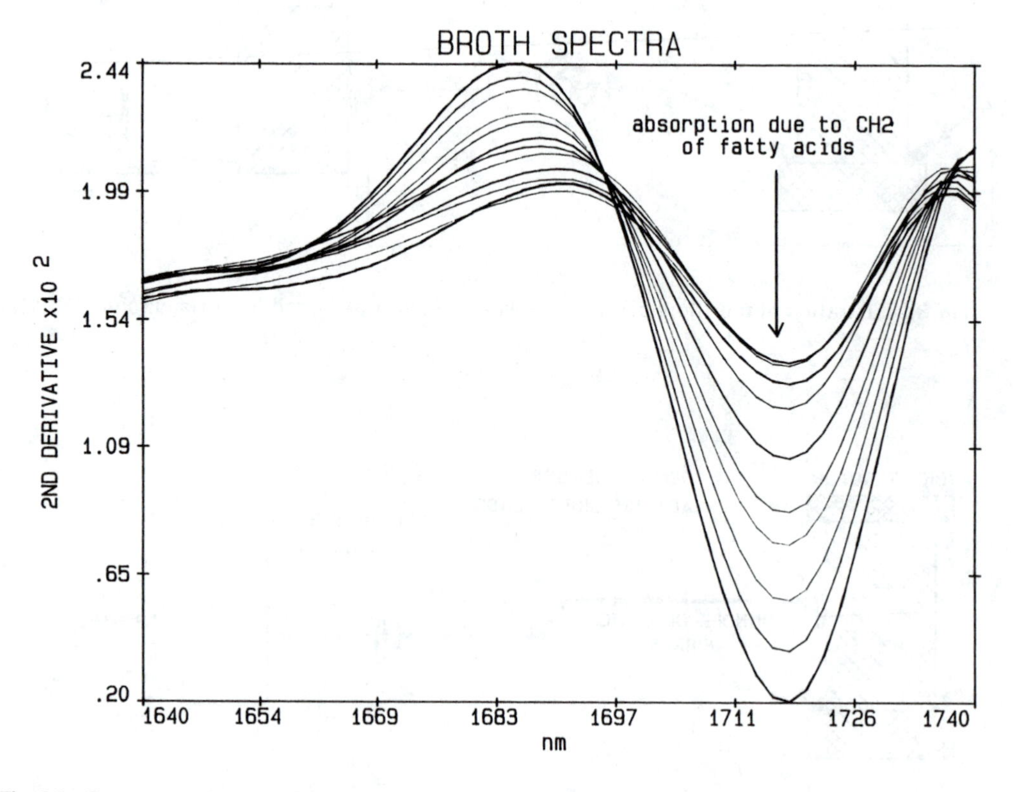

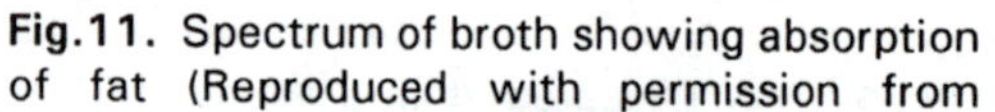
Fig.11. Spectrum of broth showing absorption of fat (Reproduced with permission from Hammond 1991. Copyright *NIR News*).

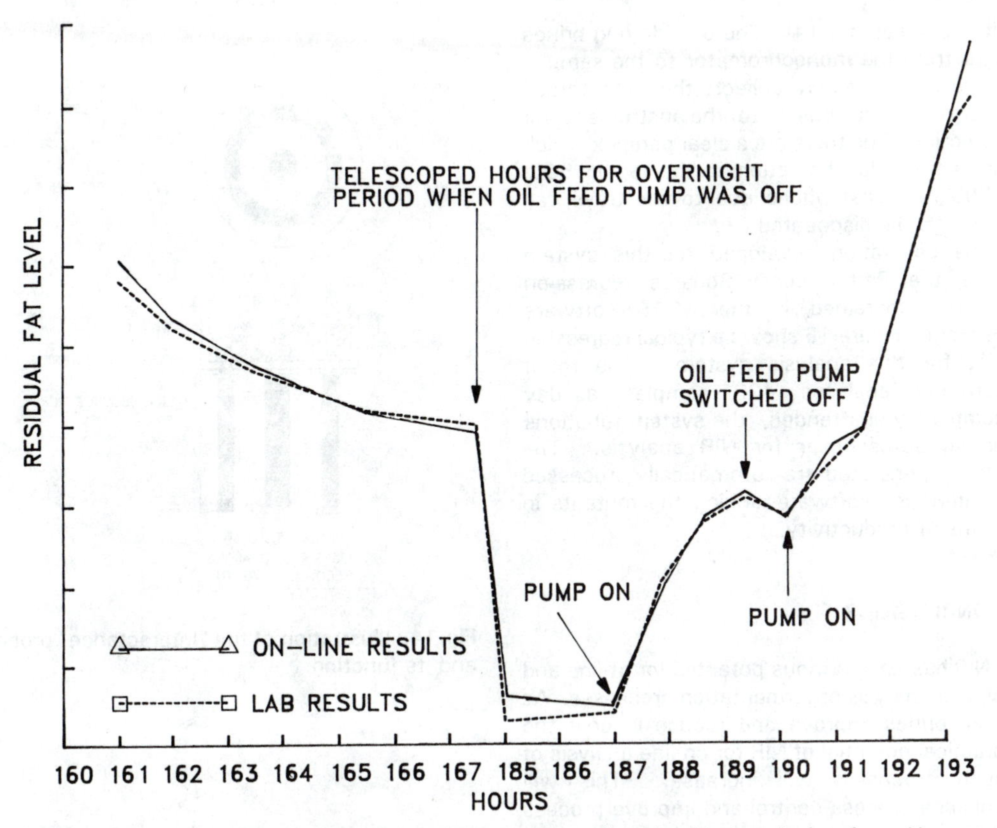

Fig.12. Plot of solvent extraction laboratory analysis against on-line flow cell analysis. (Reproduced with permission from Hammond 1991. Copyright *NIR News*).

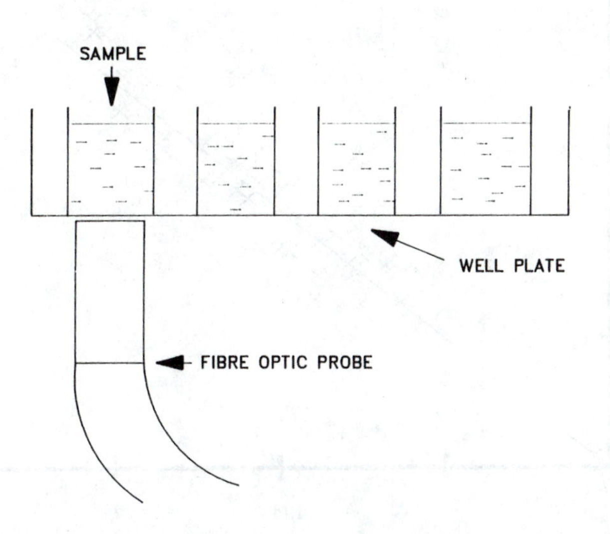

Fig.13. Diagram of the probe scanning a well plate. (Reproduced with permission from Hammond 1991. Copyright *NIR News*).

fibre bundles (fig 14). The outside ring brings light from the monochromator to the sample, while the inner core collects the "interacted" energy and returns it to the instrument for detection. The trays are a clear perspex which has very little absorption in the region 500-1300 nm, absorptions due to the trays can therefore be discounted.

The calibration developed for this system uses the Partial Least Squares regression module contained in the NSAS software package. Figure 15 shows a typical regression plot for this analysis system. The robot currently analyses 400 samples a day completely unattended, the system functions as an autosampler for NIR analysis. The results generated are automatically processed by in-house software-ranking the mutants in terms of productivity.

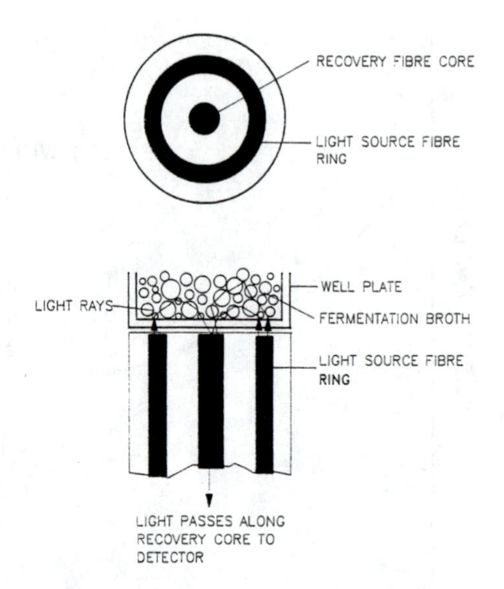

Fig.14. Illustration of the "Interactance" probe and its function.

CONCLUSIONS

NIR has an enormous potential for at-line and on-line analysis of fermentation processes. As fibre optics improve and reduce in cost the practical potential of NIR for on-line analysis of such processes will increase. This will enhance process control and improve process performance.

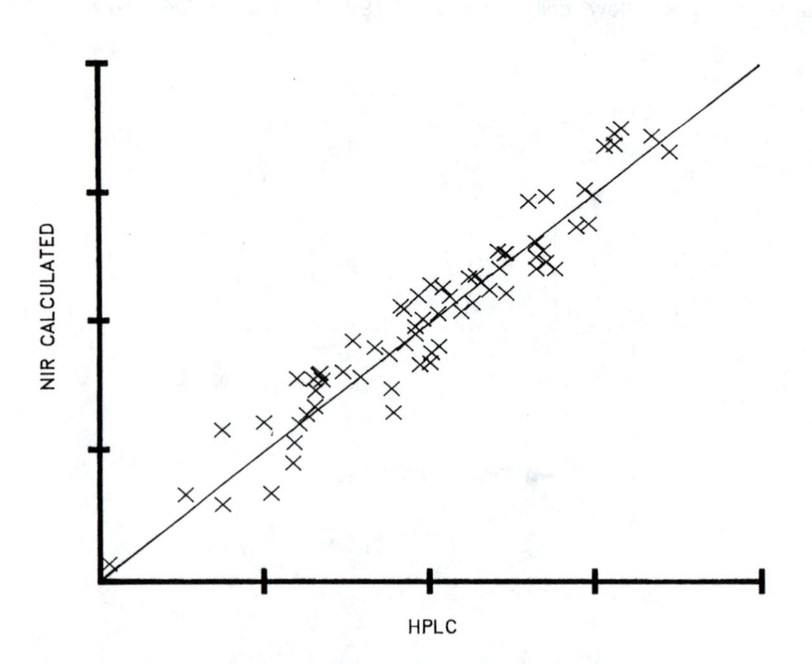

Fig.15. Correlation plot of HPLC against NIR calculated potency of wells, robot system. (Reproduced with permission from Hammond 1991. Copyright *NIR News*).

ACKNOWLEDGEMENT

Some of the information and figures used in this paper were first presented at the Fourth International Conference for Near Infrared Spectroscopy, University of Aberdeen, Scotland, August 1991.

BIBLIOGRAPHY

1. KAFFKA, K.J., NORRIS, K.H. and ROSZA-KILL, M. Determining fat, protein and water content of pastry products by the NIR technique, *Acta Alimentaria*, vol. 11 No.2 pp 199-217, 1982.

2. MURRAY, I and HALL, P.A. Animal feed evaluation by use of NIR reflectance spectro computer. *Analytical Procedure*, Volume 20 pp. 75-79, February 1983.

3. NAES, T. and ISAKSSON, T. Nonlinear Calibration by Locally Weighted Regression. *Proceedings 3rd International Conference for NIR Spectroscopy 1991, Vol. 1 page 145.*

4. Hammond, S.V. NIR Analysis of Antibiotic Fermentations. *Proceedings of the Fourth International Conference for NIR Spectroscopy*, Aberdeen 1991.

Fiber Optical Methods for Bioprocess Monitoring - Promises and Problems.

Jerome S. Schultz
Center for Biotechnology and Bioengineering, University of Pittsburgh, Pittsburgh, Pennsylvania 15261 (FAX: 412 648 7762)

Recently the intense research interest in biosensors has led to many achievements in the development of fiber optic biosensors for a variety of biochemicals. By coupling enzymes, antibodies, and membrane receptors to fiber optic devices, biosensors with a high degree of selectivity and sensitivity have been demonstrated. However, there are some major technical problems that need to be solved before these devices can be used for routine monitoring of biochemical processes. The primary problem to be solved before these devices can be used routinely is the development of techniques that will allow the sensors to be calibrated in situ, or to utilize those properties of sensors that are refractory to modest changes in amounts or activity of the reagents used in the sensor. Some current efforts to solve this problem are reviewed.

Biosensors, with their high degree of sensitivity and selectivity for biochemicals within complex biological fluids, have obvious potential to fulfill many critical biomedical and clinical needs. However, as the power and generality of biosensors has become to be appreciated, their tremendous promise has emerged in diverse other fields such as the food industry, agriculture, the environment and biotechnology, in particular, bioprocessing.

Many investigators have coupled biological recognition elements (enzymes, antibodies, membrane receptors, binding proteins, DNA or RNA segments, etc) to optical fiber sensors to produce biosensors with a high degree of selectivity so that they can be used to measure individual biochemicals within complex biological fluids (For recent reviews see Sepaniak, et al, 1988; Wolfbeis, 1991; and Wise and Wingard, 1991).

Part of the reason for the explosive interest in fiber optic biosensors is the potential information content obtainable from measurements in the optical domain which is extraordinary in comparision to other detection methods - e.g. electrochemical, capacitance, piezoelectric, quartz crystals, etc. By utilizing the specific signatures of compounds -- absorption bands, fluorescence characteristics, reflectivity, refractive index changes -- and coupling these properties to the specificity of biomolecular interactions, exquisitely finely tuned analytical devices have been constructed for substances as diverse as glucose, penicillin, lactate, ethanol, and aromatic hydrocarbons. And thus the promise of fiber optic biosensors for bioprocessing.

A wide variety of optically based analytical methods have been adapted for biosensor applications. These optical methods utilize either the intrinsic or extrinsic modes of operation of fiber optic wave guides. In the intrinsic mode of operation, a portion of the cladding is removed from the fiber at the distal end. In the decladded region, there is energy exchange between the fiber and the surrounding sample fluid through an evanescent standing wave. The intensity of illumination in this region falls off exponentially with distance with the depth of interaction on the order of the wavelength of light. Thus, in this mode of operation, the optical fiber "senses" primarily surface events. In the extrinsic mode of operation, light eminates from the cleaved end of the optical fiber, and a zone of fluid is illuminated at the distal end of the fiber. Responsive reagents are placed in the illuminated region so as to provide optical signals that are related to the concentration of the analyte of interest, Fig. 1.

Yet, relatively few of these devices have made

1054–7487/92/0334$06.00/0

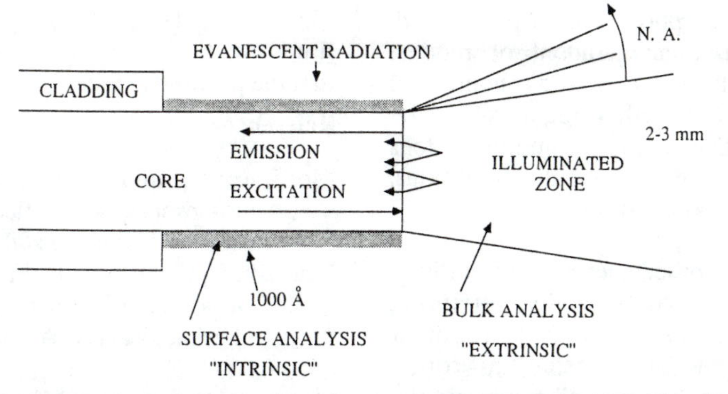

Figure 1. Schematic drawing of the modes of operation of fiber optic devices in conjunction with biosensors. In the evanescent (intrinsic) mode, cladding is stripped from the core and the reagents are placed on the surface of the fiber. The terminal end of the fiber is blackened or mirrored. The range of interactions with the surrounding fluid is on the order of the wavelength of light. In the extrinsic mode, the cladding is intact and the illumination region is outside the cleaved end of the fiber providing a useful analytical path length of about 2-3 mm. The width of the field of illumination is determined by the numerical aperture (N.A.) of the fiber.

their way into bioprocessing practice. Two key problems which have prevented their widespread use in industrial bioprocess monitoring and control are calibration and sterilization. Industrial expectations for monitoring devices are that they should be robust enough to be used on-line in the process stream. For bioreactor use this poses difficult problems. First, current industrial practice requires steam sterilization of fermentors and auxiliary equipment. In this regard, it is not the optical detector components that pose the problem, as sterilizable fluorimeter probes are commercially available. It is the biochemical and membrane elements of the sensor that are highly fragile to steam sterilization. Perhaps this factor could be overcome by creative materials engineering, if a solution to the other critical problem - reliable calibration techniques - could be developed.

The calibration of an analytical device is often achieved by simply periodically analyzing a series of samples with "known" concentrations with the procedure, and using this standardization curve to determine the concentration of "unknowns." Recently, this approach has been used with biosensors in the "flow injection analysis" (FIA) mode (Ruzicka and Hansen, 1988). Using FIA, biosensors can be used in a kinetic regime, that is, without the sensor and sample reaching equilibrium. FIA is being used in bioprocess monitoring, especially for process streams where representative samples can be withdrawn periodically. A more desirable mode of operation is to obtain continuous information by placing biosensors directly inside process units. In these situations, one cannot arbitrarily change the concentration of the analyte for calibration purposes, but at most can withdraw samples for analysis by other laboratory techniques, and use this information for "one point" calibration of the sensor.

Thus there is a need to develop fiber optic biosensors that can be calibrated by one-point methods. One of the major sources of variation for optical methods is the intensity of the light source, another is change in concentration of reagents within the sensor (e.g. chromophore, enzyme, substrate). Although a general solution to these problems have not emerged as yet, a number of approaches are being developed that will lead to more robust instruments.

An early approach to achieve independence of light intensity and chromophore variations was demonstrated in a fiber optic pH sensor (Peterson, et. al. 1980) that utilized the ratio of absorbance at two wavelengths, the isobestic point and the absorption peak of phenol red. The calibration is independent of modest variation in light intensity and phenol red concentration. The method requires that the absorption spectrum of the chromophore does not change due to environmental conditions, e.g. temperature, ionic strength.

For fluorescence based methods, in addition to light source and concentration variations, temperature, quenching due to oxygen or other interferences, energy transfer to other chromophores, and solvents all can affect the signal intensity. One approach that may overcome these effects is to devise sensor systems that utilize fluorescent lifetimes as an indicator, rather than fluorescent intensity (Petrea, et al, 1988).

Another approach which can be generally applied to test the integrity of the biosensor is to place an inert and noninterfering chromophore within the analytical chamber. The presence of this component can be measured at a different set of wavelengths, and an apparent loss of this component will indicate a breakdown in the sensor (Meadows and Schultz, 1988).

The problem of developing methodologies to check and calibrate fiber optic biosensors in situ is one that is worthy of wider attention so that the promise of biosensors can be achieved.

References

Meadows D, and Schultz JS. Fiber optic biosensors based on fluorescence energy transfer. Talanta **1988**, 35, 145-150.

Peterson JI, Goldstein SR, Fitzgerald RV, and Buckhold DK. Fiber optic pH probe for physiological use. Anal. Chem. **1980**, 52, 864-869.

Petrea RD, Sepaniak MJ, and Vo-Dinh T. Fiber optic time resolved fluorimetry for immunoassays. Talanta **1988**, 35, 139.

Ruzicka J, and Hansen EH. Flow injection analysis. **1988**, J. Wiley, New York, 498 pp.

Sepaniak MJ, Tromberg BJ, and Vo-Dinh T. Fiber optic affinity sensors in chemical analysis. Prog. Anal. Spectros. **1988**, 11, 481-509.

Wolfbeis OS Ed. Fiber Optic Chemical Sensors and Biosensors. Volume I (413pp) and Volume II (358 pp). CRC Press, Boca Raton, **1991**.

Wise DL and Wingard LB, Eds. Biosensors with Fiberoptics. Humana Press, Clifton, 370 pp, **1991**.

Acknowledgement

This work was supported in part by an AFSOR grant to the Pittsburgh Materials Research Center.

Modeling of Microbial Processes. The Status of the Cybernetic Approach.

Doraiswami Ramkrishna, Jeffrey V. Straight and **Atul Narang**
School of Chemical Engineering, Purdue University
West Lafayette, Indiana 47907
Allan E. Konopka, Department of Biological Sciences
Purdue University, West Lafayette, IN 47907

The present status of our cybernetic approach to modelling metabolic regulation in microbial processes is summarized briefly reviewing its accomplishments to date. This approach views regulation of the synthesis rates and activities of enzymes catalyzing different biochemical reactions as an optimal response of the organism based on economic utilization of its internal resources. In its current state of evolution, the influence of various intermediates on different metabolic fluxes in a pathway is the direct concern of cybernetic models consequently making them very promising for numerous applications to biotechnology.

Simple kinetic models have sufficed for the description of microbial processes in circumstances when environmental changes do not call for significant metabolic regulation. A classic example of this is the ability of the Monod model to describe batch growth in a culture in which a single rate limiting substrate is present. More complex environmental changes, however, provoke regulatory response of a kind that falls outside the scope of simple kinetic models. Thus the phenomenon of diauxic growth in the presence of mixed, substitutable substrates discovered by Monod (1942) cannot be described by his model. Similarly, steady state biomass concentrations at low dilution rates in a continuous culture cannot be described without incorporating the effects of maintenance. Although a simple way to account for maintenance (Herbert et al., 1956; Pirt, 1982) can overcome the limitations of the Monod model in describing steady state continuous culture data, *transient* behavior falls completely outside the scope of such a model. In general, transient behavior of microbial cultures is complicated by the manifestation of metabolic inertia, which produces dynamic variations not describable by kinetic models neglecting metabolic regulation. When multiple limitations exist in the microbial environment, the behavioral complexity is further enhanced by the regulation of "fluxes" of intermediates along specific directions in the different metabolic pathways. The presence of growth intermediates can also induce regulated behavior of a very complex nature. It is evident that engineering strategies for controlling microbial processes are critically dependent on quantitative understanding of metabolic regulatory processes.

The Cybernetic Approach

The cybernetic approach to modelling metabolic regulation is based on attributing to the regulatory process an optimal goal broadly in terms of economic utilization of existing internal resources. More specifically, enzymes for different reactions competing for cellular resources (either to consume a common reactant or produce a common product) are synthesized by expending cellular resources at a rate depending on the "returns" from each reaction as measured by its *rate*. Each reaction rate is modified by the relative enzyme level (prevailing enzyme level relative to its maximum attainable value) multiplied by its *activity* (a cybernetic variable denoted v). The synthesis rate of a specific enzyme is modified by a second cybernetic variable, u which represents the fractional allocation of resources (e.g., amino acid pool; alternatively, metabolic intermediates and energy) to the synthesis of that enzyme.

The two sets of cybernetic variables $\{u_i,v_i\}$, where i symbolized the ith competing reaction, are computed simply by using microeconomic principles. The cybernetic variable u_i is given by the *matching law* (Kompala et al., 1986) equating it to the ith reaction rate divided by the *total reaction rate*. The cybernetic variable v_i is given by the *proportional law* which compares the ith reaction rate to the maximum of the different reaction rates involved.

In the foregoing, the term "reaction" must be understood as a "lumped reaction" among "lumped species" which are simplifying assumptions characteristic of models of microbial growth. Similarly, it is preferable to assign this lumped status also to the term "enzyme."

The early cybernetic models successfully predicted the diauxie growth phenomenon observed by Monod for bacterial growth in the presence of multiple, substitutable substrates (Kompala et al., 1986; Ramkrishna et al., 1987). Subsequently, progressive refinements in cybernetic models have led to successful predictions of a large variety of experimental observations in fed batch and continuous cultures with single or mixed (carbon) substrate feeds. Thus with inclusion of maintenance effects, cybernetics models have been able to describe metabolic switches between maintenance and growth processes when cultures under "starvation" conditions with minimal amount of a particular carbon substrate are interrupted by replenishments with the same or an alternative carbon substrate; these models also described steady state continuous cultures fed by one or more carbon substrates (Turner et al., 1988, 1989).

Metabolic inertial effects are manifest in dynamic phenomena associated with continuous cultures. Kinetic models which describe steady states at different dilution rates are unable to predict transients following switches between dilution rates. The cybernetic models of Baloo and Ramkrishna (1991a, 1991b), by incorporating a lumped resource for the protein synthesis system, successfully predict "overshoots" and "undershoots" appearing in transients following "step-up" and "step-down" in dilution rates. More complex transient behavior in lactose-limited cultures is described by the cybernetic models of Straight and Ramkrishna (1990) which featured regulation of transport and growth processes.

Transport effects also appeared in the work of Alexander and Ramkrishna (1991) who have used cybernetic models to describe regulation in iron-limited bacterial cultures in which a metabolite (siderophore) is produced by the cells to scavenge iron from the abiotic phase.

Metabolic Pathways

Recent work by Straight in his doctoral dissertation (1991) has extended the cybernetic modelling concept to entire metabolic pathways. He recognizes different component "units" of the reaction pathway designated as *linear segments*, *diverging branch points*, *converging branch points*, and *cyclic processes*, which are reproduced in Fig. 1. Straight regards the different units as "uncoupled" from a cybernetic viewpoint which means that cybernetic variables are calculated for each unit based on a *local* optimality criterion. This optimality criterion depends on the nature of the unit. Where a single end product is involved, as in the case of linear segments and converging branch points Straight proposes maximizing formation of the end product, while for multiple end products as in a diverging branch point, the mathematical product of all the end products is proposed to be maximized. A similar procedure is used for cyclic pathways. The entire approach is presented in terms of clearly stated postulates.

To provide a sample of the nature of the thinking in the above approach consider the linear segment. Regulation will prefer activation and synthesis of the enzymes for the "upstream" reactions when intermediates have not accumulated. Upon accumulation of intermediates, however, the preference will switch to "downstream" reactions. As another example, for a diverging branchpoint the accumulation of a product from a particular branch will lead to its automatic regulation favoring those whose end products have yet to accumulate thus incorporating the effect of feedback repression and inhibition. The cybernetic laws for enzyme synthesis and activation have remained essentially the same as in prior work because of the manner in which microeconomic principles have been incorporated.

It is important to realize that no phenomenological constants other than those characterizing the kinetics of the different reactions in the pathway are present in the approach. In other words, regulation does not add any adjustable parameters to the kinetic model.

Yet another important feature of Straight's approach is the the description of growth in the presence of *complementary* substrates without regard to whether they are "interactive" or "non-interactive", an issue that has been of interest in the past (Bader, 1982, Baltzis and Fredrickson, 1988). This was accomplished by invoking the concept of a diverging branch point and the cybernetic formulation associated with it.

Straight (1991) has shown that his cybernetic model can predict batch, steady state and transient continuous culture behaviors under carbon-limited, nitrogen-limited, and under both carbon-and-nitrogen-limited conditions. Fig. 2 demonstrates successful predictions of steady state continuous culture data for *Escherichia coli* under carbon-limited conditions while Fig. 3 shows the contrasting trend of steady state cell mass concentrations under nitrogen-limited conditions. Straight also predicts several transient continuous culture data in both carbon- and nitrogen-limited cultures following shift-ups and shift-downs in dilution rate.

Growth in the Presence of Intermediates

Narang (1991) has performed experiments with growth of *E coli* in the presence of the metabolic intermediates pyruvate and fumarate,

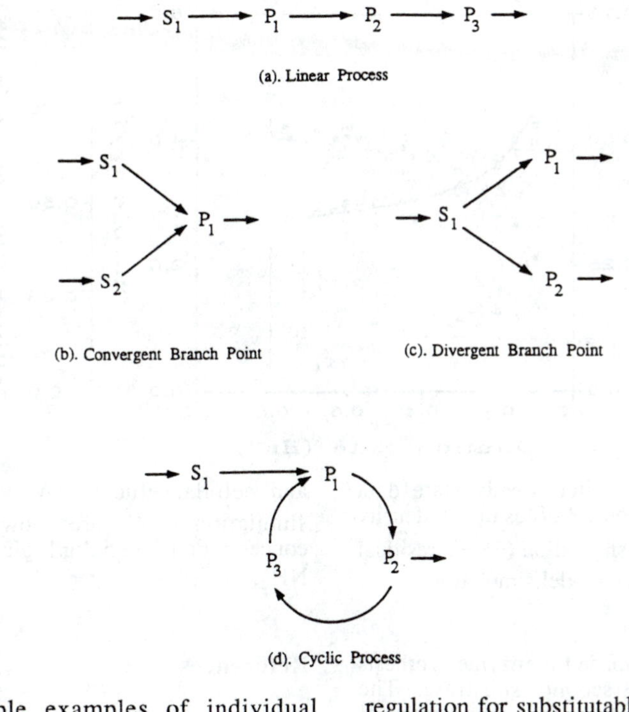

Figure 1. Simple examples of individual structural units commonly within metabolic pathways. Superscripts "s" and "c" denote regulation for substitutable anc complementary processes respectively.

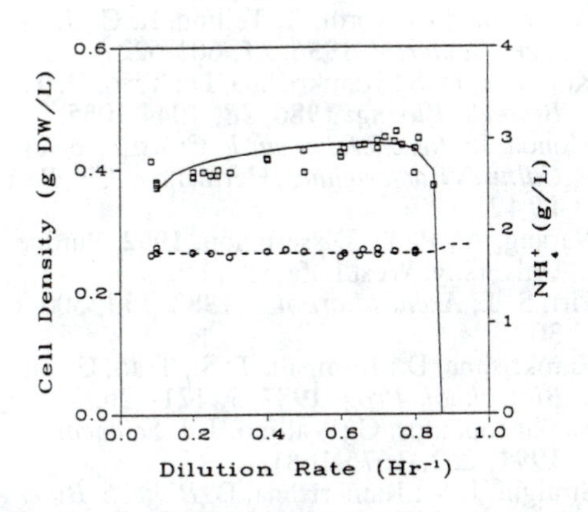

Figure 2. Carbon-limited steady state data versus model simulations. Profiles of cell density ( ) versus the model simulation (———) and NH_4^+ (O) versus the model simulation (----) are shown. Average feed concentrations are 0.98 g/L glucose and 1.85 g/L NH_4^+.

individually and with glucose. His experiments conclusively show that intermediates affect growth behavior in ways that significantly depart from the diauxie phenomenon. Thus although the growth rate on fumarate alone is greater than that on pyruvate alone when the two are supplied together, the organisms consume both substrates *simultaneously*. Interestingly, the growth rate on the two substrates exceeds that on either single substrate, and is equal to that on glucose alone. Furthermore if either of fumarate or pyruvate is present with glucose, then again simultaneous utilization of glucose and the second substrate occurs. These observations clearly challenge the matching law which was applicable to the case of diauxic growth.

Narang's investigation of this phenomenon has led to some interesting concepts of regulation that are as amenable to the cybernetic modelling approach as the diauxie itself. This explanation is built on the supply of growth precursors which are viewed to limit the rate of biomass synthesis. In meeting this precursor demand the substrate with the larger growth rate continues to be the "preferred" substrate. If the preferred substrate is able to meet the precursor demand there is no utilization of the "less preferred" substrate (constituting a case of exclusive utilization of the preferred substrate). On the other hand, if the preferred substrate does not meet the precursor

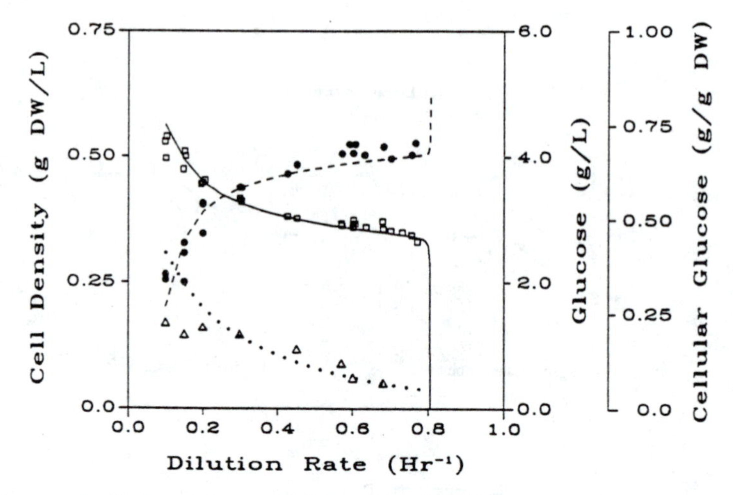

Figure 3. Nitrogen-limited steady state data versus model simulations. Profiles of cell density ( ) versus the model simulation (——), residual glucose (O) versus the model simulation (-----), and cellular glucose (Δ) versus the model simulation (······) are shown. Average feed concentrations are 5.0 g/L glucose and 66 mg/L NH_4^+.

demand, provision is made for enzyme synthesis for growth on the second substrate. The methodology submits to the calculation of cybernetic variables as in the matching law.

While the above investigation is still in progress, there are several other issues of generalization that are of interest to the analysis of metabolic pathways. Energetic considerations in the economic dispensation of catabolic pathways in meeting the demands of the anabolic machinery occupies a prominent place in future developments. Thus cybernetic coupling between the different metabolic units of Straight (1991) could become an important issue to investigate.

It is evident at this stage that cybernetic modelling has begun to address some of the important issues concerning the manipulation of metabolic fluxes by introducing intermediates of metabolism in the cells' environment. Thus what Galazzo and Bailey (1990), and Stephanopoulos and Vallino (1991) refer to as "metabolic engineering" falls naturally within the scope of cybernetic modelling. In this connection, the application of cybernetic modelling to bacterial bioremediation processes is under current investigation with the objective of promoting uptake of aromatic nutrients by invading the cellular environment with various growth intermediates.

Cybernetic modelling has provided us with a new way of *thinking* about regulatory processes, one which leads to *mathematical* models. This is akin to the manner in which chemical kinetics, a way of thinking about chemical reactions, leads to mathematical expressions for reaction rates.

References

Alexander, M.; Ramkrishna, D. *Biotech. Bioeng*. **1991**, *38*, 637-652.

Bader, F.; **1982**, In *Microbial Population Dynamics*; Bazin, M.J., Ed; CRC Press, Inc., Boca Raton, Florida, **1982**, pp 1-32.

Baltzis, B.; Fredrickson, A. G. *Biotech. Bioeng*. **1989**, *31*, 75-86.

Baloo, D.; Ramkrishna, D. *Biotech. Bioeng*. **1991b**, *38*, 1353-1363.

Galazzo, J. L.; Bailey, J. E. *Enzyme Microb. Technol*. **1990**, *12*, 162-172.

Herbert, D.; Elsworth, R; Telling, R. C., *J. Gen. Bacteriol*. **1956**, *14*, 601-622.

Kompala, D. S.; Ramkrishna, D.; Tsao, G. T. *Biotech. Bioeng*. **1986**, *28*, 1044-1055.

Monod, J.; *Rescherches sur la Croissance des Cultures Bacteriennes*, Hermann et Cie, Paris, **1942.**

Narang, A.; Ph.D. Dissertation, **1992**, Purdue University, West Lafayette, IN.

Pirt, S. J.; *Arch. Microbiol*. **1982**, *133*, 300-302.

Ramkrishna, D.; Kompala, D.S.; Tsao, G. T.; *Biotechnol. Prog*. **1987**, *3*, 121-126.

Stephanopoulos, G.; Vallino, J. J. *Science*, **1991**, 252, 1675-1681.

Straight, J. V.; Ramkrishna, D. *Biotech. Bioeng*. **1990**, *37*, 895-909.

Straight, J. V.; PhD. Dissertation, **1991**, Purdue University, West Lafayette IN.

Turner, B. G.; Ramkrishna, D.; Jansen, N. B., *Biotech. Bioeng*. **1988**, *32*, 46-54.

Turner, B. G.; Ramkrishna, D.; Jansen, N. B., *Biotech. Bioeng*. **1989**, *34*, 252-261.

Quantitative Studies of EGF/EGF-Receptor Phenomena in Cell Proliferation

Douglas Lauffenburger*, **Yong-Ho Khang, Gail Sudlow,** and **Kim Forsten,**
University of Illinois at Urbana-Champaign, Urbana, Illinois 61801
(FAX: 217-244-8068)
H. Steve Wiley,
University of Utah, Salt Lake City, Utah 84132

We describe our recent experimental and theoretical efforts aimed at increased quantitative understanding of the regulation of cell proliferation in culture by epidermal growth factor (EGF). Having previously shown that we can predict proliferative responses of fibroblast cell lines to EGF using a mathematical model that includes binding and trafficking dynamics of EGF and the EGF-receptor (EGFR) along with the dependence of proliferation on EGF/EGFR complex levels, we now explore some key features of EGF/EGFR phenomena. In particular, we provide new quantitative information on cell surface binding, intracellular recycling, and control of autocrine growth factor systems.

Control of cell proliferation is important for both bioprocessing and health care applications of biotechnology. Animal cell bioreactors offer the potential for reliable production of complex proteins, and the rational development of optimal growth media -- and cell lines optimized for desired growth properties -- remains a major goal. Improved wound healing therapies on one hand, and avenues for cancer prevention and treatment on the other, also depend on fundamental understanding of cell growth regulation. Central to all these problems is the role of peptide growth factors, by means of specific interactions with their corresponding receptors, in governing cell mitogenic behavior.

Dramatic advances in information concerning molecular properties of growth factors, growth factor receptors, and associated mitogenic signal transduction pathways have occurred in recent years (Sporn and Roberts, 1990). However, this information has been primarily descriptive and qualitative in nature. There is growing recognition that the promise of molecular cell biology will resist fulfillment unless it can be turned into a quantitative science, accessible to engineering analysis and design (Maddox, 1992).

Moving cell biology to a sound quantitative foundation has been the chief objective of our research program. Considering that receptor-mediated responses underlie most animal cell behavioral functions, our ambition is to develop mechanism-based mathematical models for such functions -- including cell proliferation, cell adhesion, and cell migration -- that can be used to analyze and design bioprocessing or health care systems. This general approach, combining principles and techniques from engineering and cell biology, has been termed "cellular bioengineering" (Lauffenburger, 1989). We are especially excited about the prospects for experimental test of such models by exploiting molecular biology methods such as site-directed protein mutation and monoclonal antibody generation to vary model parameters. The system of interest in this paper, that of cell proliferation control by EGF interactions with EGFR, is especially well-suited for attack by this approach.

Dependence of Cell Proliferation on EGF/EGFR Complex Levels

The point of departure for our attempts to develop quantitative understanding of EGF-dependent cell growth is the finding by Knauer *et al.* (1984) that the rate of DNA synthesis in fibroblasts is a simple, apparently linear, function of the number of EGF/EGFR complexes. The proportionality constant for DNA synthesis rate versus complex number can be thought of as an "intrinsic receptor/ligand complex signalling coefficient". Given this parameter for a given cell signal transduction pathway, the overall mitogenic signal can be influenced by receptor/ligand binding and trafficking dynamics, which govern the number of signalling complexes present. Hence, changes in parameters characterizing binding and trafficking

1054–7487/92/0341$06.00/0

events can affect signalling complex levels and thus the DNA synthesis rate.

To be more precise, processes leading to receptor downregulation and ligand depletion can serve to attenuate a mitogenic signal. Wells *et al.* (1990) recently found that substantially reduced EGF concentrations can stimulate growth of a fibroblast cell line transfected with a mutant EGFR (the Δ973 receptor, truncated after 973 amino acid residues of the 1186 residue wild-type receptor) possessing a diminished endocytotic internalization rate constant, compared to cells transfected with wild-type EGFR. Starbuck and Lauffenburger (1992) demonstrated that this reduction could be quantitatively predicted by a mathematical model incorporating the Knauer *et al.* relation and experimentally measured EGF/EGFR binding and trafficking parameters. The Δ973 receptor obtained less severe EGFR downregulation and EGF depletion than the wild-type receptor, minimizing signal attenuation. Based on this success, Starbuck and Lauffenburger went on to predict effects of other parameter changes such as receptor/ligand binding affinity and recycling rate constant. Our latest work has now been directed at gaining experimental data on molecular aspects of EGF and EGFR which might be used to vary these parameters.

EGFR Cytoplasmic Domain Mutations Alter EGF/EGFR Surface Binding Affinity

The often-observed appearance of multiple affinity forms of EGFR for surface binding of EGF has been hypothesized to involve interactions of an EGFR cytoplasmic domain with membrane-associated components (Mayo *et al.*, 1989), perhaps actin microfilaments (Rijken *et al.*, 1991). We have examined this hypothesis by measuring equilibrium and kinetic properties of EGF/EGFR cell surface binding using site-directed EGFR mutants. A key result is that the Δ1022 truncation mutant exhibits loss of the multiple-affinity forms; it yields a linear Scatchard plot for equilibrium binding instead of the concave-upward plot typically found for the wild-type EGFR (see Figure 1). All truncations removing additional cytoplasmic residues likewise give linear behavior, indicating that the receptor may interact with a membrane-associated component by means of a domain between residues 1022 and 1186. This possibility is supported by the transient dissociation data. The wild-type EGFR shows biphasic kinetics corresponding to fast- and slow-dissociating forms; the truncated receptor mutants obtain a significantly reduced slow-dissociating phase.

Hence, although one means for modifying EGF/EGFR binding affinity is through site-directed ligand mutants (Campion *et al.*, 1991), mutation of receptor cytoplasmic domains may also provide a useful avenue. It should be emphasized that different structural features are likely to be responsible for different aspects of receptor behavior. The 1022-1186 domain was found here to be involved in ligand binding affinity, while efficient endocytotic internalization has been previously shown to depend on the 973-991 domain (Chen *et al.*, 1989) and on tyrosine kinase activity (Wiley *et al.*, 1991).

EGF/EGFR Recycling Depends on Cytoplasmic Domain Interactions with Endosomal Components

Receptor and ligand recycling offers another event for modulating the number of signalling EGF/EGFR complexes. For a greater recycling rate constant, receptor downregulation and ligand depletion might be reduced, leading to a larger overall mitogenic signal. Therefore, we have begun to investigate recycling behavior using quantitative pulse/chase experiments to measure the recycling rate constant. Two major findings have emerged so far. First, the recycling rate constant is increased compared to the wild-type receptor when almost the entire cytoplasmic tail is eliminated via a Δ647 truncation. This suggests that a cytoplasmic receptor domain could be interacting with a membrane-associated component in the intracellular endosomal sorting compartment; this component would direct the receptor/ligand complexes to the lysosomal degradation compartment and away from the recycling pathway. Second, the recycling rate constant is higher for wild-type EGFR when greater number of EGF/EGFR complexes are internalized (see Figure 2). A possible explanation for this result is that the available endosomal sorting components, which are hypothesized to direct complexes into the degradation pathway and away from the recycling pathway, would be stoichiometrically overwhelmed at high complex levels.

Control of EGFR-Mediated Autocrine Proliferation Responses

Autocrine growth factor secretion offers yet another means for affecting cell proliferation. For example, A431 epidermoid carcinoma cells are known to secrete transforming growth factor type alpha (TGFa), which binds to the EGF-receptor to generate a mitogenic signal. Intentional introduction of autocrine EGF secretion capability into a desired cell line can easily be envisioned for bioprocessing purposes.

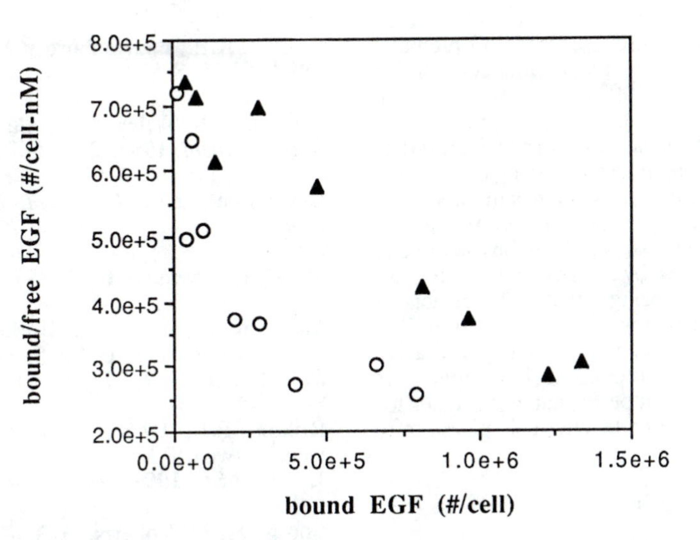

Fig. 1. Scatchard plots for equilibrium binding of EGF to EGFR surface receptors on B82 fibroblasts at low temperature preventing membrane trafficking. Binding for wild-type EGFR (open circles) exhibits nonlinearity while that for the Δ1022 truncation mutant EGFR (filled triangles) is linear.

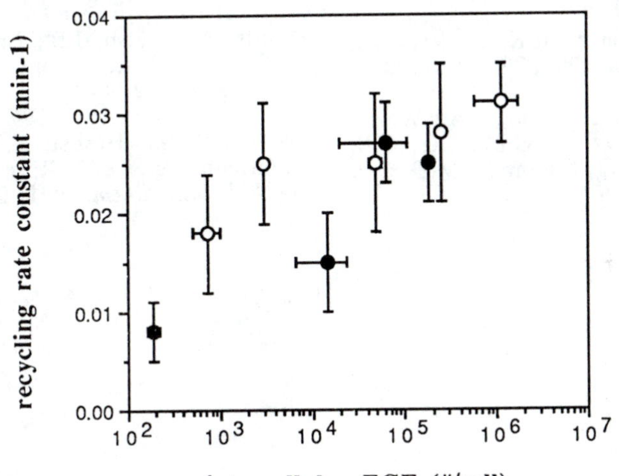

Fig. 2. EGF/EGFR recycling rate constant in B82 fibroblasts as a function of the number of internalized complexes. Open circles are wild-type EGFR, filled circles are mutant EGFR lacking tyrosine kinase activity.

In this case -- as well as in naturally-occuring malignant transformations -- an ability to suppress cell proliferation at will is desirable. A seemingly straightforward approach for this is the exogenous addition of factors in the extracellular medium which can inhibit receptor/ligand binding. Ligand "decoys", such as soluble receptors or anti-ligand antibodies are one class of inhibitors, as are receptor "blockers" such as anti-receptor antibodies or mutant, non-stimulatory ligands.

We have begun to examine possible inhibition protocols theoretically, using a mathematical model for decoy and blocking strategies. Competitive interplay between transport and binding processes makes prediction of satisfactory inhibition schemes problematic without this sort of analysis. Indeed, Forsten and Lauffenburger (1992) have predicted that ligand decoys may need to be present at concentrations in the range of four to eight orders of magnitude greater than the binding equilibrium dissociation constant in order to ensure more than 99% inhibition. Our more recent computations predict that receptor blockers may be effective at concentrations lower by a factor of 10 to 100, all kinetic parameters being equal. The effects of cell density for these two schemes are dissimilar, as well. Thus, we conclude that quantitative modeling of autocrine control strategies is necessary for efficient performance, whether for *in vitro* cell culture or *in vivo* cell physiology applications.

Acknowledgements

The authors would like to express their gratitude to the NSF Biotechnology Program for its support of this work.

References

Campion, S.R.; Matsunami, R.K.; Engler, D.A.; Niyogi, S.K. *Biochem.* **1991,** *29*, 9988-9993.

Chen, W.S.; Lazar, C.S.; Lund, K.A.; Welsh, J.B.; Chang, C.P.; Walton, G.M.; Der, C.J.; Wiley, H.S.; Gill, G.N.; Rosenfeld, M.G. *Cell* **1989,** 59, 33-43.

Forsten, K.E.; Lauffenburger, D.A. *Biophys. J.* **1992,** *61*, 518-529.

Knauer, D.J.; Wiley, H.S.; Cunningham, D.D. *J. Biol. Chem.* **1984,** *259*, 5623-5631.

Lauffenburger, D.A. *Chem. Eng. Ed.* **1989,** *Fall*, 208-213.

Maddox, J. *Nature* **1992,** *355*, 201.

Mayo, K.H.; Nunez, M.; Burke, C.; Starbuck, C.; Lauffenburger, D.A.; Savage, C.R. Jr. *J. Biol. Chem.* **1989,** *264*, 17838-17844.

Rijken, P.J.; Hage, W.J.; Van Bergen en Henegouwen, P.M.P.; Verkleij, A.J.; Boonstra, J. *J. Cell Sci.* **1991**, *100*, 491-499.

Sporn, M.B.; Roberts, A.B. *Peptide Growth Factors and their Receptors*; Springer-Verlag: New York, NY, 1990.

Starbuck, C.; Lauffenburger, D.A. *Biotech. Prog.* **1992** (in press).

Wells, A.; Welsh, J.B.; Lazar, C.S.; Wiley, H.S.; Gill, G.N.; Rosenfeld, M.G. *Science* **1990,** *247*, 962-964.

Wiley, H.S.; Herbst, J.J.; Walsh, B.J.; Lauffenburger, D.A.; Rosenfeld, M.G.; Gill, G.N. *J. Biol. Chem.* **1991,** 266, 11083-11094.

Prediction and Experimental Investigation of the Effect of Altered Peptide Chain Assembly Rate on Antibody Secretion Using a Structured Model of Antibody Synthesis and Secretion

Michael C. Flickinger*, **Theodora A. Bibila, and Kirsten Kitchin,** Department of Biochemistry and Institute for Advanced Studies in Biological Process Technology, University of Minnesota, St. Paul, Minnesota, 55108 U.S.A. FAX: 612-625-1700

We have investigated monoclonal antibody (MAb) synthesis, heavy (H) and light (L) chain assembly, interorganelle transport and secretion using a structured kinetic model and experimental results obtained from a murine hybridoma (9.2.27). Simulation results suggest that the assembly rate of the H and L chains can significantly affect MAb secretion rate during rapid growth. To alter chain assembly rate, over-expression of a thiol:disulfide exchange enzyme (PDI,ERp59) involved in catalyzing disulfide bond formation between the H and L MAb chains is being studied in 9.2.27 cells in comparison to endogenous levels in the non-MAb secreting parent myeloma P3X63-Ag653.

The ability to enhance the productivity of multichain glycoprotein synthesis and secretion *in vitro* from myeloma or hybridoma cells would be a significant advance for large scale production of therapeutic immunoglobulins, antibodies with altered structure or antibody fragments. The ability to manipulate the secretion rate of lymphoid cells, however, depends on an understanding of not only the regulation of peptide synthesis, but also chain assembly, modification, and interorganelle transport. Although a variety of lymphoid cell expression systems have been reported for antibody production (Morrison and Oi, 1989, Hendricks *et al.*, 1989, Traunecker *et al.*, 1991), alteration of the rate of post-translational steps in the secretory pathway such as the rate of interchain disulfide bond formation (assembly), stress-induced chaperone proteins, and the reducing environment in the endoplasmic reticulum (ER) where assembly of secretory proteins occurs have only recently begun to be investigated (Hwang and Sinskey, 1991).

In order to investigate the growth rate dependence and interaction between the rates of the numerous steps in the pathway of synthesis of a monoclonal antibody (MAb), a structured kinetic model has been developed (Bibila and Flickinger, 1991a, 1991b, 1992a, 1992b, Flickinger and Bibila, 1992).

Experimental data was obtained using the 9.2.27 murine hybridoma (Morgen *et al.*, 1981) on the growth rate dependence of the stability and level of monoclonal antibody heavy (H) and light (L) chain mRNA. Simulation results suggest that the specific growth rate, the H and L chain assembly rate and the heavy and light chain gene dosage can significantly affect the rate of monoclonal antibody (MAb) secretion (Bibila and Flickinger, 1992a, 1992b).

During rapid exponential hybridoma growth in batch or fed-batch culture, the level and stability (half-life) of MAb H and L chain mRNA is at a maximum (Bibila and Flickinger, 1991a, Leno *et al.* 1991a, 1991b, 1992) (Figs. 1,2). However, increasing the level of MAb mRNA does not result in a significant increase in secretion rate indicating that there is most likely a post-transcriptional rate-limiting step. Steady state and transient analysis of this pathway (Bibila and Flickinger, 1992a, 1992b) indicates that the step of MAb assembly in the ER is a good candidate for a rate-limiting step in rapidly growing hybridoma cells. Increasing the overall assembly rate constant, K_A, (Bibila and Flickinger, 1992a) might, therefore increase the antibody secretion rate and yield (Bibila and Flickinger, 1992b).

In contrast to rapid exponential growth, translation of the H and L chains is most likely rate-limiting in slowly growing or stationary phase cells (Fig.3) where mRNA levels are also

1054–7487/92/0345$06.00/0

decreasing (Bibila and Flickinger, 1991, Leno *et al.*, 1992a, 1992b). Recently, evidence has been reported to suggest that translation rates of MAb can be stimulated in slowly growing cells (Flickinger *et al.*, 1992).

As an initial step to investigate alteration of the assembly rate of the H and L MAb chains during rapid hybridoma growth, over-expression of assembly mediating proteins is being investigated. Two proteins retained in the ER that are believed to be involved in MAb assembly are heavy chain binding protein (BiP, grp78) and a thiol:disulfide exchange enzyme, protein disulfide isomerase (PDI, ERp59) (Roth and Koshland, 1981, Lewis *et al.*, 1986, Bassuk and Berg, 1989, Gething and Sambrook, 1992).

BiP is a glucose starvation or stress-induced chaperone which binds to the C_H1 region of H chains during assembly and is released along with ATP hydrolysis following completion of assembly of the intact H_2L_2 MAb (Hendershot *et al.*, 1987). BiP over-expression studies (Dorner *et al.*, 1987), however, suggest that elevated levels of this chaperone in the ER sequester assembly intermediates reducing secretion rates. BiP-H chain complexes in the ER may also be involved in the signaling pathway for induction of ER stress proteins (Lenny and Green, 1991).

PDI is believed to play a role in catalyzing

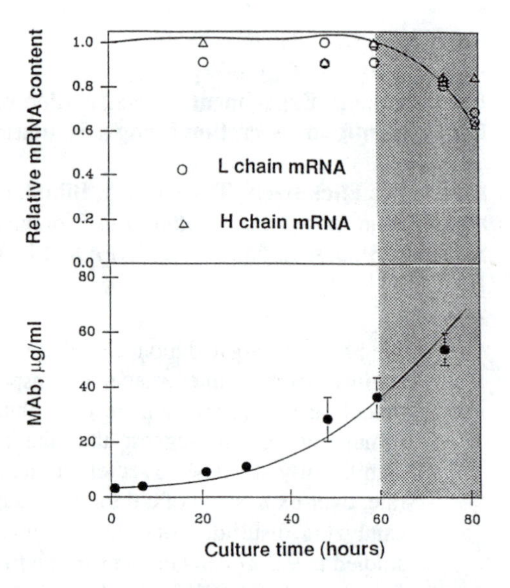

Fig. 2 Experimentally determined relative intracellular heavy (H) and light (L) chain mRNA levels and monoclonal antibody accumulation during 1 liter batch spinner culture. Solid lines represent model simulations. Dot screen indicates stationary phase. (Adapted from Bibila and Flickinger, 1991a. Copyright 1991 John Wiley & Sons.)

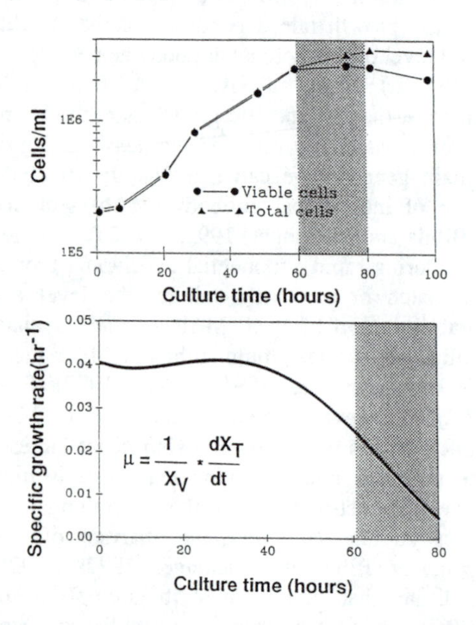

Fig. 1 Growth of the murine 9.2.27 hybridoma in a 1 liter spinner culture. Dot screen indicates period of stationary phase or slow growth where the cell population remains at a viability of >90%. Solid line indicates variation in the specific growth rate determined from the experimental data in the top panel. (Reproduced with permission from Bibila and Flickinger, 1991a. Copyright 1991 John Wiley & Sons.)

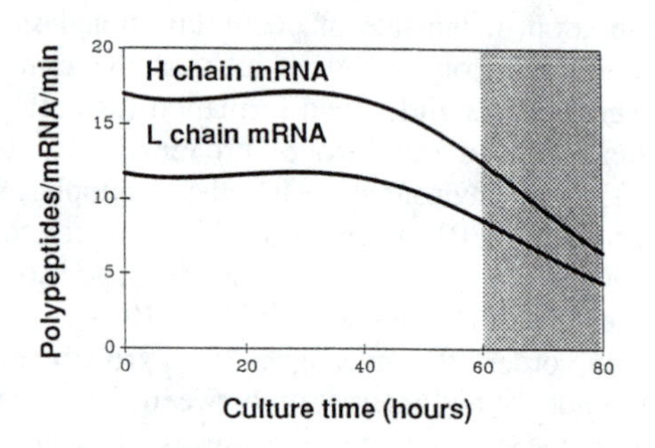

Fig. 3 Variation of heavy (H) and light (L) chain translation rate with batch culture time. Dot screen indicates stationary phase as in Figs. 1 and 2. (Adapted from Bibila and Flickinger, 1991a. Copyright 1991 John Wiley & Sons.)

disulfide bond formation during MAb assembly in the ER (Roth and Pierce, 1987, Freedman, 1984, Gething and Sambrook, 1992). The expression of this multifunctional enzyme has previously been reported to be correlated with immunoglobulin secretion in hybridoma and non-secreting myeloma cells (Roth and Koshland, 1981). It therefore is an attractive initial candidate ER protein whose level and activity in lymphoid cells during rapid growth may affect MAb chain assembly rate.

Our approach is to initially investigate the variation of endogenous PDI activity during batch growth of the 9.2.27 hybridoma in relation to MAb secretion rate, which has previously been shown to be independent of growth rate (Flickinger *et al.*, 1990). Studies were carried out in both serum-containing and serum-free media since it has been reported that insulin negatively regulates PDI expression at the transcriptional level (Nieto *et al.*, 1990) and that *in vitro* estrogen inhibits purified PDI activity (Tsibris *et al.*, 1989). Enzyme activity was determined based on the ability of PDI to catalyze isomerization of the disulfide bonds in randomly reoxidized ribonuclease. Acceleration of refolding with the regaining of ribonuclease activity was measured by the release of [5-^{3}H]CTP during degradation of [5-^{3}H]CTP-labelled RNA (Myllyla and Oikarinen, 1983). This assay was chosen over the spectrophotometric assay (Ibbetson and Freedman, 1976) as it was convenient for multiple samples in order to obtain time-course and growth-rate-dependence data.

The comparison of endogenous PDI levels between the 9.2.27 hybridoma and the non-Ig secreting parent myeloma P3X63-Ag653 was used in order to determine whether a positive correlation exists in this hybridoma between MAb secretion and PDI activity.

Following this initial work, expression vectors were developed in order to over-express this enzyme from a 2.3 kb rat cDNA (Edman *et al.*, 1985) in clones selected from transfected 9.2.27 hybridoma cells. Several vectors are being investigated for stable PDI expression based upon pMAMneo (Clontech). This vector drives expression from a glucocorticoid inducible promoter which can be substituted with a Moloney sarcoma virus 5' LTR.

Materials and Methods

The 9.2.27 hybridoma cell line was grown in spinner culture in either DMEM supplemented with 5% FBS or in a protein-free medium, PFHM II (Gibco) as previously described (Flickinger *et al.*, 1992). The P3X63-Ag653 murine myeloma cell line which is a non-Ig secreting parent of the 9.2.27 hybridoma (Kearney *et al.*, 1979) was grown in T25 flasks at 37°C, 5% CO_2, in RPMI 1640 supplemented with 10% heat inactivated FBS. Cell viability was determined using 0.4% erythrosin B and a hemocytometer. Secreted MAb was determined using an ELISA as previously described (Flickinger, *et al.*, 1990, Abrams *et al.*, 1984).

Protein disulfide isomerase (PDI) assays were performed starting with 2 x 10^6 viable cells, removed from the culture, separated by centrifugation at 600 x g, washed twice with 10 ml of PBS and stored frozen at -20°C. For analysis, frozen cell pellets were resuspended in 500 μl of 0.05 M Tris-HCl, pH 7.5 and the cells lysed by three cycles of freeze-thawing. A crude post-nuclear supernatant from the cell lysate containing 1-2 μg protein was assayed with 0.5 μg of randomly reoxidized (scrambled) ribonuclease A (sRNAse) in 0.05 M Tris-HCl, pH 7.5, 2 mM EDTA and incubated at 30°C for 30 mins in a total volume of 100μl (Myllyla and Oikarinen, 1983). [5-^{3}H]CTP RNA was prepared by pulsing 9.2.27 cells with 10 μCi [5-^{3}H]CTP/ml for 12 hrs. followed by isolation of total cytoplasmic RNA using the RNAzol B kit (Tel-Test). PDI activity levels were expressed as dpm per mg of total cell protein. Activity levels were corrected for the reappearance in RNAse activity observed when sRNAse was incubated in the absence of a cell lysate.

Results and Discussion

Growth Rate Dependence of MAb Secretion Rate: the Effect of H and L Chain Assembly Rate

Using a steady analysis of the pathway of MAb synthesis and secretion (Bibila and Flickinger, 1992b), the effects of variation in the H and L chain assembly rate constant, K_A, was simulated as a function of hybridoma specific growth rate (Fig. 4). For each specific growth rate, slow assembly of H and L chains (low assembly rate constant) resulted in very low specific MAb secretion rates. The specific secretion rate increased with increasing assembly rate until a plateau was reached. At these low assembly rates, it is likely that only a fraction of the H and L chains available for assembly in the ER are actually assembled. As the assembly rate increases, the specific secretion rate increases until some other post-transcriptional step becomes rate-limiting. As hybridoma growth rate increases, the value of the assembly rate constant at which a plateau is reached increases along with the specific MAb secretion rate (Fig. 4). This suggests that only during rapid growth would an increase in assembly rate significantly affect the specific MAb secretion rate (Bibila and Flickinger, 1992a).

Simulation analysis indicated the existence of an inflection point at an assembly rate constant value of 2-3 x 10^{-10} ((molecules/cell)•h)$^{-1}$ (Fig. 4). For assembly rate constant values greater than that corresponding to the inflection point, there is a positive correlation between secretion rate and growth rate. For assembly rate constant values lower than that corresponding to the inflection point, there is a negative correlation between secretion rate and growth rate. Near the inflection point, specific MAb secretion rates are independent of growth rate.

The specific MAb secretion rate determined experimentally for the 9.2.27 hybridoma using a total cell recycle reactor (Flickinger *et al.*, 1990) was found to be between 3-5 x 10^6 molecules/cell • h over a wide range of growth rates (Fig. 5). The steady state solution of the pathway simulation (developed independently from the experimental data) predicts that the specific MAb secretion rate of 4-5 x 10^6 molecules/cell • h should be near the inflection point (Fig. 4) and hence secretion rate should be independent of growth rate.

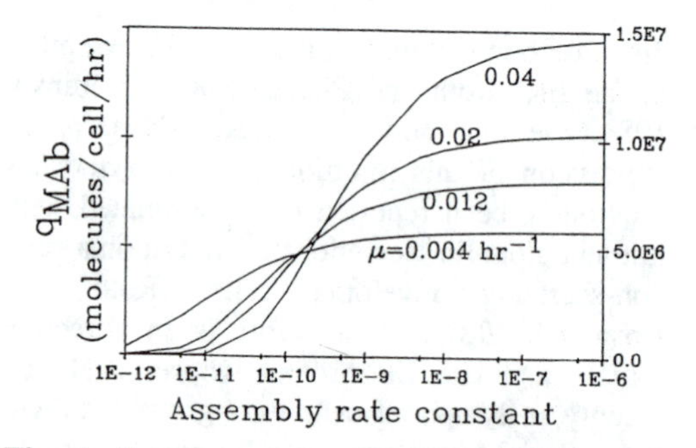

Fig. 4 Steady state specific antibody secretion rate as a function of the assembly rate constant, K_A, ((molecules/cell) • h)$^{-1}$ for specific growth rates from 0.004 h^{-1} to the maximum specific growth rate, 0.04 h^{-1}. (Adapted from Bibila and Flickinger, 1992a. Copyright 1992 John Wiley & Sons.)

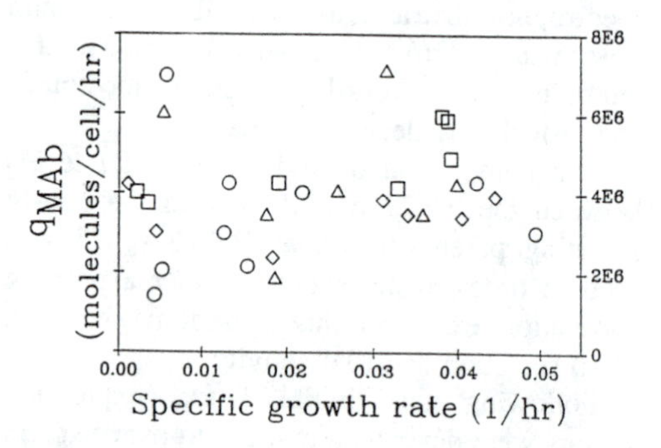

Fig. 5 Specific antibody secretion rate versus specific growth rate during cultivation of the 9.2.27 hybridoma in a 1 liter total cell recycle perfusion reactor (Flickinger *et al.*, 1990). Symbols represent reactor runs at different L-glutamine provision rates at a constant dilution rate of 0.05 h^{-1}. (Reproduced from Bibila and Flickinger, 1992a. Copyright 1992 John Wiley & Sons.)

Prediction of the Effect of Alteration of Assembly Rate by Genetic Manipulation

The model was used to simulate how an assembly rate limitation could be overcome in rapidly growing (maximum specific growth rate $\mu = 0.04$ h^{-1}) 9.2.27 cells using transient analysis (Bibila and Flickinger, 1992b). By using the model to simulate over-expression of an assembly-mediated protein (such as PDI) stably (transfected) and expressed from a controllable promoter, the affect on MAb secretion rate and yield could be predicted (Fig. 6). An exponential increase in PDI expression was used following promoter induction at t = 0 with enzyme accumulating within a physiologically relevant range throughout exponential growth. In this simulation, increases in PDI levels result in proportional increases in the assembly rate

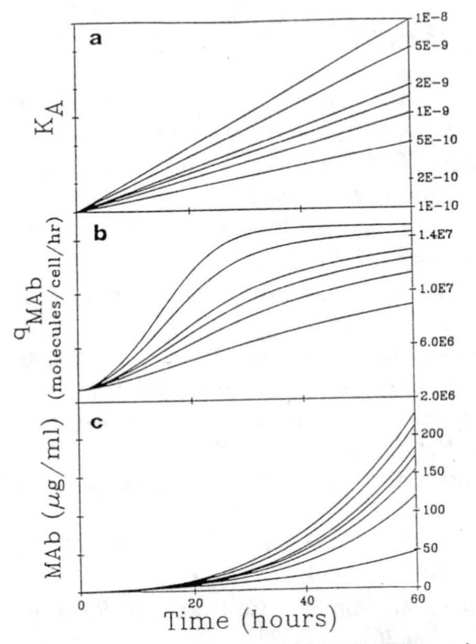

Fig. 6 Profiles of (a) exponentially increasing assembly rate constant, (b) specific antibody secretion rate, and (c) extracellular antibody accumulation during the course of the exponential growth phase following induction of PDI expression at t=0. In (a) through (c) profiles from top to bottom represent exponential increases of K_A with time to final values equal to 1, 5, 10, 15, 20, 50, and 100 times the initial steady state K_A value of 10^{-10}. (Reproduced from Bibila and Flickinger, 1992b. Copyright 1992 John Wiley & Sons.)

constant, K_A, and an increase in the secretion rate and final yield of MAb of 3 to 5 fold (Fig. 6).

Experimental Determination of Endogenous PDI Levels During Batch Cultivation of the 9.2.27 Hybridoma

Batch time-course analysis of endogenous PDI activity levels was determined in both 5% FBS and protein free media. In either media, PDI activity levels (dpm/mg total protein) were constant during rapid exponential growth and stationary phase. Enzyme levels decreased at the end of stationary growth as the fraction of viable cells decreased.

The levels of PDI detected in rapidly growing 9.2.27 hybridoma cells secreting MAb, however were not significantly different from the levels detected in the parent P3X63-Ag653 myeloma cell line. This result differs from that reported by Roth and Koshland (1981) where levels of extractable disulfide interchange enzyme (PDI) were found to be significantly higher in Ig secreting lymphoid cell lines than in the corresponding non-Ig secreting lymphoma and thymoma cell lines.

PDI over-expression studies are currently in progress. If significant increases in MAb secretion rate in the 9.2.27 hybridoma occur, the usefulness of the structured kinetic modeling approach will be verified. In addition to demonstrating the potential for manipulation of a post-transcriptional step to increase the specific secretion rate of a multi chain glycoprotein such as a monoclonal antibody, this approach may be useful for enhancing the secretion rate and yield of other proteins containing a large number of inter-chain or intra-chain disulfide bonds using lymphoid expression systems.

Acknowledgements

The authors are grateful to Nancy Goebel for ELISA assays, adaptation of the 9.2.27 hybridoma to PFHM II and maintenance of P3X63-Ag653 cells. We are also grateful to Eli Lilly for permission to use the 9.2.27 hybridoma and to W.J. Rutter and J.C. Edman for the rat PDI cDNA.

References

Abrams, P.G.; Ochs, J.J.; Giardina, S.L.; Morgan, A.C.; Wilburn, S.B.; Wilt, A.R.; Oldham, R.K.; Foon, K.A. *J. Immunol.*, **1984**, *132*, 1611-1613.

Bassuk, J.A.; Berg, R.A. *Matrix* **1989**, *9*, 244-258.

Bibila, T.A.; Flickinger, M.C. *Biotechnol. Bioeng.* **1991a**, *37*, 210-226.

Bibila, T.A.; Flickinger, M.C. *Biotechnol. Bioeng.* **1991b**, *38*, 767-780.

Bibila, T.A.; Flickinger, M.C. *Biotechnol. Bioeng.* **1992a**, *39*, 251-261.

Bibila, T.A.; Flickinger, M.C. *Biotechnol. Bioeng.* **1992b**, *39*, 262-272.

Edman, J.C.; Ellis, L.; Blander, R.W.; Roth, R.A.; Rutter, W.J. *Nature* **1985**, *317*, 267-270.

Flickinger, M.C.; Bibila, T.A.. In *Frontiers in Bioprocessing II*; Todd, P.W.; Sikdar, S.K.; Bier, M., Eds.; ACS Conference Proceedings Series, American Chemical Society, Washington, D.C., **1992**, pp. 241-257.

Flickinger, M.C.; Goebel, N.K.; Bohn, M.A. *Bioproc. Eng.* **1990**, *5*, 155-164.

Flickinger, M.C.; Goebel, N.K.; Bibila, T.A.; Boyce-Jacino, S.; *J. Biotechnol.* **1992**, (in press).

Freedman, R.B. *Trends Biochem. Sci.* **1984**, *9*, 438-441.

Gething, M.J.; Sambrook, J. *Nature* **1992**, *355*, 33-45.

Hendershot, L.; Bole, D.; Kohler, G.; Kearney, J.F. *J. Cell Biol.* **1987**, *104*, 761-767.

Hendrick, M.B.; Luchette, C.A.; Banker, M,J. *BIO/TECHNOLOGY* **1989**, *7,* 1271-1274.

Hwang, C.; Sinskey, A.J. In *Production of Biologicals from Animal Cells in Culture*; Spier, R.E.; Griffiths, J.B.; Meignier, B., Eds.; Butterworth-Heinemann, London, **1991**, pp 548-568.

Ibbetson, A.L.; Freedman, R.B. *Biochem. J.* **1976**, 159, 377-384.

Kearney, J.F.; Radbruch, A.; Liesegang, B.; Rajewsky, K. *J. Immunol.* **1979**, *123*, 1548-1550.

Leno, M.; Merten, O-W.; Hache, J. *Enz. Microb. Technol.* **1991a**, *14*, 135-140.

Leno, M.; Merten, O-W.; Vuillier, F.; Hache, J. *J. Biotechnol.* **1991b**, *20*, 301-312.

Leno, M.; Merten, O-W.; Hache, J. *Biotechnol. Bioeng.* **1992**, *39* in press.

Lenny, N.; Green, M. *J. Biol. Chem.* **1991**, *266*, 20532-20537.

Lewis, M.J.; Mazzarella, R.A.; Green, M. *Arch. Biochem. Biophys.* **1986**, 389-403.

Morgan, A.C.; Galloway, D.R.; Reisfeld, R.A. *Hybridoma* **1981**, *1*, 27-36.

Morrison, S.L.; Oi, V.T. *Ad. Immunol.*, **1989**, *44*, 65-92.

Mylla, R.; Oikarinen, J. *J. Biochem. Biophys. Methods* **1983**, *7*, 115-121.

Nieto, A.; Mira, E.; Castano, J.G. *Biochem. J.* **1990** *267*, 317-323.

Roth, R.A.; and Koshland, M.E. *Biochemistry* **1981**, *20*, 6594-6599.

Roth, R.A.; Pierce, S.B. *Biochemistry* **1987**, *26*, 4179-4282.

Traunecker, A.; Oliveri, F.; Karjalainen, K. *TIBTECH* **1991**, *9*, 109-113.

Tsibris, J.C.M.; Hunt, L.T.; Ballejo, G.; Barker, W.C.; Toney, L.J.; Spellacy, W.N. *J. Biol. Chem.*, **1989**, *264*, 13967-13970.

Transmembrane Transport Systems Studied on Actively Growing Cells

Stig Benthin, Jens Nielsen, and John Villadsen
Department of Biotechnology, The Technical University of Denmark
2800 Lyngby, Denmark (Fax 42 88 41 48)

Phosphotransferase systems (PTS) in *Lactococcus cremoris* are studied *in vivo* by experiments in which glucose, galactose, lactose and fructose are transported into the cells and subsequently metabolized. It is shown that galactose cannot support growth of the investigated strain and that one of the two operative PTS for fructose gives an intracellular fructose substrate which cannot support growth. This gives rise to oscillatory growth and product formation.

Sugars are taken up by mutually interacting transport systems, and the fate of an intracellular sugar complex depends on the transport system and on the other sugars present. Here the transport of glucose, galactose, lactose, and fructose into *Lactococcus cremoris* is discussed based on results from pulse additions of a single sugar or mixtures of sugars. No intracellular variables are followed, but biomass and extracellular sugar and metabolite concentrations are accurately measured on-line. Thus the focus is on the over-all response of actively growing cells, and substrate transport cannot be decoupled from the subsequent metabolism of sugar-phosphates. *In vitro* experiments by other investigators amply illustrate the transport process alone, but for studies of cell growth and metabolite distribution the combined transport and metabolism is of prime importance. Our experimental methods are described elsewhere (Benthin et al., 1991) and the nitrogen uptake of *Lactococcus cremoris* from yeast extract/casein peptone (=YECP) is studied in (Benthin, 1992). A comparison of glucose uptake in *Lactococcus cremoris*, *Saccharomyces cerevisiae* and *E. coli* with particular emphasis on the anomeric specificity of the transport proteins is given in (Benthin et al., 1992).

Properties of phosphotransferase systems (PTS)

The PTS is a complex protein system found exclusively in procaryotes and responsible for transport and simultaneous phosphorylation of several sugars. Phosphorylation takes place without involvement of ATP, and the activated phosphate group P comes from phosphoenol-pyruvate (PEP) via a chain of four proteins of which two (I and HPr) are general proteins which shuttle P to the sugar specific enzymes II and III. The flux of P through enzymes I and HPr determines the total sugar uptake rate, whereas the relative uptake of sugars from a mixture is determined by the II/III enzymes.

A number of sugar specific PTS systems have been found in lactococci. Among these are: Man-PTS, Lac-PTS, Gal-PTS and Fru-PTS of which the first three are constitutive. Man-PTS is quite unspecific, allowing transport of glucose, mannose, fructose and several N-sugars (Glucosamine, N-acetyl-glucose (or mannose) amine) which are used for ATP production and growth. There is no special PTS for glucose which must enter by Man-PTS. In figure 1 the transport of fructose through Man-PTS and Fru-PTS is illustrated. Since the intracellular fructose species is Fru-1P when fructose enters via Fru-PTS subsequent cell growth can only result if fructose 1,6 diphosphatase (FDPase) is expressed by the organism. Our results show that this is not the case for the strain of *Lactococcus* studied here, and fructose only contributes to growth when it enters via the Man-PTS. Galactose can enter via Lac-PTS or Gal-PTS and is phosphorylated to Gal-6P, but in the absence of FDPase Gal-6P cannot contribute to growth. Since our strain of lactococci has never been able to grow on galactose alone a Gal-permease system (which would give the growth precursor Gal-1P) must also be absent.

When lactose is transported via Lac-PTS, hydrolysis of the phosphorylated Lac-6P gives Gal-6P and intracellular Glc which is rapidly phosphorylated to Glc-6P. Since Gal-6P and Glc are metabolized at different rates in our strain the lactose uptake also exhibits some interesting features.

Experimental results

Only a few characteristic experimental results from a very large experimental investigation (Benthin, 1992) can be shown here to illustrate the function of each PTS and of the interactions between the different PTS. Unless otherwise stated a pulse of sugar (0.5-1.5 g/l) is added to a severely glucose limited steady state chemostat with D ~ 0.1 h^{-1} corresponding to an extracellular glucose concentration of ~ 5 mg /l. Only 3-4% of the sugar is incorporated into biomass, the remainder is converted to lactic acid or to byproducts (formic acid, acetic acid and ethanol), but only for certain sugars and then only if their concentration is below 8-9 mg/l. During all the pulse experiments the metabolism was homofermentative.

Man-PTS

Fig. 2a shows sequential metabolism of glucose and fructose added as one pulse. Glucose is metabolized virtually as if fructose was not present ($r_{max,Glc}$ = 1.5 h^{-1}, saturation constant K_{Glc} = 1 mg/l) and afterwards fructose is slowly metabolized ($r_{max,Fru}$ = 0.65 h^{-1}, K_{Fru} = 160 mg/l). The same high K_{Fru} = 160 mg/l is found by pulse addition of fructose alone. Since the organism was cultivated for days on glucose before the pulse experiment, fructose enters via Man-PTS which consequently has a very low affinity for fructose. Fig. 2b shows uptake of fructose via Fru PTS on a culture grown on fructose. Now $r_{max,Fru}$ = 1.6 h^{-1}, K_{Fru} ~ 1-3 mg/l and there is not even a trace of extra biomass formed during the pulse. These two experiments highlight the difference between the transport systems for fructose.

Experiments with pulses of mannose and of mannose + glucose to the glucose limited chemostat show that mannose is taken up with the same maximum specific rate as glucose, but the affinity of Man-PTS for glucose is higher (K_{man} ~ 10 mg/l) and for an equimolar pulse of α-glucose and mannose only 35% of the mannose is consumed when all glucose is gone.

Man-PTS is the only PTS with two polypeptide chains in the protein (II^{man}) which spans the cytoplasmic membrane (Meadow et al., 1990). This may perhaps explain why transport of the two glucose anomers through Man-PTS is found to occur by two different sites which do not compete with each other. A relative flux of α-D glucose f = 36% of the total (α + β) flux was found (Benthin

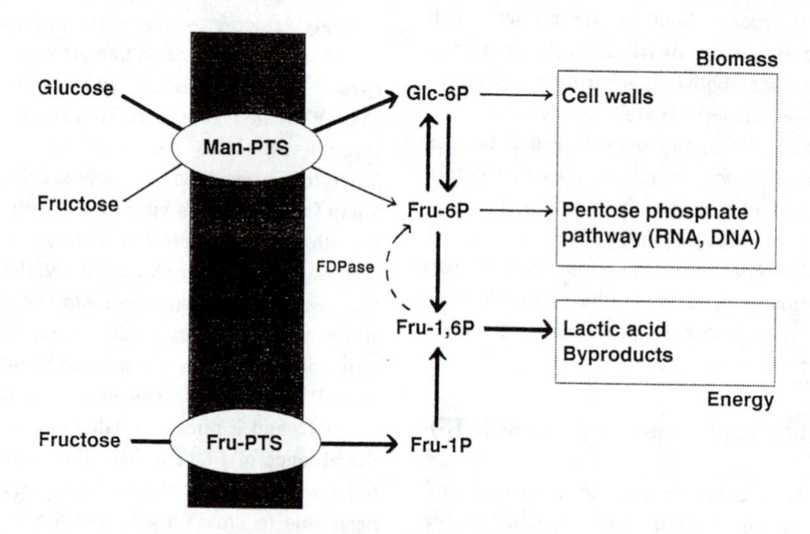

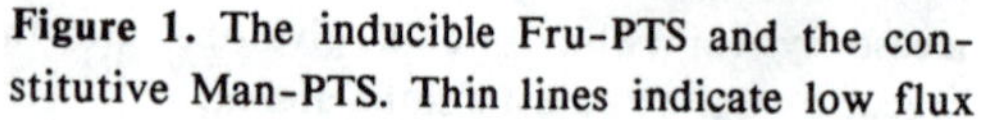
Figure 1. The inducible Fru-PTS and the constitutive Man-PTS. Thin lines indicate low flux reactions. Glucose uptake via Man-PTS is also shown.

et al., 1992) for *Lactococcus cremoris* acclimatized to growth on glucose in mutarotational equilibrium. Interestingly, the bacteria can regulate the relative flux through the two sites. After prolonged growth on mannose ($D = 0.10\ h^{-1}$) the first added pulse of glucose is metabolized with a relative flux of α glucose (f ~ 50%) which is more like that corresponding to the mutarotation equilibrium composition of mannose (68% α-anomer). After several pulses f approaches 36% and the change of f towards 36% is approximately exponential (time constant = μ^{-1}). When the culture is preconditioned with a feed of sucrose + invertase which acclimatizes the cells to α-glucose (+ β-fructose) the first pulse of glucose shows that f ~ 78%, but after 3-4 pulses (total 4 hours) the Man-PTS is reorganized to take up an equilibrium $\alpha + \beta$ glucose pulse, almost with f = 36%.

Lac-PTS

Compared with the unspecific nature of Man-PTS only galactose and lactose can be transported via the Lac-PTS. Pulse experiments with uptake of either α-lactose or β-lactose show that both anomers are transported through the same site (relative affinity $K_{\alpha lac}/K_{\beta lac} = 0.63$). Fig. 3 a-b show an interesting feature of the metabolism of a lactose pulse added to a glucose limited chemostat (D = 0 for t > 0). Whereas lactose rapidly disappears from the medium the total reducing sugar decreases more slowly. The acid production rate Apr is first constant (saturation of the general PTS enzymes). It drops rapidly when the glucose moiety of lactose has been metabolized. Fig. 3b shows the fate of the galactose moiety (calculated from lactose and reducing sugar measurements). At t = 0 exactly half the lactose is (as expected) galactose and free galactose accumulates in the medium until the glucose is gone, whereafter it seeps back through the Lac-PTS and generates lactic acid without growth of cell mass.

The Man-PTS and the Lac-PTS clearly influence each other. With a feed of lactose and any of the Man-PTS sugars lactose uptake is either completely (lactose + mannose) or severely (lactose + glucosamine) inhibited until the Man-PTS sugar transport is finished. The two sugar specific PTS may compete for the same limited phosphorylation machinery in the general proteins or Man-PTS may inhibit Lac-PTS.

When a batch fermentation is started with 15 g/l galactose and with a small, constant lactose addition (0.28 g lactose · H_2O/h), galactose is quantitatively converted to lactic acid and the biomass concentration is as high as when glucose is used as energy source. Hence the glucose moiety of lactose serves as biomass precursor while the metabolism of galactose supplies energy for production of much more biomass (2.2 g/l) than could be supported by the lactose alone (only 0.5 g/l). The conclusion is that a cultivation of lactococci on galactose alone is carbon (but not energy) limited even when YECP is in large excess. The role of the sugar substrate is twofold: to supply energy and to supply small amounts of essential biomass precursors, Glc 6P/Fru 6P.

Fru-PTS

Galactose and fructose are primarily gluconeogenic substrates, but since fructose can also enter the cell via Man-PTS (Fig. 1), an extremely complicated growth pattern results. First it is noted that all growth experiments on fructose alone (batch, steady state or pulse) have an erratic behaviour with poor reproducibility between apparently identical experiments. Batch growth on fructose alone is possible, but μ is very low and the biomass yield only 1/3 of that obtained with glucose. Addition of even a small amount of glucose stabilizes the growth and gives the same biomass yield (~ 0.18 g/g sugar) and growth rate as on glucose. This is the same situation as described for lactose supported growth on galactose.

Steady state experiments on fructose limited medium give the unusual result that for 0.05 < D (h^{-1}) < 0.45 the biomass concentration x increases almost linearly from 0.3 to 1 g/l while the specific acid production rate increases linearly from 1 to 3 h^{-1}. At all D values x is much smaller than when glucose is the substrate, but apparently the cell allocates an increasing fraction of the fructose to the Man-PTS when D increases.

Transient experiments in a chemostat give even more curious results. In Fig. 4 a fructose batch is switched to continuous operation ($D = 0.0978\ h^{-1}$) when the fructose concentration has decreased to 10 mg/l (x = 0.5 g/l). The large oscillations for the first 6 hours show that the organism vacillates between two steady states. Finally, after 40 hours

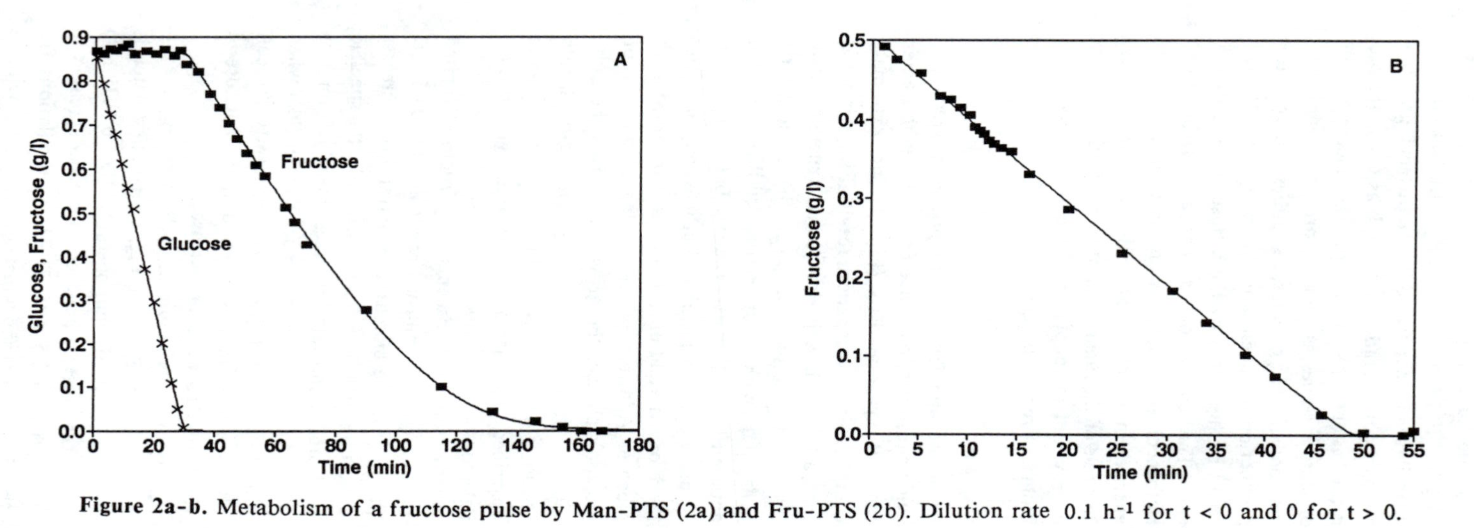

Figure 2a-b. Metabolism of a fructose pulse by Man-PTS (2a) and Fru-PTS (2b). Dilution rate $0.1\ h^{-1}$ for $t < 0$ and 0 for $t > 0$.

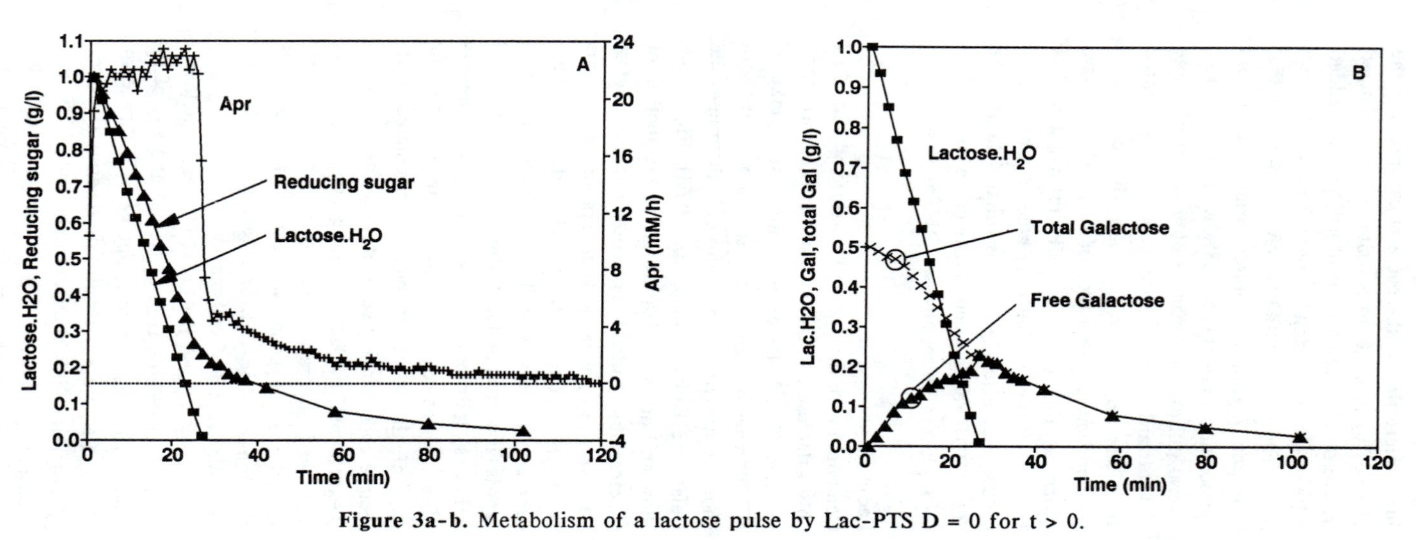

Figure 3a-b. Metabolism of a lactose pulse by Lac-PTS D = 0 for $t > 0$.

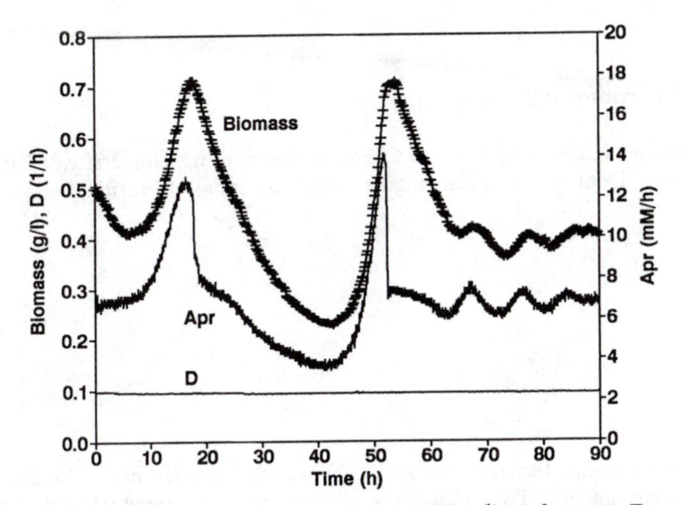

Figure 4. Start-up of a continuous fermentation ($D = 0.0978\ h^{-1}$) on fructose. For $t < 0$ the culture has grown in batch mode.

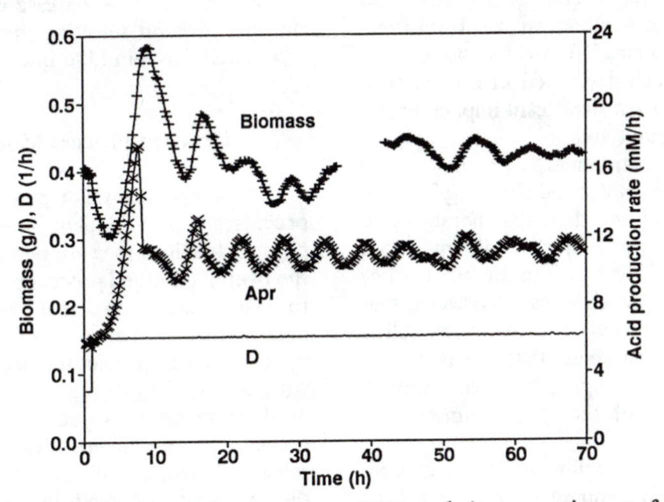

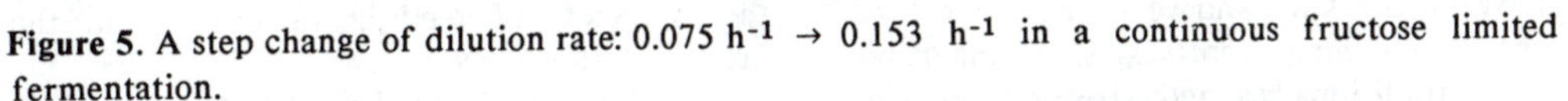

Figure 5. A step change of dilution rate: $0.075\ h^{-1} \rightarrow 0.153\ h^{-1}$ in a continuous fructose limited fermentation.

with damped oscillations a steady state with $x = 0.4$ g/l is reached. When a pulse of glucose is added to the chemostat after 120 h the maximum rate of glucose uptake is only 25% ($0.46\ h^{-1}$) of its normal value. Clearly the Man-PTS is repressed by Fru-PTS after prolonged growth on fructose.

Changes of dilution rate in a fructose limited chemostat invariably leads to long oscillatory periods. Fig. 5 illustrates persistent oscillations of biomass (around $x = 0.42$ g/l) after a change of D from $0.075\ h^{-1}$ to $0.153\ h^{-1}$. In a repetition of the experiment a different steady state of $x = 0.6$ g/l is obtained in about 25 h after several pronounced oscillations. Clearly no simple model describes the complex interactions of Fru-PTS and Man-PTS which result in the strange fructose uptake kinetics. Many, and specifically, targeted experiments must be made before any reasonably comprehensive conclusion can be made.

References

Benthin, S.; Nielsen, J.; Villadsen, J.; *Anal Chim Acta* **1991**, *247* 45-50

Benthin, S.; Nielsen, J.; Villadsen, J.; *Biotechn. Bioengr.* **1992** (accepted)

Benthin, S. Ph.D. thesis, Department of Biotechnology, DTH, **1992**

Meadow, N.D.; Fox, D.K.; Roseman, S. *Ann. Rev. Biochem.* **1990**, **59** 497-542

Modelling For Environmental Technology

Gerasimos Lyberatos, ICEHT and Dept. of Chemical Engineering, University of Patras, Greece,
Spyros A. Svoronos, Dept. of Chemical Engineering, University of Florida, U.S.A.

Wastewater treatment technologies and, more generally, Environmental Technology has gained little attention by biochemical engineers, primarily because, for reasons of tradition they have focussed their attention to traditional fermentation industries. It is clear though that the general approach developed in Biochemical Engineering can have a significant impact on this increasingly important area.

Wastewater treatment processes are characterized by a complex and time-varying feed medium as well as a insufficiently characterized mixed microbial population. It therefore becomes a rather difficult task to identify the key microbial and physicochemical processes that govern the overall process performance, as well as the key process variables that need to be accounted for in developing a process model, descriptive enough of the key parameters of interest.

In this work we will focus on the process of anaerobic digestion and we will consider three different modelling approaches. Anaerobic digestion, traditionally a method used for treating sludge produced in wastewater treatment plants, has gained much attention in the recent years as a method for energy generation, in the form of methane,from especially grown crops. The process involves the concerted action of at least five different groups of bacteria mediating the bioconversion of organic particulate matter to carbon dioxide and methane (biogas). The cellulolytic bacteria hydrolyze complex carbohydrates to simple sugars, like glucose. The acetogenic bacteria convert propionic and butyric acids to acetic acid. The acetoclastic methanogens produce methane and CO_2 from acetic acid, whereas the hydrogen utilizing methanogens convert H_2 and CO_2 from earlier steps to CH_4. The modelling approach to be followed does (and should) depend on the intended model use. In the sequel we will consider two model types that are based on off-line measurements, and one on-line identified process model. In each case we will emphasize the logic behind selecting the particular model type, having in mind the intended use.

Detailed Process Model

A model of this type includes all the key processes and variables, both biological and physicochemical, that are perceived to influence the overall observed process performance. It has to be based on real observations and measurements, whereas its quality will be judged by its ability to predict the process performance <u>without any data fitting.</u> Such a model can be used, in principle, to successfully simulate the process. Consequently, it can be of use for reactor design, as well as for the assessment of the influence of feed characteristics on the process performance.

Previous detailed models for anaerobic digesters have been developed by and Andrews and Graef (1970) assuming that methanogenesis from acetate is the rate limiting step. The model was based on this assumption and neglected methanogenesis from hydrogen utilizing bacteria. More complex models were developed by Heyes and Hall (1981), Mosey (1983), Bryers (1985) and Smith et al (1988). Smith's model assumed a fixed ratio of fatty acids (acetic/ propionic /butyric) and neglected hydrogen methanogenesis as well. Kaspar and Whyrmann (1978), observed that the VFA kinetics depended on the hydrogen partial pressure. Mosey (1983) proposed that the distribution among the various fatty acids depends on the ratio of NADH to NAD^+, which in turn depends on the pH and the hydrogen partial pressure. This proposal stems directly

1054–7487/92/0356$06.00/0

from consideration of the metabolic pathways that lead to the three VFA products, and has a sound biochemical basis.

Our model integrates the regulatory mechanism proposed by Mosey et al with the physicochemical properties of the liquid and gas phases, to give a complete CSTR model. Biomass is assumed to have the empirical formula $C_5H_9O_3N$ throughout. With glucose as the starting substrate two types of reactions are considered. The first group consists of the dissimilatory bioreactions:

Acidogenesis:

$$C_6H_{12}O_6 + H_2O \rightarrow 2CH_3COOH + 2CO_2 + 4H_2 + 4ATP$$

$$C_6H_{12}O_6 + 2H_2 \rightarrow 2CH_3CH_2COOH + 2H_2O + 2ATP$$

$$C_6H_{12}O_6 \rightarrow CH_3CH_2CH_2COOH + 2CO_2 + 2H_2 + 2ATP$$

Acetogenesis:

$$CH_3CH_2COOH + 2H_2O \rightarrow CH_3COOH + CO_2 + 3H_2 + ATP$$

$$CH_3CH_2CH_2COOH + 2H_2O \rightarrow 2CH_3COOH + 2H_2 + 2ATP$$

Methanogenesis:

$$CH_3COOH \rightarrow CH_4 + CO_2 + 0.25ATP$$

$$CO_2 + 4H_2 \rightarrow CH_4 + 2H_2O + ATP$$

The second is the group of assimilatory steps for the bacteria that mediate the process bioconversions:

$$5C_6H_{12}O_6 + 6NH_3 \rightarrow 6C_5H_9O_3N + 12H_2O$$

$$3CH_3CH_2COOH + CO_2 + 2NH_3 \rightarrow 2C_5H_9O_3N + 2H_2O + H_2$$

$$CH_3CH_2CH_2COOH + CO_2 + NH_3 \rightarrow C_5H_9O_3N + H_2O$$

$$5CH_3COOH + 2NH_3 \rightarrow 2C_5H_9O_3N + 4H_2O$$

$$5CO_2 + 10H_2 + NH_3 \rightarrow C_5H_9O_3N + 7H_2O$$

The kinetic model stoichiometry is fully defined by the above reactions. Five different independent growth rates are assumed, one for each bacterial type. In developing the growth rate expressions the following assumptions were made. Acidogenesis leads to glucose conversion to the three fatty acids in proportions that depend on the $NADH/NAD^+$ ratio (as in Mosey,1983), which in turn depends on the pH and the hydrogen partial pressure by:

$$\frac{[NADH]}{[NAD^+]} = 10^{(\log P_{H_2} + pH + \frac{E^o}{29.5})}$$

where P_{H2} is the hydrogen partial pressure, and E^o the standard redox potential for the $NADH/NAD^+$ couple (-113 mV). Acetogenesis from propionic and butyric acids is assumed to be pH dependent and inhibited by hydrogen. Acetoclastic methanogenesis is assumed to be VFA inhibited (by each VFA type independently). Hydrogen methanogenesis is pH dependent with an optimum at pH 7.1. The exact rate expressions were presented in Pullammanappallil et al (1991) and will appear in a forthcoming publication. The dissociation reactions for the fatty acids were considered, since it is the unionized forms that are inhibiting. The HCO_3^- concentration was related to the pH and the total inprganic carbon. An overall charge balance:

$$[Z] + [H^+] = [A^-] + [P^-] + [B^-] + [HCO_3^-]$$

then gives a fifth order equation, which may be solved for the pH. The CO_2 transfer rate between phases was considered and material balances for total cations (Z) and total inorganic carbon were developed. The gas phase was assumed completely mixed and always at a total pressure of 1 atm. The total number of differential equations from material balances was 14 (five biomass concentrations, glucose, acetate, propionate, butyrate, H_2, CO_2, CH_4, Z and total inorganic carbon). Detailed model equations may be found in Pullammannappallil et al (1991). Kinetic parameters were determined from independent literature data.

Model predictions for a 6 l glucose fed experimental digester, operated at 20 d retention time were compared with steady-state measurements. The results are given in table 1. As we can see an excellent agreement is seen validating the process model.

Simplified Process Model

A model of the type presented in the previous

Table 1. Model validation.

	Experimental	Model
Methane Production Rate(l/l/d)	0.75-0.78	0.68
Methane %	57	54
pH	7.1-7.2	7.19
Acetic Acid (mg-COD/l)	30 - 75	34
Propionic Acid (")	12 - 40	33
Butyric Acid (")	< 10	4

section has in principle two types of problems. The first is that it contains a large number of parameters that need to be determined with specifically designe experiments. The second is that it is extremely cumbersome to work with, especially when complex calculations are needed. An example case is that of determining an optimal control strategy. In order to determine the optimal manipulation of the dilution rate of a digester in response to entry of a methanogen inhibitor, the following much simplified model was used:

$$\frac{dx}{dt} = -D\,x + \mu(s,I)x$$

$$\frac{ds}{dt} = D\,(s_o - s) - \frac{1}{Y_s}\mu(s,I)x$$

$$\frac{dI}{dt} = D\,(I_o - I)$$

where

$$\mu(s,I) = \frac{\mu_{max}}{(1 + \frac{K_s}{s} + \frac{s}{K_{IP}})(\frac{1}{1+\beta\,I})}$$

In this model s is the total VFA (assuming instantaneous acidogenesis) and I is the inhibitor concentration. The methane production rate was assumed to be:

$$Q_{CH_4} = Y\,V\,\mu(s,I)x$$

The optimal control problem (singular) for maximization of the performance measure:

$$J[D(t)] = \int_0^{t_f} Q_{CH_4}(t)\,dt$$

was solved. A typical result is given in Fig.1. The optimal dilution rate is 0 until the singular arc is reached, and then it is varied according to an analytically determined feedback law (Pullammannappallil et al,1991a) over the singular arc. Details may be found in that paper. The same optimal control problem could in principle be solved with the complex model described in the previous section. This would be a very difficult task and, most importantly, it would not give an analytical feedback law. To the extent that the simplified model is inaccurate, this procedure leads to a suboptimal control, which however, is expected to be close enough to the optimum predicted by the detailed model.

On-line Modelling

The off-line identifiable models of the two previous sections are excellent for process design and simulation (the first) and for optimal control (the second). In order, however, to secure optimal setting of process parameters, despite temporal variations in the process, an on-line identified model needs to be used. Then following the approach introduced by Svoronos and Lyberatos (1987), the optimal settings at all times can be secured. An on-line identified model has to be a simple dynamic, discrete-time input-output model with a <u>small number</u> readily identified on-line parameters. A suitable model for the on-line optimization of the methane production rate of a continuous digester with respect to temperature was developed by Harmon

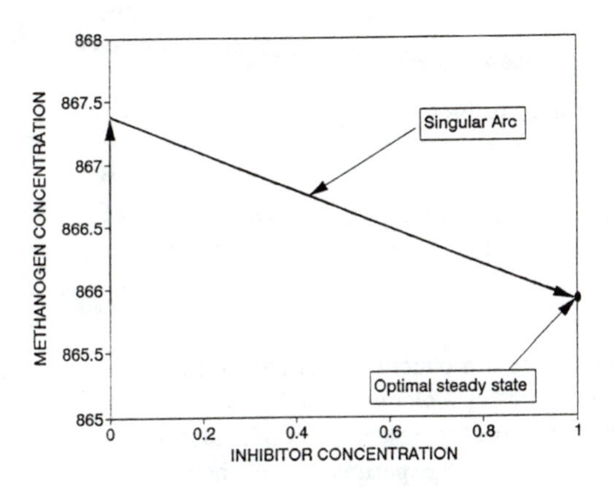

Figure 1. Optimal state trajectory for an anaerobic digester as determined by the simplified model.

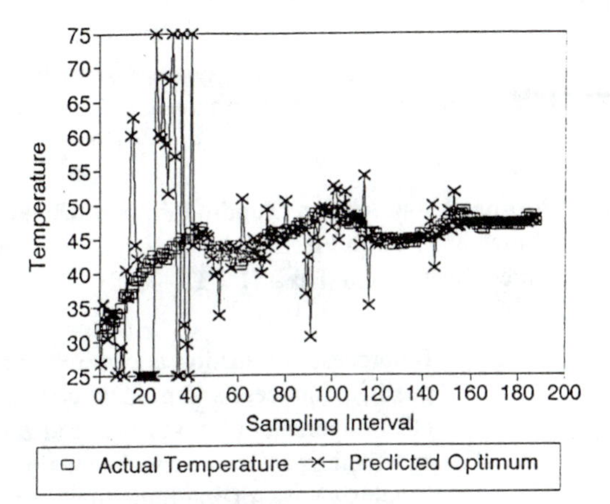

Figure 2. Convergence of the adaptive steady state optimization algorithm to the optimum temperature that maximizes methane production.

et al (1991). Details may be found in J.Harmon's PhD thesis (1990). The algorithm is based on the following model:

$$Q_{CH_4}(i) = \theta_1 Q_{CH_4}(i-1) + \theta_2 T(i-1) + \theta_3 T^2(i-$$

where the θ_i's are on-line identified parameters at each (discrete) sampling interval i. Rasults of the application of this adaptive optimization algorithm on a lab-scale anaerobic digester are shown in Fig 2. The temperature was at each step adjusted in the direction of the estimated optimum ($-\theta_2/2\theta_3$) by an amount that depended on a filtered a posteriori estimation error. The algorithm as can be seen drove the temperature to an optimum of 47.5°C, increasing the methane production rate by 64% over its initial value.

Conclusion

We have examined three different model types of anaerobic digestion and have seen that each model type can prove very useful for the design, control and optimization of the process, depending on the particular use. Modelling can certainly improve wastewater treatment processes significantly.

References

Andrews,J.F. and Graef S.P., in *ANAEROBIC BIOLOGICAL TEATMENT PROCESS*, Am. Chem. Soc., Wash.DC,1970,126-162

Bryers,J.D.,*Biot.&Bioen.*,1985,27,638-649

Harmon,J.,*PhD Thesis*,1990, U.Florida

Heyes,R.H. and Hall R.J., *Biot.Letters*, 1981,3,431-436

Kaspar,H.F. and Wuhrmann,K., *Mic.Ecol.*,1978,4,241-248

Mosey,F.E., *Wat.Sci.Tech.*,1983,15,209-232

Smith P.H., Bordeaux , F.M.Goto, M., Shoralipour,A. ,Wilke,A., Andrews,J.F., Ide,S.E. and Barnett,M.W. ,in *METHANE FROM BIOMASS*, Ed.Smith W.H. and Frank.J.R.,Elsevier App.Sci., London,1988

Pullammanappallil,P.,Owens,J,M,Svoronos,S.A., Lyberatos,G.,Chynoweth,D.P.,*AICHE Ann.Meeting*,San Francisco,Ca.,1991

Pullammanappallil,P.,Svoronos,S.A., Lyberatos,G., *Pr.ACC*,Boston,1991

Bioprocess Automation is an Excellent Tool for Research in Biology

Bernhard Sonnleitner, Claudio Filippini, Ulrike Hahnemann, Georg Locher, Thomas Münch and **Armin Fiechter**, Institute for Biotechnology, ETH Zürich Hönggerberg, CH-8093 Zürich, Switzerland (Fax +41 1 371 0658)

Bioprocess automation is certainly beneficial in a production environment: the quality of process and product can be improved. In biology research, however, other important objectives can be effectively pursued: representativity of analyses and therefrom drawn conclusions, reproducibility of experiments and, hence, conclusiveness of interpretations, accuracy of measurement and precision of control activities. All these aspects are decisive to avoid artifacts. Examples focus on recent advances in perceiving bioprocesses in vivo.

Modern biotechnological research is committed to serveral aims:

➩ quantify activities, yields and rates
➩ elucidate mechanisms of regulation and gene expression
➩ find ways to force biosystems in a desired stable physiological state
➩ determine the relaxation times which are relevant for metabolic regulation
➩ process control as well as long term and high productivity cultivations.

Any method of choice to approach these goals would preferably neither disturb the biosystem under investigation nor destroy it. Of course, non-invasive techniques are the most excellent tools.

'On-line' is synonymous to 'fully automated'. Although the available methods and devices do by far not permit to measure all the interesting variables on-line, some important success has been made in recent years. There exist non-invasive sensors and analytical instruments that are based on electro-magnetical principles (eg optics and spectrometers, spin resonance, capacitance, potentiometry). But there are many sensors and instruments that consume sample aliquots irreversibly (eg automated wet chemical, enzymatic or immune reactions either directly or following physico-chemical separation techniques). In the latter case, special care must be taken that the sampling and sample preparation procedures do not interfere with the culture.

In the following, a few examples are used to highlight decisive aspects of bioprocess analysis:

Reproducibility

Bio(techno)logical experiments can be reproduced to a very high degree. Figure 1 shows the mean values of 2 time trajectories and respective extreme values (min & max) ever found in a series of 8 individual batch cultures. Any deviation from the small bands is a sound indication for malfunction of either the technical or the biological cultivation system in use.

Time Resolution

Many biosystems are more dynamic than generally expected. For instance, a slow-growing *Bacillus subtilis* ($\mu_{max} \approx 0.5\ h^{-1}$) oscillates reproducibly with a period of approx 6 min in the initial phases of a batch culture (Fig 2). This can be nicely monitored by on-line sensors that deliver continuous signals. But on-line capillary-GC is not yet able to cope with this frequency (run time to analyze intermediates and products: ≥ 7 min).

Earning New Insights

Employing high frequency and high resolu-

1054–7487/92/0360$06.00/0

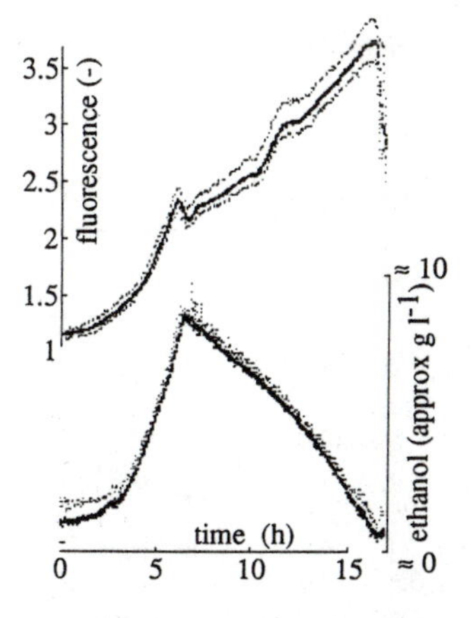

Fig 1: Time courses of fluorescence and ethanol (measured with MS) during a series of 8 batch cultures with *Saccharomyces cerevisiae*. The central trajectories represent the mean values, the lines forming the confidence band show the absolute minima and maxima found in the entire series (ethanol concentration is approximate because the on-line MS determines just the partial pressure).

tion of measurements is necessary to get new and reliable information about the temporal development of a cultivation as sketched out in all figures. Transient experimentation in a chemostat (eg batch, D-shift, pulse) can be thorougly analyzed. The high degree of reproducibility permits to discriminate between possible origins of diverse effects: the 2 batch series shown in Fig 3 differed only with respect to the method of medium preparation. Glucose was sterilized either separately or together with the mineral salts of the defined synthetic medium. The obvious differences in the cultures' responses are reason enough to reconsider and improve the design of most culture media and to include all up-stream process operations into — of course fully automated — bioprocess control.

Figure 4 surveys the oscillations of a spontaneously synchronizing yeast culture. The online DNA measurement clearly proves that the oscillations found in many state variables are correlated with the cell cycle. The oscillations

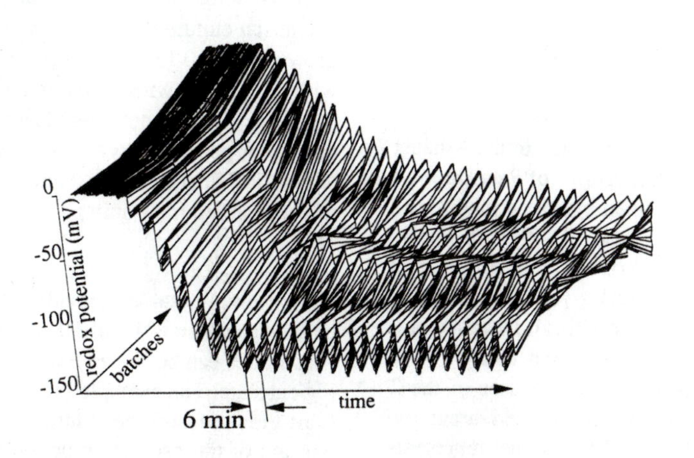

Fig 2: Four batch cultures of ***Bacillus subtilis***: first 3 h of redox traces shown. Other variables oscillate similarly. But there are various biosystems with even shorter relaxation times of dynamics.

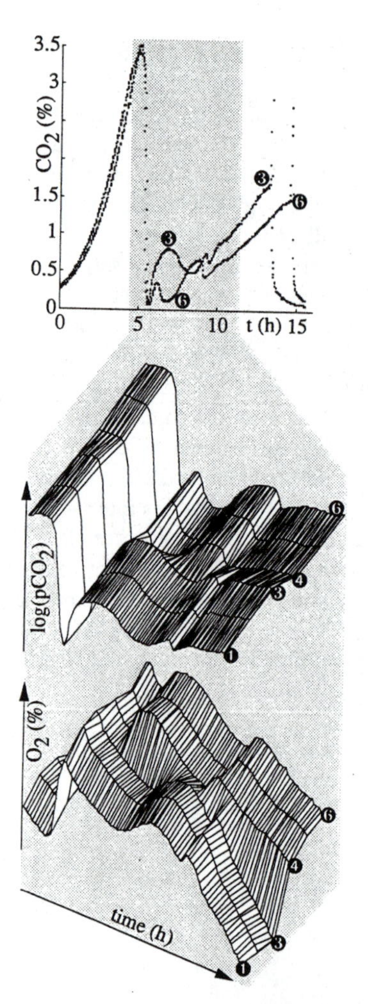

Fig 3: Top: time course of CO_2 in the exhaust gas during 2 batch cultivations of *Saccharomyces cerevisiae* on defined synthetic medium (3 %). Trace ❸ is from a series (❶-❸) with glucose been sterilized separately from the mineral salts, trace ❻ from a series (❹-❻) where the complete medium was sterilized. Distance between dots = 1 min. Middle and bottom mesh plots show the most different phases of the 2 series as indicated by the shaded area: the "white wall" in the log(pCO_2)-signal represents the final depletion of glucose from the medium. This transient is less pronounced in the signal for O_2 in exhaust gas because of the rapid change in physiology there (RQ >> 1 → RQ ≈ 0.5). The following more or less smooth peaks, that distinguish the 2 series, reflect the utilization of previously excreted pyruvic and acetic acid, respectively. Distance of mesh-lines = 3 min.

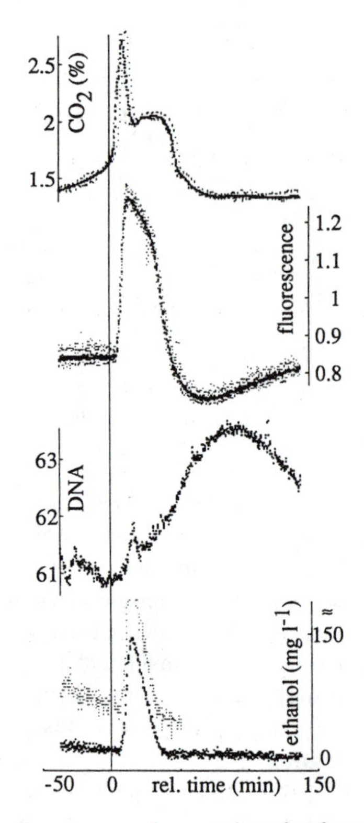

Fig 4: Spontaneously synchronized growth. Mean values (and maximal deviations therefrom) of some state variables in a synchronous chemostat culture of *Saccharomyces cerevisiae* grown at D = 0.13 h^{-1}. Ethanol was measured with MS, DNA with a continuous on-line fluorimetric method upon specific staining (reflecting DNA per culture volume; delay: dead time of this flow analysis is < 10 min; DNA and fluorescence in arbitrary units).

are extremely stable, provided the control of culture 'parameters' is sufficiently precise.

Cultures can be forced into a desired physiological state. The example in Fig 5 shows that a faint elevation of the dilution rate results in phasing of the oscillation period of synchronously growing yeast. This manipulation of the cell cycle duration is a most powerful tool.

Accuracy, Precision & Validation

Accuracy of measurements is mostly a question of sound and frequent (re)calibration. If such procedures are implemented on-line (ie

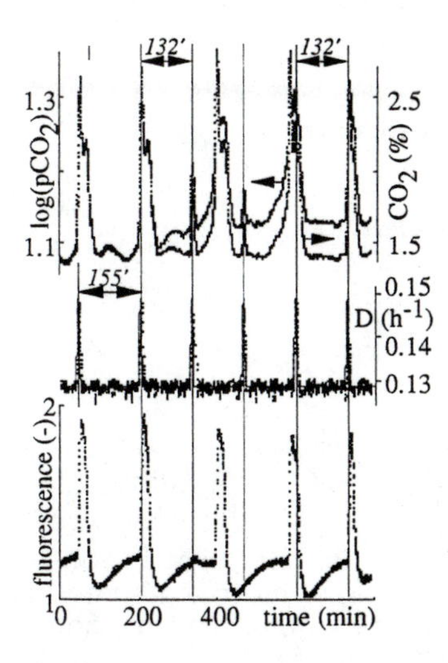

Fig 5: Forced synchronous growth. The dilution rate of a synchronized chemostat culture of yeast is repeatedly elevated by 10% at various time intervals (first: 155 min, then 132 min). In a certain permissive window, the culture responds by reducing the cell cycle duration (visualized by fluorescence and CO_2 in the exhaust gas, cf Fig 4).

automated) the signals reliability can be significantly improved. Reasonable control has to be automated anyway, but cycle time, resolution and type of algorithms of controllers certainly influence the precision of control loops. However, also systematic errors can persistently occur when only single sensors are installed or the calibration method is inappropriate. An elegant but expensive solution is the use of parallel and even better the use of alternative sensors or analytical methods. The respective signals can be compared on-line with each other or with software sensors and permit even an on-line validation.

Automated systems work around the clock with constant quality. They also operate reliably with a much higher frequency than humans.This results in substantial time savings and really complete documentations.

From the above said it can certainly be imagined that complete sequences of investigation or optimization experiments can be operated automatically; obviously, this exploits time efficiently and renders results independent of the shape of personnel in favour of objective facts.

Adaptive Control of Continuous Yeast Fermentation.

S.Bay Jørgensen, H.E.Møller and **M.Y.Andersen***

Institut for Kemiteknik
Technical University of Denmark
DK 2800, Lyngby

* Novo-Nordisk
Novo Alle'
Gladsaxe, Denmark

The dynamics and control of a continuous fermentation with Saccharomyces cerevisiae is investigated. A sensor for reducible gas concentration in the exhaust gases is used in an adaptive control loop for manipulating the substrate flow rate. A model based control design with a LQ-regulator is used. At the lowest setpoint level the process is exposed to a large disturbance in substrate feed concentration and to a small setpoint disturbance. In order to investigate the process behaviour around the critical dilution rate a sequence of setpoint changes is carried out. At each setpoint a closed loop identification experiment is carried out to enable identification of the process dynamics. The adaptive LQ-regulator performs well experimentally. Some interesting features are indicated concerning the adaptability of the microorganism and the possibilty of multiple steady states for the fermentation process around the critical dilution rate.

The purpose of this paper is to study the process statics and dynamics around the critical dilution rate. Thus attempting to reveal precisely at what dilution rate is maximal productivity achieved and what type of process dynamics occur near the critical dilution rate. The latter issue could shed light upon the difficulty of controlling the process around the critical dilution rate. Such a study is interesting in its own right since the fermentation processes are nonlinear. Nonlinear behaviour reveals itself when, e.g. the operating point is changed and the microorganism adapts its metabolism to the new conditions. These effects depend upon the location in the operating region and the size of the change. The rate of such metabolic changes is most interesting from a control point of view, since they may require tens of hours of otherwise steady operation to occur, and thus any control design should be able to handle such slow process changes as well as faster changes in fermentor operation. Another impact of these metabolic changes is however expected to be on the productivity, which of course will vary with the location within the operating region. Since fermentation processes are inherently nonlinear a control design must be able to deal with the timevarying aspects in an efficient manner. For this purpose an adaptive model based control design is used.

Fermentor system: The experiments are performed in a 20 l laboratory fermentor at Novo Nordisk A/S. The fermentor is equipped with standard measurements of pH, dissolved oxygen tension (DOT), and temperature. Oxygen uptake rate and carbon dioxide production rate are calculated from mass spectrometer measurements. The

1054–7487/92/0364$06.00/0

fermentor pressure is 50 kPascal gauge. The substrate feed rate and concentration can be controlled independently, through manipulation of feed flowrates from two weighed vessels. One of the feed streams contain 40 g glucose/l and the other 10 g glucose/l.

Computer Control system: The fermentor is normally controlled by the company's process computer (FOXBORO 300) and for practical reasons pH and temperature controls are performed by this computer. The FOXBORO computer also provides data from the mass spectrometer. The remaining control task is in principle divided between a low and high level task each undertaken by a computer. Only the low level computer is interfaced to the process. This computer handles PID control of dissolved oxygen and substrate feeding. The MODAC (MOdular Data Acquisition and Control) software package (Brabrand et al.,1990) runs under OS-2 on the high level computer and the IX (Instrumentation eXpert) software package runs on the low level computer. Both are multitasking, real-time, command based software systems developed in-house. The IX system, which runs under DOS, is specially designed to meet high requirements in timing of operations (e.g. the FIA system). The model based controller MIMOSC (Multi Input Multi Output Selftuning Controller) (Brabrand et al.,1990) is implemented in MODAC as a task. MIMOSC performs process identification and model based control using the identified model.

<u>Cultivation procedure:</u> The microorganism used in this study is Saccharomyces cerevisiae provided by Novo Nordisk (ATCC 26785). The microorganism is stored freeze dried and cultivated 3-5 days at 30 °C on agar slants (5-10 cm^2 surface) before use. The slants are then stored at 5 °C. A standard synthetic glucose-limited medium as given in Egli,(1980) was used with one modification. Citric acid was added (0.4 g/l) instead of EDTA in order to obtain pH ≈ 4 to avoid precipitation. The experiments were started as batch fermentations with inoculation directly from one agar slant. The glucose concentration of the batch medium was 10 g/l. In the continuous fermentations the standard glucose concentration in the feed was 10 g/l. The fermentations were run at 30°C, pH 4.5, 1 vvm air, 0.5 bar gauge pressure. Agitation was 300 rpm during batch operation and variable during continuous operation. The fermentor was switched to continuous operation either directly or with a fed-batch phase in between. The working volume was 8.15 l.

Adaptive Control: An adaptive Linear Quadratic controller is used to control the reducible gas concentration by manipulating the feed flow rate. The controller is implemented in MODAC as described in more detail in Brabrand and Jørgensen (1990). Fig.2 shows a block diagram of the adaptive LQ controller. Process identification: The approach to process identification is to perturb the input variables with a Pseudo Random Binary Sequence (PRBS). Proper filtering of the input-output data is done to avoid aliasing and to remove low frequency trends. Linear input-output models are used for controller design, the model structure and parameters are determined using the experimental data obtained in suitably designed experiments. Due to the lack of stable open loop operation around the critical dilution rate, as shown in fig. 1, identification experiments are carried out in closed loop. The reducible gas signal is controlled at the setpoint using the feed flow rate as manipulated variable. Perturbations are imposed on the reducible gas reference signal to the controller. Fig.2 shows a block diagram of the closed-loop identification set-up. A

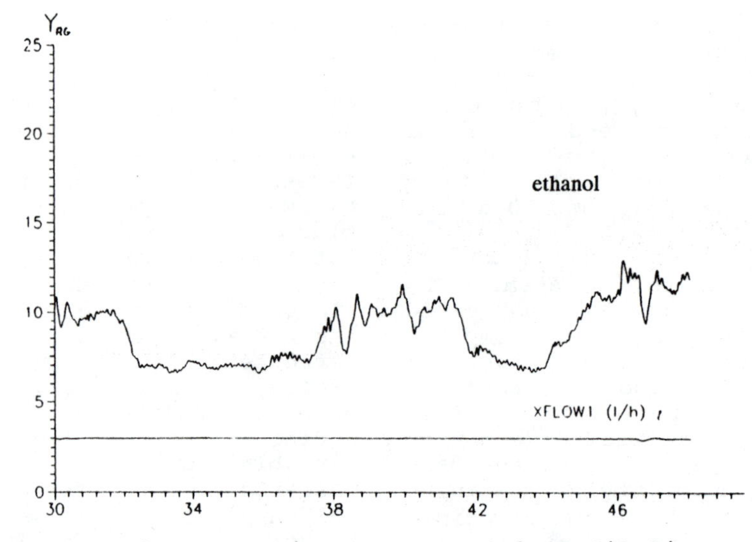

Fig.1: Open loop experiment near critical dilution rate.

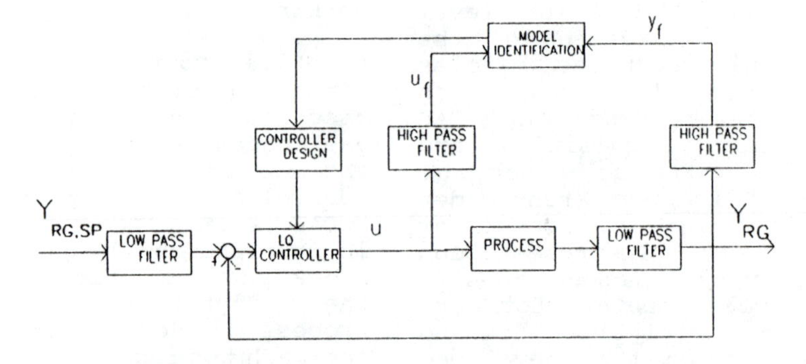

Fig.2: Closed-loop adaptive identification set-up. u:feed flow rate, y: ethanol measurement.

perturbation experiment is performed with the ethanol reference signal at 10 which corresponds to a liquid concentration of 40-50 mg/l. A PRBS sequence with a basic period of 6 min but sampled every min has been generated off-line. The sampling time for the controller is 1 min. The input-output data are shown in fig.3. A third order ARX model with a time delay of one min was used with this model structure 96 % of the sensor variance was explained. It is clearly seen how the adaptation of the model parameters with time makes the setpoint be followed more closely by the output with the delay dictated by the model structure.

The setpoint tracking ability of the adaptive LQ regulator is shown in Figures 4 and 5. In Figure 4 an experiment was conducted with a sequence of setpoint changes. The setpoint changes were introduced with a model following first order filter with a time constant of 45 minutes, such that a new setpoint should be reached in 3 hours. This is seen to be well accomplished for the approaches to Y_{rg}=50 and 90. The setpoint for the transition from 50 was 90, this transient was however interrupted at Y_{rg}= 66.5, due to a temporary sign changes in the estimated low frequency (static) gain. After 15 hours a new setpoint change towards 90 was initiated. It should be noted that at each

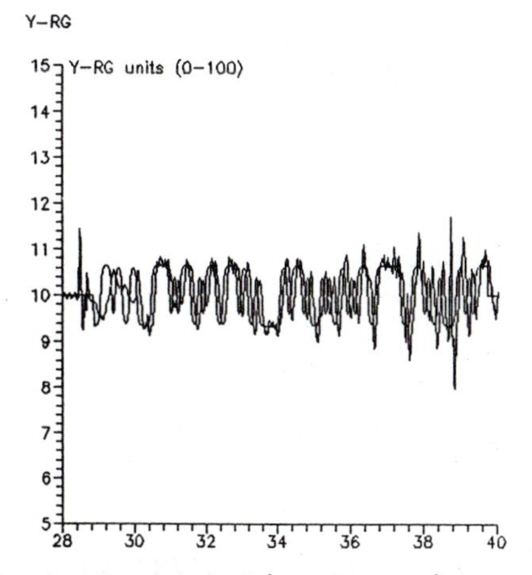

Fig.3: Perturbation experiment of 200691 around Y_{rg}=10.

setpoint a pertubation experiment was carried out. The detailed response of the setpoint change from 10 to 50 is shown in figure 5. It is clearly seen that the noise level is higher around 50 than around 10. The results demonstrate that adaptive control indeed renders it possible to perform experiments at any desired ethanol concentration near the critical dilution rate. At each setpoint the substrate flowrate is seen first to increase to make the transition, but then there is a long period during which the flowrate slowly decreases. This indicates that the microorganisms adapt to the new ethanol level by manufacturing more enzymes for ethanol production and hence can maintain the ethanol level at a lower dilution rate. Thus the substrate conversion becomes less efficient from a biomass production point of view.

In an attempt to shed some light upon the the low frequency characteristics of ethanol versus dilution rate a sketch is shown in Figure 6. The basis for the plot are the data obtained around the end of the steady reducible gas data at each setpoint. It is also clearly seen that this does not imply that the dilution rate has reached its new steady value. In the experiment in figure 4 and 5 around 5 to 10 volume changes were accomplished at each setpoint. Even with this uncertainty about the precise static relationship figure 6 shows that the highest dilution rate is achieved for Y_{rg}=10, and that there is a tendency towards lower dilution rates up to around Y_{rg}=50. From an operational viewpoint this plot indicates that there may be several steady states possible for the same dilution rate dependent upon the prior history of the

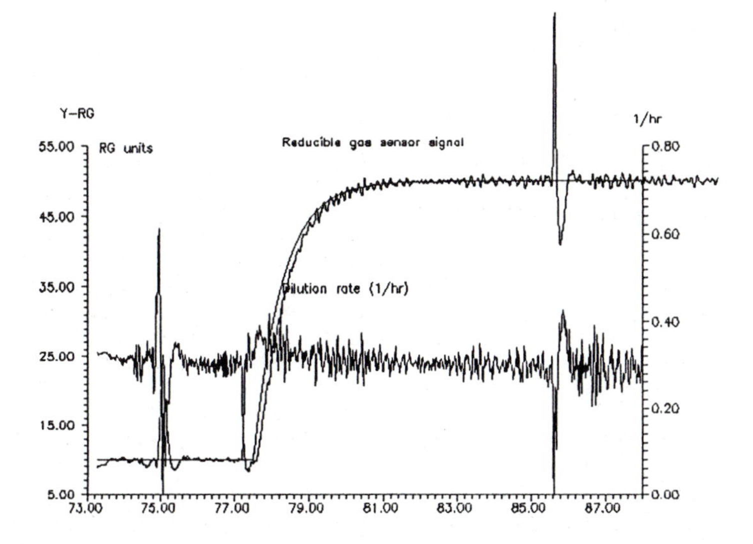

Fig.4: Setpoint tracking transient from Y_{rg}=10 to 50.

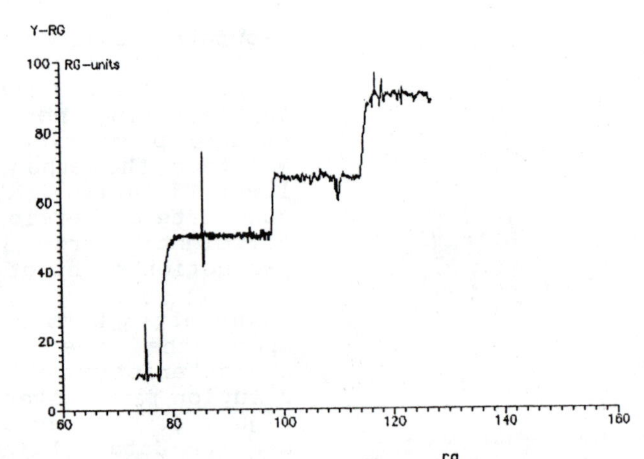

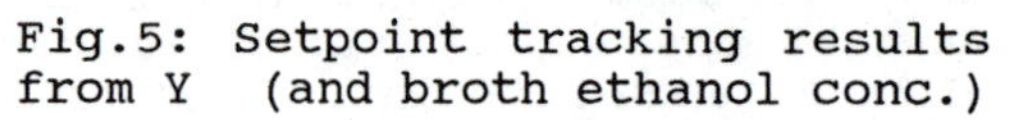

Fig.5: Setpoint tracking results from Y_{rg} (and broth ethanol conc.) =10 (44 mg/l) to 50 (334 mg/l), to 66.5 (541 mg/l), to 90 (741 mg/l).

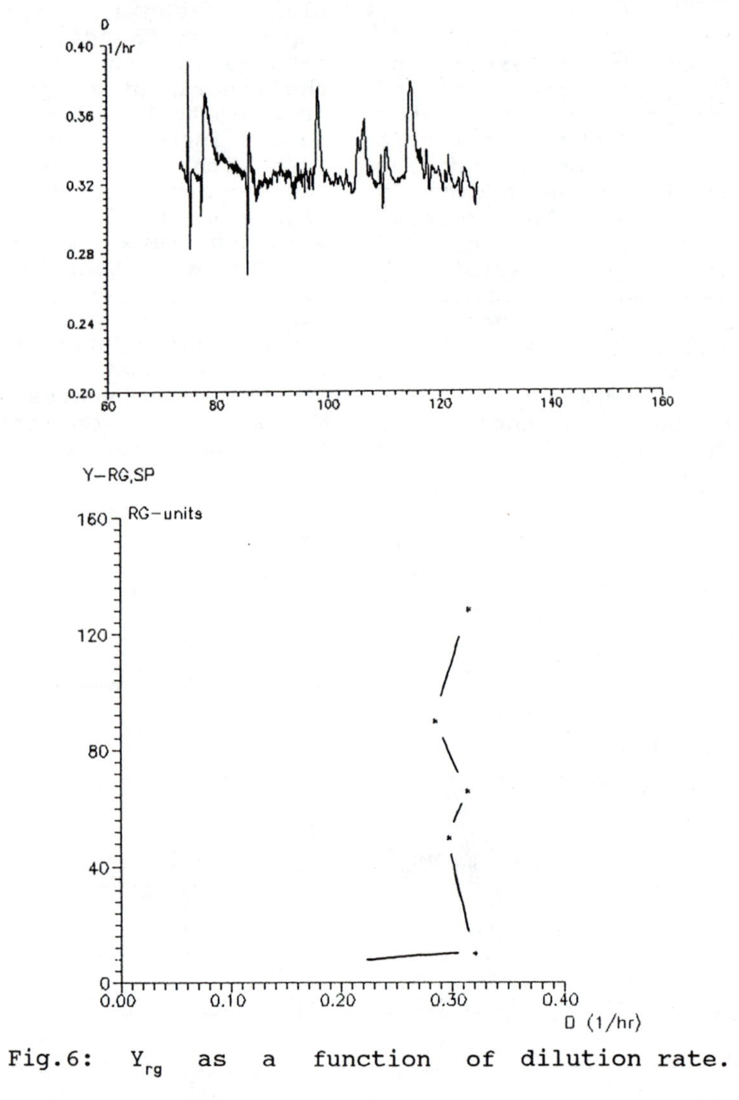

Fig.6: Y_{rg} as a function of dilution rate.

microorganism. This information may be valuable for attempts to model the microorganism behaviour. At least the result indicates that structured biomass models seem to be necessary to catch this type of behaviour. For practical application of the adaptive regulator a deadzone around the measured output should be utilized within which estimation is stopped. Adaptation is then only active when variations larger than the deadzone are measured. During adaptation addition of a PRBS to the input or the setpoint will ensure sufficient conditions for identification.

CONCLUSIONS: The control of continuous fermentation of Saccharomyces cerevisiae near the critical dilution rate has turned out to be a challenging problem due to the varying process characteristics when a family of organisms is operated under conditions near the shift between oxidative and fermentative metabolism. The timevarying characteristics of continuous fermentation with Saccharomyces cerevisiae render it necessary to perform closed loop identification around the critical dilution rate.

REFERENCES

Andersen,H.W.,Rasmussen,K.H. & Jørgensen,S.B., (1991). Advances in process identification, Chemical Process Control IV , eds. W.H. Ray and Y.Arkun, AIChE, New York. p.43??

Andersen,M.Y.,(1990). Multivariable identification of a continuous yeast fermentation. Ph.D. thesis, Department of Chemical Engineering, Technical University of Denmark, Lyngby.

Brabrand,H.,Jensen,N. & Jørgensen,S.B., (1990). MIMOSC - a tool for real-time multivariable identification and adaptive control of chemical processes. Nordic CACE symposium, November 15-16, Lyngby, Denmark.

Egli,T.,(1980). Wachstum von methanol assimilierenden hefen, Ph.D. thesis. ETH, Zürich

Møller, H.E.,(1992). **Modelling, Identification and Control of Yeast Fermentation**. Ph.D. thesis - in preparation. Department of Chemical Engineering, Technival University of Denmark.

Fermentation Diagnosis and Control: Balancing Old Tools with New Concepts

Gregory Stephanopoulos, Department of Chemical Engineering, Massachusetts Institute of Technology, Cambridge, Massachusetts, 02139 U.S.A.

Two examples are presented illustrating two distinct approaches to the subject of fermentation diagnosis and control. The first applies concepts of metabolic activity control to the development of a yield or productivity maximization control strategy based on respiration measurements. It exemplifies the use of mechanistic information in estimation and control. The second example uses projection and pattern recognition to discern, in a lower dimensional space, common features and differences among runs as well as their probable cause(s). It demonstrates how black-box methods can offer valuable insights where other methods usually fail. While some processes can benefit from the application of tested mechanistic methods, for some others only statistically-based methods have a reasonable likelihood of success. In the absence of a hybrid-type methodology in sight, or rules for the use of either one, selection between the two approaches will be a matter of personal judgement for some years to come.

State of the Art and Future Trends

Fermentation diagnosis has been the objective of a great deal of biotechnological research in the past 25 years. Even when the expressed research objective was different, the need to understand the state of a fermentation and the fundamental mechanism by which it evolved to its current state has been the driving force of much of the research activity in this area.

Despite this activity, fermentation diagnosis still remains a rather elusive goal. In contrast to chemical processes, which can be fully defined by the values of the pressure, temperature and extent(s) of principal chemical reaction(s) that take place, fermentations, and biological processes in general, are far more complex systems for which a specific set of culture and process variables that reflect in an inclusive but non-overlapping manner the underlying key mechanisms, is not easy to define. Typically, several process (i.e. fermentor) variables (concentrations, T, P, pH) are combined with physiological variables (such as specific rates, yields, intracellular compositions, etc.) to define, sometimes in a redundant manner, the state of the system. Macroscopic and elemental balances along with some models are written for the state variables aiding in the determination of unmeasured variables through the application of estimation/filtering techniques (Stephanopoulos and Park, 1991). Although system identification is complete at this point, our understanding of the true state of the process in the context of an underlying mechanism may be quite inadequate. This certainly imparts on our ability to confer a value judgement on the present state of the process and its direction, and limits the possibilities to exercise effective action by feedback control.

In light of these uncertainties with fermentation diagnosis, control applications have focused primarily on those situations where the observability requirements of the control scheme were satisfied. The decreasing number of control applications as one moves from regulatory controls to controls of metabolic activity and optimal controls reflects primarily the increasing uncertainty regarding the diagnosis of the state of the process and increasing model requirements for this purpose.

The above summary of the present state of fermentation diagnosis and control reflects the predominantly mechanistic

1054–7487/92/0370$06.00/0

approach to control taken to-date, in parallel with similar applications in the chemical processing industry. In recent years, however, an increasing body of literature is pointing to some exciting new directions in this area, such as diagnosis through projection methods and pattern recognition, expert systems (Konstantinov and Yoshida, 1992), neural networks (Breusegen, et al., 1991) and fuzzy controls (Yamada et al., 1991). These methods are applied to the entirety of data available, make little, if any, use of models or mechanisms, and their results can find use as post-mortem analytical tool or in on-line feedback control.

The two methodologies discussed above are illustrated in the two examples presented in the sequel. The first deals with the fermentation of an auxotrophic mutant of Corynebacterium glutamicum. A rather complete picture of the biosynthetic pathways and their metabolic control suggested the use of a mechanistic, RQ-based, approach to the diagnosis of the fermentation state. The subsequent control strategies of feed flow for yield and productivity maximization were met with success. The second example is an industrial enzyme fermentation with a rather involved and poorly understood regulation and bioreaction network. The projection of the plethora of data collected during production runs allowed the rational classification of runs and identification of the causal factors of the observed differences.

Metabolic Activity Control Applied to the Optimization of the Lysine Fermentation

Production of L-lysine depends on direct fermentation of carbohydrates by auxotrophic and regulatory mutants of Corynebacteria and Brevibacteria species in batch culture. As processes intensive in raw material use, process yield, in addition to productivity, is a critical measure of performance and economic viability. Although some improvements in productivity were achieved by classical mutation and selection methods, results on yield improvement, requiring redirection of carbon flows in the biosynthetic pathways, were mixed. In the work reported herein, environmental controls were employed to effect performance improvement by manipulating the metabolism to favor the desired pathway.

The strain utilized in this work was C. glutamicum ATCC 21253, auxotrophic for both L-homoserine (or L-threonine plus L-methionine) and L-leucine. L-homoserine auxotrophy results in L-lysine overproduction in cultures with low levels of L-threonine due to the bypass of concerted aspartate kinase inhibition by L-threonine plus L-lysine. Materials and Methods are described in (Kiss and Stephanopoulos, 1991).

Careful examination of batch culture dynamics revealed that although the yield and productivity of lysine rose sharply following the depletion of threonine, lysine biosynthesis decayed rapidly thereafter. Active lysine biosynthesis was correlated with elevated values of the respiratory quotient at all times. Furthermore, continuous culture experiments revealed that the specific productivity exhibits a maximum as function of the specific growth rate while yield decreases as μ increases.

The above results suggest that for yield maximization a low specific growth rate should be sought which, however, is not too low as to impair biomass activity. For productivity maximization, on the other hand, the maximum amount of biomass should be established as quickly as possible implying the need of a high specific growth rate, but not higher than the point where the specific productivity begins to decline.

These strategies were implemented in the control schemes discussed below.

The control strategy for maximizing productivity is shown below.

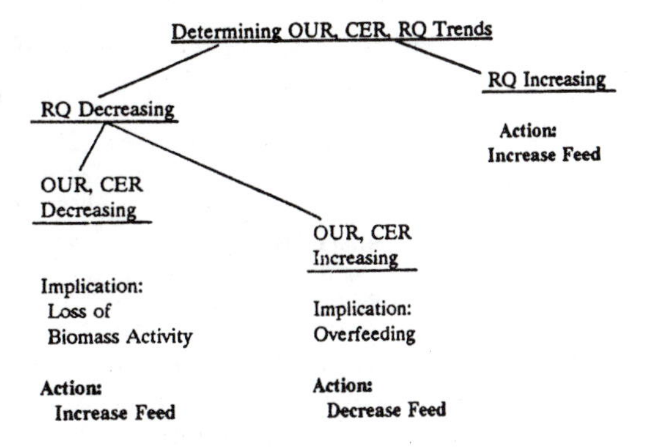

An increasing RQ implies increasing L-lysine

production; consequently, an even higher μ should be sought by further increasing the feed rate. A decreasing RQ could result from a loss of biomass activity or inhibition of the L-lysine biosynthetic pathway due to overfeeding. These two possibilities can be resolved by looking at the individual trends of OUR and CER with the respective control action taken as indicated.

The control algorithm for yield maximization is shown below.

While RQ Indicative of Lysine Production

Determine Trends in OUR and CER

OUR, CER Decreasing	OUR, CER Increasing
Implication: Biomass Activity Declining	Implication: Overfeeding
Action: **Increase Feed**	**Action:** **Decrease Feed**

Since the emphasis is on minimizing growth rate, the only two metabolic conditions which should arise in the application of this strategy are the case of declining metabolic activity due to insufficient nutrients, and the case of excessive nutrient addition leading to more growth than required to maintain maximum conversion efficiency.

The above control strategy succeeded in achieving instantaneous yields between 30-40 mol % with an overall fermentation yield of 24 mol %. A summary of the obtained results with the two strategies is given in Table 1.

It can be seen in the above Table that both strategies achieved superior levels of metabolic activity when compared to batch culture. This was achieved by correctly diagnosing the fermentation state from respiratory data and the application of simple controls for the feed. It should be also noted that the application of the productivity maximization strategy to a fermentation with high biomass concentration (30 g/L) yielded product titer of 78 g/L at a volumetric productivity of 2 g/L.h The implementation of the controls required no models or sophisticated state estimation methodologies although further improvements can be certainly obtained through the use of such methods.

Classification and Diagnosis of Industiral Fermentations by Pattern Recognition and Data Projection Methods

The second example comes from an industrial fermentation of Bacillus subtilis for the production of enzymes. Although some general features of this fermentation, such as glucose repression of protease expression, are well established, the picture of growth, production, inhibition and regulation in a complex medium is far from complete. Other than serving as markers for the initiation of certain control actions, (i.e. feed, aeration, etc.), or for the calculation of some simple quantities, (RQ), very little use of the collected data is otherwise made.

Table 1. Effect of Maximization Control Strategies on Fermentation Performance

experiment	batch base case	yield max. strategy	prod. max. strategy
final L-lysine hydrochloride concn, g/L	23	25 (1.09)	48 (2.09)
average volumetric productivity, g/(L-h)	0.378	0.479 (1.27)	0.925 (2.45)
overall fermentation yield, mol/mol	0.15	0.24 (1.60)	0.22 (1.47)

(Quantities in parenthesis indicate the ratio relative to the batch fermentation)

It is our contention that industrial data and observations often contain sufficient information and discriminating power to establish unique fingerprints of the particular runs and aid in the discrimination among different runs. The problem is that key features of a process are non-obvious or frequently hidden under a large volume of irrelevant or other data with little information content. A central question then is how to unmask the salient features of a run in order to use them for the diagnosis of the process.

Several methods have been employed for this purpose recently, neural networks and direct classification schemes among them. In this work we present some results from the application of a projection method as means of pattern recognition. We opted for a projection method in order to reduce the dimensionality of the data matrix, for high dimensionality contributes to diffusion of discriminating power and loss of resolution. Of course, dimensionality reduction is accompanied by loss of information and, in general, the greater the reduction of dimensions the less the amount of information retained. It can be shown that information loss is minimized if the data matrix (defined as the collection of column vectors containing the time points of all measured variables) is projected to the (hyper) plane of the principal eigenvectors of the covariance of the data matrix. The sum of the principal eigenvalues of the projection space is a measure of the fidelity of transformation. Ideally one would like to project all available data to a two-dimensional plane with minimal information loss.

Such a projection carried out for 18 process variables measured hourly over a period of 135 hours is shown in Figure 1. Three features are obvious in Figure 1. First, that a similar pattern characterizes all 4 runs depicted in the Figure, consisting of a growth phase (the common lower cloud) and production phase. Second, that runs of similar performance (marked as "standard set") give similar projection on the plane. Third, the projections of runs of different performance are well separated from the standard set thus allowing their discrimination from standard runs and between themselves. Similar results are obtained when on-line variables only are used indicating that the PCA projection method can have applications for on-line process diagnosis.

The patterns of Figure 1 provide no information regarding the possible causes of

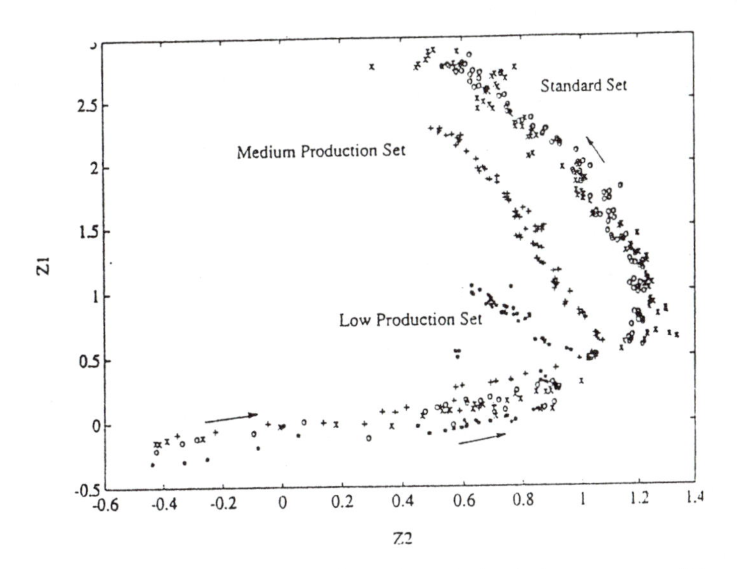

Figure 1. Data projection to a plane spanned by the principal eigenvectors of the covariance of the data matrix.

the observed differences. Such causal factors can be identified by dividing the process variables into distinct groups such as control-related variables, operator-related variables, physiological state variables, and the like. If, upon elimination of one such group and reprojection of the reduced data matrix, the differences of Figure 1 disappear, this is a strong indication that a variable in the group removed may be responsible for the difference. The process can be repeated with the individual variables of the group to identify the one most probably responsible for the deviation. This method applied to the above example revealed the supply of oxygen and the premature initiation of glucose feed as the reasons for substandard performance.

In closing, statistically-based methodologies can offer powerful insights into the diagnosis and control of industrial fermentations. Just the same, traditional mechanistic approaches can extract world-class performance from rather inconspicuous strains. The choice between the two methodologies will be a matter of personal judgement for some years to come.

REFERENCES

Breusegem, V.V., Thibault, J., Cheruy, A., The Canadian Journal of Chem. Eng., 1991, Vol. 69, 481-487.

Kiss, R.D., Stephanopoulos, G., Biotechnology Progress, 1991, Vol. 7, 501-509.

Konstantinov, K.B., Yoshida, T., Biotechnology & Bioengineering, 1992, Vol. 39 (in press).

Stephanopoulos, G., Park, S., In Biotechnology, Vol. 4, Measuring, Modelling and Control, Rehm, H.J., Reed, G.Ed, VCH, Weinheim, 1991, pp. 225-249.

Yamada, Y., Haneda, K., Muriyama, S., Shiomi, S., J. Chem. Eng. Jpn., 1991, Vol. 24, 94-99.

Development of Real-Time Expert System Applications for the On-Line Analysis of Fermentation Data

Joseph S. Alford Jr., Gary L. Fowler, Richard E. Higgs, Jr., Don L. Clapp, and Floyd M. Huber, Eli Lilly and Company, Indianapolis, Indiana, 46285 (FAX: 317 276-5499)

A knowledge-and-model based expert system approach for the operation and control of industrial fermentations at Eli Lilly & Company is described. The system includes an interface to historical data files which simplifies validation and reduces application development time. Applications are being developed for on-line data validation, real-time process analysis, and remote alarming.

One of the goals of Fermentation Process Research and Development at Eli Lilly and Company (Indianapolis, IN) is to increase the level of bioreactor automation. Among the strategies to meet this goal is one focused on the convenient capturing of knowledge of process experts and making it continuously available to automated control systems. One of the best commercially available tools for pursuing this strategy appears to be real-time expert systems. This subject has been previously discussed by Aynsley (1990), Cooney (1988) and Hitzmann (1992).

After examining available products and consulting with groups outside Lilly, the G2 real time expert system (Gensym Corporation, Cambridge, MA) was selected. Project objectives included:

* Integrating G2 software into the in-house distributed control system (DCS)
* Creating a test environment for application development
* Developing successful applications
* Comparing expert system and DCS approaches to incorporating and using "knowledge"

Integration of G2 into Lilly's Fermentation DCS

Most of Lilly's fermentors are interfaced to an in-house developed process control and data historian system (Alford, 1981). G2 is currently networked with this system in a way that provides two way system communications but one way data flow to G2. A schematic of the system is shown in Figure 1. Data is initially collected, filtered and compressed by HP1000 process control minicomputers and then sent to an HP9000 host computer. The data historian on the host computer records live data to disk and also forwards it to any G2 system on the network. Alternatively, archived data can be retrieved from disk and sent to G2 as if it were live data.

Both development and runtime G2 workstations have been added to the network. Completed applications will reside in dedicated runtime workstations in order to minimize cost and risk of system problems when compared to running in a development environment.

Lilly experience has been that wide variation exists in the ease of interfacing different technologies from different vendors. In this case, the interface was accomplished with eight person days of formal Gensym schools, about three person weeks of Lilly computer scientist effort, one person day of Gensym consulting and use of Gensym's interface product GSI. Approximately 1800 lines of new user written code was required for the interface.

Development Environment

A testing utility that directs an archived fermentation data set to a knowledge base was created for developing G2 applications. This offers a convenient alternative to writing test data sets or simulation equations that would, at best, be a subset of what could occur in the plant. The utility accesses historical lots and plays them back to G2 at up to 100 times the normal rate of real-time data acquisition. These data sets provide real-life operational information, containing all the variability, range, and variety of conditions normally encountered in fermentation processes. Therefore, they are especially useful in operational qualification of the application and validation of the system. A flowchart of typical knowledge base development information flow is illustrated in Figure 2.

1054–7487/92/0375$06.00/0

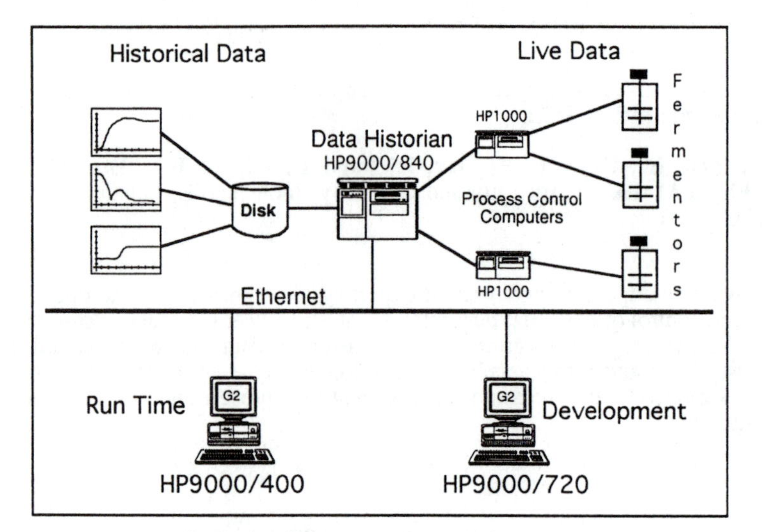

Fig. 1. System Architecture

Applications

The first G2 applications involved the use of fermentation respiration rate data. Considerations included verification of carbon dioxide evolution (CE) values as they became available from mass spectrometer measurements, the recognition of trends in the CE profile, and a consideration of relationships to other process variables. Factors leading to a focus on respiration rate applications included:

* CE is one of the most important indicators of culture metabolism. Fermentation scientists and operators often make important process changes based on CE, their general knowledge of the fermentation's CE profile and their ability to intuitively filter erroneous data.
* For processes run under the same conditions, CE profiles are consistent qualitatively but not quantitatively. The absolute magnitudes of peak values and the times at which peaks occur differ among runs.
* Data needed to calculate CE is not continuously available. Some of the information is normally acquired from expensive multiplexed analytical instruments, so intervals between data points can be several minutes. While CE data are normally highly accurate, spurious data points can result from a host of possible process transient, hardware, and calibration causes. Given the relatively low number of CE points and the need to know the validity of each one, it is more appropriate to test individual data points against a set of rules than to apply conventional DCS filtering algorithms as is normally done for continuous fermentation measurements. Filtering algorithms still include, but reduce the impact of, bad data points.
* CE relationships to other process variables (e.g. DO2) are known, which can help in verifying CE data.

Critical Nutrient Initiator Knowledge Base

The Critical Nutrient Initiator knowledge base indicates the appropriate time to begin delivery of a critical nutrient feed. It was postulated that the culture's optimal productivity depends on the time of initiating the delivery of the nutrient feed and that the optimal time for initiating feed delivery is within a narrow window near the peak of the CE profile. The goal of this knowledge-based application is to consistently identify the profile's peak as soon as it occurs.

This application fits respiration rate values to simple exponential and polynomial models. Each new value is analyzed using expected values based on extrapolation of the model and process knowledge expressed in G2 "if/then" statements. When the profile reaches a peak, the operator is informed by G2's message board that it is time to begin the delivery of a critical nutrient. Figure 3 illustrates the performance of the knowledge base compared with the performance of DCS-initiated logic. The G2-assisted process resulted in reduced variation of feed initiations as well as triggering the feed closer to optimum conditions.

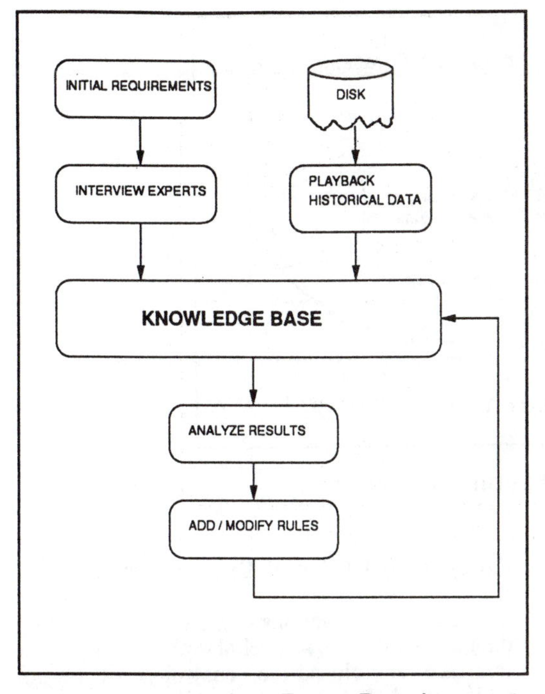

Fig. 2. Knowledge Base Development Flow Chart

Inoculation Forecaster Knowledge Base

The approach used in developing the Critical Nutrient Initiator was easily extended to another fermentation application. The Inoculation Forecaster knowledge base refits an algorithmic model each time a new CE value from a "seed fermentation" is received, checked, and accepted as valid. By extrapolating the curve to a target value, the time of inoculum transfer is determined. Figure 4 shows a G2 analyzed data set illustrating use of the Inoculation Forecaster. By accurately forecasting the inoculation time, the need for alternative methods, such as manual sampling of tanks, is reduced. In addition, operators can better plan their work activities so as to be available when conditions are optimum for inoculation.

The application can also provide early warning of contamination for slow growing cultures when the specific growth rate calculated from the incoming CE data exceeds that characterizing the specific microorganism.

Pilot Plant Remote Alarming System

The Fermentation Pilot Plant Remote Alarming System is used to monitor pilot plant operations during unattended hours. Use of a real-time expert system was considered key to a successful transition from around the clock manned operations to single shift operations. Using DCS generated alarm messages as one of several

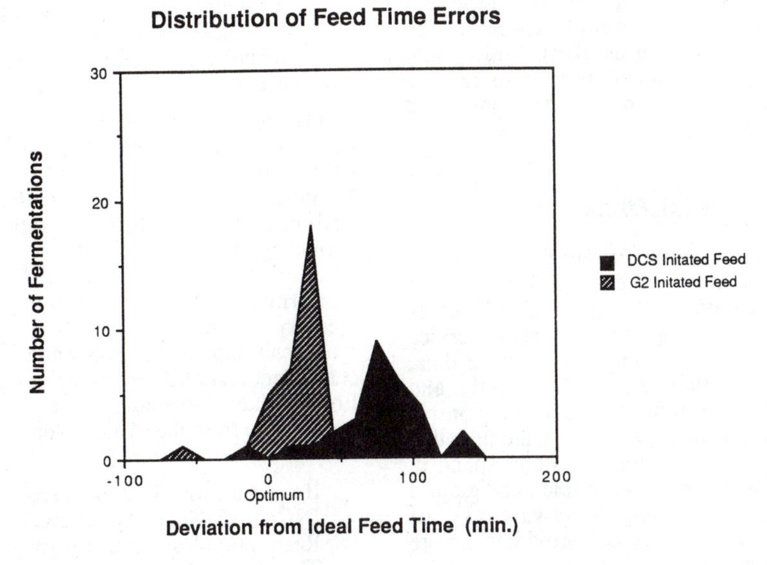

Fig. 3. Distribution of Feed Time Errors for G2 and DCS Implementation

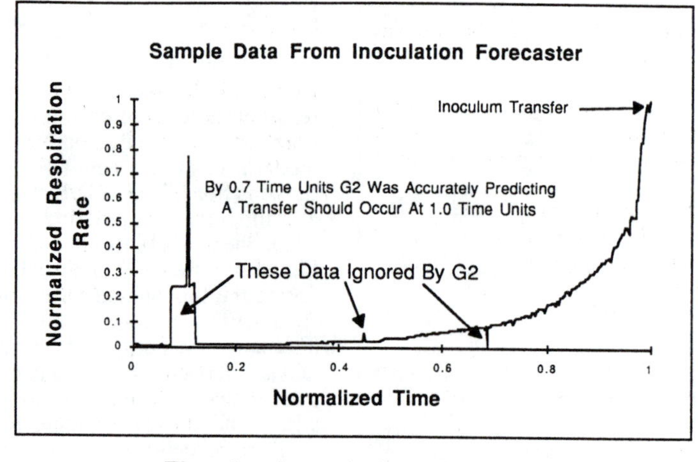

Fig. 4. Inoculation Forcaster

inputs, this system performs a high level of situation analysis, identifies existing and developing problems and notifies the appropriate personnel using an alphanumeric paging system.

One of the goals of the project has been to use high level situation analysis to minimize the occurrence of nuisance alarms, something historically difficult to maintain with DCS systems. Also helpful in reducing nuisance alarms has been the admittedly low tolerance of users to which a 2 AM non-critical pager call could be directed.

A friendly operator interface, making effective use of color, menus, graphics, icons and a mouse, permits representatives of individual laboratories to access the system locally and easily change information such as alarm limits, tank designation, designation of person to receive initial call, designation of backup personnel, and length of time between repeat calls.

Fermentation Production Advisor

The Fermentation Production Advisor monitors, in real time, data collected from production fermentors. This application involves capturing the knowledge that Technical Service personnel use in evaluating fermentation data, then performing this task automatically and continuously on incoming process control information. Goals include reducing the time that Technical Service personnel presently spend in manually reviewing process variable trend graphs, providing more immediate awareness of fermentation abnormalities, and providing a more consistent determination of when to harvest a fermentor.

Comparing DCS and Expert Systems

The primary functions of a DCS are providing data acquisition, low level alarming, an operator interface, and closed loop control. They are not well suited, however, to handling knowledge based applications. DCS limitations can include:

* lack of tools (including inference engines) to efficiently represent different kinds of knowledge and process large numbers of interrelated rules
* limited memory
* cryptic application code
* knowledge difficult to find (embedded in other application code)
* difficulty of modifying the knowledge base on-line

G2 provides an alternative for developing and running applications that overcome these limitations:

* It uses a tool set (including inference engines) designed for efficient organization and use of knowledge bases.
* It runs on a variety of different hardware platforms, accommodating user preference and memory requirements.
* Technologists can conveniently incorporate, centralize, and augment process knowledge in an English-like knowledge base. For example, a sample rule from the Critical Nutrient Initiator is:

> If the Air Flow Rate of a Fermentor is INCREASING and the New Value of the Respiration Rate of that Fermentor is HIGH, Then conclude that the New Value of the Respiration Rate is BAD.

Other rules, put into the knowledge base in no particular order, define "INCREASING" and "HIGH."

Engineers and scientists have found that knowledge expressed in this way is easy to find, read, and understand since it is similar to natural language.

* On-line changes are also easily made. The sample rule listed above could be added or modified on-line without resetting or recompiling the knowledge base.

Future Work

Future work will focus on closing the loop (two way communication of information). This includes:

* permitting G-2 conclusions to automatically trigger DCS control logic
* integrating G-2 messages and alarms into DCS operator displays in order to reduce the number of CRT workstations that operators would otherwise need to interact with.

References

Alford, J.S., Jr., Evolution of the Fermentation Computer System at Eli Lilly and Co., ICCAFT 3, Manchester, England, **1981**.

Aynsley, M., A.G. Hofland, G.A. Montague, D. Peel and A.J. Morris, A Real-Time Knowledge Based System for the Operation and Control of a Fermentation Plant. Proceedings of the **1990** American Control Conference.

Cooney, C.L., G.M. O'Conner, and F. Sanchez-Riera, An Expert System for Intelligent Supervisory Control of Fermentation Processes, Proceedings, 8th International Biotechnology Symposium, **1988.**

Hitzmann B., A. Lubbert, and K. Schugerl, An Expert System Approach for the Control of a Bioprocess. I: Knowledge Representation and Processing, Biotechnology and Bioengineering, **1992**, Vol 39, pp 33-43

Automatic Discovery of Protein Motifs

Douglas L. Brutlag and **Tod M. Klingler**
Department of Biochemistry and Section on Medical Informatics
Stanford University School of Medicine, Stanford, CA 94305-5307

We have developed a novel representation of protein motifs that permits the rapid discovery of structural features in sets of protein sequences with a common structure or function. Many popular methods for representing protein motifs (consensus sequences, weight matrices, profiles, etc.) emphasize conservation of amino acids at specific sites in the sequence. Our method looks for correlations between amino acid variations at distinct sites. Correlations between the residues represent side-chain side-chain interactions and give insight into the structural properties of the motifs. Structural correlations can be used in database search to discover other proteins bearing similar relationships. This database search is significantly more sensitive than methods depending only upon conserved residues.

Introduction

Most methods for representing protein motifs emphasize the amino acids conserved during evolution. These conserved residues are often of critical importance in the structure or function of the protein. Dictionaries of such conserved motifs have been compiled and are extremely valuable in discovering structural and functional attributes of novel protein sequences (Bairoch, 1991).

Many protein motifs are not conserved sufficiently to be represented as a consensus sequence (Dodd and Egan, 1987). When motifs are highly variable in sequence, one usually employs probabilistic methods such as weight matrices or profiles (Staden, 1984; Gribskov, McLachlan and Eisenberg, 1987). Weight matrices give the likelihood of finding each amino acid at each position in the motif based on a large set of examples. The likelihood of finding each amino acid is calculated relative to the overall frequency of finding that residue in the proteins or in the database being examined. Like the consensus sequence method, weight matrices emphasize residues conserved in evolution.

Often, one only has a few examples of a particular protein structure, which are insufficient to determine all the likelihoods required in a weight matrix accurately. In these cases, the weight matrix generated by a few examples can be multiplied by the likelihoods of each amino acid replacing another during evolution (these replacement matrices are known as PAM matrices, Schwartz and Dayhoff, 1979). The resulting likelihood matrix is known as a protein profile (Gribskov, McLachlan and Eisenberg, 1987). Profiles, again, give a measure of evolutionary sequence similarity rather than insights into structural attributes of the protein.

Methods

We have undertaken a novel approach that emphasizes sequence variation within proteins of common structure or function. Along with the amino acids conserved at any one site, we have looked specifically for conserved correlations between amino acids at two distinct sites in a motif. We felt that if the structure of a motif was highly conserved independent of the sequence, then correlated changes at two distinct sites may reflect conserved amino acid side-chain interactions. If correlations between positions were conserved in evolution despite sequence variation, testing for conserved correlations as well as conserved positions would be a more discriminating method for detecting structural motifs.

In order to represent both correlated changes and conserved residues simultaneously we have used a probabilistic method known as a belief network (Neapolitan, 1990; Pearl, 1988). Belief networks are general graphic representations of Baysian conditional probabilities. Nodes in the belief network shown in Figure 1 are of two kinds. The top node, labeled "Motif?", represents a decision node with two possible values, yes or no. The evidence nodes in these networks represent the amino acid residues that occur in five positions along a protein motif. Each has 20 possible values. The arcs drawn

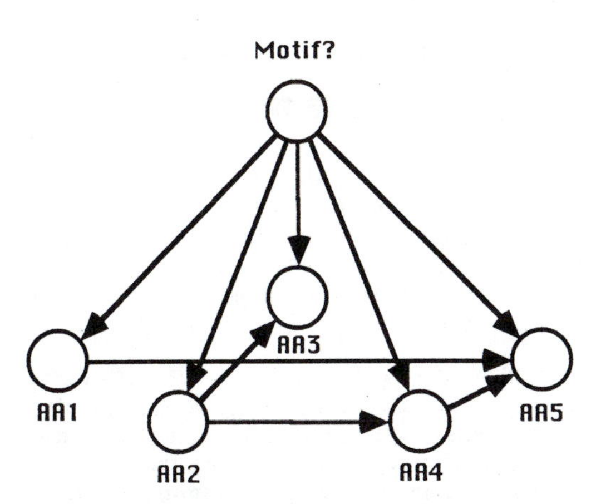

Figure 1. A generic belief network graphically illustrating the relationships between a decision node labeled Motif? and the evidence nodes (AAn). The decision node has two possible values and the evidence nodes have twenty possible values, one for each amino acid. The arc between any two nodes represents the conditional probabilities relating values for the two nodes. In this belief network there are conditional probabilities for each position in the motif depending on the knowledge of the motif and there are also dependencies between amino acids at different positions. These later conditional probabilities represent correlations between amino acids at those two positions observed among many examples of the protein motif.

between the decision node and the evidence nodes represent the conditional probabilities of each amino acid given a value for the decision node. Values for these conditional probabilities are obtained from training sets of sequences and, like weight matrices, they represent the conservation of residues at each position in the motif.

Arcs between two amino acid nodes represent the probability of amino acid correlations between two positions in the training set. We calculate probabilities of amino acid correlation using χ^2 statistics and check them using a Monte Carlo simulation. Comparisons between all pairs of nodes are determined and those arcs whose probability lies below 0.01 are retained in the belief network. Normally arcs between two positions would represent 400 different possible correlations (as there are 20 different amino acids at each position) and thus would require over 2,000 examples of a motif to determine the each correlation significantly. With limited numbers of examples, we generally can not compare individual amino acid residues at each position. Instead, we classify amino acids into a few structural or functional types and look for correlations between types of amino acids. Our initial amino acid classes include nonpolar (ILMVAGPFWY), polar but uncharged (CNQST), basic (HKR) and acidic (DE) classes. With four classes we can readily detect correlations between types of amino acids with as few as 80 examples.

Once correlations between positions are discovered, belief networks like the one shown in Figure 1 can be used to search for motifs as well. Given values for all the amino acids in the chain, one can infer the posterior probability that a sequence is the motif in question (Lauritzen and Spiegelhalter, 1988). Applying this probabilistic inference procedure to each position in a protein sequence database allows one to perform search.

Results : The Helix-Turn-Helix Motif

We have examined 80 prokaryotic helix-turn-helix containing proteins for correlations among the 22 positions comprising the helix-turn-helix motif. We discovered seven pairs of positions none of which were highly conserved, but which were strongly correlated with each other. When a database search was conducted for sequences giving a high score with either a weight matrix (finds conserved residues) or a belief network (finds conserved residues and correlations simultaneously), the number of false positives was reduced at all threshold levels. Especially significant were the relative scores for sequences clearly unrelated to a helix-turn-helix motif. The relative scores for such sequences for the belief network were 40 to 100 fold lower than for the weight matrix. This increased discrimination makes the belief network more useful in database search.

Amino Acid Interactions in α-Helices

We examined the protein sequences of 234 α-helical segments taken from a unique subset of the Brookhaven Structure Library. This subset contained only the highest resolution member of each protein superfamily. We initially looked for sequence correlations in eight amino acid segments (two complete α-helical turns) contained within these segments.

Figure 2 shows half of the highly significant correlations that were observed within α-helical segments. This belief network is displayed as a helical wheel, with the dotted line showing the path of the protein peptide backbone and the solid lines showing the correlated positions. The arcs represented in Figure 2 indicated a strong dependence of amino acids at positions $i+3$ and $i+4$ on the amino acid at position i. The correlation indicated that if amino acid i were

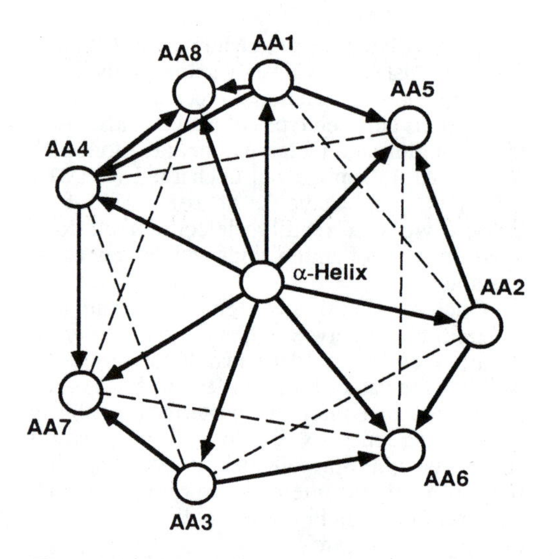

Figure 2. This belief network displays some the highly significant correlations observed between the amino acid side-chains in α-helical segments in the Brookhaven Database. The network is depicted as a helical wheel. The dotted lines show the path of the peptide main chain. The solid arrows around the circumference show the strong correlations observed between amino acid side chains that are adjacent in space. Finally the solid spokes represent the dependence of the amino acid composition on the helical nature of the sequence.

hydrophobic, then the amino acids at $i+3$ and $i+4$ were also more likely to be hydrophobic and more unlikely to be hydrophilic. This is the well-known hydrophobic patch that characterizes many α-helices.

The other half of the significant correlations not displayed here, shows a dependence of amino acids at $i+2$ and $i+5$ as well. The residues at these positions have hydrophobicity opposite that at position i. This observation is consistent with the amphipathic nature of most of the α-helices in the Brookhaven database.

Similar networks have been made for β-strand sequences taken from the Brookhaven Structure database. They also suggest the amphipathic nature of β-strands.

While these observations do not provide any novel insights into secondary structure, they show that the probabilistic representation can rediscover well-known principles of protein structure. Moreover, the quantitative relationships contained in the belief networks should allow us to discriminate regions of secondary structure with more precision than previously possible.

Phe-His Correlation at C-Termini of α-Helices

Finally we searched for correlations between the amino acids found near the ends of α-helices. We examined the first and last 5 amino acid residues from both the N- and C-terminal regions as determined by the DSSP program of Kabsch and Sander (1983). Most of the correlations we observed were of the types mentioned above for α-helical segments in general. However, among the 234 C-terminal ends of α-helices, we discovered a small but statistically significant correlation between an aromatic amino acid four residues before the terminus and a basic amino acid at the terminal position. When helices displaying this arrangement of amino acids at their C-terminus were examined, we discovered five occurrences of the amino acid pattern Phe-Xaa-Xaa-Xaa-His among the 234 helical segments examined (Figure 3). When we searched for the amino acid pattern above in all the sequences in the Brookhaven database it occurred at the C-terminal ends of α-helixes (± one residue) a total of twelve times. This pattern occurred in no other helical region.

The occurrence of the Phe-Xaa-Xaa-Xaa-His sequence has been known to stabilize α-helical segments (Shoemaker *et al.*, 1990). The nature of the interaction appears to be one of an ring hydrogen of histidine interacts with the aromatic ring of phenylalanine. This interaction

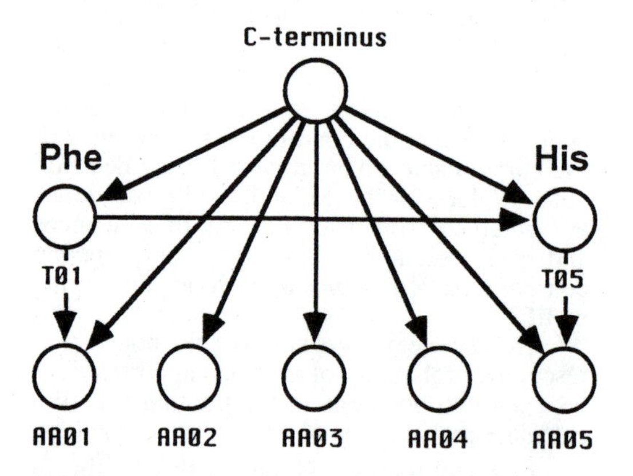

Figure 3. This belief network shows the observed the correlation between amino acids within five residues from the C-terminus of α-helices. Of the 234 α-helices examined in a high resolution unique subset of the Brookhaven Structural database, twelve displayed the pattern Phe-Xaa-Xaa-Xaa-His. This sequence pattern was found at no other position in α-helical segments in the database.

requires the histidine side-chain to bend towards the preceding phenylalanine. Our results suggest that this interaction may terminate α-helical segments.

This final result shows that the search for correlations in protein sequences can lead to the discovery of novel side-chain side-chain interactions in the structures of proteins. The application of probabilistic inference using belief networks that represent general secondary structure elements such as α-helices, β-strands, and C-terminal segments of α-helices should lead to more accurate predictions of protein secondary structure.

References

Bairoch, A. *Nucleic Acids Res.*, **1991,** *19*, 2241-2245.

Dodd, I. B.; Egan, J. B. *J. Mol. Biol*, **1987,** *194*, 557-564.

Gribskov, M.; McLachlan, A. D.; Eisenberg, D. *Proc. Natl. Acad. Sci. USA*, **1987,** *84*, 4355-4358.

Kabsch, W.; Sander, C. *Biopolymers*, **1983,** *22*, 2577-637.

Lauritzen, S. L.; Spiegelhalter, D. J. *Journal of the Royal Statistical Society*, **1988,** *50 B*, 157-224.

Neapolitan, R. E. *Probabilistic Reasoning in Expert Systems: Theory and Algorithms;* John Wiley and Sons: New York, NY, **1990**;

Pearl, J. *Probabilistic Reasoning in Intelligent Systems: Networks of Plausible Inference;* Morgan Kaufmann Publishers, Inc.: San Mateo, CA, **1988**;

Schwartz, R. M.; Dayhoff, M. O. *Atlas of Protein Structure,* **1979,** *5*, 353-358.

Shoemaker, K. R.; Fairman, R.; Schultz, D. A.; Robertson, A. D.; York, E. J.; Stewart, J. M.; Baldwin, R. L. *Biopolymers*, **1990,** *29*, 1-11.

Staden, R. *Nucleic Acids Research*, **1984,** *12*, 505-519.

The National Center for Biotechnology Information: Its Mandate and Current Activities

David Landsman, Jonathan A. Kans, Gregory D. Schuler and James M. Ostell.
National Center for Biotechnology Information,
National Library of Medicine,
National Institutes of Health,
Building 38A, Room 8S806
Bethesda, Maryland 20894 U.S.A.
(301)480-9241 (fax)

The creation of the National Center for Biotechnology Information in the National Library of Medicine in 1988, has resulted in a new approach to bioinformatics with special emphasis on molecular sequence data structure and analysis. A sampling of the current research at the Center will be reviewed, including such projects as ASN.1 specifications for biosequence data, the GenInfo® Backbone database, and ***Entrez***, a CD-ROM retrieval system developed at the NCBI which integrates access to nucleotide and protein sequence databases and relevant MEDLINE® citations.

The National Center for Biotechnology Information (NCBI) was created by Public Law 100-607 in November, 1988, and placed within the National Library of Medicine (NLM). The establishment of the Center reflects the importance of developing new information technologies to facilitate the understanding of the molecular processes that control health and disease. The Center has been given the responsibility to:

- create automated systems for storing and analyzing knowledge about molecular biology, biochemistry, and genetics
- perform research into advanced methods of computer-based information processing for the vast number of biologically important molecules and compounds
- enable biotechnology researchers and medical care personnel to use the systems and methods developed
- coordinate efforts to gather biotechnology information worldwide.

A key component for furthering the understanding of molecular biology data is basic research in areas of genomic organization and macromolecular evolution, structure and function. Thus, a major activity of the NCBI is an intramural research program with senior investigators and staff fellows committed to a computational and analytic approach for investigating these problems within an interdisciplinary framework of mathematics, biology, and computer science. Research is critically dependent upon having a body of knowledge about protein and DNA sequences and structures, as well as physical and genetic mapping data. Therefore, databases in molecular biology are not only essential archival resources but active research tools.

Consequently, the NCBI has begun building a database of molecular sequence data, named the GenInfo® sequence database. One component, the GenInfo® Backbone (GIBB), developed in collaboration with the Division of Library Operations in the NLM (the builders of

MEDLINE®), involves locating and extracting all new sequences from the published literature. This can be done efficiently at the NLM by tapping into the stream of over 325,000 articles per year indexed by MEDLINE® to identify sequence-containing articles. Sequences appearing in these articles are then evaluated, entered and annotated by professional staff of sequence indexers. The detailed design coupled with the continued support of this database is essential for the proficient scientific endeavor of many research molecular biologists.

In order to effortlessly integrate numerous databases about different sequence information (i.e. multiple alignments, motifs), a well-defined data structure is a prerequisite. Thus, the NCBI has begun developing an Abstract Syntax Notation 1 (ASN.1) specification for biological sequence data. ASN.1 is a mechanism that allows computer systems to reliably exchange arbitrary data structures. ASN.1 does not describe the content, meaning, or structure of the data, only the way in which it is specified and encoded. These properties make it an excellent choice for a standard way of encoding scientific data. Since ASN.1 does not specify content, specifications can be created as new concepts need to be represented. In addition, since it is an International Standards Organization (ISO) standard, the new specification can take advantage of various tools built to work with ASN.1 in general. It removes from scientists the role of specifying *ad hoc* file formats and focuses them instead on specifying the content and structure of data necessary to convey scientific meaning. To this end, the NCBI has created explicit definitions for biosequence objects using ASN.1. Messages passed in ASN.1 will be used to provide a hardware- and software-independent version of the information contained in the GenInfo® Backbone's relational database.

NCBI has also built tools for producing ASN.1 versions of the GenBank® DNA sequence database and the PIR® amino acid sequence database. ASN.1 versions of other databases, such as SWISS-PROT, an amino acid sequence database, and the EMBL sequence database are also under developement in collaboration with their producers. The NCBI plans to distribute ASN.1 versions of GenInfo®, as well as other sequence databases, on CD-ROMs along with information retrieval and sequence searching tools for PCs and Macintosh computers.

The database will be distributed on CD-ROM in at least 3 forms. A CD-ROM in ASN.1 format will be supplied to those developers and individuals who require this ISO standard for data manipulation, analysis and presentation. NCBI-developed software tools for accessing the data are freely available to the public. A GenBank® flat file format will be distributed for current users of that format. An ***Entrez*** CD-ROM will be distributed by the NCBI with an analysis tool.

Entrez : Sequences is a CD-ROM retrieval system developed at the NCBI which integrates access to nucleotide and protein sequence databases and relevant MEDLINE® citations. ***Entrez*** eliminates futile searches by having the list of available indexed terms appear as the user types the query. Queries are constructed by direct manipulation with a mouse, and Boolean query refinement is independent of term entry. A key feature is neighboring, which uses text and sequence similarity algorithms developed at the NCBI to identify related records within a database. Neighbors within, and links between databases are pre-computed and stored on the CD-ROM. The ability to traverse the literature and molecular sequences via neighbors and links provides a powerful yet intuitive way of retrieving the information in those databases. The database records are stored using the NCBI-designed ASN.1 specification for sequence information. A data access library allows third party vendors to adapt their packages to use the data files on the CD-ROM. ***Entrez*** is written with

VIBRANT, the NCBI user interface development library, and currently runs on the Macintosh and on PCs under Microsoft Windows, and is under developement to run on UNIX machines under Motif. The data access routines run on a variety of additional platforms, including UNIX and VAX/VMS.

A New General Approach for Searching Functional Regions Using Fractal Representation of Nucleotide and Amino Acid Sequences

Hwa A. Lim,* and **Victor V. Solovyev**

Supercomputer Computations Research Institute, Florida State University, B-186, Tallahassee, Florida, 32306-4052

(FAX: (904) 644-0098, internet: hlim@scri.fsu.edu)

A new approach combining the technique of fractal geometry and intrinsic characteristics of nucleic and amino acid sequences is described. The approach, fractal representation of sets of sequences (FRS), is used in particular to study the functional regions of genomes. Global properties of sequences are also plotted to obtain information about 5', exon, intron, 3'-regions in eukaryotic genes.

Biological sequence databases represent a major tool in today's biological research. Since the beginning of the international initiative of the Human Genome Project, the amount of data is accumulating at an ever increasing rate. The data contains all the present knowledge on primary sequences either at the nucleic acid or amino acid level. As the size of the database grows, the problem of handling, analyzing and classifying them becomes more demanding. Thus, development of advanced technique for handling, analyzing and classifying sequences has become one of the main challenges of the Human Genome Project (Woodhead et al., 1988; Bell et al., 1990).

Several studies have shown that hidden in nucleic and amino acid sequences are some distinctive patterns (Trifonov, 1990). These patterns arise from for example, functional or evolutionary reasons. Evolutionary advantageous attributes or functionally significant regions are known to be conserved in the course of evolution. These conserved regions must therefore be quite distinct from the rest of the sequence so that they can effectively carry out their basic functions.

Quite independently but in parallel, another field of study, fractal geometry, is also under intensive research (Mandelbrot, 1982; Barnsley, 1988). Fractal geometry is known to be a useful tool for extracting patterns and the approach has been used in diverse fields (Avnir, 1991).

Thus, it seems natural that fractal geometry will also be an important tool in studying patterns in *nucleic* sequences. The first attempt to do so was carried out by Jeffrey (Jeffrey, 1990). The latter used a method from chaotic dynamics. The new approach reported here, nicknamed fractal representation of sets of sequences (FRS), extends the method to a point where not only it can provide visual global properties of sequences so that coding regions in newly sequenced genes can be predicted, but also to where it can be used to obtain information about splicing sites and some regularities in 5'- and 3'-regions of eukaryotic genes (Solovyev et al., 1991; 1992). In addition, the approach is also extended to handle amino acid sequences and genetic texts of any number of letters in the alphabet (Solovyev, 1991).

The Methodology of FRS

There are four chemically different types of nucleotides (A, C, T, G) and there are twenty chemically distinct types of amino acids (A, V, L, I, C, M, P, F, Y, W, D, N, E, Q, H, S, T, K, R, G). Since the numbers are different, the FRS representations of nucleic and amino acid sequences are quite different. A general approach will first be formulated, and then the two individual cases of nucleic and amino acid will be addressed in turn.

For a genetic text of A letters (monomers) in the alphabet, it is useful to factor A into $A = v \times h$. Then the letters can be plotted consecutively (as they appear in the sequence) as dots in a regular enclosing

1054–7487/92/0387$06.00/0

geometrical shape. In this case, $v > 1, h > 1$ and they determine the number of cells in a vertical column and the number of cells in a horizontal row, respectively. The plotting routine is actually a set of iterative equations to calculate the co-ordinates of dots corresponding to the letter in question from the coordinates of the dot for the letter just plotted. The transformation is described by a set of equations

$$x \mapsto x_j + x/v, \tag{1a}$$

$$y \mapsto y_j + y/h, \tag{1b}$$

where $j = 1, \ldots, A$. The iteractive process is applied to each of the letter (monomers) in the sequence until the sequence is exhausted. In this way, a map of dots is created as a result.

Nucleic Sequences

Since there are four ($A = 4$) distinct nucleotides, a natural choice will be a regular four-sided geometrical enclosing area, i.e., a square ($h = 2, v = 2$). Each vertex of the square is associated with one of the nucleotides. This is shown in Figs. 1a & b. In creating the FRS, Equation (1) translates to

$$x \mapsto x_j + x/2, \tag{2a}$$

$$y \mapsto y_j + y/2. \tag{2b}$$

In this work, however, a slightly modified procedure (*a la* Jeffrey (Jeffrey, 1990)) is followed:

1. Pick an initial point and put a dot on the point. Mathematically, any point within the square will suffice, but in this case, the centroid of the square will be taken since it is equidistant from any of the four vertices;
2. Plot the first nucleotide in the nucleic sequence as a dot at the midpoint of the vector joining the initial dot (centroid of square) to the vertex associated with the nucleotide;
3. Plot subsequent nucleotides in the nucleic sequence as dots at the midpoint of the vector joining the previous dot to the vertex associated with the nucleotide in question;
4. Repeat step 3 until all the nucleotides in the nucleic sequence are exhausted.

Fig. 2 presents an example of FRS for human intron sequences, plotted as described above.

Amino Acid Sequences

Since there are twenty ($A = 20$) amino acids, a rectangle with length $5a_x$ and height $4a_y$, where a_x, a_y are scaling factors, is taken. The rectangle is divided into 20 ($h = 5, v = 4$) $a_x \times a_y$ cells, each associated with one of the twenty amino acids (Fig. 3). If the amino acids are numbered consecutively from 1 to 20 in the following order: (A, V, L, I, C, M, P, F, Y, W, D, N, E, Q, H, S, T, K, R, G), i.e., $j = 1$ corresponds to A, $j = 2$ corresponds to V and so forth, then the correspondence between a cell and an amino acid is determined by the following equation

$$x_j = \mathrm{Int}\left(\left(j - 1\right)/4 + 1\right), \tag{3a}$$

$$y_j = j - 4 \times \left(x_j - 1\right), \tag{3b}$$

where $x_j = 1, \ldots, 5$ and $y_j = 1, \ldots, 4$, and

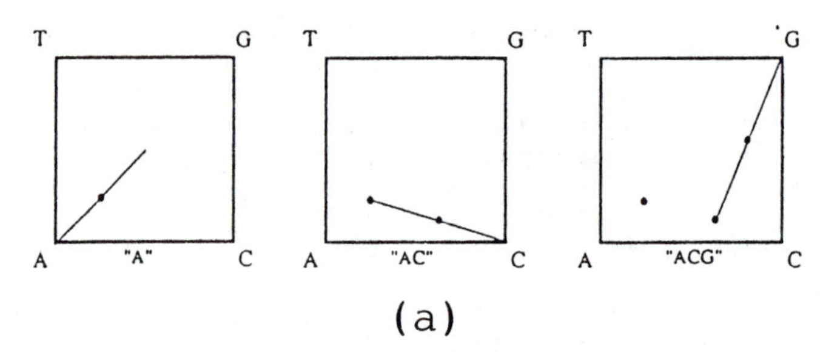

Fig. 1. (a) The representation of the sequence "ACG" as a map of dots in a primitive square. Each corner of the square is associated with one of the four nucleotides: A, T, G, C; (b) The primitive square is subdivided into cells associated with mononucleotides and dinucleotides. The process can be continued to get cells associated with trinucleotides, tetranucleotides and so on. In general, a cell of length 2^{-k} units of the length of the primitive square is associated with a polynucleotide of length k nucleotides.

(x_j, y_j) are the coordinates of the lower left vertex of the rectangular cell corresponding to amino acid of type j. In creating the FRS, Equation (1) translates to

$$x \mapsto x_j + x/5, \qquad (4a)$$

$$y \mapsto y_j + y/4, \qquad (4b)$$

or simply the following steps (for convenience, set $a_x = a_y = a$):

1. Plot the first amino acid in the sequence as a dot in the center of the rectangular cell associated with the amino acid;
2. Plot the next amino acid by projecting *conformally* the location of the dot just plotted (dot corresponding to the amino acid in the sequence which precedes the one in question) in the $5a_x \times 4a_y$ rectangle onto the $a_x \times a_y$ rectangular cell associated with the amino acid in question. Other dots, which are plotted earlier in the iterative process, are untouched;
3. Repeat step 2 until the amino acid sequence is exhausted.

Fig. 3 presents an example of such a plot for α-globin family.

Extraction of Graphics Patterns

It is obvious that a short sequence will only create a very sparse map of dots and thus cannot reveal any structures of the sequence. Therefore, for the purpose of FRS, a set of sequences are plotted to give better statistics

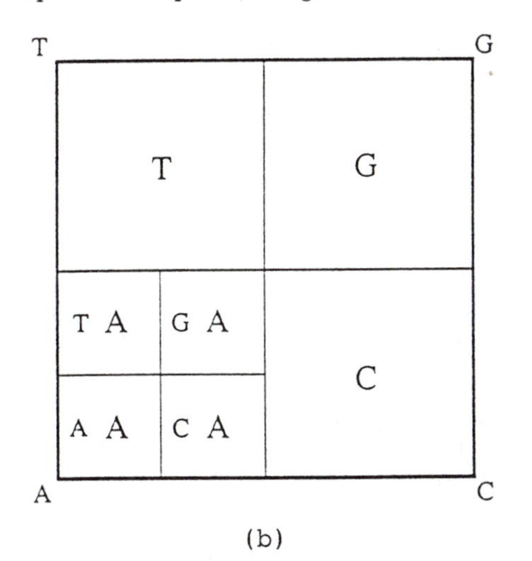

Fig. 1b

Hwa A. Lim & Victor V. Solovyev
9th Intl. Biotechnology Symposium

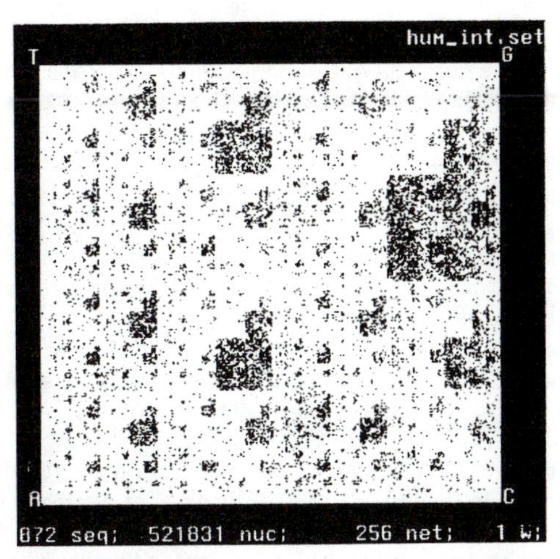

Fig. 2. The FRS of introns of human genes. Differences in the content of oligonucleotides in this set are clearly seen

and to make more transparent the existence of any possible hidden structures.

The fractal representation of a set of genetic texts (FRS) must satisfy two necessary conditions:

1. A set of random sequences should generate a plot of uniform dot density;
2. Dots of identical oligomeres of genetic texts (oligopeptides, oligonucleotides) should concentrate in same regions of FRS plots.

In the general case of genetic text of A letters in the alphabet, the second condition

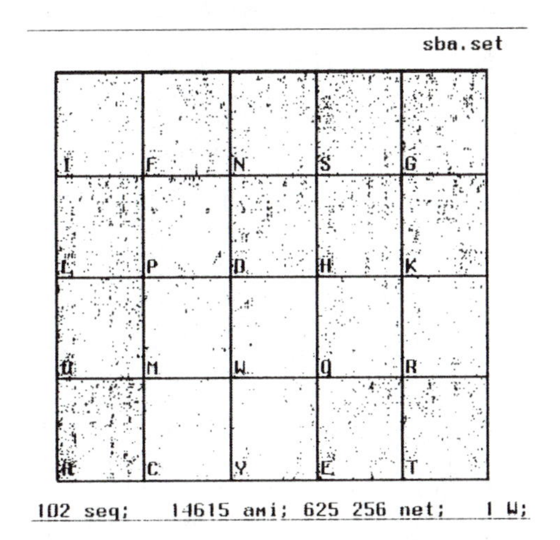

Fig. 3. A fractal representation of α- and β-globins. Note that the 20 amino acids are numbered consistently in a specific order during the creation of the FRS.

implies that the enclosing geometrical shape should be subdivided. For the case of nucleotide, the condition has an important corollary: sequences with k-terminal identical nucleotides will always be in a subsquare of length 2^{-k} units of the length of the primitive square (Fig. 1b). Thus, by subdividing the primitive squares into smaller cells, hidden structures of oligonucleotides of length k can be brought out more clearly.

Capitalizing on the fact that evolutionary useful or functionally significant portions of a sequence has certain characteristic features, one can discern these portions from the rest. This scenario is very reminiscent of cases in disciplines (astrophysics, for example) where useful data is superimposed on some background noise (white noise). The way to sift out useful data from the noise is to provide a threshold cut-off. Similarly, in genetic text a constant cut-off can be imposed, i.e., cells of dot density lower than a certain number are eliminated. Or in a statistically more rigorous way, one can define a set of random sequences S^r (sequences where all the nucleotides or amino acids occur with statistically equal probability). The random sequences are used as the background (or standard) to define a standard deviation

$$\sigma = \sqrt{\frac{1}{(n^2-1)} \sum_{i,j=1}^{n} \left(\rho_{ij}^2 - \bar{\rho}^2\right)}, \quad (5)$$

where ρ_{ij} is the density of dots in the (i,j)-th cell, $\bar{\rho}$ is the average random (background) dot density and n is the number of rows (or columns) the primitive enclosing geometrical shape is subdivided into, i.e., $n \times n$ cells. In this case, a threshold condition of $\rho_{ij} > \bar{\rho} + m \times \sigma$ can be imposed. Computationally, masks M_n^m are defined to perform the task of cut-off (Solovyev et al., 1991).

Results and Conclusions

Figs. 2-4 give some representative results. The figures show that the FRS are very distinct for different gene regions and can thus be a very powerful for their recognition, especially in newly sequenced genes. Thus FRS can lead to 1) further understanding of characteristics defining functional differences in genomic regions; and 2) development of a new approach of pattern recognition. Both of these are important in the realization of the genome project.

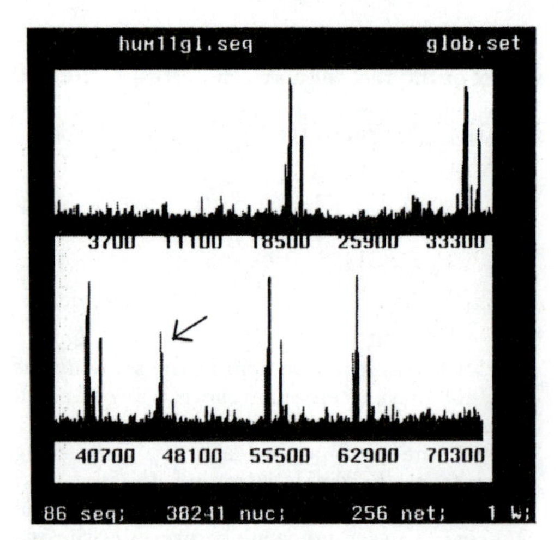

Fig. 4. Search for globin genes using the fractal image of globin family (without human genes) in human chromosome 11. The five dominant peaks correspond to the positions of globin genes and the smaller peaks correspond to the positions of pseudogenes (indicated by an arrow). The exon structure is also clearly seen.

Acknowledgments

This work is partially supported by a grant from TRDA under Contract Number TRDA 205 and by SCRI which is partially funded by the US DOE under Contract Number DE-FC05-85ER250000. Some of this work was performed with S.V. Korolev of the Engelhardt Institute of Molecular Biology, Moscow. The computational part of the project is made possible by the time allocated by SCRI on Silicon Iris 4D/240GTX workstation.

References

Avnir D. *Chemistry & Industry* **1991**, *24*, 912-916.

Biotechnology and the Human Genome: Innovations and Impact; Woodhead, A.D.; Barnhart, B.J., Eds.; Basic Life Sciences, Vol. 46; Plenum Press, New York City, New York, 1988.

Barnsley M.F. *Fractals Everywhere*; Academic Press, Inc, San Diego, California, 1988.

Computers and DNA; Bell, G.I.; Marr, T.G., Eds.; Santa Fe Institute Studies in the Sciences of Complexity, Vol. VII; Addison-Wesley Publishing Co.: New York City, New York, 1990.

Jeffrey, H.J. *Nucleic Acids Res.* **1990**, *18*, 2163-2170.

Mandelbrot B. *The Fractal Geometry of Nature*; W.H. Freeman and Co., San Francisco, California, 1982.

Solovyev, V.V.; Korolev, S.V.; Tumanyan, V.G.; Lim, H.A. *Doklady Biochemistry.* **1991**, *319*, 1496-1499.

Solovyev, V.V. *Graphical methods of representing and analysis of DNA and protein sequences*, Inst. Cytology & Genetics, Russian Acad. Sc., Novosibirsk **1991**.

Solovyev, V.V.; Korolev, S.V.; Lim, H.A. *Intl. J. Genomic Res.* **1992**, *In Press.*

Trifonov, E.N. In *Structures and Methods*; Ramaswamy H.; Sarma M., Ed.; Human Genome Initiative and DNA Recombination, Vol. 1; Adenin Press, New York City, New York, 1990; 69-77.

Computational Analysis of Metabolic Pathways

Michael L. Mavrovouniotis, Systems Research Center and Chemical Engineering Dept., University of Maryland, College Park, MD 20742
(FAX: 301 405-6707; Email: mlmavro@src.umd.edu)

The analysis of biochemical systems must often be carried out in the absence of detailed information on enzyme kinetics, concentrations of intermediates, regulation, etc. Thermodynamic arguments can be employed in such cases, through a group-contribution technique which estimates the Gibbs energies of bioreactions; the ABC approach, which estimates properties by examining alternative conjugate forms of compounds, is another useful computational thermodynamic method. Understanding the behavior of a biochemical reaction system can also be significantly aided by the identification of extreme points of the space of pathways. These are minimal pathways which define the range of permissible yields, selectivities, and other properties of pathways.

The metabolic pathways that determine the behavior of a bioprocess often involve a known set of metabolites and a set of biochemical reactions with known stoichiometries. The metabolites include the initial raw materials, products, and byproducts of the process, and metabolic intermediates which do not occur extracellularly. Individual bioreactions in the metabolic network may involve intermediates, but the overall net transformation accomplished by the bioprocess does not. In effect, the rates or extents of individual bioreactions are in such proportions such that the total production of each intermediate by some bioreactions cancels out an equal consumption of the intermediate by other bioreactions.

The detailed quantitative analysis of the metabolic pathways relevant to a bioprocess is normally hindered by the absence of detailed data on bioreaction equilibria, enzyme kinetics and inhibition, regulation of enzyme amounts, and transport processes. Thus, approaches which contribute to the analysis of metabolic pathways without requiring complete kinetic and other data are extremely important.

In this paper, we present methods for the qualitative computational analysis of metabolic pathways, along two directions. In the first part of this paper, we outline estimation methods that can be used to determine the Gibbs energy of formation of biochemical compounds in aqueous solution, and hence the equilibrium constant of any bioreaction. In the second part of the paper, we show how the behavior of fluxes or reaction extents in a metabolic network can be bounded by certain metabolic pathways which represent extreme behaviors in terms of raw material requirements and product yield and selectivities. These methods do not provide a prediction of what pathways will actually operate in the system (or what the relative fluxes of different pathways will be), they can be used to define the overall space of possible pathway activities.

Thermodynamics of Biotransformations

While the kinetics of a bioreaction depend on the source of the enzyme, reaction equilibria are independent of the enzyme involved. The equilibria depend on the stoichiometry of the reaction; if activity coefficients are taken into account, the equilibria also depend on other solutes (metabolic intermediates, inorganic ions, pH, etc.) which affect the activity coefficients.

Definitions. The relevant thermodynamic parameters are the standard Gibbs energy of reaction, $\Delta G^{o\prime}$ and the standard equilibrium constant, K', related through the equation:

$$\Delta G^{o\prime} = - RT \ln K' \quad (1)$$

The standard Gibbs energy of a bioreaction is related to the standard Gibbs energies of formation of its reactants and products. If V_i is the stoichiometric coefficient of compound S_i (with $V_i > 0$ for products and $V_i < 0$ for reactants), and $\Delta G_i^{o\prime}$ is the Gibbs energy of formation of S_i, then:

$$\Delta G^{o\prime} = \sum_i V_i \, \Delta G_i^{o\prime} \quad (2)$$

Significance of Thermodynamics. These thermodynamic parameters determine whether a biotransformation is feasible at all and whether it will occur reversibly or irreversibly (Lehninger, 1986, Chapter 4; Rawn, 1983, Chapters 1 and 12). Below, we give two examples of the many contexts in which thermodynamic information is useful.

The range of metabolite concentrations over which a bioreaction may occur (in a given

1054–7487/92/0392$06.00/0

direction) is determined by its equilibrium constant. One can also determine whether a given *set* of bioreactions may occur simultaneously, and under *what metabolite concentrations* this may happen.

The bioenergetic significance of an overall transformation which is described in terms of its *net overall* stoichiometry (rather than all the enzymatic steps that it entails) can be assessed from its Gibbs energy. If the transformation appears to entail an unfavorable Gibbs energy change, then one can determine the number of moles of ATP that must participate (converted to ADP or AMP) in the transformation in order to overcome the thermodynamic limitation. If the transformation is favorable to start with, then the Gibbs energy can be used to determine the maximum production of ATP from ADP which may be accomplished by the transformation. Of course, this analysis can only provide limits, as the extent of actual ATP participation depends on the specific enzymes involved in the pathway.

Group Contributions. We have developed a group-contribution technique for determining the Gibbs energy of formation of a metabolite in aqueous solution, $\Delta G_i^{\circ\prime}$ (Mavrovouniotis, 1990b and 1991). Through this group-contribution method and Eqs. 1-2, one can determine the Gibbs energy and equilibrium constant of a bioreaction, starting only from the bioreaction stoichiometry and the structures of the participating metabolites. In the group-contribution method, one views a compound as composed of functional groups and sums amounts contributed by each group to the overall value of the standard Gibbs energy of formation.

The method provides a set of functional groups, along with the contribution of each group to the standard Gibbs energy of formation, which has been determined from Gibbs energy data through multiple linear regression. A few corrections for special group-interactions are also provided. To estimate the property of a particular compound, one decomposes the aqueous-solution form of the compound into groups, and sums the contributions of the groups. Results for cyclic (Mavrovouniotis, 1990b) and acyclic (Mavrovouniotis, 1991) compounds have been computed.

The ABC Technique. One of the difficulties encountered in the estimation of properties of biochemical compounds is that many compounds are resonance hybrids, i.e., cannot be accurately described by a single syntactic formula with integer-order bonds. The *conjugates* of a compound are alternative formal arrangements of the valence electrons; a compound that is strongly influenced by conjugation cannot be adequately represented by any single structural formula and therefore cannot be properly decomposed into groups (**Fig. 1**).

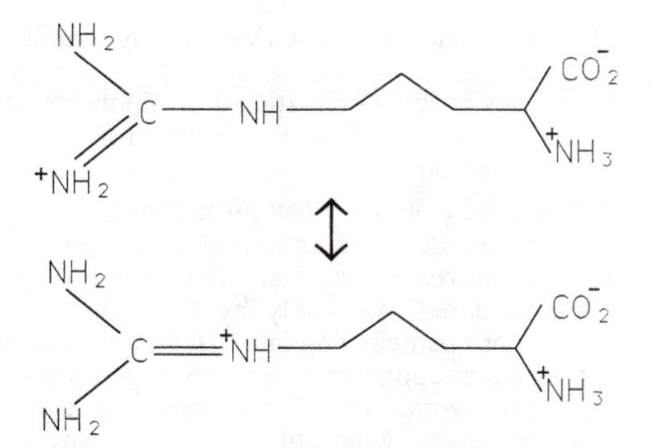

Figure 1. Two of the conjugate forms of the aminoacid arginine.

An alternative property-estimation framework under development uses contributions of Atoms and Bonds for properties of Conjugate forms (ABC), and deriving the properties of the compound from properties of its conjugates (Mavrovouniotis, 1990a, Mavrovouniotis *et al.*, 1992). By examining conjugate forms, which play an important role for many biochemical compounds, ABC models molecules much more accurately. The approach has been enhanced by approximate quantum-chemical analysis (Mavrovouniotis, 1990a). The fundamental basis of the ABC approach is discussed below.

In conjugation, a compound is viewed as a *hybrid* of a number of *conjugates* (**Fig. 1**). Each conjugate is an arrangement of atoms which are connected, in pairs, by single, double, or triple bonds; each electron pair belongs to a specific bond that connects a specific pair of atoms; each charge present on a conjugate must be integer. The hybrid cannot be represented in this way, because its electronic structure includes fractional-order bonds and fractional charges.

The basic premise of the ABC approach is that the properties of the hybrid compound can be estimated by examining the properties of each conjugate form of the compound. The properties of each conjugate are estimated from the *contributions of individual atoms and individual bonds*, rather than larger functional groups. Thus, in ABC, the intramolecular interactions among groups are not captured through a large variety of groups (which would require a large number of parameters), but through a variety of conjugates, whose properties can in fact be captured with just a small number of contributions from atoms and bonds.

A necessary task in the ABC framework is the manipulation of computational representations of molecular structures, in order to generate all possible conjugate forms of the structure. To accomplish this task, atoms, bonds, molecules, and conjugates are represented as interconnected objects within an Object-Oriented Programming environment. The generation and analysis of conjugates entail computational manipulation of these objects and their interconnections (Mavrovouniotis *et al.*, 1992).

Space of Pathway Stoichiometries

Ordinarily, the term *pathway* denotes merely a *set* of bioreactions, i.e., the pathway is considered defined solely by the list of the bioreactions participating in it. However, a fixed set of bioreactions may lead to a number of different overall transformations. The net transformation accomplished by a set of bioreactions is determined by the bioreaction fluxes (or total extents, if an analysis of time-dependence is not needed); for stoichiometry purposes, the relative proportions of bioreactions fluxes or extents are sufficient.

Consider the following glycolytic reactions which involve Fructose 1,6-Diphosphate (FruDP), Glyceraldehyde Phosphate (GAP), and Dihydroxyacetone Phosphate (DHAP) and are catalyzed by *Fructose Diphosphate Aldolase* (r_1) and *Triose Phosphate Isomerase* (r_2):

$$\text{FruDP} \rightarrow \text{GAP} + \text{DHAP} \quad (r_1)$$

$$\text{DHAP} \rightarrow \text{GAP} \quad (r_2)$$

If the fluxes of the two steps are approximately equal (i.e., 1:1 proportion), then the net transformation will be:

$$\text{FruDP} \rightarrow 2\ \text{GAP} \quad (r_3)$$

and for this system DHAP becomes effectively an intermediate with no net production or consumption. If, on the other hand, the fluxes are in a ratio of 1:2, the net transformation is:

$$\text{FruDP+DHAP} \rightarrow 3\ \text{GAP} \quad (r_4)$$

while a ratio of 1:–1 (with step r_2 actually occurring in the reverse from the nominally shown direction) makes GAP an intermediate and DHAP a net product:

$$\text{FruDP} \rightarrow 2\ \text{DHAP} \quad (r_5)$$

Thus, the proportions of fluxes induce qualitative differences among the pathways that give r_3, r_4, and r_5, since the roles of DHAP and GAP vary from metabolic intermediate to net product. We are interested in defining the space of possible net transformations and bioreaction fluxes for large systems of bioreactions.

Definitions. We will assume from now on that each individual bioreaction is either (a) irreversible or (b) reversible but with the sign of its net rate known. Thus, every reaction has a predetermined net direction. The basic ideas discussed below also apply to reactions with undetermined net directions, but this assumption simplifies the exposition of the concepts.

An individual pathway must represent a completely determined net transformation: A *pathway* contains a set of bioreactions with fixed proportions of fluxes or extents indicating the relative participations of the bioreactions in the pathway. In general, the behavior of a bioreaction system will be determined by a number of interacting pathways. Each pathway will contain a subset of the original set of bioreactions; the bioreactions not contained in the pathway can be viewed as having a coefficient of zero. Thus, a pathway can be described by a vector which has a dimension equal to the number of bioreactions; the components of the vector must be non-negative, so that each bioreactions is either used in the forward direction (positive component) or not used at all (zero component). Since we are interested only in proportions, the vector is significant only up to a positive multiplicative constant. Thus, a pathway is only a *direction* (or a line passing through the origin) in this space.

As was mentioned in the introduction, some of the metabolites are designated as metabolic intermediates which occur internally in the bioreaction network but do not occur in the net stoichiometry of the biotransformation. Given this distinction, we consider a pathway *feasible* if the net reaction it accomplishes does not involve intermediates. To state this fact algebraically, let A be a matrix with each entry a_{ij} representing the stoichiometric coefficient of intermediate j in bioreaction i. Let the non-negative vector σ represent a pathway P, with the component σ_i denoting the (relative) flux or extent of bioreaction i. The pathway P is feasible iff $\Sigma_i a_{ij}\sigma_i=0$ for all j (or, in matrix notation, $A\sigma=0$).

A feasible pathway P is *minimal* if there is no feasible pathway which involves a proper subset of the bioreactions present in P. In other words, the omission of any one or more bioreactions of the pathway P leads to a set of bioreactions which cannot give rise to a feasible pathway. Mathematically P (represented by σ in the vector space of reaction extents) is minimal iff there is no pathway P' (with the vector σ') such that $A\sigma'=0$, $\sigma'_i=0$ and $\sigma_i\neq0$ for some i, and $\sigma'_i=0$ for all i for which $\sigma_i=0$.

Cone of Feasible Pathways. We will now discuss the utility and significance of the above definitions. We focus on a qualitative overview, omitting rigorous mathematical statements or proofs, formal algorithms, and practical computational considerations. Some of these issues surrounding minimal pathways and their construction are discussed by Mavrovouniotis and Stephanopoulos (1992) and Mavrovouniotis (1992). Algebraic expressions of pathways in this section are to be understood as relations involving the corresponding σ-vectors.

Let P_1, P_2, ..., P_n be a set of feasible pathways. Any linear combination $\Sigma_k\lambda_kP_k$, with λ_k positive real numbers, is also a feasible pathway. In effect, any interpolation of feasible pathways is also a feasible pathway.

Given any feasible pathway P, there exists a

decomposition of P into minimal pathways, $P=\Sigma_k\lambda_k P_k$, with P_k distinct minimal pathways and λ_k non-negative real numbers; in general, the decomposition is not unique. The decomposition of a minimal pathway P_m is unique, with $\lambda_m=0$ for all $k\neq m$. Hence, a minimal pathway cannot be written as a combination of other minimal pathways.

In the vector space given above, the feasible pathways form a *convex cone*, whose tip coincides with the origin. The set of one-dimensional edges of the cone is precisely the set of minimal pathways. The interior of the convex cone contains precisely all feasible pathways, which can be written as non-negative linear combinations of minimal pathways.

Many features of feasible pathways are present, to an extreme degree, in one or more minimal pathways. Let f be a property whose domain is the set of feasible pathways, such that:

(I) $f(\lambda P)=f(P)$ for any positive real λ

(II) $\min(f(P_1), f(P_2)) \leq f(P_1+P_2)$

(III) $f(P_1+P_2) \leq \max(f(P_1), f(P_2))$

The conditions imposed on f guarantee that it does not depend on the arbitrary multiplicative factor implicit in all pathways, and that, for a pathway P which is an interpolation between P_1 and P_2 the property f(P) lies between $f(P_1)$ and $f(P_2)$; the property does not have to follow a linear or other quantitative interpolation function.

It can be shown that both the minimum and maximum value of f occur at minimal pathways. Each of the extrema is unique, or else it occurs precisely at a set of minimal pathways and all their possible interpolations.

A variety of important quantities can be so characterized. They include the yield of any product over any substrate and the selectivity (molar ratio) of any product over any other product. Often, an appropriate normalization must be introduced; for example, the Gibbs energy of reaction per mole of a product is a suitable property. Some of the biochemical arguments and results presented by Mavrovouniotis *et al.* (1990, 1992) stem from properties that have their extrema at minimal pathways. Their example on the synthesis of lysine, leads to a classification of pathways based on the molar yield of lysine over glucose; the molar yield of lysine over glucose can exceed 0.67 only for a specific class of pathways.

The consideration of minimal pathways is thus important in the analysis and design of processes for desired bioproducts, as it points to limitations that affect the feasibility of a process and the selection of appropriate pathways and strains.

References

Lehninger, A.E. *Principles of Biochemistry.* Worth, New York, **1986**.

Mavrovouniotis, M.L. *Ind. and Eng. Chem. Res.* **1990a**, *29*, 1943-1953.

Mavrovouniotis, M.L. *Biotech. Bioeng.* **1990b**, *36,* 1070-1082.

Mavrovouniotis, M.L., Stephanopoulos, G., and Stephanopoulos, G. *Biotech. Bioeng.* **1990**, *36,* 1119-1132.

Mavrovouniotis, M.L. *J. Biol. Chem.* **1991**, *266,* 14440-14445.

Mavrovouniotis, M.L., Stephanopoulos, G., and Stephanopoulos, G. *Comp. Chem. Eng.* **1992**, in press.

Mavrovouniotis, M.L., Prickett, S., and Constantinou, L. *Proceedings of the European Symposium on Computer-Aided Process Engineering (ESCAPE-1),* Pergamon Press, **1992**.

Mavrovouniotis, M.L., and Stephanopoulos, G. *Ind. and Eng. Chem. Res.* **1992**, in press.

Mavrovouniotis, M.L. *Ind. and Eng. Chem. Res.* **1992**, in press.

Rawn, J.D. *Biochemistry.* Harper and Row, Philadelphia, **1983**.

Policy Issues in Biotechnology Symposium VIII

J. E. Rollings: *Chair*

Intellectual Property: Session A
C. E. Van Horn and G. J. Mossinghoff: *Co-Chairs*

Safety and Containment: Session B
J. Van Houten: *Chair*

Regulatory Issues in Biotechnology and Bioprocessing: Session C
A. Moreira: *Chair*

A Patent Practitioner's View of Selected Legislative and Judicial Developments in the Protection of Biotechnology Inventions

Robert A. Armitage, Vice President, Corporate Patents and Trademarks, The Upjohn Company, 301 Henrietta Street, Kalamazoo, Michigan 49001 (FAX: 616-385-6897)

Seemingly unprecedented social, economic and public policy issues have focused attention on the patent statutes and their application to biotechnology. Industry groups have besieged the Congress with proposals to fundamentally alter patentability considerations for some biotechnology inventions, at the same time as the courts have produced surprising results in attempts to apply existing legal principles to this new technological subject matter. The most explosive biotechnology patent issue of all, the patentability of life itself, has bred a down-on-the-farm anxiety over patented farm animals, hatching a variety of responsive legislative proposals.

Should Automatic Patentability Be Accorded to the Use of Patentable Cell Lines?

Two bills in the current Congress, H.R. 1417 and S.654 (as originally introduced), would add the following provision to the U.S. patent laws concerning the existing non-obviousness requirement for patentability:

"When a process of making or using a machine, manufacture, or composition of matter is sought to be patented in the same application as such machine, manufacture, or composition of matter, such process shall not be considered as obvious ... if, such machine, manufacture, or composition of matter is novel ... and non-obvious ... . If the patentability of such process depends upon such machine, manufacture, or composition of matter, then a single patent shall issue on the application."

After introduction of S. 654, it was reported to the Senate in the amended form, as follows:

"Notwithstanding any other provision of this section, a claimed process of making or using a machine, manufacture, or composition of matter is not obvious under this section if --

"(1) the machine, manufacture, or composition of matter is novel ... and non-obvious ...; and

"(2)(A) the machine, manufacture or composition of matter, and the claimed process invention at the time it was made, were owned by the same person or subject to an obligation of assignment to the same person; and

"(B) claims to the process and to the machine, manufacture, or composition of matter, are entitled to the same effective filing date, and appear in the same patent or in different patents which are owned by the same person and are set to expire on the same date."

The intent of either version of the legislation is to make the patenting of "simple" processes, i.e., the conventional expression of a protein by a recombinant cell line, patentable -- so long as the cell line itself is patentable.

The need for legislation of this type has sharply divided the biotechnology industry. Its proponents insist certainty in the law is needed in order for the industry to have sufficient incentives to undertake high-risk research. Other factions in the biotechnology industry suggest that the existing patent law already *routinely* provides such process patentability.

Although the truth of the matter may well lie somewhere between these extreme views, no compelling circumstances have yet presented themselves that would fully support the extreme recourse of a statutory amendment.

1054–7487/92/0398$06.00/0

Should Farmers Be Accorded a Compulsory License or Other Special Exemptions from Patent Infringement for Conventional Breeding of Recombinantly-Produced Farm Animals?

In both the 100th and 101st Congresses, efforts were made to made to exempt farmers from patent infringement where the farmers were conventionally breeding patented farm animals: once the farmer had obtained (from the inventor or a licensee) a breeding pair of patented animals, that farmer would be free to indefinitely produce offspring without any liability to the inventor.

Because of opposition from biotechnology industry groups, an innovative alternative was advanced under which farmers could acquire patented animals from inventors either under "qualified sales" (that included the right to reproduce successive generations) or non-qualified sales (where reproduction would presumably incur liability to the inventor).

The following specific proposal was advanced by industry groups:

"§273. *Qualified Sales of Transgenic Farm Animals.*

"(a) [Effect of Qualified Sale.] A person shall not be liable for infringement [of a patent] based on such person's conventional breeding of an individual transgenic farm animal claimed in a patent, whenever --

"(1) the animal itself; sperm, egg, or embryo which was used to produce the animal itself; a conventionally bred ancestor of the animal; or the sperm, egg, or embryo which was used to produce the conventionally bred ancestor was purchased by such person through a qualified sale made pursuant to this section, and

"(2) the breeding is undertaken by such person solely pursuant to conventional farming practices.

"(b) [Requirements for Qualified Sale.] In order for a sale to be a qualified sale pursuant to this section, the following conditions must be met at the time of sale --

"(1) the seller must be expressly authorized by the patentee to make the sale as a qualified sale pursuant to this section,

"(2) the purchaser must be a person whose occupation or business is farming, and

"(3) the authorized seller must make an express indication that the sale is a qualified sale or the circumstances of the sale must otherwise indicate the authorized seller's intention that the sale is a qualified sale under this section.

"(c) [Presumption.] Where the authorized seller does not give notice to the purchaser that a transgenic farm animal is patented, through a writing provided at the time of sale indicating the number of the patent, it shall be presumed that the seller's intention is that the sale is a qualified sale under such patent, notwithstanding the absence of any express indication by the seller.

"(d) [Definitions.] For the purposes of this section:

"(1) the term "farming" shall mean using and conventionally breeding domesticated farm animals for the production of food or fiber and shall exclude the use of such animals for the production or sale of living reproductive materials, including sperm, eggs, and embryos.

"(2) the term "conventionally bred" shall mean bred through each successive generation using only traditional selection and breeding practices and shall exclude techniques involving gene manipulation or other biotechnological means of achieving conception, gestation or reproduction.

"(3) the term "conventional farming practices" shall mean the normal and customary practices of persons engaged in farming, including conventional breeding practices, but shall exclude breeding practices involving biotechnological means.

"(4) the term "transgenic farm animal" shall mean a domesticated farm animal whose germ cells contain genes originally derived from an animal of another species."

This alternative legislation was intended to preserve the full incentives of the patent system for inventors who discover and patent transgenic farm animals. The proposal recognized and preserved the principle that the exercise of the patent property right and the confiscation of that property right through a statutory compulsory licensing scheme were antithetical to one another. Accordingly, the proposal provided voluntary, not compulsory, participation and provided a simple means for patent owners and farmers to make sales of living, self-reproducing inventions with certainty as to the rights of both parties to the transaction.

The rights of farmers to engage in conventional farming practices without creating potential and uncertain liabilities to patent owners on account of the conventional breeding of patented transgenic farm animals through one or more generations were recognized. The mechanism employed to meet the interests of patentee-sellers and farmer-purchasers was a statutorily-defined "qualified sale." The decision to engage in a qualified sale would be in the first instance that of the inventor who must authorize such a sale. The inventor would be free to engage in only non-qualified sales or in both qualified and non-qualified sales based upon ordinary business and commercial concerns. Likewise, farmers would be free to purchase patented animals under sales of both types.

In many farming activities, breeding is a natural, inherent and unavoidable consequence of animal husbandry. In these situations, farmers introducing patented animals into a herd might benefit significantly from a statutorily qualified sale under which no infringement liability would be possible for second and subsequent generations of animals resulting from conventional breeding. In other farming activities, no breeding will take place and farmers may significantly benefit from purchasing the same patented animal through a non-qualified sale, presumably at a lower cost. A key feature of the proposal was the element of choice for both the farmer and for the patent holder.

When a sale of a patented transgenic farm animal takes place, the proposal contemplated that the seller would make clear the nature of the transaction as either a non-qualified sale or a qualified sale. Where the authorized seller provides no indication that the sale is a qualified sale and provides no written notice of the patent involved, the sale would be presumed to be a qualified sale. This presumption could be avoided only if the authorized seller were to provide in writing at the time of sale notice of the patent involved. Where a patent notice were properly provided, then the sale would be considered a qualified sale only where the seller would provide some express indication that the sale was intended as a qualified sale.

While a presumption of the above type would be unprecedented in the patent law, imposing the presumption could provide a balance between the interests of the farmer and those of the patent holder. The farmer has a strong interest in avoiding potential infringement arising from uncontrollable reproduction of purchased animals. The patent holder has a strong interest in avoiding unlicensed and unlimited reproduction of the patented invention. The patent owner is afforded the opportunity to escape the presumption by providing notice of the patent. It affords the farmer the opportunity to seek a qualified sale, or forego the purchase of the animal altogether, once the existence of the patent becomes known.

With the defeat of the chief congressional proponent of the legislation designed to clarify the rights of farmers vis-a-vis inventors, no legislative proposals have yet emerged from the 102nd Congress regarding transgenic farm animals. Nonetheless, the issue remains of potential importance as the prospect of transgenic farm animals looms on the horizon. A far more insidious alternative has been proposed: a "moratorium" on the patenting of all claims to transgenic animals. Such a moratorium would represent an unprecedented denial of the incentives of the patent system to an important field of technology.

To date, the Congress has taken no serious action on any of the various responses to animal patents.

Is the "Cell Line" Deposit Requirement A Vestigial One After *Amgen v. Chugai*?

The United States Supreme Court recently decided not to decide what certainly ranks as one of the most important biotechnology patent decisions of all time. In *Chugai v. Amgen* the court was faced with consideration of the impact of the "best mode" requirement for disclosure of the invention in a patent application. This opportunity arose in the context of a patent claiming the recombinant machinery (CHO cells) for making erythropoietin. No fewer than five organizations filed *amicus curiae* briefs, urging the Court to consider the issues involved. Organizations requesting court review included The American Foundation for AIDS Research, the AIDS Action Committee of Massachusetts, the AIDS Action Council, Whitman-Walker Clinic, American Type Culture Collection,

Council for Responsible Genetics, Professors Roger Sperry, Gerald Holton, Everett Mendelsohn, and Dorothy Nelkin, and the Taxpayers Assets Project of the Center for the Study of Responsive Law.

One of the most cogent briefs (ATCC) supported Supreme Court view for the following reasons:

"The decision below [that deposit of the cell culture was unnecessary] has already begun to impede the purposes of the patent regime's deposit rules by introducing uncertainty and disunity into deposit requirements. The Federal Circuit has upheld a patent despite a finding that its genetically engineered cell line cannot be reliably reproduced on the basis of the written patent disclosure alone, and despite the patentee's failure to place samples of the cell line in a public depository. Genetic engineers, unlike all other microbiologists working the biotechnology field now know that they might be able to get a patent while also keeping their cell lines as trade secrets. These inventors thus have no incentive to make bioculture deposits in the future. That bodes ill for the future of both basic and applied research in genetic engineering, medicine, and other areas of biotechnology."

If Amgen is able to dominantly claim *every* CHO cell line capable of producing EPO without providing a reproducible means for obtaining the *best* available to it at the time it sought its patent, what now does the requirement to disclose *best mode* mean? As with almost every patent case, the Court of Appeals for the Federal Circuit will have the final word. That final word in *Chugai v. Amgen* may mean fewer deposits of cell lines.

Conclusions

Legal issues in biotechnology have created an abundance of legal disputes -- the resolution of which will occupy biotechnology patent lawyers for years to come. With the prospect of granting patents on "obvious" processes made non-obvious by legislation, the widespread commercial application of biotechnology to the farmer in the form of patented plants and animals, and the possibility of more patents issuing with a minimalist "best mode" disclosure (and broader retained trade secret technology), inventors and their lawyers should be endowed with an abundance of patents with which to wage the biotechnology patent wars of this decade and the next.

Current Developments on Patentability of Biotechnology in Europe and Japan

Waddell A. Biggart
Sughrue, Mion, Zinn, Macpeak & Seas, Washington, D.C. 20037
(FAX:202 293 7860)

Biotechnology developments and current biotechnology related issues in Europe and Japan are described. In particular, the European Commission draft Directive on Biotechnology and the UPOV Convention on protection of plants are discussed. Also, the patentability of biotechnology developments in Japan is summarized. Further, particularly important appellate decisions and litigated cases in Europe and Japan are described.

Europe

Biotechnology patents can be obtained in many European countries. Rather than file multiple national patent applications, a single regional patent application can be filed in the European Patent Office ("EPO") under the European Patent Convention ("EPC"). If granted, a European patent provides national patent rights in European countries which are members of the EPC and which were designated when the European application was initially filed.

Of course, differences exist in the patent systems in European countries. To achieve a greater degree of harmonization, the European Commission ("EC") pursuant to the European Economic Community ("EEC") treaty has issued a draft Directive on Biotechnology. The EC draft Directive on Biotechnology once adopted by the EEC member states will have an important impact on the future of European intellectual property rights in biotechnology. This Directive was issued because of the differences in the legal protection of biotechnological inventions presently existing as to members of the EEC and in view of changes which will exist in Europe post 1992. The Directive relates to the issuance of patents on biotechnology inventions and was first proposed in 1988 for implementation by the member states by 1991. Revisions of the initial draft have been proposed and are being considered.

The Directive substantively provides that microorganisms, plant and animal parts and plants and animals other than those considered varieties would be considered patentable subject matter. The Directive defines microorganisms as including microbiological entities capable of replication, *e.g.*, bacteria, fungi, viruses, mycoplasmae, rickettsiae, algae, protozoa and cells, and defines self-replicable matter as matter possessing the genetic material necessary to direct its own replication, *e.g.*, seeds, plasmids, protoplasts, replicons and tissue cultures. Moreover, the Directive states that biological processes involving intervention by man are patentable.

The Directive also defines the rights which accrue to the patentee on issuance and states that use of a patented invention for purposes other than private or experimental use is an infringement. Further, the Directive provisions provide the progeny of replicable materials are included in the patent rights arising, unless unavoidable for commercial use other than multiplication or propagation.

Where essential to repeat a biotechnology invention, the Directive includes procedures for deposition of a microorganism or other self-replicable material which is not publicly available and which cannot be described in the patent application in a manner sufficient for one skilled in the art to reproduce the invention. Consistent with present European practice, the Directive requires a deposit of such a replicable

1054–7487/92/0402$06.00/0

material on or before the application is filed and provides third party access to the deposit 16 months after the date of filing or priority date if the application is based on an application filed in another country or at anytime to anyone having rights to access under the national laws of a member country.

Plant varieties created by classical hybridization techniques or by plant science developments on which patent issuance is proscribed can be protected in various European countries by issuance of a plant breeder's right. The requirements of the International Union for the Protection of New Varieties of Plants ("UPOV") are in general adhered to by European countries providing plant breeder's rights. The UPOV convention describes a plant variety as a group of plants defined by characteristics that are the expression of a given genotype(s) and that are distinguished from other groups of plants.

In order for a plant breeder's right certificate to issue, the variety must be shown to be (i) new, (ii) distinct, (iii) uniform and (iv) stable. Examination is done in European countries which grant such rights. Upon issuance of a plant breeder's right, rights exist as to propagating material of the protected variety and as to harvested material. In general, the rights provided include the requirement of authorization by the rights holder, *inter alia*, to reproduce, market, import or export the protected variety. Private and non-commercial use and experimental use do not constitute a violation of the breeder's rights.

The latest revision of UPOV, completed in March 1991, has been signed by a number of European countries, and notably, eliminates the limitation previously placed on signatories preventing issuance of a patent and a plant breeder's right on the same development. The United States has also signed this latest revision of the UPOV convention but the revised convention has not yet been ratified by the U.S. Senate nor have U.S. plant variety protection laws been amended for conformity with the revised convention provisions.

A number of important decisions impacting biotechnology on European patent applications have been issued recently by the Technical Board of Appeal of the European Patent Office ("EPO"). While there are specific exceptions to the patentability of certain types of inventions, such as those involving treatment of or diagnosis of disease in humans and animals and on plant and animal varieties under the EPC, generally the Technical Board of Appeal has considered these exceptions to be narrow in appealed cases thereby expanding the scope of subject matter protectable by issuance of a European patent. In the mid-1980's, the Technical Board of Appeal confirmed the patentability of plants that were not plant varieties under the EPC and the most notable decision recently issued by the Technical Board of Appeal involved a reversal of the EPO Examining Division's rejection of claims to a transgenic animal (the Harvard Mouse Decision) on the basis that the EPC prohibited issuance of patents on animal varieties. The Technical Board of Appeals held the claims were not directed to an animal variety. The EPO has now decided to issue a patent on the Harvard Mouse and it is expected that numerous oppositions to the grant of a European patent on this development will be filed.

Procedurally, the EPO now strongly recommends that applications disclosing 4 or more amino acids in a sequence or 10 or more nucleotide bases in a sequence include this information also in computer readable format. The underlying rationale for this recommendation is to facilitate computer searching and evaluation of the patentability of subject matter involving DNA/RNA and amino acid sequences.

National country developments in biotechnology include the revocation of Genentech's United Kingdom patent on tissue plasminogen activator ("TPA"), the provision of patent protection throughout the unified Germanies, the elimination of the ban on patenting of chemical and pharmaceuticals in Austria and continued opposition to the patenting of biotechnology in Scandinavia.

Japan

In Japan, biotechnology subject matter is broadly protectable, with the Japanese Patent Office ("JPO") considering microorganisms, cell lines, plants and animals to be patentable subject matter. The JPO has already issued patents on plants and on animals. Moreover, plant breeder's rights are also available in Japan and Japan follows the requirements of UPOV.

A Japanese court has held that Genentech's TPA patent was infringed by competitors. However, unlike the situation in most countries, under Japanese law a court can only consider whether a patent is infringed. The question of validity of a Japanese patent over the prior art cannot be considered during infringement litigation. The question of validity of a Japanese patent is considered by the JPO and, presently, revocation proceedings against Genentech's TPA patent are pending in the JPO seeking a holding that the Genentech TPA patent is invalid.

Summary

In summary, activity in seeking and enforcing intellectual property rights continues to accelerate in Europe, Japan and the United States as more biotechnology developments are made.

New Biotechnology Medicines in Private Development

Gerald J. Mossinghoff,
Pharmaceutical Manufacturers Association, Washington, D.C. 20005
(202) 835-3400

Research and Development in Biotechnology

The research-based pharmaceutical industry is continually exploring new frontiers in the development of improved treatments for devastating diseases. Biotechnology offers great promise in bringing powerful new medicines to patients. A survey of "Biotechnology Medicines in Development" conducted by the Pharmaceutical Manufacturers Association indicates that in 1991, 132 new medicines were in human clinical trials, or, having completed those trials, were at the Food and Drug Administration for marketing approval. A similar survey conducted in 1988 established that there were 81 new therapies in development. The 1991 figures represent a 63% increase over the past four years.

Biotechnology is thus playing an essential role in the discovery and development of treatments for many life-threatening diseases. Fifty percent of the medicines in development are being tested for cancer or cancer-related conditions, seven more than in 1990 and 26 more than in 1988. Another 17 medicines are in testing for AIDS or HIV-related conditions, nine more than in 1988.

A favorable research and development atmosphere in this country has resulted in a large number of additional new products in the biotechnology sector and has stimulated continued research in industrial laboratories. Treatment of hemophilia, skin ulcers and certain cancers should improve substantially in the next several years as products now in clinical trials win marketing approval. Other promising products for the treatment of cystic fibrosis, HIV infection, and herpes are also being pursued through biotechnology techniques.

U.S. Industry Leadership

The U.S. Patent and Trademark Office issued 3,378 biotechnology patents in 1990. Well over one-third -- 1,321 -- were health-care patents, and the vast majority -- 67 percent -- of those were of U.S. origin.

The U.S. pharmaceutical industry is unquestionably the leader in biotechnology research, with the largest share of health-care patents. Corporations received 47 percent of the health-care patents that use the advanced biotechnology technique of genetic engineering. Of the 78 genetic-engineering health-care patents issued to corporations, 50 were issued to member companies of the Pharmaceutical Manufacturers Association.

The United States is the country of origin for 890 of the 1,321 biotechnology health-care patents, representing 67 percent of the total. The European Community is a distant second with 15 percent, followed by Japan with 13 percent. An even greater share of genetic-engineering health-care patents are of U.S. origin. Of the 169 genetic-engineering health-care patents, 82 percent, or 138 patents, were of U.S. origin; Japan was second with 11 percent and the European Community third with five percent.

Patent Office Examination and Resources

However, the very success of biotechnology may jeopardize the U.S. industry's premier status

1054–7487/92/0405$06.00/0

unless the Patent Office gains the scientific and support staff, physical resources and strategies to handle the proliferation of biotechnology inventions that have deluged the Office during the past few years.

Statistics compiled by the Patent Office illustrate the severity of the application backlog. In FY 1990 a little over 9300 biotechnology patent applications were filed. The FY 1991 PTO estimate was 10,000; the actual inventory as of June 1991 was over 19,500.

A major effort is underway at the Patent Office to improve the processing time of patent applications, but budget constraints have limited the number of biotechnology patent examiners that could be added to the staff. Commissioner of Patents and Trademarks, Harry Manbeck, has announced initiatives to improve the processing of biotechnology patent applications. These include the formation of a new patent examining group devoted exclusively to the examination of biotechnology inventions, recruitment of additional expert examiners and staff support, and retention of experienced examiners. But these initiatives can only be realized with adequate appropriations from Congress. It is vitally important that the biotechnology industry, and pharmaceutical companies in general, urge Congress to provide these additional resources.

FDA Review and Resources

Many new therapies are awaiting approval at FDA, and more are rapidly entering the initial phases of clinical testing. There currently are 21 medicines at FDA awaiting approval compared with 15 in 1988, 48 in Phase I compared with 24 four years ago and 46 in Phase II versus 21 in 1988.

Despite this expanding activity in pharmaceutical biotechnology research, we are seeing some disturbing trends. Only three biotechnology medicines have been approved since PMA's last survey 16 months ago.

At the current rate of innovation in pharmaceutical biotechnology research, a bottleneck is rapidly developing at FDA. Only 14 biotechnology medicines have been approved in the past decade, an average of 1.6 approvals per year.

FDA estimates that it had 2,600 active investigational new drug (IND) applications on file in 1991 and that this number will increase to 3,250 during 1992. INDs are requests from researchers to FDA for permission to test experimental medicines in human subjects. In addition, FDA estimates that there will be 8,800 active IND amendments in 1992, up from 7,400 in 1991. Most amendments are requests for permission to run human tests of a drug for additional therapeutic indications.

The majority of biotechnology medicines are reviewed at FDA under a product licensing application (PLA) submitted by the developer. FDA anticipates that 2,400 PLA reviews will be in progress in 1992, versus 2,160 in 1991. Although FDA also reviews applications for other biological substances, such as blood products and some diagnostics, most of this increase is attributable to the rapid growth of pharmaceutical biotechnology research.

FDA required an average review time of 21.4 months (measured from the filing of the complete application with FDA authorities to date of FDA marketing approval) to approve biotechnology drugs for their original and additional therapeutic indications.

A quick glance into the future reveals a biotechnology research pipeline at the brink of a bottleneck. Since 1988, 51 new biotechnology drugs have entered clinical trials. There are now 132 biotechnology medicines in development compared with 81 in 1988.

No medicines in the monoclonal antibody category have been approved since June 1986. PMA's survey results show that 58 monoclonal antibody products are in development, eight of which are at the FDA for review for such life-threatening diseases as heart and liver transplant rejection, cancer, septic shock, graft vs. host disease and sepsis.

Fundamental changes are necessary if FDA is to handle the industry's burgeoning capacity for developing new medicines through biotechnology. More staff and resources must be devoted to the review of PLAs. Once again, it is in the best interests of biotechnology companies to support increased FDA resources and scientific expertise both within the Administration and before Congressional committees. In addition, consistent with recommendations made by the Department of Health and Human Services Advisory Committee on the FDA, FDA should better manage its role in realizing the potential of this new area of pharmaceutical research -- for the people whose lives depend on the treatments and cures biotechnology can bring.

PMA Support of Improvements in U.S. Patent System

A strong U.S. patent system is essential to ensure continued progress in biotechnology drug development. PMA and its member companies are on record in support of needed fine-tuning of patent protection. PMA endorses enactment of the Biotechnology Patent Protection Act, introduced by Representative Rick Boucher (D. 9th-VA), to afford greater protection for end products made with patented intermediaries. The Boucher bill will have particular relevance to biotechnology inventions and is needed to close an unintended loophole in U.S. law by which a competitor can make or use a patented intermediate in a foreign country and import into this country the final but unpatented product. Private industry must have adequate incentives to conduct the extremely costly and time-consuming research and development needed to bring new drugs to patients, and the U.S. patent system represents the strongest single incentive for such research.

Examining Patent Applications for Biotechnology-Related Inventions in the United States

Charles E. Van Horn
Patent Policy and Projects Administrator
Patent and Trademark Office, Washington, D.C. 20231
(703) 305-9054

Application Processing

The Patent and Trademark Office (PTO) has experienced a significant growth in the number of patent applications filed in the field of biotechnology since the Supreme Court announced in Diamond v. Chakrabarty, 447 U.S. 303 (1980), that existing patent law was intended to embrace products of human ingenuity, whether inanimate or alive. The annual rate of filing patent applications directed to biotechnology inventions has risen from about 3000 in 1982 to over 9700 in 1991.

In response to a growing inventory of biotechnology patent applications, a new patent examining group [Group 180] dedicated to biotechnology was established in 1988 as one of sixteen patent examining groups in the PTO. The original staff of 67 patent examiners in this group has grown to a level of 150, with plans to hire an additional 50 examiners before the end of September 1992. The growth in staff has enabled the PTO to reduce the average period of time from filing to final disposition of a patent application in the biotechnology area from a high of 26.7 months to a current average of 25.0 months. The average period of time from filing to first action by an examiner has been reduced from 14.7 to 8.5 months.

Patent applications are normally taken up for action according to the date of filing. Although several procedures exist for taking applications out of turn or accelerating the examination of patent applications in all fields of technology, very few applicants in the biotechnology field take advantage of these procedures.

Criteria for Patentability

Development of patent law over the past decade in the field of biotechnology has confirmed, for the most part, that courts are not moving beyond established and time-honored principles and analytic approaches when interpreting the patent statutes with respect to biotechnology inventions. The answers to key questions of claim scope and infringement are important because they are likely to influence the pace of biotechnology research and commercial development of new products in such fields as agriculture, drugs, and diagnostics.

In order to warrant the grant of a patent, each invention must fall within the scope of subject matter that can be patented. The Supreme Court in the Chakrabarty decision stated that the patent statute should be construed broadly so as to embrace new technologies. Patent applicants in the United States enjoy the most user friendly patent system in the world when it comes to biotechnology inventions. Patents are routinely granted in the U.S. to man-made organisms, to man-made human, animal and fused cell lines, plants and plant parts, animals, and processes of treating disease conditions in humans and animals. Even products which exist in nature can be the proper subject of a patent grant when the product is given a new form, quality or property that the product did not possess in its natural state.

In order to be patentable, an invention must be described in a patent application in a manner that permits a person skilled in the art to which the invention pertains to make and use the invention. The scope of the information contained in the description must be commensurate with the scope of protection granted in the patent.

The description of some biotech inventions can involve requirements that are unique to biotechnology. An invention may require access to certain biological material which is incapable of description by words alone. The patent law has adapted to this situation by recognizing a system of depositories for biological material that can be used to supplement the written description of a patent application. Samples of biological material in a depository must be accessible to the public upon grant of the patent. The PTO has also developed regulations for the submission and description of nucleotide and amino acid sequence information in patent applications. A computer program called Patentin is available from the PTO to facilitate compliance with the sequence regulations and expedite the preparation of patent applications containing sequence disclosures.

A further requirement of the application is that the best mode contemplated by the inventor at the time of filing the patent application must be described. The requirement does not permit inventors to disclose only what they know to be a second best embodiment, retaining the best for themselves as a trade secret. However, recent decisions by the Court of Appeals for the Federal Circuit relating to the best mode requirements for inventions involving monoclonal antibodies and mammalian host cells have emphasized that the disclosure of the best mode must be adequate, but need not permit an exact duplication.

Before a patent is granted, the invention must meet three criteria- it must be useful, novel and non-obvious. The purpose of the utility requirement is to ensure that the public obtains a quid pro quo in the form of an invention with substantial utility, where specific benefit exists in currently available form, before an applicant is granted exclusive rights in the form of a patent grant. The asserted utility must be believable on its face to persons skilled in the art in view of contemporary knowledge in the art at the time the application is filed.

Novelty requires that the invention is different in some respect from what has been made available to the public at the time the invention is made. In the U.S., an inventor can disclose his or her invention to the public and not lose a right to a patent in the U.S. if a patent application is filed within one year of the first public disclosure (one year grace period). That same act of public disclosure, however, does prevent the inventor from obtaining patent protection in many foreign countries that do not have a grace period.

The final requirement for patentability is that the invention must be nonobvious. The legal conclusion is based on factual determinations of the scope and content of the prior art, the differences between the claimed invention and the prior art, the level of skill in the art at the time the invention was made, and objective evidence of nonobviousness such as unexpected results, satisfaction of a long felt need, or commercial success of the invention.

One issue that arises frequently in biotechnology is whether a process of making or using a patentable product is nonobvious. Judicial precedent in this area has not been consistent and has led to significant uncertainty in assessing the patentability of this type of process claim. Process claims can be particularly important to a patent owner, particularly where protection on the final product may not be available.

Patent Term Extension

In 1984, the Drug Price Competition and Patent Term Restoration Act was enacted to compensate the patent owner for a period of patent life lost during the premarket regulatory review process conducted by the Food and Drug Administration. It provides for an extension of up to five years of the term of a patent claiming a drug product, a medical device or a food or color additive.

The Act was amended in 1988 to add animal drugs and veterinary biologics to the list of products that can form the basis for patent term extension. One difference between the Act of 1988 and the Act of 1984 is that animal drug products that are primarily manufactured through biotechnology are excluded from eligibility for patent term extension. This provision was the result of an argument by the biotech industry that patents available for biotechnology-derived drugs do not provide the same kind of market protection as is available to traditional, chemically derived drugs.

The number of patent applications for patent term extension tend to be small because of the small number of products actually approved by the FDA, and because eligibility for patent term extension generally must be based on the first approval of the product for commercial marketing and use. From 1984 to March 15, 1992, 209 applications for patent term extension have been filed - 166 (80%) were based on approval of a human drug product, 32 (15%) based on a medical device, 4 on a food additive, and 6 based on an animal drug product. Of the 137 extensions of patent term granted, 117 (85%) are based on human drug products, 12 on medical devices, 3 on food additives, and 5 on animal drug products. There are 36 applications for patent term extension pending before the PTO.

Advisory Commission on Patent Law Reform

The Commission, chaired by Commissioner of Patents and Trademarks Harry F. Manbeck, Jr., is scheduled to deliver its report to the Secretary of Commerce in August 1992. The Commission has been studying the patent system in the U.S. and considering the advisability of recommending changes that are in the best interests of the United States. Topics that are being considered that would require fundamental changes in the U.S. patent system are: granting patents on the basis of first-to-file a patent application; mandatory publication of the content of a patent application within twenty-four months of filing; changing the patent term from the current 17 years from the date of patent grant to 20 years from the date of first filing a patent application for the invention; permitting an assignee of patent rights to file an application in the name of the inventors rather than requiring the inventors to file the application; and changing the reexamination process to permit active, but controlled participation by parties other than the patent owner.

The recommendations that are made by this Commission could form the basis for legislative proposals to change U.S. law. The Commission recommendations also will be considered by the U.S. delegation negotiating a patent law harmonization treaty under the guidance of the World Intellectual Property Organization.

Containment Considerations for Bioremediation

Howard M. Hohl* and **Thomas B. Lewis**
Celgene Corporation, Warren, NJ 07059
FAX: 908/271-4184
Katherine Devine, Devo Enterprises, Inc., Washington, DC 20003

An up-to-date analysis of regulations, particularly related to containment issues, for pollution prevention, control, and remediation, is presented. The Federal Government has recently taken action to reduce barriers for the application of biotechnology to major environmental problems. The US EPA derives its authority for regulating biotechnology in hazardous waste treatment applications from the Toxic Substances Control Act and the Resource Conservation and Recovery Act.

An increasing level of activity involving industry, academia, and the US EPA began in 1990 with a coordinated effort directed towards expanding the responsible use of biotechnology for the prevention and remediation of environmental contamination. This coordinated thrust was an initiative of William K. Reilly, EPA Administrator, who publicly (Reilly, 1991) has encouraged the use of biological solutions, where appropriate, to address environmental needs. With his endorsement, a Bioremediation Action Committee was formed to serve as an effective force of Federal, State, industry and academic interest to achieve the full potential of environmental biotechnology (Skinner, 1991).

Earlier this year the White House Council on Competitiveness issued a new government policy on biotechnology based products (Federal Register, 1992), and at that time President Bush stated, "I am announcing major new ground rules for the regulation of biotechnology...the United States leads the world in biotechnology and I intend to keep it that way." The New York Times reported (New York Times, 1992) that the thrust of the policy is that genetically engineered products "should not be assumed to be inherently dangerous and that regulation should be based on evidence that the risk presented by introduction of an organism in a particular environment used for a particular type of application is unreasonable."

Thus, at the Federal level, there has been considerable action in recent times to reduce barriers to the application of biotechnology. This paper addresses the situation for the prevention, control, and remediation of environmental contaminants using biotechnology-based products and processes. It is a changing situation, and this paper documents the situation as of March 1, 1992 with an update scheduled for the presentation at the Ninth International Biotechnology Symposium in August 1992.

Three Main Applications Impacted by Regulation

The regulatory barriers are to be considered in the context of three application areas in which biotechnology-based products and processes are used for effective pollution treatment and clean-up:

- Pollution Prevention -- some of the most significant environmental benefits of advanced biotechnology (Hoyle, 1991) are coming through new, more effective and safer ways to eliminate toxic and noxious substances within manufacturing processes before emission via process streams, waste water, solid waste and vapors.

- Remediation -- specially selected or adapted microbes today can decompose approximately 100 pollutants typically found in hazardous waste sites, rendering them harmless (Brox and Hanify, 1992).

- Oil Spills -- the clean-up of oil contaminated waterways and soils using naturally occurring microbes and appropriate nutrient products has been demonstrated (Berkey, et al., 1991).

In summary, during the last few years, the use of biological methods for hazardous waste treatment and clean-up has been successfully documented in field tests and at full scale as the preferred approach (Devine, 1991).

This paper addresses two main subjects that relate to regulations, and more specifically, containment for the three applications areas:

- Containment issues related to the use of genetically engineered microbes that degrade or sequester hazardous compounds;
- Containment as it relates to hazardous waste site remediation.

In addition, there are several other subjects relating to regulations that potentially impact the use of biotechnology-based products in environmental applications, and they are briefly described at the end of the paper.

EPA's Authority is Via TSCA and RCRA

The Toxic Substance Control Act (TSCA) is the primary statutory authority for the EPA's biotechnology regulatory efforts applicable to hazardous waste treatment. Under TSCA, the EPA regulates genetically engineered microbes prior to manufacture or release through its Premanufacture Notification (PMN) process. The EPA uses Section 5 of TSCA to require submission of a PMN before a "new chemical substance" can be manufactured or released into the environment for commercial purposes. In a 1986 policy statement, EPA interpreted TSCA to include certain genetically engineered microbes (those having genetic material from more than one genus) to be new chemical substances, thus, subjecting companies to filing a PMN with the Agency (EPA, 1986) if they intend to release such microbes to the environment. For uses involving genetically engineered microbes in contained systems, the EPA continues to operate under this policy, and requires a PMN at the commercial use stage. In the case of a non-contained situation, the EPA uses a voluntary policy to request PMN filing at the R/D stage (assumes commercial intent), and requires a PMN at the commercial stage with a complete set of environmental data.

It is anticipated that the EPA will announce a proposed rule specifically for genetically engineered microbes under TSCA in 1992, now that the White House Council on Competitiveness has issued its basic policy on the subject of biotechnology-based products. Such action should provide a clearer direction for companies who are either currently developing or planning to develop genetically engineered products for environmental applications. The EPA first began drafting a proposed biotechnology rule for public comment in 1986. In the most recent public available draft (Federal Register, 1991), the Agency announced steps to reduce the burden on the regulated community for select situations of contained use.

The Resource Conservation and Recovery Act (RCRA) also provides oversight capabilities for EPA in certain areas. RCRA was enacted to minimize the amount of hazardous waste generated, and to require proper hazardous waste identification and management. As amended by the Hazardous and Solid Waste Amendments (HSWA) of 1984, RCRA provides EPA with authority to regulate hazardous waste treatment by imposing restrictions on the types of treatments that may be used and by requiring treatment facilities to obtain permits.

In January 1992 the EPA solicited comments (Federal Register, 1992) on a proposed rule relating to treatment standards for such facilities. This containment issue is particularly applicable for ex-situ approaches which generally enhance the performance of bioremediation. The EPA's proposed rule relates to the types of structures used for containment, treatment and storage. This solicitation is an example of the EPA stepping forward and requesting information related principally to bioremediation applications. In this case, the Agency was requesting comment on whether less stringent requirements for containment can be considered adequately protective. Comments were due at the same time as this paper.

Over the years there have been a series of rules that have been promulgated under RCRA that are collectively referred to as the Land Disposal Restrictions (LDRs). With these regulations, the EPA stipulates that hazardous wastes must be treated to specified treatment standards, wither by a specific technology or to a concentration level. EPA considers a technology to be demonstrated for a particular waste if the technology currently is in commercial operation for treatment of that waste. At this time, EPA has identified treatment standards for several hundred wastes considered hazardous and has specified the use of biodegradation for a few of these wastes (Federal Register, 1990). Recently, the EPA introduced a mechanism (US EPA, 1992) that recognizes the limiting aspects of this regulation with regard to new technologies, and established what it terms a treatability variance, whereby a company can provide the Agency with data that describes the capability of a site specific alternate treatment method. Thus, this mechanism is available to companies developing biology-based approaches to degrade certain hazardous wastes for which a standard is already in place.

Other Significant Regulations Relating to Containment

In December 1990 the US Department of Transportation proposed a broad regulation on packaging and shipping. This rule defined an etiologic agent as a viable microorganism or its toxin, which "causes or may cause disease...", and stated that such an agent would be subject to certain packaging and shipping requirements (Federal Register, 1990). Earlier that year the Public Health Service also issued a proposed rule which contained a similar definition and imposed packaging and shipping requirements. The definition in that proposal encompassed all microbes, and extended regulatory oversight to any microbial product that is transported. Based on comment received, DOT issued a delay in this portion of the regulation. Also, the Federal Plant Pest Act regulates the import and interstate transport of microorganisms that are classified as "plant pests". If a specific microbe were classed as a plant pest, and were to be transported for use in a biotreatment or bioremediation application, it would be subject to a permit from the US Department of Agriculture under this Federal Act.

References

Berkey, E., et al., In *Environmental Biotechnology for Waste Treatment*, Sayler, G.S., Fox, R., and Blackburn, J.W., Eds., Environmental Science Research Series, Vol. 41, Plenum Press, NY, **1991**; pp 85-90.

Brox, G.H., and Hanify, D.E., Biotreatment News, **1992**, 2, 6.

Hoyle, R., Biotechnology, **1991**, 9, 1316-7.

New York Times, February 25, **1992**, pg. A1.

Reilly, W.K., Speech to Bioremediation Action Committee, Arlington, VA, June 14, **1991**.

Skinner, J.H., In *Environmental Biotechnology for Waste Treatment*, Sayler, G.S., Fox, R., and Blackburn, J.W., Eds., Environmental Science Research Series, Vol. 41, Plenum Press, NY, **1991**; pp 61-70.

Bioremediation Case Study Collection: 1991 Augmentation of the Alternative Treatment Technology Information Center, Devine, K., US EPA Report, Washington, DC, February **1991**.

Coordinated Framework for Regulation of Biotechnology, EPA Office of Science and Technology, June 26, **1986**.

Exercise of Federal Oversite Within Scope of Statutory Authority; Planned Introduction of Biotechnology Products into the Environment, Federal Register, Vol. 57, No. 39, February 27, **1992**.

Land Disposal Restrictions for Newly Listed

Wastes and Contaminated Debris; Federal Register, Vol. 57, No. 6, January 9, **1992**.
Land Disposal Restrictions for Thirds Scheduled Waste, Federal Register, 40 CFR Pt 148, June 1, **1990**.
Microbial Products of Biotechnology: Proposed Regulation Under TSCA; Federal Register, Vol. 51, No. 123, June 21, **1991**.
Performance Oriented Packaging Standards; Federal Register, 49 CFR Pt 107, December 21, **1990**.
Regional Guide: Issuing Site-Specific Treatability Variances for Contaminated Soils and Debris from Land Disposal Restrictions; US EPA, Publication No. 9380. 3-OBFS; January **1992**.

Containment of Fermentations: Comprehensive Assessment and Integrated Control

Joseph Van Houten, Ph.D., The R. W. Johnson Pharmaceutical Research Institute, Raritan, New Jersey 08869 (FAX: 908-707-9211)

Containment is a concept that has been applied widely throughout the fermentation industry to provide for the physical separation of recombinant DNA-containing biological agents, primarily bacteria, fungi, and mammalian cells, from the workplace and surrounding environment. The intent is to prevent employee illness and contamination of the work area and surrounding location with biological agents, their products and other hazardous materials associated with the process. Decisions regarding the appropriate level of containment require a coordinated, comprehensive assessment of all the process hazards followed by an action plan that integrates the individual control decisions made for each hazard.

The requirement to contain fermentations stems from the desire to protect workers from exposure to hazardous materials and to prevent contamination of the air, water and soil surrounding the facility. Employee and environmental safety can only be assured when the hazards of the process have been adequately identified and assessed and an integrated control plan has been designed and implemented.

In general, there are three broad categories of hazard associated with fermentation processes--the biological agent, its products and by-products, and process hazards of a chemical or physical nature.

Biological Hazards

Assessment of the biological hazard begins with comparing the genus and species designation to classification schemes prepared and published by recognized experts, such as the U.S. Centers for Disease Control. Although somewhat dated, CDC's Classification of Etiological Agents on the Basis of Hazard (U. S. Centers for Disease Control, 1974) is a useful starting point for assessing the hazard of the host biological agent. This document divides harmful bacteria, viruses and fungi into four groups. Class 1 is for those that have not been shown to be pathogenic for man. Class 2 is for agents of moderate risk that are spread primarily through ingestion or inoculation. Class 3 agents are those for which transmission via inhalation of aerosols has been firmly established. Class 4 is reserved for those agents that cause lethal infections in man and for which effective treatment is unavailable. Once the hazardous nature of the organism is established, attention should be focused on the scope of work, including the volume of material and whether the organism will be handled in liquid or solid culture medium or in vivo as part of animal experimentation. Successful identification of these two elements establishes the magnitude of risk that needs to be controlled.

Containment of Small Scale Biological Hazards

If one is working with small volumes of culture medium, typically less than 10 liters, or performing in vivo studies, the biohazard risk can be adequately controlled using a combination of facilities and work practices. Standard guidelines are those found in Biosafety in Microbiological and Biomedical Laboratories, published by the U.S. Centers for Disease Control and National Institutes of Health. Described in this publication are four small scale containment levels--Biosafety Level 1 (BL-1) through Biosafety Level 4 (BL-4)--and four in vivo containment levels--Animal Biosafety Level 1 (ABL-1) through Animal Biosafety Level 4 (ABL-4). Facilities and practices recommended for BL-1/ABL-1 are predicated upon the fact that organisms requiring this level of containment are non-infectious for humans but may cause allergic or similar reactions. BL-2/ABL-2 is appropriate for organisms that are transmitted primarily via ingestion, inoculation or mucous membrane contact. BL-3/ABL-3 is for organisms whose transmission via the inhalation route has been firmly established and BL-4/ABL-4 stipulates a completely closed system for handling lethal pathogens.

Containment of Large Scale Biological Hazards

When the anticipated volume exceeds 10 liters of culture, the best guidance for containment is found in

1054–7487/92/0415$06.00/0

the Guidelines for Research Involving Recombinant DNA Molecules published by the National Institutes of Health. Appendix K of this document contains recommendations for four levels of physical containment--Good Large Scale Practice (GLSP), Biosafety Level 1-Large Scale (BL1-LS), Biosafety Level 2-Large Scale (BL2-LS) and Biosafety Level 3-Large Scale (BL3-LS).

Good Large Scale Practice (GLSP)

The GLSP level of physical containment is recommended for large scale research or production involving viable, non-pathogenic, and non-toxigenic biological agents that have an extended history of safe large scale use. Likewise, it is recommended for organisms that have built-in environmental limitations that permit optimum growth in the large scale setting but limited survival without adverse consequences in the environment. The basic elements of the GLSP level of physical containment are:

1. Formulate and implement institutional codes of practice for safety of personnel and adequate control of hygiene and safety measures.
2. Provide adequate written instructions and training of personnel to keep the workplace clean and orderly and to keep exposure to biological, chemical or physical agents at a level that does not adversely affect the health and safety of employees.
3. Provide changing and handwashing facilities as well as protective clothing, appropriate to the risk, to be worn during work.
4. Prohibit eating, drinking, smoking, mouth pipetting and applying cosmetics in the workplace.
5. Implement an internal accident reporting system.
6. Control the generation of aerosols with appropriate procedures so as to minimize the release of organisms during sampling from a system, addition of materials to a system, transfer of cultivated cells, and removal of material, products and effluents from a system.
7. Implement emergency plans to handle large losses of culture.

Biosafety Level 1--Large Scale (BL1-LS)

For those organisms that do not qualify for GLSP, the BL1-LS level is recommended for large scale research or production of viable biological agents that require BL1 containment at the laboratory scale. Elements of BL1-LS containment include:

1. A closed system to reduce the potential for escape of organisms.
2. All microorganisms must be inactivated using a validated procedure prior to removing them from the closed system.
3. Collection of samples from the closed system, addition of materials to the closed system, or transfer of materials between closed systems must be accomplished in a manner that minimizes the creation of aerosols, preferably using engineering controls (e.g., sealed sampling devices with filters).
4. All exhaust gases must be treated to minimize the release of viable microorganisms.
5. The system must be sterilized prior to its being opened for maintenance.
6. Assure that emergency plans are developed to handle large scale spills.

Biosafety Level 2--Large Scale (BL2-LS)

The BL2-LS level is intended for biological agents that require BL2 containment at the laboratory scale. Key features of this level include:

1. Implementation of all BL1-LS recommendations with the additional requirement that aerosols must be prevented during sampling, addition of materials and transfer of materials between vessels.
2. Rotating seals must be designed to prevent leakage or the entire operation must be enclosed in a containment device (e.g., biological safety cabinet).
3. Devices must be installed to monitor the integrity of containment during operation.
4. The closed system must be tested with a non-recombinant organism prior to use.
5. The system should be permanently identified and labeled with a biohazard symbol.

Biosafety Level 3--Large Scale (BL3-LS)

The last level of large scale physical containment for which pre-established guidelines exist is BL3-LS. This level is required for organisms that require BL3 containment at the laboratory scale. Highlights of this level include:

1. Implement all BL1-LS and BL2-LS recommendations.
2. Operate at as low a pressure as possible to maintain the integrity of containment.
3. The closed system must be located in an area with the following features:
   a. Access through an air lock.
   b. Surfaces that are readily decontaminated.
   c. All penetrations to the area must be sealed to permit decontamination with gas.
   d. Utilities, services, piping and wiring are sealed to prevent contamination.
   e. Handwashing facilities are available and have foot, elbow or automatic valves.
   f. A shower is provided near the controlled area.
   g. There is controlled air flow into the facility (negative pressure with respect to surrounding areas).

4. All practices outlined for BL3 laboratory work apply.

No provisions have been made in the Guidelines for large-scale research or production of viable organisms that require BL4 containment at the laboratory scale. If necessary, such requirements will be developed by the National Institutes of Health on an individual basis.

Product Hazards

The vast majority of contained fermentations that utilize recombinant DNA-containing biological agents are conducted to produce a pharmaceutical active ingredient. Today, many of these products are immune system modulators that heretofore have been obtainable only in minute quantities. As the potency and availability of these materials increase, concern for employee safety extends beyond the potentially infectious nature of the biological agent to the pharmacological activity of the product. Exposure to such products may result in a number of undesirable effects. Among the most common are evidence of therapeutic or side effects, interaction with or counteraction of current medication, potentiating current medication, exacerbating existing disease, and increasing tolerance of or inducing sensitization to a future medication.

Containment of Product Hazards

Realizing that worker exposure to a pharmaceutical active ingredient can be considered to be neither benign nor beneficial, Sargent and Kirk (Sargent, E.V.; Kirk, G.D., 1988) introduced the concept of exposure control limits (ECL) for such materials. They are modeled after the threshold limit values (TLV) established for chemicals by the American Conference of Governmental Industrial Hygienists (ACGIH) and represent a concentration that a worker can be exposed to for 8 hours per day and 40 hours per week for a working lifetime without experiencing adverse effects. Calculation of a numerical ECL requires that a "no effect" level (NOEL) be determined, preferably in humans but definitely in animal models. Early ECLs were established in the range of 1-5 mg/m^3. As the potency of compounds increases, the determination of a NOEL becomes increasingly difficult. This makes the task of calculating a numerical ECL elusive at best and in many cases impossible. In recent years, this has led to the introduction of performance-based exposure control limits (PB-ECL) that focus on physical containment requirements similar to those prescribed at BL1-LS through BL3-LS. PB-ECL1 is reserved for those compounds that are not life-threatening and whose symptoms are reversible, not incapacitating and easily managed. A partially closed system in which open handling is restricted, with virtual isolation of the process, is the recommended containment measure for this level. PB-ECL2 is recommended for those compounds that may be life-threatening at achievable levels. All effects of exposure may not be reversible and some permanent damage may result. The symptoms may be incapacitating and medical intervention may be needed. Appropriate controls include a completely closed system in which there is no open handling of the compound. PB-ECL3 is for those compounds that are life-threatening, whose effects are not reversible, and which require heroic measures to mitigate exposure. Controls include: no open handling, total containment with highly restricted access, a preference for remote operations and clean-in-place systems.

Process Hazards

Once the compound synthesis phase (e.g., fermentation) of the operation has been completed, the product of interest must be isolated and purified. This may require handling hazardous chemicals, working with radioactive materials, using specialized machinery with movable parts, and performing procedures that utilize heating, cooling or repetitive motion.

Control of Process Hazards

Controlling exposure to hazardous chemicals employs many of the same concepts used to control exposure to the final product. One benefit is that for most commercially available chemicals, the hazardous nature is known and controlling it becomes more straight forward. This may include using a partially closed or completely contained system or protective clothing and equipment such as uniforms, respirators, face shields and gloves. With respect to downstream processing equipment, an effective machine guarding program should be implemented to assure that potential pinch points are adequately protected. If the process involves repetitive motion, an ergonomic evaluation should be initiated to minimize the potential for these types of injuries. Finally, where radioactive materials are used, an effective radiation safety program that addresses facilities and procedures for employee protection must be implemented.

Comprehensive Assessment and Integrated Control

Throughout this paper, we have explored a variety of potential hazards that can be associated with biotechnological processes. In order to assure employee safety and environmental protection, all potential hazards must be identified and a coordinated control system developed. The preferred method for accomplishing this is to subject all operations to the rigors of a formal process hazards review. This can be accomplished by using simple, straight forward techniques such as "What If?" for relatively uncomplicated processes. At each step in the process or reaction, "What If?" questions are asked and the answers are considered in evaluating the effects of failure of components or errors in procedures. For slightly more complex processes, the checklist method provides a more organized approach. When analysis of a small portion of a large process or specific item of equipment, such as a reactor, is needed, a failure mode and effect analysis can be used. While this method may not evaluate operating procedure errors or omissions or the possibility or probability of operator error, it does assess the consequences of component failures on the process. In a hazard and operability (HAZOP) study, every part of the process is examined to discover how deviations from the intended design can occur and how these deviations can cause hazards. Finally, the most rigorous method is the fault tree analysis. In this method, a specific undesired process event, such as an explosion, is postulated and placed at the top of a tree, from which branches representing all possible precursor events or causes are extended. When basic causes are reached, failure rates are estimated or obtained.

Regardless of the method, the team that conducts this review should include a:

1. Biological Safety Specialist
2. Industrial Hygienist
3. Safety Specialist
4. Environmental Engineer
5. Biochemical Engineer
6. Fermentation Microbiologist
7. Scientists/operators who perform the work

Implementation of this approach will assure that there is comprehensive assessment of the process hazards and complete integration of the controls that need to be established for each individual hazard. The end result is a workplace that is safe and healthy for all biotechnology workers and a process that minimally impacts the environment surrounding the fermentation facility.

References

Federal Register, **May 7, 1986**, Vol. 51, No. 88, pp. 16958-16985.

Federal Register, **July 18, 1991**, Vol. 56, No. 138, pp. 33174-33183.

Sargent, E.V.; Kirk, G.D., American Industrial Hygiene Association Journal, **1988**, Vol. 49, p. 309.

U. S. Centers for Disease Control, Classification of Etiological Agents on the Basis of Hazard, **1974**, Fourth Edition.

U. S. Centers for Disease Control and National Institutes of Health, Biosafety in Microbiological and Biomedical Laboratories, **1988**, Second Edition.

Agricultural Biotechnology: A U.S. Department of Agriculture Perspective

Terry L. Medley, Animal and Plant Health Inspection Service, U.S. Department of Agriculture, Hyattsville, MD 20782 (FAX: 301–436–8724)

A national implementation strategy for biotechnology in agriculture should focus on solving important scientific and agricultural problems, effectively using the funds and institutional structure available to support research and training researchers in new scientific areas. However, to be a comprehensive and truly meaningful national strategy, it must place equal importance on efficiently transferring technology (3). The national strategy must assure successful technology transfer for expanded development and commercialization of the products of biotechnology. In February of 1991, the President's Council on Competitiveness issued a report on National Biotechnology Policy (hereinafter, the report). The report describes the competitive status of American biotechnology and outlines the goals of Administration policy to support free market development of biotechnology products. It examines three critical policy areas: Support for Science and Technology; Risk-based Regulation of Health and Safety; and Access to Capital and Financial Resource; and Protection of Intellectual Property Rights (8).

I will discuss some of the current regulatory policies and issues at the U.S. Department of Agriculture (USDA) with regard to the commercialization of agricultural biotechnology products. I will focus on the facilitation role of regulations in technology transfer rather than the normally attributed role of constraining or impeding technology development or commercialization. Biotechnology offers new ideas and techniques applicable to agriculture. Biotechnology, as a tool, offers tremendous potential to improve agricultural productivity; decreases our dependence on synthetic chemicals; and enhances our ability to produce food on marginal lands (9). Crucial to enhanced productivity, efficiency, and environmental acceptability is innovation. The successful safe transfer and utilization of agricultural products of biotechnology should be the result of this innovation.

In order to facilitate safe technology transfer and commercialization of agricultural biotechnology products, the regulations must be based upon logical reasoning rather than empirical methodologies. The philosophy at USDA concerning the regulation of agricultural biotechnology products is based upon logical reasoning and several principles contained in two National Research Council studies on introduction of genetically engineered organisms into the environment (4,5). Two of these key principles are: (1) There is no evidence that unique hazards exist either in the use of rDNA techniques or in the transfer of genes between unrelated organisms; and (2) The risks associated with the introduction into the environment of rDNA-engineered organisms are the same in kind as those associated with the introduction into the environment of unmodified organisms and organisms modified by other genetic techniques. Utilization of these principles enables USDA to ask the most appropriate questions about the biology of the organisms being reviewed (1).

Achieving the desired goals of expanded development, safe technology transfer, and commercialization of agricultural biotechnology products requires that a high priority be placed on utilization of appropriate "oversight" structures. "Oversight" refers to the application of appropriate laws, regulations, guidelines, or accepted standards of practice to control the use of a product based on the degree of risk or uncertainty associated with it. In the area of regulations and the implementation of mandatory review requirements, it is of paramount importance that these requirements be balanced and commensurate with risk (2).

If structured and administered properly,

regulations can facilitate rather than impede expanded development, safe technology transfer, and commercialization of agricultural biotechnology products. Regulations should prevent, or at least mitigate risks, and not inhibit innovation and product development. Development of regulations which neither overregulate nor underregulate is a most formidable task for any national authority (7).

Specifically, the national authority must ensure that the regulatory structure adequately considers health and environmental safety standards as biotechnology is transferred from the laboratory to the field to the marketplace. The exact nature of regulatory structures implemented will have a direct impact on the potential contribution to the country's economy. It will also directly impact the competitiveness of the country's agricultural producers in both domestic and world markets (6). For agricultural biotechnology, as in many other high technology industries, national regulatory structures are a critical determinant of the time and the cost to bring a product to market. The cost of testing to meet regulatory requirements, the potential for delay in regulatory approval, and the uncertainty associated with possible imposition of extensive restrictions or outright disapproval of new agricultural biotechnology research or product could present substantial barriers to product development.

At USDA, we have sought to avoid unnecessary burdens on biotechnology and eliminate unneeded regulatory requirements.

In the establishment of risk based regulations, the technique or process by which an agricultural product is modified should not be the sole litmus test for determination of risk. Although knowledge about the process used is useful in assessing the characteristics of the modified organism, use of new molecular techniques does not establish A priori risk. The goal at USDA is to ensure that regulations and guidelines affecting biotechnology are based solely on the potential risks and are carefully constructed and monitored to avoid excessive restrictions that curtail the benefits of biotechnology to society.

References

1. McCammon, S. and Medley, T., The Molecular and Cellular Biology of the Potato, Certification for the Planned Introduction of Transgenic Plants into the Environment, CAB International. 1990.
2. Miller, Henry I., Burris, R.H., Vidaver, A.H. and Wivel, N.A., Risk-Based Oversight of Experiments in the Environment, Science, Vol. 250. 1991.
3. National Research Council, Committee on a National Strategy for Biotechnology in Agriculture. Agriculture Biotechnology Strategies for National Competitiveness. National Academy Press. Washington, DC. 1987.
4. National Research Council, Committee on Scientific Evaluation of the Introduction of Genetically Modified Organisms and Plants into the Environment. Field Testing Genetically Modified Organisms: Framework for Decisions. National Academy Press, Washington, DC. 1989.
5. National Research Council, Committee on the Introduction of Genetically Engineered Organisms into the Environment. Introduction of Recombinant DNA-Engineered Organisms into the Environment: Key Issues. National Academy Press, Washington, DC. 1987.
6. Office of Science and Technical Policy, Proposal for a Coordinated Framework for Regulation of Biotechnology, 49 Fed. Reg. 50856 (December 31, 1984).
7. Shapiro, S.A., Biotechnology and the Design of Regulation, Ecology Law Quarterly, Volume 17, Number 1, Boalt Hall School of Law, University of California, Berkeley. 1990.
8. The President's Council on Competitiveness Biotechnology Working Group Report on National Biotechnology Policy. Washington, DC. 1991.
9. United States Department of Agriculture, Agricultural Research Service, Solving Agricultural Problems with Biotechnology, Program Aid 1445. Washington, DC. 1990.

Environmental Engineering and Biology Symposium IX

A. S. Menawat and S. Matsui: *Co-Chairs*

Bioremediation: Session A
A. S. Menawat and M. Reuss: *Co-Chairs*

Hydrometallurgy: Session B
B. Volesky and J. Remacle: *Co-Chairs*

Waste Treatment: Session C
R. L. Irvine and S. Matsui: *Co-Chairs*

Bioremediation: How Does the Environment Modulate Microbial Gene Evolution?

A. M. Chakrabarty, Department of Microbiology and Immunology, University of Illinois College of Medicine, Chicago, IL 60612 (FAX: 312–996–6415)

Bioremediation has become almost a household word these days, implying the use of biological agents to control problems of environmental pollution. A prominent example of bioremediation, often cited in popular press and scientific magazines, is that of Exxon Valdez oil spill where indigenous microorganisms were supplied with an oleophilic fertilizer and allowed to proliferate and consume the spilled oil (Pritchard & Costa, 1991). Numerous laboratory and field trials are being conducted using natural (and often indigenous) or adapted microorganisms to allow degradation and ultimate removal of a host of toxic pollutants from the environment (Keeler, 1991). The emphasis so far has, however, been on the use of natural microorganisms, rather than genetically manipulated ones, because of adverse public perception on the release of genetically engineered microorganisms as well as various regulatory constraints on their use (Bakst, 1991). On the other hand, mixed cultures are known to degrade pollutants, including crude oil, at a slow rate because of interactions among themselves (Friello et al., 1976). Also, many contaminated sites contain more than one pollutant, and it has been shown (Haugland et al., 1990) that a single culture having the appropriate genetic competence to degrade a mixture of chemicals would be much more efficient than a mixture of cultures having the same genetic capability. Thus, genetically improved single cultures are considered better adapted to degrade single or mixture of pollutants than mixed cultures. Even when indigenous microorganisms may utilize bulk of some ingredients of mixed pollutants, they may produce enough toxic intermediates so as not to greatly reduce the overall toxicity or teratogenicity of the biotreated sample (Mueller et al., 1991).

Another major reason for using genetic selection for the decontamination of polluted environments is the fact that in many cases, natural microorganisms have not evolved the genetic competence to utilize a synthetic compound. For example, the half-life of chlorinated dioxins or a number of highly chlorinated compounds is of the order of years, meaning that natural microorganisms have no efficient enzyme systems to break these compounds down. To generate new degradative capability against a newly-made synthetic compound, a microorganism must evolve the appropriate genes encoding enzymes that would have high affinity for the target chemical or its intermediate products as substrates. Natural evolution may sometime take years, depending upon the chemical structure, the solubility of the compound, and the nature of the environment. Genetic selection may play a critical role in enhancing the process of evolution of new biodegradative genes in natural microorganisms for utilization of such compounds.

How does one enhance the process of natural evolution for biodegradative purposes? In order to address this question, it is imperative that we understand how new genes encoding new types of catabolic enzymes evolve in nature. An interesting example of natural evolution of new genes is the microbial degradation of chlorinated aromatic compounds. These compounds have been released into the environment in large amounts in the form of herbicides and pesticides, or as industrially useful compounds such as PCBs (polychlorinated biphenyls). Because of the presence of large number of chlorine substituents, natural microorganisms that can rapidly degrade the nonchlorinated parent compounds are unable to mineralize these compounds to any significant extent. Catechol is a central intermediate of aromatic degradation (Wheelis & Ornston, 1972), and it is interesting that many chlorinated aromatic compounds, having somewhat lower number of substituent chlorine, can be degraded by natural microorganisms because of

1054–7487/92/0422$06.00/0

the evolution of new degradative genes in such organisms. This is illustrated in Fig. 1. Catechol, derived from benzoate (and other aromatic compounds) is utilized by a large number of bacterial genera, including the genus Pseudomonas. In P. putida, catechol is degraded by a set of chromosomal genes catA, catB, catC etc leading to the formation of ß-ketoadipate which is finally converted to succinate and other members of the tricarboxylic acid cycle for oxidative metabolism. The enzymes such as pyrocatechase I or cycloisomerase I (muconate lactonizing enzyme I) have high affinity for their substrate catechol or cis, cis-muconate, but have very little affinity for the chlorinated catechols (Schlomann et al., 1990). Consequently, catechol degrading microorganisms are unable to utilize chlorinated catechols. It is, however, possible to isolate from nature members of Pseudomonas species or Alcaligenes eutrophus capable of degrading chlorinated benzoic or phenoxyacetic acids or benzenes such as 3-chlorobenzoic acid (3-Cba; Chatterjee et al., 1981), 2,4-dichlorophenoxyacetic acid (2,4-D; Don & Pemberton, 1981) or 1,2,4-trichlorobenzene (1,2,4-Tcb; Van der Meer et al., 1991a). These compounds are metabolized to their corresponding chlorocatechols, viz. 3-chlorocatechol, 3,5-dichlorocatechol or 3,4,6-trichlorocatechol, which are further degraded via chloromuconates and dienelactones (Reineke & Knackmuss, 1988). Thus the pathways of catechol and chlorocatechol degradation are quite similar, as shown in Fig. 1, where catA, clcA, tfdC and tcbC all encode type I or type II 1,2-dioxygenases acting on catechol or chlorinated catechols, catB, clcB, tfdD and tcbD encode cycloisomerases acting on cis, cis-muconate or its chlorinated derivatives and catD, clcD, tfdE and tcbE encode hydrolases having affinity for enol lactone or dienelactones. Even though the pathways are largely similar, the evolved enzymes active on chlorinated catechols or their intermediate metabolites have strikingly altered enzyme specificities. Thus, catA gene product (type I dioxygenase) has little activity towards chlorinated catechol, while clcA gene product (type II dioxygenase) has high activity towards 3-chlorocatechol with a lower activity for 3,5-dichlorocatechol. In contrast, tcbC gene product preferentially acts on 3,-4-dichlorocatechol whereas tfdC has a higher affinity for 3,5-dichlorocatechol (Schlomann et al., 1990; van der Meer et al., 1991a). A similar specialization of specificity in the evolved enzymes can be discerned in subsequent steps such as in the cycloisomerization reactions.

How do microorganisms evolve new degradative genes with new substrate specificities? It is interesting to note that while the catechol degradative cat genes are chromosomal, the chlorocatechol degradative clc, tfd or tcb genes are all plasmid-borne. Recruitment of genes in nature is favored in the form of plasmids because such genes can then be transmitted to other natural microorganisms to allow degradation of the chlorinated compounds as substrates, as shown by mesocosm studies in presence of chlorobenzoates (Fulthorpe & Wyndham, 1991). Since a complete pathway of chlorobenzoate, chlorophenoxyacetate or chlorobenzene degradation requires initial oxygenases to generate chlorocatechol, it is also interesting to note that genes for such oxygenases are often recruited by the plasmid through transposable elements, either as separate gene clusters flanked by IS elements or as part of a composite transposon (van der Meer et al., 1991b; Nakatsu et al., 1991). Molecular cloning and DNA sequencing studies, as well as purification of some of the catechol and chlorocatechol degradative enzymes and their amino acid sequence data, have provided considerable insight regarding the mode of evolution of the chlorocatechol degradative genes. It should be pointed out that the host strains with the three plasmids pAC27, pJP4 and pP51 harboring the clcABD, tfdCDEF and tcbCDEF gene clusters (Fig. 1) were isolated in three different continents viz. in the United States, Australia and Europe, and the plasmids appear to be different with respect to their host range, inc property, and substrate profiles, although pAC27 and pP51 are of similar size. If these three plasmids evolved independently in natural microflora in response to the environmental release of the primary substrates for which they encode appropriate degradative enzymes, viz, 3-Cba, 2,4-D and 1,2,4-Tcb, then one could ask if pathway genes would show any similarity with one another as well as with any presumptive ancestral pathway such as the chromosomal cat pathway. Indeed, the organi-

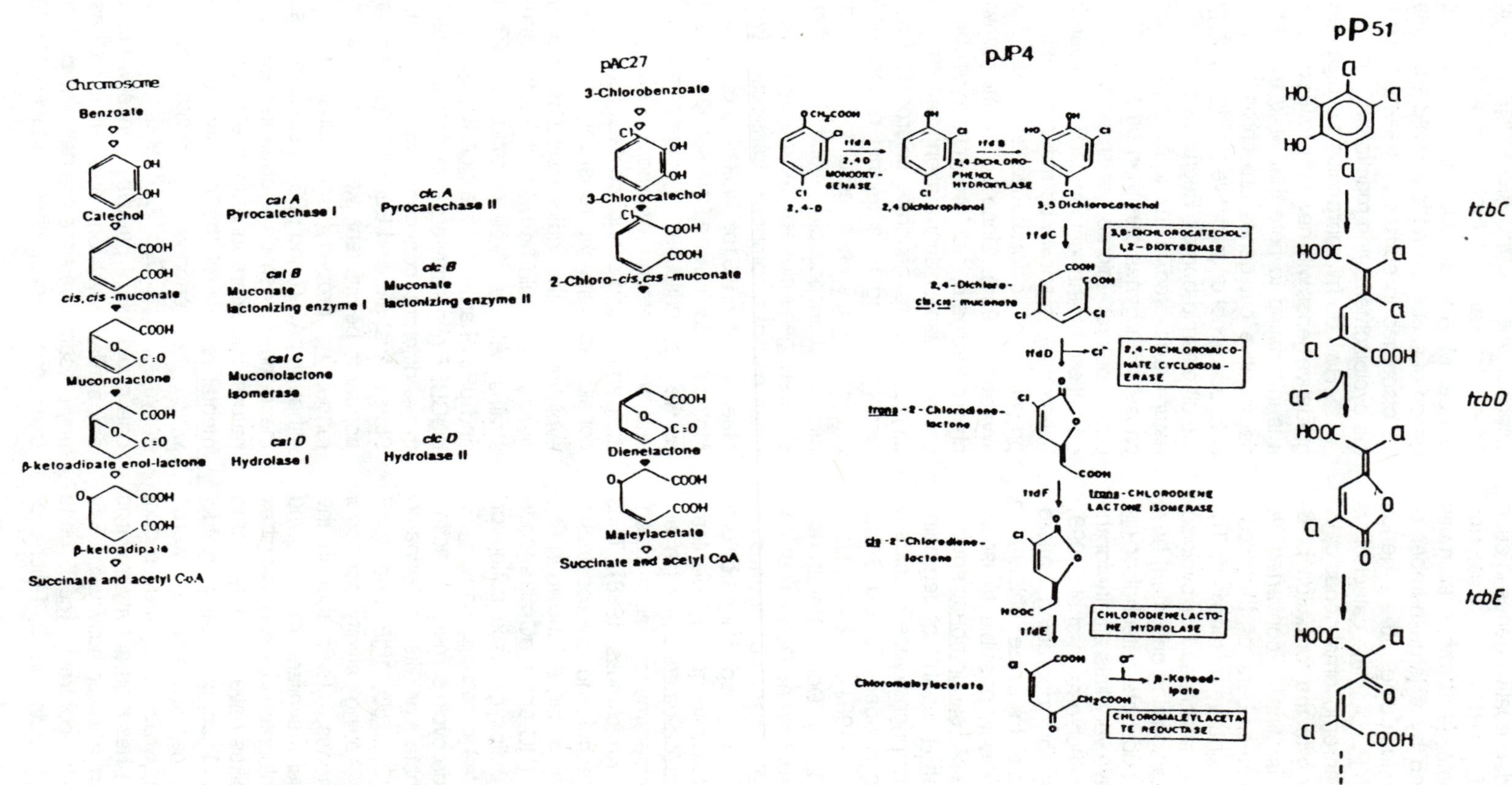

Figure 1. The pathways of the degradation of catechol, 3-chlorocatechol, 3,5-dichlorocatechol and 3,4,6-trichlorocatechol encoded by the chromosomal genes and by plasmids pAC27, pJP4 and pP51, as indicated. The genes in each step of the pathways are indicated, as are the enzymes in the first three pathways. Type II enzymes involved in 3-chlorocatechol and 3,5-dichlorocatechol utilization are also involved in 3,4,6-trichlorocatechol dissimilation.

zation of the catBC operon which is regulated by the divergently transcribed positive regulator catR shows interesting similarity with that of the clcABD operon and its regulator gene clcR (Rothmel et al., 1991a) as well as the organization of the tcbCDEF operon and its regulator tcbR (van der Meer et al., 1991c). In all these cases, the positive regulator genes are transcribed divergently from the operons they control in such a way that their promoters overlap (Rothmel et al., 1991a; van der Meer et al., 1991c; Rothmel et al., 1991b). In contrast, the tfdCDEF operon is controlled by a negative regulator tfdR, which acts as an activator in presence of inducers such as 2,4-D, 2,4-dichlorophenol or 4-chlorocatechol, but which maps several kilobases upstream of the tfdCDEF operon (Kaphammer et al., 1990). The regulator of the catBC operon, CatR, shows significant amino acid sequence homology with the other two regulatory proteins ClcR and TfdS (Rothmel et al., 1991a), while TcbR shows significant homology with parts of CatR, ClcR, TfdS etc (van der Meer et al., 1991c). Similarly, the structural genes of the four operons catBC, clcABD, tfdCDEF and tcbCDEF exhibit extensive homology. The parent CatA shows a good deal of homology to ClcA (Neidle et al., 1988; Frantz et al., 1987), while ClcA shows 50 to 60% homology with TcbC or TfdC (van der Meer et al., 1991a; Perkins et al., 1990, Ghosal & You, 1988). Similarly, CatB shows 40 to 50% homology with ClcB, TfdD and TcbD (Aldrich et al., 1987; Frantz & Chakrabarty, 1987; van der Meer et al., 1991a; Perkins et al., 1990; Ghosal & You, 1988), while ClcD shows about 50% homology with TfdE and TcbE (van der Meer et al., 1991a; Perkins et al., 1990).

In addition to nucleotide sequence identity among the evolved genes, there is striking organizational similarity in the clc, tfd and tcb operons. For example, not only are the genes organized in the same order as the pathway steps, but there is a 4 bp overlap between the stop codon of clcA and the start codon of clcB (Frantz & Chakrabarty, 1987), which also exists between the stop codon of tfdC or tcbC and the start codon of tfdD or tcbD (Perkins et al., 1990; van der Meer et al., 1991c). Most striking is the presence of an additional open reading frame, termed ORF3, between the clcB and clcD genes and between the tcbD and tcbE genes. These ORF3 copies have no known function in chlorocatechol degradation but are present in the two operons, exhibiting more than 50% homology (van der Meer et al, 1991c). Thus the evolved gene clusters exhibit among themselves a great deal of organizational and nucleotide sequence similarity, and to the parent cat genes that allow degradation of non-chlorinated catechol.

We have discussed so far the evolution of genes that took place over a long period of time in response to the release of chlorinated compounds. In nature, such compounds are often present with other biodegradable lignocellulosic materials, so that the urgency of gene recruitment and evolution of biodegradative genes against such chlorinated compounds is not apparent. We have described (Rothmel et al., 1991a) another system where microorganisms from dump sites were subjected to strong selection in a chemostat for the utilization of a recalcitrant compound 2,4,5-trichlorophenoxyacetic acid (2,4,5-T) which was supplied as the only major source of carbon and energy. Since the need to evolve a degradative pathway for the utilization of this compound was intense for the survival of the microorganisms present in the chemostat, it is interesting to note that cloned 2,4,5-T degradative (tft) genes evolved in a strain of P. cepacia AC1100 showed little DNA homology to other members of the genus Pseudomonas (Tomasek et al, 1989; Haugland et al, 1991). Recent sequencing results with a cluster of tft genes (D. Daubaras, unpublished observations) show homologies with genes encoding glutathione reductase and glutathione S-transferase, which are often associated with xenobiotic detoxification in non-Pseudomonas type of bacteria or eukaryotic organisms. It appears that a very stressed environment such as the chemostatic environment with a single recalcitrant carbon source favors gene recruitment from any available source without regard to DNA sequence similarity for enhanced evolution of a catabolic pathway. The nature of the environment therefore dictates to a large extent the mode of evolution of new degradative genes in microorganisms.

In conclusion, even though bioremediation with natural or indigenous microorganisms is being promoted for detoxification of environmental pollutants, the application of genetic selection and manipulation provides additional and highly effective tools for enhanced removal of such compounds from the environment. An understanding of how the environment modulates microbial gene expression and allows assortment and recruitment of new genes may allow in the future development of organisms highly effective in completely degrading such recalcitrant compounds as chlorinated dioxins and PCBs.

References

Aldrich, T.L.; Frantz, B.; Gill, J.F.; Kilbane, J.J.; Chakrabarty, A.M. Gene 1987, 52, 185-195.

Bakst, J.S. J. Indust. Microbiol. 1991, 8, 13-22.

Chatterjee, D.K.; Kellogg, S.T.; Hamada, S.; Chakrabarty, A.M. J. Bacteriol. 1981, 146, 639-646.

Don, R.H.; Pemberton, J.M. J. Bacteriol. 1981, 145, 681-686.

Frantz, B.; Aldrich, T.L.; Chakrabarty, A.M. Biotech. Adv. 1987, 5, 85-99.

Frantz, B.; Chakrabarty, A.M. Proc. Natl. Acad. Sci. USA 1987, 84, 4460-4464.

Friello, D.A.; Mylroie, J.R.; Chakrabarty, A.M. In Proceedings of Third International Biodegradation Symposium; Sharpley, J.M. Ed; Applied Science Publication, Essex, England, 1976, pp 205-214.

Fulthorpe, R.R.; Wyndham, R.C. Appl. Environ. Microbiol. 1991, 57, 1546-1553.

Ghosal, D.; You, I.S. Mol. Gen. Genet. 1988, 211, 113-120.

Haugland, R.A.; Schlem, D.J.; Lyons, R.P.; Sferra, P.R.; Chakrabarty, A.M. Appl. Environ. Microbiol. 1990, 56, 1357-1362.

Haugland, R.A.; Sangodkar, U.M.X.; Sferra, P.R.; Chakrabarty, A.M. Gene 1991, 100, 65-73.

Kaphammer, B.; Kukor, J.J.; Olsen, R.H. J. Bacteriol. 1990, 172, 2280-2286.

Keeler, R. R & D Mag. 1991, 33, 34-40.

Mueller, J.G.; Middaugh, D.P.; Lantz, S.E.; Chapman, P.J. Appl. Environ. Microbiol. 1991, 57, 1277-1285.

Nakatsu, C.; Ng, J.; Singh, R.; Straus, N.; Wyndham, C. Proc. Natl. Acad. Sci. USA 1991, 88, 8312-8316.

Neidle, E.L.; Hartnett, C.; Bonitz, S.; Ornston, L.N. J. Bacteriol. 1988, 170, 4874-4880.

Perkins, E.J.; Gordon, M.P.; Caceres, O.; Lurquin, P.F. J. Bacteriol. 1990, 172, 2351-2359.

Pritchard, P.H.; Costa, C.F. Environ. Sci. Technol. 1991, 25, 372-383.

Reineke, W.; Knackmuss, H. J. Annu. Rev. Microbiol. 1988, 42, 263-287.

Rothmel, R.K.; Haugland, R.A.; Sangodkar, U.M.X.; Coco, W.M.; Chakrabarty, A.M. In Biodeterioration and Biodegradation 8; Rossmore, H.W. Ed; Elsevier Applied Science, London, 1991a, pp 276-291.

Rothmel, R.K.; Shinabarger, D.L.; Parsek, M.R.; Aldrich, T.L.; Chakrabarty, A.M. J. Bacteriol. 1991b, 173, 4717-4724.

Schlomann, M.; Pieper, D.H.; Knackmuss, H.-J. In Pseudomonas: Biotransformation, Pathogenesis, and Evolving Biotechnology; Silver, S.; Chakrabarty, A.M.; Iglewski, B.; Kaplan, S. Ed; American Society for Microbiology, Washington, D.C. 1990, pp 185-196.

Tomasek, P.H.; Frantz, B; Sangodkar, U.M.X., Haugland, R.A.; Chakrabarty, A.M. Gene 1989, 76, 227-238.

van der Meer, J.R.; Eggen, R.I.L.; Zehnder, A.J.B.; de Vos, W.M. J. Bacteriol. 1991a, 173, 2425-2434.

van der Meer, J.R.; Zehnder, A.J.B.; de Vos, W.M. J. Bacteriol. 1991b, 173, 7077-7083.

van der Meer, J.R.; Frijters, A.C.J.; Leveau, J.H.J.; Eggen, R.I.L.; Zehnder, A.J.B.; de Vos, W.M. J. Bacteriol. 1991c, 173, 3700-3708.

Wheelis, M.L.; Ornston, L.N. J. Bacteriol. 1972, 109, 790-795.

Alternate Respiration as a Mode for Bioremediation

Anil S. Menawat, Department of Chemical Engineering and Molecular & Cellular Biology Program, Tulane University, New Orleans, LA 70118-5698
(FAX: (504) 865-6744)

Many facultative aerobes can use inorganic substrates instead of oxygen as terminal electron acceptors. The inducer and the oxygen can influence the formation and activity of the enzymes catalyzing the reduction of these inorganic electron acceptors in several ways. Several of these enzymes are also capable of lending gratuitous activities that are useful in bioremediation. In this study we present evidence of chlorate reduction activities in *Escherichia coli, Paracoccus denitrificans*, and *Alcagenus denitrificans* with various inducers.

Denitrification refers to the reduction of one or both ionic nitrogen oxides -- nitrate, NO_3^-, and nitrite, NO_2^-. The major products of which are the gaseous nitrogen oxides -- nitric oxide, NO, and nitrous oxide, N_2O. Such reactions occur in many of the *Enterobacteriaceae*, bacilli and clostridia, and in soil and marine sediments under anaerobic conditions. More and more applications of denitrification are emerging in bioremediation, although interest in denitrifying bacteria exists for several other reasons. There is of course the great potential for nitrogen removal from high-nitrogen hazardous materials, even explosives. However, most denitrifying bacteria are not very specific in the substrate requirements and can, therefore, lend various gratuitous activities. Such gratuitous activities arise for several reasons and offer tremendous opportunities for biodegradation. In this paper we present evidence of chlorate reduction by three denitrifying bacteria with diverse activity levels depending on the inducer used.

Many bacteria can use nitrate instead of oxygen as a terminal electron acceptor. This process generally occurs only under anaerobic conditions, and is dominated by two dissimilatory processes: respiratory denitrification and dissimilatory nitrate reduction to ammonium (Knowles, 1982). Assimilatory nitrate reduction could also occur, however. In dissimilatory processes, the cells do not consume nitrogen oxide for cellular metabolism but employ it as an electron acceptor. Nitrate, therefore, competes with oxygen. Consequently, oxygen being the preferred electron acceptor inhibits the dissimilatory nitrate reduction.

Regulation of Nitrate Reductase

Nitrate reduction is an inducible process, which occurs only in the absence of oxygen. This is evident from the free energy considerations

$$NADH + H^+ + NO_3 \rightarrow NAD^+ + NO_2^- + H_2O \qquad \Delta G° = -142.5 \text{ kJ}$$

$$NADH + H^+ + {}^1/_2\, O_2 \rightarrow NAD^+ + H_2O \qquad \Delta G° = -216.5 \text{ kJ}$$

Use of oxygen instead of nitrate gives off greater energy (Stouthamer, 1988). In general, anaerobic conditions are sufficient for formation of nitrate reductase. Nitrate reductase synthesis temporarily derrepresses when aerobic cultures shift to anaerobic conditions (Showe & DeMoss, 1968; De Groot & Stouthamer, 1970); but, the presence of nitrate can typically increase the amount of enzyme. Calder *et al*(1980) found that nitrate reductase induces to higher levels with azide than with nitrate, and is inducible by nitrite also. No suitable explanations for the observed behavior of

1054–7487/92/0427$06.00/0

induction are yet available although several hypotheses exist in literature. Only consensus, however, is that formation of nitrate reductase is very complex and much more work is necessary to unravel this mystery.

Nitrate reductase is a membrane-bound enzyme that is responsible for the reduction of nitrate. Distinct enzymes are present that act on nitrate and chlorate, but differ in substrate specificity and in their sensitivity towards inhibitors. Nitrate reductases from various organisms are very similar in subunit structure, molecular weights of the subunits and other properties (Stouthamer, 1988). They consist of two polypeptides -- subunits A and B -- of about 150 and 60 kDa respectively with each present in equimolar amounts. Subunit B is responsible for binding the enzyme to the membrane through a subunit C that is a cytochrome b of about 20 kDa. Cytochrome is essential for the functioning of the enzyme.

Forget (1971) found that the isolated nitrate reductase can reduce chlorate as effectively as nitrate. Toxic chlorite is the reduced product. However, intact cells reduced chlorate only at one tenth the rate of nitrate. In medium containing 4.5 mM chlorate and 0.5 mM nitrate, the *P. denitrificans* preferentially reduced nitrate over chlorate (John, 1977). John's explanation of this was that (i) nitrate reductase orientation in the plasma membrane is such that nitrate approaches the enzyme from the cytosol, (ii) a nitrate carrier permits nitrate entry across plasma membrane, and (iii) the carrier is unable to transport chlorate. Upon destroying the plasma integrity by triton X-100, chlorate is able to gain entry easily and undergoes reduction at similar rates as nitrate.

In this study we attempt to use induction of nitrate reductase to find interesting gratuitous activities. We use three organisms -- *Escherichia coli*, a nonrespiring denitrifier, and *Paracoccus denitrificans* and *Alcagenus denitrificans*, respiring denitrifiers.

Method

Escherichia coli (K12), *Paracoccus denitrificans* (ATCC 19367) and *Alcagenus denitrificans* were grown aerobically in 500 ml flasks with magnetic stirring at 34°C. There were four identical flasks for each culture containing defined media: sodium L-glutamate, 3.75 g l^{-1}; sodium succinate · 6 H_2O, 2.7 g l^{-1}; sodium molybdate, 0.1 mM; $(NH_4)_2SO_4$, 2 g l^{-1}; $K_2HPO_4 \cdot 3 H_2O$, 7 g l^{-1}; KH_2PO_4, 3 g l^{-1}; $MgSO_4 \cdot 7 H_2O$, 200 mg l^{-1}; $CaCl_2 \cdot 2 H_2O$, 53 mg l^{-1}; $MnSO_4$, 7.5 mg l^{-1}. After about 20 hours of growth when optical density rose to about 2 AU at 540 nm, the cells were harvested and resuspended in fresh media with 0.02 mM iron citrate. Inducers -- KNO_3, NaN_3 and $KClO_3$ -- were added to three flasks while none added to the fourth one for control. These fresh cultures were incubated at same conditions as above except under anaerobic environment. Samples were drawn at every two-hours interval to assay for nitrate and chlorate reduction.

Nitrate and chlorate reductions were measured by a kinetic method in the cell free extracts (micro-centrifuged at 13,000 rpm after cell disruption for 10 min). The oxidation rate of the reduced benzyl viologen was monitored for 180 min at 600 nm from which the endogenous rate collected over 60 min is subtracted. 25 µl of the supernatant was added to the sample cuvette while none to the reference one. Benzyl viologen was reduced by addition of 20 µl of freshly prepared solution of 25 mM $Na_2S_2O_4$ and 52 mM $NaHCO_3$. A 3.5 ml aliquot containing 0.4 mM benzyl viologen, 0.1 mM EDTA and 49 mM sodium phosphate (pH = 7.2) was poured in each cuvette that were then purged with nitrogen for 1.5 min. Oxidation was initiated under totally anaerobic conditions by adding either 15 µl of $NaNO_3$ (2.5 M) or 5 µl of $NaClO_3$ (2.5 M) to both cuvettes. Special care was necessary to maintain absolute anoxic environment.

Discussion

The focus of this study was to understand the role of an inducer in nitrate and chlorate reduction. In Figs. 1 to 3, we present the relative rates of nitrate and chlorate reduction for the three cultures used. The reported rates are deviations from the control with no inducer. In control, the shift from aerobic to anaerobic conditions initiates the reductase synthesis. The data spans over six hours with periodic sampling at every two-hours interval.

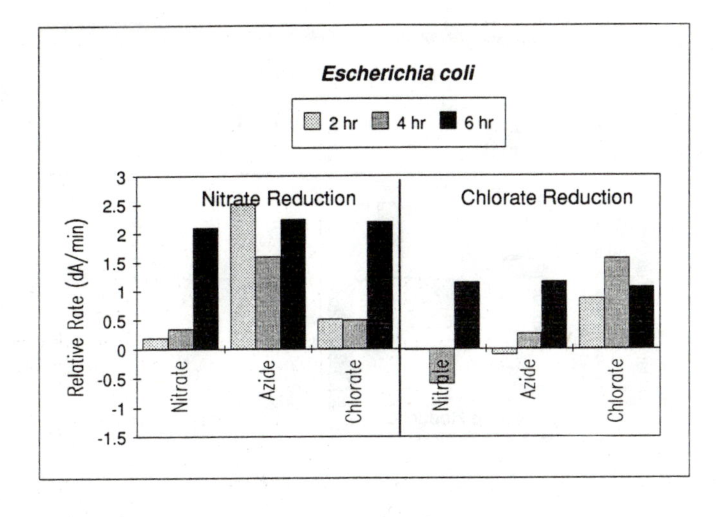

Fig. 1. Relative Rates of nitrate and chlorate reduction by *Escherichia coli* with nitrate, azide or chlorate as inducers. Relative rate is the control rate without any external inducer subtracted from the induced rate.

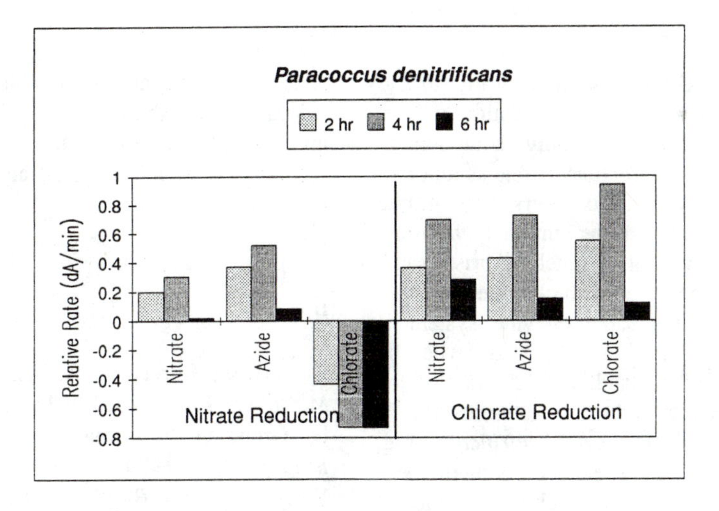

Fig. 2. Relative Rates of nitrate and chlorate reduction by *Paracocuss denitrificans* with nitrate, azide or chlorate as inducers. Relative rate is the control rate without any external inducer subtracted from the induced rate.

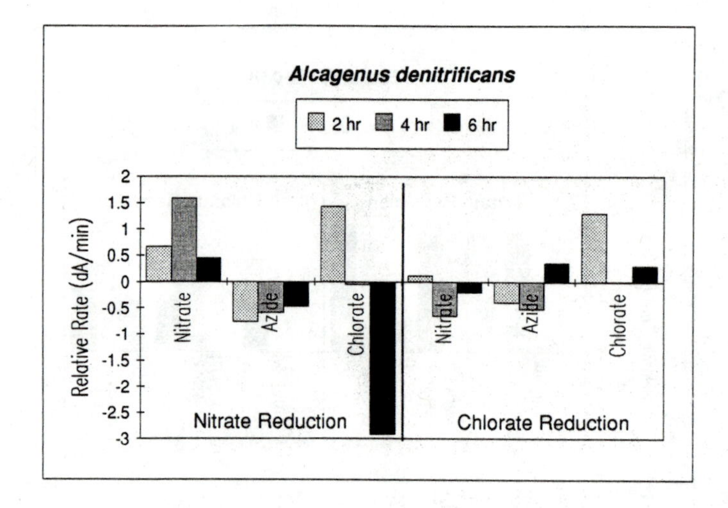

Fig. 3. Relative Rates of nitrate and chlorate reduction by *Alcagenus denitrificans* with nitrate, azide or chlorate as inducers. Relative rate is the control rate without any external inducer subtracted from the induced rate.

For *E. coli*, chlorate along with nitrate and azide provides greater nitrate reduction activity over control (cf. Fig. 1). Initially, azide confers a ten-fold higher activity than nitrate as reported by others. However, in six hours, both nitrate and chlorate catch up to the similar high nitrate reduction activity. Not to much surprise, same phenomenon is observed with chlorate reduction confirming that the same enzyme system is exhibiting a dual role as suggested earlier by others. On the other hand, with chlorate as the inducer, the nitrate reductase activities for *P. denitrificans* (Fig. 2) and *A. denitrificans* (Fig. 3) are below control levels. For both of these cultures, there are high initial chlorate reduction activities that suggest a role of toxic chlorite in killing the nitrate reduction while slowly attenuating the chlorate reduction as well.

P. denitrificans exhibits almost identical trends in nitrate and chlorate reduction with nitrate and azide as inducers, as seen in Fig 2. *A. denitrificans* exhibits similar trends (see Fig. 3) for nitrate reduction as *Paracoccus*. However, it's behavior as a chlorate reducer is different. Nitrate and azide are not suitable inducers while chlorate may show some promise. This is important to recognize for *Alcagenus* is the second most abundant genus among denitrifiers found in soil.

References

Calder, K.; Burke, K. A.; Lascelles, J. *Arch. Microbiol.* **1980**, *126*, 149-153.

De Groot, G. N.; Stouthamer, A. H. *Biochim. Biophys. Acta.* **1970**, *208*, 414-427.

Forget, P. *Eur. J. Biochem.* **1971**, *18*, 442-450.

John, P. *J. Gen. Microbiol.* **1977**, *98*, 231-238.

Knowles, R. *Microbiol. Rev.* **1982**, *46(1)*, 43-70.

Showe, M. K.; DeMoss, J. A. *J. Bacteriol.* **1968**, *95*, 1305-1313.

Stouthamer, A. H. In *Biology of Anaerobic Microorganisms*; Zehnder, A. J. B., Ed; John Wiley & Sons, New York, NY, **1988**, pp 245-304.

Anaerobic Treatment of Wastewater from the Pulp and Paper Industry

Hermann Sahm*, **Ulrich Ney, and Siegfried M. Schoberth**, Institut für Biotechnologie, Forschungszentrum Jülich, F.R.G. (FAX: 0049/2461/61 2710)

Sulfite evaporator condensate (SEC) from pulp and paper industries can be successfully treated by anaerobic mixed enrichment culture. Besides the following methanogens: *Methanobacterium spec.*, *Methanosarcina spec.*, and *Methanothrix spec.* the sulfate-reducing bacterium *Delsulfovibrio vulgaris* was detected which degrades furfural into acetate and CO_2. A consortium of *D. furfuralis, M. bryantii*, and *M. barkeri* degraded furfural in continuous culture (fixed bed loop reactor) to 92 %, while the conversion of acetate was only 67 %. The conversion of acetate could be improved to 89 % by the following consortium: *M. barkeri, M. arboriphilus, M. concilii,* and *D. furfuralis*. This defined culture degraded all the constituents of SEC anaerobically at an efficiency of almost 90 % compared to an enrichment culure under identical conditions.

In the last few years anaerobic microbial treatment of industrial wastewater has found a great interest. Few years ago we showed that sulfite evaporator condensate (SEC) from pulp and paper industries can be successfully treated by anaerobic mixed enrichment culture (Brune et al. 1982). Although in this wastewater acetic acid is the main compound (Tab. 1), furfural is the substance most toxic towards anaerobic bacteria. Recently we studied the composition of this enrichment culture using microbiological and microscopical techniques. Besides the following methanogens: *Methanobacterium spec.*, *Methanosarcina spec.*, and *Methanothrix spec.* the sulfate-reducing bacterium *Desulfovibrio vulgaris* was detected which degrades furfural into acetate and CO_2 (Brune et al. 1983).

Table 1. Composition of sulfite evaporator condensate (SEC)

Compound	Concentration	Total Carbon
Acetic acid	300-600 mM	81.6-78.7 %
Furfural	21 - 49 mM	14.3-16.1 %
Methanol	30 - 80 mM	4.1- 5.2 %
Total sulfur*	20 - 37 mM	2.7- 2.4 %
COD**	24000-60000 mg/l	
pH	2	

* Sulfate and sulfite

** Chemical oxygen demand

This furfural isolate could degrade furfural as sole source of carbon and energy in a defined mineral-vitamin sulfate medium. Optimal growth on furfural occurred at 37°C and pH 6.8. Furfural, furfurylalcohol and 2-furoic acid were fermented to acetate with the concurrent reduction of sulfate to sulfide. Sulfite or nitrate could replace sulfate as electron acceptor. During metabolism of (carbonyl-^{13}C)-furfural, 2-^{13}C-furfuryl-alcohol and 2-^{13}C-furoic acid could be detected as intermediates and the ^{13}C-labelled carbonyl-group was converted to $^{13}CO_2$; this indicates intermediary decarboxylation of the heteroaromatic compound (Folkerts et al. 1989). Here we report on defined microbial consortia that degraded very efficiently an effluent from the pulp and paper industry (SEC).

Degradation of furfural by a defined diculture

Recently we tested, if furfural can be degraded completely to biogas by defined bacterial consortium. It had first to be proven that the electron acceptor sulfate could be replaced with a hydrogenotrophic methanogen as external "electron acceptor" (cf. equ. 1 and 2). *M. bryantii* strain MoHG was chosen as H_2 accepting organism because its immunotype was abundant throughout the longterm study of the enrichment culture in an anaerobic bioreactor treating SEC (Macario et al. 1989). Figure 1 shows that a diculture consisting of

1054–7487/92/0431$06.00/0

D. furfuralis and *M. bryantii* did carry out the conversion of furfural to methane and acetic acid (equ. 1 and 2), in the absence of sulfate.

$$C_5H_4O_2 + 4\,H_2O \rightarrow 2\,CH_3COO^- + 2\,H^+ + CO_2 + 2\,H_2 \quad (1)$$
$$4\,H_2 + CO_2 \rightarrow CH_4 + 2\,H_2O \quad (2)$$
$$CH_3COO^- + H^+ \rightarrow CH_4 + CO_2 \quad (3)$$

In this experiment, only small amounts of furfural could be added stepwise, first since furfural was found to be very toxic to methanogenic bacteria, and second since the growth rate of *D. furfuralis* has a narrow maximum (0.1 h^{-1}) at furfural concentrations between 4 and 5 mM (Folkerts et al. 1989). The diculture produced about 1.85 mol acetic acid per mol furfural with simultaneous production of 0.42 mol methane according to equ. 1 and 2. The H_2 partial pressure in the culture headspace was below the detection limit of the method applied (5 ppm); this indicates that interspecies H_2 transfer was very effective.

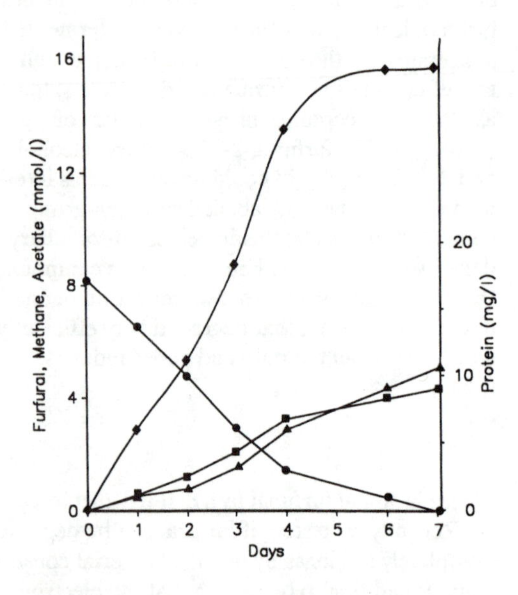

FIG. 1. Anaerobic degradation of furfural by a defined mixed culture of *D. furfuralis* and *M. bryantii* without sulfate in the medium. (●) Furfural, (◆) Acetate, (▲) Methane, (■) Protein. Experiment was performed in batch-culture as described by stepwise adding furfural. The total amount of furfural added at each point is given in the graph.

Degradation of furfural into biogas by a triculture

To degrade the acetate left over by the diculture (Fig. 1), an acetotrophic methanogen was added as third member. Both *Methanosarcina* and *Methanosaeta* ("*Methanothrix*") would have been able to convert acetate to methane and carbon dioxide, and morpho- and immunotypes of strains of both genera have been identified in SEC digesting reactors (Brune et al. 1982; Macario et al. 1989). *M. barkeri* DSM 804 was chosen not only because it is able to degrade methanol (which *Methanosaeta* is not), but also because its growth rate (0,083 h^{-1}) is higher than that of *Methanosaeta* (0,041 h^{-1}). Its inhibitory constants (K_i) for furfural were 1.83 ± 0,08 mM on methanol and 2.05 ± 0.27 mM on acetate. Compared to those of *Methanosaeta* of 2.4 ± 0.26 mM these values were lower but they were slightly more favorable than those of *M. barkeri* DSM 800 (1.25 ± 0.13 mM on methanol; 1.74 ± 0.09 mM on acetate) and of *M. barkeri* DSM 1538 (1.41 ± 0.07 mM on methanol; 1.88 ± 0.13 mM on acetate).

Growth of this triculture consisting of *M. bryantii*, *M. barkeri* and *D. furfuralis* on furfural in batch culture is shown in Fig. 2. 4.5 mM furfural were completely degraded within 7 days. The formation of methane was accompanied by intermediate formation of acetic acid reaching a maximum value of 6 mM at day 4. Then acetate concentration decreased markedly with simultaneous increase in methane production to final concentrations of 8.9 mM. Thus acetate was not completely degraded after 7 days, as already found with enrichment cultures (Brune et al. 1982).

Studies in fixed bed loop reactors

It has previously been found (Aivasidis 1985; Aivasidis and Wandrey 1988) that the efficiency and kinetics of immobilized enrichment cultures are far superior over those used in conventional stirred tank reactors. Therefore, we set up similar experiments with continuously operated fixed-bed loop (FBL) reactors (Ney et al. 1989).

In addition to the strains used in the experiments described above, *M. arboriphilus* and *M. concilii* were used. *M. arboriphilus* whose immunocounts increased considerably during treatment of SEC by enrichment cultures (Macario et al. 1989) replaced *M. bryantii*. *Methanosaeta* was added for two reasons: first, because both morpho- and immunotypes of *Methanosaeta* are found along with those of *Methanosarcina* (Macario et al.

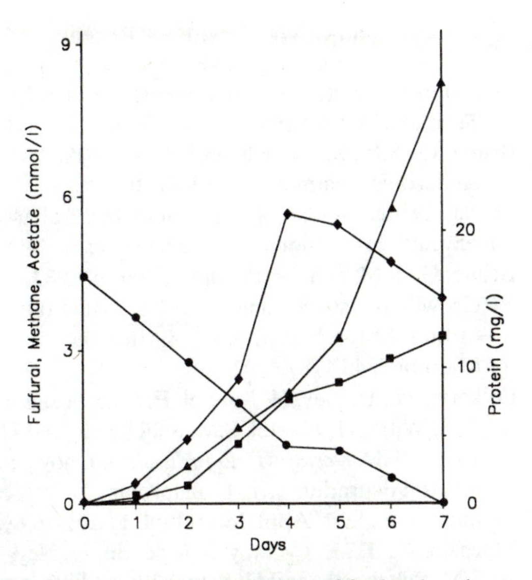

FIG. 2. Anaerobic degradation of furfural and acetate in batch-culture by *D. furfuralis*, *M. bryantii* and *M. barkeri* without sulfate as electron-acceptor in the medium. (●) Furfural, (◆) Acetate, (▲) Methane, (■) Protein.

1989) and second *Methanosaeta* has more favorable kinetic constants for acetate degradation at low concentration than *Methanosarcina* (Patel 1984; Patel 1990).

Table 2 shows that the stationary concentration of furfural was almost nil under all eperimental conditions. Acetate (A-D) and methanol (C, D) had been added to simulate more closely the composition of SEC. Methanol was completely degraded in all experiments, and acetate was reduced by almost 90 % (B-D). In the absence of small amounts of sulfate or sulfite, acetate reduction was less effective (A), as was the efficiency of methane production (yield CH_4 in Tab. 2).

It improved when small concentrations of sulfate were added which may have served as additional sulfur supplement for biosynthesis (B). The same triculture was tested for anaerobic treatment of synthetic (C). After a lag of 33 days, gas production increased and the retention time dropped to an intermediate level between 25 and 30 h.

The determination of biomass demonstrated that 65 g cellular dry weight/l was held back in immobilized form between and on the outer surface of the sinterglass fillings, 31 g/l within the inner pores of the glass rings, and only 0.9 g/l in the free liquid phase. From these data it was

Table 2. Degradation of acetate, methanol and furfural by defined cocultures in continuously operated FBL reactors. The organisms used were: *D. furfuralis* (A-D); *M. bryantii* (A-C); *M. barkeri* (A-D); *M. arboriphilus* and *M. concilii* (D). pH was 6.7. The reactors had a volume of 530 ml.

Parameter	Reactor			
Reactor	A	B	C	D
Retention time (h)	21	15.2	46.6	25.6
Sulfate added (mM)	0	10	15	10
Sulfite added (mM)	0	0	0	15
$Furfural_{in}$ (mM)	10	10	24	24
$Furfural_{out}$ (mM)	0.8	0.01	0	0
$Acetate_{in}$ (mM)	100	100	150	146
$Acetate_{out}$ (mM)	40	17	30	21
$Methanol_{in}$ (mM)	0	0	50	50
$Methanol_{out}$ (mM)	0	0	0	0
Gas_{out} ($l x l^{-1} x d^{-1}$)	4.4	7.7	8.2	8.6
CH_4 (%)	57	51	52	54
Acetate conv.* (%)	67	86	85	89
Yield CH_4** (%)	90	97	99	98

* $acetate_{out}$ per total substrate $acetat_{in}$ including acetate from furfural conversion.

** yield per total of substrates converted.

calculated that 96-98 % of the biomass was immobilized. The specific activities of methane formation from acetate, methanol, furfural and H_2/CO_2 were (in μmol CH_4/g dry weight/h; potential activities, and actual in parentheses) acetate 1411 (1336), methanol 376 (150), furfural 284 (56), H_2/CO_2 246 (56).

In a final step, we investigated the degradation of SEC from a paper mill by a defined tetracultures constisting of *D. furfuralis*, *M. arboriphilus*, *M. barkeri*, and *M. concilii*. The culture was started on model wastewater and then changed to SEC from a paper mill. SEC with a COD of 36,500 mg/l was degraded at an average hydraulic retention time of 17.9 h and a space time yield of 39,000 mg COD/l/d. The organic compounds of SEC were removed to 84 % and the turnover of acetate, methanol and furfural was equivalent to the production of 25.4 mmol CH_4/l/h. The specific activity of this defined tetraculture was comparable to that of an enrichment culture (Table 3).

Table 3. Continuous degradation of SEC (425 mM acetate, 75 mM methanol, 28 mM furfural, 28 mM total sulfur) from pulp and paper industries by a defined coculture of *Methanosarcina barkeri*, *Methanobrevibacter arboriphilus*, *Methanosaeta concilii*, and *Desulfovibrio furfuralis* (A) in comparison to degradation by an enrichment culture (B).

Reactor	A	B
Retention time (h)	17.9	16
COD (g/l)	36.5	36.7
Gas_{out} ($1x1^{-1}xd^{-1}$)	27.7	31.3
Org. space load. (g $CODx1^{-1}xd^{-1}$)	47.6	55
Spec. act. (g $CODxbiomass^{-1}xd^{-1}$)	3.3	3.8
Space time yield (g $CODx1^{-1}xd^{-1}$)	39	45
Carbon balance (biogas) (%)	94.8	91.5

Based on these results now the Siemens company builds a big plant for the treatment of the waste water from a pulp and paper industry in Germany.

These studies indicate that starter cultures for anaerobic treatment of industrial effluents might be developed. Defined bacterial cultures can be used as a "booster" to reduce long lag phases encountered at start up with an unadapted sewage sludge when treating industrial waste water. This looks also promising with respect to treat other industrial effluents by defined anaerobic cultures.

References

Aivasidis, A. 1985. Anaerobic treatment of sulfite evaporator condensate in a fixed-bed loop reactor. Water Sci. Technol. 17:207-221

Aivasidis, A. and C. Wandrey. 1988. Recent developments in process and reactor design for anaerobic wastewater treat-ment. Water Sci. Technol. 20:211-218.

Brune, G., S.M. Schoberth, and H.Sahm. 1982. Anaerobic treatment of an industrial wastewater containing acetic acid, furfural and sulphite. Proc. Biochem. 17:20-36.

Brune, G., S.M. Schoberth, and H. Sahm. 1983. Growth of a strictly anaerobic bacterium on furfural (2-furaldehyde). Appl. Environ. Microbiol. 46:1187-1192.

Folkerts, M., U. Ney, H. Kneifel, E. Stackebrandt, E.G. Witte, H. Förstel, S.M. Schoberth, and H. Sahm. 1989. *Desulfovibrio furfuralis* sp. nov., a furfural degrading strictly anaerobic bacterium. Syst. Appl. Microbiol. 11:161-169.

Macario, A.J.L., E. Conway de Macario, U. Ney, S.M. Schoberth, and H. Sahm. 1989. Shifts in methanogenic subpopulations measured with antibody probes in a fixed-bed loop anaerobic bioreactor treating sulfite evaporator condensate. Appl. Environ. Microbiol. 55:1996-2001.

Ney, U.; Schoberth, S.M.; Sahm, H. 1989. Anaerobic degradation of sulfite evaporator condensate by defined bacterial mixed cultures. In: D. Behrens and A.J. Driesel (eds.), DECHEMA Biotechnology Conferences, Vol. 3, Part B, pp. 889-892, Verlag Chemie, Weinheim; VCH Publishers, New York.

Patel, G.B. 1984. Characterization and nutritional properties of *Methanothrix concilii* sp. nov., a mesophilic aceticlastic methanogen. Can. J. Microbiol. 30:1383-1396

Patel, G.B.; and G.D. Sprott. 1990. *Methanosaeta concilii* gen. nov., sp. nov. ("*Methanothrix concilii*") and *Methanosaeta thermoacetophila* nom. rev., comb. nov. Int. J. Syst. Bact. 40:79-82

Cometabolism in Microbial Consortia under Transient State Operating Conditions

Geoffrey Hamer, Institute of Aquatic Sciences and Water Pollution Control, Swiss Federal Institute of Technology Zürich, CH-8600 Dübendorf, Switzerland [FAX: (44) 1 823 5028]

Cometabolism plays an important rôle in the biodegradation of pollutants both in natural aquatic and terrestrial environments and in engineered biotreatment processes. Gross pollution of the environment has resulted from the indiscriminate disposal and both the intentional and accidental release of products, residues and wastes. Retrospective remediation of such pollution, irrespective of whether it is performed either *in situ* by accelerating the biodegradative activity of indigenous microbes or in on-site bioreactors using enriched microbial cultures will be mediated by microbial consortia operating, because of the heterogeneities in such systems, under gradient and transient state conditions.

In a strict sense, cometabolism can be defined as the microbial transformation of a compound, which is unable to support cell replication, in the requisite presence of another transformable cosubstrate that supports cell replication. Such a definition excludes situations that merely arise from the non-specific nature of certain monooxygenases that transform non-growth substrates in the absence of a co-substrate, but appropriately extends the original definition of cooxidation to include dehalogenations, condensations and rearrangements. What remains unclear in the strict definition of cometabolism is whether energy derived from the oxidation of a non-growth substrate can be used to satisfy energy requirements for growth on the co-substrate, thereby enhancing the biomass yield coefficient for the co-substrate.

Traditionally, undue emphasis has been placed on the properties exhibited by pure microbial monocultures, which are, of course, laboratory artifacts. This has created numerous difficulties with respect to the interpretation of the functioning of mixed cultures. Until recently, interpretation has been very largely based on concepts for binary microbial interactions proposed by Bungay and Bungay (1968).

In the field of biodegradation, the most widely studied types of microbial consortia are those involved in the biodegradation of apparently recalcitrant compounds where the several strains present in the consortium are each responsible for either a step or steps in the overall biodegradative pathway, but no single strain in able to mediate the overall process by itself (Slater and Bull, 1982). However, of equal importance and interest in biodegradation are consortia that comprize only one primary substrate utilizing strain, that also exhibits cometabolic activity, and between two and five ancillary, none primary substrate utilizing, strains that enhance the growth of the primary substrate utilizing strain by reducing its fastidiousness.

The best known examples of the latter type of consortia are methanotrophic consortia of the types discussed by Wilkinson *et al.* (1974) and by Linton and Buckee (1977). A third type of consortium is one where cooperative metabolism occurs with one strain mediating not only a step in the main biochemical pathway, but also exhibiting cometabolic activity towards other incidentally present carbon substrates. A particularly interesting consortium of this type is the nitrifying consortium comprizing *Nitrosomonas* and *Nitrobacter* spp., where *Nitrosomonas* spp. can exhibit cometabolic activity. (Ensley, 1991).

When considering the functioning of microbially mediated processes, three features

1054–7487/92/0435$06.00/0

that frequently evade evaluation are that microbial systems are never at a steady state, but change with time, that gradients exist in all microbial systems, creating conditions at reaction sites that are markedly different from conditions measured in the system as a whole, and that where either the physiological state or the physiological status, within a particular state, of microbes differs, their dynamic response to either imposed or incidental changes will also differ.

Bioreactors for Remediation Processes

For the bioremediation of polluted soils, sediments or groundwaters the concept of a bioreactor can be formulated in two distinct ways; for *in situ* operations, the bioreactor comprizes the whole of the polluted zone or compartment that is subject to restoration, whilst for on-site operations, it is represented by a contained isolated volume of material undergoing treatment. In both types of system, significant macro- and microscopic scale heterogeneities are unavoidable. Here, only the latter type of system, where efforts to reduce gross macroscopic scale heterogeneities are frequently made will be considered.

When considering on-site bioremediation of polluted soil, three essentially different types of bioreactor have been proposed:

1. static beds and heaps of soil that can be intermittently irrigated, aerated and mixed;

2. rotating cylindrical and horizontal screw-mixed moving bed systems that are continuously aerated and mixed and intermittently irrigated;

3. high and low aspect ratio bioreactors in which the soil undergoing treatment is present as an aqueous slurry subjected to aeration and agitation.

In principle, bioreactors used for remediation should be operated so as to minimize the concentrations of residual pollutants remaining after treatment. Although optimization of the total cost of converting a maximum quantity of pollutants into natural non-noxious end products, with acceptable residual concentrations of pollutants remaining after treatment, is critical as far as process economics are concerned, questions of system reliability and safety and operational flexibility must not be ignored. Further, the question of the conversion of pollutants into either hazardous or toxic intermediates with a physical state that permits their migration from the soil undergoing treatment must be avoided (Castro & Belser, 1990).

The fundamental reasons for agitating bioreactor charges are to enhance process rates and to minimize segregation. The former reason assumes that it is physical transport, rather than biological, processes that first become rate limiting. The flow characteristics of particulate and granular matter in mixed solid phase reaction systems are complex and remarkably little is known concerning the dynamics of granular mass. What is clear is that its properties are neither those of a fluid nor of a solid (Brown & Richards, 1959).

A major problem concerning microbial growth in both moving granular bed and in liquid slurry type bioreactors, where density differences between the liquid and solid phases are significant is the physical effects of solids movement on the growing microbes. As far as on-site soil bioremediation processes are concerned, the development of semi-mechanistic process descriptions remains a matter for the future, largely because it is not yet possible to describe the important physical, biological and molecular processes occurring under unsteady state operating conditions that affect biodegradative activity. Therefore, to describe bioreactor operation, one can best resort to the generalized approach to regime analysis, proposed by van de Vusse (1962).

Stress and Microbes

The quantitative evaluation of microbial activity has been seriously retarded by the erroneous assumption that microbes can be regarded as particles of constant elemental composition and properties. In other words, the fluctuating macromolecular composition of microbial cells, which very largely determine culture dynamics, has largely been disregarded. Further, the enormous intrinsic flexibility of microbes in their response to changes in the availability and speciation of essential nutrients has also been neglected.

The response of a growing bacterium to perturbations due to an external stress is first, to cease the current mode of growth and metabolism; second, to protect the cell against any adverse consequences due to the initial response; third, to develop strategies that can deal with the changing conditions; fourth, to implement such strategies. In such a sequence the specific effects involve both gene expression and enzyme activity and, of course, the timing of the individual stages is strain, physiological state, nature of stress and stress intensity dependent. Even so, most bacteria have evolved mechanisms that allow rapid growth under favourable conditions, but also permit survival under conditions that are unfavourable for growth. The proteins necessary for balanced growth are usually present in requisite quantites, whilst those required for protection, repair or growth under adverse conditions are usually only present in very low concentrations, but are synthesized when needed.

The most extensively studied response to a potentially adverse environmental change is the seemingly universal heat shock response. The heat shock response involves rapid transient synthesis of a specific set of stress proteins. However, various other changes or agents also invoke a stress response. Of particular relevance in the context of bioremediation are stress protein induction by substrate and nutrient starvation, anaerobiosis, hydrogen peroxide challenge, water potential effects, acid (pH) and heavy metal effects. The concept of a generalized stress cycle has been discussed by Hamer and Heitzer (1991).

Microbial growth can occur in environments that exhibit a wide range of water availabilities. For soil environments, water potential is the most appropriate basis for describing water availability. Water potential is both a stress and a selective factor. For growth in a soil matrix unicellular microbes require a continuum of water-filled pores, whereas most fila-mentous fungi have the ability to bridge air-filled pores. This is evidenced by the domination of the latter at increasing negative water potentials and has clear implications for soil treatment systems where one seeks to enhance microbial activity in order to accelerate pollutant biodegradation. Furthermore, the effects of variations in water potential because of either intermittent irrigation or segregation can be expected to affect not only the physiological status of individual strains within microbial consortia, but also intra-consortia interactions.

Methanotrophic Consortia

Some years ago, McCarty (1988) suggested that methanotrophic bacteria would, because of their cometabolic capacities, offer marked potential for the bioremediation of chemically polluted sites. Subsequently, this concept was developed for both *in situ* groundwater remediation (Halden & Chase, 1991; Halden, 1991) and on-site bioreactor remediation of groundwater (Arvin, 1991). However, none of these studies have recognized the important fact that when methanotrophic bacteria are present in unprotected environments, they function exclusively as predominant members of methane-utilizing consortia and that the other members of such consortia markedly extend the performance characteristics of the consortia, both in the presence of methane alone and when other carbonaceous compounds are also present.

When evaluating the potential of various microbes for application in bioremediation, it is important to consider the elemental (biogeochemical) cycle as a whole. As far as methanotrophic consortia are concerned, the rôles of both a consortium with a type I methanotrophic bacterium as the primary substrate utilizer and a consortium with a type II methanotrophic bacterium as the primary substrate utilizer have been examined as far as their rôles in both the carbon and the nitrogen cycles are concerned (Hamer, 1992).

It is interesting to note that various strains of the genera that represent the ancillary strains in methanotrophic consortia are hydrocarbon-utilizers, e.g., *Pseudomonas, Flavobacterium, Acinetobacter, Moraxella, Mycobacterium* and

Nocardia spp. Clearly, if in methanotrophic consortium, one or more of the ancillary strains utilizes a range of pollutant hydrocarbons, possible competition for such hydrocarbons as carbon energy substrates and as cometabolizable carbon compounds that conceivably contribute energy for methanotrophic growth can be envisaged.

Concluding Remarks

Modern biodegradation research has emphasized the elucidation of metabolic pathways, but failed to recognize most of the important aspects of microbial physiology that are also involved. This has resulted in a surfeit of kinetically barren metabolic maps. However, Kacser (1988) has pointed out that the individual chemical transformations which constitute activity in microbial cells are primarily organized in a kinetic manner. Therefore, when considering the biodegradation of an individual pollutant it is necessary to examine the system as a whole rather than examine individual enzymes and the particular reactions they mediate. As all the enzymes comprizing a specific pathway are coupled to one another by the metabolites they share and the rates and fluxes through particular enzymes are significantly affected by adjacent enzymes, the concept of a single rate-determining enzyme in any pathway is invalid.

Knowledge concerning both the dynamic response of cultures and the manifestations of unsteady state growth is meager. Several illuminating investigations on phenotypic variability were published more than 30 years ago, but unfortunately, no results concerning transient state behaviour during changes in growth rate were provided. However, what is most remarkable is that, even now, neither intra- nor extra-cellular culture response data generated under transient conditions are widely available. What is clearly evident is that a better understanding of both biodegreadation kinetics and dynamics would undoubtedly enhance the possibilities for more effective and efficient bioremediation irrespective of either the type of process or mode of operation employed.

References

Arvin, E. *Wat. Res.*, **1991**, *25*, 873-881.

Brown, R. L., Richards, J. C. *Trans. Instn. Chem. Engrs.*, **1959**, *37*, 108-119.

Bungay, H. R., Bungay, M. L. *Adv. Appl. Microbiol.*, **1968**, *10*, 269-291.

Castro, C. E., Belser, N. O. *Environ. Toxicol. Chem.*, **1990**, *9*, 707-714.

Ensley, B. D. *Annu. Rev. Microbiol.*, **1991**, *45*, 283-299.

Halden, K. *Trans. Instn. Chem. Engrs. B*, **1991**, *69*, 173-179.

Halden, K., Chase, H.A. *Trans. Instn. Chem. Engrs. A*, **1991**, *69*, 181-183.

Hamer, G. *Proc. 4th World Congress of Chemical Engineering*, **1992**, in press.

Hamer, G., Heitzer, A. In *Environmental Biotechnology for Waste Treatment*, Sayler, G.S., Fox, R., Blackburn, J. W., Eds; Plenum Press, New York, N.Y., 1991, pp 233-248.

Kacser, H. In *Physiological Models in Microbiology*; Bazin, M. J., Prosser, J. I., Eds.; CRS Press, Boca Raton, Fl. 1988, Vol. 1; pp 2-23.

Linton, J. D., Buckee, J. C. *J. Gen. Microbiol.*, **1977**, *101*, 219-225.

McCarty, P. L. In *Environmental Biotechnology*; Omenn, G.S., Ed.; Plenum Press, New York, N.Y., 1988, pp. 143-162.

Slater, J. H., Bull, A. T. *Phil. Trans. Roy. Soc. Lond. B*, **1982**, *297*, 575-597.

van de Vusse, J. G. *Chem. Engng. Sci.* **1962**, *17*, 507-521.

Wilkinson, T. G., Topiwala, H.H., Hamer, G. *Biotechnol. Bioengng.*, **1974**, *16*, 41-59.

Dynamic phenomena in phenol degradation

Matthias Reuss, Institut fuer Bioverfahrenstechnik, Universitaet Stuttgart, Boeblingerstr. 72, D-7000 Stuttgart 1
Peter Goetz, Institut fuer Biotechnologie, Technische Universitaet Berlin

Disturbances in substrate inlet concentrations pose a major problem in industrial waste water treatment plants. Quite often the dynamic response of the system to these changes in the input conditions are not understood and mathematical models for quantitative analysis of these phenomena are not available. This lack of knowledge is a severe hindrance to a proper analysis of the stability of coexistences in mixed cultures as well as to design the neccessary and sufficient control strategies. It is the objective of this contribution to elaborate and critically discuss different approaches of structured models for the quantitative description of the dynamic behaviour of *Pseudomonas putida* growing on carbon and energy source phenol and serving as a model system for the above mentioned problems.

Mathematical modelling of microbial populations have evolved into a valuable tool for bioreactor design, predictions of operating conditions and analysis of process control. As far as the application of modelling to biodegradation processes is concerned the state of the art, however, lags behind its technical demand. With a few exceptions it seems that most studies of biodegradation processes in continuous cultures are based on models which have been formulated from steady state kinetics and have not been very successful in describing dynamic behaviour. The quantitative description of the systems behaviour under unsteady state dynamic conditions is of particular importance because such systems are usually subject to fast variations in the inlet concentrations (shock loads). In contrast to the practical importance relatively little work has been done to throw light on the complex dynamic responses to such disturbances as far as experimentally proofed models are concerned. It should be stressed that predictions of stability of coexistence in mixed cultures which is a very important task in the reaction engineering analysis of biodegradation processes are also of limited value if dynamic phenomena are not properly taken into account.
The contribution aimes at presenting and discussing some experimental observations of the dynamic response in continuous cultures of *Pseudomonas putida* growing on phenol as the energy and carbon source. The system is dynamically disturbed with the aid of a shift in the substrate inlet concentration. Attempts of modelling the observed phenomena are presented and model discrimination is performed to evaluate quantitatively the different postulates.

Modelling the Steady state Kinetics

From extensive experimental investigations in batch, fed-batch and single as as well as two-stage continuous fermentations with substrate addition into the second stage, it was possible to verify a Haldane type of substrate inhibition kinetics for growth rate in the form:

$$\mu = \mu_{max} \frac{c_S}{K_s + c_S + c_S^2/K_I}$$

The comparison of measured data and computed growth rates are illustrated in Fig. 1. From the experiences during these basic studies we are able to strongly recommend application of the two stage system for studying the steady state kinetics in the range of inhibitoric substrate concentrations. If the growth kinetics exhibits a maximum as those of Fig. 1 the biomass, substrate concentrations and specific growth rates can be iteratively predicted from the balance equations for the two stage system, in which the population growths under substrate limitation in the first stage and with the aid of substrate addition in the region of inhibition in the second stage (Fig. 2).
For illustrative purpose the region of the three steady states is represented as shown in the phase plane of Fig. 3. It must be emphasized that this phase plane is quite similar to the one predicted for a system with immobilized cells for the same kinetics. This may serve as an additional example that a two-stage system has some more interesting properties for studying the kinetic behaviour of complex systems with inhibitory kinetics in a better defined environment which is a prerequisite for more reliable models.

1054–7487/92/0439$06.00/0

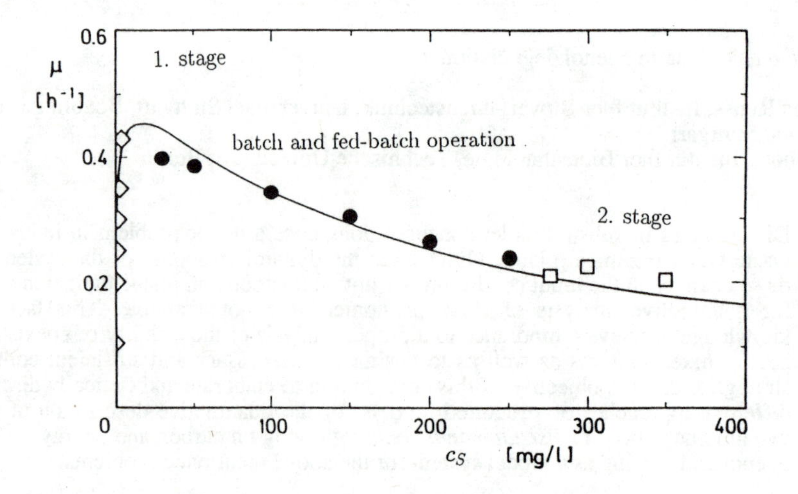

Fig. 1 Substrate inhibition kinetics from steady state measurements (*Pseudomonas putida* growing on phenol).

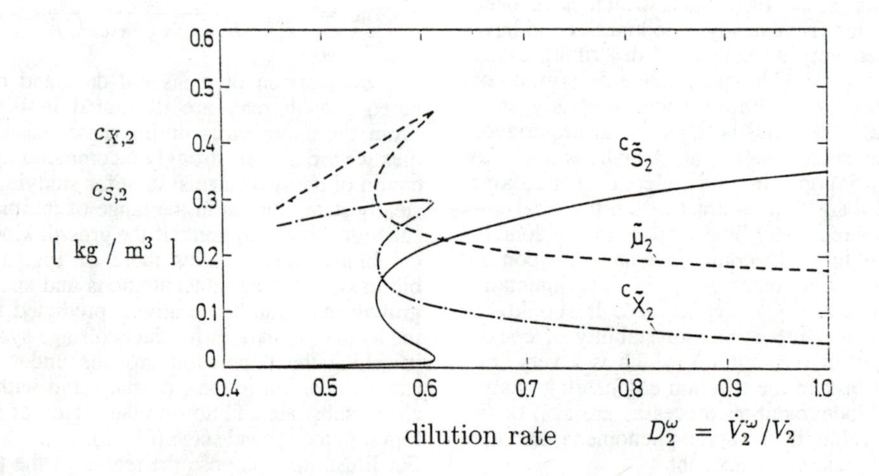

Fig. 2 Steady state conditions for substrate inhibition kinetics in a second stage of a two stage system with feeding additional substrate into the second vessel.

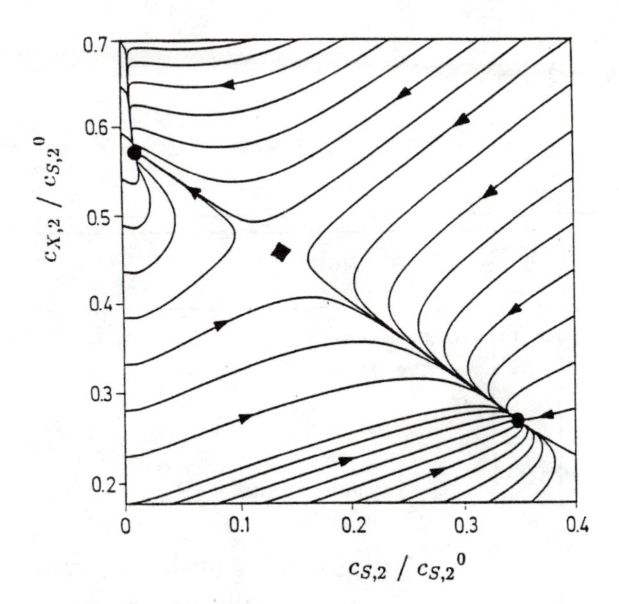

Fig. 3 Phase plane for substrate inhibition in the second stage in the region of the three steady states illustrated in Fig. 2.

Transients in Continuous Cultures

In order to investigate the dynamic behaviour of the continuous culture, inlet substrate shifts were performed. Starting from steady state, this concentration was changed from 0.2 g/l to 3 g/l for one hour and then set back to 0.2 g/l. The most important results from this experiment are summarized in Fig. 4. In contrast to the integration of the balance equations making use of the steady state kinetics (solid lines) the predicted washout could not be observed during the experiment. A more impressive presentation of the experimental observations is illustrated in Fig. 6 in which the estimated growth rates are plotted versus measured substrate concentrations. Tanner (1985) has collected extensive data from literature demonstrating that such hysteresis effects are quite common during transient periods. For quantitative description Tanner suggested some empirical correlations in which either saturation constant K_S or yield coefficient $Y_{X/S}$ varied with biomass and substrate concentrations respectively. Such empirical approaches are, however, always questionable because the fundamental intracellular dynamics are masked through the unstructured formulation of the model. The necessity for introducing a structured approach can be most impressive illustrated with the aid of the phase plane in Fig. 5, showing the biomass-substrate trajectory during the dynamic experiment. The intersection of the trajectory is a proof that the dynamic situation manifested in the two dimensional phase plane is a projection from a higher dimensional state.

The proposed modelling is based on the assumption that the observed dynamic phenomena are brought about by the ribosome synthesis. The simple interpretation of this postulate is that the increase of growth rate as a result of increasing substrate concentrations demands a corresponding change in the rate of protein biosynthesis and consequently in the amount of ribosomes. This idea has been suggested by Jerusalimski (1967) and various modifications have been summarized in the book of Roels (1983). Three different approaches are suggested in the present contribution to take into account the retardation in the change of protein biosynthesis due to the time window necessary for the step up of ribosome synthesis: (1) A first order delay, (2) an additional discrete time delay and (3) a distributed delay with the aid of polymerization kinetics for ribosome synthesis.

The first approach includes altogether 16 different variations. This enormous number results from the various assumptions regarding

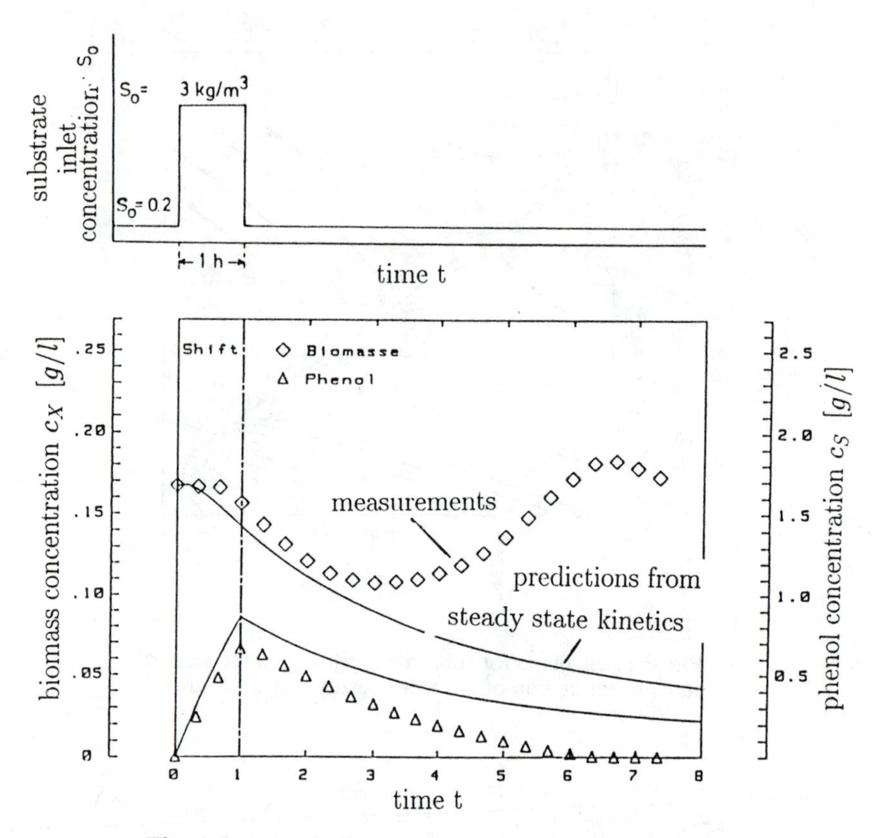

Fig. 4 Comparison between measurements and predictions from steady state kinetics during response of the system to a shift in the substrate inlet concentration.

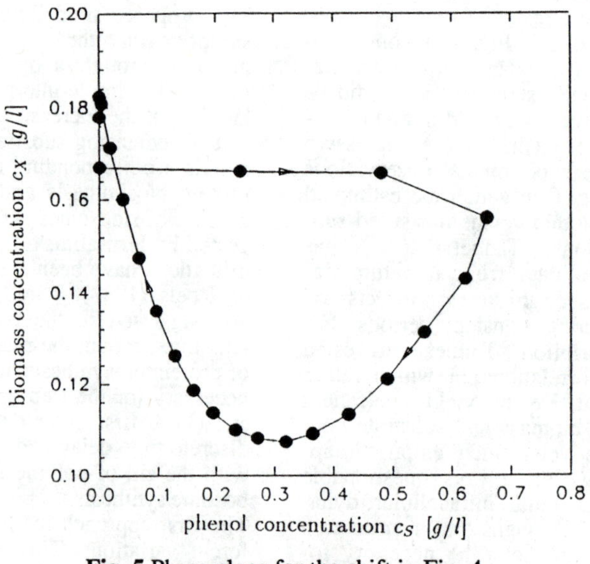

Fig. 5 Phase plane for the shift in Fig. 4.

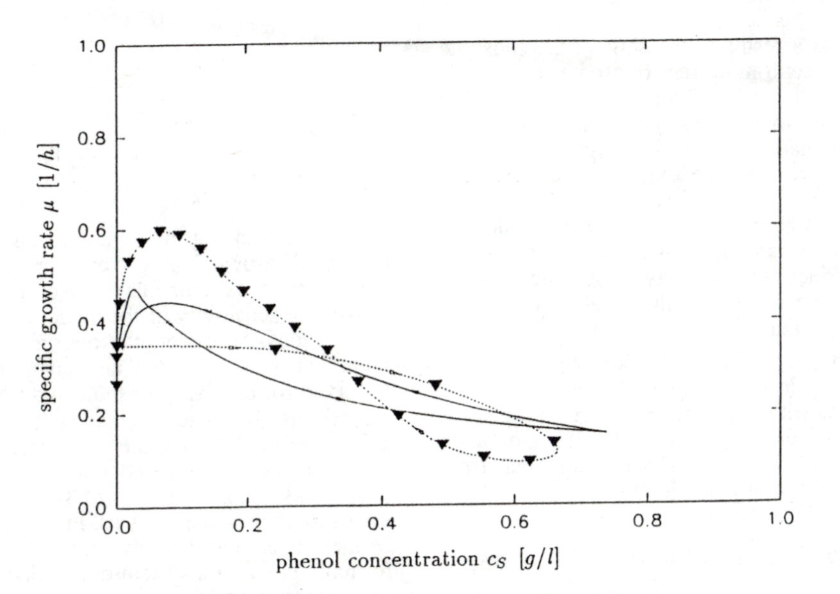

Fig. 6 Hysteresis in the growth rate. Comparison between estimates from measured biomass concentrations and predictions from the structured model including polymerization kinetics for ribosome synthesis.

the formulation of the kinetic equation for the ribosome synthesis. If growth rate is expressed by:

$$\mu = k\, R \frac{c_S}{K_M + c_S + c_S{}^2/K_N}$$

with R = actual concentration of ribosomes and the intracellular material balance equation for the ribosomes given by:

$$\frac{dR}{dt} = r_R - \mu R$$

then synthesis rate r_R may be modeled in the following way:

$$r_R = r_R{}^* + \frac{1}{\tau}(R^* - R)$$

with R^* ribosome concentration at steady state which in turn is a function of the steady state growth rate μ^*. Next we have 4 different possibilities for writing the steady state rate equation:

$r_R{}^* = \mu^* R^*$ or μR^* or $\mu^* R$ or μR

where the symbol (*) denotes the steady state conditions. Obviously these different model approaches reflect the problem of perception of the microorganisms. Furthermore we may consider another 4 possibilities for stopping further synthesis if the set point R^* is approached. The sixteen model variants have been exposed to a process of model discrimination in which the following important contraints in addition to the least square optimization have been considered: (1) The same model with the same set of parameter should be able to fit the steady state and dynamic behaviour, (2) model predictions of the key variable R should result in a reasonable fit with the measured RNA-concentrations and (3) evidence concerning the realism of parameters. By equally considering the different contraints the first model approach must be rejected.

Based upon the available informations regarding the synthesis of ribosomes it was next assumed that an additional time lag may play an important role. The time lag is interpreted as a maturation time Θ, which is the time span between starting additional rRNA synthesis and finishing the first ribosome. As a result of the analysis the material balance for the intracellular ribosome concentration is given by:

$$\frac{dR}{dt} = \left[\tilde{\mu}\, \tilde{R}\, e^{\tilde{\mu}\theta} + \frac{1}{\tau}(\tilde{R}^* - \tilde{R})\right] e^{-\int_{t-\theta}^{t} \mu(t)dt} - \mu R$$

The symbol (~) denotes the system state at time (t-Θ). The comparison of measured and computed data of biomass and substrate concentration shows a reasonable agreement. However, as a result of the shift in the substrate inlet concentration the ribosome content R displays stable oscillations, a phenomenon which is known to exist in systems of differential equations with time lags but which is not supported by the experimental observations.

Finally an attempt is untertaken to model the dynamic behaviour with the aid of polymerization kinetics which may be considered as an appropriate model for the creation of ribosomes from their various subunits (monomers).

Assuming that the final ribosome of concentration P_n in the product of n polymerization reactions with monomer P_1 the resulting set of material balance equations for the monomer, the semifinished products and the final ribosome concentrations is given by:

$$dP_1/dt = r_I - k_P P_1 \sum_{i=1}^{n-1} \frac{P_i}{i} - \mu P_1$$

$$dP_2/dt = k_P P_1 (\frac{2}{1} P_1 - P_2) - \mu P_2$$

$$dP_i/dt = k_P P_1 (\frac{i}{i-1} P_{i-1} - P_i) - \mu P_i$$

$$dP_n/dt = k_P P_1 (\frac{n}{n-1} P_{n-1} - P_n) - \mu P_n$$

The growth rate is afterwards calculated from:

$$\mu(P_n) = (\mu_{max}/A) P_n \frac{c_S}{K_M + c_S + c_S^2/K_N}$$

It must be emphasized that due to the model structure the final concentration P_n which determines the actual growth rate differs from the measured RNA-concentration which is given by:

$$RNA = \sum_{i=1}^{n} P_i^*$$

The solution of the entire system of material balance equations including:

$$dc_X/dt = (\mu(P_n) - D) c_X$$

and

$$dc_S/dt = D (c_S^0 - c_S) - \mu(P_n) c_X Y_{XS} - m_S c_X$$

is illustrated in Fig. 6 and compared with the estimated growth rates from the measured data. Further experiments are required to improve the sensitivity of the model to dynamic stimulations. However, the reasonable agreement between measured and computed data observed for biomass, substrate and RNA-concentrations gives rise to the conjective that the suggested model structure is principally suitable for description of complex dynamic behaviour as a consequence of shock loads in inlet substrate concentrations. Finally it should be mentioned that the polymerization model is formally equal to a system of coupled first order differential equations leading to a distribution of time lags. This is very similar to an approach which has been suggested by Wang et al. (1984) in a quite different model structure resulting in a so called 'impulse transfer function' which consists of a sum of exponential distributions.

References

Jerusamlimsky, N.D. cited in: Romanowsky, J.M.; Stepanova, M.V.; Chernavsky, D.S. *Kinetische Modelle in der Biophysik* VEB Gustav Fischer Verlag, Jena **1974.**

Roels, J.A. *Energetics and kinetics in biotechnology.* Elsevier Biomedical Press, Amsterdam **1983.**

Tanner, R.D. In *Proc. VII. International Biotechnology Symposium, New Dehli,* **1984,** pp 103-117.

Wang, N.M.; Stephanopoulos, G. *Biotechnol. Bioeng. Symp.* **1984,** *14,* 635-656.

Strain Selection for High Temperature Oxidation of Mineral Sulfides in Reactors

Paul R. Norris* and Jonathan P. Owen,
Department of Biological Sciences, University of Warwick, Coventry CV4 7AL, U.K.
(FAX 0203 523701)

There are few examples of commercial use of mineral-oxidizing bacteria in bioreactors for extraction of metals from mineral sulfides. Slow reaction rates and incomplete mineral dissolution can undermine the feasibility and competitiveness of biohydrometallurgical processes. Therefore, it is worthwhile establishing which bacteria and conditions promote the maximum rate and efficiency of metal extraction.

There is a wide variety of acidophilic mineral-oxidizing bacteria (Norris, 1990). The mesophile *Thiobacillus ferrooxidans* has been the most extensively studied and it is being utilized in the first commercial bioreactors processing mineral sulfides. However, there is much laboratory evidence that maximum rates and yields of metal extraction can be enhanced at elevated temperatures (Brierley and Brierley, 1986; Norris, 1990) and a pilot plant has successfully utilized moderately thermophilic bacteria (Spencer et al., 1989). The data presented here suggest that the maximum potential of the extreme thermophiles remains to be demonstrated and that this will require characterization of novel isolates.

The relative efficiency of mineral sulfide-oxidizing acidophiles in metal extraction can be influenced by the temperature at which they can grow; their characteristics of iron oxidation (affinities for ferrous iron and reaction to potential end-product inhibition by ferric iron of the oxidation); their sensitivity to toxic metals released from minerals; and their robustness in relation to agitation with mineral particles in reactors.

Effect of Temperature on Mineral Dissolution by Extreme Thermophiles

The oxidation of pyrite, nickeliferous pyrrhotite, chalcopyrite and arsenopyrite at about 70°C by *Sulfolobus* strain BC has been described (Norris and Parrot, 1986; Le Roux and Wakerley, 1987; Lindström and Gunneriusson, 1990). A slightly higher optimum temperature has been found for *Metallosphaera sedula* (Huber et al., 1989). Pyrite oxidation by *Acidianus brierleyi* in air-lift reactors has also been described (Larsson et al., 1990). Mineral oxidation by these three strains and a culture capable of growth at higher temperatures has been compared (Fig. 1). An advantage of utilizing the most extreme thermophiles exists where the yield of target metal can be approximately proportional to the process temperature, as in

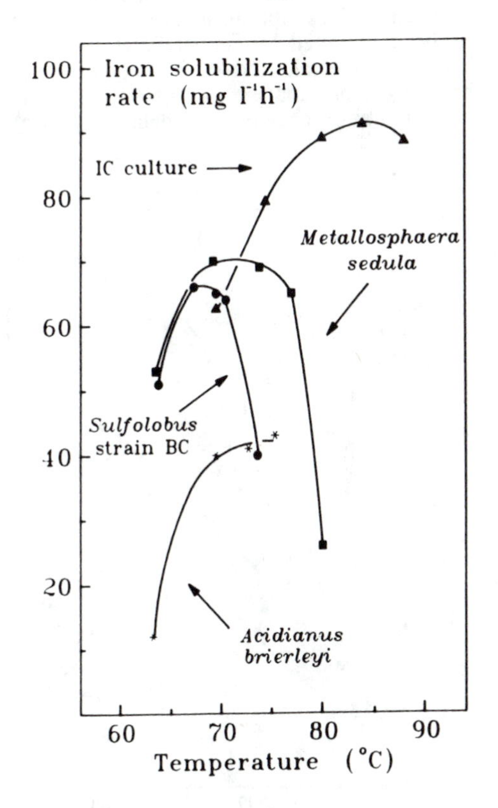

Fig. 1. The effect of temperature on pyrite solubilization during autotrophic growth of extreme thermophiles. The mineral (1% w/v, -75 μm diameter particles containing 40% w/w iron) was in a simple mineral salts medium (initially pH 2) in air-lift reactors gassed with 1% (v/v) CO_2 in air. The IC culture, an 80°C pyrite-enrichment culture of *Sulfolobus*-like bacteria from a geothermal site has not been confirmed as a pure culture.

1054–7487/92/0445$06.00/0

the solubilization of copper from chalcopyrite by bacteria (Norris, 1990 and 1992).

The maximum specific rates of iron oxidation (a key reaction in mineral sulfide dissolution) and affinities for iron differ among the thermophiles (Norris, 1992). However, these differences appear less likely to influence the mineral dissolution rate than the temperature or inhibition by metal ions which might be required in high concentration in solution to facilitate the metal recovery.

Inhibition by Copper

Sulfolobus strain BC has been grown on chalcopyrite with copper in solution at about 30 g/l (Norris and Parrott, 1986; Le Roux and Wakerley, 1987), although the extent to which this concentration of copper inhibited the bacteria was not made clear. *M. sedula* has been inhibited by copper between 1 and 5 g/l (Huber et al., 1989). A comparison of the effect of copper on growth of the thermophiles on chalcopyrite (data not shown) confirmed the relative sensitivity of *M. sedula*. It was also found difficult to adapt *A. brierleyi* or the high temperature IC culture (see Fig. 1 legend) to grow when the copper concentration was adjusted to about 10 g/l by adding copper sulfate. In contrast, chalcopyrite-grown *Sulfolobus* strain BC required little or no adaptation for growth with copper at 20 g/l. Repeated serial culture of this strain with 25 g copper/l provided an adapted culture. The pattern of chalcopyrite dissolution was complex with accumulation of ferrous iron in solution preceding copper release (Fig. 2). The separation of iron and copper release was not marked in the unadapted culture (Fig. 2A) where initial iron release was curtailed by a transient, inhibitory rise in the pH (to pH 2.3) before sulfide/sulfur oxidation generated a more favourable pH for growth. In other experiments (data not shown), the lag before extensive iron release was greatly reduced by pH control which prevented an initial reduction in acidity. The initial pH rise was for some reason only relatively slight in the

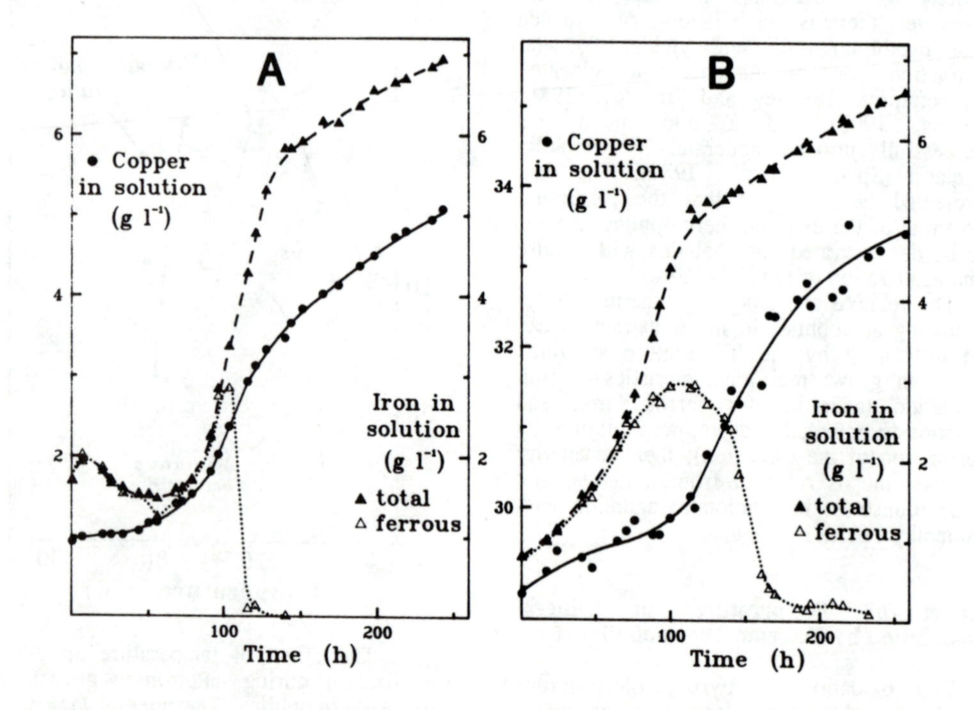

Fig. 2. The solubilization of copper from a chalcopyrite concentrate during autotrophic growth of *Sulfolobus* strain BC. Cultures at 68°C were gassed with 1% (v/v) CO_2 in air in air-lift reactors. The inocula were from cultures which previously were not adapted (A) or were adapted (B) to grow in the presence of 25 g copper/l. The mineral (5% w/v, 75-250 μm diameter particles containing 20% w/w copper) was in a simple mineral salts medium (initially pH 1.5 but 1.8 following mineral addition). In the adapted culture (B), the medium was supplemented with copper sulfate before inoculation.

copper-adapted culture (Fig. 2B) with the lag phase in this case primarily a result of inhibition by copper indicating that adaptation was not complete. The maximum rate of copper solubilization by the adapted culture with 30 g copper/l (Fig. 2B) was just over 60% of that by the un-adapted culture at low copper levels (Fig. 2A). After further adaptation through serial cultures with higher copper concentrations, rapid mineral dissolution by *Sulfolobus* strain BC occurred in the presence of over 40 g copper/l, at which concentration growth of the un-adapted culture was not possible.

Inhibition at high concentrations of mineral

Bacterial activity is restricted at high mineral concentrations in reactors by simple physical attrition of cells and possibly by mass transfer limitations (O_2 or CO_2), or by accumulation of an inhibitor released from the mineral or produced in solution. Whatever the mechanism, the tolerances of cultures can differ. A pyrite-enrichment culture from a geothermal site (a mixed culture dominated by an un-characterized *Sulfolobus*-like strain referred to as strain N) was relatively un-affected by increasing the pyrite concentration to levels which extended the lag phase (Fig. 3) and finally prevented the growth of *Sulfolobus* strain BC at 15% (w/v) mineral (data not shown).

Summary

A comparison of activity of the extreme thermophiles in relation to factors that affect their capacity for growth-associated dissolution of pyrite and chalcopyrite (Table 1) shows that an ideal strain would at least possess a combination of the best features of the bacteria so far described. It seems likely that sampling of more sites followed by varied culture-enrichment conditions would reveal a greater variety of, and more-useful, strains. Selective enrichment conditions could be designed and cultures adapted with regard to potential applications and target minerals. For example, arsenate/arsenite-resistant cultures could be required for processing of auriferous arsenopyrites if a high temperature process were to be seen as advantageous in comparison with the low temperature bacterial processes currently in operation.

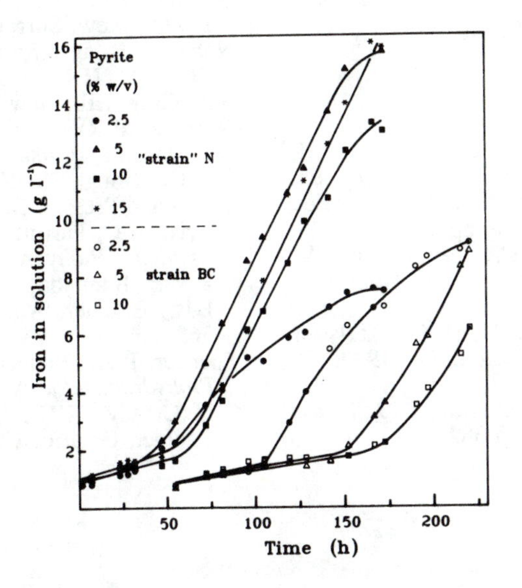

Fig. 3. The effect of mineral concentration on pyrite solubilization at 68°C during autotrophic growth of *Sulfolobus* strain BC and strain N (see text). Cultures were grown under 1% (v/v) CO_2 in air in shaken flasks (orbital shaker, 150 rpm) .

Table 1. The relative mineral-oxidizing capacities of extreme thermophiles.

	Sulfolobus strain BC	*Acidianus brierleyi*	*M. sedula*	High Temperature IC Culture	Strain N
Yield (FeS_2)[a]	+++[d]	+	+	?[e]	+++
Yield ($CuFeS_2$)[b]	+	+	++	+++	+
Copper tolerance	+++	++	+	++	?
Solids tolerance[c]	++	++	++	?	+++

[a] The yield (iron in solution) from pyrite here is primarily a measure of bacterial tolerance of the increasing acidity during the mineral dissolution in batch culture
[b] The yield (copper in solution) from chalcopyrite is primarily a measure of the upper temperature limit for growth (see text)
[c] Observed as the relative capacities to withstand agitation in air-lift reactors with increasing concentrations of pyrite
[d] The culture performing the best in relation to a given feature is scored +++, while ++ and + indicate decreasing capacities
[e] ? indicates not yet tested

Ackowledgements. Parts of this work were supported by the SERC Biotechnology Directorate, CRA Advanced Technical Development, Australia, and the EC Raw Materials Programme.

References

Brierley, J.A; Brierley, C.L. In *Thermophiles: General, Molecular, and Applied Microbiology*; Brock, T.D., Ed.; Wiley, New York, **1986**, pp 279-305.

Huber, G.; Spinnler, C.; Gambacorta, A.; Stetter, K.O. *System. Appl. Microbiol.* **1989**, *12*, 38-47.

Larsson, L.; Olsson, G.; Holst, O.; Karlsson, H.T. *Appl. Environ. Microbiol.* **1990**, *56*, 697-701.

Lindström, E.B.; Gunneriusson, L. *J. Indust. Microbiol.* **1990**, *5*, 375-382

Le Roux, N.W.; Wakerley, D.S. In *Biohydrometallurgy, Proceedings of International Symposium*; Norris, P.R.; Kelly, D.P., Eds.; Science and Technology Letters, Kew, Surrey, **1988**, pp 305-317.

Norris, P.R. In *Microbial Mineral Recovery*; Ehrlich, H.L.; Brierley, C.L., Eds.; McGraw-Hill, New York, **1990**, pp 3-27.

Norris, P.R. In *The Archaebacteria: Biochemistry and Biotechnology*; Lunt, G.G.; Hough, D.W.; Danson, M.J., Eds.; Portland Press, Colchester, **1992**, in press.

Norris, P.R.; Parrott, L. In *Fundamental and Applied Biohydrometallurgy*; Lawrence, R.W.; Brannion, R.M.R.; Ebner, H.G., Eds.; Elsevier, Amsterdam, **1986**, pp 355-365.

Spencer, P.A.; Budden, J.R.; Sneyd, R. In *Biohydrometallurgy 1989*; Salley, J.; McCready, R.G.L.; Wichlacz, P., Eds.; Canmet, Ontario, **1989**, pp 231-242.

An Integrated Approach to Biohydrometallurgical Metal Extraction, Recovery, and Waste Treatment

P.B. Marchant*, R.W. Lawrence, D. Warkentin, P. Elson, M. Rowley, The Triton Group, 538 - 999 Canada Place, World Trade Centre, Vancouver B.C. Canada (FAX 604 688 3639)

This paper provides a brief description of biologically assisted metal extraction and CN decomposition processes (which have been reported in detail previously) and a more detailed description of the Bio-sulphide process for the treatment of sulphate and metal laden waste streams. Finally, an integrated process flowsheet is described that incorporates several biological process steps incorporating both metals extraction and waste treatment.

METALS EXTRACTION

Base Metals

Fig. 1 shows a conceptual biooxidation refining circuit within an integrated process flowsheet. Fig. 1 shows acidic biooxidation of sulphides to solubilize base metals, for example copper, from a complex sulphide concentrate. The primary bioleach product enters a dewatering/wash step where the sulphate solution is separated from the residue. The washed residue is either a final concentrate (for example, after removal of the zinc from a copper-zinc concentrate) or is repulped for further processing, as shown in Fig. 1. The solution is then refined by conventional hydrometallurgical processes and the resultant barren sulphate solution can either be recycled as required, and/or bled to a biological treatment step. Waste heat from the biooxidation step can be employed to heat solutions during refining for diluent removal and solution evaporative concentration.

Precious Metals

Most sulphides can be oxidized by bacteria (Lawrence, 1974; Pinches, 1972) and consequently highly complex sulphide concentrates containing refractory precious metals can be pretreated using biooxidation prior to conventional cyanidation for precious metals recovery. Fig. 1 shows a secondary biooxidation step to pretreat refractory sulphides prior to cyanidation. There has been considerable attention focused on bioleaching of refractory gold concentrates (Lawrence and Bruynesteyn, 1983; Marchant, 1986; Marchant and Lawrence, 1986) and therefore the details of the process are not described herein.

BIO-SULPHIDE PROCESS

Sulphate wastes, such as those produced during sulphide biooxidation processes shown in Fig. 1, can present a serious environmental problem and must be neutralized for pH control and metals removal. The Bio-sulphide process was developed to treat complex sulphate solutions and acid mine wastes typically encountered during sulphide mineral processing. In the Bio-sulphide process, the chemical process of metal removal and the biological process of sulphate reduction and pH control are separated. The result is a more robust and flexible process to treat highly complex sulphate waste solutions.

Briefly, the Bio-sulphide process consists of the following stages:

- Metal sulphide precipitation by treatment with H_2S,

1054–7487/92/0449$06.00/0

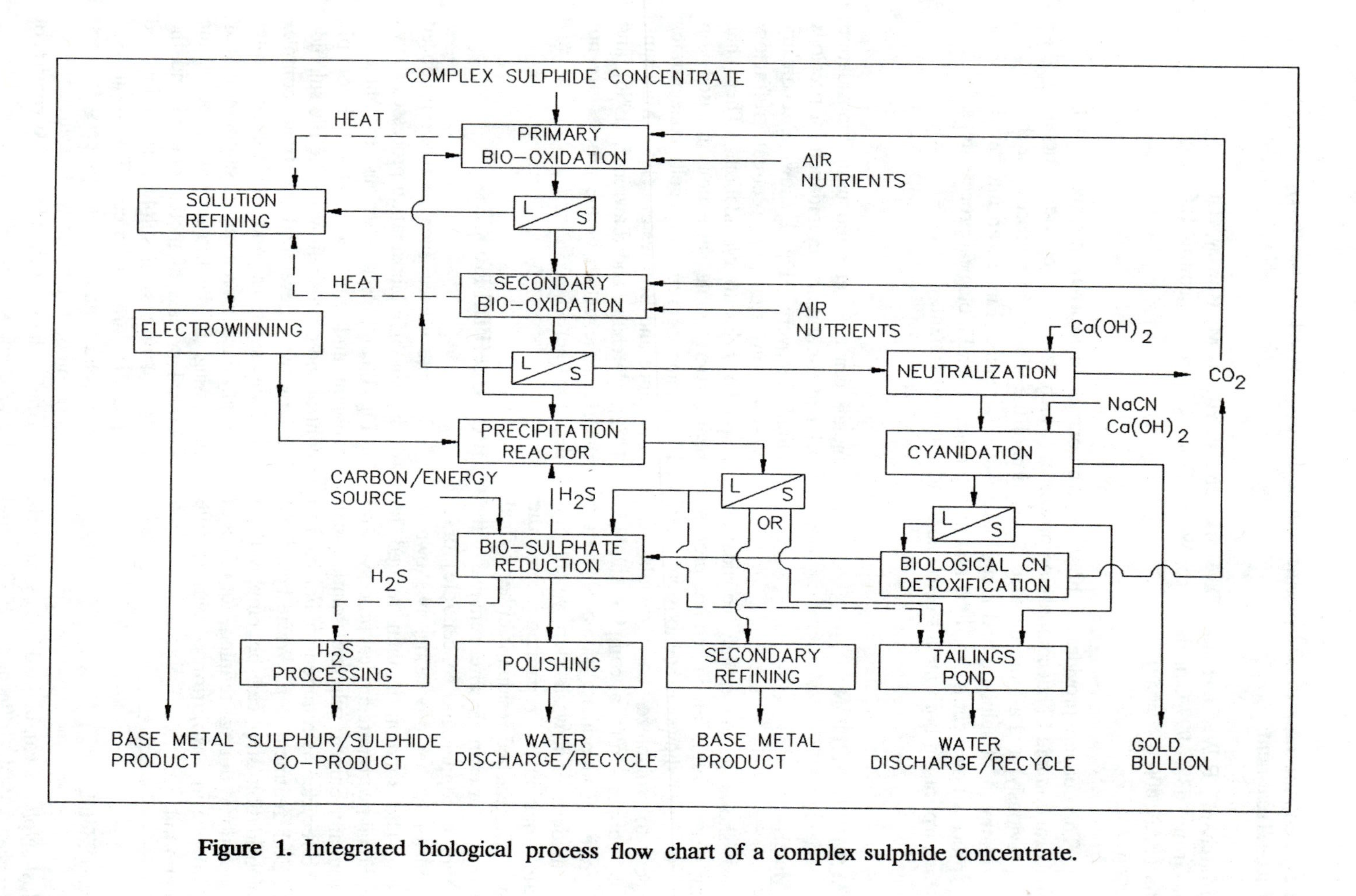

Figure 1. Integrated biological process flow chart of a complex sulphide concentrate.

- Anaerobic sulphate reduction and H_2S generation,
- H_2S stripping and concentration
- Aerobic polishing, and
- Excess H_2S processing to saleable co-products.

In the first stage, the H_2S stripped from the anaerobic reactor is used to precipitate metals from the waste stream as metal sulphides while the pH is adjusted. In the second stage, sulphate is reduced to H_2S by anaerobic bacterial action when an energy/carbon source is added. In the third stage, H_2S is stripped from the reactor vessel, concentrated, and transported to the precipitation stage. In the fourth stage, residual organic compounds are biodegraded aerobically. In the fifth stage, excess H_2S is converted to elemental sulphur or other saleable co-products.

The biological generation of H_2S is accomplished in a separate vessel and the resultant H_2S transferred to the precipitation vessel where the metals are removed. The separation of the chemical and biological stages allows maximum efficiency of both reactions to be achieved. The biological reactor can be operated at optimum conditions, including pH levels and retention time. The chemical precipitation reactor is not limited by the retention time of the biological reactor since, in most cases, only a fraction of the waste stream reports to the biological reactor. Also, the precipitated metal sulphides are much easier to remove when not mixed with the biomass. This process also has a great deal of flexibility, and for many applications not all of the process stages are required. In some cases, the aerobic or the sulphur production stages may not be required. This flexibility increases the scope of potential applications of the Bio-sulphide process.

Metal Sulphide Precipitation

In this stage, H_2S produced in the anaerobic stage is sparged through the feed waste solution. The H_2S and the metal cations in the solution form insoluble metal sulphide compounds which precipitate. Discharge solution from the anaerobic stage can be added to this stage to adjust the pH and enhance the precipitation rate of metal sulphides by increasing the S^{2-} solubility and decreasing the solubility of the resulting precipitates.

Anaerobic Stage

In the anaerobic stage sewage sludge, molasses, CO_2 and H_2, or producer gas can be used as the carbon/energy source by sulphate reducing bacteria to produce H_2S, carbonate and acetic acid. The production of alkalinity in this stage makes the direct neutralization of acidic water possible.

Stripping Stage

The H_2S formed in solution during the anaerobic stage is stripped with an inert carrier gas such as N_2. The mixture is then transported to the precipitation stage and any excess H_2S can be utilized in the sulphur production stage, if necessary, or recycled as required.

BIOLOGICAL CYANIDE DETOXIFICATION

Biological cyanide detoxification processes take advantage of naturally occurring biochemical mechanisms to convert and/or metabolize cyanide species to inert compounds and fix heavy metals derived from the decomposition of soluble complex cyanide species. The rate of metabolism will depend on a number of variables such as substrate, mass transfer, solution chemistry, temperature, nutrients, pH, and the SCN/CN ratio. The relative toxicity of metals in solution varies greatly depending on the biomass and the operating conditions. Generally, however, metallo-cyanide complexes are biodegradable. Thiocyanates are degraded rapidly by conversion to CO_2, ammonia, and sulphate.

Processing Alternatives

For solutions, the most common processing equipment anticipated on a large scale include fixed film and rotating biological contactors, aerated filters, and continuous stirred tank reactors (CSTR). For slurries, to improve substrate availability and degradation kinetics, equipment is limited to splash towers and CSTR's.

In the integrated process shown in Fig. 1, CN and SCN decomposition is achieved in an aerobic biological step. The effluent, which may contain metals, alkalinity, sulphate, ammonia and organics, can be added to the anaerobic reactor for the Bio-sulphide process.

Comparison with Alternatives

The biological processing alternative is technically competitive with the more traditional alternatives. Note that it does not provide any benefit of cyanide recovery. However, an important long term benefit of biological processing is the complete removal of thiocyanate species.

Limitations

Biological cyanide degradation is presently limited to relatively dilute solutions with a simple solution chemistry and relatively high SCN/CN ratios. Improved toxicity tolerance is a function of biomass acclimation and can be accelerated by standard microbiological methods.

INTEGRATED PROCESSING

Fig. 1 shows the integration of the unit processes. The synergistic process proposed incorporates base metal bioleach extraction from a complex sulphide concentrate followed by secondary biooxidation to enhance precious metals recovery by cyanidation. The resulting sulphate waste solutions are treated in a Bio-sulphide system to produce recyclable metal sulphides, or stable metal sulphides for disposal, process water and water for discharge to the environment, while cyanidation effluents are treated aerobically.

Process synergism is demonstrated in the following areas:

- combined treatment of sulphate solutions produced during biooxidation and wastes from base metal solution refining in a common chemical/biological process producing H_2S and a stable metal sulphide solids product,
- CO_2 generated during aerobic processing and secondary biooxidation residue neutralization can be used as a carbon source for both biooxidation stages,
- carbonate rich, biological treatment effluent can be recycled for pH control during metal precipitation and/or neutralization of bioleach residue prior to cyanidation,
- the Bio-sulphide process can treat effluent from biodegradation of complex metallo-cyanide solutions to remove soluble metals, organics and sulphates. Use of this effluent greatly enhances sulphate reduction efficiency and allows cyanide process water to be recycle for processing or neutralization.
- Use of cooling water from bioleach processing to heat base metal solutions for refining purposes and evaporative concentration of both base metal refining solutions and carbonate recycle water.
- H_2S from the bio-sulphide process can be used for selective metals removal from internal streams in bioleaching or refining.

It is apparent that biological processing has a wide range of application in sulphide processing. It has been shown here that biological processes can be integrated to assist in the extraction and refining of both base and precious metals from sulphide minerals and concurrently produce saleable co-products, (eg. metal sulphides), reusable or discharge quality water, and terminal destruction or removal of potential environmental contaminants.

References

Lawrence, R.W. (1974). Bacterial extraction of metals from sulphide concentrates. Ph.D. Thesis, University College, Cardiff, Wales, December.

Lawrence, R.W., and A. Bruynesteyn (1983). Biological preoxidation to enhance gold and silver recovery from refractory pyritic ores and concentrates. CIM Bulletin, 76 (857), 107-110, September.

Marchant, P.B. (1986). Continuous biological tank leaching of a refractory arsenical sulphide concentrate to enhance gold extraction by cyanidation. M.A.Sc. Thesis, University of British Columbia, B.C., May.

Marchant, P.B., and R.W. Lawrence (1986). Flowsheet design, process control and operating strategies in the biooxidation of refractory gold ores. 25th Conf. of Metall., CIMM, Toronto, August.

Pinches, A. (1972). The use of microorganisms for the recovery of metals from mineral materials. Ph.D. Thesis, University College, Cardiff, Wales, December.

Complex Utilization of Polymetallic Sulphide Ores by Means of Combined Bacterial and Chemical Leaching

Stoyan N. Groudev*, University of Mining and Geology, Sofia 1156, Bulgaria
(FAX: (359) 2- 621- 042)

Veneta I. Groudeva, University of Sofia, Sofia 1421, Bulgaria

Pyritic ores containing zinc, lead and copper as the relevant primary sulphide minerals, as well as gold and silver finely disseminated in the sulphide matrix, were leached in percolation columns. Firstly, zinc and copper were leached by means of chemolithotrophic bacteria in acidic sulphate-bearing medium. The ores were then leached by means of $FeCl_3$ - HCl solutions both in the presence and in the absence of bacteria to solubilize the lead. Finally, the ores were leached by thiourea or amino acids in the presence of oxidants to solubilize the gold and silver.

Low-grade lead-zinc ores, mixed oxide-sulphide lead-zinc ores as well as pyritic lead-zinc-copper ores with fine intergrowth of the separate sulphides are refractory to flotation. It has been suggested that percolation leaching could be potentially useful for processing such ores (Dutrizac and MacDonald, 1977; Dutrizac, 1979; Roy Chaudhury and Das, 1987). The leaching by chemolithotrophic bacteria or by ferric ions in sulphate-bearing media results in solubilization of zinc and copper, whereas the lead remains in the leached ore as the insoluble lead sulphate. The precious metals which may be presented in the ores remain in the leached ore together with the lead. Chemical leaching by ferric chloride could be more promising because, apart from the zinc and copper, lead is also dissolved as lead chloride. The precious metals could be then efficiently leached by different methods because they are liberated from the sulphide matrix as a result of the oxidative pretreatment of the ore.

In this study two pyritic ores containing zinc, lead and copper as the relevant primary sulphide minerals, as well as gold and silver finely disseminated in the sulphide matrix, were firstly leached in percolation columns by means of chemolithotrophic bacteria in acidic sulphate-bearing medium to solubilize zinc and copper. The ores were then leached by $FeCl_3$ solution in the presence or in the absence of bacteria to solubilize lead. Finally, the ores were leached by thiourea or amino acids in the presence of oxidants to solubilize the gold and silver.

Materials and Methods

The chemical analysis of the ores used in this study is shown in Table 1.

The sulphides sphalerite, galena and chalcopyrite were the main minerals containing zinc, lead and copper, respectively. They were presented as disseminations in the pyrite and as interstital fillings between pyrite grains. Silver and gold occurred as submicroscopic disseminations in the sulphide minerals. Most gold was associated with pyrite, and the galena was the principal silver carrier in the ores. Quartz and feldspars were the main minerals of the host rocks.

The ores were leached in PVC columns with 2200 mm effective length and 105 mm internal diameter. Each column was charged with 30 kg

1054–7487/92/0454$06.00/0

Table 1. **Data about the chemical analysis of the ores used in this study.**

Element	Content	
	Ore No 1	Ore No 2
Zinc	1.34 %	2.82 %
Lead	1.04 %	1.91 %
Copper	0.25 %	0.19 %
Iron	4.82 %	8.02 %
Sulphur	6.12 %	9.92 %
Gold	2.4 g/t	1.7 g/t
Silver	82 g/t	161 g/t

of ore crushed to minus 10 mm. Leach solutions containing chemolithotrophic bacteria (in concentrations higher than 10^8 bacteria/ml), iron ions (from 5 - 7 g/l, mainly in the trivalent state as ferric sulphate), some essential nutrients, sulphuric acid and dissolved oxygen were pumped to the tops of the columns at flow rates in the range of 50 - 100 l/t. 24 h. The pH of the solutions was in the range of 1.7 - 1.9 and the Eh was in the range of 600 - 640 mV. The column effluents were supplemented with $(NH_4)_2SO_4$ and KH_2PO_4 to concentrations of about 0.5 and 0.25 g/l, respectively, and were treated in a BACFOX unit (Groudev, 1981) in which bacteria oxidized the ferrous ions to the ferric state under conditions of an intensive aeration. The pH and iron content of the solutions were adjusted to the desired levels and the solutions were pumped through the ores repeatedly. An interrupted irrigation of the ores was applied and the ratio of the total duration of the irrigation periods to that of the rest periods was about 1:3. The leaching was carried out at 30°C for 180 days.

Mixed enrichment cultures of chemolithotrophic bacteria preliminary adapted to the ores being leached were used in these experiments. The mixed cultures were obtained by inoculation of ore suspensions in 9K iron-free nutrient medium with samples from acid mine drainage waters containing natural mixed populations of such bacteria. The samples were taken from the same mines from which the ores had been mined. The inoculated suspensions were incubated bated at 30°C. The mixed enrichment cultures were adapted to the ores by consecutive transfers on suspensions with increasing pulp density.

Sterile controls were also set up. The ores in these tests were leached by means of 0.1 M $Fe_2(SO_4)_3$ - 0.1 M H_2SO_4 solution.

The leached ores were then treated for recovering lead. The leaching was carried out by means of solutions containing $FeCl_3$ - HCl in different concentrations, at 30°C for periods from 30 - 90 days. NaCl was also added in some tests. Finally, the ores were treated for recovering gold and silver. Two different methods were applied. In some tests the ores were leached by solution containing 0.5 mol/l thiourea and 3 g/l oxone (a reagent produced by Du Pont de Nemours) at pH 0.5. In the other tests the ores were leached by solutions containing 5 g/l of gold-complexing amino acids of microbial origin and 5 g/l $KMnO_4$ at pH 9.5. The leaching of the precious metals was carried out at 30°C for 90 days.

All microbial and chemical procedures used in this study have been described elsewhere (Groudev, 1990).

Results and Discussions

Zinc and copper were leached simultaneously from the ores by means if chemolithotrophic bacteria in acidic sulphate-bearing medium. Zinc was leached at much higher rates than copper but the final extractions of these metals were similar (Table 2). Galena was partially oxidized to the insoluble lead sulphate during the leaching. The pyrite was, however, relatively slightly attacked during the efficient leaching of sphalerite and chalcopyrite. A more efficient pyrite oxidation

Table 2. **Data about the bacterial leaching of the polymetallic ores in acidic sulphate-bearing medium.**

Variable	Ore No 1	Ore No 2
Metal extraction %		
- Zn	84.2	74.7
- Cu	81.1	73.0
Mean rate of leaching, g/24 h		
- Zn	1.128	2.106
- Cu	0.250	0.139
Mean rate of leaching, %/24 h		
- Zn	0.281	0.249
- Cu	0.270	0.243
Sulphuric acid consumption, kg/t ore	10.46	6.11
Galena oxidized, %	62	53
Pyrite oxidized, %	18	24

proceeded only after the major parts of the remaining sulphides were oxidize. The observed order of leaching may be related to the galvanic corrosion mechanism as well as to the abundance of the various sulphides in the ores.

The bacteria maintained the dissolved iron in the trivalent state. It must be noted, however, that part of the dissolved ferric iron precipitated in the ore mass, mainly as jarosites. Elemental sulphur was also found in the columns.

The mixed enrichment cultures used to inoculate the ores in the columns contained Thiobacillus ferrooxidans, T. thiooxidans, T. acidophilus, Leptospirillum ferrooxidans and some acidophilic heterotrophic bacteria (mainly such related to the genus Acidiphilium). T. ferrooxidans was the prevalent microorganism in these cultures. Most chemolithotrophic bacteria were firmly attached to the ore particles and were found in highest numbers (over 10^8 bacteria/g of ore) in the upper sections of the ore mass.

The chemical leaching of the ore with 0.1 M $Fe_2(SO_4)_3$ - 0.1 M H_2SO_4 solution resulted in zinc and copper extraction similar to those obtained by bacterial leaching. However, the irrigation rate and acid consumption during the chemical leaching were much higher than those during the bacterial leaching.

After the bacterial leaching of zinc and copper, the ores were then leached by means of $FeCl_3$ solutions to solubilize the lead. In some tests the pH of solution was maintained to less than 2.0 by using a mixture of hydrochloric and sulphuric acids. The chemolithotrophic bacteria grew in such media containing both chloride and sulphate ions and maintained a high ferric to ferrous ions ratio in the percolating solutions. The bacteria were able to grow at chloride ion concentrations as high as 15 g/l. Galena was solubilized as $PbCl_2$ but part of this compound precipitated in the columns. Part of the lead sulphate was also converted to $PbCl_2$. 86.2 and 74.4% of the lead were solubilized in this way in 30 days from the two ores, respectively. A chemical leaching by using 1 M $FeCl_3$ - 1 M HCl gave slightly better results than the leaching in the presence of bacteria. Small parts of the remaining other metals were also solubilized as the relevant chlorides.

The leaching of gold and silver by thiourea or amino acids in the presence of oxidants from the pretreated ores was much more efficient than the leaching of these metals from untreated ore samples. Extractions in the range of about 65 - 80% were obtained, whereas the extractions from the untreated ores were lower than 25%.

The utilization of metal values from such polymetallic sulphide ores by using three consecutive leaching procedures seems technically feasible and economically attractive, especially for low-grade ores dumped as wastes during mining operations.

References

Dutrizac, J.E., MacDonald, R.J.C.; *Metall. Soc. CIM,* Annual Volume, **1977**, 1-9.

Dutrizac, J.E.; *CMM Bull,* October, **1979**, 1-10.

Groudev, S.N.; *Comp. rend. Acad. bulg. Sci.,* **1981**, 34, 1437-1440.

Groudev, S.N.; *Microbiological Transformations of Mineral Raw Materials,* Doctor of Biological Science Thesis, University of Mining and Geology, Sofia, **1990** (in Bulgarian).

Roy Chaudhury, G., Das, R.P.; *Int. J. Min. Process,* **1987**, 21, 57-64.

Heavy Metal Trapping by Gram Negative and Gram Positive Bacteria

Jean Remacle*, Microbial Ecology, Department of Botany, University of Liège, Sart Tilman, B 4000 Liège, Belgium (FAX:32 41 563840)

The aim of this presentation is to exemplify the potentialities of the microorganisms for recovering dissolved metals. The researches of the laboratory are mainly focused on the properties of bacteria. Gram negative and Gram positive bacteria are currently investigated more specially strains belonging to the genera *Alcaligenes* (Gram negative) and *Bacillus* (Gram positive).

Metallic ions interact with bacterial cells at three levels: close to the cell surface, in contact with the cell surface and inside the cell. More precisely metal immobilization could result from three main microbial processes: the biosorption, the bioaccumulation and the precipitation by metabolism by-products (Remacle, 1988).
Biosorption is conceived as a physico-chemical process depending on the surface characteristics of the microbial cell. Biosorption is observed in living, dead as well as in resting cells.
Bioaccumulation occurs in living cells and is induced by the presence of the dissolved metal. It is generally related to the resistance or tolerance level of the microorganism against the metal toxicity.
The third process leads to metal insolubilization as carbonate, sulphide, phosphate, hydroxide or oxalate mediated by metabolism by-products that do not depend on the presence of the metal.

Biosorption

Alcaligenes eutrophus CH34 was isolated from a sedimentation pond of a zinc factory (Houba and Remacle, 1980). Its behaviour is dependent on the cadmium concentrations in the culture medium. The strain maintained the cadmium concentration in the cytoplasm below the lethal threshold by efflux mechanisms when it was cultivated at low cadmium concentrations *i.e.* below 1 ppm (Nies and Silver, 1989). On the other hand in presence of high cadmium concentrations (up to the Minimal Inhibitory Concentration) the strain immobilized cadmium in the cell wall (Remacle *et al.*,1986). Cadmium could account up to 18% of the cell dry weight and was mainly located in the cell envelopes. This feature discriminated the cadmium-resistant strain CH34 from the cadmium-sensitive one AE104 (Hambuckers-Berhin and Remacle, 1987). The ability of bacterial cell to trap metals can be related to the cell surface characteristics. When suspended in low acidic conditions (pH 2 to 4) the net surface charge of bacterial cell is generally electronegative which results in cation affinity. It is due to the presence of carboxyl, amino and phosphoryl groups located in the cell wall (Remacle, 1990 a) and it is interesting to observed that the chemical composition of cell wall could be modified when microorganisms are cultivated in presence of toxic metals as observed in cultures of *Alcaligenes eutrophus* CH34 (Hambuckers-Berhin and Remacle,

1054–7487/92/0458$06.00/0

1988) and of *Bacillus coagulans* (Infantino-Masuy et al , 1988).
The comparison of the Cd-R and Cd-S strains of *A.eutrophus* CH34 led to the following conclusions (Hambuckers-Berhin and Remacle, 1990):

- the amount of the envelopes was the highest in the cells of the Cd-R strain
- the amount of envelopes of the Cd-R cells was related to the contamination level.

Since the shape and the size of the cells remained unchanged even at the highest cadmium concentrations it could be inferred that cadmium provoked the thickening of the envelopes in the Cd-R cells. Moreover it could be conceived that the envelope thickening improved the efficiency of this barrier against the entry of cadmium inside the cell bearing in mind that envelopes could account for 40 to 50 % of the cell dry weight. The cell wall is quantitatively the most important constituent of the cell envelopes. It is composed of an outer membrane covering a layer of peptidoglycan. The proportion of peptidoglycan significantly increased when Cd-R strain was cultivated in presence of 250 ppm of cadmium. This increase could be also observed in the Cd-S strain at a lower cadmium concentration, 33 ppm. Moreover the Cd-R peptidoglycan showed a higher cadmium affinity than the Cd-S peptidoglycan. At the same level of contamination, 33 and 60 ppm of cadmium, the Cd-R peptidoglycan immobilized 5 to 10 times more cadmium than the Cd-S peptidoglycan. It could be thus concluded that cadmium sequestration by the cell wall could be controlled by two factors, the thickness of the peptidoglycan and its cadmium affinity. The cadmium affinity of the peptidoglycan could depend on the number of the available sites and more precisely on the number of free carbolixic groups. The presence of cadmium in the culture solution provoked a shift in the amino acids composition of the peptidoglycan. A higher proportion of glutamic acid and the presence of an unusual amino acid, the aspartic acid were observed. These amino acids harboured free carboxilic groups that could trap cadmium. It was also showed that free amine groups were more numerous in cadmium loaded peptidoglycan which led to conclude that the number of cross-links between diaminopimelic acid and adjacent glycopeptide strands was lowered and that a higher number of sites was consequently available for cadmium fixation. However a stoichiometric approach indicated that the amount of carbolixic groups was not enough for the binding of cadmium trapped in peptidoglycan. It could be assumed that nucleation sites could promote metal trapping (Beveridge and Murray, 1976; Tsezos and Voleski, 1982). When considering the amount of cadmium sequestered in the wall it could be concluded that peptidoglycan only accounted for a small part. 33 and 28 % of the total cadmium in the wall were immobilized respectively by Cd-R and Cd-S peptidoglycan in the best conditions. It means that the outer membrane do play a significant role in the cadmium trapping mainly by the presence of phosphoryl and carbolixic groups located in the lipopolysaccharides, one of the constituents of the outer membrane.

The chemical characteristics of cadmium loaded wall of Gram positive bacteria were also different from cadmium free wall as observed in *Bacillus coagulans* CD17L (Infantino-Masuy *et al.*, 1988). Moreover the metal specificity of the wall was changed as shown by the selectivity coefficients of the Cu-Cd exchanges. These coefficients were calculated for two types of walls, walls of cells cultivated in the presence of 15 ppm of cadmium and wall of cells cultivated in a cadmium free medium. The partition of copper and cadmium in the walls proved that the Cd-wall displayed a higher cadmium affinity than the control.

It appears that the metal affinity of microbial cells could be modified by cultivating bacteria in metal contaminated solution. However it is interesting to note that metal trapping ability of microorganisms could be changed either by monitoring the culture conditions or by genetic engineering (Remacle, 1990 b).

Bioaccumulation

Several microorganisms are able to grow in the presence of toxic metal by keeping the intracellular concentration of the toxic form of the metal below the lethal threshold. Metal tolerant bacteria produce intracellular molecules that can complex the toxic metal. The well-known metallothioneins are characterized by their affinity for several heavy metals such as copper, zinc, cadmium, lead, mercury. The metal affinity of the metallothioneins is due to the presence of sulfhydryl groups borne by cysteine that accounts for *ca* 30 % of the metallothionein molecule. A metal protein was synthesized when *A.eutrophus* CH34 was grown in heavily cadmium or zinc contaminated cultures (Remacle and Vercheval, 1991). The zinc protein was quite different from typical metallothioneins since it harboured few sulhydryl groups (cysteine 2.1 %) but significant amounts of carboxilic groups located on acidic amino acids (glutamic and aspartic acids 36.9 %). These carboxilic groups were assumed to contribue the zinc immobilization. However this protein had lower affinity than metallothionein, 4 g at. of zinc per mole instead of 7 whereas the molecular mass was of the same order of magnitude, *ca* 20,000 Da.

Immobilization by metabolism by-products

The precipitation of metallic ions as sulphide, carbonate, hydroxide, oxalate could be the consequences of the metabolic activities of the microorganisms. *A.eutrophus* CH34 and *A.denitrificans* provoked an alkalization of the culture that promoted the formation of cadmium carbonate. In these conditions up to 90 % of cadmium was inactivated as cadmium carbonate. (Remacle *et al.*, 1992) The alkalization process is not yet understood since the pH increase cannot be only explained by the consumption of the substrate, the lactic acid. Sulphate-reducers were also very effective for the removal of heavy metals by provoking precipitation of metal sulphides (Crine *et al.*, 1989).

Conclusion

Living and dead microorganisms as well as microbial constituents offer several potentialities for recovering metals in solution at low concentrations.
The choice of the microbial process will depend on
- the kind of metal, precious, rare or toxic,
- the type of solution where metals are dissolved and specially the metal speciation,
- the efficiency, the cost of the microbial process by comparison with the conventional physico-chemical processes.

References

Beveridge, T.J.; Murray, R.G.E. *J. Bacteriol.* **1976**, 127, 1502-1518.
Crine, M.; Schlitz, M.; Salmon, T. In *Microbial Processes for the Heavy Metal Recovery;* Hecq, W.; Crine, M. Ed; Ass. Univ. Environ., **1989**, pp. 79-77.
Hambuckers-Berhin, F.; Remacle, J. In *Heavy Metal in the Environment*; Linberg, S.E.; Hutchinson, T.C. Ed; C.E.P. Publ., U.K., **1987**, pp 244-246.
Hambuckers-Berhin, F.; Remacle, J. FEMS *Microb. Ecol.* **1990**, 73, 309-316

Houba, C.; Remacle, J. *Microbial Ecol.* **1980**, 6, 55-69.
Infantino-Masuy, B.; Hambuckers-Berhin, F.; Remacle, J. In *8th International Biotechnology Symposium;* Durand, G.; Bobichon, L.; Florent, J.; Ed; Société Française de Microbiologie, Paris, **1988**, pp 230.
Remacle, J. In *8th International Biotechnology Symposium;* Durand, G.; Bobichon, L.; Florent, J. Ed; Société Française de Microbiologie, Paris, ***1988***, pp 1187-1197.
Remacle, J. a In *Biosorption of Heavy Metals;* Volesky, B. Ed; CRC Press, U.S.A., **1990**, pp 83-92.
Remacle, J. b In *Biosorption of Heavy Metals;* Volesky, B. Ed; CRC Press, USA, **1990**, pp 293-304.
Remacle, J.; Vercheval, C. *Can. J. Microbiol.* ***1991***, 37, 875-877.
Remacle, J.; Houba, C.; Hambuckers-Berhin, F. In *Perspectives in Microbial Ecology;* Megusar, F.; Ganthar, M. Ed; IVth International Symposium of Microbial Ecology, **1986**, pp 668-672.
Remacle, J.; Muguruza, I.; Fransolet, M. *Wat. Res.* **1992** (in press).
Tsezos, M.; Volesky, B. *Biotechnol. Bioeng.* ***1982***, 24, 385-401.

Removal of Heavy Metals by Biosorption

B. Volesky
Department of Chemical Engineering, McGill University, and B.V. SORBEX, Inc.
Montreal, Canada H3A 2A7 (FAX: 1-514-398.6678)

Review is presented of recent new data obtained in search for new biosorbent materials. Selected results from batch equilibrium experiments demonstrated the superb performance in cadmium and lead removal by brown marine algae *Sargassum natans* and *Ascophylum nodosum* were capable of sequestering more than 20% of their biomass dry weight in the metal. While mucoralean filamentous fungi were performing well as broad-spectrum biosorbents, the common yeast *Saccharomyces cerevisiae* is a mediocre sorbent. Dynamic sorption column tests yielded essential process scale-up parameters confirming the sorbent application potential of suitably granulated algal biomass.

Biosorption has been defined as the property of certain types of microbial biomass to bind and concentrate heavy metals. The chemical composition of the microbial cell, and its cell wall in particular, is mostly responsible for the metal biosorbent property of dead cells. In fact, the concentration of metals in the biomass is sometimes even more pronounced with the dead cell material which has lost its "active defence" mechanism against toxic heavy metal ions. The mechanism of metal binding by the biomass is still relatively little understood (Beveridge and Murray , 1980; Kuyucak and Volesky, 1989c; Kuyucak and Volesky, 1989d; Tsezos and Volesky, 1982a; Tsezos and Volesky, 1982b). Due to the complexity of the biomass there are numerous possibilities based on varied binding sites where the metal can be sequestered. Depending on the type of metal and its solution chemistry as well as on the type of biomass and the prevailing sorption conditions, some materials can sequester the metallic ions selectively while others bind them without distinguishing. The preference, however, is definitely toward "heavy metals" reflecting the size of their ionic radii (Tobin, et al., 1984).

The phenomenon of biosorption can find a broad potential application in removing toxic heavy metals particularly from industrial solutions and effluents and its background as well as applied aspects have been recently comprehensively reviewed (Volesky, 1990). Being very similar to well established ion exchange resins, the new family of biosorbent materials represents a basis for extremely cost-effective alternative process for metal removal. It is important to emphasize that current investigations indicate the possibility of regenerating the biosorbent material allowing thus not only its multiple reuse (Kuyucak and Volesky, 1989a; Kuyucak and Volesky, 1989b; Volesky and Tsezos, 1981) but also the recovery of the metal. This is particularly important for the environmentally based applications requiring the ultimate metal removal and low costs of the process.

Studies of heavy metal biosorption conducted so far have been mostly exploratory in nature and limited by the large number of variable factors increasing the scope of any in-depth investigation. Two key aspects which have to be taken into consideration in conjunction with the metal uptake capacity of the sorbent for a specific sorbate are the characteristics of the solution system and the characteristics of the sorbent. The chemical properties of the sorbent material include the types of chemical functional groups which are present in the material binding the metallic ions, the degree of ionization of the sorbent surface and ash content (Faust and Aly, 1987). The physical properties of the sorbent include its specific surface area, the pore size and its distribution (Slejko, 1985).

This contribution presents selected results of recent investigations of the biosorbent heavy metal uptake by common biomass which is readily available in nature or from industrial sources. Important aspects of biomass immobilization and dynamic sorption testing are also discussed.

Materials and Methods

The samples of brown algae *Ascophyllum nodosum*, *Fucus vesiculosus* and *Sargassum natans* were harvested from the Atlantic ocean. Dried ground biomass of green alga *Halimeda opuntia* (Australia) was repeatedly washed with 10% acetic acid until no more CO_2 gas evolved

1054–7487/92/0462$06.00/0

and there was no further change in the solution pH which remained acidic. The carbonate-free biomass was subsequently washed with distilled water to neutrality and dried. Dry samples of industrial *Penicillium chrysogenum* were supplied by the courtesy of Hindustan Antibiotics, Ltd., (Pimpri, Pune, India) and by the Research Institute of Antibiotics, (Roztoky by Prague, Czechoslovakia). Filamentous fungi were cultivated in shake flasks in the laboratory (May, 1984; Treen-Sears, et al., 1984) The biomass materials were not purified nor otherwise treated for biosorption. Crosslinked *P. chrysogenum*, was used prepared according to Jilek, et al. (1971).

Living and dried biomass samples of "baker's yeast" *Saccharomyces cerevisiae* (strains 1452-L6F + 8% 1453-L65 as) and "brewer's yeast", supplied respectively by Lallemand, Inc. and Molson Breweries, both in Montreal, Canada, were subcultured on the medium containing 10 g glucose, 5 g peptone, 3 g malt extract, and 3 g yeast extract in 1L of distilled water. The biomass from 500 mL Erlenmeyer culture flasks was washed and then used for the equilibrium and kinetics sorption experiments (Volesky, 1990).

Results

Algae: Cd, Pb and Cr Biosorption

The equilibrium (batch) uptake of cadmium by different types of biomass and by recommended ion-exchange resins was compared in sorption isotherms (Fig. 1). The metal biosorbent uptake around 100 mgCd/g at the equilibrium final concentration of 100 mg/L was observed for native *A. nodosum* and *Sargassum natans*. In comparison, the commercial ion exchange resins tested performed at approximately 50% and 25% of this value, respectively. All three brown marine algae absorbed impressively high quantities of lead as demonstrated by equilibriu sorption isotherms in Fig. 2. For instance, at the equilibrium concentration of 100 mgPb/L, *F. vesiculosus* and *A. nodosum* sequestered 150 and 180 mgPb/g, respectively, while *S. natans* at 220 mgPb/g at that point did not even reach its sorption saturation potential. Under the same conditions, commercial ionex Duolite GT-73 performed at less than 30% of that value.

Removal of Cr^{+3} by *Halimeda opuntia.*

The chromium uptake capacity of *H. opuntia* was reported earlier at approximately 380 mgCr/g biomass (Kuyucak and Volesky, 1988). However, the high content of calcium carbonate in the biomass (70 - 85 %) of this marine alga (Johansen, 1981) was apparently responsible for skewing the investigation results because of the difficulty in controlling the terminal pH of the

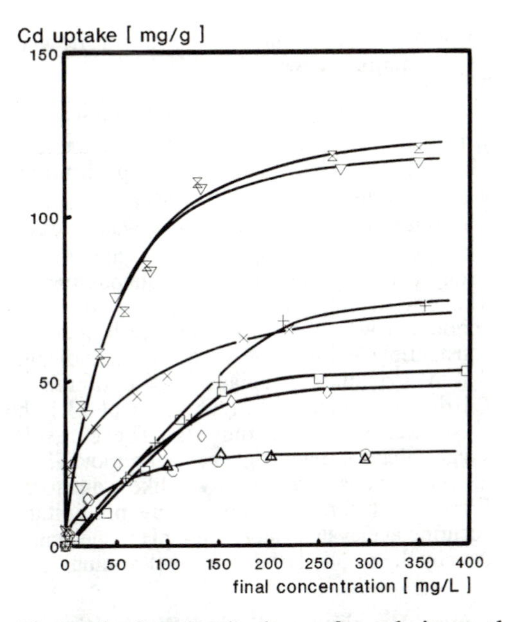

Figure 1. Sorption isotherms for cadmium and selected sorbents:
+*Fucus vesiculosus* ○*Rhizopus arrhizus*
□*Halimeda opuntia* ▽*Sargassum natans*
◇*Penicillium chrysogenum* × Duolite GT-73
⧖*Ascophylum nodosum* △Amberlite IRA-400
Equilibrium pH 3.5, temperature 26°C.

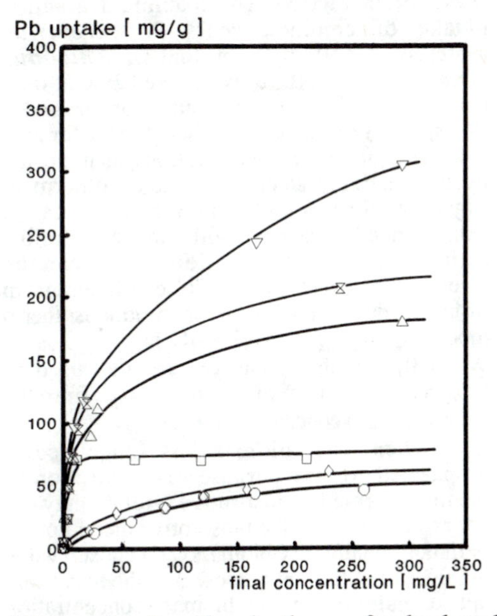

Figure 2. Sorption isotherms for lead and selected sorbents.
△*Fucus vesiculosus* ○*Rhizopus arrhizus*
⧖*Ascophylum nodosum* ▽*Sargassum natans*
◇*Penicillium chrysogenum* □ Duolite GT-73
Equilibrium pH 3.5, temperature 26°C.

sorption system. Indeed, in the follow-up experiments it was established that the equilibrium pH tended to repeatedly rise in excess of pH 7 even after multiple adjustments with acid which invariably resulted in the release of gaseous CO_2. During the preliminary washing of the biomass with 10% acetic acid, it was observed that the CO_2 gas was released from the biomass that otherwise had a very strong buffering capacity in the aqueous system. The equilibrium sorption capacity test was conducted with the carbonate-free biomass at constant pH and the new set of results indicated that the maximum Cr uptake at the optimum pH 4.1 did not exceed 40 mgCr/g. At pH 3.1 the q_{max} was below 12 mgCr/g. These results suggest that the much higher metal removal from the solution observed earlier was likely due to the filtration of chromium-containing precipitates occuring at elevated equilibrium pH values rather than by the sorption of dissolved chromium ions.

Fungi: Uranium and Zinc Biosorption

It has been noted that, in general, biosorbent affinity has been often higher for the uranyl ion than for other metals. Several types of common microbial biomass types propagated in the laboratory fermenters have been examined for the uptake of the uranyl ion and zinc. Mucoralean fungi are particularly good uranium sorbents (*Rhizopi* in Fig. 3). The maximum uranium uptake of common yeast *S. cerevisiae* represented only 63% of that of *Rhizopus arrhizus*. *S. cerevisiae* sequestered zinc at only about 60-70% of its capacity for uranium. *Rhizopus nigricans* had a higher affinity for zinc at lower equilibrium concentrations, as indicated by the desirably steeper sorption isotherm in Figure 4. Stronger affinity for the metal is demonstrated by steeper isotherm slopes at low residual metal concentrations This can be quantified by expressing the "b" coefficient from fitting the data to the Lanqmuir sorption isotherm model: $\mathbf{q = q_{max}\, b\, C_f/(1 + bC_f)}$
When fitting, the model can be also used to predict the maximum metal uptake $\mathbf{q_{max}}$ from the residual (final) concentration $\mathbf{C_f}$.

When the biomass is specifically propagated for the purpose of biosorption, a quantity termed "biosorptive yield" may be defined as a ratio of the biosorptive metal uptake in mmol Metal/g (dry biomass) at a chosen value of 90% (for definition purposes) of the maximum uptake multiplied by the biomass concentration (g/L) at the fermentation process harvest time. For every liter of fermentation broth, due to the very high uranium uptake by the mucoralean fungi, both *Rhizopi* featured high uranium biosorbent yield, 1.10 and 0.77 mmolU/L, respectively. *S. cerevisiae* showed higher

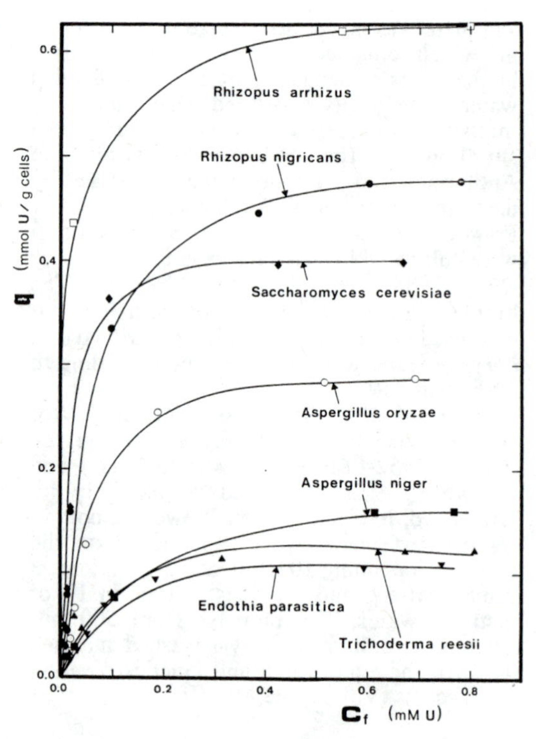

Figure 3. Sorption isotherms for uranium and selected fungal biomass.

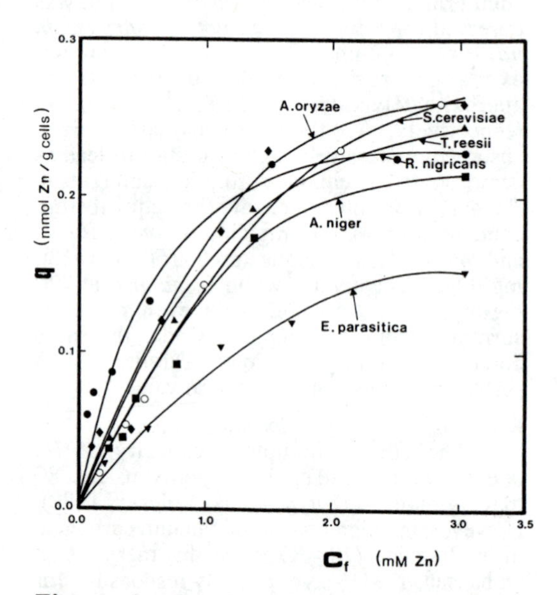

Figure 4. Sorption isotherms for zinc and selected fungal biomass.

biosorbent yields for both uranium and zinc than *R. niqricans*.

Effect of Culture Medium on Biosorption:

Rhizopus niqricans was propagated on 3 different media and harvested after the same culture period. Biomass of the culture grown on glucose-peptone medium exhibited the highest maximum uranium uptake of 0.475 mmol U/g, and biosorptive yield 2.82 mmolU/L which is 50% higher than that obtained with sucrose-grown biomass. Potato-dextrose medium resulted in biomass with the lowest biosorptive yield and maximum metal uptake of 0.34 mmolU/g, indicating the low suitability of this type of medium for producing a good biosorbent.

Asperqillus oryzae was also grown in the same series of media. The glucose–peptone medium again resulted in the best biosorbent (1.36 mmol U/L). Biomass propagated in the sucrose medium sequestered less uranium than that from potato-dextrose but the final biomass concentration in the former medium was twice that observed in the latter.

Biosorption Column Tests

Stiffening and granulation of the biosorbent "powder" is essential for the study of its sorption performance under dynamic continuous-flow conditions in a sorption column. In order to achieve the optimal volumetric process efficiency, biomas crosslinking may be preferable to its entrappment in a gel matrix. Preliminary tests with *A. nodosum* sorbing Cd revealed that formaldehyde-croslinked biomass possessed both the desirable mechanical properties and high metal uptake. The column sorption tests focused on deriving the key process parameters, namely, the Critical Bed Depth, the Saturation Uptake Capacity and the Bed Service Time. Essential for the design and process scale-up procedures, these parameters have been used in determining the useful relationship between the sorbent usage rate and the bed contact time presented as a resulting plot in Fig. 5. The packed bed column of granulated *A. nodosum* biosorbent (FCAN) reduced the Cd concentration to 1.5 ppb which is lower than the concentration specified for potable water.

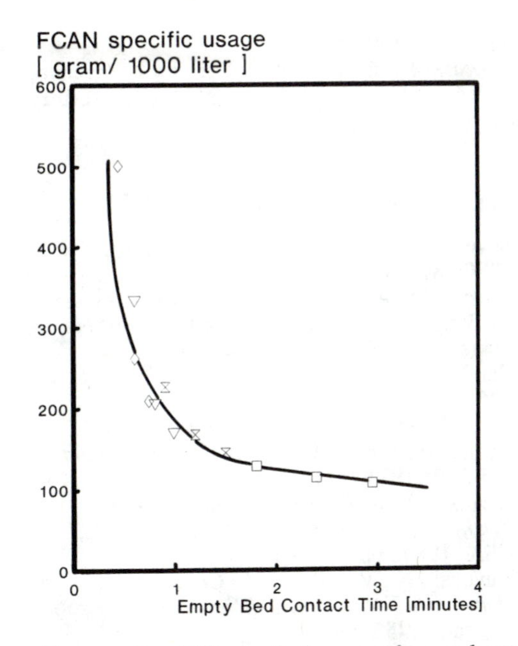

Figure 5. Effect of the sorption column residence time on the specific biosorbent usage: Formaldehyde-reinforced *A. nodosum* removing cadmium from an aqueous solution at flow-rates:
□ 2.4 L/h.cm^2 ▽ 7.2 L/h.cm^2
⧖ 4.8 L/h.cm^2 ◇ 9.6 L/h.cm^2

Discussion

The cross-section of common biomass types examined demonstrated varied capabilities in sequestering heavy metals. While uranium is often readily sorbed, biosorption of other heavy metals may vary. Some filamentous fungi, particularly from the family *Mucorales* have a good sorbent potential for a spectrum of heavy metals. The common yeast *Saccharomyces cerevisiae* is a mediocre sorbent. An outstanding promise related to at least two (Pb and Cd) of the "big three" most toxic heavy metals has been demonstrated for *Sagassum natans*, *Ascophylum nodosum* and *Fucus vesiculosus*, all belonging among the brown marine algae. It has been indicated that the alginate in the cell wall of certain types of algal biomass is responsible for their ion exchange properties (Paslans-Hurlburt, et al., 1976) of different specificity depending on the mannuronic/glucuronic acid ratio (Haug and Smidsord, 1965). While *A. nodosum* is rich in alginate (Bold and Wynne, 1978), the other two algae examined here are not, rising the possibility of a different binding mechanism based perhaps on fucoidan, another anionic and sulfated polysaccharide present in their cells.

The reinforcement and granulation of the biosorbent material, essential for its process application, can improve or diminish its metal-binding performance. Derived from the experience with formaldehyde cross-linking of cellulose (Walker, 1964), cross-linking of *A. nodosum* biomass devised in this work resulted in desirable properties of a durable granular biosorbent. It can be considered a success and can serve as a basis for further tests with the new family of biosorbents.

Acknowledgement

Experimental contributions by I. Prasetyo, H. May and Dr. Z.R. Holan are gratefully acknowledged.

References

Beveridge, T.J.; Murray , R.G.E. *J. Bacteriol* **1980**, 876-887.

Faust, S.D.; Aly, O.M. *Adsorption Processes for Water Treatment*; Butterworths Publishers, Stoneham, UK, **1987**; pp 14-163.

Haug, A.; Smidsrod, O. *Acta Chem Scan* **1965**, *19*, 341-351.

Jilek, R.; Prochazka, H.; Kuhr, I.; Fuska, J.; Nemec, P.; Katzer, J. *Czechoslovakian Patent 155 830* **1971**.

Johansen, H.W. *Corraline Algae, A First Synthesis*, CRC Press, Inc. Boca Raton FL, USA, **1981**; pp 111-118.

Kuyucak, N.; Volesky, B. In *Conference on Non-Waste Technologies (Proceedings)*; Espoo, Finland, **1988**; pp 1-13.

Kuyucak, N.; Volesky, B. *Biotechnol. Bioeng.* **1989a**, *33*, 815-822.

Kuyucak, N.; Volesky, B. *Biorecovery* **1989b**, *1*, 155-163.

Kuyucak, N.; Volesky, B. *Biotechnol. Bioeng.* **1989c**, *33*, 823-831.

Kuyucak, N.; Volesky, B. *Biorecovery* **1989d**, *1*, 164-172.

Paskins-Hurlburt, A.J.; Tanaka, Y.; Skoryna, S.C. *Bot. Marina* **1976**, *19*, 59-60.

Percival, E.; McDowell, R.H. *Chemistry and Enzymology of Marine Algal Polysaccharides*, Academic Press, London, UK, **1967**; pp 99-124.

Slejko, F.L. *Adsorption Technology*; Marcel Dekker, Inc., New York, **1985**; pp 1-68.

Tobin, J.M.; Cooper, D.G.; Neufeld, R.J. *Appl.Envir.Microbiol.* **1984**, *47*, 821-824.

Treen-Sears, M.; Martin, S.M.; Volesky, B. *Appl.Envir.Microbiol.* **1984**, *48*, 137-141.

Tsezos, M.; Volesky, B. *Biotechnol. Bioeng.* **1982a**, *24*, 955-969.

Tsezos, M.; Volesky, B. *Biotechnol. Bioeng.* **1982b**, *24*, 385-401.

Volesky, B., ed. *Biosorption of Heavy Metals*; CRC Press, Inc., Boca Raton, FL, USA, **1990**.

Walker, J.F. *Formaldehyde, 3rd ed.*; Reinhold Publ. Corp., London, UK, **1964**; pp 264-281.

Genetic Engineering Approach for Wastewater Treatment

Edward D. Schroeder, Department of Civil & Environmental Engineering, University of California, Davis, Davis, CA 95619

Microbial wastewater treatment systems have, to a large extent, evolved to their present status. Although microbial treatment systems are constructed in a number of configurations (e.g. suspended growth, attached growth, and fluidized bed) and operated under a variety of conditions (e.g. batch, continuous flow, aerobic, anaerobic) the use of heterogeneous populations of microorganisms is standard practice. Use of pure cultures, or even limited mixtures of species, is occasionally reported in laboratory experiments, but the practice is nearly unheard of in field application. In recent years there has been an increasing interest in developing systems in which the microbial populations are designed, or engineered. Much of this interest has been based on the assumption that better overall process performance could be achieved through designing organisms to carry out specific reactions [Crawford, 1988, Rawson and Kinzy, 1987]. Another objective has been the development of a "superbug" that would degrade recalcitrant organics such as chlorinated aliphatics, dioxins, and polychlorinated biphenyls. In general, there has been little success achieved, although efforts continue in a number of laboratories. The complexity of developing genetically engineered microorganisms that compete well in a diverse environment has proven to be a difficult problem [Pierce, 1984, Neidhardt, et al., 1990]. Attempts to achieve similar effects through bioaugmentation have generally failed [Wilderer et al., 1991, Koe and Ang, 1992] Thus the purpose of this paper is to consider the possible roles of genetic engineering in wastewater treatment, the advantages and limitations of such applications, and the process configurations that appear to be most suitable for this approach.

Conventional Microbial Populations

Microbial populations of conventional wastewater treatment systems include a wide range of eucaryotic and procaryotic organisms but the dominant group is the bacteria. Fungi are important in some types of treatment process and may become significant factors in soil bioremediation and microbial air cleaning systems. Algae contribute to the performance of systems such as oxidation ponds but appear to offer little to genetically engineered treatment processes. Protozoans are normally present, as are multicellular microorganisms such as rotifers and analids, but the importance of these organisms in the treatment process community is unclear and their possible contributions to genetically engineered wastewater treatment systems is not apparent.

Mixed populations of microorganisms are difficult to avoid in wastewater treatment and offer significant advantages in terms of treatment objectives. Because wastewater flow rates are usually relatively large management of the microbial population in terms of species present would be extremely difficult. Wastewaters are normally open to the environment and therefore easily contaminated with organisms from the air, soil, and other sources. Thus treatment systems are constantly open to infection by opportunistic organisms. Thus if a treatment process is to make use of genetically engineering the genetically engineered organisms must have a competitive advantage over naturally occurring species or a species control procedure must be incorporated into the system. Two types of species control procedure might be considered: feed stream sterilization and chemical addition. Feed stream sterilization is difficult to manage and extremely expensive in large, continuous feed operations, as is the case in most wastewater treatment operations. Chemical addition is also expensive and adds a new wastewater component that must be dealt with prior to discharge.

A significant advantage of the use of naturally occurring mixed microbial populations is the process stability provided. Species predominance shifts with changes in feed characteristics and operating conditions. Thus if a new compound appears in the feed stream that is not degraded by the predominant microbial groups, the population makeup is likely to shift in such a way that the biooxidation of the new compound begins in a relatively short period of time. Introduction of a toxic compound in the feed stream may result in a decrease or loss of microbial activity. However, unless the toxic material is fed continuously, recovery of the microbial population is very rapid because of the open nature of the systems. If the physical or chemical environment changes the population predominance can be expected to shift also such that the most competitive species predominate.

A second advantage of mixed microbial populations is that essentially complete removal of metabolizable compounds is possible. Single species populations can be expected to partially oxidize some compounds or to produce intermediates that are undesirable in the final discharge stream.

Advantages of GEMs

The principal advantages of the utilization of genetically engineered organisms appear to be (1) biotransformation of compounds that are recalcitrant using naturally occurring species, (2) enhancement of the biooxidation rate in a particular application, and (3) providing a needed biochemical reaction in a selected environment.

Recalcitrant Compounds: Because of the widespread problems with toxic and hazardous organic chemicals, such as solvents, pesticides, herbicides, PCBs, and dioxins, that are either very difficult to degrade or have not been observed to be degraded microbially, considerable interest has been developing genetically engineered microorganisms (GEMs) for this purpose. A major drawback of this approach is that somewhere in nature there must be found an organism that can provide the necessary genetic information. The information required required may be a step in a biochemical reaction sequence, a factor that detoxifies the target compound, or a permease that allows the target compound to be transported across the cell membrane. Once found, the genetic information must be incorporated into the genetic compliment of an organism that is competitive in the reaction environment of the treatment system. An example would be the transfer of the gene responsible for cellulase synthesis. Incorporating this gene onto a plasmid or the chromosome of a rapidly growing facultative species that competes well in aerobic biological treatment processes could result in a much improved capacity for treatment of cellulosic compounds. There has been considerable interest in the development of a "superbug" incorporating the genes responsible for the production of ligninases in white rot fungus, such as *Phanaerochaete chrysosporium*, into bacterial groups such as *Pseudomonas*.

Problems that have been observed with GEMs in wastewater treatment studies. If the required genetic sequence is encoded onto a plasmid the desired activity rapidly disappears unless a selective pressure is maintained. Competitiveness of GEMs for readily degradable substrates is low compared to "wild" organisms. Biodegradation of many of the compounds of interest (e.g. chlorinated solvents) is by cometabolism reactions that do not support growth. Ability to carry out these reactions appears to be related to consistent environmental conditions and the presence of a specific cosubstrate (e.g. toluene or methane). Some hope that stable systems can be developed is given by the work of Fujita et al. [1991] who reported that the development of P. putida strains capable of simultaneous salicylate and phenol degradation with good stability over 300 generations. Fujita et al. also introduced the gene onto plasmid of the floc forming bacteria *P. lemoignei* 551 with the thought that this would increase ecological stability in mixed cultures.

Rate Enhancement: The possibility exists that biodegradation rates of specific compounds can be enhanced by transferring capabilities for selected reactions from slower growing, less competitive species to more rapidly growing groups. Other methods of increasing biodegradation rates exist if the target compound supports growth. For example, increasing the solids retention time will result in increased cell mass concentrations and increased removal rates. In general, this approach will be considerably less expensive and troublesome than the use of GEMs. Where the target compound degradation does not support growth the use of GEMs may be appropriate. The GEMs must remain competitive in the mixed culture and this requires that the degradation rates and affinities for primary substrates remain high following introduction of the new gene or genes.

Biotransformations in Selected Environments: Removal or transformation of pollutants in conventional treatment systems is often restricted to particular environments. An example is denitrification which is carried out by a number of facultative anaerobes under anoxic or low oxygen concentration conditions. Development of a strain that respires preferentially on nitrate, or denitrifies but does respire aerobically might be quite useful. Development of methods that would allow control of gene expression in large processes would also be useful. Most wastewaters vary considerably with time in both makeup and concentration of the pollutant species. Selective pressures based on the presence of particular chemical species are difficult, if not impossible, to maintain. Rapid, and automatic development of populations capable of degrading xenobiotic compounds is often required. If this requires a build up of the concentration of a particular species success is unlikely, at least in continuous flow processes. A capacity to turn on reaction capacities in predominant species very quickly is required if response to time varying inputs is to be satisfactory.

Constraints On Genetic Engineering

Four general types of constraints on applications of genetic engineering in wastewater treatment appear to be significant: (1) risks to ecosystems and human health associated with introduction of GEMs into the "wild" environment [Allbergo and Lee, 1991, Selvartnam and Gealt, 1992] and (2) problems associated with maintaining cultures with the desired characteristics, which have been discussed above, (3) problems with production of satisfactory conversions, and (4) probable high cost. The risk question appears every time proposals to use GEMs in a noncontrolled environment are made. To date the problems associated with such uses have been quite the opposite - maintenance of the populations has been difficult. However, the irrevocability of the releases of GEMs into the environment make the concerns reasonable and the development of risk assessment procedures prudent. Optimism that use of GEMs in uncontrolled environments will be possible is reasonable considering that natural mutations that would carry out the desired reactions appear to be possible but have not developed. This would lead to the conclusion that long term stability in uncontrolled environments is unlikely.

Whether production of satisfactory conversions will be possible on a general scale is a difficult question to answer. The Monod equation, commonly used to predict process performance, has a limiting concentration at which activity ceases.

$$C_{min} = \frac{K_m k_d}{Yk - k_d} \quad (1)$$

Where C_{min} is the limiting concentration, K_m is the saturation constant, k_d is the maintenance energy constant, Y is the yield coefficient, and k is the rate constant. Whether the Monod model is realistic at low concentrations is questionable and applicability of the model to compounds removed through cometabolism seems unlikely. The predicted C_{min} is extremely sensitive to values of the constants but those available in the literature are highly variable. Production of effluents with target compound concentrations below detection limits has been reported and removals of target compounds where initial concentrations were below 50 mg/l provide reasons to be optimistic. However, evaluation of limitations will need to be on a case by case basis for the foreseeable future.

Probable High Costs: At present reliable costs estimates for applying GEMs to wastewater treatment are not available. However, the difficulty of developing and maintaining GEM cultures will probably limit their use to situations where volumes to be treated are small and the nature of the wastewater being treated allows considerable investment in both process development and process operation..

Potential Applications of GEMs

The contraints related to application of GEMs in wastewater treatment suggest that general applicability for municipal and industrial wastewaters is unlikely. Convential populations can be manipulated to provide high rates of removal, satisfactory pollutant conversion, acceptable process stability. Additionally, application of GEMs to the treatment of high volume wastewaters is not likely because of the costs involved and the problems with assessing and mitigating ecological risks. This leaves small volume wastewaters containing significant concentrations toxic or hazardous chemicals. At present such pollutants in these wastewaters are often concentrated on activated carbon and incinerated or disposed of in hazardous waste landfill. In some cases volatile components are emitted to the atmosphere in stripping operations. The low volumes involved and the high cost treatment methods now employed make treatment using GEMs more feasible, particularly because destruction of the compounds is the result.

The need to provide a controlled environment, maximize population stability, allow time for population development, and focus on small focus on small volume wastewaters suggests the appropriateness of batch systems. Use of batch systems allows application of the same type of controls that are used in the production of insulin, human growth factor and other commercial products. Additionally, batch systems can be managed in such a way that GEM populations are grown under the most stable conditions; probably those with minimal competition from wild organisms. If the GEM populations need to be developed the batch operation can be continued until growth has occurred and the reactions have been completed. Finally, with small wastewater volumes storage is less of a problem. Wastewaters requiring treatment by these methods could be collected, stored and treated in sequence. Wastewaters from diverse sources but containing similar compounds could be treated together at a central location.

In most cases wastewaters will contain a number of compounds that are relatively easy to degrade in addition to those that are recalcitrant and require treatment by GEMs. A possiblity exists that the GEMs will not metablolize all of the biodegradable compounds (particularly if only one GEMs strain is used) but will not compete well in a mixed culture. Multistage treatment may be required in which conventional microorganisms are used to degrade the majority of the materials and GEMs are used in a second stage to degrade recalcitrant compounds. If the recalcitrant compounds are broken down by cometabolism a primary substrate may be required in the second stage.

Suspended growth, attached growth, or immobilized cell processes can all be used in GEM processes. The choice of a process configuration can be made on the basis of system performance and relative population stability. Attached growth processes appear to have advantages at low nutrient concentrations and because discharge requirements for toxic compounds are often in the mg/L range this configuration may have advantages over the others.

References

Allbergo, N.: Lee, W.E. "Risk assessment & engineered organisms in remediation," Environmental Engineering: Proceedings of The 1991 Specialty Conference, ASCE, P. A. Krenkel, ed., Reno, NV, 1991, p. 198.

Fujita, M., M. Ike, Hashimoto,S. "Feasibility of wastewater treatment using genetically engineered microorganisms," *Water Research*, 1991, 25, 979.

Koe, L.C.C.: Ang, F.G. Bioaugmentation of anaerobic digestion with a biocatalytic addition: the bacterial nature of the biocatalytic addition," *Water Research*, 1992, 26, 389.

Neidhardt, F. C.:Ingraham, J.L.: Schaechter, M *Physiology of the Bacterial Cell,* Sinauer Associates, Inc., Sunderland, MA.

Olson, R. "Description and development of microorganisms for the degradation of substituted aromatic compounds; past, present, future," in Biotechnology Applications in Hazardous Waste Treatment, G. Lewandowski, Ed: Engineering Foundation, New York, 1988.

Pierce, G. E. "Diversity of microbial degradation and its implications in genetic engineering," *Impact of Applied Genetics in Pollution Control,* C.F. Kulpa, R. L. Irvine, S.A. Sojka, eds., Univ. of Notre Dame, Notre Dame, IN, 1984, p. 20.

Rawson, J.R.Y.: Kinzy, T.G. "The potential for genetic enhancement of bact-erial processes used in waste treatment," Presented at the 194th National Meeting, American Chemical Society, New Orleans, LA, August 30-Sept. 4, 1987.

Selvartnam, S.: Gealt, M.A. "Recombinant plasmid gene transfer in amended soil," *Water Research*, 1992, 26, 39.

Wilderer, P. A.: Rubio, M.A.: Davids, L. "Impact of the addition of pure cultures on the performance of mixed culture reactors," *Water Research*, 1992, 25, 1307.

Use of Biologically Based Periodic Processes in Waste Treatment

Robert L. Irvine* and Robert Chozick and James P. Earley
Center for Bioengineering and Pollution Control, Univ. of Notre Dame, Notre Dame, IN 46556
(FAX: 219 239 6306 or 219 239 8007)

In 1980, the first study on biodegradation of hazardous organics in the Sequencing Batch Reactor (SBR) was conducted at the University of Notre Dame. This study led to the construction of a full-scale SBR for the biological treatment of leachates from Hooker Chemical Company's Hyde Park Landfill. The success of the SBR, and the advantages of periodic processes demonstrated in the early 1980's, provided the impetus for research into other periodic processes. In the late 1980's, the Sequencing Batch Biofilm Reactor (SBBR), the Granular Activated Carbon - SBBR, and Soil Slurry - SBR, periodic processes with applications in municipal and industrial wastewater treatment, contaminated groundwater and soil remediation, and hazardous waste remediation, were developed.

Research and development in the broad area of hazardous waste management has been conducted at the University of Notre Dame since 1980 and has been applied to the construction of full scale bioreclamation facilities. During the early 1980's, suspended growth SBRs were used in the treatment of leachates and industrial discharges which contained high levels of hazardous materials. In 1985, focus shifted to on-site and *in situ* bioreclamation of contaminated soils (e.g., those containing petroleum hydrocarbons, phthalates, trinitrotoluene, etc.). During the past three years, the emphasis has been expanded to include projects that involve the biological treatment of lightly contaminated groundwaters (e.g., those containing benzene, toluene, trichloroethylene, etc.).

Virtually all of these research efforts have involved the use of periodic processes, including conventional suspended growth SBRs, Sequencing Batch Biofilm Reactors (SBBRs), Granular Activated Carbon - Sequencing Batch Biofilm Reactors (GAC-SBBRs), and Soil Slurry-Sequencing Batch Reactors (SS-SBRs). A review of these technologies is presented.

Sequencing Batch Reactors

The conventional, suspended growth SBR has been shown to be a cost effective and energy efficient means of degrading hazardous organics (Irvine and Ketchum, 1989). It is uniquely suited for the selection and enrichment of the desired microbial population because of the ease with which a diverse array of operating strategies and selective pressures can be implemented. This convenience stems from the time-oriented nature of the process which allows manipulation of maximum microbial growth rates and regulation of reactor oxygen levels.

Laboratory-scale SBRs have been thoroughly investigated and operating strategies for a wide variety of treatment objectives, including, for example, nitrogen removal (Palis and Irvine, 1985) and phosphorus removal (Ketchum *et. al.*, 1987) have been developed. Using the experience gained on laboratory-scale SBRs, full-scale SBRs have been designed and successfully used in municipal wastewater treatment (Irvine *et. al.*, 1983) as well as for the treatment of hazardous wastes (Irvine *et. al.*, 1984)

Sequencing Batch Biofilm Reactor

The SBBR offers all of the features of the SBR and additional advantages offered by using bubble-free aeration and biofilm operation (Chozick and Irvine, 1991). Bubble-free aeration allows for biological degradation of volatile wastewater contaminants while minimizing

1054–7487/92/0471$06.00/0

fugitive emissions, while biofilm operation offers higher efficiency for treatment of wastewaters with low levels of organic contaminants.

A laboratory scale SBBR (see Fig. 1) using silicone membrane aeration and a 12-hour operating cycle (i.e., 1 hour fill, 10 hour react, and 1 hour draw) was used to treat a BTEX-contaminated groundwater containing between 0.6 and 3.5 mg/L of total BTEX. The effluent BTEX concentration from this reactor was typically below 10 µg/L. A typical concentration profile of total BTEX in the SBBR is presented in Fig. 2.

Later, a synthetic wastewater containing approximately 115 mg/L of total BTEX was introduced to the SBBR to examine the effect of higher BTEX concentrations on reactor performance. The effluent BTEX concentration was typically below 2 µg/L during this phase of the study, and, although reduced dissolved oxygen concentrations were observed in the SBBR, the results provided above indicate that the overall performance of the reactor was not adversely affected.

Accumulation of biomass in the SBBR over extended operating periods caused a steady decline in dissolved oxygen concentrations in the reactor and eventually led to reduced reactor efficiency. The performance of the reactor was improved by the removal of excess biomass after a decline in performance was observed, indicating that a regular biomass wasting strategy is critical to long-term operation.

Volatile losses from the reactor and from the membrane aeration system accounted for approximately 0.02% and 0.3%, respectively, of the total VOCs treated, demonstrating the effectiveness of the silicone membrane aeration system for maintaining low volatile emissions from the reactor.

An extension of the SBBR, the Granular Activated Carbon - Sequencing Batch Biofilm Reactor (GAC-SBBR), is also currently under investigation (Chozick and Irvine, 1991). Car-

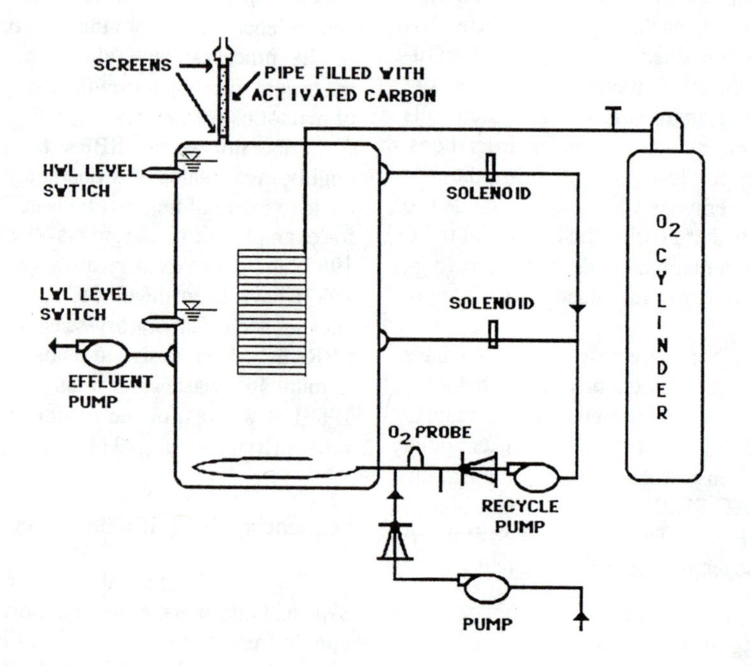

Fig. 1. Bench - Scale Sequencing Batch Biofilm Reactor

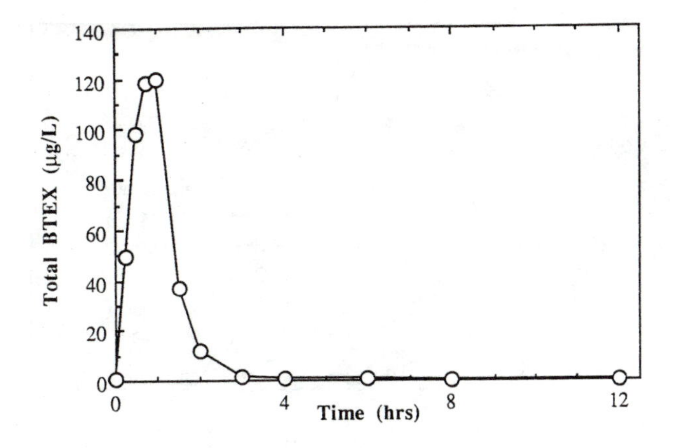

Fig. 2. BTEX Concentration Profile in the SBBR

bon-enhanced biological systems offer the advantage of biological regeneration of activated carbon, are less sensitive to the introduction of toxic compounds, and are able to treat wastewaters which contain nonbiodegradable components (Ying *et. al.*, 1987).

Soil Slurry - Sequencing Batch Reactors

SS-SBRs were used to evaluate the potential for bioremediation of contaminated soil containing bis-(2-ethylhexyl) phthalate (BEHP) and petroleum hydrocarbons. This study was undertaken to demonstrate to the New Jersey Department of Environmental Protection that BEHP is biodegradable.

The phthalate contaminated soil was collected from the plasticizer separation treatment unit at a manufacturing plant in northern New Jersey. In the first stage of this study, a Radiological/Consortia Study, using ^{14}C-labeled BEHP, demonstrated that the indigenous microbial populations could mineralize BEHP. Approximately 50-60% of the BEHP was mineralized to CO_2, while 40-50% was converted to cell mass.

In the second stage of this study, soil slurry reactors were used to determine nutrient requirements and BEHP and total petroleum hydrocarbon (TPH) removal rates for untreated (unacclimated), contaminated soil and a 50/50 blend of treated (acclimated) soil and untreated, contaminated soil.

BEHP concentration profiles during the second phase of this study are presented in Fig 3. Mixes A, B, and C represent 50/50 mixtures of untreated soil with soil from 3 slurry reactors which received different nutrient concentrations and seed cultures during an earlier study. Clearly, a lag in the degradation of BEHP can be avoided by adding nutrients to this contaminated soil, but degradation to low levels will take place even in unacclimated soils. The results for TPH (not presented) were similar to those for BEHP.

Summary

When the Sequencing Batch Reactor was initially conceived, it was considered a step back from the current state-of-the-art technology, continuous flow stirred tank reactors (CFSTRs). Today, a wide range of periodically operated systems based on the original concept of the SBR, with applications in municipal and industrial wastewater treatment, treatment of contaminated groundwater and soil, and treatment of hazardous wastes, have demonstrated that periodic processes are a part of true state-of-the-art technologies.

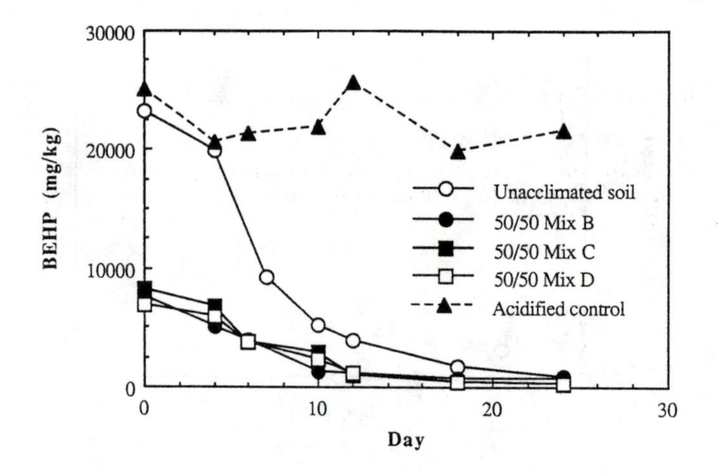

Fig. 3. BEHP Degradation in Soil Slurry Reactors

References

Chozick, R. and Irvine, R. L., "Preliminary Studies on the Granular Activated Carbon - Sequencing Batch Biofilm Reactor," *Environmental Progress*, **1991**, *10*, 282.

Irvine, R. L. and Ketchum, L. H., Jr., "Sequencing Batch Reactors for Biological Wastewater Treatment," *Critical Reviews in Environmental Control*, **1989**, *18*, 255.

Irvine, R. L., Ketchum, L. H., Jr., Breyfogle, R. E., and Barth, E. F., "Municipal Application of Sequencing Batch Treatment at Culver, Indiana," *J. Water Pollut. Control Fed.*, **1983**, *55*, 484.

Irvine, R. L., Sojka, S. A., and Colaruotolo, J. F., "Enhanced Biological Treatment of Leachates for Industrial Landfills," *Hazardous Waste*, **1984**, *1*, 123.

Ketchum, L. H., Jr., Irvine, R. L., Breyfogle, R. E., and Manning, J. F., Jr., "A Comparison of Biological and Chemical Phosphorus Removals in Continuous and Sequencing Batch Reactors," *J. Water Pollut. Control Fed.*, **1987**, *59*, 13.

Palis, J. C. and Irvine, R. L.,"Nitrogen Removal in a Low Loaded Single Tank Sequencing Batch Reactor," *J. Water Pollut. Control Fed.*, **1985**, *57*, 82.

Ying, W., Bonk, R. R., and Sojka, S. A., "Treatment of a Landfill Leachate in Powdered Activated Carbon Enhanced Sequencing Batch Bioreactors," *Environmental Progress*, **1987**, *6*, 1.

Sequencing Batch Biofilm Reactor Technology

Peter A. Wilderer, Institute of Water Quality and Waste Management, Technical University of Munich, Am Coulombwall, D-8046 Garching, Germany
FAX: 49-89-3209 3718

Sequencing Batch Biofilm Reactor (SBBR) technology profits from advantages of both, biofilm systems and periodic processes. High biomass of slow growing microorganisms can be maintained in the reactor, and biological process sequences can be controlled. Nitrogen and phosphate removal was achieved in laboratory trials, and granular activated carbon was biologically regenerated, when fixed bed biofilm reactors were operated in a fill and draw mode. Efficiency and stability of the processes were impressively high.

Studies on the performance of activated sludge reactors has revealed the importance of periodic changes of key milieu conditions. Frequent alternation of factors such as availability and lack of carbon sources, electron donors and electron acceptors allows simultaneous enrichment of a variety of microbial strains in multi-species biocommunities, and exploitation of the metabolic capabilties of the various strains. The bacteria are able to maintain high metabolic activity even when low substrate concentrations are to be established in the reactor effluent.

Realizing these relationships Irvine and his coworkers (1989) developed the Sequencing Batch Activated Sludge Reactor (SBASR). The SBASR allows easy and accurate control of the periodic changes of the various process conditions. Researchers and engineers have demonstrated efficiency and reliability of SBASR operations in pilot and full scale.

The SBASR technology - activated sludge technology in general - reaches application limits, when the quality of the wastewater constituents is poor, and the mass of substrates to be removed is little. Activated sludge cannot be built-up, and the reactor system fails. For that reason, treatment of hazardous wastes, and biological nutrient removal of the effluent of existing biological treatment plants is often difficult to achieve by means of activated sludge reactors.

The SBBR technology provides new ways to overcome these problems. Immobilization of microorganisms at the packing materials in the form of biofilms allows maintanance of high biomass and biomass-water interfaces, no matter how quickly and extensively the microorganisms can actually grow. Periodic changes of selected milieu conditions allows enrichment and control of different types of microorganisms in one biofilm culture.

Process description

A SBBR process cycle is characterized by a sequence of process phases, each pre-defined in time, duration and intensity.

Fig. 1 provides an overview of a SBBR process strategy to achieve nitrification and denitrification of secondary effluent. As a result of pretreatment of the wastewater in either activated sludge or trickling filter plants, biodegradable organic wastewater incredients are already low in concentration in the influent of the SBBR. Activated sludges cannot be built-up effectively under those dilute conditions, but accumulation of biomass in the form of biofilms is possible.

The SBBR process cycle begins with a fill phase during which the incoming wastewater trickles over the packing material. Once the reactor has been filled up to a certain level, an aerator is turned on to maintain aerobic conditions in the reactor. As a result, nitrification may proceed. Once completed, the aerator is switched off. Hydrodynamic sheer and mixing is maintained by means of recirculation. Either raw wastewater or carbon sources such as methanol or ethanol is pumped into the reactor to drive the denitrification process. Finally, the reactor is drained and kept wet until a new batch of wastewater is available.

1054–7487/92/0475$06.00/0

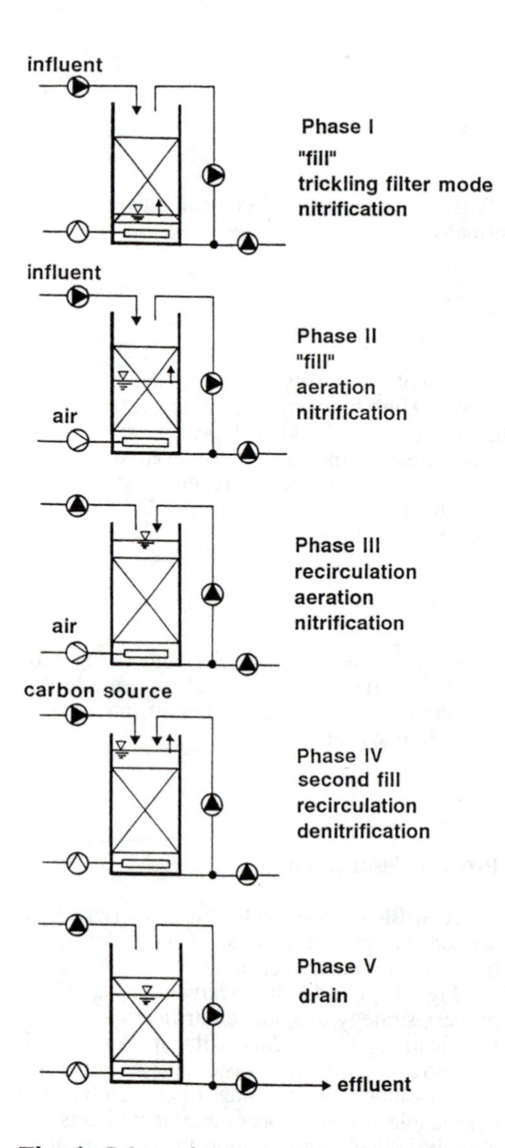

Fig. 1. Schematic representation of the SBBR process designed to achieve nitrogen removal

Results

Laboratory experiments have been conducted to study the applicability of the SBBR technology. The results obtained so far meet all expectations.

Ganzales-Martinez (1990) tried to achieve biological phosphate removal, nitrification and denitrification. Fig. 2. contains a typical result of his work. The reactor was rapidly filled, recirculated but not aerated at first to establish anaerobic conditions. The aerator was turned on - in this example 5.5 hours after start of the SBBR cycle - and operated for 6.5 hours. Finally, the reactor was completely drained and refilled.

After a start-up period of 12 weeks nitrification (99 per cent), denitrification (about 50 per cent), and phosphate removal (about 85 per cent) could be achieved. Shock loading had little or no effects.

The results of comparative studies (SBBR, continuous flow fixed film reactor) are summarized in a publication of Wilderer et al. (1992). As a result of the uniform distribution of the miroorganisms the SBBR responded to shock loading more flexible than the continuous flow reactor.

Fast fill strategies were recognized as important to develop well settleable activated sludge flocs. Rubio (1978) demonstrated with his experiments that the fast fill strategy was counter-productive when applied to SBBR. In case, the reactor received high quality organic substrates a thick biofilm developed (Fig. 3), causing clogging of the packing and subsequent system failure. Extracellular polymeric substances (EPS) accumulated, and made the biofilm very voluminous. Reaction kinetics were negatively affected by diffusion limitations. Biofilms which developed in a slowly filled reactor were thin, compact and active.

A very different application of the SBBR was investigated by Jaar (1992). He used granular activated carbon as packing material, loaded the reactor with hazardous substances, and studied subsequent responses.

Fig. 4 summarizes results obtained when the reactor was loaded with the herbizid 3-chlorobenzoate (3-CB). During the first two hours of each process cycle adsorption was the dominant process. In the bulk fluid the concentration of 3-CB decreased rapidly. Biodegradation of 3-CB contributed to the removal process. As a result, Cl^- appeared in the bulk fluid and eventually reached almost the stoichiometric value for complete mineralization. Because of the storage capacity of the activated carbon the influent concentration could be increased by a factor of about 5 without overwhelming the system. The long term loss of activated carbon adsorption potential was in the range of only 10 per cent.

Conclusions

The experiments conducted in laboratory scale have provided very encouraging results. The SBBR appears to be an interesting and most promising solution to a variety of problems. The metabolic capacity of slow growing

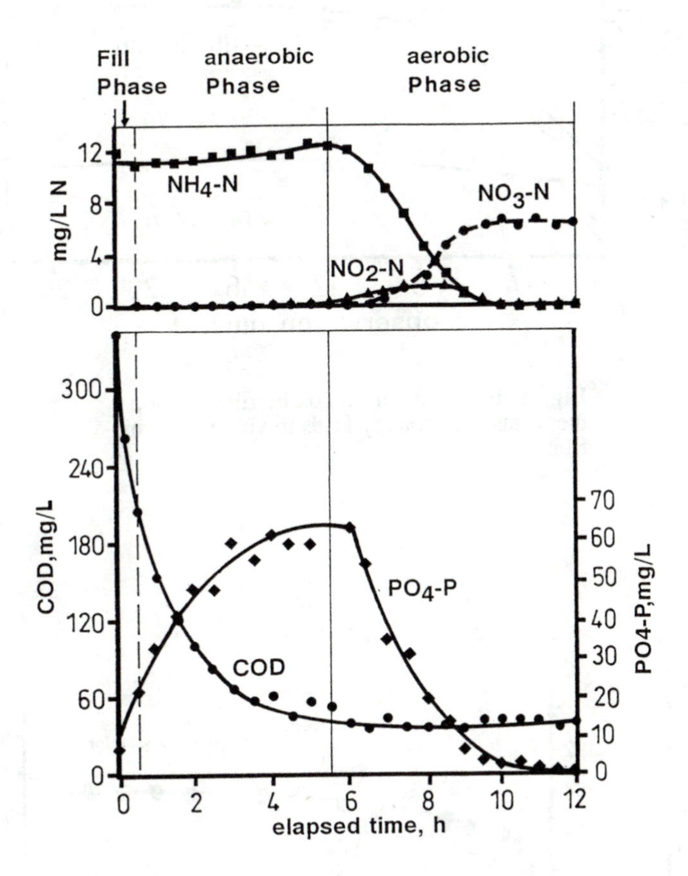

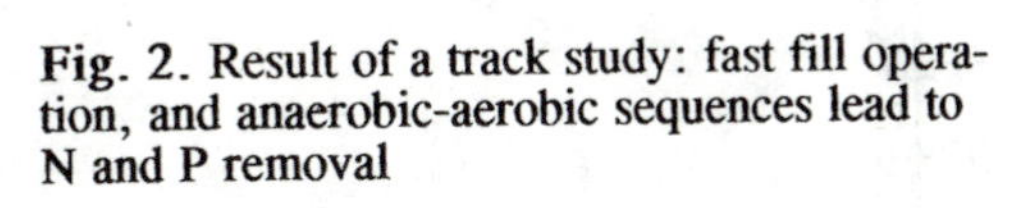

Fig. 2. Result of a track study: fast fill operation, and anaerobic-aerobic sequences lead to N and P removal

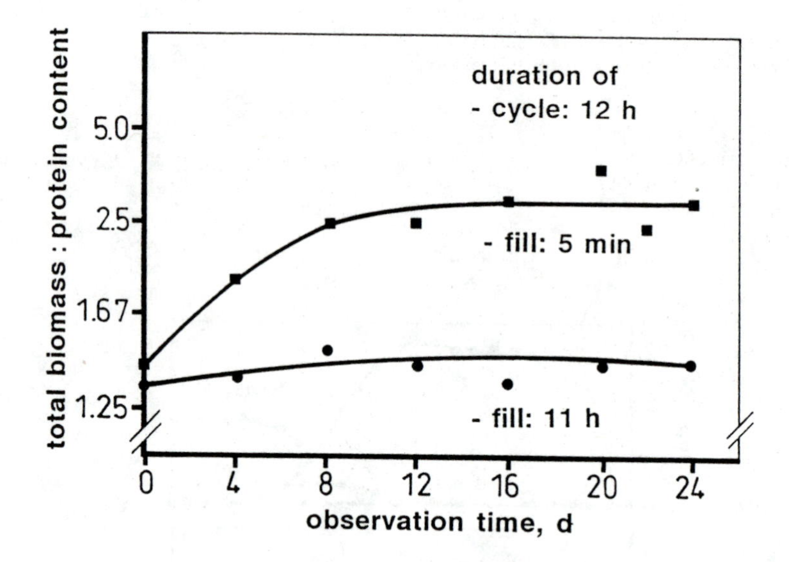

Fig. 3. Effect of fill rate on biofilm development: fast fill strategy leads to voluminous biofilms

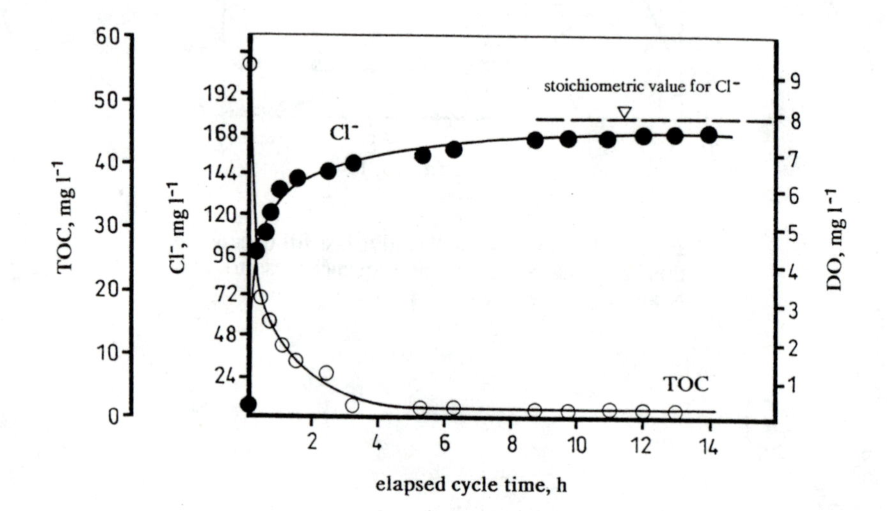

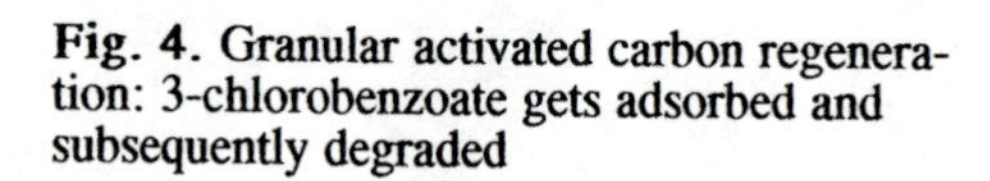

Fig. 4. Granular activated carbon regeneration: 3-chlorobenzoate gets adsorbed and subsequently degraded

microorganisms in multi-species biocommunities can be effectively and very reliably exploited. Of special interest are applications such as

- nitrification/denitrification of secondary effluents,
- treatment of landfill leachates,
- treatment of extraction fluids from soil remediation sites

In critical cases, the adsorption capacity of the packing material can be exploited to further enhance the efficacy of the SBBR. Pilot and full scale experiments are currently being executed to demonstrate applicability of the SBBR technology in practice, and to define proper design criteria.

References

Gonzales-Martinez, S., Wilderer, P.A., Wat. Sci. Techn., **1990**, 23, 1405-1416

Irvine, R.L., Ketchum L.H., Crit. Rev. Env. Contr., **1989**, 18, 255-294

Jaar, M.A.A., Wilderer, P.A., Wat. Sci. Techn., **1992**, in press

Rubio, L., Wilderer, P.A., Env.Letters, **1987**, 8, 87-94

Wilderer, P.A., Röske, I., Ueberschaer, A. Davids, L., Biofouling, **1992**, in press

Utilization of Landfills for Bioremediation of Hazardous Substances

Frederick G. Pohland, Department of Civil Engineering, University of Pittsburgh, Pittsburgh, Pennsylvania 15261 (FAX: 412-624-0135)

Landfills serve an essential role as ultimate receptors for residential, commercial and industrial wastes. Therefore, understanding the fundamental attenuating mechanisms determining the potential fate and transport of codisposed waste materials is critical to developing acceptable landfill management strategies. Results from landfill investigations with codisposed wastes are used to disclose various attenuating mechanisms, and to demonstrate the capacity of landfills for in situ bioremediation of hazardous substances.

Municipal landfills exist as dynamic, microbially-mediated and operationally-influenced treatment systems, with inherent capacities to attenuate a variety of waste constituents. In the absence of impeding or inhibiting influences, this capacity is driven by availability and sufficiency of moisture and nutrients, and is reflected by leachate and gas characteristics as the landfill matures. Normal fluctuations in these characteristics may be made more temporally and spatially predictable if the landfill is managed as a controlled bioreactor with accelerated stabilization and enhanced attenuation of waste constituents.

In the case of municipal landfills, the input waste contains various post-consumer products, including toxic or hazardous components derived from the household or small quantity commercial or industrial generators. The associated consequences can be demonstrated on the basis of a few fundamental principles and attenuating mechanisms, and by results of landfill simulations with leachate containment and recycle.

Landfill Simulations

Previous investigations (Pohland, 1975; Pohland, 1980) established the sequential nature of landfill stabilization, and emphasized the importance of the acid formation and methane fermentation phases as reflected by changes in leachate and gas characteristics. Using 5 pilot-scale landfill containment columns with leachate and gas management (Fig. 1), and loaded with municipal solid waste (MSW) and hazardous substances (Table 1), the effects on landfill evolution could be illustrated (Figs. 2, 3 and 4). Elevated leachate COD and TVA concentrations at low pH (and ORP), coupled with low gas production, indicated an early onset of acid formation and the development of a strong and chemically aggressive leachate transport phase. Consequently, rapid release of both inorganic and organic hazardous constituents occurred, as exemplified by Ni and Zn (Fig. 5) and dibromomethane (DBM) and trichloroethene (TCE) (Figs. 6 and 7).

With the onset of methane fermentation, decreased leachate COD and TVA, increased pH, persistently negative ORP and associated reduction of sulfates to sulfides, metal sulfide or hydroxide precipitation, and reduction of the mercuric ion to metallic mercury occurred. Moreover, the waste matrix provided a sink for heavy metals transported with leachate recycle from other regions of the landfill, and opportunities for physical adsorption, ion exchange, chemisorption upon complexation with insoluble ligands, and mechanical filtration and containment in transiently stagnant void volumes or pooled liquid were enhanced. Likewise, the complexation of metals by soluble ligands such as the humic-like substances (Pohland and Gould, 1986), although initially enhancing mobility, eventually resulted in sorption opportunity through formation of relatively hydrophobic interactions.

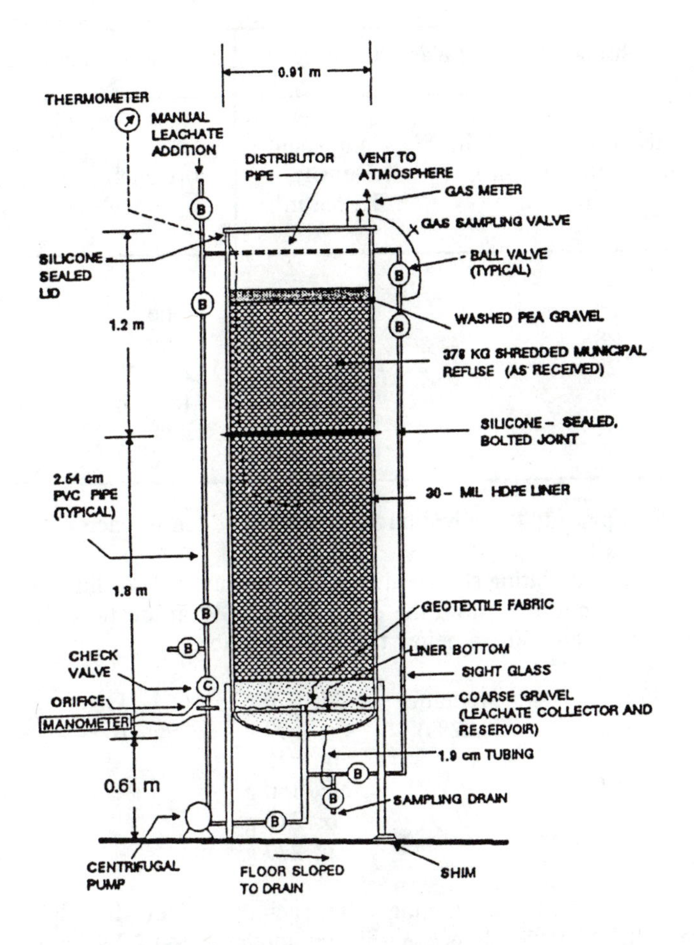

Fig. 1. Simulated landfill column with leachate recycle.

Table 1. Loading and Operational Characteristics of Simulated Landfills with Leachate Recycle

Simulated Landfill Column Identity	Shredded MSW/Waste Loading		Admixed Hazardous Constituents[a]	
	Initial Loading Height (above underdrain), cm	Compacted Density, Kg/cu m[a]	Inorganic[b]	Organic[c]
1 CR	29	313	None	None
6 OR	28	317	None	Yes
7 OLR	29	309	Low	Yes
9 OHR	29	313	Medium	Yes
10 OHR	31	293	High	Yes

[a]For shredded MSW as placed; 42 individual 9-kg batches added to each column (378 kg total).

[b]Added as augmented metal plating sludge at low (L), medium (M) or high (H) loadings divided into three equal portions and placed on the compacted MSW at locations 30 cm above the underdrain, at mid-depth, and 30 cm below the MSW surface.

Low: (0.11%)		Medium: (0.22%)		High: (0.44%)	
Cd;	35 g	Cd;	70 g	Cd;	140 g
Cr;	45 g	Cr;	90 g	Cr;	180 g
Hg;	20 g	Hg;	40 g	Hg;	80 g
Ni;	75 g	Ni;	150 g	Ni;	300 g
Pb;	105 g	Pb;	210 g	Pb;	420 g
Zn;	135 g	Zn;	270 g	Zn;	540 g

[c]Added as an organic (O) cocktail containing 120 g each of bis-2-ethylhexylphthalate, 1,4-dichlorobenzene, 1,2,4-trichlorobenzene, dibromomethane, γ-1,2,3,4,5,6-hexachlorocyclohexane, hexachlorobenzene, 2,4-dichlorophenol, 2-nitrophenol, naphthalene, nitrobenzene and trichloroethene, and 30 g dieldrin all placed as a mixture at the surface of the first 30 cm of compacted MSW in each column; 1.35 kg total or 0.36%.

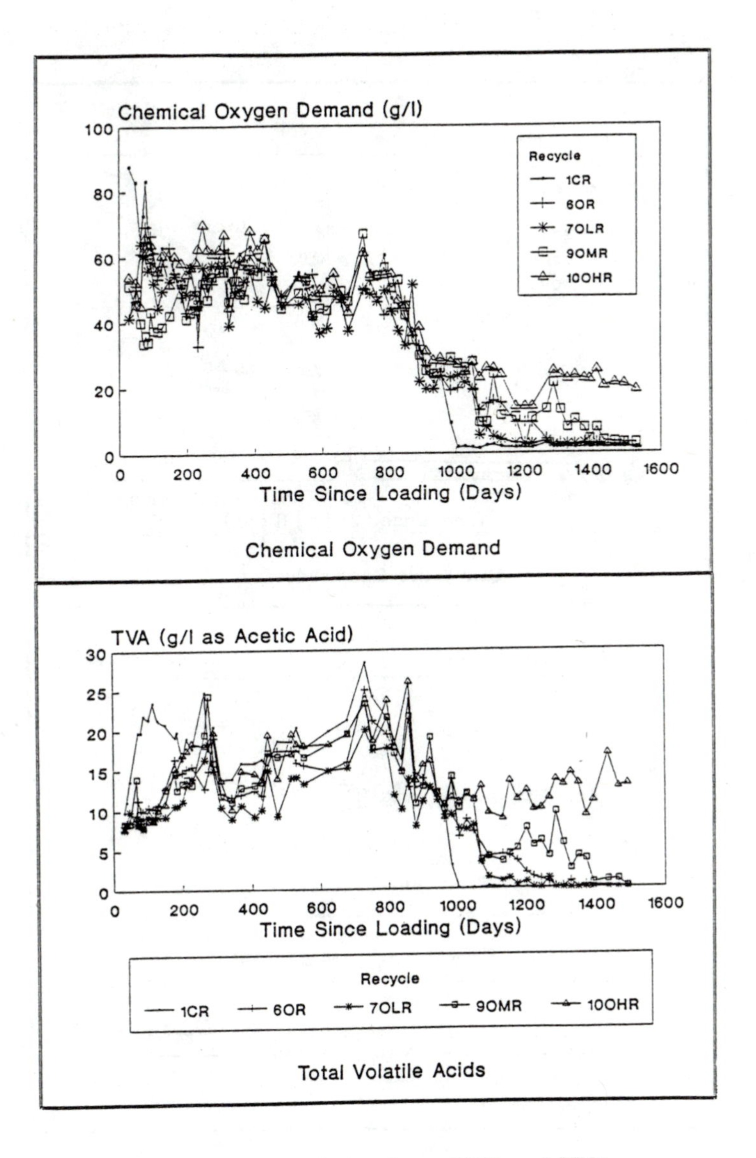

Fig. 2. Changes in leachate COD and TVA.

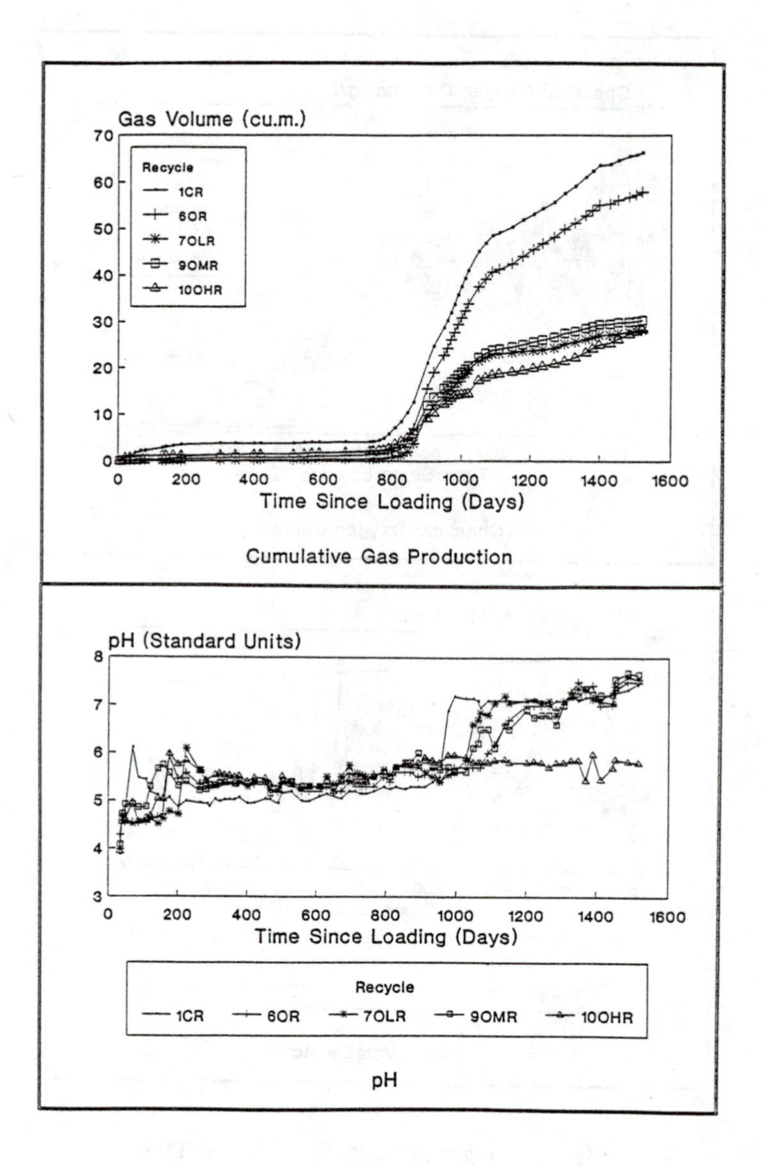

Fig. 3. Changes in gas production and leachate pH.

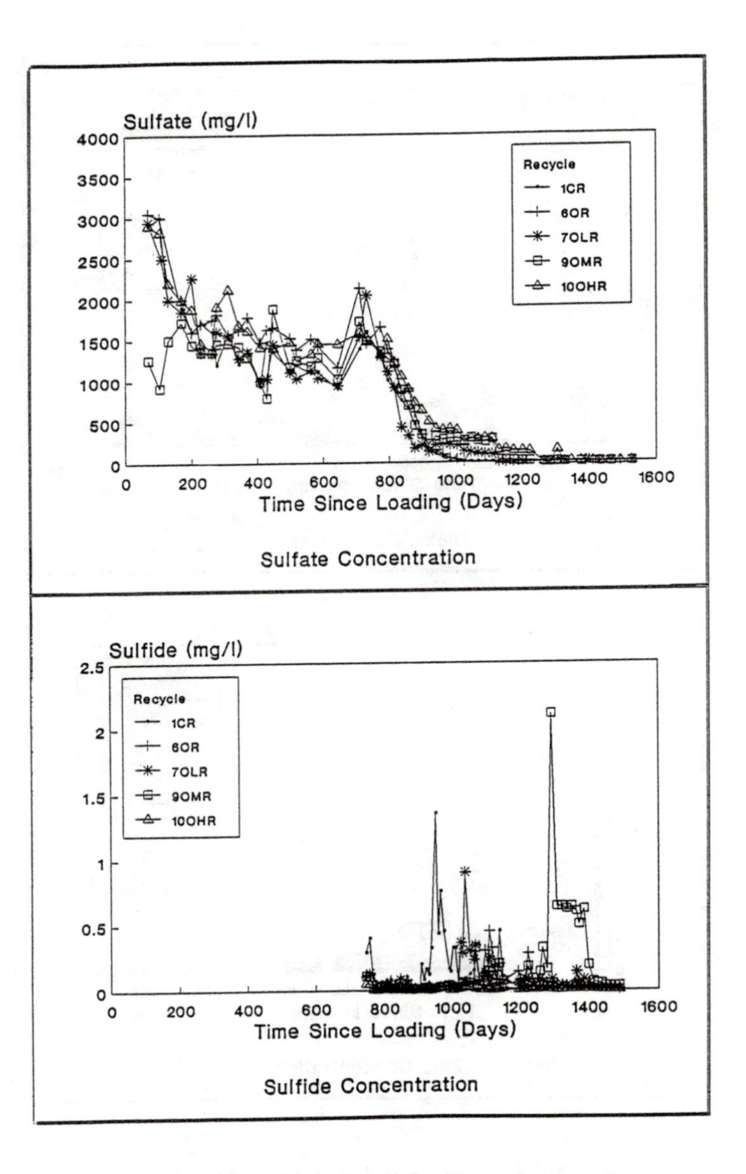

Fig. 4. Changes in leachate sulfate and sulfide.

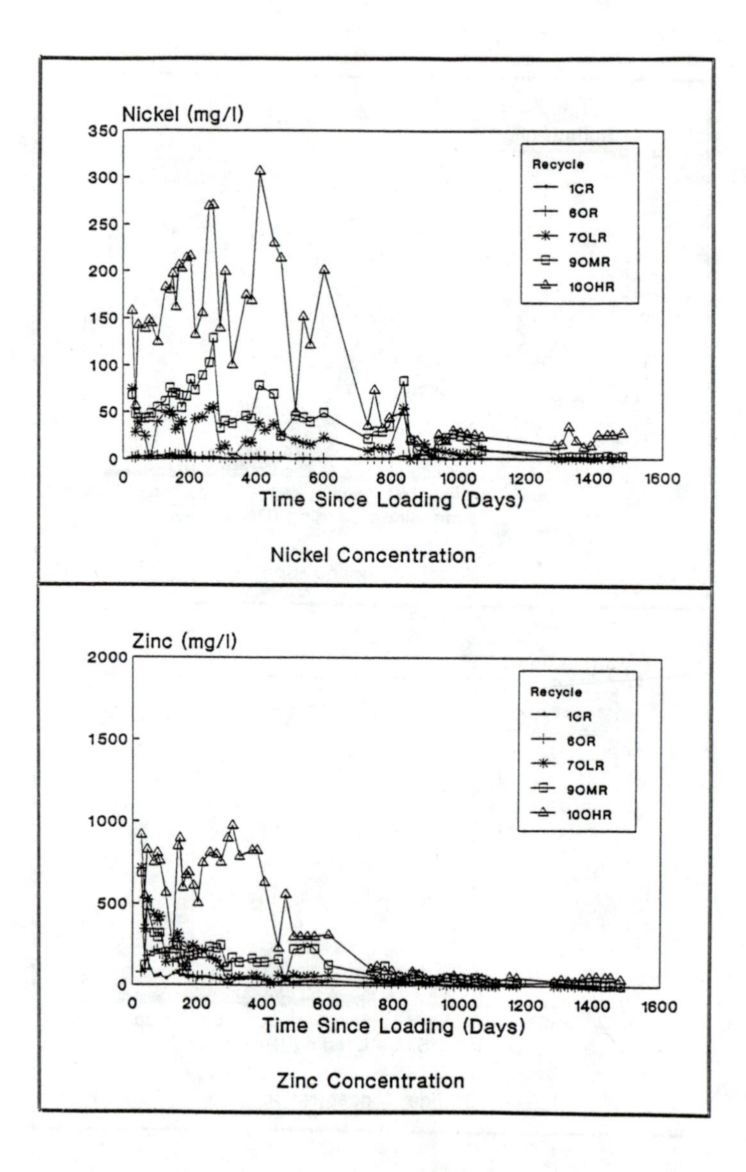

Fig. 5. Changes in leachate Ni and Zn.

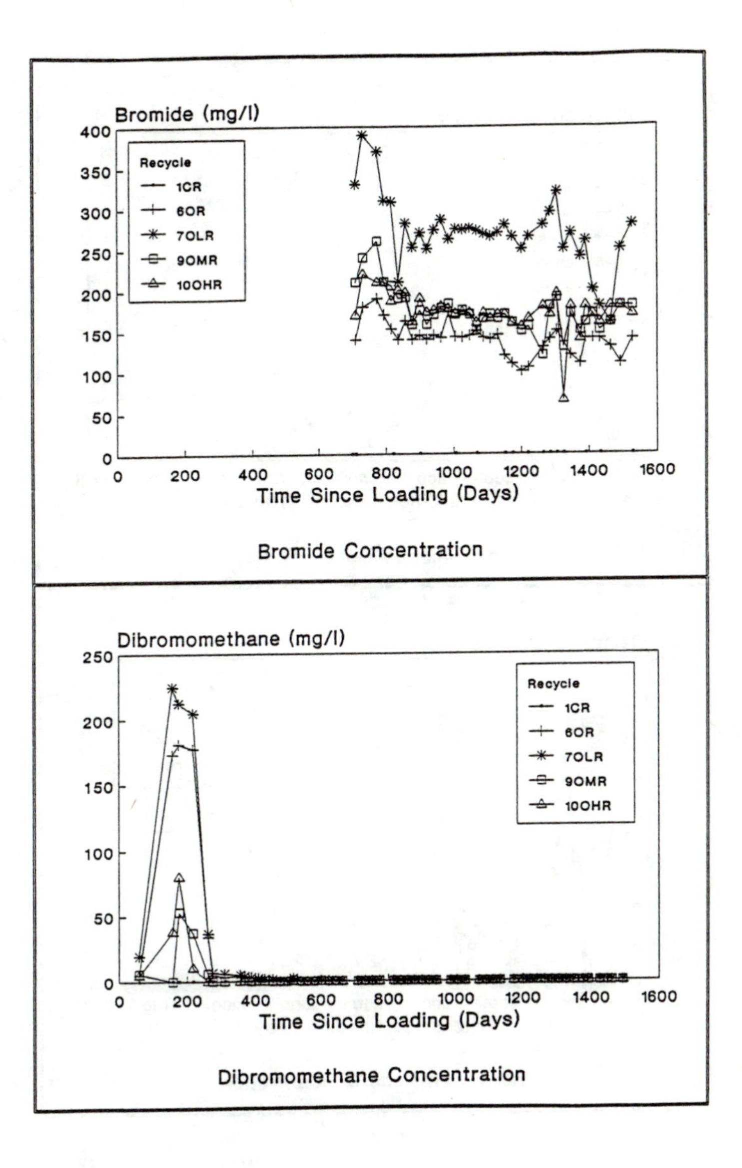

Fig. 6. Changes in leachate Br and DBM.

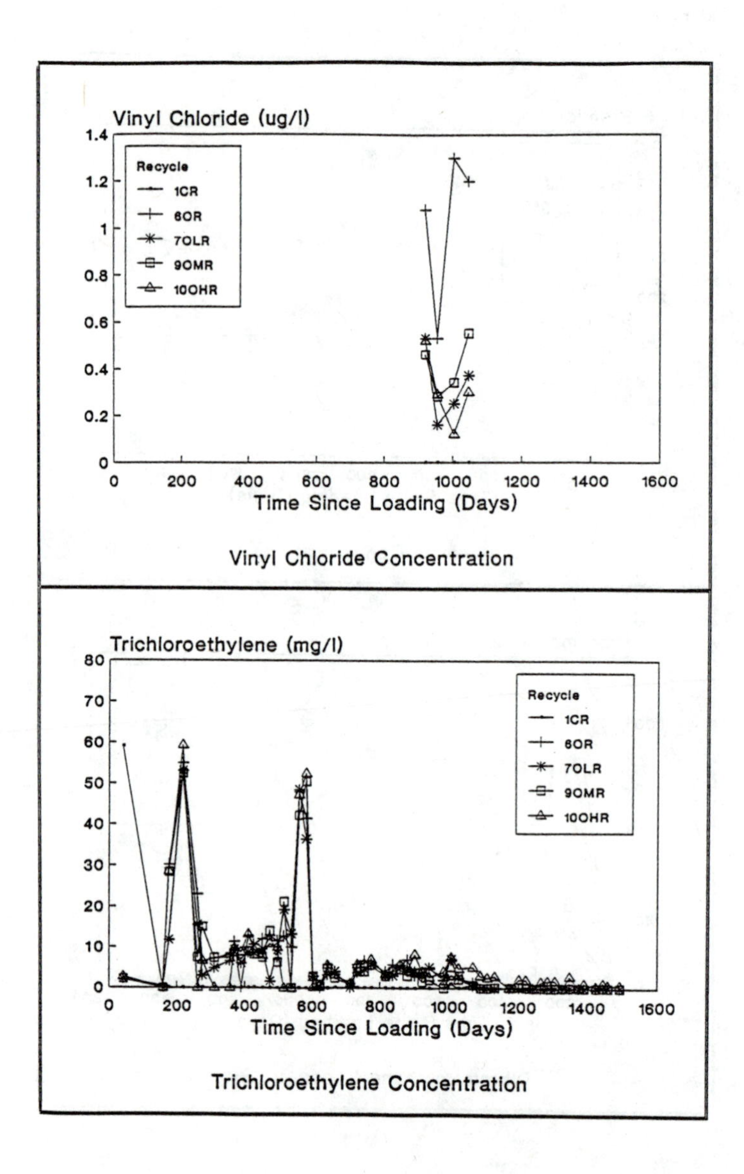

Fig. 7. Changes in gas-phase VC and leachate TCE.

Hydraulic residence times of 200 to 400 days, and the extended opportunity for contact and microbial acclimation provided by leachate recycle, promoted attenuation of hazardous organic constituents as evidenced by the decrease in DBM with accumulation of bromide, and appearance of gas-phase vinyl chloride as leachate TCE was being attenuated. In addition, the overall potential for release or attenuation of a particular organic compound to or from the gas, liquid and solid phases was largely dependent on solubility, the possible mobilizing influences of other leachate constituents on the more hydrophobic compounds, and physical sorption in the waste matrix. Only the soluble, less hydrophobic and more refractory codisposed organic compounds eluted from the waste mass, and these emerged in the approximate order of increasing affinity for the MSW constituents. The pattern of attenuation of the mobile, hydrophilic compounds by microbially-mediated transformations was influenced by the phase and extent of landfill stabilization and the enhancement provided by leachate recycle.

SUMMARY AND CONCLUSION

Municipal landfills possess finite capacities to attenuate toxic and hazardous waste constituents codisposed with MSW at loadings potentially encountered in conventional practice. Codisposed inorganic heavy metals are attenuated by in situ microbially-mediated physicochemical processes of reduction, precipitation, sorption and waste matrix capture. Codisposed organic waste constituents fractionate according to their physical and chemical properties and those of the waste matrix, and are retained or transformed primarily by sorption and/or bioconversion, with the sequential generation and conversion of identifiable reaction products. Therefore, controlled landfills employing leachate recycle and final removal can provide significant reductions in hazardous substances and in the potential for adverse health and environmental risks.

REFERENCES

Pohland, F.G. (1975) "Accelerated Solid Waste Stabilization and Leachate Treatment by Leachate Recycle through Sanitary Landfills". Progress in Water Technology, Vol. 7, 3/4, 753.

Pohland, F.G. (1980) "Leachate Recycle as Landfill Management Option". Journal of Environmental Division, ASCE, Vol. 106, EEG, 1057.

Pohland, F.G. and Gould, J.P. (1986) "Co-Disposal of Municipal Refuse and Industrial Waste Sludge in Landfills". Water Science and Technology, Vol. 18, No. 12, 177.

A New Approach for the Control of Micropollutants in Water----Biological Detection of Hydrophobic DNA Toxic Substances

Saburo Matsui
Kyoto University, Laboratory for Control of Environmental Micropollutants, Otsu City, Japan, 520
(FAX:+81 775 24 9869)

Bacillus subtilis rec-assay was applied to water samples from Lake Biwa and the Yodo River which provide municipal drinking water sources for 14 million people in the basin of Japan. DNA toxicity of the samples increased toward downstream indicating strong potential of the toxicity almost equivalent to that of MNNG, 4NQO etc.. Micropollution of hydrophobic DNA toxic substances needs more attention for the safety of ecosystems as well as drinking water. The rec-assay can be applied to water samples which contain a mixture of DNA toxic and non-DNA toxic substances. New waste water technology is necessary to focus on hydrophobic DNA toxic micropollutants.

DNA toxicity means any type of interaction between chemicals and DNA molecules as well as enzymes for DNA synthesis, followed by interference of DNA duplication, breakage of DNA molecules and alkylation of DNA bases, etc.. The possible relationship between mutation and cancer development is well known as the initiation process. However, recent findings in cancer research also show another important relationship between a deletion of cancer suppressor genes in certain regions of chromosomes and the development of cancer cells. The deletion is caused not only by mutation but also by other DNA interaction with chemicals most of which are metabolically modified and activated before injuring DNA molecules.

There are three different points of view on water quality in terms of DNA toxicity. The first is drinking water quality for man: the second is toxicity for aquatic animals : the third is food chain accumulation of DNA toxic substances. DNA toxicity in drinking water must be evaluated together with the possible intake of the toxic substances through food and air. If contamination of DNA toxic substances in drinking water increases compared with other routes of intake, water purification technology must decrease the toxicity. However, difficulties remain in the second and the third points of view. Many unknown areas are still remaining for chronic toxicity including DNA toxicity. Many biologically recalcitrant substances are subject to food chain accumulation and magnification followed by the influence of DNA toxicity on the ecosystem.

We have developed the Bacillus subtilis rec-assay which can detect wide spectrums of DNA toxicity including, intercalation, and breakage of DNA molecules, alkylation of DNA bases, etc.(Matsui, 1984, 1988, 1989). The important advantage of the rec-assay compared to other microbial mutation tests including the Ames methods, is the capability of evaluation of DNA toxicity together with non DNA toxicity of test samples. Environmental samples are always a mixture of numerous chemicals including cytotoxic effects on the test microbes, which often makes hindrance of evaluation of mutation tests. The application of the Ames methods to the environmental samples always faces so called "killing effect" that means no dose-mutation relationship obtained by death of the test bacteria due to the cytotoxic effect of the mixture.

In this paper, we applied the rec-assay to an environlmental survey of Lake Biwa and the Yodo River, in Japan, to evaluate DNA toxicity of hydrophobic substances. XAD-2 resins were used for concentration of those hydrophobic micropollutants in water. The reason why XAD-2 resins was used is that bioaccumulation of micropollutants in water is the major concern for DNA toxic substances. Lake Biwa and the Yodo River have a

1054–7487/92/0490$06.00/0

large basin area (7,281 km^2) where about 14 million people live on the highly developed water complex. There is upstream Lake Biwa which is a large natural reservoir with relatively good water quality for a drinking purpose as well as many other water utilization downstream. However, in the middle of the basin there are many urban areas developed including Kyoto City of 1.2 million population, which discharge urban and industrial wastewaters with agricultural run off.

There are big cities located very downstream such as Osaka, Kobe, etc. which take municipal water sources from the Yodo River. Deterioration of water quality along Lake Biwa and the Yodo River has become one of the most important environmental issues in Japan.

Materials and Method

Bacillus subtilis strain H17(arg-,trp-,recE+), was used as a recE function proficient strain. A derivative of this strain, M45(arg-,trp-,recE-) , which was very sensitive to gamma rays, ultra violet rays, and pharmaceuticals, was used as the recE function defective strain(Sadaie and Kada,1976).

DNA Toxicity Evaluation

A typical result for a DNA damaging substance is depicted on the left hand side of Fig. 1. One way of evaluating DNA toxicity potential is to compare the concentrations of a test substance at 50% survival turbidity for strains H17(Rec+) and M45(Rec-), i.e.: R50= RC50Rec+/RC50Rec-.

When R50 becomes large, the test substance shows stronger potential of DNA damage. The important advantage of rec-assay is to provide cytotoxicity of test samples to the bacteria in terms of RC50Rec+ which is equivalent to LC50 of the wild type of Bacillus subtilis.

Another way of evaluating the DNA damaging potential of a substance is measuring the area enclosed between the survival lines of Rec+ and Rec-. In order to obtain the precise area, a mathematical treatment can be introduced such as transformation of data by the Probit analysis. After conversion to the Probit coordinate, the two survival lines are obtained. The right hand side of Fig.1 shows the Probit conversion. The enclosed area between the converted two lines is obtained by integration and designated as S-probit.

The standard substances were selected for comparison to assess the DNA toxicity potential of chemicals. The results of the Bacillus subtilis liquid /microsome rec-assay for the standard substances are shown in Table 1.

Since we use the XAD-2 resins to concentrate and extract DNA damaging substances from environmental water samples, we can quantitatively evaluate the DNA damaging potential of the samples by introducing the following new indicator:Rec-volume = S-probit/(the volume of sample water in liter applied

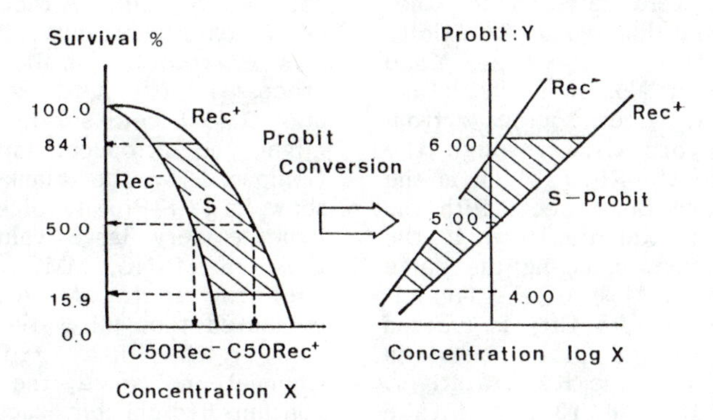

Figure 1. Survival lines for H17(Rec+) and M45(Rec-) tested with a typical DNA damaging substance. R50 is obtained from the comparison of CR50 Rec+ and CR50Rec-. Conversion to the Probit coordinate is shown on the right side, where the S-probit is shown.(Reproduced with permission from Matsui, 1989b. Copyright 1988 IAWPRC)

Table 1 Results of the Bacillus subtilis/microsome Rec-assay for Standard Substances

Compound	S9 Act.	R50 = CR50 Rec+/CR50Rec-* mg/L mg/L	S-probit
	-	$1.01 = 4.70\times10^{-1}/4.65\times10^{-1}$	0.039
KM	+	$0.99 = 6.01\times10^{-1}/6.10\times10^{-1}$	0.029
DMSO	-	$1.11 = 1.74\times10/1.57\times10$	0.054
	+	$0.99 = 1.57\times10/1.58\times10$	0.010
MNNG	-	$24.8 = 1.76\times10^{2}/7.09$	2.901
	+	$7.01 = 2.96\times10^{2}/4.22\times10$	1.390
MMC	-	$7.01 = 2.44\times10^{-2}/3.48\times10^{-3}$	1.790
	+	$5.28 = 2.97\times10^{-2}/5.62\times10^{-3}$	1.177
EMS	-	$9.08 = 2.47\times10^{3}/2.72\times10^{2}$	1.716
	+	$23.2 = 8.45\times10^{3}/3.64\times10^{2}$	2.496
4NQO	-	$54.6 = 4.48\times10^{-1}/8.21\times10^{-3}$	3.703
B(a)P	+	$26.1 = 7.16\times10^{2}/2.74\times10$	1.092
2AAF	+	$65.7 = 4.29\times10^{2}/6.53$	2.608
DMN	+	$3.48 = 6.75\times10^{4}/1.94\times10^{4}$	1.624

*The concentration indicated in this table is the concentration of the test substance with the bacteria in the interaction period.

to XAD-2 resin, by which 50% survival of Rec- is obtained).

Results of Survey and Discussion

Thirteen points were selected for the survey of micropollution of DNA toxicity along Lake Biwa and the Yodo River.Sampling point No. 1 is the intake point of municipal water sources from Lake Biwa to Kyoto City through the Sosui channel. The Uji River which is the main outlet of Lake Biwa meets with the Kizu River and the Katsura River in the middle of the stream, changing the name into the Yodo River. Most of the municipal wastewater of Kyoto City is treated and discharged to the place between points No. 6 and 7 in the Katsura River. Points No. 11, 12 and 13 are intake points for the municipal water supply of respectively, Osaka Prefectural Government, Kobe City and Osaka City in the Yodo River.

Data of the rec-assay as well as the conventional water analysis were shown in Table 2. The blank in Table 2 means results of Rec-assay of distilled water of 18L prepared with the same procedures of XAD-2 concentration, elution, and redissolving in DMSO. As indicated in Table 1, DMSO solution itself shows a very small value of S-probit. Although occurrence of DNA toxic substances in distilled water was extremely small, but during the processes of XAD resin concentration and DMSO redissolving, contamination might be introduced into blank solutions. Compared to the blank value, all data show large S-Probit values, some of which indicate very large values equivalent to those of MNNG, MMC and 4NQO. Based upon the scale of S-Probit, it can be evaluated that all water samples except No. 1, the intake of Biwako- sosui Channel and No. 2, the Seta River Weir, contain hydrophobic substances of very strong potential of DNA toxicity. DNA toxicity of those water samples could be attributed to numerous substances which we have not identified yet.

The concentration factor of Rec- at 50% survival which is equivalent to LC50,

Table 2 Results of the Rec-assay for Water Samples of Lake Biwa and the Yodo River

Point	Date	Temp.	E260	TOC	RC50Rec	RC50Rec-	S-probit	Rec-volume
		°C		mg/L	C.F.	C.F.		
1	11/05/91	16.4	.035	5.9	64.69	37.82	.47	20.54
2	11/05/91	16.7	.039	8.4	113.70	54.08	.65	19.89
3	13/11/90	13.0	.048	6.9	355.60	54.87	1.62	49.30
ibid.	06/06/91	18.7	.049	4.9	365.40	82.90	1.29	25.89
4	06/06/91	21.5	.081	6.9	55.72	14.59	1.16	132.97
5	13/11/90	14.0	.064	8.9	35.72	4.17	1.87	746.34
6	15/11/90	15.5	.020	3.3	214.40	33.36	1.62	80.74
7	15/11/90	16.7	.088	8.2	56.79	6.45	1.89	488.68
8	15/11/90	16.0	.083	7.0	67.91	6.20	2.08	559.14
9	13/11/90	14.0	.062	4.8	148.20	27.96	1.45	86.31
10	13/11/90	13.8	.062	3.3	100.80	13.46	1.75	216.57
11	08/06/91	23.4	.074	4.7	272.10	22.91	2.15	156.34
12	06/01/91	5.0	.063	5.9	122.60	4.16	2.94	1178.45
13	06/01/91	4.0	.051	6.0	237.30	4.11	3.52	1427.53
ibid.	08/06/91	24.0	.055	3.8	174.60	12.79	2.27	295.80
Blank					315.20	290.70	.07	

is only 6.45 while the concentration factor of Rec+ is 56.79 giving the value of R50, 8.80 which is almost equivalent to the value of EMS. The scale of S-probit can show the potential of DNA toxicity, but cannot indicate how much volume of sample water was used to concentrate hydrophobic micropollutants. The new index Rec-volume includes such information on volume of sample water used. Interesting results are the comparison of Rec-volume values among sampling points No. 6, 7, and 8, where the point No. 6, Kuze Bridge shows slightly contaminated conditions, while No. 7, Kuga Bridge and 8, Miyamae Bridge, clearly indicate with the sharp increase of Rec-volume values, the influence of the discharge of treated sewage, industrial wastewater, untreated gray water and urban run off from most areas of Kyoto City. After the confluence of three tributaries, the value of Rec-volume decreased probably due to dilution by the Kizu River. However, the Rec-volume values sharply increased again toward downstream due to discharge of industrial wastewater as well as untreated sewage.

The Rec-volume values exceeded 1400 which was almost 70 times higher than the values of points No. 1 and 2. Seasonal changes of Rec-volume values were also observed with results of No. 3 and 13. Further study is necessary in this respect. The water purification processes taken by water works of Osaka City, Kobe City and Osaka Prefectural Government have not included the activate carbon treatment. Those municipalities recently announced that they would introduce the activated carbon process for the near future. This decision is supported by the results of this research due to increased contamination of DNA toxicity toward downstream in the Yodo River. When the results of rec-assay were compared to other indexes of conventional water quality, no straight relationships was observed. In order to protect the ecosystems of the lake and rivers from DNA toxicity, new waste water technology is necessary to be developed.

References

Kada, T.; Tutikawa, T.; Sadaie Y. Mutation Research, **1972**, 16, 165-174.

Matsui, S. In Freshwater Biological Monitoring; Pascoe, D.;R.W. Edwards, R. W. Eds; Advances in Water Pollution Control, Pergamon Press, London, **1984**, pp 143-152.

Matsui,S. Toxicity Assessment, **1988**, 3. 173-193.

Matsui, S. In Bacillus subtilis: Molecular Biology and Industrial Application; Maruo, B.; Yosikawa, H. Eds. Elsevier, Kodansha Ltd.,Tokyo, **1989a**, pp 241-260.

Matsui, S., Yamamoto, R. and Yamada, H.(1989). Water Science & Technology, **1989b**, 21, 875-887.

Agriculture and Food Biotechnology Symposium X

M. R. Ladisch and C. Rolz: *Co-Chairs*

Impact of Transgenic Plants: Session A
G. Kishore and J. Schell: *Co-Chairs*

Biomass Utilization: Session B
M. R. Ladisch and C. Rolz: *Co-Chairs*

Food Flavors and Additives: Session C
S. Gendel and P. Schreier: *Co-Chairs*

Characterization of Genes Affecting Plant Growth and Development

Jeff Schell*, **Csaba Koncz, Angelo Spena, Rick Walden,** and **Klaus Palme**
Max-Planck-Institute for Plant Breeding Research, D-5000 Köln 30, FRG
(Fax: 0221-5062213)

Plant development is characterized by the fact that the fate of most cells is not fixed at some relatively early stage during the development of the embryo. Cell differentiation and cell fate determination in most plants are ongoing processes throughout development. As a consequence, most organs consist of cells that are potentially "totipotent", since their fate is not irreversibly fixed. This observation has been well recognized and forms the basis of both experimental and commercial plant tissue culture techniques. Different regimes of so-called growth hormones - such as auxins and cytokinins -, of energy sources (e.g. sucrose) and of various ions can be used to trigger somatic embryogenesis or organogenesis in a wide variety of plant species.

A number of soil bacteria have evolved the capacity to exploit this property of plants to their own advantage. Well known examples are pathogens such as *Agrobacterium tumefaciens, A. rhizogenes* and *Pseudomonas savastanoi.*

Indeed these bacteria have evolved mechanisms allowing them to specifically modify the growth and differentiation of various plant organs to suit their own growth requirements. It has been rewarding to study the structure and function of the procaryotic genes that allow these bacteria to specificallly interfere with the "normal" development of their eucaryotic hosts. It was our assumption that the mechanisms by which these procaryotic genes control plant growth and development would be representative of the mechanisms active in the plants themselves. In this, as well as in other ways, these procaryotic genes could be regarded as "oncogenes".

This approach, which was initiated in the early seventies, has thus far provided us with insights in the following mechanisms of control of plant growth and differentiation:

I. Deregulated synthesis of plant growth factors leads to dedifferentiation and tumorous growth

A. tumefaciens induces so-called "crown-gall" tumours on a variety of mostly dicotyledonous host plants. The dedifferentiated growth of the infected tissues is the direct consequence of the deregulated synthesis of auxins and cytokinins by enzymes coded for by three genes (*iaa*M, *iaa*H and *ipt*) carried by a DNA fragment (T-DNA) which is transferred from a plasmid (Ti-plasmid), carried by the pathogenic bacteria, into the genome of the infected plant cells.

II. The activity of growth factors can be modulated by the synthesis of specific antagonists

The T-DNA segments of *A. tumefaciens* strains not only harbour the *iaa*M, *iaa*H and *ipt* genes coding for enzymes catalyzing the synthesis of auxins and cytokinins, but in addition these T-DNAs carry genes, such as gene 5, the function of which is to modulate the activity of the growth factors produced by the major oncogenes.

Indeed, gene 5 was recently shown to be responsible for the synthesis, in transformed plant cells, of an auxin-analogue: indole-lactic acid (ILA). Transgenic tobacco plants expressing gene 5 under control of the 35S RNA promoter of the plant virus CaMV, produce ILA and develop without readily observable alterations. Their seedlings however tolerate levels of exogenously applied auxins that are toxic to isogenic non-transgenic tobacco seedlings. This protection against toxic levels of the growth hormone auxin, might well result from the observation that indole-lactic acid (ILA)

1054–7487/92/0496$06.00/0

competes with active auxins such as indole-acetic acid (IAA) for binding to auxin binding proteins that act as auxin-receptors. T-DNAs therefore not only introduce genes in plant cells forcing them to synthesize growth-factors in a deregulated fashion but also introduce a linked gene (gene 5) coding for the synthesis of a growth factor antagonist. The expression of these procaryotic "pathogenesis-genes" in the transformed host cells is fine tuned since it has been shown that the promoter of gene 5 becomes active in the presence of auxins but is repressed in the presence of both auxin and its antagonist ILA.

III. Plant growth factors can not only act extracellularly after transport to target cells but also intracellularly in a cell specific fashion, by activation of intracellular pools of inactive conjugates

Agrobacterium rhizogenes is a pathogen that induces the formation of adventitious roots (called "hairy roots") on a number of plant organs that would not otherwise make such roots. Also, in this case, the abnormal growth was shown to be due to the transfer and expression in plant cells of a set of genes carried on a transferable T-DNA fragment harboured by a plasmid (Ri-plasmid) in *A. rhizogenes*.

The procaryotic genes responsible for the abnormal growth were called *rol* (for root locus). It has been demonstrated that the *rol*B gene, in combination with either the *rol*C gene or the *rol*A gene, was sufficient to induce root growth in several plants and that these genes acted in a cell specific fashion. Indeed, only cells containing these *rol* genes were able to grow as transformed roots. It was therefore thought unlikely that these genes would somehow be involved in the synthesis of growth hormones since these were expected to act extracellularly also on non-transformed cells. In fact it was found that the *rol*C gene codes for an enzyme that releases active cytokinins from inactive intracellular cytokinin glucosides, whereas *rol*B was similarly shown to code for an enzyme capable of hydrolizing inactive auxin-glucosides, thus releasing active auxins in the transformed cells.

IV. Use of gene tagging to identify genes involved in phytohormone perception/regulation

In order to test whether or not plants normally make use of mechanisms similar to those evolved by these soil bacteria, we initiated a search for plant cell mutants that would be able to grow and differentiate in the absence of extracellularly supplied auxins. In order to rapidly identify and clone genes involved in conveying auxin-independent growth, specially designed T-DNA vectors were used to activate and tag genes that are normally silent in tobacco callus cultures which require auxins for growth. Because the inserted tag is designed to stimulate the transcription of genes, the mutants are expected to be dominant.

At least 4 different classes of dominant auxin-independent mutants were thus obtained. Calli from these mutants grow well in the absence of extracellularly supplied auxins but can be regenerated to form fertile plants. Protoplasts derived from the leaves from these mutants were shown to be able to form calli on media devoid of auxins.

Currently work is underway to characterise the genes responsible for directing auxin independent growth and we are interested in testing whether these mutants will suppress or enhance the abnormal growth of transgenic plants expressing the oncogenes from pathogenic or symbiotic gram negative soil bacteria.

V. Isolation of plant genes encoding auxin binding proteins

A further approach to study the mechanism of action of phytohormones consists in the isolation and functional characterization of potential receptors involved in phytohormone perception and transduction. We have concentrated on the isolation of auxin binding proteins that might act as auxin-receptors. Previous binding and inhibitor studies had shown that plant cells contain different auxin binding sites located in the endoplasmatic reticulum or the tonoplast or the plasma membrane.

An auxin binding protein (ERabp1) was

recently purified and the corresponding genes from *Zea mays* and *Arabidopsis* were cloned and their primary structure deduced. ERabp1 codes for a protein located in the lumen of the ER (with a KDEL sequence at its C-terminus), but also possibly associated with the plasma membrane. Indirect evidence suggests that this auxin binding protein may be a receptor for the auxin signal. Other auxin binding proteins, located at the plasma membrane, may well be involved as influx and efflux carriers in the transport of auxins in plant tissues.

To identify and isolate different auxin binding proteins located on plant cell plasma membranes, we have synthesized auxin-derived photoaffinity probes (5-azido-7-^{3}H-indol-3-acetic acid) to covalently label auxin binding proteins. Three different proteins were labelled called pm23, p58 and p60.

p60 is apparently a member of a gene family and might well be synthesized as a larger precursor protein having a N-terminal extension. The predicted amino acid sequence of a cDNA clone that corresponds to a p60 protein shares peptide homology with a number of eucaryotic ß-D-glucosidases. Indeed, p60 was shown to have ß-D-glucoside glucohydrolase activity (E.C.3.2.1.21). Interestingly a conserved central region of p60 shares similarity with a region of the *A. rhizogenes rol*C gene which encodes a cytokinin-N-glucoside glucosidase. Preliminary evidence indeed suggests that p60 can hydrolyze cytokinin-glucosides. Whereas p60-like proteins were detected both in plasma membrane enriched and cytosolic fractions, this is apparently not the case for pm23 which was detected only in fractions highly enriched for plasma membranes. Its labeling by 5-azido-IAA was efficiently and specifically inhibited by TIBA (2,3,5 triiodobenzoic acid), a potent polar auxin transport inhibitor and by the phytotropin naphthylphthalamic acid (NPA) indicating that pm23 could well be part of an auxin efflux carrier. A cDNA clone corresponding to pm23 from maize coleoptiles was isolated and sequenced. No extensive homology to ERabp1 and no similarity to sequences in data bases were detected.

The p58 protein has not been thus far subjected to further studies.

In summary: the extensive analysis of auxin binding proteins has not revealed any typical eucaryotic transmembrane receptor protein. The question therefore arises whether such classical receptors are involved in auxin perception and response.

Relevant Bibliography

I. Review articles:

Kado, C. *Crit. Rev. in Plant Sci.* **1991**, *10*, 1-32.
Walden, R.; Schell, J. *Eur. J. Biochem.* **1990**, *192*, 563-576.
Kondorosi, A.; Kondorosi, E.; John, M.; Schmidt, J.; Schell, J. In *Genetic Engineering, Vol. 13*; Setlow, J.K. Ed; Plenum Press, New York, **1991**, pp. 115-136.

II. Function of T-DNA gene 5:

Körber, H.; Strizhov, N.; Staiger, D.; Feldwisch, J.; Olsson, O.; Sandberg, G.; Palme, K.; Schell, J.; Koncz, C. *EMBO J.* **1991**, *10*, 3983-3991.

III. Function of *A. rhizogenes rol*B and *rol*C genes:

Spena, A.; Schmülling, T.; Koncz, C.; Schell, J. *EMBO J.* **1987**, *6*, 3891-3899.
Schmülling, T.; Schell, J.; Spena, A. *EMBO J.* 1988, *7*, 2621-2629.
Estruch, J.J.; Schell, J.; Spena, A. *EMBO J.* **1991**, *10*, 3125-3128.
Estruch, J.J.; Chriqui, D.; Grossman, K.; Schell, J.; Spena, A. *EMBO J.* **1991**, *10*, 2889-2895.

IV. Tagging of genes involved in auxin mechanism of action:

Walden, R.; Hayashi, H.; Schell, J. *Plant J.* **1991**, *1*, 281-288.

V. Auxin binding proteins as potential receptors:

1. For review see:

Davies, P.J. In *Plant Hormones and Their Role in Plant Growth and Development; Davies, P.J.*

Ed; Martinus Nijhoff Publishers, Kluwer Academic Publishers Group, Dordrecht, **1987**, pp. 1-23.

2. Auxin binding sites:

Dohrmann, U.; Hertel, R.; Kowalik, W. *Planta* **1978**, *140*, 97-106.

3. ERabp1:

Hesse, T.; Feldwisch, J.; Balshüsemann, D.; Bauw, G.; Puype, M.; Vandekerckhove, J.; Löbler, M.; Klämbt, D.; Schell, J.; Palme, K. *EMBO J.* **1989**, *8*, 2453-2461.

4. Is ERabp1 an auxin receptor?

Barbier-Brygoo, H.; Ephritikhine, G.; Klämbt, D.; Maurel, C.; Palme, K.; Schell, J; Guern, J. *Plant J.* **1991**, *1*, 83-93.

5. Photoaffinity labeling of auxin binding proteins:

Hicks, G.; Rayle, D.; Lomax, T. *Science* **1989**, *245*, 52-54.

Feldwisch, J.; Zettl, R.; Hesse, F.; Schell, J.; Palme, K. *Proc. Natl. Acad. Sci. U.S.A.* **1992**, *89*, 475-479.

Moore, I.; Feldwisch, J.; Campos, N.; Zettl, R.; Brzobohaty, B.; Bakó, L.; Schell, J.; Palme, K. *Biochem. Soc. Trans.* **1992**, *20*, 70-73.

Choice of Equipment Sizes for SSC Plants

Carlos Rolz*, **Silvia Robles**, Central American Research Institute for Industry (ICAITI), P.O.Box 1552, Guatemala 01901, Central America
(FAX:502 2/317470)
Joan Mata-Alvarez, Dpt. Enginyeria Quimica, Universitat de Barcelona, Marti i Franques 1, E-08028 Barcelona, Spain

A flowsheet option is described for a process to enrich green banana flour by fungal solid state culture. Equipment sizing is done by considering the process divided into semicontinuous trains and true batch units for the case of single product compaigns with non-overlapping product batch cycles. Total capital investment required for a fixed quantity of final product in a time horizon has been expressed as a design of a multiproduct batch plant and optimized by the Simplex algorithm

Solid substrate culture of fungi can be employed to produce protein enriched flours for monogastric animal feeding. The process is of interest to tropical countries for upgrading quality rejects and byproducts from cash fruit crops like banana, coffee and sugar (Raimbault, 1988). Green banana fruit rejects, with 21% of its weight of skin, are rich in starch. Pretreated flour can be enriched to a flour containing above 20% of true fungal protein, biologically available to growing chicks.
The flowsheet design alternative (Fig.1) contains the following sections: a) raw material preparation, b) protein enrichment in a SSC bioreactor and c) product stabilization. The flour preparation comprises continuos fruit washing and low pressure extrussion. The resulting mash is partially dryed **batchwise** on trays. The material is then continuosly milled, heat treated to gelatinize the starch, cooled, mixed with the inoculum and charged to the bioreactor. Fungal growth is done **batchwise**. The protein upgraded moist product is continuosly dryed and milled/agglomerated to a coarse flour which is easily incorporated into animal feed formulations.

The process has **two** batch operated units and **three** semicontinuos trains. In a multiproduct plant (Rippin, 1983), **one** product only is processed at a time and **all** potential products follow the same flowsheet sequence. Optimal design of multiproduct batch plants has been addressed as an equipment sizing problem (Loonkar and Robinson, 1970; Sparrow et al., 1975; Grossmann and Sargent, 1979 and Knopf et al., 1982), in which capital investment is optimized by varying the throughput rates of the semicontinuos trains within a production estimate in a time horizon. The optimization is

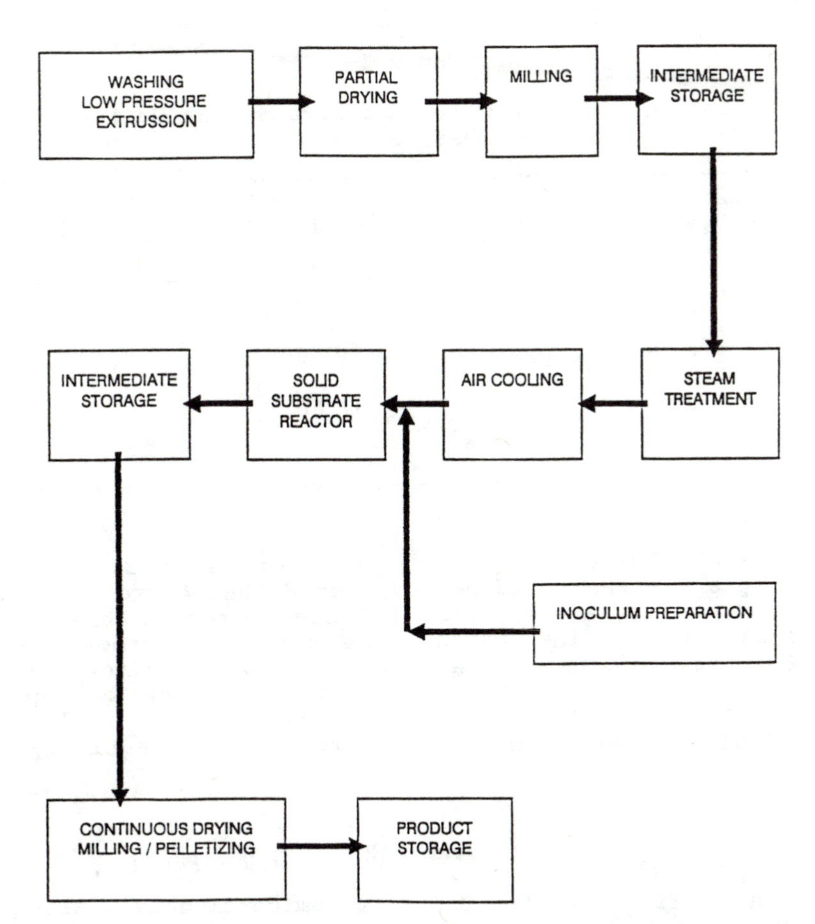

Figure 1. Flowsheet option for the protein enrichment of green banana fruit rejects by SSC of fungi. The batch operated blocks are the ones identified by: **partial drying** and **solid substrate reactor**. The analysis will not take into account intermediate and product storage.

important for this process as about 40% of the total manufacturing cost is represented by capital investment; raw materials only account for 13%, but energy expenditures are high, about 31% (ICAITI, 1986).

Problem Formulation.

The objective function to be minimized is:

$$I = \sum_i m_i a_i (V_i)^{\alpha_i} + \sum_k n_k b_k (R_k)^{\beta_k}$$

where **I** is total capital investment, a_i, b_k, α_i, β_k are power law cost coefficients for m_i batch and n_k continuos units respectively, the V_i's are the volumes (in m^3) of the batch units and the R_k's the throughput rates of the semicontinuos trains in kg per h. The cost coefficient data is shown in Tables 1 and 2.

The operating times for the batch units are: 30 (t_1) and 44 h (t_2) for the tray dryer and the bioreactor respectively. The yearly production requirements are set at 440,000 kg of dry product (**Q**). The production horizon (**T**) is set at 7920 h. The batch time, T_b, is

Table 1

Cost Coefficients for Batch Equipment Units (US$, 1991)

Equipment	ai	αi
Tray dryer	21 600	0.51
Bioreactor	57 600	0.60

calculated as shown on Fig. 2, where the ϕ's are the residence times of the material in the semicontinuos units. The batch size, **B**, is calculted as follows:

$$QT_b/B = T$$

The calculation procedure is shown in Table 3. It has been assumed that there is a 17% loss of dry matter in the bioreactor due to fungal growth, and that 137 and 125 kg of dry matter per cubic meter can be charged to the tray dryer and the SSC bioreactor, respectively.

Table 2

Cost Coefficients for Units in Semicontinuous Trains (US$, 1991)

Equipment	b_k	β_k
Train 1		
Band washer	1330	0.23
Low pressure extruder	510	0.60
Conveyor	890	0.23
Train 2		
Conveyor	890	0.23
Diskmill	110	0.60
Conveyor	590	0.23
Thermal screw conveyor	1450	0.53
Cooler screw conveyor	1200	0.53
Conveyor	890	0.23
Train 3		
Conveyor	890	0.23
Rotary dryer	1970	0.64
Diskmill/Pelletizer	190	0.60
Conveyor	890	0.23

Table 3

Calculation Procedure

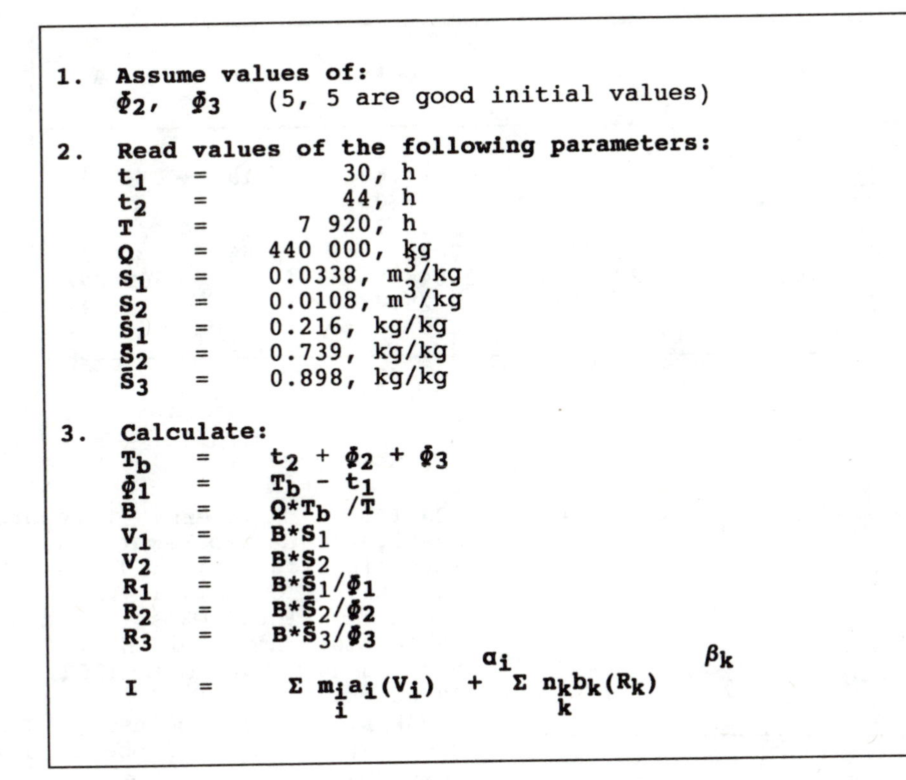

1. **Assume values of:**
   Φ_2, Φ_3 (5, 5 are good initial values)

2. **Read values of the following parameters:**
   t_1 = 30, h
   t_2 = 44, h
   T = 7 920, h
   Q = 440 000, kg
   S_1 = 0.0338, m^3/kg
   S_2 = 0.0108, m^3/kg
   $\bar{S}_1$ = 0.216, kg/kg
   $\bar{S}_2$ = 0.739, kg/kg
   $\bar{S}_3$ = 0.898, kg/kg

3. **Calculate:**
   $T_b = t_2 + \Phi_2 + \Phi_3$
   $\Phi_1 = T_b - t_1$
   $B = Q*T_b / T$
   $V_1 = B*S_1$
   $V_2 = B*S_2$
   $R_1 = B*\bar{S}_1/\Phi_1$
   $R_2 = B*\bar{S}_2/\Phi_2$
   $R_3 = B*\bar{S}_3/\Phi_3$

$$I = \sum_i m_i a_i (V_i)^{\alpha_i} + \sum_k n_k b_k (R_k)^{\beta_k}$$

Problem Solution

Proper initial values for Φ_2 and Φ_3 were found by solving the problem employing the tkSolver plus software. These were employed to start the Simplex-Nelder-Mead optimization algorithm in order to search for the optimum. This was found in 55 iterations employing the tkLibrary version of the Simplex. Results are shown in Table 4.

The optimization has been done only in two dimensions, as the residence time for the first semicontinuos train (Φ_1) is set by the problem formulation as shown in Fig. 2. The search has been unconstrained because there is only one product being produced.

Future work will include products of differerent protein content, as required by other animals, pigs for example. Also the possibility of employing other raw materials will be considered. These will be more complex cases were multipurpose (Rippin, 1983) flowsheet options will be analyzed with the objective of optimizing production scheduling.

Table 4

Calculation Results

	1	2	Optimum	3	4
Φ_2	3	5	4.68	10	15
Φ_3	3	5	7.03	10	15
T_b	50	54	55.7	64	74
B	2778	3000	3095	3556	4111
I	936 271	907 478	902 045	924 412	969 099
Φ_1	20	24	25.7	34	44

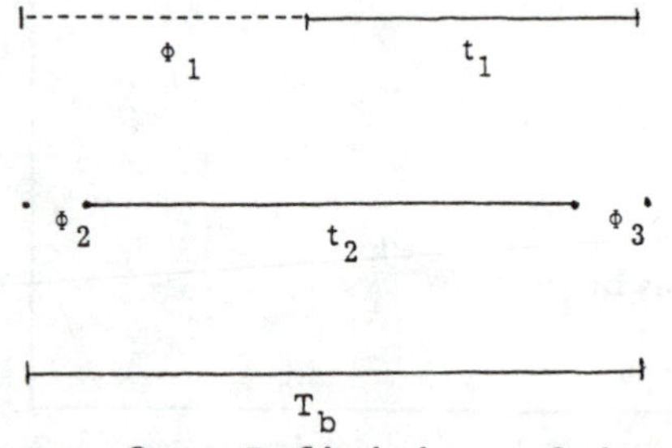

Figure 2. Definition of batch time (T_b). Relationship between t_i's and the Φ_k's. The independent variables are only Φ_2 and Φ_3.

References

Grossmann, I.E.; Sargent, R.W.H. Ind. Eng. Chem Process Des. Develop. **1979**, 18, 343-348

ICAITI. Internal Report, Instituto Centro Americano de Investigación y Tecnología Industrial, Guatemala **1986**

Knopf, F.C.,;Okos, M.R.; Reklaitis, G.V. Ind. Eng. Chem. Process Des. Develop. **1982**, 21, 79-86

Loonkar, Y.R.; Robinson, J.D. Ind. Eng. Chem. Process Des. Develop. **1970**, 9, 625-629

Raimbault M. In Solid State Fermentation in Bioconversion of Agroindustrial Raw Materials; Raimbault, M. Ed; ORSTOM, France, **1988**, pp 5-12

Rippin, D.W.T. Computers chem. Engng **1983**, 7, 137-156

Sparrow, R.E.; Forder, G.J.; Rippin, D.W.T. Ind. Eng. Chem. Process Des. Develop. **1975**, 14, 197-203

Testimony of a Bioprocess--Total Conversion of Biomass

Tarun K. Ghose, and Purnendu Ghosh Biochemical Engineering Research Center, Indian Institute of Technology, Hauz Khas, New Delhi, India

The conversion of rice straw into ethanol and co-products is presented as a case study on the integration of bioprocess engineering, biotechnology, and agronomic factors into a viable manufacturing technology. Since biomass, in general, represents a low to moderate density raw material, the staging of the process steps which result in a concentrated final ethanol product are an important consideration. Novel approaches to biomass fractionation into hexose and pentose based fermentable substrates using enzyme hydrolysis are presented. An integrated microbial conversion/ethanol recovery system will also be discussed based on the results of on-going large scale trials in India.

Although there is no lack of cellulosic substances in the world, it does not necessarily mean that all such materials are logistically convenient for acceptance and technologically suitable for enzymatic conversion. Moreover, both long and short fibres of the most woody materials are already in use for pulp and paper making. Even waste paper is being recycled.

Any cellulosic substance considered for its rendering into chemicals through an environmentally acceptable process must be cheap in terms of collection, transport, handling and storage and use of a technology easily adapted to such substances. The proper choice of all these factors largely depend on local conditions.

A totally integrated biotechnology of rice straw conversion into ethanol and co-products developed at the IIT-Delhi which is currently undergoing large scale trials will be reported. It will deal with: (a) ethanol refining of rice straw to segregate cellulose from pentose sugars and lignin, (b) preparation of highly active mixed cellulose enzymes, (c) a novel reactor system allowing rapid product formation involving enzymatic hydrolysis of cellulose to sugars followed by microbial conversion of the latter into ethanol and its simultaneous flash separation employing a programmed recompression of ethanol vapours and condensation, and (d) concentration of ethanol via alternate approaches.

Earlier studies reported by Seeley (1976), Wilke and associates (1977, 1983a, 1983b), Ramalingham and Finn (1979), Viikari et al. (1980), Ladisch et al. (1983), Ghose and associates (1982, 1983, 1984, 1985, 1986) constitute the background of this report.

Process mass balances of cellulose, ethanol and water and the effects of concentrations of non-fermentables on ethanol production based on rice straw cellulose will be discussed.

Specifics of operation of a novel bioreactor employing refined rice straw cellulose as substrate reflect the unique features of the process. Cost estimates are based on the Indian market price of input and output components as of September 1991. The major advantages of the process include: (a) a very high quality ethanol with a yield coefficient of 0.52 (stoichiometric coefficient of cellulose to ethanol is 0.568); (b) energy expenditure for flash vacuum cycling operation is low (0.006 MJ per mole ethanol) with an average ethanol concentration of 13.9% (w/v); (c) more than 90% of the cellulose (100% cellulose basis) is rendered into ethanol; and (d) the process does not generate any appreciable quantity of liquid effluent.

References

Seeley, D. B. *Biotech. Bioeng. Symp. Ser.* **1976,** *6*, 285.

Cysewski, G. R.; Wilke, C. R. *Biotechnol. Bioeng.* **1977,** *19*, 1125.

Ramalingham, A; Finn, R. K. *Biotechnol. Bioeng.*,**1977,** *19*, 583.

Viikari, L.; Nybergh, P.; Linko, M. *6th Interna-*

1054–7487/92/0505$06.00/0

tional Fermentation Symposium; 6th International Fermentation Symposium; London, Ontario; Abstract, p. 80.
Ghosh, V. K.; Ghose, T. K.; Gopalakrishna, K. S. *Biotechnol. Bioeng.* **1982,** *24*(1), 241.
Ghose, T. K.; Selvam, P.; Ghosh, P. *Biotechnol. Bioeng.* **1983,** *25*, 2577.
Maiorella, B.; Blanch, H. W.; Wilke, C. R. *Proc. Biochem.* **1983,** *18*(4), 5.
Wilke, C. R.; Maiorella, B.; Sciamanna, A.; Tangnu, S. K.; Wiley, D.; Wong, H. **1983** Noyes Data Corp., Park Ridge, NJ, USA.
Ladisch, M. R.; Lin, K. W.; Voloch, M.; Tsao, G. T. *Enz. Microbial Technol.* **1983,** *5*, 82.
Ghose, T. K.; Roychoudhury, P. K.; Ghosh, P. *Biotechnol. Bioeng.* **1984,** *26*, 377.
Panda, T.; Ghose, T. K.; Bisaria, V. A. *Biotechnol. Bioeng.* **1985,** *27*, 1353.
Roychoudhury, P. K.; Ghose, T. K. *Biotechnol. Bioeng.* **1986,** *28*, 972.

Genetic Engineering of Novel Bacteria for the Conversion of Plant Polysaccharides into Ethanol

Lonnie Ingram, Department of Microbiology and Cell Science, University of Florida, Gainesville, Florida 32611 (FAX: 904 392 8479)

Nature has provided a marvelous diversity of pathways and metabolic capabilities, grouping these variously to produce the contemporary flora of microorganisms. In some cases, organisms such as *Saccharomyces cerevisiae* can be used directly for the commercial production of useful chemicals such as ethanol. However, in most cases, further genetic modifications are required to optimize biological conversion processes for commercial utility. The tools of molecular genetics now allow the construction of hybrid organisms in which diverse pathways can be combined using the background knowledge of microbial physiology and biochemistry as a guide. We have employed this approach to construct a series of new bacterial strains which are capable of converting all of the sugars found in the polymers of woody biomass (hexose and pentose sugars) into ethanol.

Approximately 1 billion gallons of ethanol are blended in gasoline each year in the United States (Lynd *et al.*, 1991). This ethanol is manufactured almost exclusively from corn starch and utilizes 1/3 of the total U.S. corn production. With the passage of the 1992 Clean Air Act, an expanded market has been established for ethanol and other oxygenates as components in reformulated gasoline. Some of this higher demand for oxygenates can be met by chemical synthesis from petrochemicals and some by increased ethanol production from corn. However, multi-billion gallon production of fuel ethanol as the dominant oxygenate will require the use of additional feedstocks.

Lignocellulosic biomass is the only renewable feedstock which is sufficiently abundant to fill the expanded need for fuel ethanol. Lignocellulose is available as waste streams from wood processing, from pulp and paper, from landfills, and as undervalued agricultural residues.

In general, the compositions of all lignocellulosic materials are similar on a dry weight basis: 50% cellulose, 20% hemicellulose, and 25% lignin and other materials (Eriksson *et al.*, 1990). Both cellulose and hemicellulose are carbohydrate polymers which can be potentially fermented. Cellulose is a ß-1,4-linked homopolymer of glucose while hemicellulose is a complex group of polymers rich in pentose and hexose sugars, partially acetylated and variously modified to include covalent bonds with lignin. As with the manufacturing of ethanol from starch (Fogarty, 1983), polymers of cellulose and hemicellulose must be solubilized and converted to sugars before fermentation. The saccharification of cellulose and hemicellulose as well as the efficient metabolism of the resulting hexose and pentose sugar mixture have been identified as primary problems in the development of commercial ethanol processes from ligocellulose.

Over the past few years, our lab has concentrated on the use of recombinant DNA techniques to develop improved organisms for fuel ethanol production (Alterthum and Ingram, 1989; Ingram *et al.*, 1987 & 1988). No organisms are known from nature which are capable of rapidly and efficiently converting all of the sugars from lignocellulose into ethanol. We have developed a series of recombinant bacteria with this ability by combining the traits of different organisms.

In fermenting bacteria, ethanol is a waste product which is generated during the disposal of excess electrons under anaerobic conditions. Pyruvate, a product of glycolysis, is first cleaved to acetaldehyde and CO_2 by pyruvate decarboxylase. Acetaldehyde is then used as a terminal electron acceptor to regenerate the cofactor, NAD^+, allowing glycolysis to continue. Only two enzymes are required for ethanol production from pyruvate and both have been cloned and sequenced from the bacterium, *Zymomonas mobilis* (Conway *et al.*, 1987a&b). Although *Z. mobilis* is an excellent organism for the production of ethanol from glucose, this organism lacks the ability to metabolize other sugars in hemicellulose polymers.

There are many bacteria which are able to metabolize all of the sugars present in cellulose and hemicellulose but produce mixtures of organic acids and ethanol during fermentation. By assembling the two *Z. mobilis* genes required for ethanol production into an artificial operon and inserting this operon into acidogenic bacteria, we have been able to redirect the central metabolism of acidogenic bacteria to produce ethanol. This has been done with a variety of Gram negative organisms, although in principal, analogous metabolic engineering can be widely applied to other organisms. The affinity (Km) of *Z. mobilis* pyruvate decarboxylase for pyruvate is lower than that of competing fermentation enzymes which produce acids. Expression of the *Z. mobilis* genes at high levels should lower the intracellular pool of pyruvate sufficiently to minimize carbon flow into organic acids in most organisms.

Ethanologenic *E. coli* strains have been most extensively studied and have been used to produce up to 68 g ethanol/liter (Guimaraes and Ingram, 1992) with volumetric productivities of almost 2 g/liter per hour (Beall *et al.*, 1991; Ohta *et al.*, 1991a). These strains efficiently ferment all of the sugars which are constituents of biomass. They have been further improved by integrating the *Z. mobilis* genes into the *E. coli* chromosome to produce a highly stable organism, eliminating the need for a plasmid (Ohta *et al.*, 1991a). Additional modifications include the disruption of minor pathways for acidic fermentation products and the insertion of defects in recombination. Acid hydrolysates of softwood hemicellulose and cellulose (Barbosa *et al.*, 1992) and of hardwood hemicellulose (Lawford and Rousseau, 1991) have been fermented with these organisms.

Analogous strains of ethanologenic *Klebsiella oxytoca* have also been developed in which the *Z. mobilis* ethanol genes have been integrated into the chromosome (Ohta *et al.*, 1991b). These organisms have some advantages over *E. coli*-based constructs in that *K. oxytoca* contains transport systems and enzymes which allow the intracellular metabolism of short oligosaccharides from cellulose and xylan (Burchhardt and Ingram, 1992; Wood *et al.*, manuscript in preparation), eliminating the need for ß-glucosidase and xylosidases.

Ethanologenic recombinants in which the alcohol production genes have been integrated into the chromosome serve as excellent hosts for the production of plasmid-encoded, recombinant proteins. Enzymes useful for hydrolysis or higher value recombinant products can be produced as co-products with ethanol. Examples investigated thus far include xylanase, endoglucanase, α-amylase, and pullulanase (Burchhardt and Ingram, 1992; Guimaraes *et al.*, manuscript in preparation).

This work has been supported by the Florida Agricultural Experiment Station (R-02202) and by grants from the U.S. Department of Agriculture, the U.S. Department of Energy, and BioEnergy International, a subsidiary of the Quadrex Corporation.

References

Alterthum, F.; Ingram, L.O. *Appl. Environ. Microbiol.*, **1989**, *55*, pp 1943-1948.

Barbosa, M. de F.S.; Beck, M.J.; Fein, J.E.; Potts, D.; Ingram, L.O. *Appl. Environ. Microbiol.*, **1992**, *58*, In Press.

Beall, D.S; Ohta, K.; Ingram, L.O. *Biotechnol. Bioengin.*, **1991**, *38*, pp 296-303.

Burchhardt, G.; Ingram, L.O. *Appl. Environ. Microbiol.*, **1992**, *58*, In Press.

Conway, T.; Osman, Y.A.; Konnan, J.I.; Hoffmann, E.M.; Ingram, L.O. *J. Bacteriol.*, **1987a**, *169*, pp 949-954.

Conway, T.; Sewell, G.W.; Osman, Y.A.; Ingram, L.O. *J. Bacteriol.*, **1987b**, *169*, pp 2591-2597.

Eriksson, K.-E. L.; Blanchette, R.A; Ander, P. In *Microbial and Enzymatic Degradation of Wood and Wood Components*. Springer-Verlag, New York, **1990**.

Fogarty, W.M. In *Microbial Enzymes and*

Biotechnology; Fogarty, W.M., Ed.; Applied Science Publishers Ltd: Essex, England, **1983**; pp 1-92.

Guimaraes, W.V.; Dudey, G.L.; Ingram, L.O. *Biotechnol. Bioengin.* **1992**. *39*, In Press.

Ingram, L.O.; Conway, T. *Appl. Environ. Microbiol.*, **1988**, *54*, pp 397-404.

Ingram, L.O.; Conway, T.; Clark, D.P.; Sewell, G.W.; Preston, J.F. *Appl. Environ. Microbiol.*, **1987**, *53*, pp 2420-2425.

Lawford, H.G.; Rousseau, J.D. 1991. *Biotechnol. Let.*, **1991**, *13*, pp 191-196.

Lynd, L.R.; Cushman, J.H.; Nichols, R.J.; Wyman; C.E. *Science*, **1991**, *251*, pp 1318-1323.

Ohta, K.; Beall, D.S.; Mejia, J.P.; Shanmugam, K.T; Ingram, L.O. *Appl. Environ. Microbiol.* **1991a**, *57*, pp 893-900.

Ohta, K.; Beall, D.S.; Mejia, J.P.; Shanmugan, K.T.; Ingram, L.O. *Appl. Environ. Microbiol.*, **1991b**, *57*, pp 2810-2815.

Intercalation in the Pretreatment of Cellulose

Michael R. Ladisch, Lori Waugh, Paul Westgate, Karen Kohlmann, Rick Hendrickson, Yiqi Yang, and Chris Ladisch, Laboratory of Renewable Resource Engineering, A.A. Potter Building, Purdue University, West Lafayette, IN 47907 (FAX: 317–494–0811)

The structural features of cellulose are known to profoundly influence the kinetics of cellulose hydrolysis. Cellulose in biomass is resistant to hydrolysis due to hydrophobic interactions between cellulose sheets, hydrogen bonding between adjacent cellulose chains, and cellulose's close association with lignin. A useful pretreatment disrupts hydrophobic and hydrogen bonds, as well as the lignin seal, in a manner which minimizes chemical change of the cellulose and formation of undesirable degradation products. The resulting polysaccharide structure must be stabilized against spontaneous recrystallization, once pretreatment conditions are removed. Otherwise the benefit of enhanced hydrolysis is lost. This work reports the intercalating effects and mechanisms of sulfate esters, and the role of water in altering the physical properties of pretreated cellulose. A mechanism is proposed which leads to a leveling off in particle size (LOPS) during enzyme hydrolysis of lignin free, microcrystalline cellulose.

Biomass materials consist principally of hemicellulose, cellulose, and lignin. Compositions are typically in the range of 25 to 35 % hemicellulose, 30 to 40 % cellulose, and 10 to 15 % lignin. Other components include inorganic compounds, proteins, and other non-cellulosic constituents. Autohydrolysis in water, or hydrolysis by dilute acids or xylanase enzymes readily removes the hemicellulose fraction and gives pentose rich hydrolysates. These treatments also remove a significant fraction of the other soluble components. The remaining cellulose is closely associated with lignin and has a crystalline structure which makes it resistant to both acid and enzyme hydrolysis.

The recalcitrant nature of cellulose continues to impede economical production of fermentable sugars from biomass materials and cellulosic residues. Many pretreatments have been proposed and tested which disrupt the crystalline structure of cellulose and make the cellulose sufficiently reactive to give 80 to 90% yields of glucose upon acid or enzyme hydrolysis. These include: cellulose solvents such as cadoxen, or FeTNa; strong mineral acids; amines; amine oxides; ammonia; alkali/alcohol solutions; modified pulping processes; and physical treatments (grinding, beating, and radiation) (Navard and Haudin, 1986; Moriyama and Saida, 1986; Ladisch, 1988; Wood and Saddler, 1988; Holtzapple et al., 1992).

A review on the cost of energy or chemicals used in various pretreatments (Fan et

1054–7487/92/0510$06.00/0

al., 1987) enabled estimation of the contribution of pretreatment to the cost of fermentable sugars (Ladisch and Svarczkopf, 1991). Substrate, and energy or chemical costs of pretreatment, in the best case, range from 5 to 20 cents per pound of sugar. This calculation does not include capital, labor or operational costs elsewhere in the process. Both the glucose and xylose, derived from the hydrolysis of the biomass, were assumed to be fermentable to ethanol at close to theoretical yield (i.e., about 0.5 lbs ethanol/lb glucose and xylose).

The overall cost (including raw material, capital and operating costs) of biomass derived sugars needs to be about 4 cents/pound at current economic conditions to be competitive with corn as a substrate for fuel ethanol production. Consequently, cellulose pretreatment costs need to be reduced to improve the prospects for converting biomass to value-added fermentation products. Starch pretreatment by liquefaction and gelatinization using principally water, enzymes, and heat, serves as a benchmark against which the cost and effect of cellulose pretreatments can be compared.

Starch, like cellulose, needs to be pretreated to make it susceptible to enzyme hydrolysis. Pretreatment of starch consists of adding a thermostable *a*-amylase to partially hydrolyze the starch and reduce the viscosity of the starch slurry. The slurry is gelatinized and liquefied at temperatures ranging from 90 to 163 C, depending on the process used (Borlum, 1980). The concentration of the starting slurry is lower for corn than for refined starch and ranges from 20 to 35 %.

Our research seeks an approach which is conceptually similar to that of starch pretreatment. Such an approach would be based on water, and minimizes the addition of chemicals which would need to be recovered or recycled.

Cellulose Structure

Cellulose, a linear polymer of *B*-1,4 anhydroglucose monomers, has a lower proportion of hydroxyl groups available for solvation than starch. This difference arises from the *B*-1,4 bonds which link the anhydroglucose monomers in cellulose compared to the *a*-1,4 bonds in linear forms of starch. The *B*-1,4 bonds result in a polymer with both inter and intramolecular hydrogen bonding and a repeat unit which is a dimer. X-ray diffraction studies show the supramolecular and crystalline structure of cellulose is organized into unit cells with hydrogen bonding believed to be confined within sheets of cellulose chains. Hydrophobic interactions perpendicular to the sheets are thought to help hold the sheets together. Native cellulose thus has a high cohesive energy which makes it water insoluble relative to starch (Nevell and Zeronian, 1985; French, 1985).

Studies on cotton cellulose and lignocellulose suggest that hydrolysis kinetics will be affected by the surface area available for reaction (Rowland and Bertoniere, 1985; Lin et al., 1985), as well as cellulose crystallinity. Surface area can be increased by particle attrition to give a higher surface area per unit volume, and/or by disrupting the supramolecular and crystalline structure to give an internal porosity within the cellulose particle which is large enough to facilitate diffusion of the enzyme into the particle. Cellulose with internal "pore" dimensions of approximately 60 angstroms or larger enables penetration of fungal cellulases with molecular weights in the range of 60 kdal (Grethlein, 1985; Stone et al., 1969). In either case, a major change in surface area will correspond to a change in the perceived crystallinity of the cellulose.

Unlike starch or cellulose, lignocellulose has an additional barrier to hydrolysis - lignin. Lignin is closely associated with both the macroscopic plant cell wall structure as well as the microscopic cellulose crystalline structure of biomass materials. Lignin imparts resistance to hydrolysis of cellulose regions with which it is associated even though these would otherwise be accessible to cellulolytic enzymes, i.e., susceptibility is reduced. Explosive depressurization of wood (Aspen) impregnated with 500 to 1000 psig steam removes the lignin (a potentially valuable co-product of biomass conversion) without radically altering the crystallinity of the resulting cellulose (Bungay, 1982; Nesse, et al., 1977). Increased susceptibility of the cellulose may help to explain the 85% conversion achieved upon subsequent enzyme hydrolysis. Similar effects were obtained for wheat straw at enzyme levels of 80 to 100 FPU/g substrate (Vallender and Eriksson, 1985).

This paper describes results for a lignin free, microcrystalline cellulose, Avicel(R). This material has well defined properties, including an average particle size of 68 to 80 microns (wet basis) and a leveling off degree of polymerization (LODP) of about 200. Microcrystalline cellulose is a useful model substance for probing the effects of different pretreatments on the physical structure of cellulose.

Hydrolysis of Sulfated Cello-oligosaccharides

Cellodextrins and sulfated cello-oligosaccharides are obtained by dissolving 100 grams of Avicel in 50 mL of H2SO4 at 4 C for 7 minutes. Water is added and the slurry is heated to 70 C. This is followed by addition of 100 mL ethanol at 4 C and activated carbon to remove color bodies. The resulting slurry is filtered. Next, 2.4 L ethanol is added to the filtrate. A white precipitate, containing cellodextrins and sulfated cello-oligosaccharides, is formed and recovered by centrifugation (Pereira et al., 1988).

Cellodextrins are only sparingly soluble at a degree of polymerization (DP) above 6 (molecular weight of 991 D) (Huebner et al., 1978). However, if the cello-oligosaccharides are sulfated to give 1 sulfate per 3 to 5 glucose units, a solubility in excess of 100 g/L is obtained even if the DP exceeds 6. This illustrates the effectiveness of an intercalating agent, e.g. sulfate, in keeping the structure open to the point that the cellulose dissolves. The derivatized cello-oligo-saccharides lose the sulfate groups when heated in water at 220 C for several minutes. The resulting decrease in pH is sufficient to cause rapid formation of water soluble cellobiose which in turn is completely converted to glucose by cellulase enzymes at pH 4.8 and 50 C in a subsequent hydrolysis step (Mobedshahi, 1987; Pereira et al., 1988).

Sulfation must be carried out in strong acid in order to swell the cellulose and make it accessible. Since the sulfation reaction is strongly inhibited by the presence of water, the cellulose must be at low moisture. A practical process thus requires a drying step for agricultural residues and other biomass feedstocks which typically contain at least 40 % moisture. The sulfuric acid must be recycled for both economic and environmental reasons. This can be readily achieved using ion exclusion (Neuman et al., 1986) or other means.

A low temperature sulfuric acid process has been developed

and piloted for saccharification of corn residue (stalks) based on the use of sulfuric acid for both pretreatment and hydrolysis with an intermediate drying step (Tsao et al., 1978; Bienkowski et al., 1984; Ladisch, 1989; Barrer, 1991). This approach is also compatible with enzyme hydrolysis, although wet processing techniques which avoid an intermediate drying step and minimize extraneous reagents are more attractive.

Enzyme Hydrolysis of Microcrystalline (Insoluble) Cellulose

Microcrystalline cellulose was sieved (dry) to give a material with an initial particle size of greater than 53 microns. A quantity of 2.5 grams was then suspended in 48 mL of 50 mM citrate buffer, pH 4.8. Cellulase enzyme (Cytolase CL, Genencor International, 90 GCU/mL), 2 mL, was added and the mixture incubated under constant agitation at 50 C. A control experiment was also run in which the enzyme solution was replaced with 2 mL of buffer. Wet particle size, particle surface characteristics, and conversion to glucose were monitored as a function of time using a particle size analyzer (Malvern 2600C), scanning electron microscope (in the Materials Engineering Department), and glucose analyzer, respectively. The cellulose samples for scanning electron microscopy were prepared by simply drying them under vacuum.

The average initial particle size of the wet cellulose (68 microns) decreased to 23 microns within 5 minutes after the reaction had been started. Part of this decrease is due to the agitation or other factors as indicated by a separate control study in which the particle size decreased from an initial value of 80 microns to 60 microns after 1 hour. The data clearly show that this enzyme preparation causes a significant decrease in cellulose particle size during the initial part of the hydrolysis reaction. Scanning electron microscopy illustrates a typical particle at the beginning of the reaction. (see Figure 1(a)).

The hydrolysis rapidly proceeds to 25% conversion during the first 180 min while the average wet particle size remains essentially unchanged after its initial drop to the 20 micron range (Figure 2(a)). In a separate run at the same conditions, the hydrolysis was monitored over a period of 6 days (Figure 2(b)) during which 82 % conversion was obtained. The average particle size at the end of this run was still 20 microns. Scanning electron micrographs indicated that some of the remaining cellulose particles had a splintered appearance (Figure 1(b)). A third hydrolysis experiment, carried out for 11 days, gave 83 % conversion to glucose. A glucose conversion of 80 to 85 % appears to represent the upper limit at this concentration of enzyme. The well known inhibitory effects of cellobiose and glucose are a likely factor in the decreasing rate of the reaction, particularly as the glucose concentration approaches 42 g/L at the end of 6 days.

The leveling off of particle size suggests the hypothesis that enzyme hydrolysis of the microcrystalline cellulose is dominated by a tunneling mechanism. We postulate that the enzyme complex attacks the cellulose by penetrating into the interior of the particle. Once a large extent of hydrolysis occurs, and a significant fraction of the particle's interior is depolymerized, a high internal porosity results. Eventually this leads to a decrease in the particle's structural integrity causing the particle to break apart. The average size would thus appear to remain essentially constant until the

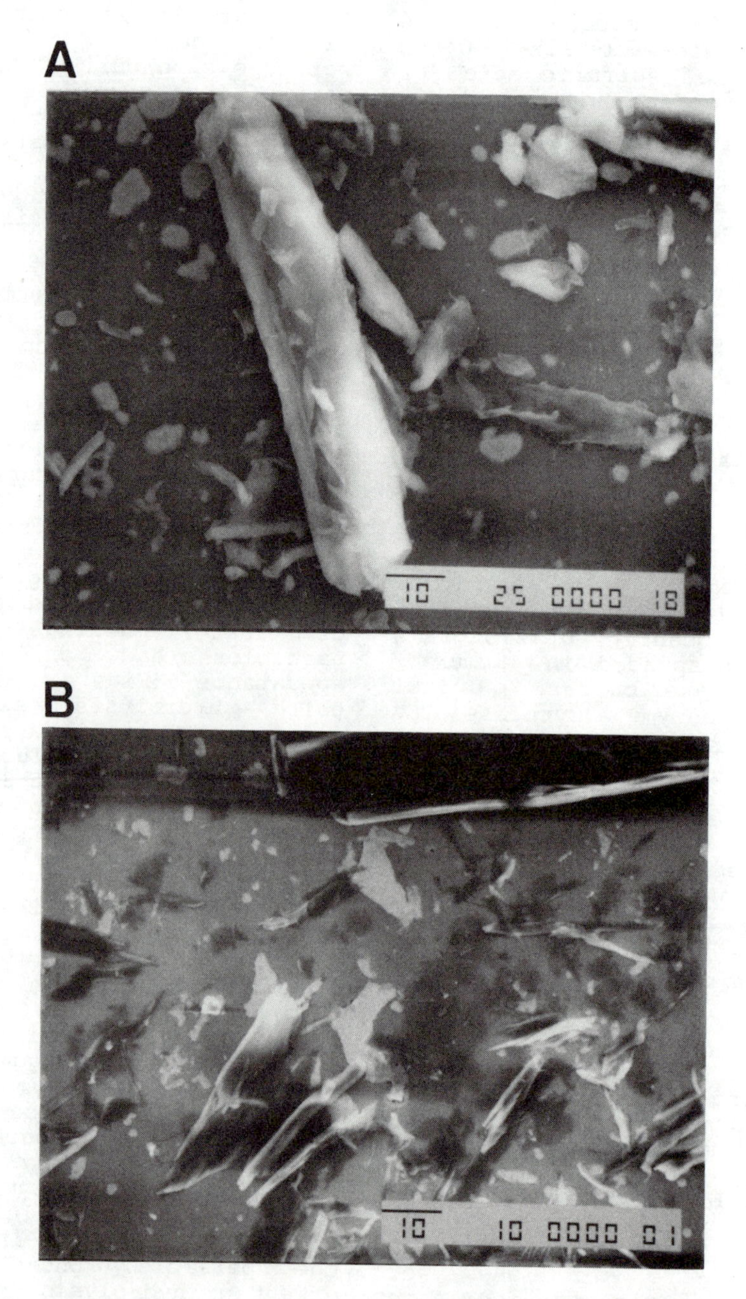

Figure 1. SEM of micro-crystalline cellulose at 800x magnification (Scale bar = 10 microns): (a) Control sample at t= 0; (b) hydrolyzed cellulose after 143 hours of incubation with enzyme at pH 4.8 and 50 C.

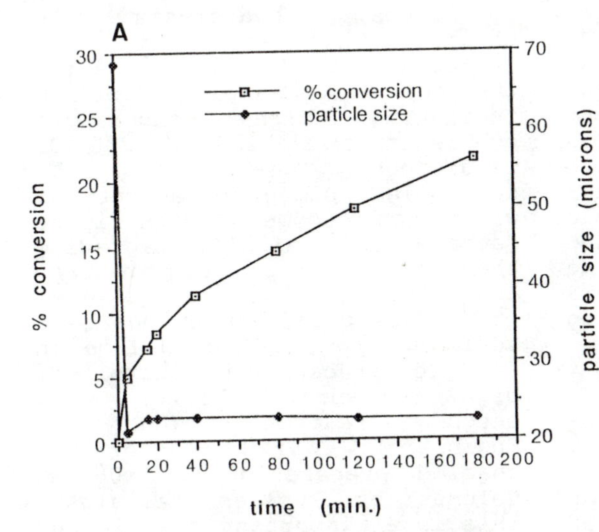

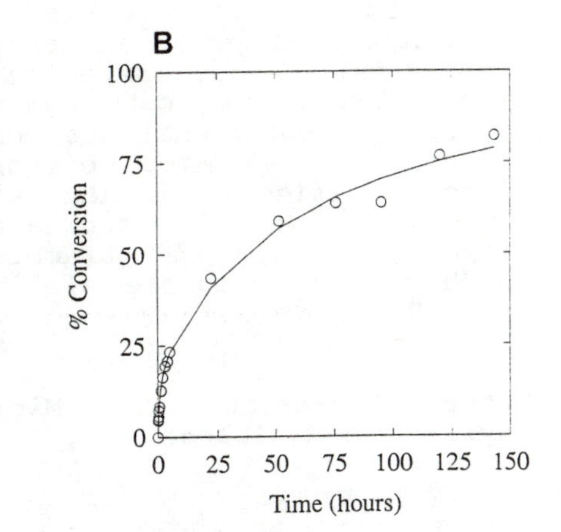

Figure 2. Hydrolysis time course showing (a) change in cellulose conversion (square symbols) and particle size (diamonds) during 180 minutes of hydrolysis; (b) conversion over 6 days.

particle disintegrates. In comparison, if hydrolysis of the solid cellulose were to occur by erosion at the outer surface, a gradual decrease in particle size would be expected. This has not been observed. Further experiments are needed to test our proposed explanation of the leveling off in particle size (abbreviated LOPS) during the enzyme hydrolysis of lignin free cellulose.

Solubility of Microcrystalline Cellulose in Sodium Hydroxide

The solubility of microcrystalline cellulose in aqueous sodium hydroxide is known and was used in the development of a procedure for the gel permeation chromatography of non-derivatized cellulose (Bao et al., 1980). We carried out further experiments on the solution properties of microcrystalline cellulose at 5% concentration in 5, 10, 20, and 30 % NaOH (Yang et al., 1991). Three temperatures were examined: ambient, 6 C and -10 C.

There was little change in the cellulose at ambient temperature, even after 3 days, although the sodium hydroxide solutions developed a yellow color.

Gelatinization in NaOH occurs at temperatures below ambient. Swelling and the onset of gelatinization increased with increasing NaOH concentration up to a maximum level of about 20 % and was lower at 30 % NaOH. At 6 C and 10 % NaOH, the cellulose began to resemble a gel after 24 hours and became opaque after 3 days. A similar result was obtained in 20% NaOH, while at 30% a slight precipitate was observed in the gel. At -10 C, an opaque gel was formed in 5, 10, and 20 % NaOH after 24 hours, although the gel in the 20 % NaOH had a yellow hue indicating that degradative reactions were likely occurring. At 30 % NaOH the gel contained a precipitate after 24 hours and large particles after 3 days.

The results show that sodium hydroxide can function as an intercalating agent. However, it must either be removed or neutralized prior to enzyme hydrolysis. These experiments also indicate that both osmotic effects (proportional to NaOH concentration) and entropic effects (reflected by increased swelling with decreasing temperature) may be parameters which affect cellulose swelling by NaOH.

Water Treatment of Microcrystalline Cellulose

Water is known to be an active agent in the cooking of pulp, and can cause undesirable gelatinization in pulp production (Aronovsky and Gortner, 1930; Mithel et al., 1957). Prior research on the role of water in cellulose transformations shows that partial cellulose dissolution may occur above 200 C between pH 7 and 8 (Mithel et al., 1957). These reports form the starting point of our research. As in other pretreatments, the goal of cellulose swelling and gelatinization using water is to increase both accessibility and susceptibility of the cellulose to enzyme hydrolysis.

The development of conditions which give reproducible swelling of microcrystalline cellulose at temperatures above 200 C was initially carried out in 5.5 mL, high pressure, stainless steel reaction tubes filled with an aqueous slurry of cellulose. The tubes were heated by submersing them in a preheated sandbath (Techne).

Heating of a 16.8 % cellulose slurry in deionized water at 190 C for 4 hours gave a swollen form of cellulose having an apparent volume of up to 4 times that of freshly suspended material. At these conditions, hydrolysis is minimal. This result led to the choice of 190 C for preheating the cellulose sample prior to initiating high temperature pretreatment at 220 to 230 C. Conditions need to be controlled since the acidity of water can become a significant factor in chemically degrading the cellulose at temperatures above 200 C.

This pretreatment has been scaled up to a 150 ml volume in a microprocessor controlled, high pressure, stirred tank batch reactor (Autoclave Engineers). This reactor has enabled preparation of 100 ml volumes of swollen cellulose. This is an important step in our research since we can now prepare sufficiently large quantities of pretreated cellulose to enable further study with respect to its physical and chemical changes, and response to enzyme hydrolysis. These bench scale runs confirm that the cellulose swells 5-10 fold over its original volume and that the swollen state is stable for periods of up to an hour. This stability is a necessary condition for further examining and characterizing the effect of aqueous cooking of cellulose as a pretreatment. We hope that this will help to foster the development of a practical process for obtaining readily saccharified cellulose in water.

Conclusions

Pretreatments using aqueous based approaches show that complete solubilization of cellulose is attainable when sulfate or sodium hydroxide is used as the intercalating agent. However, added reagents are needed to achieve this effect. Consequently, an alternate approach using water alone is being studied. Water has been found to cause major swelling when cellulose is cooked for short periods of time at temperatures above 200 C.

Hydrolysis of untreated microcrystalline cellulose gives in excess of 80 % conversion to glucose when incubated with a commercially available fungal cellulase system at pH 4.8 and 50 C for 6 days. The cellulose particle size decreases to 20 microns during the first few minutes of the reaction, and then remains at this size. This phenomenon of leveling off of particle size (LOPS) suggests the hypothesis that the enzyme system penetrates the particle, and disrupts the internal cellulose structure.

Acknowledgments

The material in this work was supported by USDA grant CSRS 90-37233-5410. We also acknowledge support from NSF grant CBT 8351916, the Purdue NASA NSCORT program, and an NSF instrument grant (CTS 8906628). We thank David McCabe of the Electron Microscope facility for his assistance in SEM of cellulose, and Jim Wallace for assistance in carrying out the experiments on sodium hydroxide swelling of cellulose. The cellulase enzyme was a gift of Genecor International, Inc. We also thank Dr. Bernie Tao and Joe Weil for helpful comments regarding this paper.

References

Aronovsky, S.I.; Gortner, R.A. *Ind. Eng. Chem.* **1930**, *22*, 264.

Bao, Y.; Bose, A.; Ladisch, M.R.; Tsao, G.T., *J. Appl. Polymer Sci.* **1980**, *25*, 263-275.

Barrer, J.W., Personal Communication, Atlanta, Ga., **1991**.

Bienkowski, P.; Ladisch, M.R.; Voloch, M.; Tsao, G.T., *Biotechnol. Bioeng. Symp. Ser.* **1984**, *14*, 512-524.

Borlum, G.B., In *Fuel from Biomass and Wastes*; Klass, C.; Emert, G. Ed., Ann Arbor Science, Ann Arbor, MI, **1980**, pp 297-310.

Bungay, H.R., *Science* **1982**, *218*, 643-646.

Fan, L.T.; Ghapuray, M.M.; and Lee, Y-H., *Biotechnol. Monograph* **1987**, 3, 21-119.

French, A.D, In *Cellulose Chemistry and its Applications*, Nevell, T.P.; Zeronian, S.H. Ed, Halsted Press, NY, **1985**, pp 84-111.

Grethlein, H.E., *Bio/technology* **1985**, *3*, 155-160.

Holtzapple, M.T.; Lundeen, J.E.; Sturgis, R.; Lewis, J.E.; Dale, B.E., , *Applied Biochemistry and Biotechnology* **1992**, *34/35*, 5-22.

Huebner, A.; Ladisch, M.R.; Tsao, G.T., *Biotechnol. Bioeng.* **1978**, *20*, 1669-1677.

Ladisch, C. M., In *Methods in Enzymology*, 160,A, **1988**, pp 11-19.

Ladisch, M.R., In *Biomass Handbook*; Hall, C.W.; Kitani, O. Ed; Gordon and Breach, London, **1989**, pp 434-451.

Ladisch, M.R.; Svarczkopf,J., ,*Bioresource Technol.* **1991**, *36*, 83-95.

Lin, J.K.; Ladisch, M.R.; Voloch, M.; Patterson, J.A.; Noller, C.H., *Biotechnol. Bioeng.* **1985**, *27*, 1427-1433.

Mithel, B.B.; Webster, G.H.; Raoson, W.H., *TAPPI* **1957**, *40*, 1.

Mobedshahi, M., M.S. Thesis, Purdue University, December, **1987**.

Moriyama, S.; and Saida, T., In *Cellulose Structure, Modification and Hydrolysis*; Young, R.A.; Rowell, R.M. Ed; John Wiley and Sons, NY, **1986**, pp 323-336.

Navard, P.; and Haudin, J. M., In *Cellulose, Structure, Modification and Hydrolysis*; Young, R.A.; Rowell, R.M. Ed; NY, **1986**, pp 247-261.

Nesse, N.; Wallack, J.; Harper, J.M., *Biotechnol. Bioeng.* **1977**, *19*, 323-336.

Neuman, R.P.; Rudge, S.R.; Ladisch, M.R., *Reactive Polymers J.* **1986**, 5, 55-61.

Nevell, T.P.; Zeronian, S.H., In *Cellulose Chemistry and its Applications*; Nevell, T.P.; Zeronian, S.H. Ed, Halsted Press, NY, **1985**, pp 15-29.

Pereira, A.N.; Mobedshahi, M.; Ladisch, M.R., In *Methods in Enzymology, 160*, A, **1988**, pp 26-38.

Rowland, S.P.; and N. R. Bertoniere, In *Cellulose Chemistry and its Applications*, Nevell, T.P.; Zeronian, S.H. Ed, Halsted Press, NY, **1985**, pp 112-137.

Stone, J.E.; Scallan, A.M.; Donefer, E.; Ahlgren, E., *Adv. Chem Ser.* **1969**, *95*, 219-241.

Tsao, G.T.; Ladisch, M.; Ladisch, C.; Hsu, T.A.; Dale, B.; Chou, T., In *Annual Reports in Fermentation Processes*, 2; Perlman, D. Ed; Academic Press, N.Y., **1978**, pp 1-21.

Vallender, L.; Eriksson, K.E., *Biotechnol. Bioeng.* **1985**, *27*, 650-659.

Wood, T.M.; and Saddler, J.N., In *Methods in Enzymology*, 160,A, **1988**, 3-11.

Yang, Y.; Lebrecht, T.; Wallace, J.; Ladisch, C.; Ladisch, M.R., *Intercalating Agents in Cellulose Pretreatment*, Poster Paper 325, *Cellulose '91*, New Orleans, LA, December 6, **1991**.

Plant Cells for the Production of Food Ingredients

D. Knorr[1,2], C. Caster[2], H. Dörnenburg[1], R. Dorn[2],
D. Havkin-Frenkel[3], A. Podstolski[2], S. Semrau[1],
P. Teichgräber[1], U. Werrmann[1] and U. Zache[1]

1 Department of Food Technology, Berlin University of Technology, Königin-Luise-Str. 22, Berlin 33, FRG

2 Department of Food Science, University of Delaware, Newark, DE 19716, USA

3 David Michael & Co., Philadelphia, PA 19154, USA

Effects of environmental conditions such as light or precursor addition on the biosynthetic potential of *Vanilla planifolia*, *Mentha canadensis* and *Mentha piperita* cultures are evaluated. Product degradation by these cultures and the application of adsorption agents is discussed. An enzymatic approach for permeabilization of the tonoplast for product release while maintaining plant cell viability is being presented and the biosynthetic potential of germinating seeds exemplified by their ability to effectively produce cold tolerant carboxypeptidases or chitinases with antimicrobial activity.

Plant cells are able to synthesize many metabolites, which currently cannot be produced by microbial cells (Reinhard and Alfermann, 1980). One of the tools for such production is the application of plant cell and tissue cultures technology (Payne et al., 1991), which has reached a point where a variety of culture types can be critically assessed as potential sources of existing and novel food ingredients (Stafford, 1991).

However, within the food industries the profile of plant cell and tissue cultures has been low. Some of the hesitations to become more actively involved in this field include the traditional use of microorganisms and whole plants for metabolite production and extraction, the lack of information regarding the biosynthesis of many metabolites in intact plants or plant cultures, slow growth of plant cells, and often low producing and instable cell lines. In addition there is a lack of activities to utilize plant cell and tissue cultures for the production of novel metabolites rather than "mimicking" traditional plant metabolites. On the other hand exciting progress is being made with root and shoot cultures especially with the availability of fast growing and productive "hairy root" lines (Payne et al., 1991).

Some additional attempts to improve cell productivity are being presented in this paper.

Another approach currently under investigation is to benefit from the biosynthetic potential of germinating plant seeds for the production of useful biochemicals. Some examples will also be given below.

PLANT CELL AND TISSUE CULTURE

Vanilla extract from cured vanilla beans contains more than 150 volatile flavor components. However, only 26 of these are present in concentrations greater than one ppm. Those that contribute most to the flavor include vanillin, vanillic acid, p-hydroxybenzaldehyde and p-hydroxybenzoic acid. While these flavor compounds are relatively abundant in cured beans, few if any are found in the mature green beans.

We are examining various environmental conditions on vanilla flavor production in Vanilla planifolia cell and tissue cultures including effects of light conditions or precursors.

Vanilla planifolia cluster cultures that had been maintained in the dark were grown under Na-light, blue light, Agro-light for complete spectrum and darkness as a control.

All lights tested (intensity = 80 µE/sec/m²) stimulated synthesis of 3-metoxy-4-hydroxybenzyl alcohol (HMBA), protocatechuic aldehyde (PROALD) and vanillin (Vn). Selection of blue light resulted in an approx. 450 % increase in vanillin concentration after 14 days of incubation over the control. Results regarding the effects of light and precursor addition on flavor component production are presented in Figure 1.

Clusters grown either in the dark or in light for 3 weeks were homogenized in 0.05 M Na-acetate buffer and 2 mM of coumaric acid, ferulic acid or vanillyl alcohol added as precursors. The following products were analyzed: PROALD, 4-OH-benzaldehyde (BA), Vn and caffeic acid (CA). Homogenates from light grown clusters had a higher ability for oxidation and chain shortening of p-coumaric acid resulting in PROALD synthesis. The reaction may proceed via caffeic acid which was found among the products. Similar activity was indentified with ferulic acid which was converted to vanillin. The homogenates were also capable of oxidation of vanillyl alcohol to vanillin. Efficiency of this reaction was significantly higher in homogenates from light grown tissues.

Menthol production via plant cell and tissue cultures is a challenging target. On the one hand the aim of production is one single component rather than a range of flavor components. On the other hand within intact plants glandular trichomes are the primary site of monoterpene biosynthesis,

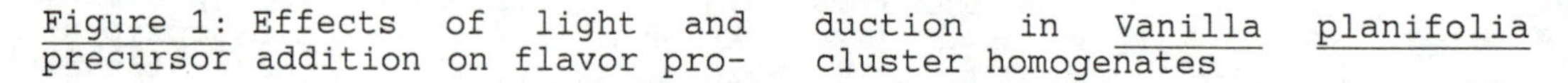

Figure 1: Effects of light and precursor addition on flavor production in Vanilla planifolia cluster homogenates

essential oils are stored in specific tissues and organs and the conversion rate from menthone to menthol is only approx. 40 %. Various biochemicals involved in the biosynthesis of monoterpenes in Mentha cultures have been tested to possibly identify bottlenecks in the biosynthesis of menthol (Table 1).
These data suggest that degradation of limonene and piperitone occured but bioconversion into tested substances was poor, that pulegone was preferably converted into isomenthone rather than menthone, that isomenthone, neomenthol and menthol were converted - at high rates - into unidentified compounds, and that the conversion of menthone was directed towards neomenthol rather than menthol.
From this it seems clear that the bottleneck was the effective conversion of pulegone to menthone and then to menthol, the pathway of menthol formation suggested in the literature. It is also interesting to note that menthylacetate a supposed end product of monoterpene biosynthesis in Mentha species is (re)converted into menthol and that menthol is degraded effectively.
Product degradation and adsorption is also an area of current consideration in our plant cell and tissue culture work. For example degradation of 90 % limonene or menthol was observed in Mentha cultures within 96 h or 6 h respectively (Table 1).
Vanillin added to Vanilla planifolia cultures was degraded effectively within 12 to 24 hours. One approach to possibly overcome product degradation is the use of adsorbents.
The use of adsorption agents has also been suggested to maintain a concentration gradient between the metabolite producing cell and the bulk of the culture medium and to maintain cell productivity.
Studies with various plant cell and tissue cultures (e.g. Vanilla planifolia, Morinda citrifolia, Mentha piperita) showed that use of adsorbents in the culture medium including activated char-

Table 1: Biotransformation of various biochemicals in Mentha piperita (A) and Mentha canadensis (B) callus cultures

Substrate	Product	Degradation (90 %)	Conversion rate (%) A	Conversion rate (%) B
Limonene	nd	96 h	nd	nd
Piperitone	Neomenthol	24 h	nd	<5
Pulegone	Isomenthone	6 h	15	94
	Menthone		<1	4
	Menthol		<1	1
Isomenthone	nd	72 h	90	90
	Piperitone		<4	<3
Menthone	Neomenthol	6 h	49	40
	Menthol	6 h	1	1
	nd		45	50
Neomenthol	nd	6 h	90	90
Menthol	nd	6-12 h	95	90
	Pulegone			<4
Menthylacetate	Menthol	6 h	48	20
	Pulegone		<1	<1
	nd		50	75

nd: not detectable

coal, miglyol or β-cyclodextrins improved secondary metabolite production and recovery.
Many products of commercial interest are stored within the producing plant cells. When no product release occurs the cells may be induced to release intracellularly stored products by permeabilization of cell membranes. Since the desired products are most often stored within the vacuoles, the procedure must permeabilize the plasma membrane and the tonoplast. Numerous chemical permeabilizing agents have been tested and cell viability is commonly lost upon release of vacuolar products (Brodelius, 1988). A method for the release of anthraquinones while maintaining the viability of Morinda citrifolia cells has been developed (Dörnenburg and Knorr, 1992) based on the interaction of lipases with the plasma membrane and tonoplast lipids. Treatment with several lipases as well as a glucoamylase with lysophospholipase activity caused product release while maintaining viability. The highest product release into the medium was found to be 32 % after 20 days of incubation (Figure 2). We assume that the lipases interacted with some of the phospholipids resulting in a "perforation" of the tonoplast. However, these phospholipids, because of their polar-apolar characteristics, attempt to bridge this gap as quickly as possible leading to restoration of an intact tonoplast.

GERMINATING SEEDS
Germinating seeds have proven to be useful sources for valuable plant metabolites. For example, cold tolerant carboxypeptidases could be extracted from germinating winter wheat (Zache, 1992), and where capable of partial hydrolysis for improvement of functionality of soy protein isolates at cold storage temperatures (0 to 10 °C). Germinating soybean seeds proved to be an effective

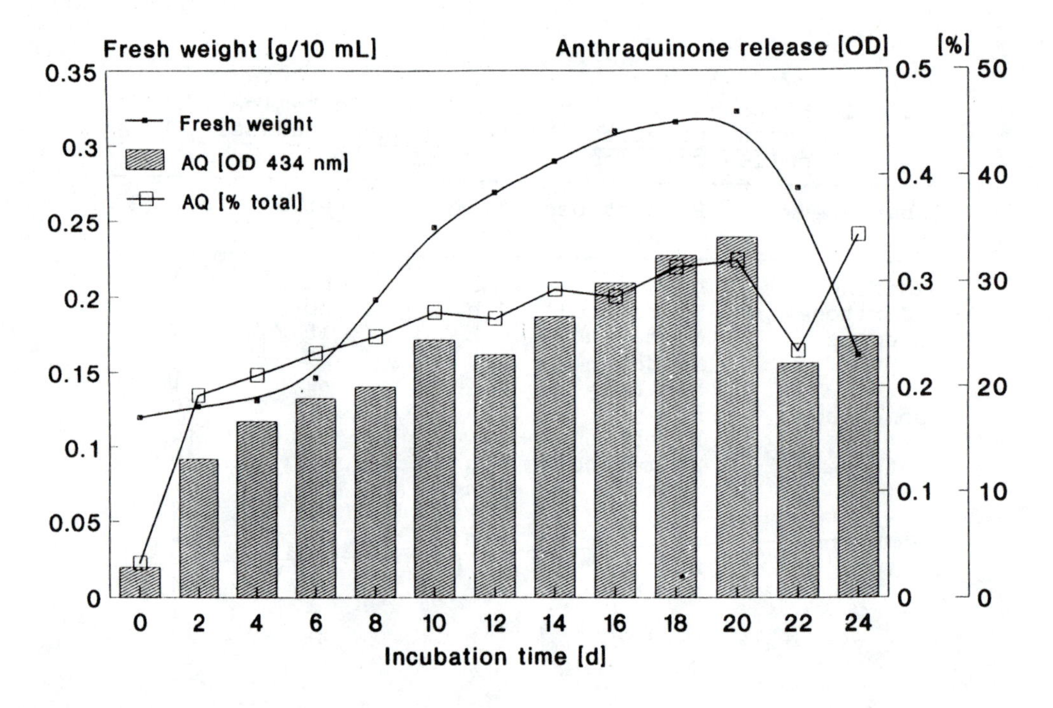

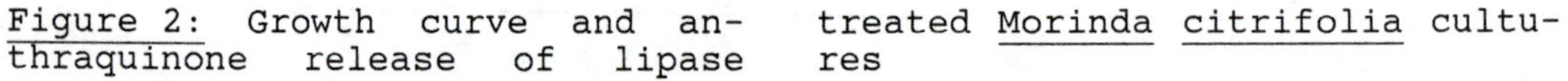
Figure 2: Growth curve and anthraquinone release of lipase treated Morinda citrifolia cultures

source for chitinase with antimicrobial activity (Teichgräber et al., 1991). Purification of plant chitinases could be performed by a simple procedure using affinity chromatography on chitin (Colantuoni et al., 1992).

ACKNOWLEDGEMENTS
Parts of this work were supported by David Michael & Co., Philadelphia, USA, the German Research Foundation and Unilever Research Laboratory, Vlaardingen, Netherlands.

REFERENCES
Brodelius, P. Appl. Microbiol. Biotechnol. 1988, 27, 561-566.
Colantuoni, D.; Popper, L. and Knorr, D. Agro-Ind. High-Tech. (in press).
Dörnenburg, H. and Knorr, D. Process Biochem. (in press).
Payne, G.; Bringi, V.; Prince, C.; Shuler, M. Plant Cell and Tissue Culture in Liquid Systems. Hanser Publisher, Munich, 1991.
Reinhard, E. and Alfermann, A.W. Adv. Biochem. Eng. 1980, 16, 49-83.
Stafford, A. Trends Food Sci. Technol. 1991, 2, 116-121.
Teichgräber, P.; Popper, L. and Knorr, D. Agro-Ind. High-Tech. 1991, 2 (3), 11-14
Zache, U. Isolation of proteases active at low temperatures from germinated wheat. PhD Thesis, Berlin Univ. Technol., Berlin, 1992.

Engineering of Lactic Acid Bacteria for Improved Production of Foods, Flavors and Additives

Willem M. de Vos. Molecular Genetics Group,Netherlands Institute
for Dairy Research (NIZO), P.O. Box 20, 6710 BA EDE, The Netherlands
(FAX: 31-8380-50400)

Lactic acid bacteria are traditionally applied in industrial milk, meat and vegetable fermentations because of their favorable fermentation properties, their capacity to produce flavors, and their antimicrobial activity. The use of molecular techniques and natural gene transfer methods has resulted in the selection and construction of lactic acid bacteria with improved industrial properties. Moreover, novel lactic acid bacteria have been obtained by genetic, metabolic and protein engineering.

This paper reviews the present state of the art and perspectives of engineering lactic acid bacteria for the improved production of foods, flavors and additives. Most attention will focus on lactococci that were the first to be characterized genetically and for which food-grade genetic modification methods have been developed. However, in most cases similar methodologies are applicable to other lactic acid bacteria including lactobacilli, leuconostocs, streptococci, and pediococci.

Genetic Tools

Initial genetic interest in lactic acid bacteria has been mainly focused on the mesophilic strains of Lactococcus lactis, because of their easy laboratory manipulation, simple metabolic pathways, and large economical importance as starter cultures for the production of cheese, quark and fermented butter.

Early studies have shown that lactococci contain a wealth of mobile DNA elements. These include a panoply of plasmids and various transposons encoding key enzymes in important metabolic pathways (McKay, 1983; Rauch & de Vos, 1992), and a great variety of bacteriophages, which frustrate industrial fermentation processes (Klaenhammer, 1987). With the use of these elements basic genetic transfer processes such as conjugation and transduction have been established, which are effective for constructing improved starter strains (Gasson, 1990).

Various lactococcal plasmids have been converted into useful vectors (de Vos, 1987). A family of patented, stable and high copy number vectors based on the replicon of pSH71 has been developed, that appeared to be functional in all transformable lactic acid bacteria, several other gram-positive bacteria and also in Escherichia coli. In combination with efficient transformation procedures based on electro-poration, and several plasmid-free genetic model strains, a variety of host-vector systems has been developed for lactic acid bacteria (de Vos, 1987). These and other vectors have been converted into expression, secretion, and integration vectors that are now used for the genetic modification of many lactic acid bacteria.

Biosafety Considerations and Food-Grade Selection Systems

Appropriate biosafety considerations are a prerequisite to the application of engineered lactic acid bacteria for the production of foods, flavors or additives. In all cases the nature of the genetic improvement should be tuned to the ultimate application of the lactic acid bacteria, which may be either in foods to be placed on the market (uncontained use) or in industrial fermentors in which they are eventually separated from the desired products (contained use). In many cases the incorporation of transmissible antibiotic-resistance genes in lactic acid bacteria is either not allowed or not preferred. Therefore, food-grade markers have been developed for selecting and maintaining the desired genetic construction. In the case of improvement of L.lactis starter cultures by

1054–7487/92/0524$06.00/0

conjugation, a natural genetic transfer mechanism, selection has been made using metabolic markers, such as plasmid-encoded lactose metabolism (Sanders, 1988) or transposon-encoded sucrose fermentation (Rauch & de Vos, 1992). A homologous food-grade system for use in self-cloning and genetic modification has been developed on the basis of the lactose genes of L.lactis which have been extensively characterized (de Vos et al., 1990). Using the small (0.3 kb) lacF gene for Enzyme III^{lac} in combination with various LacF-deficient L.lactis strains, an effective system has been developed and applied in lactococci to allow both the selection of improved strains and the stable maintenance of recombinant plasmids during industrial fermentation processes on lactose-containing media (de Vos, 1988).

Phage-resistant strains

A variety of bacteriophage-insensitivity mechanisms have been identified in L.lactis (Klaenhammer, 1987). Commercial successes have been realized by stacking into one L.lactis strain plasmids that encoded different bacteriophage-insensitivity mechanisms (Sanders, 1988; Klaenhammer, 1987). In these experiments conjugation has been used to transfer plasmids and counterselection has been based on the capacity of the recipient strain to ferment lactose. In this way a food-grade selection was realized and genetic probing has been used to verify the successful transfer of bacteriophage-insensitivity plasmids. Various starter culture producers now provide these improved strains, illustrating the application potential of advanced knowledge of plasmid biology.

Strains with improved proteolytic capacity

Lactic acid bacteria are nutritionally fastidious and contain complex proteolytic systems to provide essential amino acids and peptides that support growth during fermentation in protein-rich environments such as milk or meat. The key enzyme in the cascade of proteolytic degradation of the milk protein casein in lactococci is a cell envelope-located serine proteinase, that belongs to the well-known family of subtilases (Siezen et al., 1991). The proteinase of the industrial strain L.lactis SK11 has been studied in detail and used to engineer its processing, specificity, and production (de Vos et al., 1990). Interestingly, proteinase-overproducing lactococci were able to grow faster in milk, indicating that the initial step in the proteolysis is a limiting factor during starter growth and that this limitation may be partially overcome by proteinase engineering (Bruinenberg et al., 1992). In addition, by protein engineering based on knowledge-based modelling, various novel proteinases have been obtained that showed improved substrate specificity and stability (Vos et al., 1991). These and other proteinases may have application in the production of novel foods,or flavors from casein.

Following the initial degradation of casein by the lactococcal proteinases, further conversion into amino acids or peptides is realized by an extensive set of peptidases that have been found in lactococci. One of these peptidases is aminopeptidase N, which has a broad substrate specificity and the capacity to debitter casein hydrolysates. Following the cloning and characterization of the pepN gene encoding aminopeptidase N, lactococci have been constructed that highly overproduce this aminopeptidase (van Alen-Boerrigter et al., 1991). In subsequent studies the food-grade marker based on the lacF complementation (de Vos, 1988) was incorporated, resulting in the first example of self-cloned lactococci that overproduce an industrially important enzyme. L.lactis strains harboring these homologous plasmids have now been exempted from the biosafety directives for contained use.

Improved Production of Diacetyl

Diacetyl is an important flavor compound present in buttermilk, fermented butter, and also included in some margarines. Small amounts of this flavor compound are formed by oxidative decarboxylation from α-acetolactate, an intermediate in the citrate metabolism of some lactic acid bacteria. The metabolic pathway of α-acetolactate production has been analyzed by NMR studies (Verhue et al., 1991). There are various approaches to increase the production of α-acetolactate and hence diacetyl, by using fermentation or metabolic engineering. One is the application of strains that have a low level of α-acetolactate decarboxylase. One such strain is present in the L.lactis culture NIZO 4/25 which is used in various industrial diacetyl production processes. Another approach is to allow for efficient citrate transport by fermenting at optimal pH. The characterization and functional expression of the L.lactis citP gene for citrate permease in E.coli membrane vesicles has allowed for the detailed analysis and optimization of this transport process (David et al., 1989). Moreover, an improved production of α-acetolactate could be realized by growing L.lactis under aerated conditions (Starrenburg & Hugenholtz, 1991). Finally, the gene for α-acetolactate synthase has been cloned and characterized, allowing its overproduction in appropriate L.lactis strains (Verrips et al., 1989). It is expected that these and other changes in the metabolic flux leading to α-acetolactate will result in further improvement of the diacetyl production in lactic acid bacteria by metabolic engineering.

Novel Natural and Engineered Nisine Analogs

Nisin is a 34-residue, heat-stable, post-translationally modified polypeptide containing several dehydrated amino acids that form thioether bridges, designated lanthionines (Gross and Morell, 1971). Nisin is widely used as a natural food preservative since it shows antimicrobial activity against various gram-positive bacteria, including Clostridium, Listeria and Bacillus spp. and most lactic acid bacteria. Nisin production and immunity, as well as the capacity to ferment sucrose are located on a 70 kb conjugative transposon, Tn5276, in L.lactis strain NIZO R5 (Rauch & de Vos, 1992).

In the course of screening naturally occurring nisin overproducers, we identified a L.lactis strain that produced a natural nisin analog, designated nisin Z, with apparent higher specific activity than the known nisin, here designated nisin A. Nisin Z has been characterized at the level of its gene, nisZ, and at the protein level by NMR, resulting in the structure shown in Fig. 1 (Mulders et al., 1991). Nisin Z differs from nisin A by the substitution His27Asn, that was caused by a single mutation. Further studies of the antimicrobial activity of nisin Z showed that the larger clearing zones obtained in bioassays could be attributed to better diffusion properties, which are relevant for its application in food products (Hugenholtz & de Veer, 1991). Similar to Tn5276 encoding nisin A production (Rauch & de Vos, 1992), nisin Z production together with the ability to ferment sucrose also appeared to be located on a transposon, that could be conjugally mobilized into other lactococci allowing the food-grade construction of industrial L.lactis strains that produce nisin Z.

Using various expression systems, wild-type and mutant nisZ genes have been (over)expressed in L.lactis strains resulting in novel, nisin analogs (Kuipers et al., 1991). This protein engineering approach is facilitated by the recently proposed three dimensional model of nisin in aqueous solution (van der Ven et al., 1991). It not only permits a detailed analysis of

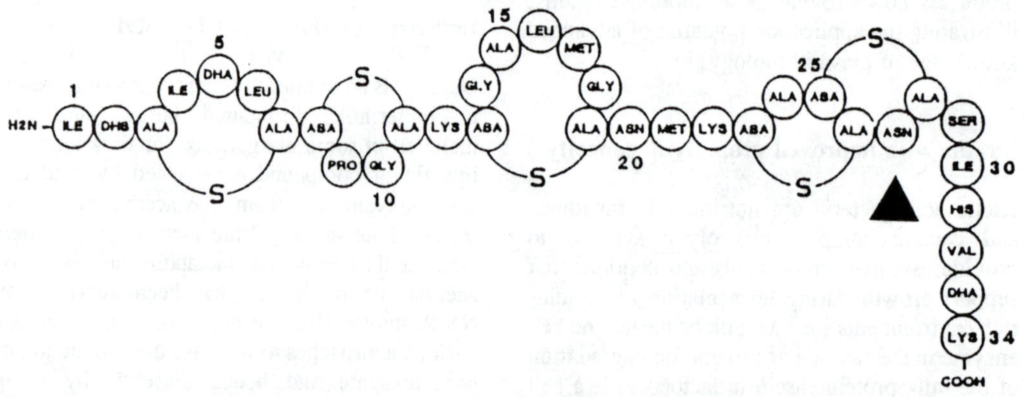

Fig. 1: Proposed structure of the natural nisin analog, nisin Z. DHA, DHB, ALA-S-AL and ALA-S-ABU indicate dehydroalanine, ß-methyldehydroalanine, lanthionine and ß-methyllanthionine, respectively. The difference with nisin A is indicated.

the structure-function relation of nisin but also allows for the design and production of nisin analogs with improved stability, increased antimicrobial activity or altered spectrum of antimicrobial activity.

References

Bruinenberg, P.G.; Vos, P., de Vos, W.M. Appl. Environ. Microbiol. **1992**, 58, 78-84.

David, S., van der Rest, M.R., Driessen, A.J.M., Simons, G; de Vos, W.M. J. Bacteriol. **1990**, 5789-5794.

De Vos, W.M. FEMS Microbiol. Rev. **1987**, 46, 281-295.

De Vos, W.M. European Patent Application **1988**, 0355036

De Vos, W.M.; Boerrigter, I; van Rooijen, R.J.; Reiche, B.; Hengstenberg, W. J. Biol. Chem. **1990**, 265:2254-22560.

De Vos, W.M., Boerrigter, I.J., Vos, P., Bruinenberg, P.J., R.J. Siezen. in: Genetics and Molecular Biology of Streptocoocci, Enterococci and Lactococci, Dunny, P.; Cleary, P.P.,; McKay, L.L., Ed.; American Society for Microbiology, Washington DC, **1990**, pp 115-119.

Gasson, M.J. FEMS Microbiol. Rev. **1990**, 87, 43-61.

Gross, E.; Morell, J.L. J. Am. Chem. Soc. **1971**, 93, 4634-4635.

Hugenholtz, J.; de Veer, G.J.C.M. in Nisin and Novel Lantibiotics; Jung, G; Sahl, H.J. Ed; ESCOM, Leiden, 1991, pp 440-448.

Kuipers, O.P.; Yap, W.M.G.; Rollema, H.S.; Beerthuyzen, M.M.; Siezen, R.J.; de Vos, W.M. in ibid, pp 250-259.

McKay, L.L. Ant. van Leeuwenh. **1983**, 49, 259-274.

Mulders, J.W.M.; Boerrigter, I.J.; Rollema, H.S., R.J. Siezen; de Vos, W.M. Eur. J. Biochem. **1991**, 201, 581-584.

Rauch, P.J.G; de Vos, W.M. J. Bacteriol. **1992**, 174, 1280-1287.

Sanders, M.E. Biochimie **1988**, 70, 411-423.

Siezen, R.J.; de Vos, W.M.; Leunissen, J.A.; Dijkstra, B.W. Protein Eng. **1991**, 4, 719-737.

Starrenburg, M.; Hugenholtz, J. Appl. Environ. Microbiol. **1991**, 57, 3535-3540.

Van Alen-Boerrigter, I.J.; Baankreis, R.; de Vos, W.M. Appl. Environ. Microbiol. **1991**, 57, 2555-2561.

Van der Ven,F.J.M.; van der Hoven,H.W.; Konings, R.N.H.; Hilbers, C.W. Eur. J. Biochem. **1991**, 202, 1181-1188.

Verhue, W.M., Tjan, S.B. Appl. Environ. Microbiol. **1991**, 75, 3371-3377.

Verrips, C.T.; Tjan, S.B.; Marugg, J.D.; Goelling, D.; and Verhue, W.M. Proc. Third Netherlands Biotechnol. Congress, Vol. 1, 1990, pp 56-63.

Vos, P.; Boerrigter,I.J.; Buist, G.; Haandrikman, A.J.; Nijhuis, M.; de Reuver, M.B.; Siezen, R.J.; Venema, G.; de Vos, W.M.; Kok, J. Protein Eng. **1991**, 4, 479-484.

Flavor Biotechnology

P. Schreier, Lehrstuhl für Lebensmittelchemie, Universität Würzburg, Am Hubland, 8700 Würzburg, Germany (FAX:+49-9369-8167)

The principal biotechnological methods and their potential application for the production of flavorings are discussed with particular emphasis to microbial biotransformations and the use of enzymes catalyzing the release of volatiles from their glycoconjugates. In addition, the importance of analytical techniques for the differentiation between biotechnically ("natural") and chemically ("nature-identical") synthesized flavor compounds is stressed.

In recent years, the increasing demand for natural flavors has stimulated a broad area of flavor research, in which biotechnical processes play an important role. Biotechnology comprises biotransformation, *de-novo* synthesis, and genetic engineering. In each of these areas, plant cells, microorganisms, and enzymes can be used. In the past, several comprehensive reviews have been provided (Parliment and Croteau, 1986; Lindsay and Willis, 1989; Mor, 1990; Schreier, 1988; 1989; 1991). Thus, in the following, our interest will be focused on the most recent trends and developments only.

Plant Tissue and Cell Culture

In the course of the symposium, Knorr's contribution is devoted to this particular topic. Thus, only a few comments will be given. The production of secondary metabolites by plant cell cultures is potentially a very promising field (Charlwood and Rhodes, 1990). Plants are able to produce complex compounds that may be difficult to synthesize in a chemical laboratory. World-wide, a great effort is being made to synthesize many different plant products in this way. Up till now, the commercial success of all these efforts has been very limited. One of the major reasons for this delay in commercialization is the low productivity of plant cell cultures. One of the highest productivities reported is that of rosmarinic acid by *Coleus blumei* cells (Petersen, 1989), but this example is more the exception that proves the rule. In general, productivities are too low to make a feasible economic process.

As an alternative the concept of so-called "precursor-stimulated flavor formation" has to be mentioned (Drawert, 1988). It consists of the use of (i) aged tissue cultures; (ii) precursor atmosphere storage; and (iii) homogenates. Among these possibilities, the latter has been published for industrial purposes concerning the production of C_6 and C_8 volatiles originating from lipid metabolism. Isomeric hexenals and the corresponding C_6 alcohols are known to be formed in plant homogenates from Z,Z-1,4-pentadienoic acids (e.g., linoleic and linolenic acid) by a sequence of enzymatic reactions, in which acyl hydrolase, lipoxygenase, hydroperoxide lyase, and alcohol dehydrogenase are involved (Hatanaka et al., 1986). Feeding suitable plant homogenates with fatty acid precursor leads to an increased production of C_6 volatiles until < 1g/kg fresh weight (Drawert et al., 1986; Goers et al., 1989). In this way, preferably E-2-hexenal is available. The same concept, i.e. using homogenates, but from mushrooms, has been recommended for the production of mushroom flavoring, i.e. (R)-1-octen-3-ol, which comprises homogenizing mushrooms (*Agaricus bisporus*; *A. bitorquis*; *A. campestris*) and, during or after homogenization, contacting the mushrooms with an aqueous medium containing a water-soluble salt of linoleic acid and oxygen (Kibler et al., 1988).

Microbiology

A great number of reviews has been provided dealing with microbial methods for the production of industrially important compounds including flavorings (cf. references cited by Winterhalter and Schreier, 1992). There are two principal ways to use microorganisms, i.e. (i) fermentation; and (ii) microbial transformation. Recently, these two fields of flavor biotechnology has been reviewed (Delest, 1991). In addition, in the course of this symposium, De Vos'contribution deals with the engineering of lactic acid bacteria for the biotechnological production of food flavors and additives. Thus, only a few additional remarks are given

1054–7487/92/0528$06.00/0

both for fermentation and biotransformation processes. The selected data outlined in Table 1 may summarize the actual situation in microbial flavor formation. As to biotransformations, particular attention has been directed to the microbial bioconversion of terpenes (Abraham, 1990). Principal types of reactions such as partial degradation, hydroxylation, epoxidation, hydration of double bonds, and reduction have been observed in a number of studies of terpene conversions using various species of microorganisms. They have also been found in our recent biotransformation studies using a common fungus, well-known in winemaking, i.e. *Botrytis cinerea.* Using α- and ß-damascone as substrates, a number of oxo- and hydroxy derivatives of this C_{13} norisoprenoid has been detected (Schoch et al., 1991; Schwab et al., 1991a). As shown from Fig. 1, the controlled conversion of 2-methyl-2-hepten-6-one by *B.c.* has resulted in the stereoselective formation of (S)-(+)-2-methyl-2-hepten-6-ol (sulcatol) (ee = 90 %) together with the formation of isomeric "sulcatol oxides" (Schwab et al., 1991b). A more practical approach is provided by a recent patent describing the production of "natural" vanillin by microbial biotransformation from eugenol and/or isoeugenol (Rabenhorst and Hopp, 1991). These bioconversion reactions are examples for the transformation of inexpensive, easily available flavor chemicals to higher-valued products.

Enzymes

Much of the available information in this field has been provided in a number of reviews (cf. references cited by Winterhalter and Schreier, 1992). Actually, just four biocatalytic systems predominate, i.e. pig liver esterase; pig pancreatic lipase (PPL); and the lipases from *Candida cylindracea* and from baker's yeast. In general, other than as products themselves, enzymes are being developed for use as reagents in chemical synthesis. Several enzymes catalyzing separate reaction steps are looked for in the growing market of agrochemicals. One may expect these to become more and more important because of questions of specificity of action needed, e.g., for particular sensory effects and legal safety of stereoisomer mixtures. Thus, from the viewpoint of enzyme technology, the ability to exert some measure of control on enzyme regioselectivity and enantioselectivity and thus produce products of desired structure and chirality would be highly desirable. Our recent work carried out on PPL-catalyzed esterification of secondary alcohols on the effect of the structure of substrate on enzyme selectivity and reactivity may illustrate one approach to this goal (Lutz et al., 1990).

Discussing the potential use of enzymes for the production of flavor compounds, it has to be considered that in plant tissues a definite amount of flavor substances is bound in non-volatile form,

Table 1. Selection of microbially produced flavor compounds

Substance	Source	Substrate	Yield	Reference
Acetaldehyde	*Pichia pastoris*	Ethanol	40 g/l	Duff and Murray,1988
Butanal	*Pichia pastoris*	1-Butanol	14 g/l	Duff and Murray,1988
3-Methyl-butanal	*Pichia pastoris*	3-Methyl-1-butanol	8 g/l	Duff and Murray,1988
Methoxy-benzaldehyde	*Ischnoderma benzoinum*	-	100 mg/l	Berger et al., 1986
6-Pentyl-α-pyrone	*Trichoderma viride*	-	170 mg/l	Collins and Halim, 1972
γ-Decalactone	*Yarrowia lipolytica*	Castor oil	5 g/l	Farbood and Willis, 1983
γ-Decalactone	*Tyromyces sambuceus*	Castor oil	2.5 g/l	Kapfer et al., 1991
Ethyl acetate	*Candida utilis*	Ethanol	4.5 g/l	Armstrong et. al., 1984
Ethyl butanoate	*Candida cylindracea*	Ethanol + But.acid	5 g/l	Duff and Murray,1988
Butyl butanoate	*Aspergillus niger*	1-Butanol+ But.acid	17 g/l	Duff and Murray,1988
Methyl anthranilate	*Pycnoporus cinnabarinus*	-	19 mg/l	Gross et al., 1990

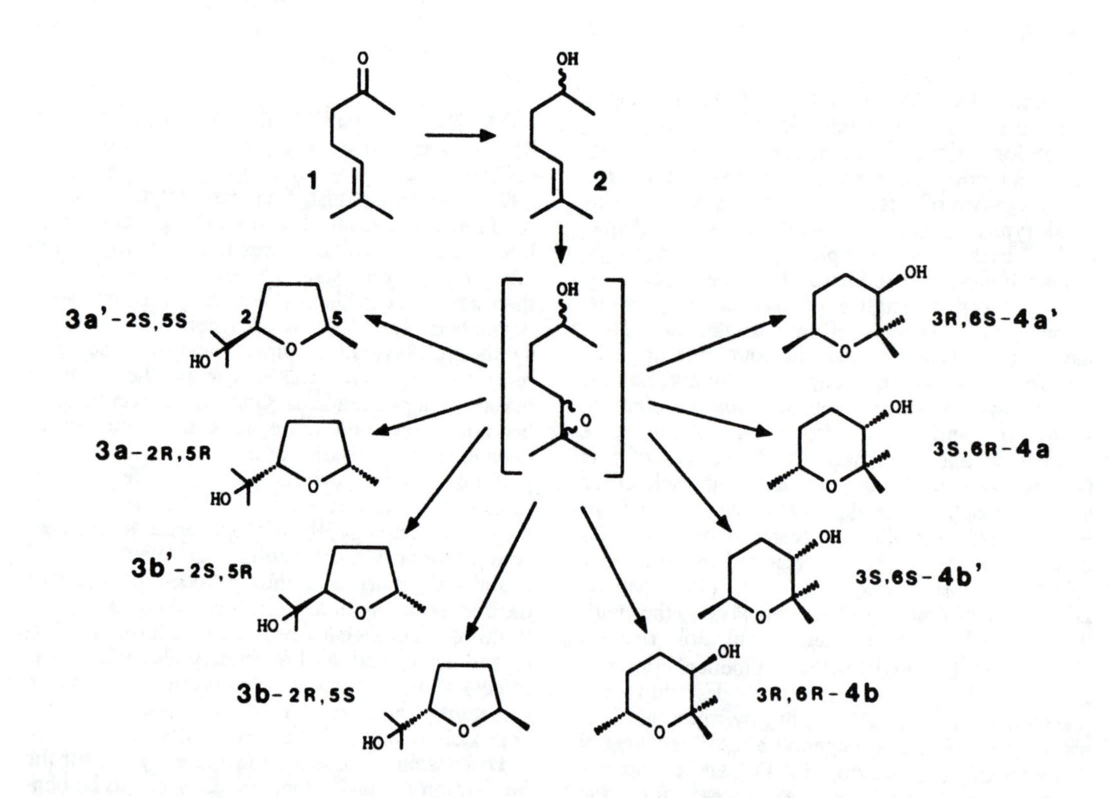

Figure 1. Structures of biotransformation products formed from 2-methyl-2-hepten-6-one **1** by *B. cinerea.* **2** sulcatol **3a** (2R,5R)- **3a'** (2S,5S)- **3b** (2R,5S)- **3b'** (2S,5R)-2-(1-hydroxy-1-methylethyl)-5-methyltetrahydrofuran; **4a** (3S,6R)- **4a'** (3R,6S)- **4b** (3R,6R)- **4b'** (3S,6S)-tetrahydro-2,2,6-trimethyl-2*H*-pyran-3-ol. (Reproduced with permission from Schwab et al., 1991. Copyright 1991 Pergamon Press.)

from which they can be liberated by use of enzymes, i.e. glycosidases (Schreier, 1991). A representative example is raspberry fruit, in which 57 aglycones have been detected; among them a number of attractive C_{13} norisoprenoids have been found (Fig. 2) (Pabst et al., 1991). Structural elucidation of their precursors occurring in ppm ranges in the fruit has revealed the structures of a number of new glycosides (Pabst, 1991). The use of glycosidases for the release of glycosidically bound volatiles from plant tissues has been proposed for industrial application (Ambid, 1988; Gunata et al., 1989).

Analytical Aspects

Due to the increasing use of "bioflavors" in foods, analytical methods to differentiate between the "nature-identical" and "natural" status of flavorings has become more and more important. Recently, the state-of-the art has been discussed in detail in the course of international meetings (Krueger, 1991; Martin, 1991). Among the different isotope techniques and chromatographic methods, the latter are of more practical relevance due to their fast applicability. As high-sophisticated chromatographic technique, the on-line coupling of multidimensional gas chromatography with mass spectrometry (MDGC-MS) using achiral (silicone)/chiral (modified cyclodextrin) column combination (Krammer et al., 1990) allows the sensitive and accurate enantiodifferentiation of chiral flavor constituents from complex natural sources without laborious sample preparation. At present, the enantiodifferentiation of γ- and δ-lactones, secondary alcohols, esters, in particular, 2-butanoates, as well as α-ionone are of high practical relevance (Werkhoff et al., 1991).

Conclusions

In general, the flavor industry has been somewhat slow in understanding or accepting biotechnology as an integral part of industrial processes, but this attitude is changing gradually.

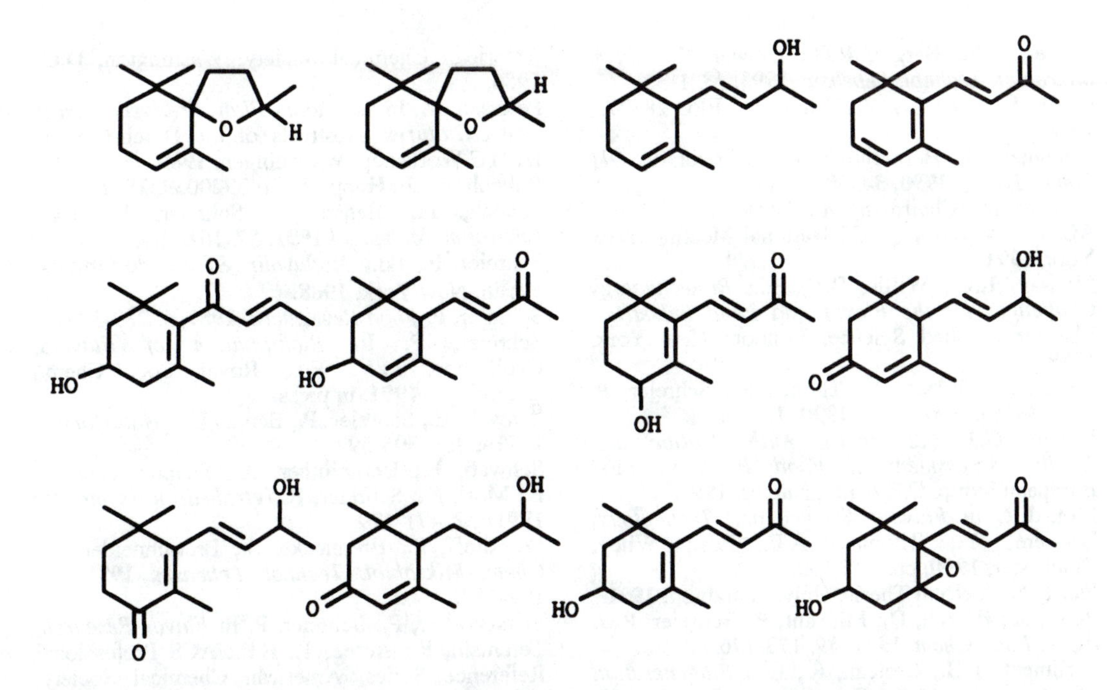

Figure 2. Structures of aglycones enzymatically (Rohapect D5L, Röhm) released from a glycosidic fraction of raspberry fruit. From left to right: theaspiranes A/B; α-ionol; 3,4-dehydro-ß-ionone; 3-hydroxy-ß-damascone; 3-hydroxy-α-ionone; 4-hydroxy-ß-ionone; 3-oxo-α-ionol; 4-oxo-ß-ionol; 3-oxo-7,8-dihydro-α-ionol; 3-hydroxy-ß-ionone; 3-hydroxy-5,6-epoxy-ß-ionone.

Thus, meanwhile, microbial flavor biotechnology has been introduced by all big flavor houses. In addition, the high interest devoted to flavor precursors by academic and industrial research groups demonstrates the importance of biotechnical processes for the increasing development of natural flavors.

Acknowledgements

I wish to express my deep appreciation to my coworkers, who are identified in the references. Current studies are gratefully supported by the Deutsche Forschungsgemeinschaft, Bonn.

References

Abraham, W.R. In *GBF Wissenschaftl. Erfolgsbericht 1990,* GBF, Ed.; GBF, Braunschweig, pp. 209-210.

Ambid, C. EPO 309339, **1988**.

Armstrong, D.W.; Martin, S.M.; Yamazaki, H. *Biotechnol. Bioeng.* **1984,** *26,* 1038-1042.

Berger, R.G.; Neuhäuser, K.; Drawert, F. *Flav. Fragr. J.* **1986,** *1,* 181-184.

Charlwood, B.V.; Rhodes, M.J.C., Eds.; *Secondary Products from Plant Tissue Culture,* Phytochemical Soc. Europe, Oxford University Press, Oxford, **1990**.

Collins, R.P.; Halim, A.F. *J. Agric. Food Chem.* **1972,** *20,* 437-439.

Delest, P. In *Authentication and Quality Assessment of Food Products,* Abstracts of the First European Seminar, Nantes, **1991**.

Drawert, F. In *Bioflavour '87,* Schreier, P.; Ed.; W. de Gruyter, Berlin, New York, **1988,** pp. 3-31.

Drawert, F.; Kler, A.; Berger, R.G. *Lebensm. Wiss. u. Technol.* **1986,** *19,* 426-430.

Duff, S.J.B.; Murray, W.D. *Biotechnol. Bioeng.***1988,** *31,* 44-49.

Farbood, M.I.; Willis, B.J. US 306091, **1983.**

Goers, S.K.; Ghossi, P.; Patterson, J.T.; Young, C.L. US 4806379, **1989.**

Gross, B.; younnet, G.; Picque, D.; Brunerie, P.; Corrieu, G.; Asther, M. *Appl. Microbiol. Biotechnol.* **1990,** *34,* 387-391.

Gunata, Z.; Bitteur, S.; Baumes, R.; Brillouet, J.M.; Tapiero, C.; Bayonove, C.; Cordonnier, R. EPO 332281, **1989**.

Hatanaka, A.; Kajiwara, T.; Sekiya, J. In *Biogeneration of Aromas,* Parliment, T.H.; Croteau, R., Eds.; ACS Symposium Series 317, American Chemical Society, Washington, **1986,** pp. 167-182.

Kapfer, G.F.; Berger, R.G.; Drawert, F. *Chem. Mikrobiol. Technol. Lebensm.* **1991,** *13,* 1-3.

Kibler, L.A.; Kratky, Z.; Tandy, J.S. EPO 288773, **1988.**

Krammer, G.; Bernreuther, A.; Schreier, P. *GIT Fachz. Labor.* **1990,** *34,* 306-311.

Krueger, D. (Chairman), *Adulteration of Flavors,* AGFD Div. Symp., ACS National Meeting, New York, **1991.**

Lindsay, R.C.; Willis, B.J., Ed.; *Biotechnology Challenges for the Flavor and Food Industries,* Elsevier Applied Science, London, New York, **1989.**

Lutz, D.; Güldner, A.; Thums, R.; Schreier, P. *Tetrahedron Asymmetry* **1990,** *1,* 783.

Martin, G.J. (Chairman), *Authentication and Quality Assessment of Food Products,* First European Symp. C.E.A.I.S., Nantes, **1991.**

Mor, J.R. In *Flavour Science and Technology;* Bessière, Y.; Thomas, A.F. Eds.; Wiley, Chichester, **1990,** pp. 135-160.

Pabst, A. Doctoral Thesis, Univ. Würzburg, **1991.**

Pabst, A.; Barron, D.; Etiévant, P.; Schreier, P. *J. Agric. Food Chem.* **1991,** *39,* 173-176.

Parliment, T.H.; Croteau, R., Ed.; *Biogeneration of Aromas,* ACS Symposium Series 317, American Chemical Society, Washington, D.C., **1986.**

Petersen, M. In *Secondary Plant Products from In Votro Cultures,* Abstracts of the Dutch-German IAPTC Workshop, Wageningen, **1989.**

Rabenhorst, J.; Hopp, R. DE 3920039, **1991.**

Schoch, E.; Benda, I.; Schreier, P. *Appl. Environm. Microbiol.* **1991,** *57,* 15-18.

Schreier, P., Ed.; *Bioflavour '87,* W. de Gruyter, Berlin, New York, **1988.**

Schreier, P. *Food Rev. Intern.* **1989,** *5,* 289-315.

Schreier, P. In *Bioformation of Flavors;* Charlwood, B.V. Ed.; Royal Soc. Chem., Cambridge, **1991,** in press.

Schwab, E.; Schreier, P.; Benda, I. *Z. Naturforsch.* **1991a,** *46c,* 395-397.

Schwab, E.; Bernreuther, A.; Puapoomchareon, P.; Mori, K.; Schreier, P. *Tetrahedron Asymmetry* **1991b,** *2,* 471-479.

Werkhoff, P.; Brennecke, S.; Bretschneider, W. *Chem. Mikrobiol. Technol. Lebensm.* **1991,** *13,* 129-152.

Winterhalter, P.; Schreier, P. In *Flavor Research,* Teranishi, R.; Acree, T., Eds.; ACS Professional Reference Series, American Chemical Society, Washington, **1992,** in press.

Application and Economics of Enzymes for the Food Industry

Bruno Sprössler, Röhm GmbH, Kirschenallee, W-6100 Darmstadt,
Fax-No. (0 61 51) 18-30 32

Over 5 000 enzymes are known, but only about 20 are produced for industrial use. Sales realized with these enzymes total roughly 700 million dollars, and circa 60 % thereof are sales to the food industry, where enzymes are used as processing aids (Uhlig, 1991). Enzyme technology is nearly as old as mankind itself, but it is due to modern biotechnology that we are able to produce enzymes with certain desired properties in an economical way. Genetic engineering allows us to reduce development and production costs even further, because microorganisms now produce the desired enzymes in larger quantities at a lower material consumption.

Most widely used on an industrial scale are hydrolases, which break down natural polymers like starch, protein and fat. Tailored enzyme products made it possible to convert the acid-based process to an enzymatic process. All isomerizations in the manufacture of high-fructose syrup are performed with enzymes. The baking industry balances the deficits of endogenous enzymes with tailored enzyme products. The fruit juice industry cannot produce clear juices without the help of enzymes; berries cannot be pressed without them.

Enzymes offer the food processor certain advantages over physical and chemical processes. They work under mild conditions and enable reactions which would otherwise require high pressures and extreme pH ranges. "Tailored" enzymes can exert an influence on the appearance, texture, flavor, color, shelf life and nutritive value of foods, but also on process parameters like viscosity, filtering rate, yield and ease of handling.

Enzymes are processing aids, i. e. they act while the foodstuffs are processed, but are normally inactive in the end product and in this state as easily digested on consumption as all other proteins. Enzymes for the food industry are plant proteases like papain and malt flour as well as chymosin obtained from animals. However, these raw material sources are limited and the properties of the enzymes fixed.

Microorganisms provide alternative sources of supply. They are available in unlimited quantities, produce enzymes with different properties and are easy to influence both qualitatively and quantitatively. It is the costs which decide on the acceptance of an enzyme-based process, which means that the benefits must outweigh the costs of enzymatic treatment.

A target-oriented development of new enzymes is impossible unless the developer is familiar with the user's processes in detail and also with the problems the user wants to solve. Usually it is not very effective to translate the results obtained by public research institutes directly into reality. Advances made in research will materialize much faster if the enzyme producer translates them into marketable products as part of an innovative supplier-customer relationship and is able to demonstrate the benefits on the spot. Let this be illustrated by a few examples:

Baking

Besides alpha- and beta-amylases, flour contains proteases, pentosanases, lipases, lipoxidases and esterases. The endogenous enzyme content depends on the type of grain, the climate and the harvesting conditions. In the past, the baking process would be adapted to the varying flour qualities. Today, with more and more baking processes being automated, the enzyme content of the flour is optimized with a view to the process parameters by adding microbiologically produced enzymes. The targets are good quality of

1054–7487/92/0533$06.00/0

Table 1: Use of Enzymes in Food Industry

Industry	Enzymes	Application
Starch	Amylases	Dextrose
	Isomerase	High-fructose corn syrup
	Glucoxyl transferase	Oligosaccharides
	Cellulase	Starch/gluten separation
Fruit	Polygalacturonase with pectinesterase;	Juice clarification
	pectin lyase,	Mash treatment
	macerating poly-	Vegetable maceration
	galacturonase	Citrus pulp mash
Baking	Amylases	Bread volume
	Proteases	Dough conditioning
	Pentosanase	High-fiber bread
		Frozen dough
	Lipoxidase	Bleaching
Brewing	Amylases	Saccharification
	Beta-glucanase	Filtration
	Proteases	Free amino compounds
	Papain	Chill-proofing
Dairy	Chymosin	Curd formation
	Proteases	Ripening
		Protein modification
	Lipase	Flavor
	Lactase	Lactose conversion
Protein	Protease	Functional hydrolysates
	Peptidase	Meat and fish processing
		Pet food
Fats	Lipases	Speciality fats

the baked goods, assured results, machinability of the dough and good dough tolerance.

Fungal amylase from Aspergillus oryzae is ideal for standardizing amylase activity. Given the right amylase supplementation, damaged starch in the dough is converted to dextrin. Endogenous beta-amylase forms maltose, which yeast converts into the CO_2 gas that leavens the dough. Proteinases change the rheological properties of the gluten. Correctly chosen proteases give extensible doughs for bread or else plastic, non-shrinking doughs for cookies and crackers. Pentosanases lower the viscosity of the pentosans taking part in dough preparation and enhance, in combination with amylases and proteases, the dough properties as well as the bread volume and freshkeeping (Ter Haseborg et al, 1988). This is particularly important with rye flours, as these contain about 5 - 8 % pentosans (Meuser et al, 1986). Oxidases for crosslinking protein chains by S-S bonds are presently subject to close scrutiny.

Fruit and Vegetable Processing

It took the pectinases to enable the manufacture of clear apple juice at reasonable costs. The viscosity of the juices is reduced by polygalacturonase plus pectin esterase or pectin lyase, the cloud particles are destabilized and can be separated with ease (Pilnik et al, 1990).

Black currants contain so much pectin that it requires the addition of pectinase to release and express the juice from the gel-like crushed fruit mass. This mash enzyme process paved the way for treatment of the apple mash. It increases the juice yield by about 15 %. The cost/benefit calculation reveals an additional gain of about 20 - 40 dollars per ton of apples.

The mechanical manufacturing process

of vegetable products is facilitated by the use of macerating polygalacturonase, which makes it possible to obtain new products of unique quality. The joint action of polygalacturonase and cellulase decreases the viscosity of carrot mash to such an extent that the carrot becomes fluid and can even be concentrated threefold to give a pasty product. Sweet peppers and onions can be processed to spices in the same way.

Wet Milling of Corn

Steeping corn kernels in water is an easy way of separating starch granules, protein, fiber and germs. The starch yield can be increased with the aid of cellulases and hemicellulases. They free the starch bound to fibrous material and protein. If phytase is additionally added to the steeped corn (Vaara, Alko Ltd., personal communication), this combination releases the phosphorus bound by phytate and makes it available for nutrition. As a result, the addition of phosphate to animal feed can be reduced, the manure contains less phosphorus and the entrophication of waters decreases.

Protein Hydrolysis

Proteases are traditionally used in cheese production and for chill-proofing of beer. Partial hydrolysis of casein improves the whipping property and the emulsibility. Further hydrolysis of the whey protein leads to dietetic nutrients with a higher resorption in the gastrointestinal tract. A suitable combination of proteases and peptidases is capable of reducing the formation of bitter peptides (Moll et al, 1990). Intense efforts are being made to replace hydrochloric acid for total hydrolysis of protein for flavorings. Proteases and peptidases are also successfully employed for the utilization of fish and poultry waste.

Chymosin hydrolyses κ-casein specifically between Phe 105 and Met 106. This animal enzyme, which is so important for cheese production, is not abundantly available. Chymosin substitutes from mucor mihei have somewhat different properties. Meanwhile it has become possible, however, to express animal genes in microorganisms and to produce animal-identical chymosin fermentatively.

Fats

In the fats industry the use of lipases is currently restricted to certain sectors, such as the production of butter and cheese flavor. Otherwise the partial hydrolysis of fat is still widely performed with acids and bases. However, some highly interesting approaches to the production of mono- and diglycerides have already been suggested. Inexpensive fats and oils can be reesterified to give high-grade products of cocoa butter quality, but the breakthrough to industrial manufacture remains to be achieved. As we deepen our knowledge of them and they become more readily available, it looks as though lipases are products with a future.

Outlook

Thanks to genetic engineering, the production of enzymes is becoming ever more economical. By replacing amino acids in the peptide chains of enzymes we can vary such properties as heat stability, reaction equilibrium, and pH sensitivity. The availability of tailoring enzymes at reasonable prices opens up new applications in the animal feed and pulp and paper sectors. If this aim is to be achieved, we must engage in more high-risk research and continue our innovative cooperation in putting the results to good economic use.

References

Christensen, F.M.; *Enzyme Technology versus Engineering Technology in Food Industry;* Biotech. and Appl. Biochem., **1989,** 11, pp 249-265.

ter Haseborg, E.; *Quality problems with high-fiber breads solved by use of hemicellulase enzymes;* Cereal Foods World, **1988,** pp 419-422

Meuser, F; Sukow, P.; In *Chemistry and Physics of Baking;* Blanshard, J.M.V. et al; R. Soc. Chem., London, 1986; pp 42-61

Moll, D.; Plainer, H.; Sprössler, B.;

Proteinases in Food Technology; ZFL, **1990,** 41, pp 829-832.

Pilnik, W.; Voragen, A.G.D.; *Use of Enzymes in Food Processing;* Food Biotechnol., **1990,** 4, pp 319-328.

Spradlin, J.E.; *Tailoring Enzyme Systems for Food Processing*; ACS Symposium Series, 389; American Chemical Society: Washington, D.C., **1989,** pp 24-43.

Uhlig, H.; *Enzyme arbeiten für uns;* Hauser: München, Wien, **1991,** pp 409-410

Some Practical Applications of Gene Probes and PCR to Food Safety

Walter E. Hill* and **Marleen M. Wekell.**
Food and Drug Administration, Bothell, WA 98115
(FAX:206 483-4996)

Traditional microbiological identification schemes have relied on the morphological, physiological, and biochemical characteristics of strains. Recombinant DNA techniques have allowed the cloning, isolation, and purification of specific DNA fragments. These pieces of DNA can be labeled and used in DNA hybridization tests to determine if a given bacterial strain harbors a particular gene. Nucleotide sequencing and in vitro laboratory synthesis of oligonucleotides have made these molecules readily available as probes and primers. The amplification of specific DNA segments (Saiki et al., 1988) allows the rapid detection of specific genes in contaminated foods or subsequent enrichments. This methodology holds the promise for the rapid identification of microbial pathogens in contaminated foods but a considerable amount of effort may have to be devoted to the preparation of samples for molecular biological analyses.

Gene Probes

Colony hybridization format - Colony hybridization has been used to detect pathogenic bacteria in foods by targeting a gene contributing to virulence (Hill et al., 1983; Hill et al., 1985; Jagow and Hill, 1986). The use of gene probes for the detection of foodborne pathogens has reviewed (Hill and Lampel , 1990).

Colonies on spread plates are tested using hybridization. By counting the number of colonies that contain a particular gene, the number of cells with that same gene present in the original food sample can be calculated. The ability to enumerate pathogenic microorganisms is essential when testing foods for which upper limits for these bacteria have been set.

Given the current state of detection technology, about 10^5 to 10^6 cells are required to yield a positive signal. When total bacterial count is high, the pathogenic cells may not be able achieve sufficient growth due to competition. To overcome this difficulty caused by high numbers of indigenous microbes, selective technigues may be applied to increase the sensitivity of the colony hybridization test (Jagow and Hill, 1988).

Widespread use of this format has been constrained by the need to use a radioactive label to achieve sufficient sensitivity and to overcome background problems caused by cell and food debris. Recent work on non-isotopic labels is promising and should lead to a broad implementation of this technique (Gicquelaia et al., 1990; Peterkin et al., 1991; Wilson et al., 1990). Because of the sample dilution that occurs during blending and limitations on the amount of liquid that can be spread on an agar plate, the colony hybridization test has a practical lower limit of detection of about 20-100 cells per gram.

Dip stick - The colony hybridization technique can provided quantitative data but it is labor-intensive. A two-probe system has been developed that uses a dipstick to remove probe-target complexes from a hybridization reaction mixture where both the probe and the target are free in solution. This technique uses a non-isotopic labeling system and is well-suited for the analysis of large numbers of food samples after completion of an appropriate enrichment scheme (Curiale et al., 1990; King et al., 1989). However, because enrichments are required, the tests results are qualitative and the bacterium sought is reported as being present or absent.

The first tests of this type for the analysis of foods used gene probes targeted to regions of ribosomal RNA. Such tests may be more sensitive because cells can contain over 10,000 copies of these molecules (Edelstein et al., 1987). A considerable amount of data on the evolution of ribosomal DNA nucleotide sequences has been collected, genus specific and, in some cases, species specific regions can be targeted by probes (Delong et al., 1989; Hogan, 1989; Romaniuk and Trust, 1989). The sensitivity of this type of test has been as low as a single cell in a 25 gram sample. The results of these hybridization tests are very similar to those obtained using conventional microbiological techniques but often provide results days earlier.

Points to consider - These DNA hybridization methods offer rapid screening techniques that allow the detection of bacteria harboring specific genes. Because most foods may be contaminated by $<10^5$ cells/gram of a bacterial pathogen, enrichment cultures must be carried out. While the mechanism of DNA hybridization reactions are fairly well understood, it is critical that growth steps be successful in increasing cell densities to a sufficient level or false negative results will be observed. Special attention may have to be paid to the resuscitation of injured cells. Also, these tests are best used on bacteria that cause infections. Species causing disease by intoxication, such as *Staphylococcus aureus* and *Clostridium botulinum* may not be viable in the sample.

Care is required in interpreting results because false positive or false negative conclusions may be reached. Pathogenicity is a multigenic phenomenon and the presence of a single gene does not mean that a particular strain can, indeed, cause disease in humans. Additional tests based on a biological system should be preformed to confirm *in vivo* pathogenicity. Hybridization assays based on ribosomal RNA may not distinguish between individual strains and hence positive results may not reflect pathogenic potential other than that usually ascribed to the species as a whole. However, finding a bacterium that is a member of the same genus as a human pathogen indicates a potential for contamination by strains that are of human health concern. The presence of such microorganisms should signal an increased level of concern for public health.

Polymerase Chain Reaction

Specific DNA segments can be amplified using oligonucleotide primers, and temperature cycling. This technique, the polymerase chain reaction (PCR), can increase the amount of a specific DNA region over one-million fold in a few hours (Saiki et al., 1985). PCR-amplified DNA can be used for gene probes, for sequencing, or for cloning. In some cases the appearance of amplified DNA can be used as an assay for the presence of particular bacteria.

Mechanism - Usually, PCR is carried out with a target DNA region of known nucleotide sequence and primers complementary to different DNA strands. The reaction mixture is heated (to around 94°C) to separate (denature) the DNA strands and then cooled to allow annealing of the primers which are present in a large excess. The annealing temperature is critical in controlling the specificity of PCR. If the temperature is too low, primers will anneal to sites that are not perfect complementary matches and the reaction will have a low specificity and possibly a high level of false positive results.

After annealing, the temperature is raised to about 72°C for efficient DNA replication by a thermostable DNA polymerase (Saiki et al., 1988). This three temperature cycle is usually repeated 25-50 times; often the increase in the target segment can be over a million fold. The preponderant product is a fragment of DNA equal to the length of the region of target DNA bounded by and including the primers. The appearance of these molecules can be observed by gel electrophoresis.

Sample preparation - While the molecular biology of this reaction is fairly well understood, the greatest challenges facing those who wish to apply this technique for detecting pathogenic bacteria in contaminated foods is that of sample preparation. The direct detection of a few cells per gram is theoretically possible but efficient recovery of template DNA and the removal of interfering substances from the food may require a substantial amount of time to develop.

Generally, extraction of DNA from the contaminated food has not proved promising. The most repeatable results have been obtained after incubation of food samples or homogenates for a few hours in an appropriate bacteriologic medium. Apparently, there are inhibiting substances in some food homogenates, such as those made from oysters (Hill et al., 1991), that are resistant to the effects of chaotropic salts and that reduce the efficiency of the *Thermus aquaticus* thermostable DNA polymerase to such as degree that no amplified DNA is observed by gel electrophoresis.

Target selection - The type of genetic region selected for amplification by PCR depends on the type of information that is required. When information on the presence or absence of a particular taxonomic group is sought, a unique region of a ribosomal RNA gene is an appropriate choice (Wesley et al., 1991). When a test for the presence or absence of a particular bacterial pathogen is needed, primers that target a gene required for causing disease are developed (Bessesen et al., 1990; Olive et al., 1988; Hill et al., 1991). The same caveats regarding the assumption of virulence based on the presence of a single gene, as discussed with respect to gene probes, are valid here as well.

Primer selection - Computer programs can assist in the selection of PCR primer sequences for use as PCR primers (Lowe et al., 1990; Rychlik and Rhoads, 1989; Rychlik et al., 1990); it is not always clear as to what features of a sequence make it an efficient primer. Primer binding sites are usually between 100 and 2000 nucleotides apart although some shorter and longer regions have been amplified. Primers are normally 18 to 25 nucleotides in length although both shorter (see below) and much longer molecules have been used. There should be limited complementarity with other primer and no substantial homology with other DNAs likely to be in the sample. The region to be amplified should also be free of substantial secondary structure, such as hairpin loops, which may impede the replicative progress of the polymerase. If a reaction using several sets of primers is conducted (multiplex PCR), one should use primers with similar thermal stabilities that bracket regions of different lengths so that the size of the PCR product will indicate which region is being amplified.

Applications - PCR has been using to detect particular species or pathogenic strains microorganisms in foods (Bessessen, et al., 1990; Border et al., 1990; Denner and Boychuk, 1991: Furrer et al., 1991; Hill et al., 1991; Lampel et al., 1991; Thomas et al., 1991; Wernars, Delfgou et al., 1991; Wernars, Heuvelman, et al., 1991). A multiplex PCR has been used to detect several microorganisms in environmental samples (Bej, DiCesare et al., 1991; Bej, McCarty et al., 1991) and for three species of pathogenic vibrios (W.E. Hill, L. Leja, and P. Trost, unpublished results).

Limits of detection - Theoretically, PCR should be able to detect a single copy of a DNA target if a sufficient number of amplification cycles are used. For example, if a single copy of a 500 base pair region is to be amplified, and the limit of detection for an agarose gel is about 10 ng per band, about 2 x 10^{10} copies of the molecule are required. This requires about 55 cycles if the reaction is 50 percent efficient but only about 35 cycles if the reaction is 90 percent efficient.

Another factor that reduces PCR detection sensitivity is the efficiency recovery of template DNA from the sample. In complex food matrices such as whole milk or homogenized oysters, DNA recovery might be only a few percent. Food substances may be co-purified with the bacterial DNA may reduce amplification efficiency (Hill et al., 1991; Keasler and Hill, manuscript submitted for publication). Increasing the number of target copies by cell growth will raise the probability of recovering target DNA as well as reducing the relative amount of food borne inhibiting substances but does increase the time required for analysis.

Dead vs. viable cells - Dead cells may contain amplifiable DNA. Therefore, a positive PCR result does not mean that culturable or even viable cells are present in the sample (Brauns et al., 1991). Using PCR to analyze foods that are expected to contain large numbers of dead cells, such as pasteurized products, will generate positive results that are of no public health significance. mRNAs have relatively short half lives and may not survive long in dead cells. Coupling reverse transcriptase with PCR makes it possible to detect RNA molecules by having them serve as the initial amplification template (Mahbubani et al., 1991). Because of relative rapid mRNA degradation, dead cells would not often yield positive results.

Strain characterization: Arbitrary Primer PCR

A popular technique for the characterization of genomes is the restriction endonuclease fragment length polymorphism (RFLP). However, this method requires the purification of several micrograms of DNA and subsequent digestion with a restriction endonuclease. Recently, techniques for applying PCR to characterize bacterial species, or possibly strains, have been described (Welsh and McClelland, 1991; Williams et al., 1990). Patterns representing data similar to that obtained by restriction endonuclease digestion can be generated using single primers of arbitrary sequence for the PCR. This obviates the need for isolation of microgram amounts of bacterial DNA because PCR generates a sufficient amount of product for analysis.

Mechanism - Under low stringency, the arbitrary sequence primers are expected to anneal to a large number of sites on the target genome. Some sites will be within a few hundred base pairs of each other on opposite strands; regions between these primers will be amplified during PCR generating a pattern of fragments readily observed after gel electrophoresis which serve as a fingerprint of a particular genome. Such a technique may be useful as a rapid method for species or strain characterization and providing data for epidemiological analyses.

References

Bej, A.K., J.L. DiCesare, L. Haft, and R.M. Atlas. **1991**. *Appl. Environ. Microbiol.* 57:1013-1017.

Bej, A.K., S.C. McCarty, and R.M. Atlas. **1991**. *Appl. Environ. Microbiol.* 57:2429-2432.

Bessesen, M.T., Q. Luo, H.A. Rotbart, M.J. Blaser, and R.T. Ellison, III. **1990**. *Appl. Environ. Microbiol.* 56:2930-2932.

Border, P.M., J.J. Howard, G.S. Plastow, and K.W. Siggens. **1990**. *Lett. in Appl. Microbiol.* 11:158-162.

Brauns, L.A., M.C. Hudson, and J.D. Oliver. **1991**. *Appl. Environ. Microbiol.* 57:2651-2655.

Curiale, M.S., M.J. Klatt, and M.A. Mozola. **1990**. *J. Assoc. Off. Anal. Chem.* 73:248-256.

Deneer, H.G., and I. Boychuk. **1991**. *Appl. Environ. Microbiol.* 57:606-609.

DeLong, E.F., G.S. Wickham and N.R. Pace. 1989. *Science* 243:1360-1363.

Edelstein, P.H., R.N. Bryan, R.K. Ennes, D.E. Kohne, and D.L. Kacian. **1987**. *J. Clin. Microbiol.* 25:1022-1026.

Furrer B., U. Candrian, Ch. Hoefelein, and J. Luethy. **1991**. *J. Appl. Bacteriol.* 70:372-379.

Gicquelais, K.G., M.M. Baldini, J. Martinez, L. Maggi, W.C. Martin, V. Prado, J.B. Kaper, and M.M. Levine. **1991**. *J. Clin. Microbiol.* 28:2485-90

Hill, W.E., S.P. Keasler, M.W. Trucksess, P. Feng, C.A. Kaysner, and K.A. Lampel. **1991**. *Appl. Environ. Microbiol.* 57:707-711.

Hill, W.E., and K.A. Lampel. In *Biotechnology and Food Safety*, D.D.Bills and S.-d. Kung (eds.) Butterworth-Heinemann, Boston, MA, 1990; pp. 139-165.

Hill, W.E., J.M. Madden, B.A. McCardell, D.B. Shah, J.A. Jagow, W.L. Payne, and B.K. Boutin. **1983**. *Appl. Environ. Microbiol.* 45:1324.

Hill, W.E., W.L. Payne, G. Zon., and S.L. Moseley. **1985**. *Appl. Environ. Microbiol.* 50:1187-1191.

Hogan, J.J. In *Nucleic acid and monoclonal antibody probes*. B. Swaminathan and G. Prakash (eds.), Marcel Dekker, Inc., New York, NY, 1989; Chap. 3.

Jagow, J.A., and W.E. Hill. **1986**. *Appl. Environ. Microbiol.* 51:441-443.

Jagow, J.A., and W.E. Hill. **1988**. *Mol. Cell. Probes* 2:189-195.

King, W., S. Raposa, J. Warshaw, A. Johnson, D. Halbert, and J.D. Klinger. **1989**. *Int. J. Food Microbiol.* 8:226-232.

Lampel, K.A., J.A. Jagow, M. Trucksess, and W.E. Hill. **1990**. *Appl. Environ. Microbiol.* 56:1536-1540.

Lowe, T., J. Sharefkin, S.Q. Yang, and C.W. Dieffenbach. **1990**. *Nucl. Acids Res.* 18:1757-1761.

Mahbubani M.H., A.K. Bej, R.D. Miller, R.M. Atlas, J.L. DiCesare, and L.A. Haft. **1991**. *Biotechniques*, 10:48-9.

Miller, V.L., J.J. Farmer, III, W.E. Hill, and S. Falkow. **1989**. *Infect. Immun.* 57:121.

Olive, D.M., A.I. Atta, and S.K. Setti. **1988**. *Mol. Cell. Probes* 2:47.

Peterkin, P.I, E.S. Idziak, and A.N. Sharpe. **1991**. *Appl. Environ. Microbiol.* 57:586-591.

Romaniuk, P.J., and T.J. Trust. **1989**. Mol. Cell. Probes 3:133-142.

Rychlik, W., and R.E. Rhoads. **1989**. *Nucleic Acids Res.* 17:8543-8551.

Rychlik, W.J. Spencer, and R.E. Rhoads. **1990.** *Nucleic Acids Res.* 18:6409-6412.

Saiki R.K., D.H. Gelfand, S. Stoffel, S.J. Scharf, R. Higuchi, G.T. Horn, K.B. Mullis, H.A. Erlich. **1988**. *Science* 239:487-91.

Saiki, R.K., S. Scharf, F. Faloona, K.B. Mullis, G.T. Horn, and H.A. Erlich. **1985**. *Science* 230:1350.

Thomas, D.J.G., R.K. King, J. Burchak, and V.P.J. Gannon. **1991**. *Appl. Environ. Microbiol.* 57:2576-2580.

Welsh, J., and M. McClelland. **1990**. *Nucleic Acids Res.* 18:7213-7218.

Wernars,K., E. Delfgou, P. S. Soentoro, and S. Notermans. **1991**. *Appl. Environ. Microbiol.* 57:1914-1919.

Wernars, K., C.J. Heuvelman, T. Chakraborty, and S.H. W. Notermans. **1991**. *J. Appl. Bacteriol.* 70:121-126.

Wesley, I.V., R.D. Wesley, M. Cardella, F.E. Dewhirst, and B.J. Paster. **1991**. *J. Clin. Microbiol.* 29:1812-1817.

Williams, J.G.K., A.R. Kubelik, K.J. Livak, J.A. Rafalski, and S.B. Tingey. **1990**. *Nucleic Acids Res.* 18:6531-6535.

Wilson, S.G., S. Chan, M. Deroo, M. Vera-Garcia, A. Johnson, D. Lane, and D.N. Halbert. **1990**. *J. Food Sci.* 55:1394-1398.

Biotechnology in Developing Countries Symposium XI

J. P. Cherry: *Chair*

Suitable Bioengineering Processes: Session A
J. P. Cherry and J. E. Allende: *Co-Chairs*

Stimulating Biotechnology for Small-Scale Farmers: Session B
J. P. Cherry and D. Alani: *Co-Chairs*

Biotechnology in Developing Countries: An Overview

John P. Cherry, Eastern Regional Research Center, Agricultural Research Service, U.S. Department of Agriculture, 600 E. Mermaid Lane, Philadelphia, PA 19118
B. Onuma Okezie, Office of International Programs, Alabama A&M University, Normal, AL 35762

Many problems need solutions in "developing," or "newly industrialized," countries. In developing countries, lifestyles include a significant involvement in food production and marketing. Hence, like developed countries the primary objectives of agricultural research are those that can be readily applied to meet the needs of the farmer, industry and consumer. However, there is less opportunity for chemical and physical intervention of biotechnology in agriculture or for medical and veterinary applications. The difficulty lies not in identifying problems, but in deciding in concept where biotechnology can quickly provide appropriate solutions. The purpose of "Symposium XI: Biotechnology in Developing Countries" is to present successful examples of developments and accomplishments that are making quality contributions to developing countries in two areas—(1) "Suitable Bioengineering Processes"; and (2) "Stimulating Biotechnology for Small-scale Farmers." These areas include discussions on educating researchers in advancing their developments of biotechnology at the farming, industrial, marketing, and political sectors.

The opening presentation, "Marine Biotechnology and the Third World: Research and Applications," by R.A. Zilinskas, analyzes areas of marine biotechnology that are particularly relevant to developing countries. Island countries and countries with long coast lines have barely drawn on their marine capital, probably for many, the Earth's richest assets. Measures are suggested whereby developing countries can build their capabilities in marine technologies. Biotechnology is defined as "a set of scientific techniques that use living organisms, or parts of organisms such as cells, to make or modify products, to improve plants or animals, or to develop organisms for specific applications." Areas for advancement in marine biotechnology include aquaculture, marine natural products, bioremediation, biofilm-bioadhesion, cell culture, biosensors and terrestrial agriculture.

In Latin American Countries, biotechnology is having an impact on most of the essential products produced. J.F. Allende notes in "Biotechnology Efforts in Latin America," that one reason for these advances is the large critical mass of scientists in many different scientific and technological disciplines collaborating in efforts to advance biotechnology research. Latin American countries have recognized the advantage of pooling resources. Additionally, a large number of international and regional agencies have come together in collaborative programs with Latin America countries. This has greatly strengthened industry-science contacts and advanced research in human resource development, multinational programs and biosafety guidelines.

"Scope and Some Applications of Biotechnology in Developing Countries—Case Study of Saudi Arabia," by D. Alani, discusses efforts to apply biotechnological techniques for inexpensive solutions to selected problems. Like most developing countries, Saudi Arabia has large amounts of bioresources that provide opportunities for bioconversion into needed products. The country also has major issues with petroleum resources (that carry a risk of

1054–7487/92/0542$06.00/0

environmental pollution) and a limited water supply. Examples of research include (1) date palm propagation and use of this fruit as a source of antibiotics and cellulosic materials; (2) use of refinery wastes, oily sludges, tank residues and oil contaminated soils; and (3) applications of microorganisms for wastewater treatment and agricultural applications.

The successes of the Republic of China, Taiwan, in land reform, agro- and food-based industries, capital intensive manufacturing and technology-based developments have led this country into the age of high-technology electronics and computers. D.G. Mou, in "Development of the Biotechnology Industry in Developing Countries: The Taiwan Experience," shows biotechnology designated as essential to this country's industry, society and environment in the 21st century. Successful programs are emerging in hepatitis B vaccine production. The establishment of a Development Center for Biotechnology is underway and planned is a new Pharmaceutical Research and Development Center to meet the biotechnology challenge. A strategy is being developed to expand and transform the country's primarily market and domestically oriented and fragmented bio-industry into one that is technologically sophisticated and globally oriented.

Studies to advance biotechnological developments in the newly industrialized country, Korea, by immobilizing hybridoma cells that produce monoclonal antibodies are presented by H.N. Chang in "Continuous Production of human Chorionic Gonadotropin (hCG) Monoclonal Antibody in Various Membrane Bioreactors." Cells were immobilized and cultivated within dual hollow fiber Ca-alginate capsules and cylindrical depth fiber bioreactors. In the dual hollow fiber bioreactor, hybridomas were successfully cultured, and monoclonal antibodies produced, for two months. Hybridomas encapsulated within Ca-alginate gels had high monoclonal antibody production. However, a long term continuous culture could not be maintained because of the diffusion limitations of nutrients and wastes. A newly developed depth filter perfusion system, based on the immobilization of hybridoma cells within a cylindrical depth filter matrix, yielded a stable continuous culture producing high amounts of monoclonal antibody. This bioreactor is projected as a process for continuous large-scale production.

Jute, a lignocellulosic bast fiber source, is a cash crop of Bangladesh, India, Nepal, Thailand and China. Due to incomplete extraction of fiber from the plant, 20-40% of jute produced is of low grade because of excess hemicellulose. The conventional industrial methods use chemical and mechanical means to prepare spinnable fibers of limited quality. Efforts are underway in various research laboratories to improve jute fibers by biotechnological processes. One such approach is described by M. Hoq in "Enzyme Biotechnology in Developing Countries." This study revealed that *Thermomyces lanuginosus* could produce pH-stable and cellulase-free thermostable xylanase on inexpensive medium. The enzyme has potential for upgrading jute and other fibers. Studies are underway to develop suitable bioprocesses for enzyme production and industrial application.

In "Building the Base for Critical Mass of Biotechnological Scientists in a Developing African Country", A.I. Robertson explains that the first use of biotechnology in Africa is to improve agriculture. Low-technology applications use tissue culture techniques; high-technology, DNA manipulations; and industry, fermentation and downstream processes. Robertson states that "practicing agriculture provides the experience that a holistic approach is needed," i.e., in a developing country, if there is one weak link in agricultural production, the result can mean that there is no crop, no viability and sudden debt or starvation. This makes risk-avoidance a high priority. Hence, efforts should build a base and work toward technological competence. Organizational infra-structures must grow and along with them research breakthroughs should induce improvements that only fine-tune already established technologies. Examples presented include gene-transfer-induced improvements in selected crops.

A.S. ElNawawy states that there are biotechnological applications that have impact on the maintenance of soil fertility, sustaining increased agricultural productivity and improving farmers' profitability in "Impact of Biotechnology in Sustaining Food and

Agriculture in Developing Countries." Examples include biofertilizers for legumes and non-legumes, organic matter recycling for composting and ensilage of mushrooms, and feed production through solid state fermentation. Biogas technology provides a way of supplying alternative fuels from available farm wastes. Earthworm biotechnology can lead to improved soil fertility. Bioremediation of contaminated soils and water is one technology that is available for removing/metabolizing toxic substances so that soil can be reused. New crops that are thermo-tolerant, salt tolerant, resist pesticides, and fix nitrogen are imminent through recombinant DNA and tissue culture techniques.

Crop production by small farmers in the Philippines and other developing countries is hampered by poor soil fertility and the high costs of chemical fertilizers. Microorganisms such as mycorrhiza have been tapped as alternatives to chemical fertilizers. R.E. dela Cruz presents five such developments in "Applications of Mycorrhizal Technologies for Agriculture and Forestry." These technologies increase the survival, growth, and yield of agricultural crops, especially in adverse sites of marginal fertility. Vesicular-Arbuscular Mycorrhizal (VAM) fungi have been shown to replace 65-85% of the chemical fertilizers required for crop growth.

J.P. Moss, in "Constraints to Production in the Semi-arid Tropics—Can Biotechnology Help the Small Farmer?" notes that the climates of the world's semi-arid tropics are harsh with inadequate, uncertain rainfalls, usually infertile soil, high incidences of pests and diseases and low availability of investment capital. Biotechnology offers the hope of introducing agricultural cultivars that can overcome these restraints. Biotechnology also offers breeders, pathologists, entomologists, and physiologists ways of speeding the breeding process (via *in vitro* regeneration, haploids, and micropropagation), increasing the precision of screening (restriction fragment length polymorphism, polymerase chain reaction and other markers) and improving the diagnosis of diseases (enzyme-linked immuno-solvent assay, monoclonal antibodies). The potential for achieving breakthroughs is greatest when scientists in developed and developing countries cooperate to make the best use of facilities and resources.

"On-farm Anaerobic Treatment of Agricultural Wastes in Some Countries in Asia," by W. Tentscher, examines the progress in the design of small-scale fixed-dome digesters and of medium and large scale biogas plants for processing agricultural wastes. China and India are among the leaders in the development of biogas technologies. Modern methods are being applied to determine structural stability of the digesters and maximum process performance. Progress is occurring in mathematical modelling of kinetic reactions and computer technologies. As a result, ability to assess the economic implications of changes in technical parameters has improved. A new pilot project in a medium scale pig farm in Thailand, employing the Hybrid Plug-flow, High-rate digester system for diluted wastewaters with suspended solids is being introduced on-farm. Biogas technology, be it at small farms, feed lots, or sewage treatment plants is gaining increasing attention. With these processes, CO_2 and mineral-organic matter can be recycled.

In "developing" or "newly industrialized" countries, biotechnology research in agriculture is growing in importance. This is in spite of the difficulties that sometimes occur in deciding, in concept, where biotechnology can quickly provide appropriate solutions to meet the immediate needs of the farmer, industry and consumer. Critical masses of scientists and engineers in many different disciplines collaborating worldwide are contributing to these solutions. The results are technological competence and risk-reduction in the application of biotechnological developments, so important to many developing countries. Hence, biotechnological breakthroughs being realized in developed countries are also occurring in "developing" or "newly industrialized" countries at a level directed to immediate needs.

References

Ninth International Biotechnology Symposium and Exposition; "Harnessing Biotechnology for the 21st Century"; August 16-21, 1992; Crystal City, Virginia, USA.

Marine Biotechnology and the Third World: Research and Applications

Raymond A. Zilinskas, Center for Public Issues in Biotechnology, Maryland Biotechnology Institute, UMBC, Catonsville, Maryland 21228 (FAX:410 455 1077)

The diversity of life in the oceans is the world's most abundant and least utilized resource. Yet, nations with large Exclusive Economic Zones have barely drawn on their marine capital. Some poor nations oversee potentially the Earth's richest assets. Marine biotechnology, "the application of scientific and engineering principles to the processing of materials by marine biological agents to provide goods and services", may be used to tap these treasures. This article reviews marine biotechnology sub-areas relevant to developing countries and suggests ways to build up capabilities in marine biotechnology. Sub-areas include aquaculture; marine natural products; bioremediation; biofilms; cell culture; biosensors; and marine R&D for terrestrial agriculture.

The tremendous diversity of life forms teeming in the oceans represents the world's most abundant, but possibly its least utilized living resource. Island countries, such as Haiti, Indonesia and Palau, and countries with long coast lines, including Chile, China, India, and Sudan, have barely drawn on their marine capital. Thus, some of Earth's most impoverished countries oversee potentially its richest assets. The question is, how can they gain from their rich but untouched resources?

The ability of marine biotechnology to advance marine-related research and generate new products is becoming widely recognized (Colwell et al., 1992). Marine biotechnology, defined above (Bull et al., 1982), has seven sub-areas; some are important to developing countries (Zilinskas et al., 1992). They are:

Aquaculture. Marine biotechnology may benefit aquaculture in two ways. Its research techniques can (1) enhance a cultured organism's growth rate, procreation proficiency, disease resistance, and ability to endure adverse environmental conditions, improving the organism's ability to grow and survive in intensive aquaculture (Renn, 1990); and (2) be used to develop vaccines against bacteria and viruses that attack aquacultured organisms (Meyer, 1991), protecting them from diseases that now periodically decimated stocks, causing enormous economic damage in Asia and Latin America.

Marine natural products chemistry. Marine organisms have evolved differently than terrestrial life. As part of their metabolism, many marine organisms secrete compounds that help them survive and, incidentally, have properties beneficial to mankind (Austin, 1989). Screening programs have discovered algae, corals, sponges and tunicates that produce compounds showing antibiotic, antitumor, anti-viral, or anti-inflammatory activity. However, fewer than 1% of marine species have been screened. Further, present screening methods have limited scope, detecting only a few bioactive properties (Boyd et al., 1988). As procedures are improved many marine organisms producing anti-parasitic, pesticidal, immune-enhancing, growth-promoting, wound healing-promoting, etc., chemicals will be discovered.

Bioremediation. Bioremediation is the use of microorganisms to break down pollutants and wastes in soil or water to harmless or less toxic end-products. Microorganisms bioremediate by feeding directly on the organic pollutant or by secreting enzymes that attack the pollutant. (Office of Technology Assessment, henceforth OTA, 1991b). The microbes used in bioremediation are usually recovered from natural sites, but their natural capability for breaking down pollutants is enhanced via R&D programs. Bioremediating microorganisms decrease in numbers after the substance they

1054–7487/92/0545$06.00/0

feed on has been destroyed. Because it causes less damage to the environment than do the present clean-up methods, bioremediation holds significant advantages over conventional techniques. Early results from the bioremediation of an oil-spill in the Prince William Strait shows that this procedure is effective, safe to humans, and environmentally benign. As its techniques are perfected, bioremediation may become the preferred approach for cleansing polluted harbors and waterways, and decontaminating sensitive coastal biota (Holloway, 1991).

Biofilm/bioadhesion. Surfaces exposed to seawater are settled by marine organisms, which soon form a crust. Organisms enmeshed in the crust produce acids, corroding piers, derricks and other structures. Encrustation also increases hull drag in ships, raising operating costs (Costerton et al., 1989). At present, paint containing heavy metals are used to coat surfaces, thereby preventing organisms from settling. However, toxic paints pose health hazards to workers and pollute seawater. Marine biotechnology research seeks to clarify the molecular basis of the settling and adhesion process; findings may be used to develop methods for preventing marine organisms from settling on ships and marine structures.

Cell culture. Cells from many species of terrestrial plants are being propagated in culture flasks, where they grow and subdivide much like bacteria. Cultured cells can be used to generate whole plants or they may be stimulated to synthesize natural products that are normally produced by the whole plant. Contrary to chemical synthesis, the production of compounds by cell culture is energy conserving and essentially non-polluting. Cell culture systems are now producing pharmaceuticals, food additives and pesticides. Marine cell culture systems, based on algae and other marine plants, are undergoing like developments; one operational system makes agarose (French Technology Survey, 1991).

Biosensors. Sensors are devices that detect a specific substance or organism. In biosensors the detecting element is a biological material; two types are of particular interest. The first, chemoreceptors, are biomolecular assemblies involved in physiological functions, such as smell and taste. One example is the crab's sensing antennule, which continually monitors water for dissolved substances ranging in chemical complexity from simple salts to pheromones. In the laboratory, antennules dissected out of the crab have displayed instantaneous, quantitative responses to various amino acids, hormones, nucleotides, drugs, and toxins (Rechnitz, 1988). Chemoreceptors will find use to detect and monitor substances found in very low concentrations in water or air.

Second, the recognition element in immunological sensors is a monoclonal antibody or DNA probe (Ho et al., 1992). These biological molecules are extremely selective; a monoclonal antibody will bind itself to only one antigen, such as a virus, a component of a bacterial cell wall, or a certain chemical. Kits based on monoclonal antibodies or DNA probes are being used to detect and identify pathogenic bacteria causing cholera, shigellosis and typhoid fever, generating results in 20 minutes. Kits to detect a variety of viruses are being tested. Previously, the culturing and identification of bacterial pathogens took 24-72 hours; that of a virus, weeks or would prove impossible.

Huge quantities of raw sewage and industrial wastes are dumped untreated in the oceans every day. Little is known about the fate of chemicals and pathogens released in the marine environment. The problem is the difficulty of identifying and monitoring substances and pathogens in the oceans, then establishing cause-effect between an agent and health event. Rugged, accurate biosensors will help investigators clarify the effects of contaminated and polluted sea water on public health.

Terrestrial agriculture. Marine biotechnology researchers are seeking to transfer valuable traits inherent to marine animals and plants to their terrestrial counterparts. For example, the winter flounder survives sub-zero temperatures that would kill most animals, including other fish. Scientists have synthesized the flounder anti-freeze gene and have inserted it in yeast and higher plants, where it was expressed (McGraw-Hill's Biotechnology Newswatch, 1991). If the trait is stable, transformed crops grown at high altitudes or northern latitudes will be protected from sudden freezes. Another example involves the world's most salt-tolerant plant, which is a microalgal species inhabiting the Dead Sea, whose water contains 29% salt. Researchers are implanting genes that code for salt tolerance in crop plants. Success would mean that rice, soya and other crops can be grown when irrigated by brackish or salt water.

Marine biotechnology and the Third World. Many developing countries have a base for capability-building in marine biotechnology consisting of institutes and universities where teaching and research in general and marine biology are done and where some biotechnology techniques have been adopted (Zilinskas et al., 1992). Existing programs may be augmented to encompass marine biotechnology R&D. For example, many Third World agricultural research laboratories are familiar with terrestrial cell and tissue culture. Similar techniques may be used to grow cells of marine plants in culture. Also, laboratories are screening terrestrial plants and microorganisms for natural products. Analogous techniques are applicable to marine organisms, although collection is more difficult.

Marine Biotechnoloby and the Third World

While it is clear that biotechnology holds immense promise for the Third World, capability-building in this field is a difficult process. Few developing countries have the in-depth R&D infrastructure to undertake wide-ranging research because they lack the trained scientific personnel, well-equipped laboratories, and dependable supplies of rare, labile biochemicals (UN Industrial Development Organization, 1986). Even fewer have the industrial capability to develop research results and exploit them (Zilinskas, 1989). In view of these problems, intergovernmental organizations (IGOs) have a vital role to play in helping bring biotechnology to the Third World.

Precedents exist in natural resource development for cooperative ventures between R&D units in developed countries and counterparts or industries in industrialized countries. Common to these projects is that both sides benefit, the developed country partner gains access to a raw material and retains certain marketing rights, while the developing country partner gains expertise, financial backing and regional marketing rights. Some of these types of arrangements have been brokered by IGOs, including UN Development Program, UN Industrial Development Organization and World Health Organization. Specifically, IGOs may assist in developing profit-based options for joint ventures between research units in developing countries and industrial firms in industrialized countries. Risks for failure or misunderstanding would be minimized by IGOs, such as the World Bank, providing expertise in drawing up contracts, making available start-up funds, and equitably assigning intellectual property rights (Zilinskas et al., 1992).

Research and applications involving genetically engineered organisms, whether terrestrial or marine, may present risks to humans and/or the environment. Risks can be minimized by applying safe procedures as specified by appropriate regulations. Guidelines for research and field testing have been formulated by national governments and various IGOs, including the European Communities and the Organization for Economic Cooperation and Development. An UN interagency working group has elaborated guidelines for field testing applicable to developing countries (United Nations Industrial Development Organization, 1991). IGO guidelines may be used as models by Third World governments for local laws. However, adapting foreign guidelines and regulations requires expertise in risk assessment and management. Local scientists and regulators will have to be trained in these subjects, possibly in countries where regulatory frameworks have been developed and the appropriate skills are available (Taylhardat et al., 1992).

While barriers exist that hinder capability building in biotechnology by developing countries, they are certain to be overcome because this field is so important. Biotechnology is in fact already stimulating economic development throughout the world. For example, the OTA estimates that American biotechnology companies employ 35,900 workers, of which 18,600 are scientists and engineers (OTA, 1988). By 1995, pharmaceuticals produced via biotechnology will be worth an estimated $25 billion per year;

agricultural biotechnology products will generate about $35 billion by 2005 (OTA, 1991a). No wonder Nobelist Abdus Salam believes that the 21st century will be known as the Biological Age.

The Biological Age encompasses the marine as well as the terrestrial environment. If wisely and correctly employed, that subset of biotechnology termed marine biotechnology presents the Third World with tools for research that may lead to increased supplies of high quality food; the sustainable, environmentally sound exploitation of natural resources; the destruction or detoxification of harmful pollutants; improved public health; and a better understanding of the role of the ocean in Earth system processes.

References

Austin, B., *J. App. Bacter.* 1989, **67**, 461-470.

Boyd, M.R.; Shoemaker, R.H.; Cragg, G.M.; Suffness, M. in *Pharmaceuticals and the Sea*; Jefford, C.W.; Rinehart, K.L.; Shield, L.S. Ed; Technomic Publishing Company, Inc., Lancaster, PA, 1988, pp. 27-44.

Bull, A.T.; Holt, G.; Lilly, M.D. *Biotechnology: International Trends and Perspectives.* Organization of Economic Cooperation and Development, Paris, 1982.

Colwell, R.R.; Lipton, D.; Zilinskas, R.A. *Marine Biotechnology in the United States and Other Countries.* Maryland Sea Grant College, 1992.

Costerton, W.; Lappin-Scott, H.M. *ASM News.* 1989, **55**, 650-654.

French Technology Survey. 1991, December, 8.

Ho, M.Y.K.; Rechnitz, G.A. in *Immunological Assays and Biosensor Technology for the 1990s*; Nakamura, R.M.; Kasahara, Y.; Rechnitz, G.A. Ed; American Society for Microbiology, Washington D.C., 1992, pp. 275-290.

Holloway, M. *Sci. Am.* 1991, **265**, 102-116.

McGraw-Hill's Biotechnology Newswatch. 1991, **11(9)**, 2.

Meyer, F.P.J. *Anim. Sci.* 1991, **69**, 4201-4208.

Office of Technology Assessment, *New Developments in Biotechnology: U.S. Investments in Biotechnology.* US Government Printing Office, Washington D.C., 1988.

Office of Technology Assessment, *Biotechnology in a Global Economy.* US Government Printing Office, Washington D.C., 1991a.

Office of Technology Assessment, *Bioremediation For Marine Oil Spills.* US Government Printing Office, Washington D.C., 1991b.

Rechnitz, G.A. *C&E News.* 1988, **66**, 24-36.

Renn, D.W. *Hydrobiol.* 1990, **204/205**, 7-13.

Taylhardat, A.R.; Zilinskas, R.A. *Bio/Technology.* 1992, **10(3)**.

United Nations Industrial Development Organization, *Capability Building in Biotechnology and Genetic Engineering in Developing Countries* by McConnell, D.J.; Riazuddin, S.; Wu, R.; Zilinskas, R.A. 1986 (Document # UNIDO/IS.608).

United Nations Industrial Development Organization, *Voluntary Code of Conduct for the Release of Organisms Into the Environment.* Vienna, Austria, 1991.

Zilinskas, R.A. *Genome. 1989, 31(2)*, 1046-1054.

Zilinskas, R.A.; Lundin, C.G. *Marine Biotechnology for the Developing Countries.* World Bank, Washington D.C., 1992.

Regional Biotechnology Efforts in Latin America

Jorge E. Allende, Departamento de Bioquímica, Facultad de Medicina, Universidad de Chile, Casilla 70086, Santiago 7, Chile (FAX: 56 2 37-6320)

Recent years have brought important advances in Latin American political and economic integration. Biotechnology has been in the forefront of the movement to integrate the efforts of the individual countries in the science and technology area. There are several reasons why the regional approach has been readily accepted in biotechnology. One of the main reasons is that biotechnology is having an impact on most of the essential products of all of the Latin American countries. Another factor is the large critical mass of scientists that are required to carry out competitive biotechnological research. Practically none of our countries has all the expertise required by large biotechnology projects; the advantage of pooling resources at the regional level has become obvious. The third factor has been the initiatives that a large number of international and regional agencies have undertaken in the past decade to put together some regional biotechnology programs.

Biotechnology has become an area in which Latin America is achieving a promising degree of regional integration. Undoubtedly the examples of the European Economic Community and the steps taken by the U.S. towards forming a North American common Market with Canada and Mexico have had a great political impact in advancing regional integration. In both examples scientific and technical collaboration has spearheaded the more difficult task of achieving political and economic integration.

There are several reasons why the regional approach has been readily accepted in Latin American biotechnology development. One of the primary reasons is the fact that biotechnology is affecting, and will affect even more in the coming years, the main sources of trade revenue of the Latin American countries. The countries of the region, with rare exceptions, are dependent on primary production-mostly agricultural. Biotechnology has already greatly impacted the world sugar market and is on its way to affecting many other fields of agriculture, animal production and aquaculture. Latin American leaders with a vision of the future have concluded that biotechnology presents a great challenge and an opportunity for the region. Biotechnology could greatly enhance the productivity of the Latin American countries and improve the quality of life of its people. On the other hand, if the opportunity is missed, the region's comparative advantages of climate and genetic diversity will soon be lost and our countries and regions will be condemned to increased dependence and poverty. Furthermore, it has become obvious that the priorities in biotechnology for the various Latin American countries are very similar: plant biotechnology for increased productivity, diagnosis and vaccines for human infectious diseases, biological control of pests that affect food production, environmental biotechnology for detoxification of water and atmosphere, and aquaculture production through advanced technologies.

A second motive that has stimulated regional integration in biotechnology is the fact that meaningful projects in this area require large transdisciplinary teams of highly trained scientists. Despite the recent efforts of some countries, the scientific communities of the Latin American nations are very small and do not cover the wide range of discipline that are necessary to mount ambitious biotechnology programs. A criterium of the scarcity of scientific research in Latin America is evident

1054–7487/92/0549$06.00/0

from the statistics of Table 1 (adapted from J.J. Brunner, 1989). Latin America, with 8% of the world population, only has 1.3 of the global number of the authors of scientific papers. This low number of scientists is even more dramatic in several key areas of biotechnology such as industrial microbiology, plant viruses, and genetics of marine organisms. Several interesting analyses of Latin American scientific productivity have been recently published (Braun et al., 1985; Inter American Development Bank, 1988; Krauskopf, 1990;).

The lack of critical mass for biotechnology projects and the coincidence in priority areas have made it evident that there is an advantage in joining forces for the purpose of collaboration in the generation and implementation of such projects. However, a third factor has been instrumental in providing an impetus in the regional integration of biotechnology efforts. This factor has been the contribution of several international agencies that have supported Latin American biotechnology programs with a regional approach. Table 2 shows some of the major international organizations that are now supporting biotechnology research in Latin America. This is undoubtedly an incomplete list but it serves our purpose to demonstrate that there is a large number of initiatives in this area. This is, of course, good because it shows the right diagnosis on the importance of the topic and on the need for international funds to permit contacts and collaborative projects in the region. The bad news is that each of these initiatives has very limited funding and that there has been very little coordination of their programs. These problems are serious because the small scientific community is rather confused and frustrated by the great limitations and the diversity and incongruity of these separate initiatives.

Nevertheless, we must admit that several of these programs have been rather important in promoting regional cooperation. I will highlight a few of these international projects that I am personally acquainted with.

The Latin American Network of Biological Sciences (RELAB)

Although this network does not deal with the biotechnology proper, its work among biologists of Latin America sets the stage for development of the contacts and mechanisms that were instrumental in the later emergence of regional biotechnology programs. As shown in Table 3, this network originated in 1975 as a regional program funded by the United Nations Development Program and executed by UNESCO. It now has 13 member countries and 6 regional societies and is sponsored by the International Council of Scientific Unions, by Unesco and by the Organization of American States. Its main contribution has been to establish the benefits of regional collaboration and to generate many scientific partnerships among the community of Latin American

TABLE 1

Economic-Technological Impact at the Beginning of the 1980's.

	Latin America %	United States %	Japan %	Federal Rep. of Germany %
Population	8.0	5.0	2.5	1.3
Gross Domestic Prod.	7.0	27.0	9.4	5.8
Manufactured Products	6.0	18.0	11.7	9.4
Capital Assets	3.0	14.7	11.1	9.6
Engineers & Scientists	2.4	17.4	12.8	3.4
R & D Resources	1.8	30.1	10.2	6.7
Scientific Authors	1.3	42.6	4.9	5.4

Percentage of participation with respect to world total (Data obtained from Brunner, 1989).

TABLE 2

Some International Agencies with Specific Biotechnology Programs in Latin America

U N D P	F A O	O. A. S.
U N E S C O	W H O - P A H O	S E L A
U N I D O	Rockefeller Foundation	C Y T E D - D (Spain)
U N U	Cartagena Agreement Countries	I C S U - C O B I O T E C H

TABLE 3

The Latin American Network of Biological Sciences (RELAB)

Sponsors: ICSU, UNESCO, OAS

Members:

National	Regional
Argentina	
Bolivia	- Pan American Association of Biochemical Societies (PAABS)
Brazil	
Chile	- Latin American Association of Physiological Sciences (ALACF)
Colombia	
Cuba	- Latin American Botany Association (ALB)
Ecuador	
Honduras	- Latin American Genetics Association (ALAG)
México	
Paraguay	- Latin American Pharmacology Association (ALF)
Perú	
Uruguay	- Iberoamerican Society of Cell Biology (SIABC)
Venezuela	

Started operating in 1975 as the: Latin American Regional Program for Postgraduate Training in Biological Sciences

Governing Council:
- Regional Executive Council with representatives of the 13 governments and 13 National Committees and the 6 Regional Members
- Technical Coordination

Presently supports:
- Training courses, Workshops, scientific society activities, and meetings to generate projects.
It also stimulates links between Latin American Scientists with the International Scientific Community.

biologists. It was this organization that requested UNDP to create a regional biotechnology network. Presently it centers its activity in the organization of training courses and symposia and in the publication of regional reviews on topics such as the usefulness of the biological sciences for Latin American development, strategies to counter the brain drain, and the mechanisms for funding biological research in the region.

The Latin American Regional Program of Biotechnology

This program started officially in 1986 and is financed by UNDP and executed by both Unesco and UNIDO. This is an interesting aspect in itself because it has succeeded in gathering the participation of several UN agencies. As seen in Table 4, this program also has a large number of Latin American countries that participate, and during the first 5 year period has had a total budget of 5 million US dollars. The most significant aspect of this program is that it has concentrated in generating and implementing multinational research projects in the fields of biotechnology dealing with plants of great economic importance in Latin America, the diagnosis of human diseases prevalent in the region, some industrial enzyme production and utilization, and the large scale production of monoclonal antibodies. These projects have been highly successful and have demonstrated the usefulness of the collaboration among the different countries.

It also has resulted in creating national biotechnology committees in the participating countries and, through its Regional Executive Council, it provides a regional forum for the discussion of policies and priorities and the exchange of information.

The International Centre for Genetic Engineering and Biotechnology (ICGEB)

The ICGEB was established under the sponsorship of UNIDO with the objective of strengthening the development and application of biotechnology in the developing world. It's scope of action is, therefore, not restricted to Latin America, especially since it has

TABLE 4

The Regional Program of Biotechnology for Latin America and the Caribbean
UNDP/UNESCO/UNIDO RLA/83/003-009

Participating Countries:

Argentina	Chile	Cuba	México
Bolivia	Colombia	Ecuador	Perú
Brazil	Costa Rica	Guatemala	Uruguay
			Venezuela

Founded in 1983 - Started Operations in 1986.

Governing Body - The Regional Executive Council with representatives of the 13 participating governments.
- Coordinators for UNESCO and UNIDO activities.

Total budget from UNDP approximately 5 million USD.

Started in 1987 - Ends in December 1992.- Second Phase is being negotiated.

Activities
- Multinational research projects.
- Fellowships, Training Courses and Symposia.
- Support of National Committees.

TABLE 5

UNIDO International Centre for Genetic Engineering and Biotechnology (ICGEB) Activities in Latin America

Latin American Countries that are Official Members of the ICGEB

Argentina*	Chile*	Cuba*	Panamá
Bolivia	Colombia	Ecuador	Perú
Brazil*	Costa Rica	México*	Trinidad & Tobago
			Venezuela*

(*) Latin American Countries that have Affiliated Centres of the ICGEB

Total number of participating countries = 43

Headquarter Laboratories in Trieste, Italy and New Delhi, India

Activities: Collaborative Research Projects, Pre-Doctoral and Postdoctoral Fellowships, Training Courses, symposia, workshops, technical services to member countries.

Total budget: US$14 million per year.

established its main laboratories in Trieste, Italy, and in New Delhi, India. However, the ICGEB has been very active in Latin America, with a large number of regional member countries (Table 5). The ICGEB has large research programs in plant biotechnology, human virology, lignocellulose utilization and molecular pathology and immunology. In addition it supports collaborative research projects at 6 Affiliated Centres in Latin America (Argentina, Brazil, Chile, Cuba, Mexico and Venezuela). It is now supporting a total of 23 projects with a total budget of $1,283,000 (US). It provides long-term pre-doctoral and postdoctoral fellowships and also supports courses and workshops both in its central laboratories and in its Affiliated Centres. Recently, the ICGEB organized a meeting in Santiago, with delegates from all the Latin American member countries, that had the objective of stimulating the Latin American participation and enhancing collaboration among the different countries of the region. In this meeting representatives of RELAB and of the UNDP/Unesco/UNIDO Regional Program, attended which is itself an affiliated network of the ICGEB.

As I stated at the beginning, there are many other international programs with biotechnology working in Latin America. It is worthwhile to mention the Biotechnology Action Council of UNESCO, which is chaired by Prof. Indra Vasil of the University of Florida. This program emphasized the awarding of short-term fellowships in the areas of plant and aquatic biotechnology. An important activity is also sponsored by the International Council of Scientific Unions, which has created a special committee—COBIOTECH—conducting a very important activity in training and in information exchange in biotechnology.

In conclusion, biotechnology has become a way for development and for the integration of the Latin American countries. There is certainly no lack of initiatives; the need now is to bring them to fruition and to achieve some coordination and complementation among participating institutions.

References

Braun, T., Glünzel, W., Schubert, A. *Scientometric Indicators. A 32-Country Comparative Evaluation of Publishing Performance and Citation Impact*. World Scientific Publ. Co., Singapore, Philadelphia, 1985.

Brunner, J.J. *Recursos Humanos para la Investigación en América Latina.* FLACSO, Santiago (Chile), **1989**.

Inter American Development Bank. *Economic and Social Progress in Latin America. Special*

Subject: Science and Technology Report. I.D.B., Washington, DC, **1988**.
Krauskopf, M. *Indicadores epistemométri cos que perfilan la productividad científica en América Latina: realidades y desafíos. In La Biologia como Instrumento de Desarrollo para América Latina;* Allende, J.E., Ed.; RELAB: Santiago (Chile), **1990**, pp 535-561.

Scope and Some Applications of Biotechnology in Developing Countries—Case Study of Saudi Arabia

Daham I. Alani, King Abdulaziz City for Science & Technology, P.O. Box 6086, Riyadh 11442, Saudi Arabia (Fax: 966-1-488-3438)

This paper attempts to describe the scope of biotechnology and its application in a developing country with its specific nature and unique local problems, i.e., Saudi Arabia. It deals with efforts made to apply biotechnological techniques including plant cell culture for development purposes through the research conducted in this field. It also reviews the current status of biotechnological applications. Moreover, the potentialities and prospects for solving some local problems are put in a future perspective.

Biotechnology is considered most convenient particularly for the developing countries because:

- the technology used is very simple and easily available,
- it provides inexpensive solutions to certain problems such as food, feed, fertilizer and energy, and
- most of the developing countries have large storages of bioresources that provide vast opportunities for bioconversion into needed products.

In the context of Saudi Arabia, a developing country with arid land, there is more justification to make use of this technology. The abundant petroleum resources though very valuable, carry a risk of environmental pollution. The refinery wastes, oily sludges, residue from tank and oil contaminated soils are considered among the most dangerous for the environment in Saudi Arabia. Saudi Arabia is also one of the leading countries producing dates and has the potentialities to become the world leader in the international trade of dates and its by-products. However, the date leaves have to be put to proper use through biotechnology to benefit from this source of cellulosic material. Water supplies in Saudi Arabia are through desalination and are therefore very costly. The wastewater can be treated by microorganisms and used for agricultural purposes.

Research Conducted in Saudi Arabia Since 1980

1. *Date palm propagation and genetic improvement via cell and tissue culture.* While the date palm is long lived and produces fruit prolifically, it produces relatively few offshoots suitable for transplanting in its lifetime. Some of the most highly valued varieties produce the lowest number of offshoots. Consequently, offshoots of the best varieties, for example Barhi and Nebut Seif, are always in short supply and very expensive (Ansley, 1989). The research conducted in Saudi Arabia began in 1985 to apply micropropagation *via* bud proliferation and embryogenesis to regenerate date palm plants. Date palm proved to be capable of high frequency regeneration *in vitro* by both embryogenesis and organogenesis under the influence of growth regulating compounds and inorganic basal media. Immediate application of the technique is in micropropagation of highly desirable cultivars of date palm. Further application of embryogenesis is for genetic improvements by genetic engineering of cell cultures.
2. *Utilization of Saudi dates and their by-products by Streptomyces spp. for the*

1054–7487/92/0555$06.00/0

production of antibiotics (oxytetracycline or OTC). The research can be summarized as follows:

- Chemical analyses of date components (date fleshy epi- and mesocarp, lipids, vitamins and elements).
- Preferential selection of the best medium and OTC-producer for biosynthetic formation of OTC.
- Investigation of biochemical aspects which occur during biosynthesis of OTC.
- Recovery and purification of OTC from the fermented mash by organic solvents and or chelating affinities of OTC to divalent cations, and determination of its purity to be used in treatment of human infectious diseases.
- Utilization of OTC and dry biomass of OTC-producer in nutrition of chicks as feed supplement.

3. *Isolation of some microorganisms capable of utilizing methanol and various hydrocarbon sources* This especially includes those capable of growing under high growth incubation temperature having the ability to utilize the experimental raw materials. Emphasis was focused on the screening of the best hydrocarbon-utilizer.
4. *Utilization of different fractions of diesel oil in fermentative production of single cell protein by Candida tropicalis.* The investigations concern the following:
   - Chemical and biochemical evaluation of single cell protein produced by microorganisms.
   - Chemical analyses of SCP for amino acids composition.
   - Chemical analyses of SCP for essential and non-essential amino acids.
   - Chemical analyses of SCP for quantitative determination of amino acids.
   - Chemical analyses of SCP for vitamins.
   - Nucleic acids analyses.
   - Utilization of SCP in growing chicks.
   - Utilization of SCP as animal feed in growing mice.

Current Status of Biotechnological Applications in Saudi Arabia

1. *Biodisposal of Oil Refinery Waste.*

Oily waste is the residue of cleaning and restoration of crude oil tanks and other process equipment in the refinery. It is a mixture of oil emulsion with naphthalenic compounds, water, waxes, and iron oxide. Biodisposal, also known as "Landfarming," is a biotechnological application that uses soil aerobic microorganisms to biodegrade the oily waste in the upper soil zone to carbon dioxide and water.

In 1982, the Arabian American Oil Company (ARAMCO) considered landfarming for disposal of waste. Therefore, a pilot landfarm was established in Ras Tanura Refinery to evaluate this treatment method and to develop techniques that would be used in Saudi refineries.

The primary objective of this pilot facility was to determine whether biodegradation of tank bottoms crude oil would occur. The secondary objectives were to determine the volume of sludge that a given area would absorb before it became saturated, the biodegradation rate, and the advantages of watering and adding bacteria (Ala'ud-Din, 1989).

The biodegradation of the sludge was completed after two months at the biodegradation rate of 475,000 liters per hectare per month. The biodegradation rate generally depends upon the factors given in Table 1.

Table 1: Factors Controlling Degradation

- Composition of wastes
- Presence of suitable microorganisms
- Contact between waste and soil microorganisms
- Appropriate depth of tilling the landfarm
- Presence of adequate oxygen
- Temperature and pH
- Presence of available inorganic nutrients
- Moisture content
- Presence or absence of toxics
- Loading rate and type of soil

The conclusions obtained from this pilot landfarm were: a) Disposal of refinery sludge by landfarming is possible; and b) Physical observations of saturation of soil indicate that an initial soil loading of 475,000 liters per hectare per month is possible according to biodegradation rates.

In 1984, the first sludge application was made and ARAMCO reached the conclusion that land treatment was the safest, most cost effective and environmentally acceptable waste management technology available for the tankage wastes generated as part of its refinery operation.

The data gathered from the Ras Tanura Refinery landfarm further indicate that this is a biotechnologically effective method of treatment and disposal of oil waste (Ala'ud-Din, 1989).

2. True-to-Type Date Palm Produced Through Tissue Culture Techniques:

As indicated earlier, research on the micropropagation of date palm began in 1985 but its application for commercial purposes has not yet started. Al-Ghamdi and collaborators (1992) at the Date Palm Research Center, King Faisal University, Saudi Arabia have been conducting research since 1985 to produce true-to-type date palm through a tissue culture technique and have completed the following:

- Production of plantlets *in vitro*
- Juvenility and leaves morphology
- Flowering and fruit set
- Physical properties
- Chemical properties
- Sugar content
- Mineral content
- Inflorescence and pollen grain evaluation

The four cultivars, namely: Deglet Noor, Medjool, Zahdi, and Thoory, produced through a tissue culture technique were compared with those produced through traditional methods at each step. No significant differences were found. It is hoped that the plants produced by the Center would soon be available for commercialization.

Future Perspective for Saudi Arabia

Saudi Arabia is now self sufficient in many food crops and has become a net exporter of some food crops. For instance, in 1984, the Kingdom ranked first among the world's date producers. The Government is strongly supportive of agricultural developments at all levels which has enabled the rapid expansion of farming activities and the creation of a sound research and development base. Biotechnological developments, which have progressed rapidly over the past ten years, are being used extensively in many countries to effect improvements in agriculture. It is, therefore, imperative that Saudi Arabia should consider the use of biotechnological application to many of the problems and potentials which face arid land farming in the Kingdom.

The Kingdom is a major producer of crude oil and natural gas and accommodates the largest toil reserves in the world. The refining industry and world-size petrochemical plants produce abundant hydrocarbon. The Saudi government is strongly committed to the industrialization of the country. The application of biotechnology has shown very encouraging results. It is, therefore, natural that Saudi Arabia should consider the biotechnological route for the utilization and processing of hydrocarbons for the development of the following:

- Methanol as a feedstock
- Enhanced Oil Recovery
- System for petroleum wastes disposal and oil spillage
- Specialty chemicals and high value-added products (KACST, 1989)

Biotechnology and genetic engineering should be used in the fields of agriculture and marine sciences.

Research and development in biotechnology should be undertaken and supported particularly in the following areas (Siddiqi, 1991):

- Crop improvement such as achieving better stress tolerance to salinity, heat and drought, diseases and pest resistance.
- Micropropagation technology for ornamentals, vegetable and fruit crops leading to the development of a micropropagation industry.
- Development of biofertilizers to improve productivity of legume crops and trees (legume inoculants), of cereals (associative nitrogen fixing organisms) and of all desert crops (phosphate dissolving organisms, e.g., microrrhizal fungi).

- Establishment of a Gene Bank through tissue culture technology which should focus upon plants of economic importance in the Gulf region.

References

Ala'ud-Din, S.H. *Oily Waste Landfarming at ARAMCO's Ras Tanura Refinery;* Proc. of the National Seminar on Genetic Engineering & Biotechnology, Riyadh, Saudi Arabia, 1987, published by KACST, Saudi Arabia, **1989**, pp 121-130.

Al-Ghamdi, A.S. *Personal Communications*; **1992**, (Not yet published).

Aynsley, J.S. *The Role of Biotechnology and Genetic Engineering in Saudi Arabian Development: The Commercial Application of Biotechnology to Date Palm and Potato*; Proc. of the National Seminar on Genetic Engineering and Biotechnology, Riyadh, Saudi Arabia, 1987, published by KACST, Saudi Arabia, **1989**, pp 57-62.

Siddiqi, B.A. *The Potentials and Opportunities in Biotechnology and Genetic Engineering and Their Utilization for the Benefit of Saudi Arabia*; KACST, Special Report, Technology Transfer and International Cooperation, 1991.

Report on the Seminar of the National Seminar on Genetic Engineering & Biotechnology; Proc. of the National Seminar on Genetic Engineering & Biotechnology, Riyadh, Saudi Arabia (1987) published by KACST, Saudi Arabia, 1989, pp. 168-173.

Development of the Biotechnology Industry in Developing Countries: The Taiwan Experience

Duen-Gang Mou, Development Center for Biotechnology, 81 Chang Hsing St., Taipei, Taiwan, Republic of China (Fax 886 2 732 5181)

Taiwan's successes in land reform in the 50's, in agriculture- and food-based industries in the 60's, in moving toward a capital intensive manufacturing industry in the 70's and in the development of a technology-based industry in the 80's led to today's development of a high-technology electronic and computer industry. For continual infrastructural growth of her industry and society as well as improvement of the well-being of her people and natural environment, the Republic of China's (R.O.C.) government on Taiwan has designated biotechnology as one of the key industries for the 21st century. Formidable tasks remain, however, in expanding and transforming a primarily market-based, domestically oriented and fragmented bio-industry into a largely technology-based, globally oriented and integrated one.

First, Lets make it clear that the word "developing" in the title of this paper is often used synonymously with the expression—"newly industrialized". Logically they fit the space between "less developed" and "developed" and between "less industrialized" and "industrialized", respectively. It is not clear though at what point one draws the line which marks the crossing from one stage to the next. In this author's view, a milestone which signals the birth of a developing or a newly industrialized country (NIC) is when intellectual property protection becomes a keen concern of business partners inside as well as outside the country. As for where lies the line which qualifies a NIC the status of a developed country, one saying was when a nation's per capita GNP had reached that of Italy [1]. This was what had happened to Japan in 1963. Now one knows the rest of the story, so that is that.

In any comparison of national economic performance over the last two decades, four economies—Hong Kong, Singapore, South Korea, and Taiwan—invariably stand out. The experiences of the city-states of Hong Kong and Singapore depend on unique geographies and histories and are thus largely irrelevant for the majority of developing countries. However, Taiwan and South Korea, with natural resources and endowments even poorer than many, both had initial conditions similar to other developing nations. Their economic experiences thus can provide valuable information toward a better understanding of the development process. This can also bring about a practical understanding of a region of enormous economic potential: the Pacific Rim countries. While we meet here, twelve Pacific Rim countries are holding their third biennial biotechnology conference in Taipei. Just last year, Hong Kong announced the opening of their new Hong Kong Institute of Biotechnology—the fourth such modern center among the four economies in a decade.

Taiwan as well as Korea show clear interest in new technologies, or more precisely, key or strategic technologies. Both countries have science parks, government sponsored industrial technology development projects and centers, and have been undertaking joint ventures with foreign companies that are active in new industries or active in exploring opportunities for growth. Backed by solid educational and intellectual achievements, both have done well in this round of global economic downturn. Of course, they can not expect to be completely competitive with world giants in such *areas* as micro-electronics,

1054–7487/92/0559$06.00/0

computers, information processing and distributing, pharmaceuticals, health delivery, and robotics. They can, however, excel in certain *lines* and make significant contributions to *portions* of highly integrated processes.

In his forward in *Models of Development,* Nobel Laureate Lawrence R. Klein stated *"Under changing world economic conditions they* (Taiwan and Korea) *could make a comeback in steel, petrochemicals, pharmaceuticals, and shipbuilding, but it is much more likely that they will succeed in the new technologies, where financial and physical capital may count for less than human capital. Their highly favorable supply of high-quality human capital makes me confident that the numerical projections to the year 2000 are sensible...I find much to admire in the successful industrial policies carried out by these two nations now approaching the line of separation between developing and developed countries."* It was against this backdrop that this author set off in preparing this presentation and try to substantiate Dr. Klein's expectation, if only partially, in the area of biotechnology industry development in Taiwan.

General Background

Taiwan's successes in land reform in the 50's, in agriculture- and food-based industry in the 60's, in moving toward a capital intensive manufacturing industry in the 70's and in the development of a technology-based industry in the 80's led to today's development of a hi-tech electronic and computer industry. In real figures, the following milestones were achieved: per capita GNP 145 USD in 1951, 7954 USD in 1990; average rate of annual economic growth, 8.91%; total employment of the manufacturing industry over 2 million, roughly 10% of the population; and a jobless rate around 2% over the entire period. Three important thresholds of economic performance are to be achieved in 1992—total GNP in excess of 200 billions USD, ranks 20th in the world; per capita GNP in excess of 10,000 USD, ranks 25th; and total goods trade 150 billions USD, ranks 14th.

For continual infrastructural growth of her industry and society as well as improvement of the well-being of her people and natural environment, the Republic of China's (R.O.C.) government on Taiwan has designated biotechnology as one of the key industries for the 21st century. To deal with this commitment in a responsible way, the government and business sectors have focused on biotechnology as one of *the eight strategic sciences and technologies* since 1982. Successful programs in hepatitis B vaccine production and the establishment of the Development Center for Biotechnology are noted. Recently, additional resources have been allocated for a 5-year plan in the building of a new Pharmaceutical Research and Development Center. Formidable tasks remain, however, in expanding and transforming a primarily market-based, domestically oriented and fragmented bio-industry into a largely technology-based, globally oriented and integrated one. A rational and incremental strategy in approaching this goal will be key to a productive and rewarding R&D program in the 90's.

Biotechnology Bases in Taiwan

Traditionally, biotechnology applications involve activities in agriculture, food, medicine and industry. Recent advances in the technique of recombining genes and fusing cells with definition and control have brought new promises to these traditional application areas. Emerging new applications such as environmental protection also demand new biotechnology to meet ever growing expectation.

With anticipated 1992 per capita income at over 10,000 USD, Taiwan is rapidly building her industrial base. Recent figures indicate a combined bio-industry output of 27 plus billions USD per annum, 17% of all industries' total. Since industrial application of the new biotechnology began only recently, the above-mentioned production outputs are all contributed by the traditional agricultural, food, pharmaceutical and specialty chemical industries, and, with a very minor share coming from the latter two. Also, this is a very domestically oriented industry, with total import and export, respectively, at 440 and 220 millions USD per annum. Due to a small domestic market, ROC's pharmaceutical companies can best be viewed as an union of plants, each of which involves little more than product formulation and re-packaging, instead of an industry with structure, capital and regulation.

Combined R&D expenditures in agricultural and medical sciences and industries totaled in excess of $300 million (US) in fiscal year 1989-90. There are universities and Academia Sinica supported by the government at the upstream, 33%; governmental, quasi-governmental/non-profit, and privately supported R&D laboratories and centers at the midstream, 50%; and private and public enterprises at the downstream, 17% of the total. Private and public funding are 14 and 86% of the total, respectively.

For rapid growth and infrastructure building into a technology-driven and high value-added manufacturing industry, one needs to take a closer look at the current base on which emphases are placed on capital might, process knowhow and marketing channel. Most of these industries are manufacturing products based on traditional microbiology and biochemistry. If categorized into a fermentation industry and a biological products industry, their combined annual sales can best be estimated at 2,500 and 100 millions USD, respectively. Of the former, Taiwan's home-grown monosodium-glutamate (MSG, a flavor enhancing amino acid) industry has a 300 millions USD/yr business from three companies and four production plants—by far the single largest multinational biotech industry in Taiwan. Nevertheless, the brewing and sugar businesses, *engine* for new biotech investment in many industrialized countries, are still state owned and monopolized. Their combined sales occupy more than 80% of this manufacturing base, but had contributed little new capitals and few channels pushing for R&D and manufacturing knowhow into new products and markets.

Development Opportunities

With rapid growth of overall industrialization, traditionally important agriculture sectors now bear more political and social significance than technological or economic ones. Restructure of governmental R&D resources with a clear national consensus is of great importance toward further upgrading of social welfare and entering a new era of life science evolution with major industrialized nations. A recent campaign for a national health insurance program is a major step in this direction. However, even from a regional perspective, Taiwan's past technology achievements and management expertise in agro-business still have a key role to play in places such as mainland China and Southeast Asia.

Successful development of high-end biotech products often requires large capital, plenty of practical experience, solid industrial infrastructures, well trained multi-disciplinary team players, well disciplined operators, seasoned management, shielded long incubation periods, and high risk. With a majority of her GNP contributed by tens of thousands small, fragmented and extremely market-driven businesses, Taiwan's near-term perspective in high-end new biotech products is at best uncertain. In the short run, one may expect new developments following the less complicated chemical industry model, i.e., the bulk manufacturing single dimension path or the specialty manufacturing two-dimension path, with emphases on capital, process and market channel. Recent acquisition of Glaxo's penicillin fermentation plant at Cambois, U.K., by the China Synthetic Rubber Co. serves as a typical example. Important points to consider in this model include raw materials availability technological knowhow, local or regional market demand, environmental protection, intellectual property protection, government incentive programs, and technology transfer.

In addition to technology acquisition abroad, functioning on an international level requires academia to teach "linking skills" to allow future R&D to consider all routes for profitable products. Also, effects of increasing population growth on the living standards need to be tempered as follows:

- improved engineering education domestically;
- reduced reliance on academia from western countries;
- education activities that include economics, management, and communication skills, as well as traditional skills of math, science, and research; and

- engineers taught to commercialize their research.

References

1. *Models of Development: A Comparative Study of Economic Growth in South Korea and Taiwan,* Lau, L. J., Editor, ICS Press, Inst. of Contemporary Studies, San Francisco, California, **1986**.

2. Lee, Shu-Jou, Republic of China's industrial development strategy and programs, Keynote speech given at the 2nd annual meeting of the Bioindustry Development Association of ROC, November 14th, 1991, as Vice Minister of the Ministry of Economic Affairs, *Bioindustry* (Taiwan), 2(4): 76-80 **(1991)**

3. Su, Yuan-Chi; Chang, William T. H., Current status of biotechnology R&D in Taiwan, *Bioindustry* (Taiwan) 1(2): 52-62 **(1990)**

4. *Indicators of Science and Technology, Republic of China 1991,* National Science Council, Taipei, Taiwan, ROC, **1991**.

5. Soong, Tai-Sen, Current industrial biotechnology development in Taiwan, *Agro-Industry Hi-Tech*, 2(2): 11-17 **(1991)**

6. Mou, D. G., Considerations in the development of exportable biotechnological products in developing countries. Paper presented at the fifth European Congress on Biotechnology, Copenhagen, July 8-13, **1990**.

Continuous Production of hCG Monoclonal Antibody in Various Membrane Bioreactors

Ho Nam Chang* and **Duk Jae Oh**
Bioprocess Engineering Research Center and Dept. of Chemical Eng., Korea Advanced Institute of Science and Technology, Daeduk Science Town, Taejon 305-701, Korea Tel: (042) 829-8811,8812 Fax: (042) 869-8800,3910

Continuous cultures of high density hybridoma cells secreting monoclonal antibodies against hCG hormone were investigated in various membrane bioreactors. In a dual hollow fiber bioreactor, hybridoma cells were successfully cultured to a high density of ca. 1.87 x 10^8 cells/mL within the extracapillary space for two months. The Mab productivity reached 205 mg/L·day. The cells encapsulated in a Ca-alginate gel reached a density of 1.5 x 10^8 cells/mL of the intra-capsule space and yielded a productivity of 744 mg/L·day and showed a stable operation.

Immobilization of hybridoma cells producing monoclonal antibodies (Mabs) has many advantages (Brunt, 1986). The commercially available immobilized cell systems include immobilization of the cells within semipermeable membranes such as hollow fibers (Evans and Miller, 1988) or capsules (Posillico, 1986), or in packed beds (Lazar et al., 1988), or on ceramic matrix (Putnam, 1987), but these systems often face difficulties in large-scale applications.

Described here is the comparison of the three membrane reactor systems for the production of Mab against hCG hormone: dual hollow fiber bioreactor (DHFBR) (Oh and Chang, 1988), Ca-alginate capsule (Nigam et al., 1988; Oh et al., 1990), and depth filter perfusion system (DFPS) (Oh, 1992). The traditional stirred cultures in a spinner flask were used as the reference for the comparison of reactor performances, and to determine the growth rate, secretion and metabolic characteristics of a hybridoma cell line.

Cell Line, Medium and Analytical Methods

Hybridoma cell line, Alps 25-3 producing the monoclonal antibody against hCG (human Chorionic Gonadotropin) β-subunit was donated by the Genetic Engineering Research Institute (Korea). The culture medium was a mixture of IMDM/F-12 (1:1) supplemented with 0.5-2.0% (v/v) bovine serum (Hyclone). HEPES and $NaHCO_3$ were supplied for the pH control. The cells were subcultivated every 3 or 4 days in a 5% CO_2 incubator. Viable or dead cells were counted by a hemocytometer (Hausser), using the dye exclusion test with 0.4% (w/v) Trypan blue in PBS. Glucose concentration was determined by an enzymatic method. Ammonia concentration was measured with ammonia assay kits (Wako Chem.). Monoclonal antibody (IgG_1 type) concentration was determined with the single radial immunodiffusion method using the goat anti-mouse IgG (Sigma).

Bioreactor Systems

The cross-sectional diagram of dual hollow fiber bioreactor (DHFBR) is shown in Fig. 1. Ten dual hollow fiber units were bundled together in a parallel assemblage. Each unit consisted of three hollow fibers inside a silicone tubule. The volume of extracapillary space (ECS) where cells grew was 2.74 cm^3. This cartridge-type DHFBR was connected to a 60 mL medium reservoir, in which a pH of 7.0 $\pm$ 0.1 was controlled with CO_2 gas and a 0.2 m NaOH solution. The culture medium was recycled between the reservoir and the DHFBR at a rate of 2.0 L/h. Oxygen was supplied to the packed cells from both the medium in lumen space and the wall of silicone tubule. Thus DHFBR receives much more oxygen than the conventional hollow fiber bioreactor.

1054–7487/92/0563$06.00/0

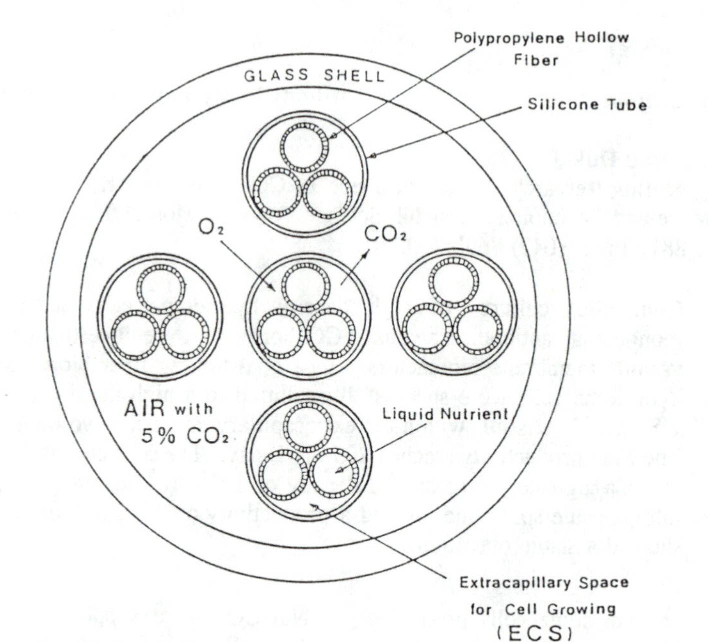

Fig. 1. Cross-sectional diagram of dual hollow fiber bioreactor. Cells are immobilized and grow in the extracapillary space. Glass shell (I.D.: 0.8 cm, length: 31 cm), silicone tubule (E.D.: 0.147 cm, O.D.: 0.196 cm) and polypropylene hollow fibers (I.D.: 0.033 cm, O.D.: 0.063 cm) are used.

The Ca-alginate capsule system is a simplified and economical encapsulation process in comparison with that of Lim and Sun, 1980. The pre-cultured cells were suspended in a 10% (w/v) gelatin and 1% $CaCl_2$ aqueous solution. The cell suspension was dropped with a spray assembly into a 0.5% (w/v) sodium-alginate and 0.05% (v/v) Tween 80 solution. The formed capsules (E.D.: 0.1 cm, thickness: 0.016 cm) were meshed, then washed with a 0.85% NaCl solution. Capsules were treated with a 1% $CaCl_2$ solution for hardening before they were introduced into a culture vessel. In a continuous culture of Ca-alginate capsule system, a 130 mL spinner flask equipped with a sedimentation column was used for separating the capsules from the culture media.

The depth filter perfusion system (DFPS) also consisted of two parts. The one was a filter matrix of cell immobilization. The other was a vessel for the control of media conditions. A reactor system with total working volume of 300 mL, and with 20 μm pore size filter was tested.

Results and Discussion

The typical profiles of batch and continuous suspension cultures of hybridoma Alps 25-3 in a 250 mL spinner flask are shown in Fig. 2 and Fig. 3. The maximum cell density and the antibody concentrations were in the ranges of 1.5-2.5 x 10^6 cells/mL and 20-40 μg/mL depending on the inoculum cell density and the serum concentration. The specific rates of cell growth, Mab production, and glucose consumption in the suspension cultures were 0.065 h^{-1}, 1.0 μg Mab/10^6 cells·h, and 0.1 mg glucose/10^6 cells·h, respectively. These specific values were used to estimate the density of immobilized cells in DHFBR. Fig. 4 shows the time courses of glucose and Mab concentrations, and cumulative productivity in a continuous culture with the DHFBR for two months. Medium feeding rate was controlled to maintain the glucose concentration at about 1.5 g/L, which meant that 50% of the supplied glucose was utilized. Antibody was produced continuously in concentrations of 100-130

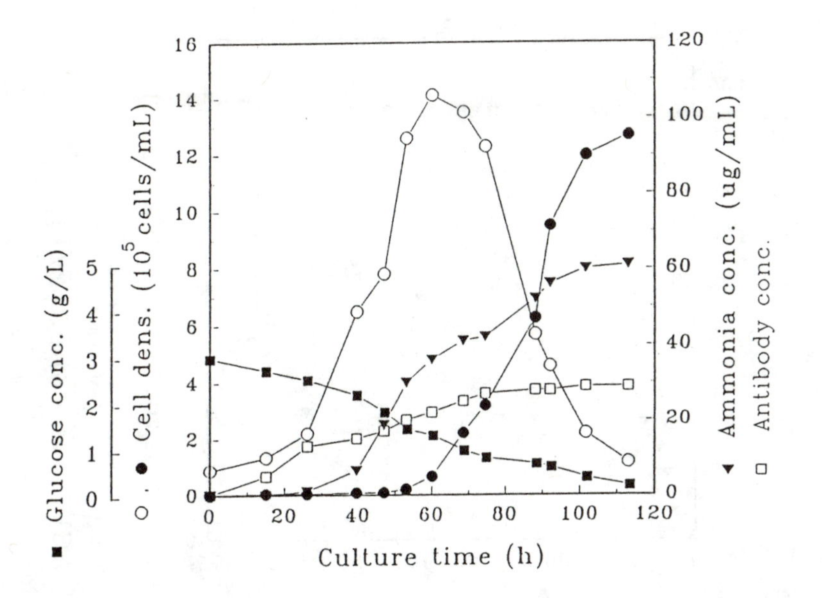

Fig. 2. Batch suspension culture of hybridoma Alps 25-3 in a 250 mL spinner flask (2% bovine serum). Hollow and filled circles mean the densities of viable and dead cells, respectively.

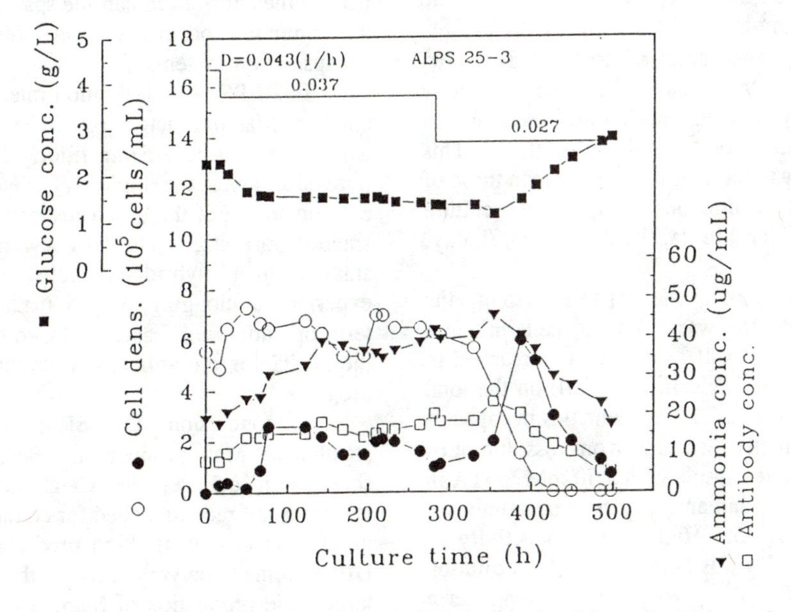

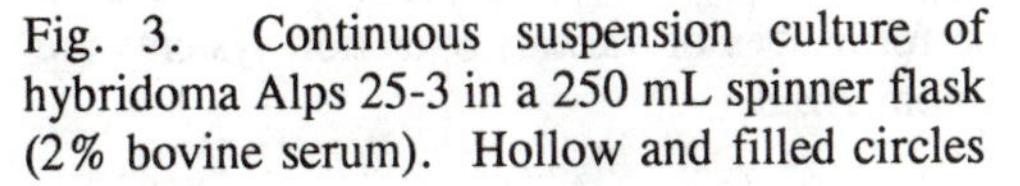
Fig. 3. Continuous suspension culture of hybridoma Alps 25-3 in a 250 mL spinner flask (2% bovine serum). Hollow and filled circles mean the densities of viable and dead cells, respectively.

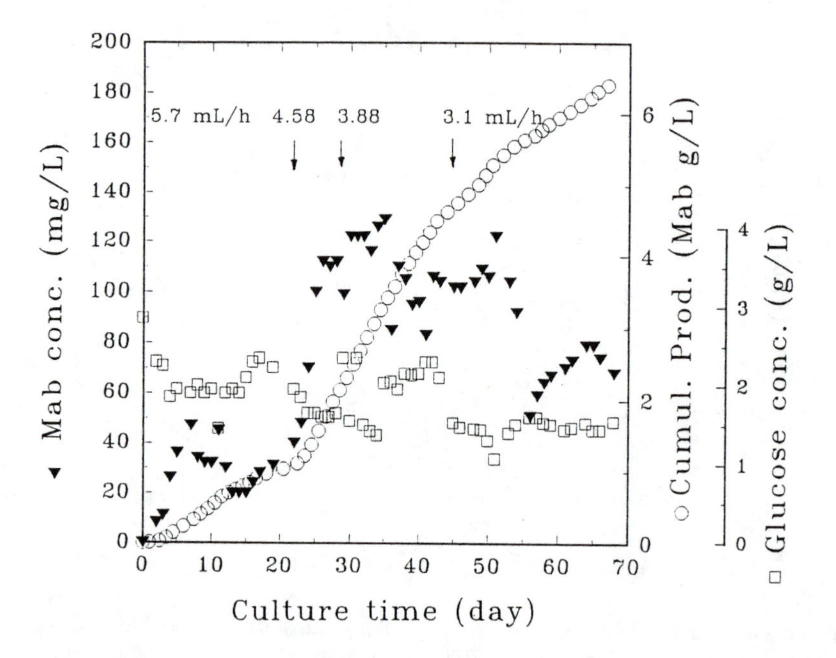

Fig. 4. Time courses of glucose, antibody concentrations, and cumulative productivity in a continuous culture of hybridoma Alps 25-3 in a dual hollow fiber bioreactor for two months. Total working volume was 60 mL. Medium with 2% bovine serum was recirculated at a rate of 2L/h.

μg/mL for 30 days. The viable cell density in the 2.74 cm^3 in the ECS of DHFBR at 28th culture day was estimated to be 1.87 x 10^8 cells/mL. At that time the antibody productivity was 205 mg/L·day based on the total reactor volume of 60 mL. This productivity is considerably higher than those of batch and continuous suspension culture systems, which are ca. 10 and 20 mg/L·day, respectively.

In the Ca-alginate capsule system, the initial cell density was 1.5 x 10^6 cells/mL in a capsule; and 1.5 x 10^8 cells/mL was reached in 7 days. It was 1.5 x 10^7 cells/mL on the total reactor volume basis. Cell densities in capsules could be directly measured from dissolution of the Ca-alginate capsule with a 50 mM EDTA in PBS. Glucose and antibody concentrations are shown in Fig. 5. High Mab productivity of 652.8 mg/L·day was obtained at a high dilution rate of 0.67 h^{-1}. However, a long term continuous culture could not be carried out because the supply of nutrients and the removal of wastes were limited, as the grown cells packed the entire intra-capsule space. This was a common occurrence in the general encapsulation system.

In DFPS, a high dilution rate of 0.72 h^{-1} and high Mab productivity could be obtained by using a 20 μm pore depth filter. The smaller pore size gives the higher efficiency of cell entrapment. But there is a restriction in using smaller pore size. The DFPS with the pores smaller than hybridoma cells of 13 μm experienced clogging, which made the long-term operation not feasible. More research on the DFPS for the animal cell culture is under progress.

In conclusion, DHFBR is suitable for small-scale Mab production. Because of its short term culture, the Ca-alginate capsule system is not recommended for continuous Mab production despite its high productivity. The DFPS seems to be very useful in the continuous large-scale production of Mab. In Table 1 the performances of various operation systems are summarized.

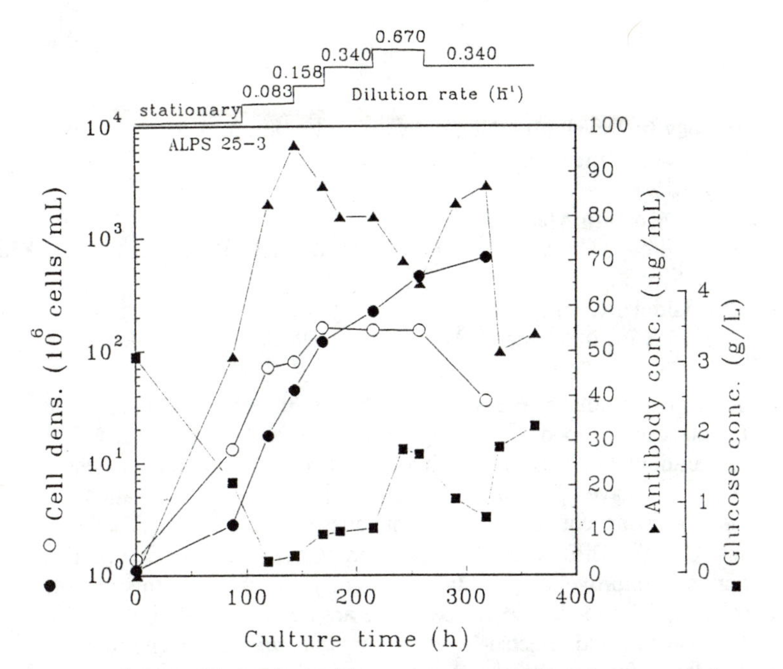

Fig. 5. Continuous cultures of hybridoma Alps 25-3 in Ca-alginate capsule system (working volume: 130 mL, 20% [v/v] capsules in a IMDM/F-12 mixture supplemented with 2% bovine serum).

Table 1. Reactor performances in the culture of hybridoma Alps 25-3

	Cells density (10^6 cells/mL)	Productivity (mg/L·day)
Batch suspension	1.5-2.5	10
Continuous suspension	1.0-2.0	20-30
DHFBR	187[a] 8.54*	205
Ca-alginate capsule	150[b] 15*	650
DFPS (20 μm pore)	60[c] 30*	744

Productivity is based on the working volume of one liter. Cell density: [a]in extracapillary space; [b]in capsules; and [c]in depth filter matrix; *based on the total working volume.

References

Brunt, J.V. *Bio/Technology* **1986**, 4, pp 506-517.

Chang, H.N.; Oh, D.J. 196th ACS National Meeting **1988**, Los Angeles, U.S.A.

Evance, T.L.; Miller, R.A. *Bio/Technology* **1988**, 6, pp 762-766.

Lazar, A.; Silberstein, L.; Mizrahi, A.; Reuveny, S. *cytotechnology* **1988**, 1, pp 331-337.

Lim, F.; Sun, A.M. *Science* **1980**, 210, pp 908-910.

Nigam, S.C.; Tsao, I.F.; Sakoda, A.; Wang, H.Y. *Biotechnol. Techniques* **1988**, 2, pp 271-276.

Oh, D.J.; Chang, H.N.; Choe, T.B. Asia-Pacific Biochem. Eng. Conference' 90 **1990**, Kyungju, Korea.

Oh, D.J. Ph.D. Thesis **1992**, KAIST, Korea

Posillico, E.G. *Bio/Technology* **1986**, 4, pp 114-117.

Putnam, J.E. Commercial production of monoclonal antibodies: a guide for scale-up; Marcel Dekker: New York, **1987**, pp 119-138.

Enzyme Biotechnology in Developing Countries: Production and Application of Thermostable Xylanase in Jute Fiber Upgradation

M. Mozammel Hoq*, Mostafa Alam
Department of Microbiology, University of Dhaka, Dhaka 1000, Bangladesh (Fax: 880 2 863060; Tel.: 880 2 505430)
Gulam Mohiuddin, Isidore Gomes
Bangaldesh Jute Research Institute, Manik Mia Avenue, Dhaka, Bangladesh

Some extracellular enzymes are being imported and used in Bangladesh in food processing, tanneries, and in pharmaceutical preparations. The hemicellulolytic and pectinolytic enzymes, however, have particular usage in jute fiber upgradation in jute—growing countries, viz. Bangladesh, India, Nepal, Thailand, and China. Efforts are Underway in various laboratories of this region to improve low quality jute fiber by enzymes. In this regard, cellulase—free xylanase and pectinase are particularly desirable. We have isolated and selected *Thermomyces lanuginosus* which produced cellulase—free xylanase and pectinase on lignocellulosics at 55°C. The xylanase production conditions have been optimized in solid state fermentation on wheat bran medium. The diluted enzyme filtrate exhibited 172 U/ml of xylanase activity at 70°C and a half-life of 4 weeks at 55°C. The enzyme demonstrated a pH maxima of 6.0 and a broad pH stability range between pH 5.0 to 11.0. Low quality jute fibers were upgraded by the enzyme.

Xylan is a biopolymer, the main component of hemicellulose, present in nature in large amounts. It is present in stalks of crop plants like wheat, rice, maize, jute, mesta, etc. The wastes from pulp and paper industries also contain large amounts of hemicellulose. It, therefore, represents a significant renewable resource which should be utilized to improve the economics of bioconversion of plant biomass to useful products. Uses of xylanases are found in the food industry (Royer and Nakas, 1989), in the cellulose pulp preparations, and plant fiber upgradation through selective removal of xylan (Paice and Jurasek, 1984).

Jute is a lignocellulosic bast fiber, the principal cash-crop of Bangladesh. It is also grown in India, Nepal, Thailand, and China. Under normal biological retting practices, the bottom part of the jute fiber remains harder due to the content of excess amounts of hemicellulose and pectin. These bottom hard parts, cut down in jute mills, are known as "jute cuttings" amounting to 20-40% of total jute production (Mohiuddin et al., 1978). These cuttings are not suitable for effective usage and pose difficulties in proper spinning. More than one million bales (one bale weighs about 180 kg) of jute cuttings are stockpiled each year in Bangladesh alone. Other jute exporting countries, viz. China, India, Nepal, and Thailand, are also facing the same problem with jute cuttings.

These jute cuttings need extensive softening prior to proper spinning. In the conventional methods, softening is done by piling jute cuttings pretreated with oil-water emulsion and kept in concrete bins for about 15 days where the temperature is raised up to 65 to 70°C (Mohiuddin et al., 1978). This conventional method makes limited improvements and are not adequate in softening the jute cuttings with required improvement of their spinnability. Hence, treatment with microbial enzyme became an alternative approach for the purpose. Most of the work dealt with the enzyme from

1054–7487/92/0568$06.00/0

mesophilic fungi, viz. *Aspergillus* spp. (Ghosh and Datta, 1983; Mohiduddin et al., 1992) and **Sclerotium rolfsii** (Ghosh et al., 1968; Haque et al., 1990). These enzyme preparations contained cellulase which affected the cellulosic structure of the fiber. Moreover, the enzymes were not thermostable.

Since jute fiber upgradation requires specific removal of hemicellulose and pectin at elevated temperatures, a cellulase-free thermostable xylanase enzyme preparation should be appropriate for the purpose.

The objective of this work was to screen thermophilic fungi from various sources including jute compost for high xylanase and pectinase activities without cellulase with a view to develop a suitable biotechnological process for upgradation of cellulosic fibers. The present report will encompass the production of xylanase from a local isolate of *Thermomyces lanuginosus* under solid state fermentation, its properties and application to jute fiber upgradation.

Experimental

T. lanuginosus was isolated from jute mill wastes and the culture was maintained on Potato dextrose agar medium (PH 5.5). Unless otherwise stated, the solid state fermentation (SSF) was carried out in 1 L Erlenmeyer flasks containing 50 g lignocellulosic substrate moistened (50-80%) with water, modified Mandel's (Steiner et al., 1987) or Czapek's basal salt solution. The sterilized medium inoculated with the spores ($\sim 2x10^6$) of *T. lanuginosus* previously grown on yeast malt extract agar (Oxoids) was cultivated at 55°C under 50-70% RH. The crude enzyme filtrate from the fermented mass was obtained by aqueous extraction (500 ml distilled water and 5 g NaCl). The xylanase and pectinase activities were measured according to Khan (1986). FP-cellulase activity was measured according to Mandels et al. (1976). The enzyme assays were made in triplicate.

Results and Discussion

Previously it was found that *T. lanuginosus* can produce cellulase-free xylanase and pectinase in submerged cultivation using hemicellulose, pectin, or jute cuttings powder as a carbon source at 55°C (Gomes et al., 1991). The fungus did not produce any cellulase in the medium containing microcrystalline cellulose. Hence, *T. lanuginosus* was selected for partial removal of pectin and xylan from low grade jute fiber and jute cuttings without affecting their cellulosic structure.

With a view to develop an inexpensive bioprocess for xylanase production, solid state fermentation (SSF) conditions were preferred. The locally available cheap agro-industrial ligno-hemicellulosic material was selected as a major component of fermentation medium for investigation. The results (Table 1) show that of all substrates, wheat bran with Mandel's basal solution supported the maximum xylanase production under the conditions employed.

Table 1. Effects of various natural lignocellulosic substrates on xylanase production by *T. lanuginosus* under solid state fermentation at 7 days cultivation.

Carbon source*	Xylanase activity (U/ml)
Wheat bran	8.90
Rice bran	2.14
Sulphite pulp	1.92
Sugarcane bagasse	3.98
AT saw dust	2.10
AT jute dust	4.36
AT rice straw	1.26

*Moistened with 50% Modified Mandel's solution. AT=Alkali-treated.

Since moisture content is a significant controlling factor of enzyme formation in SSF, the effects of the amount of moisture content (20-80%) in wheat bran was studied. The optimum initial moisture content for the enzyme production was 80% at 55°C (Fig. 1). Kitpreehavanich et al. (1984) obtained maximum enzyme production at moisture content of 65% at 45°C with *Humicula lanuginosa* in wheat bran.

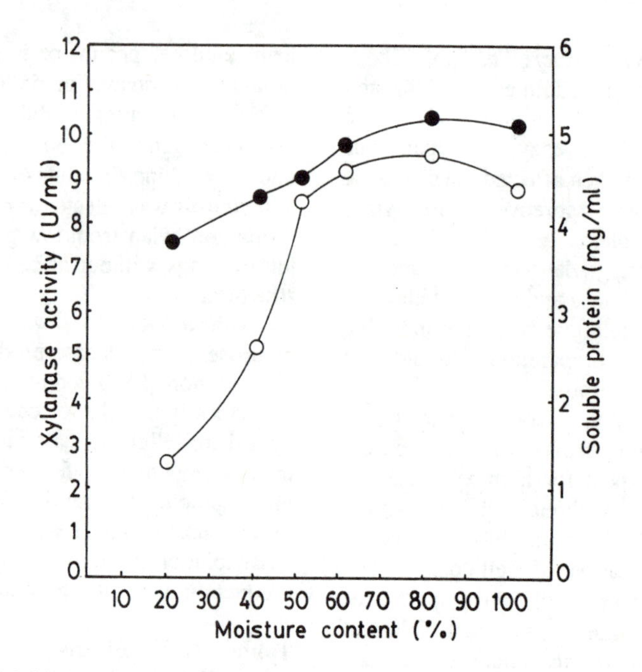

Fig. 1. Effect of initial moisture content of wheat bran on the production of xylanase and soluble protein by *T. lanuginosus*. O=xylanase activity; ●=soluble protein.

To improve the enzyme production on wheat bran by *T. lanuginosus*, the effects of Czapek's basal or Mandel's basal solution, surfactant, peptone, or induction with xylan were examined. No increase of the enzyme activity was observed with both of the basal solution supplements (Table 2). A positive effect was observed with Tween-80 and peptone. Xylan, however, appreciably induced the enzyme production. Further optimization is required with varied concentrations of xylan and their sources. Kitpreehavanich et al. (1984) obtained 2-fold increases of enzyme formation with 0.6% xylan. Supplementation of wheat bran with calcium, magnesium, and potassium salts for the production of enzyme also did not increase the xylanase activity (results not shown). These results, however, suggested that wheat bran with water only supported comparable xylanase production and could be used for the development of an inexpensive bioprocess for xylanase production.

Table 2. Supplementation of wheat bran with different nutrients for xylanase production of *T. lanuginosus* at 7 days cultivation.

Supplements (80% v/w)	Xylanase activity (U/ml)
Water	9.25
Xylan suspension (0.7%)	11.03
Czapek's basal solution	8.76
Mandel's basal solution	8.84
Tween-80 solution (0.1%)	9.74
Peptone solution (1%)	9.49

Some properties of xylanase in crude filtrate were studied. The temperature optimum and thermal stability of the xylanase are shown in Fig. 2. The optimum temperature for the enzyme activity for 20 min incubation was 70°C, indicating its thermophilic nature. The thermostability was tested by keeping the enzyme solution for 30 min at varied

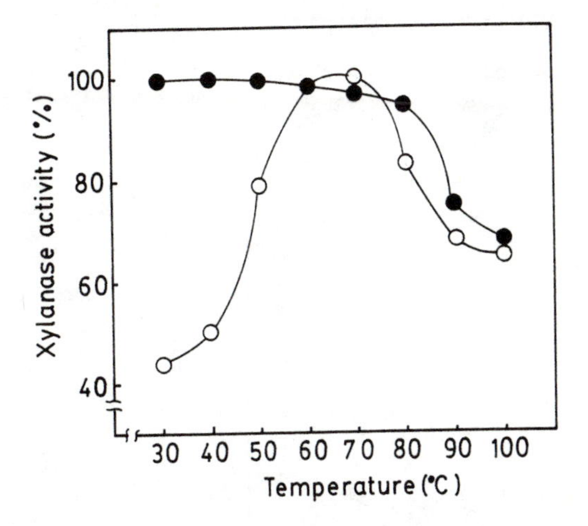

Fig. 2. Temperature optimum and thermostability of xylanase of *T. lanuginosus*. O=temperature optimum; ● =thermal stability.

temperatures and residual activity was determined at the optimum temperature of 70°C. The enzyme was stable up to 70°C demonstrating its sufficient thermostability.

Above 70°C, the enzyme activity decreased gradually, however, the activity did not decrease by 50% of its original activity even at 100°C. Storage stability of the crude enzyme filtrate at different temperatures indicate that the enzyme may be preserved at chilled temperatures (10°C) to 4°C for prolonged periods (Fig. 3). The half-life of the crude enzyme filtrate at 55°C appeared to be not less than one month.

Fig. 4 shows the pH profile and pH stability of the xylanase. The optimum pH for the enzyme activity was pH 6.0. The pH stability of the enzyme was examined by keeping the enzyme at each pH (2.0-12.0) at 4°C for 24 h. The pH stability exhibited between 5.0 and 11.0 indicated that the enzyme is alkali tolerant and thus may also be used for decomposition of industrial alkaline effluent (Kelly and Fogarty, 1976).

For improvement of jute cuttings or low quality jute fiber, the materials were moistened with the liquid enzyme preparation and kept for 24 h at 65°C. The hemicellulose content, fiber strength and "feel-and-touch" test of the fibers

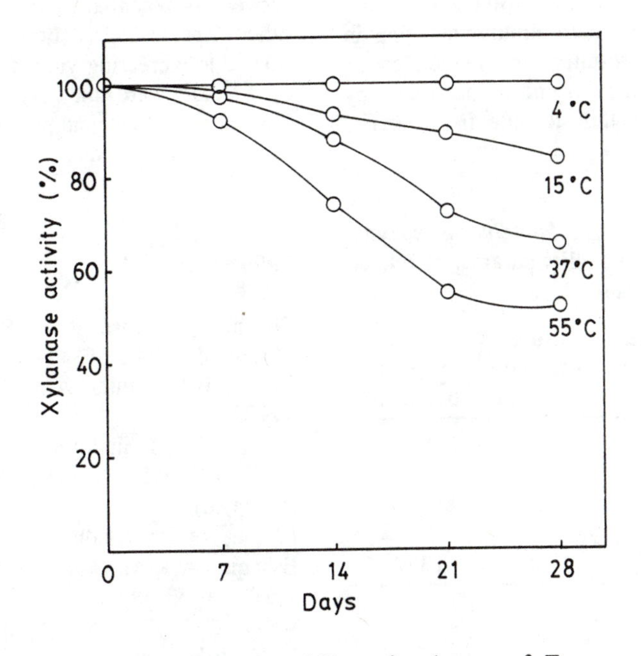

Fig. 3. Storage stability of xylanase of *T. lanuginosus* at different temperatures.

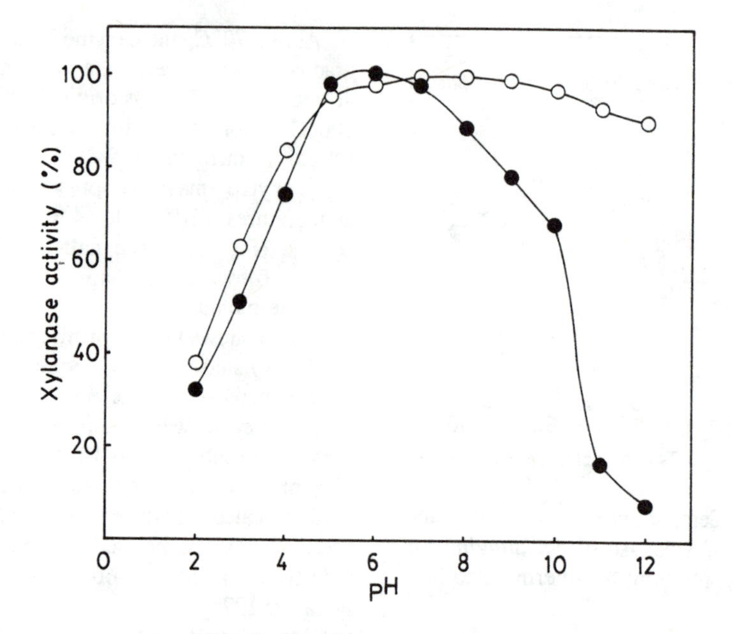

Fig. 4. pH profile and pH stability of crude xylanase of *T. lanuginosus*. ● =pH profile; O=pH stability.

before and after enzyme treatment (Table 3) suggest that the enzyme preparation has selectively removed hemicellulose bringing in physical changes, resulting in upgradation of the jute fibers. This enzyme preparation may be also used for cellulosic pulp in the paper industry.

Table 3. Effect of *T. lanuginosus* enzyme preparation on some quality parameters related to jute fiber upgradation.

Parameters of jute cuttings	Before enzyme treatment	After enzyme treatment
Feel and touch	Hard	Soft
Fiber strength (Pressly index)	4.2	4.1
Hemicellulose (%)	20.0	13.0

Conclusion

This study revealed that the ability of *T. lanuginosus* to grow and produce cellulase-free thermostable xylanase on inexpensive medium and the wide range of pH stability of xylanase proved its potentiality to be used in jute and other fibers upgradation. Further studies related to increasing yield of enzymes are being continued towards the development of a suitable bioprocess for enzyme production as well as its industrial applications.

References

Ghosh, B.L.; Bose, R.G.; Basu, S.N. *Can. J. Microbiol.* **1968**, 14, 459-466.

Ghosh, B.L.; Dutta, A.K. J. *Text Inst.* **1983**, 74, 83-91.

Gomes, I.; Mohiuddin, G. Saha, R.K.; Hoq, M.M. *World J. Microbiol. Biotechnol.* (accepted).

Haque, M.S.; Asaduzzaman, M.; Alam, S.; Eshaque, A.K.M. *Bangladesh J. Jute Fib. Res.* **1990**, 14, 53-58.

Kelly, C.T.; Fogarty, W.M. *Proc. Biochem.* **1976**, *July-August*, 3-9.

Khan, A.W. *Enzym. Microbiol. Technol.* **1986**, 8, 373-378.

Kitpreehavanich, V.; Hayashi, M.; Nagai, S. *J.*

Ferment. Technol. **1984**, 62, 63-69.
Majumder, S.; Ghosh, B.S.; Kundu, A.B. *Indian J. Text. Res.* **1987**, 12, 213-216.
Mandels, M.: Andreotii, R.; Roche, C. *Biotechnol. Bioeng. Symp.* **1976**, 6, 21-33.
Mohiuddin, G.; Akhter, S.; Mia, M.M. *Bangladesh J. Jute Fib. Res.* **1978**, 3, 28-34.
Mohiuddin, G.; Talukder, S.H.; Lutfar, L.B.; Sobhan, M.A.; Kabir, K.M. *J. Text. Inst.* **1992** (submitted).
Paice, M.G.; Jurasek L. *J. Wood Chem. Technol.* **1984**, 4, 187-198.
Royer, J.C.; Nakas, J.P. *Enzym. Microbiol. Technol.* **1989**, 11, 405-410.
Steiner, W.; Lafferty, R.M.; Esterbauer, H. *Biotechnol. Bioeng.* **1987**, 30, 169-178.

Building the Base for a Critical Mass of Biotechnological Scientists in a Developing African Country

Alan Robertson, Crop Science Department, University of Zimbabwe, PO Box 167MP, Harare, Zimbabwe. Fax: 263-4-732828; e-mail: lanRob@zimbix.uz.zw.

The first use of biotechnology in Africa tends to be in the improvement of agriculture. Low-techonolgy applications use tissue culture techniques; high-techonology applications use DNA manipulations; industrial applications use fermentation and downstream processes. Practicing agriculture provides the experience that a holistic approach is needed: one weak link and there is no crop; no viability, you are suddenly in debt or starving. **The latter makes risk-avoidance a high priority**. In *this* context, building a critical mass involves much more than providing the buildings, the equipment, and the scientists. The following suggestions might help the critical mass function more effectively. 1) Build on established breeding. 2) Provide a career path for scientists. 3) Locate in a viable economy, or insulate with high incentives. 4) Identify client funders like farmers' associations, commercial feedstock users, donor agencies. 5) Inform national decision-makers early. 6) Create space for indigenous scientists to practice decision-making. 7) Identify "genuine" leadership. 8) Use ingenuity to problem-solve in the jungles of bureaucracy as well as in the lab. 9) Establish links with officials: this is new technology and can give rise to misunderstandings. 10) Be at the right place at the right time!

In Africa there are something like 400 biotechnologists (Edroma, 1989), scattered haphazardly across disciplines from medicine to brewing, and critical masses of plant biotech scientists in Kenya, in Zimbabwe and in South Africa. Others are growing in Morocoo, Egypt and Tanzania (Okonkwo, 1992). I shall confine my observations to Zimbabwe; and restrict them to the agricultural scene which of political necessity is the most frequented field for both low and high-tech applications. Over the past eight years it has been my concern to try to build a team that will work towards the capacity to double the yields and halve the inputs for all our major crops, starting with the so-called "orphan crops" that have so far received little research interest or funding. **Table 1** shows the emerging team and their target crops.

From the contents of the table you may discern the pyramidical structure of our ambitions. There is no future in hand-me-down technological packages that purport to do one job and often break down quickly and lie idle forever. **We want to build a base and work our way up to the peak of technological competence.** The pyramid we are climbing is illustrated in **Figure 1**.

From any level of this pyramid value-added fermentation and downstream processes can enhance the productivity, the profitability, the import substitution capacity and the job generation from the crop. From even the most modest levels like tissue culture propagations small businesses can emerge, based on good science and employing a variety of skill levels. These are the motives for indulging in biotech capacity building: to be of service to development, whether the motives derive from political, social or moral imperatives. Mine is moral: I believe we (I hold a British passport as well as a Zimbabwean one) owe it to the Third World to help solve pressing problems with the most sophisticated technology available, which is fortunately transferable,

1054–7487/92/0574$06.00/0

Table 1: The critical mass of scientists and their level of training, their chosen crops and the target technologies that they are developing.

Scientist	Level	Crop and Target Technology
Edouard Sakina	PhD	**cassava**: virus-elimination
Victor Masona	PhD	**coffee**: cloning, transformaton
Khola Magotsi	PhD	**cowpea**: cold tolerance
Wendy Martin	PhD	**potato**: virus-elim, transformation
Andrew Connolly	PhD	**sweet potato**: virus-e., transform.
Rachel Chikwamba	PhD	**sorghum**: transform., regeneration
Jerry Ndamba	MPhil	**gopo tree**: cloning, moluscicide
Tsi-tsi Ndowora	PhD	funngi, plant protection
John Makoni	3rd Year	**sweet potato**: gene-gun
Prescilla Dembetembe	3rd year	**cassava,** rapid multiplication
Jacob Nyagweta	2cd Year	**brassicas**: regeneration
Tichafa Munyikwa	MSc	**tomato**: transposons
Thabani Dube	MSc	**petunia**: antisense RNA
Tichaona Mangwende	MSc	**oil-seed rape**: transformation
Modern Muzondo	MSc	**cassava**: non-sterile fermentation
Baldwin Chipangura	Bsc	**Pyrethrum**: tissue culture imports

Pop!
Custom-built crops
Multi-gene inheritanced traits
Tissue-specific pharmaceutical crops
Antisense RNA controls, RAPD gene-fishing
RFLP and transposon tagged accelerated breeding
Inter-specific gene transfers:single trait resistances etc
Embryo rescue:wide crosses plus embryogenesis in recalcitrants
Intra-specific improvements:somaclonal selection and anther dihaploids
Plant tissue culture:clean exotic imports; virus-elimination; rapid propagation
Modern plant breeding; exotic cultivar selections; high input chemical agronomy
Traditional agriculture:folk selections; low-input agronomy, indigenous genepool

Figure 1. The various levels of sophistication in agricultural biotechnology that will give access to value added products to higher profits, to more jobs and to import substitution or self-sufficiency.

applicable and affordable. The biggest bottleneck is organizational. Can we get it together? Galbraith (1986) points out that to move from the primitive feudal and merchant capitalist era to the modern power structure **the main ingredient is improved organizational capacity.**

To allow this organizational infrastructure to grow the following suggestions may help:

1) Build on established breeding. It is possible to by-pass the traditional breeder and develop virus-resistant this and caterpillar-resistant that, but the result can only be a one of a kind cultivar that could be a six-year wonder but that will lead to no net gain in capacity to be self-sufficient in improving other cultivars and crops.

Hence in Zimbabwe we have taken care to build programs in conjunction with the well-established traditional crops of maize, cotton, potato, tobacco, and more recent developments with soyas, groundnuts, winter wheat and barley. **Thus any gene-transfer induced improvements will only fine-tune already established cultivars.** This tuning of course can make a dramatic and disproportionate difference to viability.

2) Provide a career path for scientists. This is essential; if we are to recruit the best minds. **In a new field, there is always a credibility gap which can only be filled by actual products,** however modest. Our own efforts at virus-elimination (and rapid multiplication of the resulting planting material) in strawberries, potato, cassava, sweet potato, coffee and hops have led to funding in five crops by indigenous businesses or producer associations. Local funding lends confidence to the scientists involved and also to potential outside donors.

3) Locate in a viable economy. Since the 1980 Independence, our economy has grown slowly but surely and our debt service ratio is fairly modest. We are usually self-sufficient in food. Droughts have been survived, although inflation has now hit us severely. There are two plans to offer the biotechnologists a decent salary. One rests on Government plans for a Scientific and Industrial Research and Development Centre where salaries are proposed to be well above University or Government. Another plan is my company Agri-Biotech (Pvt) Ltd which hopes to offer support and eventually an independent research lab based on the commercialization of tissue culture and gene-transfer research yielding reasonable royalties. At least one of these plans will perhaps insulate scientists from further inflation problems.

7) Leadership that is honest and sincere clearly will have the advantage in a sensitive and long-term project, and it is worth looking for these qualities in the volunteers.

8) I should only add as a footnote that outwitting the bureaucrats takes more patience and brain-teasing than the simple problems found at the bench. Human nature is infinitely more variable and obtuse than DNA and enzyme kinetics.

9) Discuss problems with officials as they arise. Many problems can be reduced this way.

10) Finally it is all in vain unless you turn up at the right place at the right time. Like Joe Montana, quarterback for the professional football team, the San Francisco 49ers, and his wide receivers, this is not just a matter of luck.

References

Edroma, E. Chairman, African Biosciences Network, Introductory address: UNCTD/ABN Conference on "Biotechnology for Semi-Arid Lands" Dakar, October 8-11, 1990.

Okonkwo, SAM, Coordinator, ABN Biotechnology Network recently formed at Nairobi Conference of ABN, Nairobi, February 17-21, 1992.

Galbraith, JK in *The Anatomy of Power,* Nicolson, NY, 1986.

Bunders, JFG et al in "*Biotechnology for small-scale farmers in developing countries: analysis and assessment procedures*", Free University, Amsterdam. June, 1990.

Impact of Biotechnology in Sustaining Food and Agriculture in Developing Countries

Amin S. ElNawawy, Biotechnology Department, Kuwait Institute for Scientific Research
P.O. Box 24885, Safat 13109, Kuwait Fax: (00965) 4830432

There are alternative biotechnology applications which have impact on the maintenance of soil fertility, sustaining increased agricultural productivity and improving farmers' profitability, e.g., biofertilizers for legumes and non legumes, organic matter recycling for composting, ensilage, and mushroom production, and feeds from solid state fermentation processes. Biogas technology provides a source of supplying alternative fuel for cooking and simultaneously making available good quality manure. Farmers should be motivated via demonstration projects to apply appropriate technology to convert and reuse their available farm wastes and other agricultural wastes. The fertility of soils can be greatly enhanced by the action of earthworms. Earthworm biotechnology can accomplish this in two main ways, e.g., increasing availability of nutrients for plant growth, and by improving soil structure. Soils contaminated with pesticide residue cause reduced productivity either in quality or quantity. *In situ* and on site bioremediation for contaminated soils and water is one of the most promising technologies to transform or mineralize the potentially toxic substances, e.g., oil spills. New types of crops that are thermotolerant, salt tolerant, pesticide resistant, and fix nitrogen, are imminent, through gene splicing, recombinant DNA and advances in tissue culture techniques. Thus, developments in plant biotechnology will help farmers in increasing their productivity.

The concept of integrating all possible biotechnological aspects into a productive agricultural system has received increasing attention in many countries during recent years. The main target is the maintenance of soil feasibility, sustaining agricultural productivity and improving farmers profitability through the efficient use of biofertilizers, soil conditioners, optimization of organic waste recycling, bioremediation of soil and water; as well as getting the benefit of advanced plant biotechnology. Farmers should be motivated via demonstration projects to apply appropriate technologies to sustain this productivity. In this article the impact of major biotechnological aspects is being highlighted.

Biofertilizers: Biofertilizers denote preparations containing living microorganisms, e.g., Rhizobia for legumes, blue green algae and Azolla for rice fields, Azospirillum for cereal crops, and Mycorrhiza as a phosphate dissolvent. Applying biofertilizers improves soil fertilizer, and minimizes the consumption of mineral fertilizer, leading to decreased environmental pollution due to the extensive use of these substances. The potential of biofertilizer systems is clear. However the effect in soil depends on environmental conditions, e.g., climate, soil composition and structure, as well as plant rotation systems.

The contribution of nutrients added through these sources would be deducted from the doses recommended for the cropping system (ElNawawy and Hamdi 1975, Hamdi 1982).

Composting: Composting is mainly used to improve the soil's physical and biological characteristics and to control soil erosion.

It is important to motivate farmers to conserve and compost their farm residues in correct ways, e.g., to select an appropriate site for composting, chopping the residues, adding a starter dose of nitrogen and phosphorus if needed.

1054–7487/92/0577$06.00/0

Co-composting of sewage sludge and HHR has physical and chemical conditions appropriate for stabilization through composting. The technical feasibility of co-composting has been demonstrated in Holland, France, Austria and Germany (Obeng & Wright 1987). Co-composting now is a viable alternative in many developing countries where there is much concern over the large amounts of garbage and poorly disposed sewage sludge. These waste materials can by recycled through co-composting to improve the urban environment soil quality and productivity.

Earthworm Biotechnology: There are three areas where earthworms can benefit farmers. They are:

*Animal and vegetable wastes and municipal sludge can be stabilized by earthworms. Vermistabilization is a new, innovative, alternative technology that is environmentally sound; is not energy-, capital- or equipment-intensive; and should not require extensive management (Loehr et al., 1985). Earthworms have also accelerated the decomposition of waste water sludge in drying beds (Mitchell et al., 1980). They promote stabilization, reduce odors and manufacture auxins, the equivalence of some pituitary hormones in plants. They also create more favorable conditions for 'useful' bacteria, such as azotobacters, actinoymcetes, and some cellulase-secreting species and they fight off some 'harmful' microorganisms, such as *E. coli, Salmonella typhimurium,* and certain other bacterial pathogens (Hartenstein et al., 1984). Also earthworms can assimilate nematodes in soil (Dash et al., 1980).

*Soil fertility can be enhanced by the action of earthworms. Earthworms accomplish this in two main ways when introduced to soil (Laird and Kroger, 1981); e.g. increasing nutrient availability for plant growth; and improving soil structure.

*Earthworms are a valuable source of protein. Their average protein content is 67% on a dry weight basis (Lawrence and Miller, 1945; Hartenstein, 1985). A ton of farm animal manure could potentially culture 83.9 kg earthworms, which, at 8.8% protein, would be 7.4 kg protein (McInroy, 1971).

Ensilage: Agriculture wastes can be conserved as green fodder through ensilage. According to analysis of the available green parts of plants, the plant materials should be enriched with certain nitrogen supplements and soluble carbohydrates (e.g., milk whey) to ensure good quality silage. Fish silage and animal product silage are liquefied during ensilage. The product may be fed either wet or dry onto a carrier, depending on the type of livestock. The principal use of this silage is as a source of good quality protein that may be acquired far more economically than fish or meat meals and that may even be more accessible and cheaper than oilseed cakes in some instances. Poultry offal silages, which have a high energy value due to their high fat content, can be included in poultry finisher rations as an energy source (Zaleski et al.,

Mushrooms Cultivation: The mushroom is cultivated on organic wastes, such as composted straw, wood wastes, and horse and chicken manures. Certain mushroom species can be grown on several lignocellulosic wastes, e.g., cardboard and paper (Chang and Miles, 1984). Other mushroom species can be grown on sawdust, cotton waste, bagasse, straw, shredded paper and banana leaves (Gupta and Langar, 1988; Commanday and Macy, 1985). One species can be harvested nine days after the crop is started.

Particular kinds of cultivated mushrooms are consumed in different parts of the world. The USA, Australia and the UK consume primarily button mushrooms. In Europe, extensive methods have been developed to grow the oyster mushroom, which is now the second most important edible mushroom grown in Europe after the button mushroom. In Asia, the shiitake (straw mushroom) and oyster mushrooms are the predominant types of cultivated mushrooms.

Biogas Technology: The process of biogas fermentation from crop residues, animal droppings and human habitant wastes, attracted the interest of many developed and developing countries. The fermentation of wastes to generate biogas, yields a number of benefits, including producing an energy source that can be stored and used more efficiently for cooking, lighting, generation of electricity and running of water pumps; creating a stabilized residue that retains the fertilizer at a value similar to that of the original material; reducing faecal pathogens and improving public health. Small farmers can establish small digestors to improve their need of energy and manures, besides improving health conditions and the quality of life in villages (NAS, 1981).

Integrated Systems for Waste Bioconversion: For optimal resources use, essentially closed systems are used whereby the wastes of one process serve as the raw material for another. Integrated systems approaches have been developed to use animal, human and agro-industrial wastes together to produce food, feed, fertilizer and fuel. The conversion processes are combined and balanced to minimize external input and maximize self-sufficiency. The selection of a particular integrated system depends on many factors, e.g., types of wastes available, local needs and priorities, and psychological and economic aspects. The adaptation of integrated systems increases resource utilization and enhances self-sufficiency. Examples of these integrated systems are: Vermico-compost, Biogas Sewage Sludge and Mushroom-vermiculture-compost.

Bioremediation of Soil and Water: In many cases, soil might be polluted with certain hazardous organic wastes, which affect soil productivity. Also, it might be necessary to remediate the treated waste water used in irrigation. Much research has been conducted on detoxifying residual pesticides, e.g., phenolic compounds, chlorinated aliphatic compounds and trichloroethylene. Immobilized mixed microbial cells can be used for toxic organic waste treatment, e.g., heavy and crude oil spills. Also, some specific microorganisms can be used as biotraps for heavy metals. For heavy metal removal, a novel means of scrubbing specific compounds out of waste systems or slurries is being developed. The process involves the use of a unique class of bacteria-made proteins called periplasmic binding proteins. These proteins, normally used to transport nutrients into the cell for purposes of metabolism, have a high affinity and specificity for their ligand; are resistant to biodegradation; and are able to bind ligands over broad ranges of pH and ionic strength. Experiments with immobilized, biologically synthesized cadmium binders are also under way (Venosa, 1987).

Plant Biotechnology: Rapid development of plant cell and tissue culture techniques for plant propagation offer great potential for the future of agriculture. A wide range of plant species has now been cultured and regenerated through plant cell, tissue or organ culture, allowing many identical plants, including trees, to be grown. The plant species that can be propagated through tissue culture include asparagus, beets, cauliflower, cabbage, lettuce, spinach, alfalfa, millet, clover, potatoes, carrots, tomatoes, citrus fruits, oil palms and many ornamental species (OTA, 1984). Introducing agriculture to desert and arid areas has been a major challenge to many nations. Genetic engineering can play a significant role in developing new cultivars of crops suited for agriculture in arid and desert areas of the world. Gene transfer technology combined with the techniques of tissue culture, protoplast fusion and regeneration of plants from a single isolated protoplast are now the efficient tools to develop and grow a complete fertile and tolerant variety of crops suited for arid and desert conditions. This is another hope for sustaining food and agriculture in the world.

REFERENCES

Chang, S.T.; Miles P.G. *Bioscience*. 1984. **34:359-362.**

Commanday, F.; Macy, J.M. Arch. Microbiol. 1985. **142:61-65.**

Dash, M.C.; Senapath, B.K.; Mishra, C.C.; OIKOS. 1980; **34:322-326.**

ElNawawy, A.S.; Hamdi Y.A. In *Nitrogen fixation by free living microorganisms*; Stewart, W.D.P. Ed; Cambridge University Press, UK. 1975. IBP Vol 6: **219-228.**

Gupta, V.K.; Langar, D.N. 1988. *Biological wastes* **23:57-64.**

Hamdi, Y.A. *FAO Soil Bull 49.* FAO. Rome 1982.

Hartenstein, R.; Kaplan, D.L.; Neuhauser, E.F. 1984. *Journal of the Water Pollution Control* **56:294-298.**

Hartenstein, R. *Advances in ecological research*. 1985. **15:379-409.**

Laird, J.M. and M. Kroger. In *Critical Reviews in Environmental Control* Straub, C.P. Ed. Boca Raton, Florida: CRC Press, Inc., 1981. pp. **189-218.**

Lawrence, R.D.,; Miller, H.R. *Nature* (London), 1945. **155:157.**

Loehr, R.C.; Neuhauser, E.F.; Malecki, M.R. *Water Research*. 1985. **19(10):1311-1317.**

McInroy, D.E. *Feedstuff*. 1971. **43:37.**

Mitchell, M.J.; Horner, S.G.; Abrams, B.I. *Journal Environmental Quality*. 1980. **9:373-378.**

Obeng, L.A.; Wright, F.W. *The co-composting of domestic solid and human wastes*. World Bank Technical paper No. 57. The World Bank, Washington, D.C. 1987. PP. 102.

OTA. *Commercial biotechnology*. An International Analysis. Washington, D.C. office of Technology Assessment. 1984.

NAS, Food, Fuel and Fertilizers from Organic Wastes, 1981. National Academy of Science. Washington, DC.

Venosa, A.D. *UNEP Industry and Environment*. 1987. 10(4), 11-12.

Zaleski, S.J.; Lawik. B; Malicki A; Szubinska. M; Tereszkeiwicz, R.; Zeleski, G.; *UNEP Industry and Environment*. 1985. 8 (4):13-14.

Applications of Mycorrhizal Technologies for Agriculture and Forestry

Reynaldo E. dela Cruz*, Nelly S. Agganagan, Jocelyn T. Zarate and Elsie B. Lorilla,
National Institutes of Biotechnology and Applied Microbiology, UPLB, College, Laguna, 4031 Philippines, Fax: 010-63-94-2721 or (63-2)8170598

Five mycorrhizal technologies effective for agriculture and forestry have been developed in the Philippines. They are MYCOGROE tablets, mycorrhizal beads, MYKOVAM 1 soil inoculant, MYKOVAM 2 granules and Direct-Seeding Block. These technologies increase the survival, growth and yield of plants especially in adverse sites of marginal fertility. MYCOGROE tablets and MYKOVAM 1 can replace from 65-85% of the chemical fertilizers required for crop growth in the field.

Crop production by small farmers in developing countries is hampered by poor soil fertility and high cost of chemical fertilizers. Microbially-based fertilizers such as mycorrhiza have been tapped as alternatives to chemical fertilizers. Five technologies using mycorrhiza are presented.

MYCOGROE tablets contain basidiospore of ectomycorrhizal fungi compressed into small tablets. The tablets are inoculated to seedlings in the nursery. Growth of inoculated seedlings are significantly increased in the nursery and field (Table 1). MYCOGROE tablets can replace from 65 to 85% of the chemical fertilizers required for growth in the field (Table 2).

Mycorrhizal beads contain vegetative mycelia of the fungi grown first in fermentors and embedded in alginate beads. When inoculated to seedlings Vesicular-Arbuscular Mycorrhizal (VAM) fungi produced in trap plants. They are effective when inoculated to tree seedlings in the nursery and field (Table 4). It is also effective together with **Rhizobium** in inducing yield of agricultural crops such as peanut and mungbean (Table 5).

Table 1. Effect of MYCOGROE tablet inoculation on growth of Eucalyptus and Pinus species in the nursery and field

MYCORRHIZAL TREATMENTS	NURSERY HEIGHT (cm)	FIELD HEIGHT (m)	FIELD VOLUME (cc)
1. **Eucalyptus deglupta**			
Inoculated	22.2 a	6.3 a	12204 a
Uninoculated	16.8 b	3.9 b	3159 b
% Increase	32%	64%	285%
Age	3 months	18 months	
2. **Pinus caribaea**			
Inoculated	41.4 a	2.5 a	968 a
Uninoculated	30.4 b	1.5 b	673 b
% Increase	36%	68%	44%
Age	7 months	3 years	

Table 2. Estimated replacement of inorganic fertilizers in the field by mycorrhiza derived from inoculation with MYCOGROE tablets in the nursery

MYCORRHIZAL TREATMENTS	EUCALYPTUS CAMALDULENSIS	EUCALYPTUS DEGLUPTA	PINUS KESIYA
	-- Gram Fertilizer Needed per Plant --		
INOCULATED	20	10	17
UNINOCULATED	60	72	68
AMOUNT OF FERTILIZER REPLACED BY MYCORRHIZA, %	65	85	75

Table 3. Height growth of E. camaldulensis and P. caribaea 8 weeks after inoculation with alginate beads containing mycelia of Pisolithus tinctorius

TREATMENTS	HEIGHT (cm)
1. **Eucalyptus camaldulensis**	
Control	2.67 b
Beads stored in water	4.67 a
Beads stored dry	4.25 a
2. **Pinus caribaea**	
Control	6.3 b
Beads stored in water	6.9 ab
Beads stored dry	7.8 a

Table 4. Height growth of some reforestation species in the nursery and field after inoculation with MYKOVAM 1 containing specific VAM fungi

VAM TREATMENTS	HEIGHT (CMS) ACACIA AURICULIFORMIS.	ACACIA MANGIUM	PTEROCARPUS VIDALIANUS
	A. NURSERY (after 4 months)		
Uninoculated	27.1 b	17.9 b	31.7 b
G. margarita	26.0 b	13.7 c	37.3 a
G. macrocarpum	27.6 b	16.3 bc	38.7 a
S. calospora	33.5 a	30.7 a	37.3 a
	B. FIELD (after 13 months)		
Uninoculated	183.6 b	98.6 b	105.4 b
G. margarita	193.2 ab	149.1 ab	131.7 a
G. macrocarpum	217.4 a	157.0 ab	124.2 a
S. calospora	206.2 ab	184.8 a	132.2 a

Table 5. Seed yield of Vigna radiata and Arachis hypogaea during a dry-season cropping as affected by fertilization and inoculation with MYKOVAM 1 and/ or Rhizobium

TREATMENTS	SEED YIELD (kg per hectare) VIGNA RADIATA	ARACHIS HYPOGAEA
R_0M_0 = No **Rhizobium** - Mycorrhiza	162 b	175 d
R_0M_1 = No **Rhizobium** + Mycorrhiza	224 b	237 c
R_1M_0 = + **Rhizobium** - Mycorrhiza	249 b	332 c
R_1M_1 = + **Rhizobium** + Mycorrhiza	449 a	640 ab
R_1P_1 = + **Rhizobium** + 30 kg P/ha	429 a	628 ab
R_1P_2 = + **Rhizobium** + 100 kg P/ha	432 a	623 ab
N_1P_1 = 30 kg N/ha + 30 kg P/ha	444 a	650 ab
N_1P_2 = 30 kg N/ha + 100 kg P/ha	455 a	682 a
N_2M_1 = 100 kg N/ha + Mycorrhiza	419 a	619 ab
N_2P_2 = 100 kg N/ha + 100 kg P/ha	421 a	580 b

MYKOVAM 2 is a granulated form of MYKOVAM 1 containing VAM fungi found effective for tree seedlings (Table 6).

Direct-Seeding Blocks (DSB) containing VAM are effective in promoting growth of reforestation species (Table 7). DSB with VAM can speed up the rate of reforestration by completely bypassing traditional nursery operations thereby allowing direct sowing of seeds in the field.

Table 6. Effect of inoculation with MYKOVAM 2 granules on height growth of some reforestation species in the nursery

TREATMENTS	HEIGHT (CMS) LEUCAENA LEUCOCEPHALA	ACACIA MANGIUM
UNINOCULATED	28.9 b	6.9 b
MYKOVAM 2	56.0 a	17.5 a

Table 7. Height growth of reforestation species in response to DSB-VAM treatments

DSB-VAM TREATMENTS	HEIGHT (CMS)		
	PARASERIANTHES FALCATARIA	**PTEROCARPUS VIDALIANUS**	**ACACIA MANGIUM**
Uninoculated	4.3 c	16.5 c	13.2 d
DSB - No Mycorrhiza	4.0 c	17.2 bc	16.7 cd
DSB + **Gl. macrocarpum**	23.8 b	21.2 ab	20.1 bc
DSB + **G. margarita**	43.1 a	22.3 a	46.0 a

Constraints to Crop Production in the Semi-Arid Tropics: Can Biotechnology Help the Small Farmer

J. P. Moss, K. K. Sharma, N. Mallikarjuna, and D. McDonald, International Crops Research Institute for the Semi-Arid Tropics (ICRISAT), Patancheru, Andhra Pradesh 502 324, India (FAX: 91–842–241239)

The small farmer of the semi arid tropics farms in a region where there are many constraints, but often he does not have the physical or financial means to overcome them. The major contribution of biotechnology will be to provide him with resistant genotypes of his crops. Many crops that he grows have not received major research emphasis, but techniques are available to produce plants resistant to bacteria, viruses, insect pests, and to abiotic stresses. In addition, diagnostics can assist in the correct identification and characterization of the constraints.

Introduction

The semi-arid tropics (SAT) is a harsh region with inadequate, uncertain rainfall for much of the year, usually infertile soil, a wide range of constraints to crop production, particularly lack of water, high incidence of pests and diseases and low capital availability. In many parts of the developing world, solutions to these problems--pesticides, fertilizers and irrigation--are not available, or the farmer does not have the finances needed. Some cultivars with resistance to specific pests and diseases have been made available to the farmers.

Biotechnology offers the hope of introducing cultivars which are resistant to the biotic constraints and which can make best use of the available moisture and low levels of fertility which occur in these soils. Biotechnology also offers help to breeders, pathologists, entomologists and physiologists, for example by speeding the breeding process (in vitro regeneration, haploids and micropropagation), by increasing the precision of screening (with Restriction Fragment Length Polymorphism (RFLP), Polymerase Chain Reaction (PCR) and other technologies), and by better diagnosis of diseases (using enzyme linked immunosorbent assay (ELISA), cDNA probes and monoclonal antibodies). However, biotechnology can be expensive, and although some developing countries can undertake research much is being funded by and done in developed countries.

Constraints can be classified into biotic--yield reducing organisms, such as weeds, pests and diseases-- and abiotic, the edaphic and climatic conditions that limit production.

Biotic Constraints

Diseases. Fungal diseases are major yield reducers, and some resistance genes have been identified in cultivated and wild germplasm, and resistant cultivars developed. Wild relatives of peanut have been used extensively at ICRISAT as sources of resistance, and necessary techniques of wide hybridization, embryo rescue and tissue culture have been developed (Moss et al 1988). Similar research can taken up for the various crop problems in the SAT countries.

Many pathogenic fungi exist in a number of forms--pathotypes or physiological races--and the ability of the pathogen to produce new races casts doubt on the use of resistance based on one gene. Marker technology--use of RFLP's, RAPDs, PCR, and isozymes--will enable the

1054–7487/92/0585$06.00/0

identification and manipulation of genes, and it can be taken up in developing countries and deserves attention.

In addition to the yield loss caused by fungal diseases, the production of mycotoxins can be a serious health hazard. Mycotoxin production is often increased in developing countries by a lack of equipment to harvest crops at optimum stage, poor transport, and a lack of proper drying and storing facilities. Aflatoxin, produced when peanuts are infected with *Aspergillus flavus*, is a serious constraint for peanut farmers. Although oil extracted by the solvent method is free from aflatoxin, contamination of seed often leads to downgrading from human use to animal feed. This leads to loss of income for the farmer, and there is still the danger of contamination of dairy products from animals fed contaminated cake. Resistance to *A. flavus* has bee identified in cultivated and wild peanuts (Mehan et al 1986, Ghewande et al 1989), and careful management can reduce the growth of *A. flavus*, but these approaches cannot eliminate aflatoxin production. There is still a need for a biotechnological approach to the reduction of formation of aflatoxin. ELISA-based methods for the detection of various aflatoxins in small quantities have been developed at ICRISAT (Ramakrishna et al 1991).

Bacterial diseases have not received as much attention by breeders as fungal diseases, partly because they tend to be more local in occurrence and do not frequently cause epidemics, though yield losses can be high. Selection of resistant somaclonal variants by exposing the cells to the bacteria or to bacterial toxins has been successful (Behnke, 1979 & 1980, Hammerschlag 1988 Biondi et al 1991, Frame et al 1991). The facilities required for this technology are readily available in many developing countries, and competitive in cost with other breeding approaches.

Viral diseases are widespread in crop plants. The phenomenon of cross protection--resistance of a plant to one strain of a virus when infected with a closely related strain--has led to the use of viral coat protein genes to confer resistance (Nejidat and Beachy 1990, Lawson et al 1990). However, this phenomenon is not universal, and transformed plants are not always resistant to related viruses (van Dun et al 1988).

An understanding of genome organization of viruses is essential prior to utilizing viral genes in transformation. Many developing countries do not have the facilities and expertise for this research most of which has been done in developed countries. However, necessary expertise for the transformation and regeneration of locally important crops is becoming more frequent in developing countries.

In any research into management of viruses through resistance breeding, correct identification of the causal virus is essential as in many cases symptoms of viral diseases resemble each other. Diagnostic techniques have relied on serological methods such as ELISA, in which polyclonal antibodies are used in most cases. However, monoclonal antibodies have advantages in detection, especially in tailoring broad specificity for the identification of viruses or groups of viruses, or narrow specificity for the identification of viral strains. ICRISAT supplies antibodies to a range of viruses of its mandate crops to assist scientists in the SAT for in viral detection and identification. Viral nucleic acid based diagnostic procedures are extremely reliable and sensitive, specially in the detection of isolates of a virus.

Virus elimination from plants through tissue culture is a proven technique for which the facilities and expertise are available in many developing countries. The greatest potential benefit is from its application to vegetatively propagated crops which can be multiplied and maintained in culture, and virus-free plants provided to farmers. This has been successfully applied to staple food crops (cassava, potato, sweet potato) and in many fruit crops (bananas, citrus).

Pests. Pests are important as yield reducers during crop growth, as storage pests after harvesting, and as vectors of viral diseases. Pesticides have controlled some pests, but many have shown remarkable ability to develop resistance. This has stimulated a search for alternative means of control, among which genes which confer resistance without having any mammalian toxicity have received attention. Among these are the B.T. gene from *Bacillus thuringensis* (Delannay et al 1989) and the trypsin inhibitor gene (Boulter et al 1989). There is hope that other genes can be found (Kladitskaya et al 1989).

Weeds:

The ultimate aim for minimising losses due to weeds must be an integrated approach combining good management with improved crops that compete well with weeds, e.g. through vigorous early growth. However, herbicides can be of value in the SAT, if care is taken not to degrade the environment.

Considerable progress has been made in devising strategies for increasing herbicide tolerance in plants (Mazur and Falco 1989). One of these strategies is the transformation of plants with foreign genes which detoxify herbicides. The herbicide Basta inhibits glutamate synthetase, leading to accumulation of ammonia. Full resistance to Basta can be obtained through the use of the *bar* gene from *Streptomyces hygroscapicus*, thus acetylating and inactivating the active component of Basta (DeGreef et al, 1989). Over expression and desensitization of the target of the herbicide may be achieved in the use of the herbicide Glyphosate (Shah et al. 1986; Della-Cioppa et al. 1987). To reduce the chance that weeds would acquire herbicide tolerance through cross-pollination, it would be desirable to integrate genes for herbicide resistance into the chloroplast DNA of the transgenic plants since, in most crop plants, chloroplast DNA is inherited maternally (Tilney-Bassett 1984).

Abiotic constraints.

Less progress has been made towards resistance to abiotic stresses. This is partly due to the fact that most of these are polygenic in character. An exception is heavy metal ion tolerance which has been successfully enhanced in plants through use of the human metallothionine gene (Misra and Gedamu 1989). Several attempts have been made to select salt tolerance at the cellular level (Stavarek and Rains, 1984). Attempts are also being made to identify the genes which are differentially expressed in the acquired cellular salt tolerance in a variety of crops. Studies have shown that a 26 KD protein (Singh, et al. 1985) is induced by NaCl in suspension cultures of tobacco; it is also expressed in whole plant roots (King et al. 1986). Similar studies to identify proteins related to other stresses, and the genes that control them, -- e.g. drought and temperature -- would be helpful in developing stress-resistant varieties through biotechnological means.

Hybrid seed production

The infrastructure necessary for hybrid seed production and utilization--a seed production industry and good distribution system--are not available in many developing countries, but hybrids deserve serious consideration. One approach to cytoplasmic male sterility (cms) is to use protoplast fusion or wide hybridization to introduce alien cytoplasm. An alternative approach has been to use a bacterial ribonuclease gene (*Barnase*), and a promoter specific for the tapetal cells which results in death of the tapetal cells and hence male sterility (Mariani et al. 1990). Diversification of cytoplasms is essential for hybrid seed production, and their characterization is an important aspect. At ICRISAT, RFLP of mtDNA has been used efficiently to classify cms lines of pearl millet and sorghum.

Conclusions

If the small farmer can be given varieties adapted to the harsh climate in which he lives, and resistant to the many constraints on crop growth, he has much to gain from biotechnology,. This can best be achieved by scientists in developed and developing countries cooperating to make best use of facilities and resources.

References

Behnke, M. *Theor Appl Genet* **1979**, *55*, 69-71.

Behnke, M. *Theor Appl Genet* **1980**, *56*, 151-152.

Biondi, S., Mirza, J., Mittempergher, L., and Bagni, N. *J Plant Phys* **1991**, *137(5)*, 631-634.

Boulter, D., Gatehouse, A.M.R. and Hilder, V. *Biotechnol Adv.*, **1989**, *7(4)*, 489-498.

De Greef, W., Delon, R., De Block, M., Leemans, J. and Botterman, J. *Bio/Technology*, **1989**, *7*, 61-64.

Delannay, X., Bradley J. LaVallee, Fuchs, R.L., Sims, S.R., Greenplate, J.T., Marrone, P.G., Dodson, R.B., Augustine, J.J., Layton, J.G., Fischhoff D.A. *Bio/Technology*, **1989**, *7(12)*, 1265-1270.

della-Cioppa, G., Bauer, S.C., Taylor, M.L., Rochester, D.E., Klein, B.V., Shah, D.M., Fraley, R.T. and Kishore, G.M. *Bio/Technology*, 1987, *5*, 579-584.

Frame, B., Yu, K.F., Christic, B.R. and Pauls, K.P. *Physiol and Mol Plant Path*, **1991**, *38(5)*, 325-340.

Ghewande, M.P., Nagaraj, G., and Reddy, P.S. In *Aflatoxin contamination of groundnut*. Proc Int Workshop, International Crops Research Center for the semi-Arid Tropics, Patancheru, A.P. 502 324 India, **1989**, pp 237-243.

Hammerschlag, F.A. *Theor. Appl. Genet* **1988**, *76(6)*, 865-869

King, G.J., Hussey, C.E. Jr., and Turner, V.A. *Plant Molecular Biol.*, **1986**, *7*, 441-450.

Kladitskaya, G.V., T.A. Vallueva, T.V. Cherendnikova, and V.V. Mosolov. *Biochemistry USSR*, **1989**, *54*, 461-467.

Lawson, C., Kaniewski, W., Haley, L., Rozman, R., Newell, C., Sanders, P., and Turner, N.E. *Bio/Technology*, **1990**, 127-134.

Mariani, C., De Beuckeleer, M., Trueltner, J., Leemans, J. and Goldberg, R.B. *Nature*, **1990**, *347*, 737-741.

Mazur, B.J., and Falco, S.L. *Ann. Rev. Plant Physiol. Plant Mol. Biol.* **1989**, *40*, 441-470.

Mehan, V.K., McDonald, D., Ramakrishna N., and Williams, J.H.. *Peanut Sci.*, **1986**, *13 (2)*, 46-50

Misra, S. and Gedamu, L. *Theor. Appl. Genet.* **1989**, *78*, 161-168.

Moss, J.P., Singh, A.K., Sastri, D.C., and Dundas, I.S. In: *Biotechnology in Tropical Crop improvement*. J.M.W. de Wet, Ed., ICRISAT, Patancheru, India. **1988**, pp. 87-95.

Nejidat, A. and Beachy, R.N. *Molecular Plant-Microbe Interactions*. **1990**, *3 (4)*, 247-251.

Ramakrishna, N., Mehan, V.K., Candlish, A.A.G., Smith, D.H., and McDonald, D. In: *Abstracts of the Second International Groundnut Workshop*. ICRISAT, Patancheru, India. **1991**, pg. 15

Shah, D.M., Horsch, R.B., Klee, H.J., Kishore, G.M., Winter, J.A., Turner, N.E., Hironaka, C.M. *Science*, **1986**, *233*, 478-481.

Singh, N.K., Handa, A.K., Hasegawa, P.M., and Bressan, R.A. *Plant Phys.*, **1985**, *79*, 126-137.

Stavarek, S.J. and Rains, D.W. In: *Salinity Tolerance in Plants: Strategies for Crop Improvement*. R.C. Staples and G.H. Toenniessen, Eds., John Wiley, New York, **1984**, pp. 321-334.

Tilney-Bassett, R.A.E. In *Chloroplast Biogenesis*. Ellis, R.J. Ed., Cambridge University Press, Cambridge, **1984**, pp. 13-50.

van Dun, C.M.P., Overduin, B., van Vloten-doting, L., and Bol, J.F. *Virology*, **1988**, *164*, 382-389.

On-farm Anaerobic Treatment of Agricultural Wastes in Some Countries in Asia

Wolfgang Tentscher, Energy Technology Division, Asian Institute of Technology, P.O. Box 2754, Bangkik 10501, Thailand, (Fax: +66/2/5245552, Tel: +66/2/5245428)

The progress in the design of small-scale fixed-dome digesters and of medium and large scale biogas plants in some countries in Asia is discussed. It became obvious that a considerable knowledge was necessary to properly design and construct biogas plants and to guarantee structural and operational stability. At the AIT, several digester types were subjected to a detailed stress pattern analysis with the result that crack proof designs could be suggested. China and India are the leading countries in the region in regard to biogas technology. High attention is devoted in R&D and dissemination to high-rate digestion processes. A new pilot project in a medium scale pig farm in Thailand, employing the HYPHI process is introduced. Comprehensive utilization of gas and effluent is aimed at to achieve attractive payback periods and internal rates of return. New substrates such as agricultural residues are the topic of many research and development projects.

Biogas technology is no home-made technology any more. After another decade of research, development and dissemination new knowledge was obtained. Modern methods were applied to determine structural stability and the highest possible process performance. Progress in mathematical formulation of kinetic reactions and in PC computer technology led to a computerized design which is a valuable and indispensable tool for decision-making. The economic implications of changes in technical parameters can be easily assessed. Due to the progress in all the areas on one and the CO_2 and green house problem in the atmosphere on the other side, biogas technology, be it in small farms or feed lots or sewage treatment plants, is gaining increasing attention. The energetic, environmental and hygienic benefits come into full play. With these anaerobic processes. CO_2 and mineral and organic matter can be recycled.

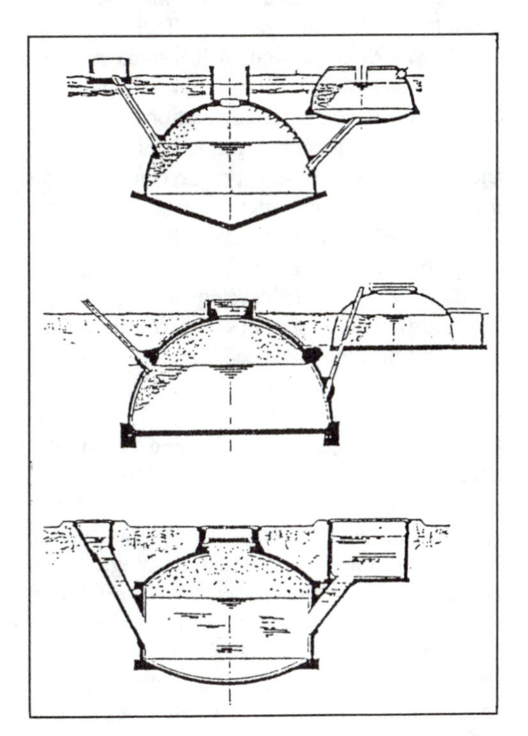

Fig. 1: Digester designs analysed: Sasse/BORDA (S/B)-design (Above), CAMARTEC (CA)-design (middle), Chinese cylindrical (CH)-design (below) (Adapted from MEESE et al. 1988).

Statical improvement of fixed-dome digesters

Fig. 1 shows the three types of fixed-dome digesters analysed. They are the Sasse/BORDA, from Germany, CAMARTEC, Tanzania and Chinese cylindrical designs.

The most important parameters influencing the stress pattern are soil properties, material strength data, digester shapes and volumes, gas pressure, slurry level, and earth cover on top of the digester (RINGKAMP et al., 1988). Six basic loads were taken into consideration and combined to 6 *load cases*: empty, water filled, under full gas pressure and with ground water pressure. It could be concluded that the CAMARTEC design is the only one that is crack proof in its original shape, since the sphere and bottom ring beams can be freely displaced and since the sphere contains the *Weak Ring* at 50 to 60° from the centre line (RINGKAMP, 1989).

The *Weak Ring* was first constructed in CAMARTEC designs and can be considered a milestone in improving the structural stability, on which AINEA (1987) reports. It is a flexible joint and effectively stops the further propagation of cracks. This could be confirmed with computer simulations.

The real interaction of soil and wall was measured with strain gauges. The results on the CAMARTEC design in Tanzania and the Chinese cylindrical design in China well confirmed the computer calculations.

As the main practical result, MEESE (1989) and RINGKAMP (1989) prepared nomograms for *dimensioning and design* of the CAMARTEC design. A recent PC program on design of fixed-dome and floating drum biogas plants BIO CALC (Biosystem 1991) includes algorithms resulting from these computations, demonstrating the advanced stage of design.

Farm-scale digesters in Asian countries

China

The total potential for biogas plants may be calculated from the potential of biogas from manures, straws and stalks which results in some 206 million digesters. The total number by the end of 1988 was 4.7 million (DENG, 1990). In 1988, about 210,000 digesters had been built. Thus, the digester construction activity was reduced by about 50% as compared with 1986. In the Sichuan province, the most commonly constructed design is still the cylindrical fixed-dome one.

By the end of 1988, 2719 medium and large scale digesters with an average digester volume of 85 m^3 and average gas production of 1.3 m^3 gas/m^3/d had been constructed in agro-industries, animal farms and cities and towns. The actual construction activity is about 300 digesters per year (HE, 1990).

Subsidies and official expenses in the biogas programme were reduced. In 1989-90, the central government allocated about 1 million Yuan for training, research, and demonstration, compared to a peak of 18 million Yuan in 1983. In addition, the local governments provided 10 times more than the central government for training, research, demonstration and "biogas industries", irrespective of the scale (DENG et al., 1990). By the mid-80s, the expenses for digester construction and by-product utilization had been borne by the farmers themselves. Subsidies, provided by the government in the past, had been cut almost completely.

The biogas development programme in China has shifted from promoting simple fermenting equipment (about 900 biogas industries produce appliances) to developing biogas projects. Strategies were developed to promote mutually complementary multi-energy systems and comprehensive utilization, to develop the technology from the domestic to the productive field, from the rural towards the urban, from just saving the farmers' fuel shortages towards effective recycling in the context of agricultural ecology, and from experimental biogas plants towards integrated energy systems (DENG et al., 1990; XU, 1988).

India

As in China, biogas technology is strongly promoted in India through a national programme reaching to grass roots levels. Environmental problems and energy shortages motivate its implementation. A total potential of about 10 million family-scale biogas plants may be assumed.

by the end of 1989, the total number of installed plants since 1981 may be around or over 1 million. The annual construction activity in 1989-90 was about 160,000 plants. By the year 2001, it is planned to set up 12 million digesters with an annual construction activity of 800,000 digesters. This ambitious goal may need a new implementation strategy. Commercialization of family-scale digesters is in development.

The construction activity of large size community and institutional biogas plants (CBF/IBP) increased to 500 in 1989. The government provides subsidies of up to 90% of the total cost. These plants of 25-140 m^3 gas/day are of KVIC type design and most of them are being fed with cattle dung. Research is going on to treat such wastes and waste waters with advanced biomethanation systems (KHANDELWAL, 1990).

A new monitoring and evaluation scheme has been developed, operating at three levels as follows: 1) Inspection by the state governments and implementing agencies, followed by, 2) DNES monitoring through its six regional offices and seven Biogas Monitoring Cells (100% basis in villages and blocks selected at random), and 3) the evaluation survey studies conducted by several independent agencies, with Regional Biogas Development and Training Centres being among them (2 to 5% sample size) (KHANDELWAL, 1990).

In 1988-89, the central subsidy was changed insofar as only a constant subsidy for capacities larger than 6 m^3 gas/d would be given. It is proposed to further reduce the limit to 4 m^3 gas/d or even 2 m^3 gas/d in the near future (KHANDELWAL, 1990). This will also discourage the owners from ordering oversized digesters.

Thailand

The total potential for family-scale biogas plants digesting cattle and pig manure is about 820,000 (TENTSCHER, 1988a). There may be actually only a very few operating digesters left over from the previous national biogas programme (1973-83).

In five northern provinces, the Chiang Mai University in cooperation with the Department of Agricultural Extension, carries out the Thai-German Biogas Project (TG-BP), a GTZ sponsored biogas dissemination project for mainly family-sized fixed-dome digesters. Improved digester designs are being promoted. Medium and large scale digesters such as the Modular Double Biogas System for Swine Manure are also included (BLOH et al., 1991). This system took the experience of the first pilot plant of this type in Thailand into consideration, a plug-flow digester designed and constructed by the AIT in cooperation with Kasetsart University (KU) (TENTSCHER, 1988b), which has operated successfully since 1987. A further development is the HYPHI (Hybrid Plug flow High-rate) digester system for diluted wastewaters with suspended solids (TENSTSCHER et al., 1991). A computer package exists to properly design the HYPHI system and its various options. The system can be designed for maximum methane production rate or minimum pay-back period and a sensitivity analysis can be done with special algorithms (TENTSCHER et al., 1991). All of these systems are under test and demonstration, and the HYPHI system is being intensively researched to confirm the kinetics and to find improvements in technical and economic aspects.

Lignocellulosic materials as new emerging substrates

Agricultural residues represent a huge potential for biogas production and would, if digestion processes operate satisfactorily and operation is easy, considerably expand, if not double, the total potential. Biogas could eventually be generated by those parts of the population which do not have enough animals to support a conventional biogas plant. Therefore it is essential to develop biogas plants for alternative feedstocks as well as mixed feeds.

Substrates such as non-edible deoiled cakes, water hyacinth, etc., banana stems and peelings, composite agricultural wastes, etc., will be included in the community and institutional biogas plants scheme in India (KHANDELWAL, 1990).

In China, the batchwise digestion of rice straw in conjunction with continuous digestion of manures is common. Dry fermentation of composted agricultural residues was also done in some parts of China (FANG, 1990).

A hybrid process or a two phase process was developed for straws in China (ZHOU et al., 1990) and for water hyacinth at AIT (TENTSCHER et al., 1991). It consists of a leach bed reactor (LB) which is loaded with lignocellulosic material, and an attached anaerobic filter (AF) or UASB. Such systems are not disturbed by floating layers. Digestion of water hyacinth may be economical if the bulk density is about 500 kg/m^3.

References

AINEA, E.,. (1987): Standardization of Biogas Units (Biogas Extension Service, Centre for Agricultural Mechanization and Rural Technology, Proc. Biogas Workshop East Africa, April 27- May 1, p. 24-37).

Biosystem (1991): BIO-CALC (Sponsored by GTZ).

BLOH, H. VON, BOON-LONG, P., SALADYANANT, S., POTIKANOND, N., PHUAGPHONG, B. (1991): Modular Double Biogas Systems for Swine Farms (Proc. Seminar on Bioprocess Engineering of On-Farm Digesters for Waste Treatment, AIT, Oct 28-29, pp. 8).

FANG, G.-Y. (1990): Development, utilization & socio-economic viability evaluation of Chinese domestic biogas digesters (International Conf. on biogas Technologies and Implementation Strategies, January, Pune, p. 56).

DENG, K., CHAI, Y. (1990): China actively promotes the development of biogas technology (Intern. Conf. on Biogas Technologies and Implementation Strategies, January, Pune, p. 59-69).

HE, L. (1990): Technical and economic evaluation on development of large and medium sized biogas projects in China (International Conf. on Biogas Technologies and Implementation Strategies, January, Pune, p. 72).

KHANDELWAL, K.C. (1990): Biogas Technology Development & Implementation Strategies-Indian experience (International Conf. on Biogas Technologies and Implementation Strategies, January, Pune, p. 1-22).

MEESE, L., FANG, G.Y., HASAN, S.M., UY, S.G., REN, L. (1988): Final Report (Report No. 2) on statical improvement of family-sized fixed-dome digesters, a computer aided design approach (GTZ-AIT, PN 88.2001.1-01.100).

RINGKAMP, M., TENTSCHER, W., SCHILLER, H. (1988): Preliminary Results on Statical Optimization of Family-sized Fixed-dome Digesters (Proc. 5th Intern. Symposium on Anaerobic Digestion, Poster-Papers. Bologna, Italy, May 21-24, p. 419-423).

TENTSCHER, W. (1988a): Biogas Development and Applications in some Countries in Asia (Chapter in the Book: Exchange of Environmental Experience, Series of Books produced by INFO-TERRA/PAC, p. 17-46).

TENTSCHER, W. (1988b): Experience with a full-scale plug-flow digester with a RMP-film as Gasholder (Proc. 5th Intern. Symposium on Anaerobic Digestion, Poster-papers. Bologna, Italy, May 21-24, p. 419-423).

TENTSCHER, W., NGO, v.T., WANCHANA, A., CHANTSAWANG, S. (1991): Engineering of a Biogas Plant System for Medium and Large Scale Pig Farms with Advanced Design Tools (The First Asia International Exhibition and Conference on New and Renewable Energy Information on Policy, Planning, Technology and Equipment, 17-20. Oct., Bangkok, Thailand 1991).

WANCHANA, A., TENTSCHER, W. CHANTSAWANG, S. (1991): The HYPHI Digester System (Proc. Seminar on Bioprocess Engineering of On-Farm Digesters for Waste Treatment, AIT, Oct. 28-29).

ZHOU, M., PEN, W. XIONG C. (1991): The Development of Biogas Fermentation Processes (In: Ten Years of Biogas Development in China (1980-1990), p. 17-27).

XU Z.-F. (1990): Biogas technology implementation in China (In: RAMANI, K.V. [Ed.]: Rural Energy Planning, Asian and Pacific Experiences, p. 283).

INDEX

Index

F

G

H

I

J

K

N

O

P

Q

R

T

U

V

W

X

Y

Z

Production: Margaret J. Brown
Indexing: Deborah H. Steiner
Acquisition: Cheryl Shanks
Cover design: Amy Hayes

Printed and bound by Victor Graphics, Baltimore, MD